2021 16th European Microwave Integrated Circuits Conference (EuMIC 2021)

London, United Kingdom
3 – 4 April 2022

IEEE Catalog Number: CFP21GAS-POD
ISBN: 978-1-6654-4722-5

Copyright © 2021, European Microwave Association (EuMA)
All Rights Reserved

*** *This is a print representation of what appears in the IEEE Digital Library. Some format issues inherent in the e-media version may also appear in this print version.*

IEEE Catalog Number:	CFP21GAS-POD
ISBN (Print-On-Demand):	978-1-6654-4722-5
ISBN (Online):	978-2-87487-064-4

Additional Copies of This Publication Are Available From:

Curran Associates, Inc
57 Morehouse Lane
Red Hook, NY 12571 USA
Phone: (845) 758-0400
Fax: (845) 758-2633
E-mail: curran@proceedings.com
Web: www.proceedings.com

EuMIC01: Large Signal and Non-Linear Characterization Techniques

Chair: Teresa M. Martín-Guerrero, Universidad de Málaga, Spain
Co-Chair: Nuno Borges Carvalho, Universidade de Aveiro, Portugal
09:00–10:40, Sunday 3rd April 2022, Room 10

1   **Load-Pull Measurement of SiGe:C HBT in BiCMOS 55nm Featuring 11dBm of Output Power at 185GHz**
C. Maye[1], Sylvie Lépilliet[1], E. Okada[1], E. Brezza[2], A. Gauthier[2], M. Margalef-Rovira[1], Daniel Gloria[2], Guillaume Ducournau[1], Christophe Gaquière[1]
[1]IEMN (UMR 8520), France; [2]STMicroelectronics, France

5   **Nonlinear Characterization of Wideband Power Amplifiers with Frequency-Dependent Match Load**
Sanket Chaudhary[1], Nuno Borges Carvalho[1], Marina Jordão[1], Marc Vanden Bossche[2], Adam Cooman[3], Sergio C. Pires[3]
[1]Universidade de Aveiro, Portugal; [2]National Instruments, Belgium; [3]Ampleon, The Netherlands

9   **Intermodulation Products of a CMOS SP6T Antenna Switch: Results Comparison Between an Experimental Test-Bench and a Corresponding Simulated Virtual Test-Bench**
M. Ben-Sassi[1], H. Saleh[1], O. Sow[1], I. Lahbib[1], Greg D. U'Ren[1], C. Hellepee[2], D. Passerieux[2], G. Neveux[2], Denis Barataud[2]
[1]X-FAB, France; [2]XLIM (UMR 7252), France

13   **A Computationally-Efficient Self-Consistent Large Signal Model for GaN HEMTs Based on ASM-HEMT**
S. Khandelwal[1], K. Kikuchi[2], H. Yamamoto[2]
[1]University of South Florida, USA; [2]Sumitomo Electric Industries, Japan

17   **Large-Signal Modeling for Nonlinear Analysis of Experimental Devices in 22nm FDSOI Technology**
Quang Huy Le[1], Dang Khoa Huynh[1], Anurag Nayak[1], Steffen Lehmann[2], Zhixing Zhao[2], Thomas Kämpfe[1], Matthias Rudolph[3]
[1]Fraunhofer IPMS, Germany; [2]GlobalFoundries, Germany; [3]BTU, Germany

EuMIC02: Silicon Based RF Solutions

Chair: Peter Magnee, NXP Semiconductors, The Netherlands
Co-Chair: Rüdiger Quay, Fraunhofer IAF, Germany
09:00–10:40, Sunday 3rd April 2022, Room 11

(NA)	**SiGe BiCMOS as Enabling Technology for Next Generation RF & THz Systems**
	Gerhard Kahmen, IHP, Germany

22	**Analysis of the Relaxed Contacted-Poly-Pitch Effect on the RF Performance of Strained-SiGe-Channel p-FETs in 22nm FDSOI Technology**
	Quang Huy Le[1], Dang Khoa Huynh[1], Steffen Lehmann[2], Zhixing Zhao[2], Thomas Kämpfe[1], Matthias Rudolph[3]
	[1]Fraunhofer IPMS, Germany; [2]GlobalFoundries, Germany; [3]BTU, Germany

26	**Design Methodology of Wide Tuning Range DGS-Based VCO for K-Band Applications in 0.18-μm CMOS Technology**
	Baichuan Chen[1], Samundra K. Thapa[1], Nusrat Jahan[2], Adel Barakat[1], Ramesh Pokharel[1]
	[1]Kyushu University, Japan; [2]CUET, Bangladesh

30	**Linearity Assessment of GaN HEMTs on Si Using Nonlinear Characterisation**
	Rana ElKashlan[1], Ahmad Khaled[1], Raul Rodriguez[1], Vamsi Putcha[1], Uthayasankaran Peralagu[1], AliReza Alian[1], Nadine Collaert[1], Piet Wambacq[2], Bertrand Parvais[2]
	[1]imec, Belgium; [2]Vrije Universiteit Brussel, Belgium

34	**Reconfigurable and Scalable Monolithic Band Reject Circuit Utilizing Phase-Change Switch Matrices**
	Tejinder Singh, Raafat R. Mansour, University of Waterloo, Canada

EuMIC03 : Transceiver MMICs

Chair: Mehmet Karaaslan, Teledyne e2v, UK
Co-Chair: Friedel Gerfers, Technische Universität Berlin, Germany
09:00–10:40, Sunday 3rd April 2022, Room 12

(NA) **mmW GaN/Si MMICs: The 3rd Generation of III/V Processes to Complement Si RFIC Solutions**
Marc Rocchi, OMMIC, France

39 **A Bidirectional 28GHz RF Transceiver Front-End with Test and Calibration Interface for 5G Phased Arrays**
Katharina Kolb[1], Julian Potschka[1], Tim Maiwald[1], Klaus Aufinger[2], Amelie Hagelauer[3], Marco Dietz[1], Robert Weigel[1]
[1]FAU Erlangen-Nürnberg, Germany; [2]Infineon Technologies, Germany; [3]Technische Universität München, Germany

43 **A 2-Channel TX and 4-Channel RX in SiGe BiCMOS for X-Band MIMO Radar Applications**
M. Kucharski, B. Błaszczuk, M. Klemm, R. Piesiewicz, SIRC, Poland

47 **Two-Element 81–86GHz SiGe Transmitter Beamformer for Backhaul Applications**
Roee Ben Yishay, Oded Katz, Danny Elad, ON Semiconductor, Israel

51 **A W-Band Single-Chip Receiver in a 60nm GaN-on-Silicon Foundry Process**
Robert Malmqvist[1], Rolf Jonsson[1], Mingquan Bao[2], Rémy Leblanc[3], Koen Buisman[4], Christian Fager[4], Kristoffer Andersson[2]
[1]FOI, Sweden; [2]Ericsson, Sweden; [3]OMMIC, France; [4]Chalmers University of Technology, Sweden

EuMIC04 : EuMIC Opening Session

Chair: John Christopher Clifton, EuMIC 2021 Chair
Co-Chairs: Shokrollah Karimian, EuMIC 2021 Co-Chair and Edward Wasige, EuMIC 2021 TPC Chair
11:20–13:00, Sunday 3rd April 2022, Room 7-9

(NA) **Welcome Address: Opening of the European Microwave Integrated Circuits Conference 2021**
John Christopher Clifton, EuMIC 2021 Chair

(NA) **III-V Nitride Semiconductors for Microwave Applications**
Christopher Snowden, ERA Foundation, UK

(NA) **High-Efficiency PAs for Broadband High-PAR Signals**
Zoya Popović, University of Colorado Boulder, USA

EuMIC05: Integrated Circuit Modelling and Design Methodology

Chair: Vadim Issakov, Technische Universität Braunschweig, Germany
Co-Chair: Matthew OKeefe, INEX Microtechnology, UK
14:20–16:00, Sunday 3rd April 2022, Room 5

55 **A 30-to-38GHz Active and Passive Combined Down-Conversion Variable Gain Mixer with Low OP_{1dB} Variation in 65-nm CMOS**
Mu-Heng Li, Chun-Nien Chen, Yunshan Wang, Huei Wang, National Taiwan University, Taiwan

59 **Analog Linearization of a 10-W GaN Power Amplifier by Baseband Feedback**
Mathani Eltayeb[1], Morten Olavsbråten[2], Gian Piero Gibiino[1], Alberto Santarelli[1]
[1]Università di Bologna, Italy; [2]NTNU, Norway

63 **Statistical Modeling of GaN HEMTs by Direct Transfer of Variations to Model Parameters**
Petros Beleniotis[1], Serguei Chevtchenko[2], Matthias Rudolph[1]
[1]BTU, Germany; [2]FBH, Germany

67 **Design of Terahertz InP pHEMT Using Machine Learning Assisted Global Optimization Techniques**
Jing Wang, Li-Yuan Xue, Bo Liu, Chong Li, University of Glasgow, UK

71 **Low-Power *Ka*- and *V*-Band Miller Compensated Amplifiers in 130-nm SiGe BiCMOS Technology**
Batuhan Sutbas[1], Herman Jalli Ng[2], Jan Wessel[1], Alexander Koelpin[3], Gerhard Kahmen[1]
[1]IHP, Germany; [2]Hochschule Karlsruhe, Germany; [3]Technische Universität Hamburg, Germany

EuMIC06: Integrated PAs for 5G, SATCOM and Vehicular Applications

Chair: TBA
Co-Chair: Alessandro Cidronali, Università di Firenze, Italy
14:20–16:00, Sunday 3rd April 2022, Room 10

(NA) **Buffer-Free GaN-on-SiC HEMT Heterostructures for Sub-6GHz and mmWave RF Devices**
Jr-Tai Chen, SweGaN, Sweden

76 **A High GBW High Power Wideband Power Amplifier for Automotive Radar Application**
Kambiz Hadipour, Dominik Amschl, Daniel Knauder, Stefano Di Martino, Infineon Technologies, Austria

80 **A Highly Rugged 39GHz 19.3dBm Power Amplifier for 5G Applications in 45nm SOI Technology**
Alice Bossuet[1], Baudouin Martineau[1], Cédric Dehos[1], Benjamin Blampey[1], Alexis Divay[1], Yvan Morandini[2]
[1]CEA-Leti, France; [2]Soitec, France

84 **44dBm Output Power and High Gain K-Band GaN Power Amplifier for Satellite Communication**
Takuma Torii, Yoshifumi Kawamura, Eigo Kuwata, Masaomi Tsuru, Mitsubishi Electric, Japan

88 **125W Solid State Power Amplifier for 17.3–20.2GHz SatCom Applications**
R. Giofrè[1], P. Colantonio[1], L. Cabria[2], M. Lopez[2]
[1]Università di Roma "Tor Vergata", Italy; [2]TTI Norte, Spain

EuMIC07: Frequency-Converting Circuits

Chair: Ingmar Kallfass, Universität Stuttgart, Germany
Co-Chair: Lars-Erik Wernersson, Lund University, Sweden
14:20–16:00, Sunday 3rd April 2022, Room 17

92   **A Ka-Band MMIC Single-Chip Frequency Converter for Telecom Satellite Applications**
Francesco Scappaviva[1], Davide Resca[1], Luca Cariani[1], Andrea Biondi[1], Francesco Vitulli[2], François Deborgies[3]
[1]MEC, Italy; [2]Thales, Italy; [3]ESA, The Netherlands

96   **A V-Band Low-Power and Compact Down-Conversion Mixer with Low LO Power in 130-nm SiGe BiCMOS Technology**
Batuhan Sutbas[1], Herman Jalli Ng[2], Jan Wessel[1], Alexander Koelpin[3], Gerhard Kahmen[1]
[1]IHP, Germany; [2]Hochschule Karlsruhe, Germany; [3]Technische Universität Hamburg, Germany

100   **A 60GHz Frequency Doubler with 3.4-dBm Output Power and 4.4% DC-to-RF-Efficiency in 130-nm SiGe BiCMOS**
Yu Zhu, Hatem Ghaleb, Vincent Rieß, Niko Joram, Frank Ellinger, Technische Universität Dresden, Germany

104   **A 14.6GHz – 19.2GHz Digitally Controlled Injection Locked Frequency Doubler in 45nm SOI CMOS**
Olli Kursu, Timo Rahkonen, Aarno Pärssinen, University of Oulu, Finland

108   **A W-Band Up-Conversion Mixer with Integrated LO Frequency Doublers in a 60nm GaN Technology**
Mingquan Bao[1], Robert Malmqvist[2], Rolf Jonsson[2], Jonas Hansryd[1], Kristoffer Andersson[1]
[1]Ericsson, Sweden; [2]FOI, Sweden

EuMIC08 : Components and Subsystems for 100GHz and Above

Chair: Ullrich Pfeiffer, Bergische Universität Wuppertal, Germany
Co-Chair: Herbert Zirath, Chalmers University of Technology, Sweden
16:40–18:20, Sunday 3rd April 2022, Room 8

(NA)   **Highly Integrated Multi-Channel D-Band Radar Transceivers in Silicon Technologies**
Vadim Issakov, Infineon Technologies, Germany

113   **SiGe BiCMOS Building Blocks for E- and D-Band Backhauling Front-Ends**
G. Amendola[1], L. Boccia[1], F. Centurelli[2], Pascal Chevalier[3], A. Fonte[4], S. Karman[5], S. Levantino[5], A. Mazzanti[6], C. Mustacchio[1], A. Pallotta[7], I. Petricli[8], C. Samori[5], F. Tesolin[5], Pasquale Tommasino[2], Antonio Traversa[4], Alessandro Trifiletti[2]
[1]Università della Calabria, Italy; [2]Università di Roma "La Sapienza", Italy; [3]STMicroelectronics, France; [4]SIAE MICROELETTRONICA, Italy; [5]Politecnico di Milano, Italy; [6]Università di Pavia, Italy; [7]STMicroelectronics, Italy; [8]Università di Pavia1, Italy

117   **A Superheterodyne 300GHz Transmit Receive Chipset for Beyond 5G Network Integration**
Iulia Dan[1], Christopher Grötsch[1], Laurenz John[2], Sandrine Wagner[2], Axel Tessmann[2], Ingmar Kallfass[1]
[1]Universität Stuttgart, Germany; [2]Fraunhofer IAF, Germany

121   **Implementation of Slow-Wave Thin-Film Microstrip Transmission Lines in a 35nm InGaAs Technology**
Athanasios Gatzastras[1], Hermann Massler[2], Arnulf Leuther[2], Sébastien Chartier[2], Ingmar Kallfass[1]
[1]Universität Stuttgart, Germany; [2]Fraunhofer IAF, Germany

125   **A 140GHz to 170GHz Active Tunable Noise Source Development in SiGe BiCMOS 55nm Technology**
Victor Fiorese[1], Joao Carlos Azevedo Goncalves[1], Simon Bouvot[1], Emmanuel Dubois[2], Christophe Gaquière[2], Guillaume Ducournau[2], François Danneville[2], Sylvie Lépilliet[2], Daniel Gloria[1]
[1]STMicroelectronics, France; [2]IEMN (UMR 8520), France

EuMIC09: High Performance LNAs

Chair: Lars-Erik Wernersson, Lund University, Sweden
Co-Chair: Ingmar Kallfass, Universität Stuttgart, Germany
16:40–18:20, Sunday 3rd April 2022, Room 9

129 **200GHz Low Noise Amplifiers in 250nm InP HBT Technology**
Utku Soylu[1], Ahmed S.H. Ahmed[1], Munkyo Seo[2], Ali Farid[1], Mark Rodwell[1]
[1]University of California at Santa Barbara, USA; [2]Sungkyunkwan University, Korea

133 **Output Power Limited Rugged GaN LNA MMIC**
Evelyne Kaule, Cristina Andrei, Matthias Rudolph, BTU, Germany

136 **A Highly Linear 79GHz Low-Noise Amplifier for Civil-Automotive Radars in 22nm FD-SOI CMOS with -6dBm iP$_{1dB}$ and 5dB NF**
Songhui Li[1], David Fritsche[1], Laszlo Szilagyi[1], Xin Xu[1], Quang Huy Le[2], Defu Wang[2], Thomas Kämpfe[2], Corrado Carta[1], Frank Ellinger[1]
[1]Technische Universität Dresden, Germany; [2]Fraunhofer IPMS, Germany

140 **Highly Linear D-Band Low-Noise Amplifier with 8.5dB Noise Figure in InP-DHBT Technology**
M. Hossain[1], Ralf Doerner[1], Hady Yacoub[1], Tom K. Johansen[2], Wolfgang Heinrich[1], Viktor Krozer[1]
[1]FBH, Germany; [2]Technical University of Denmark, Denmark

144 **C-Band Low-Noise Amplifier MMIC with an Average Noise Temperature of 44.5K and 24.8mW Power Consumption**
Felix Heinz, Fabian Thome, Arnulf Leuther, Oliver Ambacher, Fraunhofer IAF, Germany

EuMIC10: Advances in Si and GaN Based Integrated PAs

Chair: Khaled Elgaid, Cardiff University, UK
Co-Chair: Rocco Giofrè, Università di Roma "Tor Vergata", Italy
16:40–18:20, Sunday 3rd April 2022, Room 10

148 **Transient Field-Plate Thermometry Demonstrated on a 20-W X-Band GaN Power Amplifier**
Simon J. Mahon, Melissa C. Gorman, Michael C. Heimlich, Macquarie University, Australia

152 **A 27dBm Ku-Band SiGe Power Amplifier Working up to 90°C with High Robustness to the 2:1 SWR**
B. Coquillas[1], Eric Kerhervé[1], S. Redois[2], A.-C. Amiaud[2], L. Roussel[2], Bruno Louis[2], E. Itcia[2], T. Merlet[2]
[1]IMS (UMR 5218), France; [2]Thales, France

156 **A 4GBaud 5Vpp Pre-Driver for GaN Based Digital PAs in 22nm FDSOI Using LDMOS**
Frowin Buballa[1], Sebastian Linnhoff[1], Thomas Hoffmann[2], Andreas Wentzel[2], Wolfgang Heinrich[2], Friedel Gerfers[1]
[1]Technische Universität Berlin, Germany; [2]FBH, Germany

160 **400-Watt S-Band Power Amplifier MMIC**
A.P. de Hek, G. van der Bent, F.E. van Vliet, TNO, The Netherlands

164 **A 41.5dBm Broadband AlGaN/GaN HEMT Balanced Power Amplifier at K-Band**
S. Samis[1], C. Friesicke[2], T. Maier[2], Rüdiger Quay[2], Arne F. Jacob[1]
[1]Technische Universität Hamburg, Germany; [2]Fraunhofer IAF, Germany

EuMIC 2021 Table of Contents

EuMIC11: Broadband Integrated Circuits

Chair: Friedel Gerfers, Technische Universität Berlin, Germany
Co-Chair: Mehmet Karaaslan, Teledyne e2v, UK
16:40–18:20, Sunday 3rd April 2022, Room 17

(NA) **Is SiGe BiCMOS an Essential Technology for 6G?**
Pascal Chevalier, STMicroelectronics, France

169 **120GBd SiGe-Based 2:1 Analog Multiplexer Module for Ultra-Broadband Transmission Systems**
C. Schmidt[1], Tobias Tannert[2], J.H. Choi[1], C. Caspar[1], D. Pech[1], S. Wünsch[1], G. Ropers[1], J. Schostak[1], V. Jungnickel[1], R. Freund[1], Markus Grözing[2], Manfred Berroth[2]
[1]Fraunhofer HHI, Germany; [2]Universität Stuttgart, Germany

173 **A 7–30GHz, 80-dBΩ Noise-Optimized, Bandpass-Like TIA in 130nm SiGe BiCMOS Technology for Quasi-Coherent Optical Receivers**
Tom K. Johansen[1], Guillermo Silva Valdecasa[1], Monika Kupska[2], Jose A. Altabas[2], Omar Gallardo[2], Michele Squartecchia[2], Jesper B. Jensen[2]
[1]Technical University of Denmark, Denmark; [2]Bifrost Communications, Denmark

177 **Multi-Phase Clock Path Circuit up to 57GHz Including 5 Bit Programmable Phase Interpolators for Time-Interleaved Broadband Data Converters in a 28nm FD-SOI CMOS Technology**
Daniel Widmann, Tobias Tannert, Xuan-Quang Du, Markus Grözing, Manfred Berroth, Universität Stuttgart, Germany

181 **A DC to 20GHz Variable Gain Amplifier with Tunable Input Matching in 22nm FDSOI Technology**
Seyyedmohsen Seyyedrezaei, Manu Viswambharan Thayyil, Corrado Carta, Frank Ellinger, Technische Universität Dresden, Germany

EuMIC12: Device Modelling and Simulation of Parasitic Phenomena

Chair: Raphaël Sommet, XLIM (UMR 7252), France
Co-Chair: Valeria Brunel, United Monolithic Semiconductors, France
09:00–10:40, Monday 4th April 2022, Room 13

185 **Noise Modeling of GaN/AlN HEMT**
Sanaul Haque[1], Frank Schnieder[2], Oliver Hilt[2], Ralf Doerner[2], Frank Brunner[2], Matthias Rudolph[2]
[1]BTU, Germany; [2]FBH, Germany

189 **Efficient TCAD Temperature-Dependent Large-Signal Simulation of a FinFET Power Amplifier**
E. Catoggio, S. Donati Guerrieri, F. Bonani, G. Ghione, Politecnico di Torino, Italy

(NA) **A TCAD Simulation Study on Gated-Anode Diodes for Microwave Applications**
Arijit Bose[1], Debaleen Biswas[1], Qiang Ma[1], Yoichi Tsuchiya[1], Hidemasa Takahashi[2], Yuji Ando[2], Akio Wakejima[1]
[1]NITech, Japan; [2]Nagoya University, Japan

197 **Trap Characterization in InAlN/GaN and AlN/GaN Based HEMTs with Fe- and C-Doped Buffers**
Emmanuel Dupouy, P. Vigneshwara Raja, Florent Gaillard, Raphaël Sommet, Jean-Christophe Nallatamby, XLIM (UMR 7252), France

201 **Mechanisms of Buffer and Surface Traps in GaN HEMTs for Low Frequency Y21 and Y22 Parameters**
Tomohiro Otsuka[1], Yutaro Yamaguchi[1], Masaomi Tsuru[1], Toshiyuki Oishi[2]
[1]Mitsubishi Electric, Japan; [2]Saga University, Japan

EuMIC13: Receiver Components

Chair: Friedel Gerfers, Technische Universität Berlin, Germany
Co-Chair: Lars-Erik Wernersson, Lund University, Sweden
09:00–10:40, Monday 4th April 2022, Room 14

205   **A Low Phase Noise Phase-Locked Loop with Short Settling Times for Automotive Radar**
Tobias T. Braun[1], Marcel van Delden[1], Christian Bredendiek[2], Jan Schoepfel[1], Nils Pohl[1]
[1]Ruhr-Universität Bochum, Germany; [2]Fraunhofer FHR, Germany

209   **A Passively-Coupled 39.5GHz Colpitts Quadrature VCO in SiGe HBT Technology**
Janis Wörmann[1], Aleksey Dyskin[2], Sébastien Chartier[3], Ingmar Kallfass[1]
[1]Universität Stuttgart, Germany; [2]Technion, Israel; [3]Fraunhofer IAF, Germany

213   **30–46GHz 1.5dB IL Negative Gate Control SPDT with 24.5dBm IP1 in 130nm CMOS**
Sumeet Londhe[1], Noam Bar-Helmer[2], Samuel Jameson[2], Eran Socher[1]
[1]Tel Aviv University, Israel; [2]Rafael, Israel

217   **A Highly Linear SiGe BiCMOS Gilbert-Cell Based Downconversion Mixer for 5G Applications**
Mir Hassan Mahmud, Abdurrahman Burak, Can Çalışkan, Tahsin Alper Ozkan, Ali Bahadir Ozdol, Melik Yazici, Yasar Gurbuz, Sabanci University, Turkey

221   **37.2-to-42.0GHz VCO with -93.4dBc/Hz Phase Noise for FMCW Radar in 22nm FDSOI**
Laszlo Szilagyi[1], Songhui Li[2], Xin Xu[2], Paolo Valerio Testa[1], Andres Seidel[2], Corrado Carta[2], Frank Ellinger[2]
[1]GlobalFoundries, Germany; [2]Technische Universität Dresden, Germany

EuMIC14: Advances in mm-Wave and High Power Integrated PA Technologies

Chair: Franco Giannini, Università di Roma "Tor Vergata", Italy
Co-Chair: Simon J. Mahon, Macquarie University, Australia
09:00–10:40, Monday 4th April 2022, Room 17

225   **A 100GHz Class-F-Like InP-DHBT PA with 25.4% PAE**
Amit Shrestha[1], Ralf Doerner[1], Hady Yacoub[1], Tom K. Johansen[2], Wolfgang Heinrich[1], Viktor Krozer[1], Matthias Rudolph[1], Andreas Wentzel[1]
[1]FBH, Germany; [2]Technical University of Denmark, Denmark

229   **A 117.5–130GHz 22.1dBm 11.5% PAE DAT Based Power Amplifier in InP 130nm HBT Technology**
Linsheng Zhang, Vinay Iyer, Jay Sheth, Linli Xie, Robert M. Weikle, Steven M. Bowers, University of Virginia, USA

233   **A 47–50GHz 3W MMIC Power Amplifier Using 100nm Gallium Nitride Technology**
Seifeddine Fakhfakh[1], Guillaume Callet[1], Estelle Byk[1], Laurent Favede[1], Aleksandra Malko[2], Sandra Riedmueller[2], Pierre Denis[2], Hervé Blanck[2], Marc Camiade[1]
[1]UMS, France; [2]UMS, Germany

237   **A D-Band Power Amplifier with 12dBm P1dB, 10% Power Added Efficiency in InP-DHBT Technology**
M. Hossain[1], T. Shivan[1], Ralf Doerner[1], S. Seifert[1], Hady Yacoub[1], Tom K. Johansen[2], Wolfgang Heinrich[1], Viktor Krozer[1]
[1]FBH, Germany; [2]Technical University of Denmark, Denmark

241   **A 28-GHz-Band GaN HEMT MMIC Doherty Power Amplifier Designed by Load Resistance Division Adjustment**
Ryo Ishikawa, Takuya Seshimo, Yoichiro Takayama, Kazuhiko Honjo, University of Electro-Communications, Japan

EuMIC15: EuMIC Posters

Chair: Mustafa Bakr, University of Oxford, UK
10:40–13:30, Monday 4th April 2022, Exhibition Hall

245   **Field-Plate Mixer**
Simon J. Mahon, Michael C. Heimlich, Macquarie University, Australia

249   **17.6dB Variable-Gain and Variable-Bandwidth Upconverter in 65nm CMOS for 60GHz Bands**
Oner Hanay, David Bierbuesse, Renato Negra, RWTH Aachen University, Germany

253   **A Derating-Rules Compliant Ka-Band GaN-on-Si Power Amplifier Designed for Highly Reliable Satellite Applications**
F. Costanzo[1], L. Pace[1], P.E. Longhi[1], W. Ciccognani[1], S. Colangeli[1], Rémy Leblanc[2], E. Limiti[1]
[1]Università di Roma "Tor Vergata", Italy; [2]OMMIC, France

257   **A 3.3 to 11.3GHz Differential LNA with Slight Imbalance Active Balun in 0.15-μm GaAs pHEMT Process for Radio Astronomical Receiver**
Ting-Hsuan Fan[1], Chau-Ching Chiong[2], Huei Wang[1]
[1]National Taiwan University, Taiwan; [2]Academia Sinica, Taiwan

261   **Benefits of AlGaN/GaN Thermal ROM Coupling with Industrial Non-Linear Transistor Model**
C. Chang, Laurent Brunel, UMS, France

(NA)   **1–6GHz 35W Balanced GaN-HEMT Power Amplifier with Innovative Quadrature Couplers**
Alexey Radchenko, Sergey Garmash, Andrei Kishchinsky, Microwave Systems, Russia

269   **A 300GHz Frequency Doubler in Transferred Substrate InP DHBT Technology**
Arsen Turhaner[1], M. Hossain[2], Mohamed Brahem[2], Tom K. Johansen[1]
[1]Technical University of Denmark, Denmark; [2]FBH, Germany

273   **55% Fractional-Bandwidth Doherty Power Amplifier in 130-nm SiGe for 5G mm-Wave Applications**
Aniello Franzese[1], Nebojsa Maletic[1], Mohamed Eissa[1], Muh-Dey Wei[2], Renato Negra[2], Andrea Malignaggi[1]
[1]IHP, Germany; [2]RWTH Aachen University, Germany

277   **Full Octave Continuously Tunable SiGe Bipolar LC-VCO in Ku-Band**
Christian Bredendiek[1], Klaus Aufinger[2], Nils Pohl[3]
[1]Fraunhofer FHR, Germany; [2]Infineon Technologies, Germany; [3]Ruhr-Universität Bochum, Germany

281   **An E-Band Bidirectional PALNA in 0.13μm SiGe BiCMOS Technology**
R. Ahamed[1], Mikko Varonen[2], Dristy Parveg[2], M. Najmussadat[1], Mikko Kantanen[2], Y. Tawfik[1], K.A.I. Halonen[1]
[1]Aalto University, Finland; [2]VTT, Finland

285   **A Ka-Band 40W Output Power and 30% PAE GaN MMIC Power Amplifier for Satellite Communication**
Keigo Nakatani, Yutaro Yamaguchi, Masaomi Tsuru, Mitsubishi Electric, Japan

289   **Probabilistic Poly Harmonic Distortion Model**
Anna Davis Manjaly, Justin King, Trinity College Dublin, Ireland

EuMIC16: Phased Array Components from S-Band up to 300GHz

Chair: Frank E. van Vliet, TNO, The Netherlands
Co-Chair: Michael Schlechtweg, Fraunhofer IAF, Germany
14:20–16:00, Monday 4th April 2022, Room 4

293   **An S-Band 34dBm Stacked-HBT Phase Driver in 0.25μm BiCMOS Technology for GaN-Based Phased-Array Radar Transmit Chain**
J. Essing, Alice Bossuet, R. Knight, A.P. de Hek, F.E. van Vliet, TNO, The Netherlands

297   **A Phase Coherent DC–25GHz 6-Bit SiGe BiCMOS Step Attenuator with IP$_{1dB}$ >20dBm**
Hamza Kandis, Abdurrahman Burak, Cengizhan Kana, Melik Yazici, Yasar Gurbuz, Sabanci University, Turkey

301   **A 26GHz to 34GHz Active Phase Shifter with Tunable Polyphase Filter for 5G Wireless Systems**
Alok Sethi, Rehman Akbar, Mikko Hietanen, Timo Rahkonen, Aarno Pärssinen, University of Oulu, Finland

305   **A 25–50GHz Digitally Controlled Phase-Shifter**
Steeven Voisin[1], Vincent Knopik[2], Eric Kerhervé[1]
[1]IMS (UMR 5218), France; [2]STMicroelectronics, France

309   **A 270–330GHz Vector Modulator Phase Shifter in 130nm SiGe BiCMOS**
Mohammad Hassan Montaseri, Sumit Pratap Singh, Markku Jokinen, Timo Rahkonen, Marko E. Leinonen, Aarno Pärssinen, University of Oulu, Finland

EuMIC17: EuMIC Closing Session

Chair: John Christopher Clifton, EuMIC 2021 Chair
Co-Chairs: Shokrollah Karimian, EuMIC 2021 Co-Chair and Edward Wasige, EuMIC 2021 TPC Chair
16:40–18:20, Monday 4th April 2022, Room 8-11

(NA)   **Awards Ceremony**
Kamal K. Samanta, EuMW 2021 Awards Chair

(NA)   **Efficient, Broadband and Linear Radio Frequency Amplifier Architectures**
Kevin Morris, University of Bristol, UK

(NA)   **6G — Known Technologies with a Twist or Maybe Not?**
Nadine Collaert, imec, Belgium

(NA)   **Closing Remarks and Invitation to EuMIC 2022**
John Christopher Clifton, EuMIC 2021 Chair

EuMIC 2021 Table of Contents

EuMIC/EuMC01: Novel Filtering Devices in Integrated Technologies

Chair: Roberto Gómez-García, Universidad de Alcalá, Spain
Co-Chair: Michael Höft, CAU, Germany
09:00–10:40, Monday 4th April 2022, Room 1

(NA) **A Millimeter-Wave Substrate Integrated Waveguide Filter in Si-BCB Technology**
Jordan Corsi[1], Giuseppe Acri[1], Maxime Moulin[1], Nicolas Zerounian[2], Anne-Sophie Grimault-Jacquin[2], Loïc Vincent[3], Guillaume Ducournau[4], Frédéric Aniel[2], Florence Podevin[1], Philippe Ferrari[1], Emmanuel Pistono[1]
[1]RFIC-Lab (EA 7520), France; [2]C2N (UMR 9001), France; [3]CIME Nanotech, France; [4]IEMN (UMR 8520), France

(NA) **A 100GHz Bandpass Filter Employing Shielded Folded Ridged Quarter-Mode SIW Resonator in CMOS Technology**
Baichuan Chen, Samundra K. Thapa, Adel Barakat, Ramesh Pokharel, Kyushu University, Japan

(NA) **SAW Resonator Band-Pass Filter on GaN/Si Operating at 8GHz**
Alina-Cristina Bunea[1], Dan Neculoiu[2], Adrian Dinescu[1]
[1]IMT Bucharest, Romania; [2]UPB, Romania

(NA) **Engineered High Resistivity Silicon Substrates in IPD Technology Used for Miniaturized Sub-6GHz Filters**
Atte Haapalinna[1], Heikki Holmberg[1], Arto Hujanen[2], Katja Parkkinen[1], Pekka Rantakari[2], Jan Saijets[2], Tauno Vähä-Heikkilä[2]
[1]Okmetic, Finland; [2]VTT, Finland

(NA) **Glass-Integrated Single- and Dual-Band Bandpass Filters**
Andrea Ashley, Dimitra Psychogiou, University of Colorado Boulder, USA

EuMIC/EuMC02: THz components

Chair: Emma MacPherson, University of Warwick, UK
Co-Chair: Oleksiy Sydoruk, Imperial College London, UK
09:00–10:40, Monday 4th April 2022, Room 4

(NA) **A SiGe Based 0.48THz Signal Source with 45GHz Tuning Range**
Jonathan Wittemeier[1], Florian Vogelsang[1], David Starke[1], Holger Rücker[2], Nils Pohl[1]
[1]Ruhr-Universität Bochum, Germany; [2]IHP, Germany

(NA) **The Effect of Surface Passivation for Sub-THz Silicon Gradient Refractive Index Lens**
Antti Lamminen[1], Aleksi Tamminen[2], Jaakko Saarilahti[1], Vladimir Ermolov[1], Pekka Pursula[1]
[1]VTT, Finland; [2]Aalto University, Finland

341 **Optoelectronic Millimeter-Wave Integrated Circuits Fabricated in Pure Silicon-Based Technologies**
Uroschanit Yodprasit, Wolfgang Winkler, Silicon Radar, Germany

(NA) **140GHz Differential Antennas in Embedded Wafer Level Ball Grid Array Technology**
Akanksha Bhutani, Elizabeth Bekker, Teng Li, Lucas Giroto de Oliveira, Thomas Zwick, KIT, Germany

(NA) **Enhancing mmWave On-Chip-Antennas Using In-Package Electromagnetic Bandgap Structures**
Dmitrii Kruglov, Oleg Iupikov, Marianna V. Ivashina, Rob Maaskant, Chalmers University of Technology, Sweden

EuMIC/EuMC03 : MMIC Power Amplifiers and Supply Modulation

Chair: Jeff Powell, Teratech Components, UK
Co-Chair: Markus Mayer, ARELIS, France
14:20–16:00, Monday 4th April 2022, Room 14

(NA) **A 6–18GHz 13W and 22% PAE GaN Power Chipset**
Mehdi Dinari, Benoît Mallet-Guy, Yves Mancuso, Thales, France

(NA) **On-Chip Power Combining with 3-Stage 75–110GHz GaN MMIC Power Amplifiers**
Shane Verploegh, Timothy Sonnenberg, Mauricio Pinto, Akim Babenko, Zoya Popović,
University of Colorado Boulder, USA

(NA) **Wideband Phase Modulator MMIC for K-Band Supply-Modulated Power Amplifier**
Linearization
Gregor Lasser[1], Connor Nogales[1], Maxwell R. Duffy[2], Zoya Popović[1]
[1]University of Colorado Boulder, USA; [2]Northrop Grumman, USA

(NA) **Compact Design of a L-Band 40W 40MHz Envelope Tracking GaN Power Amplifier for**
Small Cells
Olivier Nonet[1], Wilfried Demenitroux[1], Frederic Ploneis[1], Denis Barataud[2],
Michel Campovecchio[2]
[1]Thales, France; [2]XLIM (UMR 7252), France

(NA) **A 600-W Enhancement-Mode GaN Multi-Level Dynamic Converter for Supply**
Modulated PAs
Connor Nogales, Zoya Popović, Gregor Lasser, University of Colorado Boulder, USA

EuMIC/EuMC04 : EuMIC/EuMC Posters

Chair: Mustafa Bakr, University of Oxford, UK
13:50–16:40, Monday 4th April 2022, Exhibition Hall

370 **Microwave Sensing Using Metal-Insulator-Metal Diodes Based on 4-nm-Thick Hafnium**
Oxide
Martino Aldrigo[1], Mircea Dragoman[1], Sergiu Iordanescu[1], Mazen Al Shanawani[2],
George Deligeorgis[3]
[1]IMT Bucharest, Romania; [2]Università di Bologna, Italy; [3]FORTH, Greece

374 **Automatic Nonlinear Nonquasi-Static Diode Model Extraction from Large-Signal**
Measurements
A. García-Luque[1], Teresa M. Martín-Guerrero[1], Alberto Santarelli[2],
Carlos Camacho-Peñalosa[1]
[1]Universidad de Málaga, Spain; [2]Università di Bologna, Italy

378 **Compact GaN RF-Switches for Power Applications**
Samira Driad[1], Charles Teyssandier[1], Laurent Caille[1], C. Chang[1], Laurent Brunel[1],
Benoit Lambert[1], Hermann Stieglauer[2], Valeria Brunel[1]
[1]UMS, France; [2]UMS, Germany

382 **Analysis of RF Stress Influence on Large-Signal Performance of 22nm FDSOI CMOS**
Transistors Utilizing Waveform Measurement
Dang Khoa Huynh[1], Quang Huy Le[1], Steffen Lehmann[2], Zhixing Zhao[2],
Germain Bossu[2], Wafa Arfaoui[2], Defu Wang[1], Thomas Kämpfe[1], Matthias Rudolph[3]
[1]Fraunhofer IPMS, Germany; [2]GlobalFoundries, Germany; [3]BTU, Germany

EuMIC/EuMC04 continues next page...

EuMIC/EuMC04 continued...

386 **Towards an Excitable Microwave Spike Generator for Future Neuromorphic Computing**
Qusay Al-Taai[1], Razvan Morariu[1], Jue Wang[1], Abdullah Al-Khalidi[1], Ali Al-Moathin[1], Bruno Romeira[2], José Figueiredo[3], Edward Wasige[1]
[1]University of Glasgow, UK; [2]INL, Portugal; [3]Universidade de Lisboa, Portugal

390 **Numerical and Experimental Investigations of Self-Mixing Effect of a Planar Gunn Diode Oscillator**
Ming Yan Zhong, David R.S. Cumming, Chong Li, University of Glasgow, UK

(NA) **An Ultra-Wideband Microstrip-to-WR15 Waveguide Transition for MMIC Applications**
Bent Walther, Marcel van Delden, Thomas Musch, Ruhr-Universität Bochum, Germany

(NA) **An Integrated Multiphysics Model for Phase-Change Material Switches**
Ines Bettoumi[1], Kateryna Kiryukhina[1], Olivier Puig[1], Pierre Blondy[2]
[1]CNES, France; [2]XLIM (UMR 7252), France

(NA) **Doherty Load Modulation Based on Non-Reciprocity**
Paul Saad[1], Han Zhou[2], Jose-Ramon Perez-Cisneros[2], Rui Hou[1], Christian Fager[2], Bo Berglund[1]
[1]Ericsson, Sweden; [2]Chalmers University of Technology, Sweden

(NA) **Adopting Supercapacitors in a Single-Stage Marx-Type Multilevel Supply Modulator**
Lukas Hüssen, Renato Negra, RWTH Aachen University, Germany

(NA) **A 30-W GaN Quasi-MMIC Doherty Power Amplifier Based on All-Distributed Inductors Load Network**
Rui-Jia Liu[1], Xiao-Wei Zhu[1], Jing Xia[2], Peng Chen[1], Chao Yu[1], Lv Zhang[3], Zhi-Yong Chen[1]
[1]Southeast University, China; [2]Jiangsu University, China; [3]Guobo Electronics, China

(NA) **A Digital Power Amplifier for 32-QAM**
Gavin T. Watkins, Toshiba, UK

(NA) **Effect of Switch Figure of Merit on Frequency-Reconfigurable Power Amplifier Performance**
Adam Der, William Sear, Taylor Barton, University of Colorado Boulder, USA

(NA) **Practical Work for Master2 Students: MMIC Distributed Amplifier Design for High Data Rate Receiver on GaAs-UMS Technology**
C. Algani[1], E. Leclerc[2]
[1]ESYCOM (UMR 9007), France; [2]UMS, France

Hub Page

2021 16th European Microwave Integrated Circuits Conference

3-4 April 2022 ■ London, UK

The European Microwave Week 2021 organisers would like to thank the following companies for their help and valued support throughout this year's event.

PLATINUM SPONSOR

SESSION LIST

EuMIC01	Large Signal and Non-Linear Characterization Techniques
EuMIC02	Silicon Based RF Solutions
EuMIC03	Transceiver MMICs
EuMIC04	EuMIC Opening Session
EuMIC05	Integrated Circuit Modelling and Design Methodology
EuMIC06	Integrated PAs for 5G, SATCOM and Vehicular Applications
EuMIC07	Frequency-Converting Circuits
EuMIC08	Components and Subsystems for 100GHz and Above
EuMIC09	High Performance LNAs
EuMIC10	Advances in Si and GaN Based Integrated PAs
EuMIC11	Broadband Integrated Circuits
EuMIC12	Device Modelling and Simulation of Parasitic Phenomena
EuMIC13	Receiver Components
EuMIC14	Advances in mm-Wave and High Power Integrated PA Technologies
EuMIC15	EuMIC Posters
EuMIC16	Phased Array Components from S-Band up to 300GHz
EuMIC17	EuMIC Closing Session
EuMIC/EuMC01	Novel Filtering Devices in Integrated Technologies
EuMIC/EuMC02	THz components
EuMIC/EuMC03	MMIC Power Amplifiers and Supply Modulation
EuMIC/EuMC04	EuMIC/EuMC Posters

WELCOME MESSAGES

Welcome to the 24ᵗʰ European Microwave Week

It is with great pleasure that we welcome you to the 24ᵗʰ European Microwave Week (EuMW), which is taking place at ExCeL in London, UK. At the time of writing this message, the world is still in the grip of the coronavirus pandemic. The pandemic has greatly affected the way we live our lives for almost two years now. Many people around the world have lost their lives and many have had their lives changed permanently by the pandemic. One impact has been how we, the human race, interact with each other. The human instinct is usually to come together to help deal with problems, to form strategies and build partnerships, and, to celebrate successes.

EuMW is one such event that is motivated by these instincts. It is for this reason the organising team for this year's event has worked long and hard to ensure we have an event where we can come together and meet, face to face, as a community to continue to develop and celebrate our area of science, engineering and technology. We feel that it is vital to achieve this goal. This is the reason why EuMW 2021 is taking place during April 2022 – we have delayed hosting the event so that it is more feasible to hold a successful in-person event whilst still respecting any national and international restrictions on social interactions and travel.

This is the third time that EuMW has been hosted in London, following on from previous highly successful events in 2001 and 2016. London is a natural venue for prestigious scientific events, being the home of such long-standing scientific institutions as the Royal Society (founded in 1660), the University of London (founded in 1836) and the Institution of Electrical Engineers (founded in 1871), as well as the home to many famous scientists, including James Clerk Maxwell, Lord Rayleigh, Charles Wheatstone, Alan Turing, etc. Our moto for this year's EuMW is 'United in Microwaves'. This reflects the traditional feeling of unity in our community, and, demonstrates how we can use this conference to re-establish and further develop this feeling of unity within our community of colleagues and fellow professionals, despite the recent problems caused by the pandemic.

EuMW 2021 continues the annual series of highly successful microwave events that started back in 1998. EuMW 2021 comprises three co-located conferences: European Microwave Conference (EuMC); European Microwave Integrated Circuits Conference (EuMIC); European Radar Conference (EuRAD). There are also many workshops and short courses associated with each of these conferences, along with several Special and Focused Sessions. Two particular highlights are Special Sessions on

the life and works of two prominent members for our community who sadly passed during 2020; Professor Peter Clarricoats and Professor Roberto Sorrentino. Peter Clarricoats was Chair of the first European Microwave Conference (EuMC), held in London in 1969, and Chair of the 9ᵗʰ EuMC in the UK in 1979; he received a EuMA Distinguished Service award in 2005. Roberto Sorrentino was a founder member of the European Microwave Association and President of EuMA from 1998 to 2009. They will both be sadly missed.

In addition, there are three Forums, covering: Defence, Security and Space; Automotive; and, Beyond 5G technologies. There is also a very large trade show – the largest RF and microwave trade show in Europe – where the leading companies from our industry exhibit their very latest technological developments. EuMW 2021 also has several activities aimed specifically at students. These include: the Tom Brazil Doctoral School of Microwaves; the European Microwave Training School; the Career Platform; and, IEEE Young Professionals. There is also the Women in Microwave event, in which both women, and men, are encouraged to participate.

We sincerely hope that you will enjoy a memorable experience in London at EuMW 2021.

NICK RIDLER
EuMW General Chair
National Physical Laboratory, UK

JOHN CUNNINGHAM
EuMW General Co-chair
University of Leeds, UK

Welcome from the President of the European Microwave Association

On behalf of the European Microwave Association (EuMA), I warmly welcome you to the 24th edition of the European Microwave Week in London! EuMA stands up for our microwave and RF community. We foster networking between scientists, engineers, decision makers and end-users. The European Microwave Week (EuMW) is our main asset and a real networking event. It's the place to get information you can't get anywhere else and to meet colleagues you don't see every day.

EuMA is continuously improving itself to support our microwave community. We recently released a 22 pages White Paper "For a Strong & Competitive European Wireless Technologies Ecosystem". A free download is available at our website www.eumwa.org. Soon EuMA will announce a new series of webinars with interesting and qualified speakers.

EuMA is very active on social media. Follow us @eumassociation on Facebook, LinkedIn, Twitter, YouTube and Instagram and discover our latest posts.

EuMA actively supports young researchers. Thereto our Innovation Team is launching the fourth edition of the *EuMA Internship Award*. Each year, up to seven prizes of 4.500 € each are awarded to selected master and PhD students to spend a period of at least 3 months abroad in one of the leading European microwave industries or institutes. Details are at our website. EuMA continues to provide grants and reduced registration fees to students and delegates from NIS countries to attend the EuMW.

EuMA offers a membership to all working in the field of microwaves. Members enjoy reduced fees for attending EuMW and EuMA-sponsored events as well as the IEEE IMS and the APMC. EuMA members have free access to our archive of publications and the on-line version of the International Journal on Microwave and Wireless Technologies.

EuMW is the premier microwave conference and exhibition event in Europe. We value the cooperation with IEEE Societies MTT, AP and ED and the GAAS° Association and our long-standing partner Horizon House / Microwave Journal as event organiser.

Preparing and hosting the EuMW is a major effort, from paper submission and review to on-site organisation. This is accomplished by a team of volunteers year by year. My special and sincere thanks go to Nick RIDLER and John CUNNINGHAM the 2021 General Chair and Co-chair; to Peter GARDNER, General TPC Chair; to Adrian CROSS, Treasurer; to Emma MCPHERSON and Yi WANG, EuMC Chair and TPC Chair; to Chris CLIFTON and Edward WASIGE, EuMIC Chair and TPC Chair; and to James WATTS and Matt RITCHIE, EuRAD Chair and TPC Chair – just to name a few on behalf of the entire team. Thank you!

The European Microwave Week is back again in UK after the successful events in 2001, 2006, 2011 and in 2016. All members of the team have been working hard to set up an outstanding technical and scientific programme and I am sure they will make your stay in London exciting, enjoyable, and a rewarding experience of Britain's hospitality. I would like to cordially invite you to EuMW 2021. Come to the wonderful city of London. Join us at EuMW2021 and discover information you won't get anywhere else. Take the opportunity to meet and talk to colleagues and friends from all over the world you don't see every day. I hope to see you in London! And most of all: Get involved in our community!

FRANK
VAN DEN BOGAART

President
European Microwave Association

EuMA is now very active on various social media. Follow us @eumassociation on Facebook, LinkedIn, Twitter and Instagram.

Welcome to the 16th European Microwave Integrated Circuits Conference

It is a great pleasure for us to welcome you to London for the 16th European Microwave Integrated Circuits (EuMIC) Conference which has been jointly organised by the GAAS® Association and EuMA since 2006. For the second year, the conference will not happen in Autumn but will be held in the Spring of 2022, on Sunday 3rd and Monday 4th April 2022.

EuMIC is the premier European technical conference for RF & microwave microelectronics as part of the European Microwave Week (EuMW).

The aim of the conference is to promote the discussion of recent developments and trends and to encourage the exchange of scientific and technical information covering a broad range of microwave, mmWave, terahertz and related topics, from materials and technologies to integrated circuits and applications that will be addressed in all of their aspects: theory, simulation, design and measurement.

Sunday is a busy day with a large offering: beyond the Opening Session, there will be twelve regular sessions as well as the traditional Foundry Session. This lively Foundry session brings together key representatives of the RF and microwave semiconductor foundries and will run together with the EuMIC Cocktail Reception, once again kindly sponsored by the GAAS® Association, to conclude the day. On Monday, there will be two regular sessions, two joint sessions with EuMC, two poster sessions, one jointly with EuMC, and the Closing Session. A number of the regular sessions will feature keynote industry talks on topical themes.

The EuMIC Opening Session will feature two keynote addresses by eminent speakers. Sir Christopher Snowden, Fellow of the Royal Society and Chair of the ERA Foundation, will speak on "III-V Nitride Semiconductors for Microwave Applications", while Zoya Popovic, Distinguished Professor, Department of Electrical, Computer and Energy Engineering at the University of Colorado, Boulder, USA, will speak on "High-Efficiency PAs for Broadband High-PAR Signals".

This year, the EuMIC Closing Session will start with the celebration of our best contributors. The EuMIC Prize for the best paper and the EuMIC Young Engineer Prize will be awarded by the EuMIC Prize Committee. For the next three years, the traditional GAAS® Association Fellowship Award will be replaced with the Tom Brazil Fellowship Award (by the GAAS® Association) in dedication to a friend and colleague who made such significant contributions to our microwave community. This award will focus on promoting and encouraging the achievements of research students and further announcements on the details will be made prior to the conference. This session will be concluded by two keynote presentations, one by Dr. Ebrahim Bushehri, CEO and Founder of Lime Microsystems (UK) on "Flexible and Open Source: The brave new world of Software Defined Radio and Open RAN", followed by Dr. Nadine Collaert, Program Director at imec on "6G – Known Technologies with a Twist or Maybe Not?"

We take this opportunity to show our appreciation to our authors for their technical contributions and for choosing to disseminate their work at EuMW and the dedication of the reviewers and TPC members who have spent their free time making the selections in order to provide the best possible programme. Workshops and Short Courses are a major offering of the EuMW and so we would also like to thank the organisers for gathering key experts to cover the latest developments. We also wish to acknowledge the support of the previous EuMIC teams, in particular Utrecht, who were always ready to advise. Finally, we would like to thank the 2021 EuMC and EuRAD teams for sharing experiences as well as to all our colleagues working in the background supporting EuMW as a whole.

We look forward to welcoming you personally in London for an exciting EuMIC!

CHRIS CLIFTON
EuMIC Chair
Sony Europe B.V.

SHOKROLLAH KARIMIAN
EuMIC Co-Chair
University of Oxford, UK

EDWARD WASIGE
EuMIC TPC Chair
University of Glasgow, UK

Welcome from the General TPC Chairs

I am delighted that we have been able to host the European Microwave Week in London again. Although much has changed since EuMW 2016, when we were last here, the UK capital city continues to be one of the great global centres for culture, the arts, sport, entertainment, shopping and tourism. I look forward to seeing you during EuMW 2021 as you enjoy our excellent conference programmes, and I hope you will be able to experience some of the other opportunities that this city offers while you are here.

As General Technical Programme Committee Chair, I would like to thank the many people who have worked through very difficult, challenging and frequently changing circumstances to generate what I am sure you agree is an excellent set of conference programmes.

First of all, of course, the paper authors and presenters. Excellent research and development work has continued in our widespread scientific and technical community despite the difficulties and complications caused by the pandemic lockdowns and other restrictions, and this has been reflected in a very high quality set of submitted papers. I must also thank the authors for their patience as circumstances forced us to change submission deadlines several times and postpone the conference.

We now share a distinction with the 2020 Olympics and the European Soccer championships, in holding our event during the year after that which appears in the name!

I also owe huge thanks to the excellent group of over 500 expert reviewers who scrutinised and provided constructive critiques on over 700 submitted papers to enable the Technical Programme Committee to select the best of them for our conferences. The delayed timing of the review period made this task more challenging this year, and I am extremely grateful to all those who fitted in their review tasks during or around their well-earned vacation periods.

For the second time, the Technical Programme Committee meeting has been conducted as a distributed virtual event, because of pandemic related travel constraints. I owe huge gratitude to the EuMW 2020 team for establishing the processes that made it possible to do this efficiently and effectively, and to the EuMW 2021 Operations Team and the EuMA Software Officers who made it work so well again this time. Of course, I would also like to acknowledge the excellent work done by the TPC chairs of the three individual conferences and our Technical Programme Committee, over 100 highly experienced experts in their fields. In their 26 sub-committees, they considered

all of the reviewed papers and selected the best, with an overall acceptance rate of approximately 65 %, and they formed them into the coherent and attractive set of 84 sessions on key topics in our discipline that make up our three conferences. Several of the sessions also feature invited talks from industry experts, to highlight the industrial context of those key topics. The programmes for the week as usual feature a set of specialist workshop and short courses.

Ours is a dynamic and constantly evolving discipline. As always, the EuMW programmes cover the important and fast developing themes, including: new applications and new passive and active technologies for the high mmWave and low THz bands; advanced manufacturing processes creating new possibilities in component design and system integration; artificial intelligence (AI), both as a tool in design and fabrication and as a new paradigm in signal processing; new applications for radar in many different aspects of our lives; and the many technologies and applications associated with communications systems, including 5G and beyond.

It has been a pleasure and a privilege to serve as General TPC Chair for EuMW 2021. I look forward to seeing you in London in April 2022.

PETER GARDNER
EuMW General TPC Chair
University of Birmingham, UK

DJURADJ BUDIMIR
EuMW General TPC Co-chair
University of Westminster, UK

International Journal of Microwave and Wireless Technologies: EuMW 2021 Special Issue

The International Journal of Microwave and Wireless Technologies was created in 2009 by the European Microwave Association (EuMA) and Cambridge University Press for the benefit of the microwave research community in Europe and overseas.

The journal is published ten times a year. It allows academic and industrial researchers to promote their work and stay connected with the most recent developments in microwave and RF technology. The journal is referenced in databases such as Scopus and Google Scholar and is indexed in the Thomson Reuters Web of Science. Following the success of previous microwave weeks, the journal will again publish a special issue dedicated to European Microwave Week 2021.

The authors of several highly ranked papers presented at the conferences will be invited to submit an extended version for publication in the journal. The special issue will be guest edited by Yi Wang, TPC chair of EuMC 2021, Edward Wasige, TPC chair of EuMIC 2021, and Matthew Ritchie, TPC chair of EuRAD 2021.

Accepted papers will be published online at http://journals.cambridge.org/MRF and can be referenced using their DOI (Digital Object Identifier). Once all submissions are received, the articles will be collated into the Special Issue, which is expected to appear in August 2022.

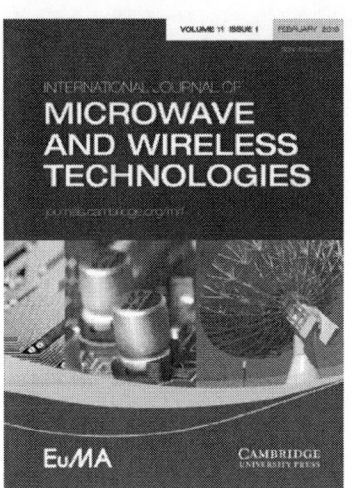

YI WANG
EuMC 2021 TPC Chair

EDWARD WASIGE
EuMIC 2021 TPC Chair

MATTHEW RITCHIE
EuRAD 2021 TPC Chair

Follow Us on Social Media

DARYNA PESINA
EuMA Social Media Officer

MOHAMMED JAHANGIR
EuMW 2021 Social Media Officer

2020 European Microwave Week in Utrecht Best Paper Prizes: EuMIC

EuMIC Prize
Sponsored by Delft University of Technology

François Deborgies
EuMIC Chair

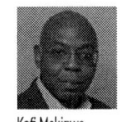

Kofi Makinwa

𝝜U Delft

Authors
Eswara Rao Bammidi[1], Ingmar Kallfass[1]
[1]Institute of Robust Power Semiconductor Systems (ILH) - University of Stuttgart

Paper Title
An Analog Costas Loop MMIC in 130 nm SiGe BiCMOS Technology for Receiver Synchronization of QPSK and BPSK Modulated Signals

EuMIC Young Engineer Prize
Sponsored by GAAS® Association

François Deborgies
EuMIC Chair

Paolo Colantonio

Authors
Tejinder Singh[1], Raafat R. Manso[1]
[1]University of Waterloo

Paper Title
Reconfigurable PCM GeTe-Based Latching 6-bit Digital Switched Capacitor Bank

Roberto Sorrentino Prize

This is a new prize, named in remembrance of Roberto Sorrentino. The prize has been initiated by Linda Di Carlo Sorrentino in cooperation with RF Microtech, the Italian EM Society (SIEm) and EuMA. Awarded every year for at least ten years, it will recognize an outstanding young professional who has distinguished technical achievements (not on a single paper) within the microwave field. The technical achievements may include technical papers in journals and/or conferences/symposia sponsored or technically sponsored by EuMA. The technical achievements may also include services as a committee member for these Journals and/or conferences/symposia. This prize focuses on the individual rather than the achievements and would preferably be in yearly alternation between university and industry.

A nominee must be a member of the EuMA and no more than 38 years of age at the time of nomination deadline (i.e. not having reached their 39th birthday). To help bridge the gender gap in the microwave community however, this deadline is postponed by one year per child for women that have had children. A nomination must be made by a EuMA member (not a student member) who has known the nominee for more than 2 years. Self-nomination is not allowed. Two references in addition to the nominator are required. A selection panel, chaired by a member of the EuMA Board of Directors, selects every year a suitable number of panel members (from 5 to 7), whose names are not public. The Chair does not vote. Because of the large financial coverage, the Jury has one member designated by RF Microtech and one by SIEm, respectively.

The annual prize comprises a certificate, a medal and a financial award of 4.000 €, contributed by Mrs Linda Di Carlo Sorrentino, RF Microtech, SIEm, and EuMA. Collectively this might sustain the prize for a longer period as it is intended to keep the amount of the prize at 4.000 € therefore increasing the number of years of availability of the prize beyond 10 years. The prize will be presented at the Opening Session of the European Microwave Week. The first prize will be presented during EuMW 2021.

The EuMW 2021 Organising Committee

Nick Ridler
General Chair
National Physical Laboratory, UK

John Cunningham
General Co-chair & Sponsors Chair
University of Leeds, UK

Peter Gardner
General TPC Chair
University of Birmingham, UK

Adrian Cross
Treasurer
University of Strathclyde, UK

Djuradj Budimir
General TPC Co-Chair & Local
Arrangements
University of Westminster ,UK

Xiaobang Shang
Operational Director & Automotive
Forum Local Arrangement Chair
National Physical Laboratory, UK

Emma MacPherson
EuMC Chair
University of Warwick, UK

Dominique Schreurs
EuMC Co-chair
University of Leuven, Belgium

Yi Wang
EuMC TPC Chair
University of Birmingham, UK

Chris Clifton
EuMIC Chair & 5G and Beyond Forum
Co-chair
Sony Europe B.V.

Shokrollah Karimian
EuMIC Co-chair
University of Oxford, UK

Edward Wasige
EuMIC TPC Chair
University of Glasgow, UK

James Watts
EuRAD Chair
Theta Technologies Ltd., UK

Stephen Harman
EuRAD Co-chair
Aveillant Ltd, UK

Matthew Ritchie
EuRAD TPC Chair
University College London, UK

Chong Li
Workshops & Short Courses Chair
University of Glasgow, UK

Qammer Abbasi
Workshops & Short Courses Co-chair
University of Glasgow, UK

Jiafeng Zhou
Publications Chair
University of Liverpool, UK

David Prinsloo
Publications liaison
ASTRON, The Netherlands

Thomas Zwick
Automotive Forum Chair
Karlsruhe Institute of Tech., Germany

Frank Gruson
Automotive Forum Co-chair
Continental AG, Germany

Martin Kunert
Automotive Forum Co-chair
Bosch, Germany

Chris Baker
DSS Forum Chair
University of Birmingham, UK

Mike Antoniou
DSS Forum Co-chair
University of Birmingham, UK

Kamal Samanta
Awards Chair
AMWT Ltd / Sony Europe B.V.

Akram Alomainy
Student Volunteers Chair
Queen Mary University of London, UK

Martin Salter
Visas Chair
National Physical Laboratory, UK

Lai Bun Lok
Doctoral School Chair
University College London, UK

Claudio Paoloni
Doctoral School Co-chair
University of Lancaster, UK

Noushin Karimian
Chair of Women in Engineering &
Career Platform
Manchester Metropolitan Univ., UK

ORGANISERS

WWW.**EUMW2021**.COM

Helen Duncan
Career Platform Co-chair
MWE Media, UK

Dilbagh Singh
Grants Chair
National Physical Laboratory, UK

Mustafa Bakr
Posters Chair
University of Oxford, UK

John Crute
Student School Chair
The Technology Academy, UK

Mohammed Jahangir
Social Media Chair
University of Birmingham, UK

Lutfi Albasha
5G and Beyond Forum Chair
American University of Sharjah, UAE

Colin Whyte
Focused & Special Sessions Chair
University of Strathclyde, UK

Daniel Stokes
Operational Officer & Attendee
Survey Officer
National Physical Laboratory, UK

Manoj Stanley
Operational Officer & Video Chair
National Physical Laboratory, UK

Ron Ginley
IMS 2022 liaison
Retired from NIST, USA

Andy Gibson
EuMW Honorary Chair
Manchester Metropolitan University,
UK

Ian Hunter
EuMW Honorary Chair
University of Leeds, UK

John Walker
EuMW Honorary Treasurer
Integra Technologies, UK

Patrice Gamand
DSS Forum - EuMA Representative
RF Consulting PG, France

Patrick Hindle
DSS Forum - Microwave Journal
Representative
Microwave Journal, USA

Renato Lombardi
5G and Beyond Forum - EuMA
Representative
Huawei Europe Ltd.

Marc van Heijningen
EuMA Conference Software Officer
TNO, The Netherlands

Cristina Andrei
EuMA Conference Software Officer
Brandenburg University of Technology
Germany

Matthias Rudolph
Electronic Submission Advisor
Brandenburg University of Technology
Germany

Daryna Pesina
EuMA Social Media Officer
O. Ya. Usikov Institute for Radiophys-
ics and Electronics NAS, Ukraine

Lorenz-Peter Schmidt
EuMW Officer
Friedrich-Alexander University
Erlangen-Nürnberg, Germany

**Annemie van
Nieuwerburgh**
EuMA Headquarters Assistant
Belgium

Michel Zoghob
Event Director
Horizon House, UK

Julie Mills
Event Manager
Horizon House, UK

Sally Garland
Hotels and Partner Programme
Connex Hotels and Events, UK

EuMA and EuMW Committees

EuMA GENERAL ASSEMBLY

Board of Directors: Frank van den Bogaart · Gilles Dambrine ·Patrice Gamand · Andrew Gibson · Willem Hol · Renato Lombardi · Luca Perregrini · Herbert Zirath · Danielle Vanhoenacker-Janvier · Thomas Zwick
EuMW Chairs: Frank van Vliet, EuMW'20 · Nick Ridler, EuMW'21 · Luca Perregrini, EuMW'22
Ordinary Members: Serge Verdeyme, Group 1 · Ingmar Kallfass, Group 2 · Alessandra Costanzo, Group 3 · Chong Li, Group 4 · Dominique Schreurs, Group 5 · Christian Fager, Group 6 · Vitaliy Zhurbenko, Group 7 · Jan Vrba, Group 8 · Bartlomiej Salski, Group 9 · Kateryna Arkhypova, Group 10 · vacancy, Group 11 · Oleg V. Stukach, Group 12 · Jasmin Grosinger, Group 13 · Nuno Borges Carvalho, Group 14 · Dick Snyder, Group 15 · Hiroshi Okazaki, Group 16 · Amr Safwat, Group 17

Founder Members: Leo Ligthart · Asher Madjar · Holger Meinel · Steve Nightengale · Roberto Sorrentino † · André Vander Vorst
IJMW Editor-in-Chief: Fransisco Medina-Mena
EuMA Honorary Secretary: Andrew F Wilson
By invitation: Wolfgang Heinrich, Past President · Jozef Modelski, MTT-S Observer · Almudena Suárez Rodríguez, Publication Officer · Lorenz-Peter Schmidt, EuMW Officer · Annemie Van Nieuwerburgh, HQ Assistant
Countries Represented: Group 1 – France, Monaco · Group 2 – Germany · Group 3 – Italy, San Marino, Vatican City · Group 4 – United Kingdom, Ireland, Gibraltar, Malta · Group 5 – Belgium, The Netherlands, Luxembourg ·

Group 6 – Iceland, Norway, Sweden · Group 7 – Denmark, Faroe Islands, Finland, Greenland · Group 8 – Bulgaria, Czech Republic, Hungary, Romania, Slovakia · Group 9 – Estonia, Latvia, Lithuania, Poland · Group 10 – Armenia, Azerbaijan, Georgia, Moldova, Ukraine · Group 11 – Albania, Bosnia and · Herzegovina, Croatia, Cyprus, FYR Macedonia, Montenegro, Greece, Israel, Serbia, Slovenia, Turkey · Group 12 – Belarus, Russia · Group 13 – Austria, Liechtenstein, Switzerland · Group 14 – Andorra, Portugal, Spain · Group 15 – North America · Group 16 – Asia-Pacific Group 17 – Africa and Middle East countries

EuMW STEERING COMMITTEE

EuMA Board of Directors
EuMW Officer: Lorenz-Peter Schmidt
GAAS° Representative: Paolo Colantonio
EuMW Chairs: Denis Barataud, EuMW 2019 · Frank van Vliet, EuMW 2020 · Nick Ridler, EuMW 2021 · Luca Perregrini, EuMW 2022 · Thomas Zwick, EuMW 2023
MTT-S Observer: Jozef Modelski
Conference Software Officers: Marc van Heijningen · Matthias Rudolph · Christina Andrei

APMC Delegate 2018 – 2020: Kamran Ghorbani
2020 Conference Chairs, TPC Chair and Treasurer: Alexander Yarovoy, TPC Chair · Wim van Cappellen, EuMC · François Deborgies, EuMIC · Mayazzurra Ruggiano, EuRAD · Ioan E. Lager, Treasurer
2021 Conference Chairs, TPC Chair and Treasurer: Peter Gardner, TPC Chair · Emma MacPherson, EuMC · Chris Clifton, EuMIC · James Watts, EuRAD · Adrian Cross, Treasurer
EuMW Operational Officers 2019 – 2022: Bernard Jarry,

2019 Oper. Officer · Marcel van der Graaf, 2020 Oper. Officer · Xiaobang Shang 2021, Oper. Officer · Lorenzo Silvestri, 2022 Oper. Officer.
HH Representatives: Ivar Bazzy, President Michel Zoghob, Event Director
By Invitation: Wolfgang Heinrich, Past President · André Vander Vorst, Secretary Emeritus/DPO · Andrew F. Wilson, Hon. Secretary · Annemie Van Nieuwerburgh, HQ Assistant

EuMW TECHNICAL PROGRAMME COMMITTEE

EuMC: Joachim Oberhammer · Catherine Algani · Shmuel Auster · Denis Barataud · John Batchelor · Pierre Blondy · Nuno Borges Carvalho · Maurizio Bozzi · Alessandra · Costanzo · Francesco Fornetti · Vincent Fusco · Alessandro Galli · Matthias Geissler · Anthony Ghiotto · Kamran Ghorbani · Roberto Gomez-Garcia · Katia Grenier · Jonas Hansryd · Yang Hao · Zhangcheng Hao · Michael Höft · Jiasheng Hong · Stavros Iezekiel · Mehmet Kaynak · Alexander Kölpin · Panos Kosmas · Giuseppe Macchiarella · Ferran Martin · Petronilo Martín Iglesias · Marion Matters-Kammerer · Markus Mayer · Francisco Medina · Giuseppina Monti · Michal Mrozowski · Bart Nauwelaers · Marco Pasian · Jose Carlos Pedro · Dirk Plettemeie · Adrian Porch · Jeff Powell · Francesc Purroy · Hendrik Rogier · Ilona Rolfes · Lorenz-Peter Schmidt · Dominique Schreurs · Xiaobang Shang · Richard Snyder · Jan Stake · Almudena Suarez Rodriguez · Oleksiy Sydoruk · Luciano Tarricone · Paul Tasker · Loh Tian Hong · Cristiano Tomassoni · Wim van Cappellen · Guy Vandenbosch · Gavin Watkins · Nils Weimann · Colin Whyte · Tudor Williams · Ke Wu

EuMIC: Alessandro Cidronali · Paolo Colantonio · Nathalie Deltimple · Khaled Elgaid · Didier Floriot · Patrice Gamand · Mike Geen · Friedel Gerfers · Franco Giannini · Ruonan Han · Vadim Issakov · Ingma Kallfass · Mehmet Karaaslan · Asher Madjar · Peter Magnee · Simon Mahon · Teresa Martin-Guerrero · Angel Mediavilla · Farid Medjdoub · Matthew O'keefe · Ekmel Özbay · Ulrich Pfeiffe · Rüdiger Quay · Michael Schlechtweg · Raphael Sommet · Joe Staudinger · Safumi Suzuki ·Frank van Vliet · Lars-Erik Wernersson · Herbert Zirath
EuRAD: Jabran Akhtar · Andre Bourdoux · Alex Charlish · Jacco de Wit · Reinhard Feger · Laurent Ferro-Famil · Marina Gashinova · Frank Gruson · Stephen Harman · Willem Hol · Maria Pilar · Jarabo Amores · Pierfrancesco Lombardo · Claire Migliaccio · Roland Oechslin · Debora Pastina · Niels Pohl · Matthew Ritchie · Mayazzurra Ruggiano · Andy Stove ·Martin Vossiek · Alex Yaravoy

EuMW 2021 Reviewers
To our reviewers: Thank you for your great work!

EUMW REVIEWERS

A: Muhammad Ali Babar Abbasi · Elena Abdo-Sánchez · Edward Ackerman · Parisa Aghdam · Janne P. Aikio · Jabran Akhtar · Lutfi Albasha · Catherine Algani · Abdullah Al-Khalidi · Rozenn Allanic · Arokiaswami Alphones · Giandomenico Amendola · Cristina Andrei · Guillaume Andrieu · Mykhaylo Andriychuk · Iltcho Angelov · Kimia Ansari · Fritz Arndt · Holger Arthaber · Klaus Aufinger
B: Damienne Bajon · Christopher Baker · Mustafa Bakr · Ali Banai · Abhishek Banerjee · Biddut Banik · Denis Barataud · Paweł Barmuta · Jan Barowski · César Barquinero · Juraj Bartolic · Mikhail Belkin · Didier Belot · Brahim Benbakhti · Johannes Benedikt · Olof Bengtsson · Tibor Berceli · Manfred Berroth · Miguel Beruete · Jan Geralt Bij de Vaate · Stéphane Bila · Hans-Ludwig Bloecher · Pierre Blondy · Luigi Boccia · Fabrizio Bonani · Marlene Bonmann · Nuno Borges Carvalho · Vicente E. Boria · Jens Bornemann · André Bourdoux · Maurizio Bozzi · Philippe Brouard · Stefan Brüggenwirth · Jack Brunning · Djuradj Budimir · Pascal Burasa
C: Pedro Miguel Cabral · Vittorio Camarchia · Marc Camiade · Juan Luis Cano · Tommaso Cappello · Christian Carlowitz · Corrado Carta · Carlos Castillo · Robert Caverly · Malgorzata Celuch · Subhradeep Chakraborty · Alexander Charlish · Nan-Wei Chen · Nickolay T. Cherpak · Kevin Chetty · Pascal Chevalier · J-C Chiao · Heungjae Choi · Kevin Chuang · Alessandro Cidronali · Lorenzo Cifola · Kevin Cinglant · Elisa Cipriani · Paolo Colantonio · Alessandra Costanzo · Luis Carlos Cotimos Nunes · Angela Coves · Diego Cristallini · Pedro Miguel Cruz
D: Thomas Dallmann · Christian Damm · Patrik Dammert · François Danneville · Jean-Yves Dauvignac · Peter de Hek · Luisa de la Fuente · Pedro de Paco · Maria Merlyne De Souza · Jacco de Wit · François Deborgies · Cyril Decroze · Jochen Dederer · Christophe Delaveaud · Sebastien Delcourt · Nicolas Delhote · Nathalie Deltimple · Yann Deval · Franz Dielacher · Jian Ding · Thanh Viet Dinh · Marco Dionigi · Massimiliano Dispenza · Tarek Djerafi · Yassen Dobrev · Ralf Doerner · Simona Donati Guerrieri · Johan Donkers · Viktor Doychinov · David Dubuc · Guillaume Ducournau · Francois Duport · Yvan Duroc
E: Abdalla Eblabla · Khaled Elgaid · Jens Engelmann · Filipe Miguel Esturrenho Barradas · Philippe Eudeline
F: SALIM FACI · Christian Fager · Francisco Falcone · Angie Fasoula · Reinhard Feger · Michael Feiginov · Mónica Fernández Barciela · Laurent Ferro-Famil · Michele Fiorini · Georg Fischer · Nelson Fonseca · Francesco Fornetti · Nicolas Fortino · Erwan Fourn · Thomas Fromenteze · Dominic Funke · Vincent Fusco
G: Michael Ernst Gadringer · Gaspare GALATI · Alessandro Galli · Patrice Gamand · Markus Gardill · Marina Gashinova · Mike Geen · Matthias Geissler · Apostolos Georgiadis · Bertrand Gerfault · Friedel Gerfers · Anthony Ghiotto · Kamran Ghorbani · Franco Giannini · Gian Piero Gibiino · Andy Gibson · Pere L. Gilabert · Raphaël

Gillard · Rocco Giofrè · Oleksandr Glubokov · Roberto Gomez-Garcia · Jose Manuel Gonzalez Perez · Jan Grahn · Katia Grenier · Jasmin Grosinger · Markus Grözing · Frank Gruson · Marco Guglielmi · Cheng Guo
H: Jonas Hansryd · Zhangcheng Hao · Stephen Harman · Ronny Harmanny · Marlene Harter · Zhongxia Simon He · Wolfgang Heinrich · Reinhold Herschel · Jeffrey Hesler · Jan Hesselbarth · Michael Höft · Willem A. Hol · Jiasheng Hong · Colin Horne · Rui Hou · Zhirun Hu · Heinz-Wilhelm Hübers · Isabelle Huynen
J: Arne F. Jacob · María-Pilar Jarabo-Amores · Olivier Jardel · Bernard Jarry · Powell Jeff · Ulf Johannsen · Ari Joki
K: Ingmar Kallfass · Karol Kalna · Despoina Kampouridou · Tomoya Kaneko · Mehmet Karaaslan · Tom Keinicke Johansen · James Kelly · Eric Kerherve · Ata Khalid · Kevin Kim · Jim Kirchgessner · Dietmar Kissinger · Tero Kiuru · Jens Klare · Toshiro Kodera · Alexander Koelpin · Hiroshi Kondoh · Panos Kosmas · Slawomir Koziel · Oleg Krasnov · Futoshi Kuroki
L: Ariana Lacorte Caniato Serrano · Ioan Lager · Jérôme Lanteri · Miguel A. G. Laso · Tuami Lasri · Rémy Leblanc · Kean Lee · Timothy Lee · Domine Leenaerts · Friedrich Lenk · Mauro Leonardi · Yoke Choy Leong · Giorgio Leuzzi · Quentin Lévesque · Ulrich Lewark · Chong Li · Changzhi Li · Ernesto Limiti · Fujiang Lin · Julien Lintignat · Ignacio Llamas · Tian Hong Loh · Lai Bun Lok · Pierfrancesco Lombardo · Errikos Lourandakis · Yunlong Lu · Fabian Lurz · Cyril Luxey
M: Tzyh-Ghuang Ma · Rui Ma · Giuseppe Macchiarella · Jan Machac · María J. Madero-Ayora · Asher Madjar · Peter Magnee · Simon J. Mahon · Mateusz Malanowski · Robert Malmqvist · Chunxu Mao · Zlatica Marinković · Paulo Marques · Ferran Martín · Marta Martínez-Vázquez · Teresa M. Martín-Guerrero · Petronilo Martín-Iglesias · Łukasz Maślikowski · Diego Masotti · David Mata-Moya · Marion K. Matters-Kammerer · OKeefe Matthew · Holger Maune · Markus Mayer · Giuseppe Mazzarella · Francisco Medina · Farid Medjdoub · Cyrille Menudier · Wolfgang Menzel · Stéphane Méric · Caterina Merla · Francisco Mesa · Claire Migliaccio · Mikael Mikael · Konstantinos Mimis · Mohamed Missous · Jaber Moghaddasi · Priyanka Mondal · Sebastien Mons · Giuseppina Monti · Gabriel Montoro · Antonio Morini · Michal Mrozowski · José-María Muñoz-Ferreras · Hiroshi Murata · Mira Naftaly
N: Jean-Christophe Nallatamby · Nasrin Nasr Esfahani · Bart Nauwelaers · Irina Nefedova · Renato Negra · Fatemeh Norouzian · Dirk Nüßler
O: Joachim Oberhammer · Roland Oechslin · Morten Olavsbråten · Arnaldo Oliveira · Abbas Omar · Ekmel Özbay
P: Cristiano Palego · Andy Panks · George Papaioannou · Youngjin Park · Marco Pasian · Daniel Pasquet · Debora Pastina · Mario Pauli · Dimitris Pavlidis · Alain Peden · José Carlos Pedro · Ana Peláez Pérez · Luca Pelliccia ·

Luca Perregrini · Nial Peters · Nikita Petrov · Oscar Antonio Peverini · Jean Marie Pham · Michael Philipakkis · Gia Ngoc Phung · Sergio Pires · Marco Pirola · Emmanuel Pistono · Emanuele Piuzzi · Florence Podevin · Nils Pohl · Jean-Luc Polleux · Zoya Popovic · Adrian Porch · Arnaud Pothier · Gaëtan Prigent · Dimitra Psychogiou · Francesc Purroy
Q: Roberto Quaglia · Rüdiger Quay · Abdul Quddious
R: Muhammad Rabbani · Antonio Raffo · Antti Räisänen · Franco Ramirez · Gustavo Rehder · Apolinar Reynoso-Hernandez · Vishal Riché · David Ricketts · Matthew Ritchie · Eric Rius · Francois Rivet · Michael Roberg · Helena Rodilla · Hendrik Rogier · Yves Rolain · Ilona Rolfes · Manuel Rosa-Zurera · Mayazzurra Ruggiano · Jorge A. Ruiz-Cruz · Anders Rydberg
S: Amr Safwat · Martin Salter · Miguel Sanchez-Soriano · Philip Sanders · Paul Sangaré · Alberto Santarelli · Fabrizio Santi · Bruno Sauviac · J. Christoph Scheytt · Michael Schlechtweg · Lorenz-Peter Schmidt · Dominique Schreurs · Hermann Schumacher · James Scott · Diebold Sebastian · Daniel Segovia-Vargas · Alexandre Serres · Xiaobang Shang · Oksana Shramkova · Manuel Sierra Castañer · Hjalti H. Sigmarsson · Christopher Silva · Talal Skaik · Richard Snyder · Eran Socher · Jacques Sombrin · Nutapong Somjit · Raphaël Sommet · Ho-Jin Song · Andreas Springer · Jan Stake · Robert Staraj · Joseph Staudinger · Andrew Stove · Almudena Suarez Rodriguez · Noriharu Suematsu · Safumi Suzuki · Oleksiy Sydoruk
T: Alexandru Takacs · Luciano Tarricone · Paul Tasker · Chakib Taybi · Murat Temiz · Manos M. Tentzeris · Cristiano Tomassoni · Ichihiko Toyoda · Pier Andrea Traverso · Jeanne Treuttel · Nikolaos Tsitsas · Stefano Turso
U: Ingrid Ullmann · Thomas Ußmüller
V: Valeria Vadalà · Issakov Vadim · Wim van Cappellen · Mark van der Heijden · Eddy van Eeuwijk · Marc van Heijningen · Wim van Rossum · Frank E. van Vliet · André Vander Vorst · Danielle Vanhoenacker-Janvier · Giorgio Vannini · Venkata Vanukuru · Andriy Vasylyev · Serge Verdeyme · Jordi Verdú · Colin Viegas · Valerie Vigneras · Ville Viikari · Jean-François Villemazet · Shelly Vishwakarma · Hubregt Visser · Michael Vogt · Andrei Vorobiev · Martin Vossiek · Haris Votsi · Tan-Phu Vuong
W: Simon Wagner · Christian Waldschmidt · Huei Wang · Yi Wang · Gavin Watkins · Robert Weigel · Nils Weimann · Li Wenda · Lars-Erik Wernersson · Colin Whyte · Werner Wiesbeck · Tudor Williams · Richard Wilson · Marian Wnuk · Kenneth Kin Wong · Sai-wai Wong · Ke Wu · Ke-Li Wu · Zhipeng Wu · Chung-Tse Michael Wu
Y: Li Yang · Alexander Yarovoy · Paul Young · Ming Yu
Z: Marco Zappatore · Jiafeng Zhou · Anding Zhu · Herbert Zirath · Farid Zubir · Thomas Zwick

SATURDAY OVERVIEW

Room	09:00 – 13:00	14:20 – 18:20
1	**WS01** Advances of Wireless Sensing in Harsh and Severe Environments	
2		
3		
4		
5		
6		**SS01** Advanced Non-linear Characterization and Design of Highly Efficient Power Amplifiers Using Load-Pull Data for sub-6GHz and mmWave Applications
7	**WS04** New Trends in Microwave and mmWave Filters	
8	**SS02** Fundamentals of Microwave PA Design	
9		
10	**WS06** Progress and Status of Gallium Nitride Monolithic Microwave Integrated Circuits	
11		
12	**WS08** Technology for RF 5G and Satcom: From Material to Packaged Demonstrators	
13	**SS04** Terahertz Technology, Instrumentation and Application	
14	**WS09** Research in Power and S-parameters Measurements at mmWave and Terahertz Frequencies	
15		
16		
17	**WS03** mmWave Plastic Waveguide High Data Rate Communications	

EuMC   EuMIC   EuRAD   Students   EuMW   Exhibitors

SUNDAY OVERVIEW

PROGRAMME – CONFERENCE SESSIONS MATRIX   WWW.EUMW2021.COM

Room	09:00 – 10:40	11:20 – 13:00	14:20 – 16:00	16:40 – 18:20	EVENING PROGRAMME
1	**SM01** R&D Trends & Challenges in RFPAs for Medium/High-Volume Products		**WM01** Optimizing Modulation Quality Measurements on Wide Bandwidth Signals - from Conformance Through R&D		
2	**WM02** Advances in Circuits and Systems for mmWave Radar and Communication in Silicon Technologies				
3					
4	**Tom Brazil Doctoral School of Microwaves**				
5			**EuMIC05** Integrated Circuit Modelling and Design Methodology		
6	**WM03** Microwave and mmWave Techniques for Sensing, Imaging and Characterisation of Biological Tissues				
7					
8		**EuMIC04** Opening Session		**EuMIC08** Components and Subsystems for 100 GHz and Above	
9				**EuMIC09** High Performance LNAs	
10	**EuMIC01** Large Signal and Non-linear Charaterization Techniques		**EuMIC06** Integrated PAs for 5G, SATCOM and Vehicular Applications	**EuMIC10** Advances in Si and GaN Based Integrated PAs	
11	**EuMIC02** Silicon Based RF Solutions		**SM02** Intuitive Microwave Filter Design with EM Simulation		
12	**EuMIC03** Transceiver MMICs		**SM03** Phase-Noise in Next-Generation Aerospace/Defense and Commercial Wireless Communications		
13	**WM04** RF On-wafer Calibration and Measurement Eco-system Workshop				
14					
15	**WM05** Microwave Equipment Based on SM EM Relays				
16	**WM06** Recent Developments in Wireless Power Transfer and Energy Harvesting				
17	**WM07** Beyond 5G: mmWave and Terahertz Techniques of 6G Research		**EuMIC07** Frequency-Converting Circuits	**EuMIC11** Boradband Integrated Circuits	**EuMIC Foundry Session** 18:30 - 20:00
ICC Capital Suite (Foyer)					**EuMIC Cocktail Reception** 18:00 - 20:00

EuMC   EuMIC   EuRAD   Students   EuMW   Exhibitors

SUNDAY 09:00 – 10:40

ROOM	Room 10	Room 11	Room 12
	EuMIC01 Large signal and Non-linear Characterization Techniques	**EuMIC02** Silicon Based RF Solutions	**EuMIC03** Transceiver MMICs
	Chair: Teresa M. Martín-Guerrero[1] Co-Chair: Nuno Borges Carvalho[2] [1]Universidad de Málaga, [2]University of Aveiro / Instituto de Telecomunicações	Chair: Peter Magnee[1] Co-Chair: Rüdiger Quay[2] [1]NXP Semiconductors, [2]IAF-Fraunhofer: Fraunhofer Institute for Applied Solid-State Physics	Chair: Mehmet Karaaslan[1] Co-Chair: Friedel Gerfers[2] [1]Teledyne E2V UK Ltd, [2]TU Berlin
09:00 – 09:20	**EuMIC01-1** Load-pull measurement of SiGe:C HBT in BiCMOS 55 nm featuring 11 dBm of output power at 185 GHz Caroline Maye[1] [1]University of Lille - IEMN	**EuMIC02-1** SiGe BiCMOS as enabling technology for next generation RF & THz Systems Gerhard Kahmen[1] **INDUSTRIAL KEYNOTE** [1]IHP - Leibniz-Institut fur innovative Mikroelektronik	**EuMIC03-1** mmW GaN/Si MMICs: The 3rd generation of III/V processes to complement Si RFIC solutions Marc Rocchi[1] **INDUSTRIAL KEYNOTE** [1]OMMIC
09:20 – 09:40	**EuMIC01-2** Nonlinear Characterization of Wideband Power Amplifiers with frequency dependent match load Sanket Chaudhary[1], Nuno Borges Carvalho[1], Marina Jordão[1], Marc Vanden Bossche[2], Adam Cooman[2], Sergio Pires[3] [1]Universidade de Aveiro / Instituto de Telecomunicações, [2]National Instruments, [3]Ampleon, The Netherlands BV	**EuMIC02-2** Analysis of the Relaxed Contacted-Poly-Pitch Effect on the RF Performance of Strained-SiGe-Channel p-FETs in 22nm FDSOI Technology Quang Huy Le[1], Dang Khoa Huynh[1], Steffen Lehmann[2], Zhixing Zhao[2], Thomas Kämpfe[1], Matthias Rudolph[3] [1]Fraunhofer Institute for Photonic Microsystems (IPMS), [2]Globalfoundries, Germany, [3]BTU, Germany	**EuMIC03-2** A Bidirectional 28 GHz RF Transceiver Front-End with Test and Calibration Interface for 5G Phased Arrays Katharina Kolb[1], Julian Potschka[1], Tim Maiwald[1], Klaus Aufinger[2], Amelie Hagelauer[1], Marco Dietz[1], Robert Weigel[1] [1]FAU Erlangen-Nuremberg, [2]Infineon Technologies AG, [3]University of Bayreuth
09:40 – 10:00	**EuMIC01-3** Intermodulation Products of a CMOS SP6T Antenna Switch: Results Comparison Between an Experimental Test-Bench and a Corresponding Simulated Virtual Test-Bench Marwen Ben Sassi[1], Hassan Saleh[1], Ousmane Sow[1], Imene Lahbib[1], Gregory D U'Ren[1], C. Hallépée[2], et al. [1]XFAB, [2]Xlim - UMR 7252 - CNRS- Limoges University	**EuMIC02-3** Design Methodology of Wide Tuning Range DGS-based VCO for K-band Applications in 0.18-µm CMOS Technology Baichuan Chen[1], Samundra Kumar Thapa[1], Adel Barakat[1], Ramesh Kumar Pokharel[1] [1]Kyushu University	**EuMIC03-3** A 2-channel TX and 4-channel RX in SiGe BiCMOS for X-band MIMO Radar Applications Maciej Kucharski[1] [1]SIRC Sp. z o.o.
10:00 – 10:20	**EuMIC01-4** A computationally-efficient self-consistent large signal model for GaN HEMTs based on ASM-HEMT Sourabh Khandelwal[1], Ken Kikuchi[2], Hiroshi Yamamoto[2] [1]Macquarie University, [2]Sumitomo Electric Industries Ltd Japan	**EuMIC02-4** Linearity Assessment of GaN HEMTs on Si using Nonlinear Characterisation Rana ElKashlan[1], Ahmad Khaled[1], Raul Rodriguez[1], Vamsi Putcha[1], Uthayasankaran Peralagu[1], AliReza Alian[1], Nadine Collaert[1], Piet Wambacq[1], Bertrand Parvais[1] [1]Imec	**EuMIC03-4** Two-Element 81-86 GHz SiGe Transmitter Beamformer for Backhaul Applications Roee Ben-Yishay[1], Oded Katz[1], Danny Elad[1] [1]ON Semiconductor
10:20 – 10:40	**EuMIC01-5** Large-Signal Modeling for Nonlinear Analysis of Experimental Devices in 22nm FDSOI Technology Quang Huy Le[1], Dang Khoa Huynh[1], Anurag Nayak[1], Steffen Lehmann[2], Zhixing Zhao[2], Thomas Kämpfe[1], Matthias Rudolph[3] [1]Fraunhofer Institute for Photonic Microsystems (IPMS), [2]Globalfoundries, [3]Brandenburg University of Technology (BTU)	**EuMIC02-5** Reconfigurable and Scalable Monolithic Band Reject Circuit Utilizing Phase-Change Switch Matrices Tejinder Singh[1], Raafat R. Mansour[1] [1]University of Waterloo	**EuMIC03-5** A W-Band Single-Chip Receiver in a 60 nm GaN-on-Silicon Foundry Process Robert Malmqvist[1], Rolf Jonsson[1], Mingquan Bao[2], Rémy Leblanc[3], Koen Buisman[4], Christian Fager[4], Kristoffer Andersson[1] [1]Swedish Defence Research Agency (FOI), [2]Ericsson Research, Ericsson AB, [3]OMMIC S.A.S, [4]Chalmers University of Technology

SUNDAY 11:20 – 13:00

ROOM   **Room 7 – 9**

EuMIC04
EuMIC Opening Session

Chair: Chris Clifton[1], EuMIC Chair

Co-chair: Shokrollah Karimian[2], EuMIC Co-chair; Edward Wasige[3], EuMIC TPC Chair

[1]Sony Europe B.V., [2]University of Oxford, UK, [3]University of Glasgow, UK

**11:20
–
11:30**

Welcome Address: Opening of the European Microwave Integrated Circuits Conference 2021

Chris Clifton[1]

[1]EuMIC Chair

**11:30
–
12:15**

III-V Nitride Semiconductors for Microwave Applications

Christopher Snowden[1]

[1]Fellow of the Royal Society, Chair of the ERA Foundation

The demand for high performance semiconductor devices for microwave applications to meet the exacting power, frequency, bandwidth and linearity requirements of 5G, radar and remote sensing has driven the development of wide-band gap semiconductors capable of delivering high powers, with high efficiencies at frequencies up to 100 GHz. This presentation will review the state-of-the-art, highlighting the development and introduction of GaN-based FETs for use in discrete and integrated circuits. A comparison with Si, SiGe and GaAs-based technologies will be made, discussing the relative merits and RF performance. Examples, of applications, and details of the technology for GaN-based devices will be given for operation in the frequency range 2 to 96 GHz. Particular emphasis will be placed on the use of GaN on Si substrate HEMTs in power amplifiers in both MIC and MMIC forms. The modelling and design of devices and circuits will be covered together with some insight into fabrication and production. Finally, future prospects for III-V nitride devices will be presented.

**12:15
–
13:00**

High-Efficiency PAs for Broadband High-PAR Signals

Zoya Popovic[1]

[1]University of Colorado, USA

Achieving power amplifiers (PAs) with high efficiency and good linearity is challenging if the amplified signals have wide instantaneous bandwidths (> 100 MHz) and high peak-to-average power ratios (PAPR > 10 dB). Examples of such signals include multi-carrier concurrent signals, both closely and widely spaced, and band-limited noise-like signals, typical of 5G and other multi-carrier aggregated signal applications. This talk will overview techniques for supply modulation of broadband signals amplified by different GaN PAs, including a 2 – 4GHz single ended hybrid PA, an X-band MMIC PA, and a K-band MMIC PA. The signals that are considered include band-limited noise with bandwidths from 10 to 250 MHz, and widely spaced multi-carrier with up to 800 MHz spacing. Both continuous and discrete supply modulation of multiple amplifier stages is demonstrated, and linearization methods discussed.

SUNDAY 14:20 – 16:00

ROOM	Room 5	Room 10	Room 17
	EuMIC05 Integrated Circuit Modelling and Design Methodology Chair: Vadim Issakov[1] Co-Chair: Matthew OKeefe[2] [1]Technische Universitaet Braunschweig, [2]INEX Microtechnoogy Ltd	**EuMIC06** Integrated PAs for 5G, SAT-COM and Vehicular Applications Chair: TBA[1] Co-Chair: Alessandro Cidronali[2] [1]TBA, [2]University of Florence	**EuMIC07** Frequency-Converting Circuits Chair: Ingmar Kallfass[1] Co-Chair: Lars-Erik Wernersson[2] [1]University of Stuttgart, [2]Lund Univeristy
14:20 – 14:40	**EuMIC05-1** A 30-to-38 GHz Active and Passive Combined Down-conversion Variable Gain Mixer with Low OP1dB Variation in 65-nm CMOS Mu Heng Li[1], Chun-Nien Chen[1], Yunshan Wang[1], Huei Wang[1] [1]National Taiwan University	**EuMIC06-1** Buffer-free GaN-on-SiC HEMT heterostructures for Sub-6GHz and mmWave RF devices Jr-tai Chen[1] **INDUSTRIAL KEYNOTE** [1]SweGaN	**EuMIC07-1** A Ka-Band MMIC Single-Chip Frequency Converter for Telecom Satellite Applications Francesco Scappaviva[1], Davide Resca[1], Andrea Biondi[1], Luca Cariani[1], Francesco Vitulli[2], François Deborgies[3] [1]MEC – Microwave Electronics for Communications, [2]Thales Alenia Space Italia, [3]ESA / ESTEC
14:40 – 15:00	**EuMIC05-2** Analog Linearization of a 10-W GaN Power Amplifier by Baseband Feedback Mathani Eltayeb[1], Morten Olavsbråten[2], Gian Piero Gibiino[1], Alberto Santarelli[1] [1]University of Bologna, [2]NTNU	**EuMIC06-2** A High GBW High Power Wideband Power Amplifier for Automotive Radar Application Kambiz Hadipour[1], Dominik Amschl[2], Daniel Knauder[2], Stefano Di Martino[2] [1]Infineon Technologies Linz GmbH & Co KG, [2]Infineon Technologies Austria	**EuMIC07-2** A V-band Low-Power and Compact Down-Conversion Mixer with Low LO Power in 130-nm SiGe BiCMOS Technology Batuhan Sütbaş[1], Herman J. Ng[2], Jan Wessel[3], Alexander Koelpin[3], Gerhard Kahmen[1] [1]IHP - Leibniz-Institut fur innovative Mikroelektronik, [2]Karlsruhe University of Applied Sciences, [3]Hamburg University of Technology
15:00 – 15:20	**EuMIC05-3** Statistical Modeling of GaN HEMTs by Direct Transfer of Variations to Model Parameters Petros Beleniotis[1], Serguei Chevtchenko[2], Matthias Rudolph[1] [1]Brandenburg University of Technology (BTU) Cottbus-Senftenberg, [2]Ferdinand-Braun-Institut, Leibniz-Institut für Höchstfrequenztechnik	**EuMIC06-3** A highly rugged 39 GHz 19.3 dBm Power Amplifier for 5G Applications in 45nm SOI Technology Alice Bossuet[1], Baudouin Martineau[1], Cedric Dehos[1], Benjamin Blampey[1], Alexis Divay[1], Yvan Morandini[2] [1]CEA - LETI, [2]SOITEC Grenoble	**EuMIC07-3** A 60 GHz Frequency Doubler with 3.4-dBm Output Power and 4.4% DC-to-RF-Efficiency in 130-nm SiGe BiCMOS Yu Zhu[1], Vincent Rieß[1], Hatem Ghaleb[1], Niko Joram[1], Frank Ellinger[1] [1]TU Dresden
15:20 – 15:40	**EuMIC05-4** Design of Terahertz InP pHEMT Using Machine Learning Assisted Global Optimization Techniques Jing Wang[1], Li-Yuan Xue[1], Bo Liu[1], Chong Li[1] [1]University of Glasgow	**EuMIC06-4** 44 dBm Output Power and High Gain K-band GaN Power Amplifier for Satellite Communication Takuma Torii[1], Yoshifumi Kawamura[1], Eigo Kuwata[1], Masaomi Tsuru[1] [1]Mitsubishi Electric Corporation	**EuMIC07-4** A 14.6 GHz - 19.2 GHz Digitally Controlled Injection Locked Frequency Doubler in 45 nm SOI CMOS Olli Kursu[1], Timo Rahkonen[1], Aarno Pärssinen[1] [1]University of Oulu
15:40 – 16:00	**EuMIC05-5** Low-Power Ka- and V-Band Miller Compensated Amplifiers in 130-nm SiGe BiCMOS Technology Batuhan Sütbaş[1], Herman J. Ng[2], Jan Wessel[3], Alexander Koelpin[3], Gerhard Kahmen[1] [1]IHP - Leibniz-Institut fur innovative Mikroelektronik, [2]Karlsruhe University of Applied Sciences, [3]Hamburg University of Technology	**EuMIC06-5** 125W Solid State Power Amplifier for 17.3-20.2GHz SatCom Applications Rocco Giofrè[1], Paolo Colantonio[1], Lorena Cabria[2], Mariano Lopez[2] [1]University of Roma Tor Vergata, [2]TTI Norte	**EuMIC07-5** A W-Band Up-Conversion Mixer with Integrated LO Frequency Doublers in a 60 nm GaN Technology Mingquan Bao[1], Robert Malmqvist[2], Rolf Jonsson[2], Jonas Hansryd[1], Kristoffer Andersson[1] [1]Ericsson AB, [2]Swedish Defence Research Agency (FOI)

SUNDAY 16:40 – 18:20

ROOM	Room 8	Room 9	Room 10	Room 17
	EuMIC08 Components and Subsystems for 100 GHz and Above Chair: Ullrich Pfeiffer[1] Co-Chair: Herbert Zirath[2] [1]University of Wuppertal, [2]Chalmers University of Technology	**EuMIC09** High Performance LNAs Chair: Lars-Erik Wernersson[1] Co-Chair: Ingmar Kallfass[2] [1]Lund Univeristy, [2]University of Stuttgart	**EuMIC10** Advances in Si and GaN Based Integrated PAs Chair: Khaled Elgaid[1] Co-Chair: Rocco Giofrè[2] [1]Cardiff University, [2]Università di Roma Tor Vergata	**EuMIC11** Broadband Integrated Circuits Chair: Friedel Gerfers[1] Co-Chair: Mehmet Karaaslan[2] [1]TU Berlin, [2]Teledyne E2V UK Ltd

16:40 – 17:00

EuMIC08-1

Highly-Integrated Multi-Channel D-Band Radar Transceivers in Silicon Technologies

Vadim Issakov[1]

INDUSTRIAL KEYNOTE

[1]Infineon Technologies

EuMIC09-1

200 GHz Low Noise Amplifiers in 250nm InP HBT Technology

Utku Soylu[1], Ahmed Samir Hamed Sayed Ahmed[1], Munkyo Seo[2], Ali Farid[1], Mark Rodwell[1]

[1]University of California, Santa Barbara, [2]Sungkyunkwan University

EuMIC10-1

Transient Field-Plate Thermometry Demonstrated on a 20-W X-Band GaN Power Amplifier

Simon J. Mahon[1], Melissa Gorman[1], Michael Heimlich[1]

[1]Macquarie University

EuMIC11-1

Is SiGe BiCMOS an essential technology for 6G?

Pascal Chevalier[1]

INDUSTRIAL KEYNOTE

[1]STMicroelectronics

17:00 – 17:20

EuMIC08-2

SiGe BiCMOS Building Blocks for E- and D-Band Backhauling Front-Ends

Giandomenico Amendola[1], Luigi Boccia[1], Francesco Centurelli[2], Pascal Chevalier[3], Alessandro Fonte[4], Saleh Karman[5], Salvatore Levantino[5], Andrea Mazzanti[6], Carmine Mustacchio[1], Andrea Pallotta[7], et al.

[1]University of Calabria, [2]Sapienza University of Rome, [3]STMicroelectronics, France, [4]SIAE Microelettronica S.p.A., [5]Politecnico di Milano, [6]University of Pavia, [7]STMicroelectronics

EuMIC09-2

Output Power Limited Rugged GaN LNA MMIC

Evelyne Kaule[1], Cristina Andrei[1], Matthias Rudolph[1]

[1]Brandenburg University of Technology Cottbus-Senftenberg

EuMIC10-2

A 27 dBm Ku-band SiGe Power Amplifier Working up to 90°C with High Robustness to the 2:1 SWR

Benjamin COQUILLAS[1], Eric Kerhervé[2], Samuel Redois[3], Anne-Charlotte AMIAUD[4], Laurent ROUSSEL[4], Bruno Louis[5], Eric ITCIA[5], Thomas Merlet[4]

[1]THALES LAS, France / University of Bordeaux, France, [2]University of Bordeaux, France, [3]THALES DMS, France, [4]THALES LAS, France, [5]THALES DMS

EuMIC11-2

120 GBd SiGe-Based 2:1 Analog Multiplexer Module for Ultra-Broadband Transmission Systems

Christian Schmidt[1], Tobias Tannert[1], Jung Han Choi[1], Christoph Caspar[1], Detlef Pech[1], Sebastian Wünsch[1], Greta Ropers[1], Jonathan Schostak[1], Volker Jungnickel[1], Ronald Freund[1], Markus Grözing[2], Manfred Berroth[2]

[1]Fraunhofer Heinrich Hertz Institute, [2]Universität Stuttgart

17:20 – 17:40

EuMIC08-3

A Superheterodyne 300 GHz Transmit Receive Chipset for Beyond 5G Network Integration

Iulia Dan[1], Christopher Grötsch[2], Laurenz John[3], Sandrine Wagner[1], Axel Tessmann[4], Ingmar Kallfass[5]

[1]Qorvo, [2]Keysight Technologies, [3]Fraunhofer IAF, Fraunhofer Institute for Applied Solid State Physics, [4]Fraunhofer IAF, Fraunhofer Institute for Applied Solid State Physics, [5]Institute of Robust Power Semiconductor Systems (ILH) - University of Stuttgart

EuMIC09-3

A highly linear 79 GHz Low-Noise Amplifier for Civil-Automotive Radars in 22 nm FD-SOI CMOS with -6 dBm iP1dB and 5 dB NF

Songhui Li[1], David Fritsche[1], Laszlo Szilagyi[1], Xin Xu[1], Quang Huy Le[1], Defu Wang[1], Thomas Kämpfe[2], Corrado Carta[1], Frank Ellinger[1]

[1]TU Dresden, [2]Fraunhofer Institute for Photonic Microsystems (IPMS)

EuMIC10-3

A 4 GBaud 5 Vpp Pre-Driver for GaN based Digital PAs in 22 nm FDSOI using LDMOS

Frowin Buballa[1], Sebastian Linnhoff[1], Thomas Hoffmann[2], Andreas Wentzel[2], Wolfgang Heinrich[2], Friedel Gerfers[1]

[1]Technische Universität Berlin, [2]Ferdinand-Braun-Institut gGmbH, Leibniz-Institut für Höchstfrequenztechnik

EuMIC11-3

A 7-30 GHz, 80-dBΩ Noise-Optimized, Bandpass-Like TIA in 130 nm SiGe BiCMOS Technology for Quasi-Coherent Optical Receivers

Tom Keinicke Johansen[1], Guillermo Silva Valdecasa[1], Monika Kupska[2], Jose Altabas[2], Omar Gallardo[2], Michele Squartecchia[2], Jesper Bevensee Jensen[2]

[1]Technical University of Denmark, [2]Bifrost Communications Aps.

17:40 – 18:00

EuMIC08-4

Implementation of Slow-Wave Thin-Film Microstrip Transmission Lines in a 35nm InGaAs Technology

Athanasios Gatzastras[1], Hermann Massler[2], Arnulf Leuther[2], Sébastien Chartier[1], Ingmar Kallfass[1]

[1]Institute of Robust Power Semiconductor Systems (ILH) - University of Stuttgart, [2]Fraunhofer IAF

EuMIC09-4

Highly Linear D-Band Low-Noise Amplifier with 8.5dB Noise Figure in InP-DHBT Technology

Maruf Hossain[1], Ralf Doerner[1], Hady Yacoub[1], Tom Keinicke Johansen[3], Wolfgang Heinrich[1], Viktor Krozer[1]

[1]Ferdinand-Braun-Institut (FBH) Leibniz-Institut für Höchstfrequenztechnik, [2]Ferdinand-Braun-Institut (FBH), [3]Technical University of Denmark (DTU)

EuMIC10-4

400-Watt S-band Power Amplifier MMIC

Peter de Hek[1], Gijs van der Bent[1], Frank E. van Vliet[1]

[1]TNO

EuMIC11-4

Multi-Phase Clock Path Circuit up to 57 GHz Including 5 bit Programmable Phase Interpolators for Time-Interleaved Broadband Data Converters in a 28 nm FD-SOI CMOS Technology

Daniel Widmann[1], Tobias Tannert[1], Xuan-Quang Du[1], Markus Grözing[1], Manfred Berroth[1]

[1]University of Stuttgart

18:00 – 18:20

EuMIC08-5

A 140 GHz to 170 GHz Active Tunable Noise Source Development in SiGe BiCMOS 55 nm Technology

Victor Fiorese[1], Joao Carlos Azevedo Goncalves[1], Simon Bouvot[1], Emmanuel Dubois[2], Christophe Gaquière[2], Guillaume Ducournau[2], François Danneville[2], Sylvie Lépilliet[2], Daniel Gloria[1]

[1]STMicroelectronics, [2]University of Lille

EuMIC09-5

C-Band Low-Noise Amplifier MMIC with an Average Noise Temperature of 44.5 K and 24.8 mW Power Consumption

Felix Heinz[1], Fabian Thome[1], Arnulf Leuther[1], Oliver Ambacher[1]

[1]Fraunhofer IAF, Fraunhofer Institute for Applied Solid State Physics

EuMIC10-5

A 41.5 dBm Broadband AlGaN/GaN HEMT Balanced Power Amplifier at K-Band

Stanislav Samis[1], Christian Friesicke[2], Thomas Maier[3], Rüdiger Quay[3], Arne F. Jacob[1]

[1]Hamburg University of Technology, [2]Fraunhofer IAF, Fraunhofer Institute for Applied Solid State Physics, [3]Fraunhofer IAF, Fraunhofer Institute for Applied Solid State Physics

EuMIC11-5

A DC to 20 GHz Variable Gain Amplifier with Tunable Input Matching in 22 nm FDSOI Technology

Seyyedmohsen Seyyedrezaei[1], Manu Viswambharan Thayyil[1], Corrado Carta[1], Frank Ellinger[1]

[1]Technische Universität Dresden, Germany

SUNDAY 18:30 – 20:00

ROOM **Room 17**

EuMIC Foundry Session

Chair: Marc Rocchi, Ommic

Co-Chair: Sunday Expo, Manchester Metropolitan University

18:30 – 20:00 Panel Session.

Panel comprising leading III-V and Silicon foundries with a lively discussion on the merits of each cutting edge technology for the next generation of communication devices.

PROGRAMME – CONFERENCE SESSIONS MATRIX

WWW.**EUMW2021**.COM

MONDAY OVERVIEW

Room	08:30	09:00–10:40	11:20–13:00	14:20–16:00	16:40–18:20	EVENING PROGRAMME
1		EuMIC/EuMC01 Novel Filtering Devices in Integrated Technologies		EuMC02 Innovative Microwave Circulators and Phase Shifters	EuMC05 Novel Structures for Power Combiners and Couplers	
2		Exhibitor Workshops		Exhibitor Workshops		
3		Exhibitor Workshops		Exhibitor Workshops		
4		EuMIC/EuMC02 THz Components		EuMIC16 Phased Array Components from S-band up to 300 GHz	EuMC06 3D to 2D Transitions and New Materials for mmWave System Integration	
5		Exhibitor Workshops		Exhibitor Workshops		
6		EuMW01 Teaching Methods for Microwave Engineering		EuMC03 Non-planar Filters I	EuMC07 Non-planar Filters II	
7					EuMC08 Digital Predistortion, PA Optimisation and MIMO Architectures	
8						
9			EuMW02 Opening Session		EuMIC17 Closing Session	
10						
11						
12					EuMC09 Metasurfaces and Frequency Selective Surfaces	
13		EuMIC12 Device Modelling and Simulation of Parasitic Phenomena		EuMC04 Active Antennas and Architectures	EuMC10 Innovative Antenna Methodology and Design	
14		EuMIC13 Receiver Components		EuMIC/EuMC03 MMIC Power Amplifiers and Supply Modulation	EuMC11 Front-End and Transceiver Modules	
15		Career Platform		Women in Microwaves (Panel to 3 pm, Visit to 6:30 pm)		
16		EuMC01 Advanced Packaging and Interconnect Technologies for Emerging Applications				
17		EuMIC14 Advances in mmWave and High Power Integrated PA Technologies		EuMW03 Special Session in Memoriam of Prof. Roberto Sorrentino	EuMC12 THz Systems and Applications	
Exhibition Hall	Tom Brazil Fellowship Award (by the GAAS® Association) Finalists Pitching The Role of Microwaves in Contributing to a Sustainable World (Venue: MicroApps)		EuMIC15 Posters	EuMIC/EuMC04 Posters		
Conference Center: Platinum Suite						EuMW Welcome Reception 18:30–22:00

▨ EuMC   ▨ EuMIC   ▨ EuRAD   ▨ Students   ▨ EuMW   ▨ Exhibitors

MONDAY 09:00 – 10:40

ROOM	Room 16	Room 1	Room 4	Room 13
	EuMC01 Advanced Packaging and Interconnect Technologies for Emerging Applications Chair: Mehmet Kaynak[1] Co-Chair: Mustafa Bakr[2] [1]IHP Microelectronics GmbH, [2]University of Oxford	**EuMIC/EuMC01** Novel Filtering Devices in Integrated Technologies Chair: Roberto Gomez-Garcia[1] Co-Chair: Michael Höft[2] [1]University of Alcala, [2]Christian-Albrechts-Universität zu Kiel	**EuMIC/EuMC02** THz components Chair: Emma MacPherson[1] Co-Chair: Oleksiy Sydoruk[2] [1]University of Warwick, [2]Imperial College London	**EuMIC12** Device Modelling and Simulation of Parasitic Phenomena Chair: Raphaël Sommet[1] Co-Chair: Valeria Brunel[2] [1]University of Limoges XLIM, [2]United Monolithic Semiconductors

ROOM	Room 16	Room 1	Room 4	Room 13
09:00 – 09:20	**EuMC01-1** Advanced Integration and Packaging of High-Power Components and Amplifiers for 5G/Beyond Industrial Applications Kamal K. Samanta[1] **INDUSTRIAL KEYNOTE** [1] AMWT Ltd, Edgware, UK	**EuMIC/EuMC01-1** A Millimeter-Wave Substrate Integrated Waveguide Filter in Si-BCB Technology Jordan Corsi[1], Giuseppe Acri[1], Maxime Moulin[1], Nicolas Zerounian[2], Anne-Sophie Grimault-Jacquin[2], Loic Vincent[2], Guillaume Ducournau[4], Frédéric Aniel[2], Florence Podevin[3], Philippe Ferrari[1], Emmanuel Pistono[1] [1]Université Grenoble Alpes, [2]Université Paris-Saclay, CNRS, Centre de Nanosciences et de Nanotechnologies, [3]Grenoble INP, [4]IEMN UMR 8520	**EuMIC/EuMC02-1** A SiGe Based 0.48 THz Signal Source with 45 GHz Tuning Range Jonathan Wittemeier[1], Florian Vogelsang[1], David Starke[1], Holger Rücker[2], Nils Pohl[1] [1]Ruhr-Universität Bochum, [2]IHP - Leibniz Institut für innovative Mikroelektronik	**EuMIC12-1** Noise Modeling of GaN/AlN HEMT Sanaul Haque[1], Frank Schnieder[2], Oliver Hilt[2], Ralf Doerner[2], Frank Brunner[2], Matthias Rudolph[1] [1]Brandenburg University of Technology, [2]Ferdinand-Braun-Institut gGmbH, Leibniz-Institut für Höchstfrequenztechnik
09:20 – 09:40	**EuMC01-2** Design and Measurement of Interconnects in Fan-Out Wafer-Level Packaging (FOWLP) for mm-Wave Applications up to 100 GHz Sherko Zinal[1], Ivan Ndip[2], Marco Rossi[2] [1]Fraunhofer Institut Zuverlässigkeit & Mikrointegration - IZM, [2]Fraunhofer IZM, Berlin	**EuMIC/EuMC01-2** A 100GHz Bandpass Filter Employing Shielded Folded Ridged Quarter-Mode SIW Resonator in CMOS Technology Baichuan Chen[1], Samundra Kumar Thapa[1], Adel Barakat[1], Ramesh Kumar Pokharel[1] [1]Kyushu University	**EuMIC/EuMC02-2** The Effect of Surface Passivation for Sub-THz Silicon Gradient Refractive Index Lens Antti Lamminen[1], Aleksi Tamminen[2], Jaakko Saarilahti[1], Vladimir Ermolov[1], Pekka Pursula[1] [1]VTT Technical Research Centre of Finland, [2]Aalto University	**EuMIC12-2** Efficient TCAD temperature-dependent Large-Signal simulation of a FinFET power amplifier Eva Catoggio[1], Simona Donati Guerrieri[1], Fabrizio Bonani[1], Giovanni Ghione[1] [1]Politecnico di Torino
09:40 – 10:00	**EuMC01-3** Demonstration of Millimeter-wave SMT Chip Scale Package using Hot-via MMICs and plastic BGA Encapsulation Alexandre Bessemoulin[1] [1]United Monolithic Semiconductors	**EuMIC/EuMC01-3** SAW Resonator Band-Pass Filter on GaN/Si operating at 8 GHz Alina Cristina Bunea[1], Dan Neculoiu[1], Adrian Dinescu[1] [1]National Institute for Research and Development in Microtechnologies - IMT Bucharest	**EuMIC/EuMC02-3** Optoelectronic Millimeter-Wave Integrated Circuits Fabricated in Pure Silicon-Based Technologies Uroschanit Yodprasit[1], Wolfgang Winkler[1] [1]Silicon Radar GmbH	**EuMIC12-3** A TCAD simulation study on gated-anode diodes for microwave applications Arijit Bose[1], Debaleen Biswas[1], Qiang Ma[1], Yoichi Tsuchiya[1], Hidemasa Takahashi[1], Yuji Ando[2], Akio Wakejima[1] [1]Nagoya Institute of Technology, [2]Nagoya University
10:00 – 10:20	**EuMC01-4** Modeling and Measurement of Double Stacked Microvia in Antenna-in-Package Module for 5G mmWave Applications Kavin Senthil Murugesan[1], Stefan Kosmider[2], Oliver Schwanitz[2], Uwe Maaß[2], Ivan Ndip[2], Andreas Ostmann[2], Klaus Dieter Lang[1] [1]TU Berlin, [2]Fraunhofer IZM, Berlin	**EuMIC/EuMC01-4** Engineered High Resistivity Silicon Substrates in IPD Technology Used for Miniaturized sub-6 GHz Filters Atte Haapalinna[1], Heikki Holmberg[1], Arto Hujanen[2], Katja Parkkinen[1], Pekka Rantakari[2], Tauno Vähä-Heikkilä[2] [1]Okmetic Oy, [2]VTT Technical Research Centre of Finland	**EuMIC/EuMC02-4** 140 GHz Differential Antennas in Embedded Wafer Level Ball Grid Array Technology Akanksha Bhutani[1], Elizabeth Bekker[1], Teng Li[1], Lucas Giroto de Oliveira[1], Thomas Zwick[1] [1]Karlsruhe Institute of Technology (KIT)	**EuMIC12-4** Trap Characterization in InAlN/GaN and AlN/GaN based HEMTs with Fe- and C-doped Buffers Emmanuel Dupouy[1], Vigneshwara Raja Paramasivan[1], Florent Gaillard[1], Raphaël Sommet[1], Jean-Christophe Nallatamby[1] [1]XLIM UMR 7252, University of Limoges/CNRS
10:20 – 10:40	**EuMC01-5** Optimised Hot-Via Transition with 20 dB Return Loss for MMIC Packaging from DC to 110 GHz Leigh Milner[1], Shyam Mehta[1], Leaonard Hall[1], Simon J. Mahon[2], Sudipta Chakraborty[3], Michael Heimlich[3] [1]DST Group, [2]Free Space Solutions, [3]Macquarie University	**EuMIC/EuMC01-5** Glass-Integrated Single- and Dual-Band Bandpass Filters Andrea Ashley[1], Dimitra Psychogiou[1] [1]University of Colorado Boulder	**EuMIC/EuMC02-5** Enhancing Mmwave on-Chip-Antennas Using in-Package Electromagnetic Bandgap Structures Dmitrii Kruglov[1], Oleg Iupikov[1], Marianna Ivashina[1], Rob Maaskant[1] [1]Chalmers University of Technology	**EuMIC12-5** Mechanisms of Buffer and Surface Traps in GaN HEMTs for Low Frequency Y21 and Y22 parameters Tomohiro Otsuka[1] [1]Mitsubishi Electric Corporation

PROGRAMME

MONDAY 09:00 – 10:40

ROOM	Room 14	Room 17	Room 6
	EuMIC13 Receiver Components Chair: Friedel Gerfers[1] Co-Chair: Lars-Erik Wernersson[2] [1]TU Berlin, [2]Lund Univeristy	**EuMIC14** Advances in mmWave and High Power Integrated PA Technologies Chair: Franco Giannini[1] Co-Chair: Simon J. Mahon[2] [1]University of Rome 'Tor Vergata', [2]Macquarie University	**EuMW01** Teaching Methods for Microwave Engineering Chair: David S Ricketts[1] Co-Chair: Bart Smolders[2] [1]North Carolina State University, [2]Eindhoven University of Technology

09:00 – 09:20	**EuMIC13-1** A Low Phase Noise Phase-Locked Loop With Short Settling Times for Automotive Radar Tobias T. Braun[1], Marcel van Delden[1], Christian Bredendiek[2], Jan Schoepfel[1], Nils Pohl[1] [1]Ruhr-University Bochum, [2]Fraunhofer FHR	**EuMIC14-1** A 100 GHz Class-F-Like InP-DHBT PA with 25.4% PAE Amit Shrestha[1], Ralf Doerner[1], Hady Yacoub[1], Tom Keinicke Johansen[2], Wolfgang Heinrich[1], Viktor Krozer[1], Matthias Rudolph[3], Andreas Wentzel[1] [1]Ferdinand-Braun-Institut gGmbH, Berlin, Germany, [2]Technical University of Denmark (DTU), Kgs. Lyngby, Denmark, [3]Ulrich L. Rohde Chair of RF and Microwave Techniques, Brandenburg University of Technology (BTU), Cottbus, Germany	**EuMW01-1** Teaching 100 remote students hands-on microwave design: Building a 16 QAM radio at home by hand David Ricketts[1], Jordan Besnoff[1] [1]North Carolina State University
09:20 – 09:40	**EuMIC13-2** A Passively-Coupled 39.5GHz Colpitts Quadrature VCO in SiGe HBT Technology Janis Wörmann[1], Aleksey Dyskin[2], Sébastien Chartier[3], Ingmar Kallfass[4] [1]Institute of Robust Power Semiconductor Systems (ILH) - University of Stuttgart, [2]Israel Institute of Technology, [3]IAF, Freiburg, [4]University of Stuttgart	**EuMIC14-2** A 117.5-130 GHz 22.1 dBm 11.5% PAE DAT Based Power Amplifier in InP 130 nm HBT Technology Linsheng Zhang[1], Vinay Iyer[1], Jay Sheth[1], Linli Xie[1], Robert M. Weikle[1], Steven Bowers[1] [1]University of Virginia	**EuMW01-2** Launching the First Massive Open Online Course (MOOC) on Microwave Engineering and Antennas A. Bart Smolders[1], Domine Leenaerts[2], Kevin Hastenberg[1], Ellen den Boer[1], Ulf Johannsen[1] [1]Eindhoven University of Technology, [2]NXP Semiconductors
09:40 – 10:00	**EuMIC13-3** 30-46 GHz 1.5dB IL Negative Gate Control SPDT with 24.5dBm IP1 in 130nm CMOS Sumeet Londhe[1], Noam Bar-Helmer[2], Samuel Jameson[2], Eran Socher[1] [1]Tel Aviv University, [2]Rafael Advanced Teachnologies	**EuMIC14-3** A 47-50GHz 3W MMIC Power Amplifier Using 100nm GaN Technology Seifeddine Fakhfakh[1], Guillaume Callet[1], Estelle Byk[1], Laurent Favede[1], Aleksandra Malko[2], Sandra Riedmüller[2], Pierre Denis[2], Hervé Blanck[2], Marc Camiade[1] [1]United Monolithic Semiconcuctors SAS, [2]United Monolithic Semiconductors GmbH	**EuMW01-3** RF Circuits Laboratory for Remote Learning and Massive Open Online Courses Carlos Mendes da Costa, Jr.[1], Lino van Mulken[1], Rainier van Dommele[1], Peter Baltus[1] [1]Eindhoven University of Technology - TU/e
10:00 – 10:20	**EuMIC13-4** A Highly Linear SiGe BiCMOS Gilbert-Cell based Downconversion Mixer for 5G Applications Mir Hassan Mahmud[1], Abdurrahman Burak[1], Can Çalışkan[1], Tahsin Alper Ozkan[1], Ali Bahadir Ozdol[1], Melik Yazici[1], Yasar Gurbuz[1] [1]Sabanci University	**EuMIC14-4** A D-Band Power Amplifier with 12 dBm OP1dB, 10 % Power Added Efficiency in InP-DHBT Technology Maruf Hossain[1], Tanjil Shivan[2], Ralf Doerner[2], Sten Seifert[2], Hady Yacoub[2], Tom Keinicke Johansen[3], Wolfgang Heinrich[2], Viktor Krozer[2] [1]Ferdinand-Braun-Institut (FBH) Leibniz-Institut für Höchstfrequenztechnik, [2]Ferdinand-Braun-Institut (FBH), [3]Technical University of Denmark (DTU)	**EuMW01-4** A Radar Kit for Hands-On Distance-Learning Markus Gardill[1], Tushar Tandon[2] [1]Brandenburgische Technische Universität Cottbus - Senftenberg, [2]Julius-Maximilians-Universität Würzburg
10:20 – 10:40	**EuMIC13-5** 37.2-to-42.0 GHz VCO with -93.4 dBc/Hz Phase Noise for FMCW Radar in 22 nm FDSOI Laszlo Szilagyi[1], Songhui Li[1], Xin Xu[1], Paolo Valerio Testa[1], Andres Seidel[1], Corrado Carta[1], Frank Ellinger[1] [1]Technische Universität Dresden	**EuMIC14-5** A 28-GHz-Band GaN HEMT MMIC Doherty Power Amplifier Designed by Load Resistance Division Adjustment Ryo Ishikawa[1], Takuya Seshimo[1], Yoichiro Takayama[1], Kazuhiko Honjo[1] [1]The University of Electro-Communications, Chofu, Tokyo, Japan	**EuMW01-5** Microwave Engineering education during COVID-19 pandemic: challenges and solutions implemented in practical work Andrés Fontana[1], Olivier Tantot[1], Nicolas Delhote[1], Serge Verdeyme[1], Denis Barataud[1], Cyril Decroze[1], Guillaume Neveux[1], Thomas Fredon[1], Guillaume Andrieu[1] [1]XLIM - Université de Limoges

MONDAY 10:40 – 13:30

Posters will be ready by 10:40. Presenters will be around their stands at 10:50 – 11:20 and 13:00 – 13:30.

Exhibition Hall

EuMIC15
EuMIC Posters

Chair: Mustafa Bakr[1]

[1]University of Oxford

EuMIC15-1
Field-Plate Mixer

Simon J. Mahon[1], Michael Heimlich[1]

[1]Macquarie University

EuMIC15-2
17.6 dB Variable-Gain and Variable-Bandwidth Upconverter in 65 nm CMOS for 60 GHz Bands

Oner Hanay[1], David Bierbüsse[1], Renato Negra[1]

[1]Chair of High Frequency Electronics, RWTH Aachen University

---

EuMIC15-3
A derating-rules compliant Ka-Band GaN-on-Si power amplifier designed for highly reliable satellite applications

Ferdinando Costanzo[1], Lorenzo Pace[1], Patrick Ettore Longhi[1], Walter Ciccognani[1], Sergio Colangeli[1], Rémy Leblanc[2], Ernesto Limiti[1]

[1]University of Rome "Tor Vergata", [2]OMMIC

EuMIC15-4
A 3.3 to 11.3 GHz Differential LNA with Slight Imbalance Active Balun in 0.15-μm GaAs pHEMT Process for Radio Astronomical Receiver

Ting-Hsuan Fan[1], Chau-Ching Chiong[2], Huei Wang[1]

[1]National Taiwan University, [2]Academia Sinica Institute of Astronomy and Astrophysics (ASIAA)

EuMIC15-5
Benefits of AlGaN/GAN thermal ROM coupling with industrial non-linear transistor model

Christophe Chang[1], Laurent Brunel[1]

[1]United Monolithic Semiconcuctors SAS

EuMIC15-6
1-6 GHz 35W Balanced GaN-HEMT Power Amplifier with Innovative Quadrature Couplers

Alexey Radchenko[1], Sergey Garmash[1], Andrei Kishchinsky[1]

[1]Microwave Systems JSC

EuMIC15-7
A 300 GHz Frequency Doubler in Transferred Substrate InP DHBT Technology

Arsen Turhaner[1], Maruf Hossain[2], Mohamed Brahem[2], Tom Keinicke Johansen[1]

[1]Technical University of Denmark, [2]Ferdinand Braun Institut (FBH)

EuMIC15-8
55% Fractional-Bandwidth Doherty Power Amplifier in 130-nm SiGe for 5G mm-Wave Applications

Aniello Franzese[1], Nebojsa Maletic[1], Mohamed H. Eissa[1], Muh-Dey Wei[3], Renato Negra[3], Andrea Malignaggi[1]

[1] IHP - Leibniz-Institut für innovative Mikroelektronik, [3]HFE RWTH-Aachen

---

EuMIC15-9
Full Octave Continuously Tunable SiGe Bipolar LC-VCO in Ku-Band

Christian Bredendiek[1], Klaus Aufinger[2], Nils Pohl[3]

[1]Fraunhofer FHR, [2]Infineon Technologies AG, [3]Ruhr-Universität Bochum

---

EuMIC15-10
An E-band Bidirectional PALNA in 0.13 μm SiGe BiCMOS Technology

Raju Ahamed[1], Mikko Varonen[2], Dristy Parveg[1], Md Najmussadat[1], Mikko Kantanen[2], Yehia Tawfik[1], Kari A. I. Halonen[1]

[1]Aalto University, [2]VTT Technical Research Centre of Finland

---

EuMIC15-11
A Ka-Band 40 W Output Power and 30 % PAE GaN MMIC Power Amplifier for Satellite Communication

Keigo Nakatani[1]

[1]Mitsubishi Electric Corporation

EuMIC15-12
Probabilistic Poly Harmonic Distortion Model

Anna Manjaly[1], Justin King[1]

[1]Trinity College Dublin

MONDAY 13:50 – 16:40

Posters will be ready by 13:40. Presenters will be around their stands at 13:50 – 14:20 and 16:00 – 16:30.

Exhibition Hall

EuMIC/EuMC04
EuMIC/EuMC Posters

Chair: Mustafa Bakr[1]

[1]University of Oxford

EuMIC/EuMC04-1

Microwave sensing using metal-insulator-metal diodes based on 4-nm-thick hafnium oxide

Martino Aldrigo[1], Mircea Dragoman[1], Sergiu Iordanescu[1], Mazen Al Shanawani[2], George Deligeorgis[3]

[1]National Institute for Research and Development in Microtechnologies (IMT), [2]University of Bologna, [3]FORTH

EuMIC/EuMC04-2

Automatic Nonlinear Nonquasi-Static Diode Model Extraction from Large-Signal Measurements

Aarón García-Luque[1], Teresa M. Martín-Guerrero[1], Alberto Santarelli[2], Carlos Camacho-Peñalosa[1]

[1]Universidad de Málaga, Andalucía Tech, [2]Università di Bologna

---

EuMIC/EuMC04-3

Compact GaN RF-Switches for Power Applications

Samira Driad[1], Charles Teyssandier[1], Laurent Caillé[1], Christophe Chang[2], Laurent Brunel[1], Benoit Lambert[1], Hermann Stieglauer[2], Valeria Brunel[1]

[1]United Monolithic Semiconductors SAS, [2]United Monolithic Semiconductors GmbH

EuMIC/EuMC04-4

Analysis of RF Stress Influence on Large-Signal Performance of 22nm FDSOI CMOS Transistors utilizing Waveform Measurement

Dang Khoa Huynh[1], Quang Huy Le[1], Steffen Lehmann[2], Zhixing Zhao[2], Germain Bossu[2], Wafa Arfaoui[2], Defu Wang[1], Thomas Kämpfe[1], Matthias Rudolph[3]

[1]Fraunhofer Institute for Photonic Microsystems (IPMS), [2]Globalfoundries, Germany, [3]Brandenburg University of Technology (BTU)

EuMIC/EuMC04-5

Towards an Excitable Microwave Spike Generator for Future Neuromorphic Computing

Qusay Raghib Ali Al-taai[1], Razvan Morariu[1], Jue Wang[1], Abdullah Al-Khalidi[1], Ali Al-Moathin[1], Bruno Romeira[2], José Figueiredo[3], Edward Wasige[1]

[1]University of Glasgow, [2]International Iberian Nanotechnology Laboratory, [3]Universidade de Lisboa, Campo Grande

---

EuMIC/EuMC04-6

Numerical and Experimental Investigations of Selfmixing Effect of a Planar Gunn Diode Oscillator

Mingyan Zhong[1]

[1]University of Glasgow

EuMIC/EuMC04-7

An Ultra-Wideband Microstrip-to-WR15 Waveguide Transition for MMIC Applications

Bent Walther[1], Marcel van Delden[1], Thomas Musch[1]

[1]Ruhr-University Bochum

EuMIC/EuMC04-8

An Integrated Multiphysics Model for Phase-Change Material Switches

Pierre Blondy[1], Ines Bettoumi[2], Kateryna Kiryukhina[3], Olivier Puig[3]

[1]Xlim - UMR 7252 - CNRS- Universite De Limoges, [2]XLIM-UMR CNRS 7252 - Université de Limoges, [3]Centre National d'Études Spatiales (CNES)

EuMIC/EuMC04-9

Doherty Load Modulation Based on Non-Reciprocity

Paul Saad[1], Han Zhou[2], Jose-Ramon Perez-Cisneros[2], Rui Hou[1], Christian Fager[2], Bo Berglund[1]

[1]Ericsson AB, [2]Chalmers University of Technology

EuMIC/EuMC04-10

Adopting Supercapacitors in a Single-Stage Marx-Type Multi-level Supply Modulator

Lukas Hüssen[1], Renato Negra[1]

[1]HFE RWTH-Aachen

EuMIC/EuMC04-11

A 30-W GaN Quasi-MMIC Doherty Power Amplifier Based on All-Distributed Inductors Load Network

Rui-Jia Liu[1], Xiao-Wei Zhu[1], Jing Xia[2], Peng Chen[1], Chao Yu[1], Lv Zhang[3], Zhi-Yong Chen[3]

[1]Southeast University, [2]Jiangsu University, [3]Guobo Electronics Corporation

---

EuMIC/EuMC04-12

A Digital Power Amplifier for 32-QAM

Gavin Watkins[1]

[1]Toshiba Europe Limited

EuMIC/EuMC04-13

Effect of Switch Figure of Merit on Frequency-Reconfigurable Power Amplifier Performance

Adam Der[1], William Sear[1], Taylor Barton[1]

[1]University of Colorado, Boulder

EuMIC/EuMC04-14

Practical Work for Master2 Students: MMIC Distributed Amplifier Design for High Data Rate Receiver on GaAs-UMS Technology

Catherine Algani[1], Eric Leclerc[2]

[1]Le Cnam, [2]UMS

MONDAY 14:20 – 16:00

ROOM	Room 1	Room 6	Room 13	Room 14
	EuMC02 Innovative Microwave Circulators and Phase Shifters	**EuMC03** Non-planar Filters I	**EuMC04** Active Antennas and Architectures	**EuMIC/EuMC03** MMIC Power Amplifiers and Supply Modulation
	Chair: Bart Nauwelaers[1]	Chair: Giuseppe Macchiarella[1]	Chair: Nils Pohl[1]	Chair: Jeff Powell[1]
	Co-Chair: Marco Pasian[2]	Co-Chair: Vicente E. Boria[2]	Co-Chair: Kevin Morris[2]	Co-Chair: Markus Mayer[2]
	[1]KU Leuven, [2]University of Pavia	[1]Politecnico di Milano[2], Universitat Politecnica de Valencia	[1]Ruhr University Bochum, [2]University of Bristol	[1]Teratech Components, [2]Arelis

14:20 – 14:40	**EuMC02-1** Microwave Ferrite Components - an Industry Perspective	**EuMC03-1** The Extracted Zero Technique	**EuMC04-1** 7.5 GHz-Band Digital Beamforming Using 1-bit Direct Digital RF Transmitter with 10GbE Optical Module	**EuMIC/EuMC03-1** A 6-18 GHz 13 W and 22% PAE GaN Power Chipset
	John Ascroft[1] **INDUSTRIAL KEYNOTE** [1] Honeywell	Simone Bastioli[1] **INDUSTRIAL KEYNOTE** [1]RS Microwave	Ryo Tamura[1], Mizuki Motoyoshi[1], Suguru Kameda[1], Noriharu Suematsu[1] [1]Research Institute of Electrical Communication, Tohoku University	Mehdi DINARI[1], Benoit MALLET-GUY[2], Yves Mancuso[2] **INDUSTRIAL KEYNOTE** [1]Thales DMS France, [2]Thales Defence Mission Systems (TDMS)

14:40 – 15:00	**EuMC02-2** Broadband Ku- and Ka-Band Circulators in LTCC Using Sintered Bulk Ferrites	**EuMC03-2** Dielectric-loaded Ku-Band Filter for High-power Space Applications based on Barrel-shaped cavities	**EuMC04-2** Quadruple-fed Aperture-coupled Microstrip Patch Antenna for On-antenna Power Combining	**EuMIC/EuMC03-2** On-chip Power Combining with 3-Stage75-110 GHz GaN MMIC Power Amplifiers
	Carsten Weil[1], Tim Hauck[1], Johannes Schur[2], Jens Müller[2] [1]AFT microwave GmbH, [2]TU Ilmenau	Paolo Vallerotonda[1], Fabrizio Cacciamani[2], Luca Pelliccia[2], Francesco Aquino[3], Cristiano Tomassoni[3], Petronilo Martín-Iglesias[4], Vittorio Tornielli di Crestvolant[4] [1]RF Microtech s.r.l / University of Perugia, [2]RF Microtech s.r.l, [3]University of Perugia, [4]ESA / ESTEC	Timothée Le Gall[1], Anthony Ghiotto[2], Stefan Varault[1], Gwenaël Morvan[1], Bruno Louis[1], Grégoire Pillet[1] [1]Thales DMS France, [2]Bordeaux INP, IMS Laboratory	Shane Verploegh[1], Timothy Sonnenberg[1], Mauricio Pinto[2], Akim Babenko[1], Zoya Popovic[1] [1]University of Colorado at Boulder, [2]Raytheon Company

15:00 – 15:20	**EuMC02-3** Quasi-Reflectionless Differential Phase Shifter with Arbitrary Prescribed Group Delay and Flat Phase Difference	**EuMC03-3** LTCC based Ka-Band Diplexer for Miniaturized Ground-Segment User Terminals	**EuMC04-3** Antenna Mutual-Coupling Mitigation With Analogue Compensation Network	**EuMIC/EuMC03-3** Wideband Phase Modulator MMIC for K-Band Supply-Modulated Power Amplifier Linearization
	Girdhari Chaudhary[1], Daehan Lee[1], Muhammad A. Chaudary[2], Yongchae Jeong[1] [1]Jeonbuk National University, [2]Ajman University	Davide Tiradossi[1], Paolo Vallerotonda[2], Luca Pelliccia[1], Stefano Moscato[3], Antonio Traversa[3], Giandomenico Cannone[3], Petar Jankovic[4], Fabrizio De Paolis[4] [1]RF Microtech s.r.l, [2]RF Microtech s.r.l / University of Perugia, [3]SIAE Microelettronica S.p.A., [4]ESA / ESTEC	Roger Green[1], Tommaso Cappello[1], Geoffrey Hilton[1], Mark Beach[1] [1]University of Bristol	Gregor Lasser[1], Connor Nogales[1], Maxwell R. Duffy[2], Zoya Popovic[1] [1]University of Colorado, Boulder, [2]Northrop Grumman Corporation

15:20 – 15:40	**EuMC02-4** A Phase Shifter Composed of Reduced-Size Rat-Race Coupler with CRLH Transmission Lines and Resonating Reactance Circuits	**EuMC03-4** Quadrature-Based Approach Used for Improved Fitting of Filter Measured S-parameters	**EuMC04-4** Conformal Antenna with Reconfigurability of Monopole-like and Broadside Patterns Realized with Polymer-Conductive Textile Composite	**EuMIC/EuMC03-4** Compact Design of a L-Band 40W 40 MHz Envelope Tracking GaN Power Amplifier for Small Cells
	Masashi Nakatsugawa[1], Fusuke Kurotani[1], Yuya Chiba[1], Tamami Maruyama[1] [1]National Institute of Technology, Hakodate College	Jedrzej Michalczyk[1], Jerzy Michalski[1] [1]SpaceForest	Roy B. V. B. Simorangkir[1], Bahare Mohamadzade[2], Ali Lalbakhsh[2], Sanjeev Kumar[3], John L. Buckley[1], Toni Bjorninen[4], Brendan O'Flynn[1] [1]Tyndall National Institute, [4]Macquarie University, [3]Tampere University	Olivier Nonet[1], Wilfried Demenitroux[1], Frederic Ploneis[1], Denis Barataud[2], Michel Campovecchio[2] [1]Thales Group, [2]Xlim - CNRS- Unversite De Limoges

15:40 – 16:00	**EuMC02-5** Simultaneous Electric and Magnetic Two-Dimensional Tuning in Nonlinear Magnetic Transmission Line	**EuMC03-5** Narrowband Extracted Pole Filters With Mixed Dielectric and Waveguide Resonators in Ku-Band	**EuMC04-5** Design of a Multi-mode Transmission System Based on Vortex Electromagnetic Wave	**EuMIC/EuMC03-5** A 600-W Enhancement-Mode GaN Multi-Level Dynamic Converter for Supply Modulated PAs
	MuhibUr Rahman[1], Ke Wu[1] [1]Polytechnique Montreal	Patrick Boe[1], Daniel Miek[1], Fynn Kamrath[1], Kennet Braasch[1], Michael Höft[1] [1]Christian-Albrechts-Universität zu Kiel	Jialin Zhang[1] [1]Beihang University (BUAA)	Connor Nogales[1], Zoya Popovic[1], Gregor Lasser[1] [1]University of Colorado at Boulder

MONDAY 14:20 - 16:00

ROOM	Room 4	Room 17
	EuMIC16	**EuMW03**
	Phased Array Components from S-band up to 300 GHz	Special Session in Memoriam of Prof. Roberto Sorrentino
	Chair: Frank E. van Vliet[1]	Chair: Cristiano Tomassoni[1]
	Co-Chair: Michael Schlechtweg[2]	Co-Chair: Maurizio Bozzi[2]
	[1]TNO, [2]Fraunhofer Institute for Solid State Physics IAF	[1]University of Perugia, [2]University of Pavia

14:20 – 14:40	**EuMIC16-1** An S-band 34dBm Stacked-HBT Phase Driver in 0.25µm BiCMOS Technology for GaN-Based Phased-Array Radar Transmit Chain Jaap Essing[1], Alice Bossuet[1], Rob Knight[1], Peter de Hek[1], Frank E. van Vliet[1] [1]TNO	**EuMW03-1** How did EuMA start? André Vander Vorst[1] [1]European Microwave Association
14:40 – 15:00	**EuMIC16-2** A Phase Coherent DC-25 GHz 6-bit SiGe BiCMOS Step Attenuator with IP1dB >20 dBm Hamza Kandis[1], Abdurrahman Burak[1], Cengizhan Kana[1], Melik Yazici[1], Yasar Gurbuz[1] [1]Sabanci University	**EuMW03-2** Remembering Roberto Sorrentino-A man inspired by knowledge and culture Wolfgang Hoefer[1] [1]University of Victoria
15:00 – 15:20	**EuMIC16-3** A 26 GHz to 34 GHz Active Phase Shifter with Tunable Polyphase Filter for 5G Wireless Systems Alok Sethi, Rehman Akbar[1], Mikko Hietanen[1], Timo Rahkonen[2], Aarno Pärssinen[1] [1]Center for Wireless Communication, University of Oulu, [2]Circuits and systems Group, University of Oulu	**EuMW03-3** My time with Roberto Richard Snyder[1] [1]RS Microwave
15:20 – 15:40	**EuMIC16-4** A 25-50 GHz Digitally Controlled Phase-Shifter Steeven Voisin[1], Vincent Knopik[1], Eric Kerhervé[2] [1]STMicroelectronics, [2] University of Bordeaux, Bordeaux INP, UMR CNRS 5218, IMS Laboratory	**EuMW03-4** Roberto Sorrentino as EuMA President Wolfgang Heinrich[1] [1]Ferdinand-Braun-Institut gGmbH, Leibniz-Institut für Höchstfrequenztechnik
15:40 – 16:00	**EuMIC16-5** A 270 – 330 GHz Vector Modulator Phase Shifter in 130nm SiGe BiCMOS Mohammad Hassan Montaseri[1], Sumit Pratap Singh[1], Markku Jokinen[1], Timo Rahkonen[1], Marko E. Leinonen[1], Aarno Parssinen[1] [1]University of Oulu	**EuMW03-5** When academic excellence gets inspired by new challenges: the growth of RF Microtech Elisa Fratticcioli

PROGRAMME

MONDAY 16:40 – 18:20

Room 8 – 11

EuMIC17
EuMIC Closing Session

Chair: Chris Clifton[1], EuMIC Chair

Co-chair: Shokrollah Karimian[2], EuMIC Co-chair; Edward Wasige[3], EuMIC TPC Chair

[1]Sony Europe B.V., [2]University of Oxford, UK, [3]University of Glasgow, UK

16:40 – 16:50 Awards Ceremony

Kamal K Samanta[1]

[1]EuMW 2021 Awards Chair

EuMIC Prize
EuMIC Young Engineer Prize
Tom Brazil Fellowship Award (by the GAAS® Association)

16:50 – 17:30 Efficient, Broadband and Linear Radio Frequency Amplifier Architectures

Kevin Morris[1]

[1]University of Bristol, UK

As communications systems have evolved to provide ever higher data rates this has resulted in the need for amplifiers that operate over wider bandwidths with signals with increasing peak to average power ratios. This creates significant challenges with power consumption. Gallium Nitride (GaN) devices have enabled the design of broadband, linear and efficient power amplifiers. GaN devices with their higher breakdown voltages and higher input and output impedances expand the design space to make use of more efficient and wideband amplifier classes. This talk will explore how broadband amplifier designs can be combined with linearisation techniques to create amplifiers that can meet both the linearity and efficiency requirements for future systems. The talk will also look at how device technology is developing with the move to emerging GaN on diamond technology.

17:30 – 18:10 6G – Known Technologies with a Twist or Maybe Not?

Nadine Collaert[1]

[1]Imec, Leuven

With 5G in full deployment, industry has kickstarted the research for the next generation of wireless communication 6G. While the vision for 6G still needs to take shape, and with CMOS scaling under pressure, more than ever System-Technology Co-optimization (STCO) will be needed to define the best blend of technologies to get benefits at the system level. In this talk, we will discuss how compound semiconductor devices and advanced packaging could play a key role in enabling ultra-fast, reliable and power-efficient connectivity

18:10 – 18:20 Closing remarks and invitation to EuMIC 2022

Chris Clifton[1], Paolo Colantonio[2]

[1]EuMIC 2021 Chair, [2]EuMIC 2022 Chair

TUESDAY OVERVIEW

Room	09:00 – 10:40	11:20 – 13:00	14:20 – 16:00	16:40 – 18:20	EVENING PROGRAMME
1		EuMC17 New Design Concepts for Microwave Filters in Planar and Hybrid Technologies	EuMC22 Advanced Implementations for Substrate-Integrated and Quasi-Planar Filters		
2	Exhibitor Workshops		Exhibitor Workshops		
3	Exhibitor Workshops		Exhibitor Workshops		
4	EuMC13 Non-planar Passive Components	EuMC18 Frequency Generation, Conversion and Nonlinear Modelling		EuMW04 Memorial Session for Professor Tatsuo Itoh	
5	Exhibitor Workshops		Exhibitor Workshops		
6	SW01 Joint Range-angle Super Resolution MIMO Radar		SW02 Radar Design from the Ground Up		
7	WW01 Technologies for 6G Front End Modules				
8					
9	EuRAD01 Opening Session	Defence, Security and Space (DSS) Forum			
10					
11					
12	Automotive Forum				
13	EuMC14 Electromagnetic Scattering and Diffraction Effects	EuMC19 3D Printing: Processes and Reliability	EuMC23 5G Communication and Beyond"	EuMC26 Novel 3D Printing Approaches for mmWave Applications	
14	EuMC15 Metamaterial Based Devices and Applications	EuMC20 Advanced High Efficiency Power Amplifier Techniques	EuMC24 Advances in Electromagnetic Modeling and Numerical Techniques	EuMC27 Measurements for 5G and 6G Systems	
15		IEEE Young Professionals Lunch	IEEE Young Professionals Session		
16					
17	EuMC16 Integrated Components for Transceivers	EuRAD02 Radar Applications	EuRAD03 Emerging Radar Applications	EuMC28 5G and mmWave Arrays	
Exhibition Hall		EuMC21 Posters	EuMC25 Posters		
North Greenwich Pier (by the O2)					The EuMW Cruise on the River Thames 19:00 – 22:00
Off-site					Automotive Forum Networking Dinner 19:00 – 22:00

■ EuMC   ■ EuMIC   ■ EuRAD   ■ Students   ■ EuMW   ■ Exhibitors

Welcome from the Workshop and Short Courses Chairs

As the first physical event in the field of microwave engineering after the outbreak of COVID-19 pandemic, EuMW 2021 committee are uniting students, academics, and industrial experts again in a less formal format at our workshops and short courses sessions. After careful considerations, we are pleased to offer an extensive and diverse programme of workshops and short courses throughout the entire week.

The wide-ranging programme of half-day and full-day workshops and short courses has been chosen to cover a range of important topics of interest to the whole EuMW community. The short courses will cover fundamental knowledge of specific areas or hands-on experience such as power amplifier designs, filter simulation, radar signal processing and AI technology for designing antennas and filters etc.; on the other hand, the workshops attracted world-leading scientists and engineers in the fields showcasing the latest developments of popular subjects such as 6G front-end developments, wireless power transfers, microwave in biomedical engineering, advanced manufacturing, and millimetre-wave and terahertz on-wafer/off-wafer S-parameters, power and load-pull measurements etc. Advances in semiconductor devices circuits based on GaN, SiC, CMOS and SiGe technologies and their applications will also be covered in a series of workshops.

We are very grateful to all the organisers, presenters and authors of workshops and short courses for their hard work and dedication before and during the conference. Each workshop and short course is individually endorsed by one or two of the conferences of EuMW. However, they are available and accessible to any scientist or engineer wishing to gain a broader perspective on microwave and RF systems and devices, or to learn about a new specialism within our broad field. Workshop organisers have been asked to provide panel sessions within their events for discussion and interaction, and we hope that you will benefit from participating in the international networking opportunity that this will present.

The workshops and short courses are mainly arranged on Saturday, Sunday and Thursday with a few on Tuesday and Wednesday. The EuMC endorsed sessions are distributed over the entire week and the EuRAD endorsed sessions are scheduled on Tuesday onwards, after the end of the main EuMIC conference sessions. We are confident that this structure will enable you to attend multiple workshops or short courses to incorporate into your schedule for the week, to enhance conference experience in London.

Slides for the workshops and short courses will be available to download from the conference's websites approximately two weeks before the conference. No hard copies of the slides will be provided. Instructions for the download process will be provided to the registered participants near the conference.

Finally, we would like to welcome you all in London in April 2022 and be United in Microwaves!

CHONG LI
Workshop, Short Courses Chair
University of Glasgow, UK

QAMMER ABBASI
Workshop, Short Courses Co-Chair
University of Glasgow, UK

SATURDAY 09:00 – 18:20

WS02
EuMIC/
EuMC

Terahertz Device, Circuit and System Fundamentals and Applications

Chair: Dimitris Pavlidis[1]

Co-Chair: Imran Mehdi[2] and Javier Mateos[3]

[1]Florida International University, [2]Jet Propulsion Laboratory, [3]University of Salamanca

Room 4

THz technology has reached a certain degree of maturity but there are still important developments necessary for implementing it to systems. At the same time, there are still needs for device and circuit studies in order to improve, frequency, power, sensitivity performance and provide integrated solutions to system requirements. The workshop will provide the opportunity to new generations of scientists and engineers to learn about the unique features of Terahertz technologies, while at the same time addressing the latest achievements in the field. THz applications to be discussed extending among from sensing and spectroscopy to communications and imaging. The workshop will bring together experts from various academic, national labs and commercial enterprises to discuss the most recent advances in their respective fields and to provide insight into what the future might hold for exploration of this frequency range. It will focus on a variety of materials such as traditional III-Vs, III-Nitrides, Silicon, Graphene and Transition metal dichalcogenides (TMDs), as well as various device concepts for efficient THz generation and detection. The operation of the components to be discussed is based on plasmonics, photoconductors, plasma waves, photomixing, Resonant Tunneling, Negative Differential Resistance, CMOS and High-Electron Mobility Transistors. Devices such as Quantum Cascade Lasers, Self-switching Diodes and Uni-Traveling-Carrier Photodiodes and nanoscale Vacuum Transistors will also be addressed. Advanced Sensing, Imaging and Communications and terrestrial, space applications will be discussed. The Workshop is intended for young scientists and engineers who are interested in learning about this emerging field, as well as individuals with a more advanced understanding of related concepts. The topics addressed include fundamental and engineering considerations together with the latest results in Terahertz technology

PROGRAMME

THz applications: from devices to space systems
Imran Mehdi[1]
[1]Jet Propulsion Laboratory

Nitride-based two- and three-terminal devices for THz applications; from diodes to transistors and Nanoscale Vacuum transistors
Dimitris Pavlidis[1]
[1]Florida International University

Terahertz characterization and applications of III-Nitride and complex oxide heterostructures
Berardi Sensale-Rodriguez[1]
[1]The University of Utah

Gated planar nanodiodes for tHz detection
Javier Mateos[1]
[1]University of Salamanca

InP HBTs for THz microsystems
Miguel Urteaga[1]
[1]Teledyne Scientific Company

Low-power consumption THz metasurface quantum-cascade VECSELs.
Benjamin Williams[1]
[1]University of California Los Angeles

THz devices and systems: from technology to applications
Guillaume Ducournau[1], J. F. Lampin[1], E. Peytavit[1] and S. Barbieri[1]
[1]CNRS - University of Lille

Terahertz communications using resonant tunneling diodes
Masayuki Fujita[1]
[1]Osaka University

Resonant-tunneling-diode THz oscillators and applications
Safumi Suzuki[1] and Masahiro Asada[1]
[1]Tokyo Institute of Technology

High-speed terahertz wireless is hot, how about its contrary?
Ruonan Han[1]
[1]MIT

Challenges and advances in terahertz antennas
Maria Alonso del Pino[1]
[1]Delft University of Technology

THz spectroscopy of agricultural samples
Marion K. Matters-Kammerer[1]
[1]Eindhoven University of Technology

SATURDAY 09:00 – 18:20

WS06
EuMIC/
EuMC

Progress and Status of Gallium Nitride Monolithic Microwave Integrated Circuits

Chair: Rüdiger Quay[1]

Co-Chair: Farid Medjdoub[2] and Ernesto Limiti[3]

[1]Fraunhofer IAF, [2]U. Lille/IEMN, [3]University of Rome

Room 10

This workshop gives an overview of the progress of important Gallium Nitride MMIC technologies available to the microwave and RF community for frequencies from 400 MHz to 200 GHz. Prominent industrial vendors of GaN MMICs have been invited and have agreed to participate. Several international speakers will give their view to the evolution of important applications such as sensing, data com, with emphasis on mmWave IC technology. Several roadmaps will be provided to enable the audience to estimate the progress of MMIC on a global scale. Further, the research progress with respect to higher frequency scaling beyond commercial technologies is addressed. The workshop thus will provide an overview on the overall status of MMIC technology, device technology, circuit design, reliability, and integration.

PROGRAMME

Self-configuring, adapting and reconfigurable GaN MMICs
Charles Campbell[1]
[1]Qorvo (USA)

Recent development of GaN power Technology applied to RF sensors
Didier Floriot[1]
[1]UMS (France, Germany)

State of the art mmWave GaN/Si MMICs
Marc Rocchi[1]
[1]Ommic (France)

Design of high performance microwave and millimeter wave GaN HPAs
Bill Pribble[1]
[1]Wolfspeed (USA)

Design of GaN power amplifier MMICs operating beyond 100 GHz
Maciej Cwiklinski[1]
[1]Fraunhofer IAF (Germany)/Rohde und Schwarz

mmWave space applications using GaN at ESA
Piero Angeletti[1]
[1]European Space Agency/ESA

Reliability evaluation, failure modes and mechanisms of scaled RF GaN high electron mobility transistors
Enrico Zanoni[1], Matteo Buffolo[1], Carlo De Santi[1], Matteo Meneghini[1] and Gaudenzio Meneghesso[1]
[1]U. Padova

Technological development towards high performance

mmWave GaN HEMTs and enhanced reliability
Farid Medjdoub[1]
[1]U. Lille/IEMN

Multifunctional front-end integration for radar/earth observation at mmWaves
Ernesto Limiti[1]
[1]U. Rome Tor Vergata

WORKSHOPS AND SHORT COURSES

WWW.**EUMW2021**.COM

SATURDAY 09:00 – 18:20

SS02
EuMIC

Fundamentals of Microwave PA Design

Chair: Paolo Colantonio[1]

Co-Chair: Franco Giannini[1]

[1]University of Roma Tor Vergata

Room 8

This short course aims to provide a comprehensive overview of all aspects related to the design of microwave power amplifier design. It is an introductory course, dedicated to graduate engineers who have moved into the field of RF design, as well as to microwave designers who aim to deeply understand the power amplifier basic concepts. This short course features a range of presentations and will provide a comprehensive overview and basic understanding on recent important progress and novel state-of-the-art achievements in semiconductor power amplifiers. Advances in semiconductor amplifiers and their applications will also be covered.

Starting from the fundamental concepts on semiconductor devices and their modelling development, the theoretical foundations of a power amplifier design are discussed. It will include fundamental concepts and state-of-the-art results on actual designs of a range of semiconductor power amplifiers using existing foundries. The load pull technique is also addressed and focused from the designer perspective.

The presentations will also cover a variety of advanced topics and will provide the attendees with a clear overview of the main streams of current and important research trends worldwide in this field, as the Doherty architecture and the load modulation power amplifier design concepts.

The short course will also focus on the major challenges, such as stability (small and large signal) and how to address these in amplifier design. Finally, accounting for the linearity issue, a basic overview on linearization techniques and their adoption to properly mitigate the amplifier distortion effects will conclude the short course.

PROGRAMME

Semiconductor devices and modelling for PAs

Iltcho Angelov[1]

[1]Chalmers University

PA basic concepts

Franco Giannini[1]

[1]University of Roma Tor Vergata

Design and model oriented load pull techniques

Marco Pirola[1]

[1]Politecnico di Torino

The Doherty power amplifier

Paolo Colantonio[1]

[1]University of Roma Tor Vergata

Load modulated PAs

Steve Cripps[1]

[1]Cardiff University

X-parameters high-power PAs modeling for system level analysis

Alessandro Cidronali[1]

[1]University of Florence

Linear and nonlinear stability analysis of power amplifiers

Giorgio Leuzzi[1]

[1]University of L'Aquila

Linearization techniques overview

Pere L. Gilabert[1]

[1]Universitat Politècnica de Catalunya UPC-Barcelona Tech.

Design of the power amplifier section of a X-band MMIC single chip front end

Davide Resca[1], Francesco Scappaviva[1]

[1]MEC srl

WORKSHOPS AND SHORT COURSES

WWW.**EUMW2021**.COM

SUNDAY 09:00 – 18:20

Advances in Circuits and Systems for mmWave Radar and Communication in Silicon Technologies

WM02
EuMIC/
EuMC

Chair: Vadim Issakov[1]

Co-Chair: Farzad Inanlou[2]

[1]TU Braunschweig, [2]Globalfoundries

Room 2

Recent developments in nano-scale CMOS, allow for MOS transistors to achieve fT and fmax in excess of several hundreds of gigahertz. This enables realization of highly integrated radar and communication systems operating at mmWave frequencies. Particularly, the frequency range around 140 GHz is an interesting candidate to become approved for licensed usage worldwide in the near future for radar and for 6G wireless communication applications. In this workshop, we discuss highly-integrated radar and communication systems operating at W-band and D-band realized in advanced nano-scale CMOS and BiCMOS technologies. The workshop offers a balanced distribution between both fields. We cover a wide range of topics starting from the technology choice for mmWave applications, a talk by

Globalfoundries. Bosch will provide a vision on a fully integrated automotive radar system-on-chip in 22nm FDSOI technology. The team of Prof. Zwick (IEEE Fellow) presents packaging and antenna solutions for D-band FMCW radar. Next, mixed-signal part and RF part of a digitally modulated PMCW 140 GHz radar transceiver is discussed by TU Dresden and Infineon, respectively. Second half of the workshop focuses of communication transceivers towards 6G. It covers system considerations, mixed-signal part (ADCs and DACs) and novel system architectures. Finally, Prof. Mark Rodwell closes the workshop with a talk on 140 GHz MIMO arrays transceiver in CMOS and InP. In this workshop we have a good mixture of industry (Globalfoundries, Bosch, Infineon) and academia (FAU Erlangen, TU Dresden, TU Berlin, TU

Braunschweig, Karlsruhe Institute of Technology). We have presentations from Europe and the USA. We will round up the workshop by a panel discussion in which we will address the challenges and future directions for circuit design for mmWave frequency in radar and communication transceivers.

PROGRAMME

Enabling silicon technologies for mmWave radar trends and requirements

Farzad Inanlou[1]

[1]Globalfoundries

An automotive radar demonstrator with a 22nm CMOS FD-SOI transceiver

Philipp Ritter[1] and Juergen Hasch[1]

[1]Robert Bosch GmbH

Millimeter-wave antenna and packaging solutions for D-band FMCW radar Systems

Akanksha Bhutani[1] and Thomas Zwick[1]

[1]Karlsruhe Institute of Technology

High-speed ADCs for D-band radar in 22 nm FDSOI CMOS

Simon Buhr[1] and Frank Ellinger[1]

[1]Dresden University of Technology (TU Dresden)

Design of a D-band PMCW radar transceiver in 45 nm RFSOI technology

Vadim Issakov[1], Vincent Lammert[2] and Michael Leyrer[3]

[1]TU Braunschweig, [2]Infineon Technologies, [3]Infineon

mmWave advanced-sampling transceiver enabling 6G data transmission with 100 Gbit/s per mobile User

Patrick James Artz[1], Julius Edler[1] and Friedel Gerfers[1]

[1]Technische Universität Berlin

High-speed DACs in 22 nm FDSOI CMOS for D-band wireless communication towards 6G

Tobias Schirmer[1] and Frank Ellinger[1]

[1]Dresden University of Technology (TU Dresden)

High-speed ADC (>20 GS/s) with high resolution (≥10 bit) for Low-IF receiver in 22nm FDSOI

Julius Edler[1], P. Artz[1], E. Wittenhagen[1], N. Lotfi[1] and F. Gerfers[1]

[1]Technische Universität Berlin

Highly-integrated mmWave transceivers for communication systems in BiCMOS technologies

Marco Dietz[1] and Robert Weigel[1]

[1]Friedrich-Alexander University of Erlangen-Nuremberg

D-band CMOS+InP and CMOS-only MIMO communication transceiver technologies

Mark Rodwell[1], Ali Farid[1], Ahmed Ahmed[1], Utku Solyu[1] and Munkyo Seo[1]

[1]UC Santa Barbara

WORKSHOPS AND SHORT COURSES

WWW.**EUMW2021**.COM

TUESDAY 09:00 – 18:20

WW01
EuMIC/
EuMC

Technologies for 6G Front End Modules

Chair: Ned Cahoon[1]

Co-Chair: Jack Pekarik[1]

[1]GlobalFoundries

Room 7

Carrier frequencies > 100GHz are attractive for next generation 6G cellular systems due to the large amount of available spectrum in the D- and G-bands that can be leveraged for high data rate communications. Operation at these high mmWave frequencies comes with many challenges, though, particularly in the demands placed on technology performance, integration and cost for the phased array front end. Losses on and off chip are very high. Transistor performance is significantly worse, with challenges in achieving acceptable gain, Pout, and PAE for the PA, acceptable gain and NF for the LNA and low insertion loss for the switch. Thermal management and antenna/FEM/transceiver integration will be particularly demanding due to the constraints of the lattice spacing at these frequencies. This workshop will delve into the candidate semiconductor and packaging technologies for the 6G beamformer FEM, and will explore the unique strengths and limitations of each for addressing these challenges.

PROGRAMME

View and trends in RF design towards 6G

Aarno Pärssinen[1]

[1]University of Oulu

SiGe BiCMOS technologies for 6G millimeter-wave

Pascal Chevalier[1]

[1]STMicroelectronics

State of the art SOI technology for mmWave FEM applications

Sameer Jain[1]

[1]GlobalFoundries

D-band circuits in 16nm FinFET: design and layout considerations

Patrick Reynaert[1] and Bart Philippe[1]

[1]KU Leuven ESAT-MICAS

InP and GaN devices for the next generation of wireless communication

Nadine Collaert[1]

[1]Imec

Chip package co-optimization: circuit-level optimization with RDL passive components for mmWave Power Amplifiers in 22nm FDSOI

Corrado Carta[1]

[1]TU Dresden

Circuits and technologies for applications above 100 GHz

Hua Wang[1]

[1]University of California - Santa Barbara

WWW.EUMW2021.COM   EXHIBITOR WORKSHOPS AND SEMINARS

Rohde & Schwarz Workshop
Tutorial Seminars and Technical Workshops

Date: Monday 4th and Wednesday 6th April 2022
Location: ICC Capital Suite - Level 3 - Room 3

TUTORIAL SEMINARS – RF TEST AND MEASUREMENT BASICS

Advances in 5G and mmWave communications are making major changes to the world of cellular and non-cellular communications. Technologies in the automotive industry and Internet of Things have had a significant impact on mmWave engineering and the design of new products. Modern communications technologies, telemetry applications, radar technologies and industrial assembly of mmWave circuits have all increased cross-disciplinary collaboration.

Nowadays, mmWave engineers also face the challenge of mastering RF signal technologies and digital communications. Every mmWave engineer needs a sound understanding of RF and mmWave testing methods to implement solutions and designs in RF and mmWave circuits.

The Rohde & Schwarz seminars covering RF test and measurement basics will familiarize you with the fundamental aspects of signal generators, power sensors, spectrum analyzers and network analyzers. You will learn the benefits of our highly flexible T&M equipment when designing RF and mmWave circuits.

MONDAY 4TH APRIL, 2022

09:30 - 10:45	Fundamentals of signal generators and oscillators (YIG vs. VCO)
11:00 - 12:15	Fundamentals of power measurements

TUESDAY, 5TH APRIL, 2022

09:30 - 10:45	Fundamentals of spectrum analysis
11:00 - 12:15	Fundamentals of phase noise testing

WEDNESDAY, 6TH APRIL, 2022

09:30 - 10:45	Fundamentals of vector network analysis
11:00 - 12:15	Calibration in vector network analysis

EXHIBITOR WORKSHOPS AND SEMINARS

WWW.EUMW2021.COM

ROHDE & SCHWARZ
Make ideas real

– Free to attend –

For more information, details and registration, visit
http://www.rohde-schwarz.com/eumw

TECHNICAL WORKSHOPS

MONDAY 4TH APRIL, 2022, 13:30 – 16:15
mmWave and THz technology beyond 5G

Workshop chair: Dr. Taro Eichler, Market Segment Manager Wireless Communications, Rohde & Schwarz

Millimeterwave (mmWave) and THz technologies are key components when looking beyond 5G and 6G. The radio spectrum between 30 GHz and 300 GHz helps resolve spectrum crunch and enables ultrabroadband mobile communications up to the terabit range. The research and development of such systems creates new challenges for frontend, mixed signal and baseband technologies and new requirements for the test and measurement industry. Since highly integrated frontends with array antennas will be implemented, advanced over-the-air testing methods with a far greater frequency range up to 500 GHz will become mandatory. Furthermore, using extreme wideband channels up to several GHz will be a challenge for broadband signal generation and signal analyzers. These require an interdisciplinary approach with close collaboration between semiconductor, assembly and signal processing experts.
This workshop gives an overview of recent developments in broadband mmWave and THz communications systems with a special focus on radio channel and OTA measurements as well as on hardware implementation issues.

13:30 - 14:00	**Space THz receivers and sources for the next generation of ESA ice cloud imager (ICI) and planetary spectrometer (SWI) instruments** Presenter: Dr. Bertrand Thomas, Co-Authors: M. Philipp, T. Stangier, M. Brandt, G. Sonnabend, P. Krause, A. Kilian, N. Wehres, M. Trasatti, M. Schmitz, and A. Walber
14:15 - 14:45	**Optoelectronic cw-THz systems and their use in wireless communication: Where are we, and what's next?** Presenter: Prof. Björn Globisch, Technical University Berlin

TUESDAY, 5TH APRIL, 2022, 13:30 – 16:15
Modern RF frontend design and testing

Workshop chair: Markus Lörner, Market Segment Manager RF and Microwave Components, Rohde & Schwarz

5G is here. The focus is now on improving systems and enhancing them with mmWave technology, which is driving the growing integration of components and the creation of more efficient designs to minimize the form factor, improve energy efficiency and drive down overall costs. Multifunction RF components such as beamformers are used in 5G mmWave technology as well as in satellite communications and defense applications. High density of RF frontends for massive MIMO systems require unprecedented energy efficiency to minimize physical size while maintaining stable temperature conditions. When we look at RF frontend design, we start with realistic simulations using target application scenarios and digital predistortion in simulation and hardware verification.
The workshop will survey the latest RF frontend technologies and requirements, focusing on improved efficiency and enhanced integration. Test and measurement experts and industry partners will provide solutions to meet demanding requirements

13:30 - 14:00	**RF measurement uncertainty: how signals interact** Presenter: Tim Fountain, Rohde & Schwarz Application Segment Manager Aerospace & Defense.
14:15 - 14:45	**From design to real RF device – connecting EDA simulation and hardware test** Presenter: Markus Lörner, Market Segment Manager RF and Microwave Components, Rohde & Schwarz
15:00 - 15:30	**Real-time frequency and time domain tuning for 5G and mmWave microwave filters** Presenter: Diamond Liu, Synmatrix
15:45 - 16:15	**VCO design and testing** Presenter: Ian Collins Richardson RFPD and Kieran Barrett, ADI
16:30 - 17:00	**Enabling mmWave 5G through antenna, IC, packaging and algorithm innovations** Presenter: Harish Krishnaswamy CTO, MixComm Inc.

SATURDAY OVERVIEW

Room	09:00 – 13:00	14:20 – 18:20
1	**WS01** Advances of Wireless Sensing in Harsh and Severe Environments	
2		
3		
4		
5		
6		**SS01** Advanced Non-linear Characterization and Design of Highly Efficient Power Amplifiers Using Load-Pull Data for sub-6GHz and mmWave Applications
7	**WS04** New Trends in Microwave and mmWave Filters	
8	**SS02** Fundamentals of Microwave PA Design	
9		
10	**WS06** Progress and Status of Gallium Nitride Monolithic Microwave Integrated Circuits	
11		
12	**WS08** Technology for RF 5G and Satcom: From Material to Packaged Demonstrators	
13	**SS04** Terahertz Technology, Instrumentation and Application	
14	**WS09** Research in Power and S-parameters Measurements at mmWave and Terahertz Frequencies	
15		
16		
17	**WS03** mmWave Plastic Waveguide High Data Rate Communications	

EuMC   EuMIC   EuRAD   Students   EuMW   Exhibitors

PROGRAMME – CONFERENCE SESSIONS MATRIX   WWW.**EUMW2021**.COM

SUNDAY OVERVIEW

Room	09:00 – 10:40	11:20 – 13:00		14:20 – 16:00	16:40 – 18:20	EVENING PROGRAMME
1	**SM01** R&D Trends & Challenges in RFPAs for Medium/High-Volume Products			**WM01** Optimizing Modulation Quality Measurements on Wide Bandwidth Signals - from Conformance Through R&D		
2	**WM02** Advances in Circuits and Systems for mmWave Radar and Communication in Silicon Technologies					
3						
4	**Tom Brazil Doctoral School of Microwaves**					
5				**EuMIC05** Integrated Circuit Modelling and Design Methodology		
6	**WM03** Microwave and mmWave Techniques for Sensing, Imaging and Characterisation of Biological Tissues					
7						
8		**EuMIC04** Opening Session			**EuMIC08** Components and Subsystems for 100 GHz and Above	
9					**EuMIC09** High Performance LNAs	
10	**EuMIC01** Large Signal and Non-linear Charaterization Techniques			**EuMIC06** Integrated PAs for 5G, SATCOM and Vehicular Applications	**EuMIC10** Advances in Si and GaN Based Integrated PAs	
11	**EuMIC02** Silicon Based RF Solutions			**SM02** Intuitive Microwave Filter Design with EM Simulation		
12	**EuMIC03** Transceiver MMICs			**SM03** Phase-Noise in Next-Generation Aerospace/Defense and Commercial Wireless Communications		
13	**WM04** RF On-wafer Calibration and Measurement Eco-system Workshop					
14						
15	**WM05** Microwave Equipment Based on SM EM Relays					
16	**WM06** Recent Developments in Wireless Power Transfer and Energy Harvesting					
17	**WM07** Beyond 5G: mmWave and Terahertz Techniques of 6G Research			**EuMIC07** Frequency-Converting Circuits	**EuMIC11** Boradband Integrated Circuits	**EuMIC Foundry Session** 18:30 – 20:00
ICC Capital Suite (Foyer)						**EuMIC Cocktail Reception** 18:00 – 20:00

EuMC   EuMIC   EuRAD   Students   EuMW   Exhibitors

MONDAY OVERVIEW

Room	08:30	09:00 – 10:40	11:20 – 13:00	14:20 – 16:00	16:40 – 18:20	EVENING PROGRAMME
1		EuMIC/EuMC01 Novel Filtering Devices in Integrated Technologies		EuMC02 Innovative Microwave Circulators and Phase Shifters	EuMC05 Novel Structures for Power Combiners and Couplers	
2		Exhibitor Workshops		Exhibitor Workshops		
3		Exhibitor Workshops		Exhibitor Workshops		
4		EuMIC/EuMC02 THz Components		EuMIC16 Phased Array Components from S-band up to 300 GHz	EuMC06 3D to 2D Transitions and New Materials for mmWave System Integration	
5		Exhibitor Workshops		Exhibitor Workshops		
6		EuMW01 Teaching Methods for Microwave Engineering		EuMC03 Non-planar Filters I	EuMC07 Non-planar Filters II	
7					EuMC08 Digital Predistortion, PA Optimisation and MIMO Architectures	
8						
9			EuMW02 Opening Session		EuMIC17 Closing Session	
10						
11						
12					EuMC09 Metasurfaces and Frequency Selective Surfaces	
13		EuMIC12 Device Modelling and Simulation of Parasitic Phenomena		EuMC04 Active Antennas and Architectures	EuMC10 Innovative Antenna Methodology and Design	
14		EuMIC13 Receiver Components		EuMIC/EuMC03 MMIC Power Amplifiers and Supply Modulation	EuMC11 Front-End and Transceiver Modules	
15		Career Platform		Women in Microwaves (Panel to 3 pm, Visit to 6:30 pm)		
16		EuMC01 Advanced Packaging and Interconnect Technologies for Emerging Applications				
17		EuMIC14 Advances in mmWave and High Power Integrated PA Technologies		EuMW03 Special Session in Memoriam of Prof. Roberto Sorrentino	EuMC12 THz Systems and Applications	
Exhibition Hall	Tom Brazil Fellowship Award (by the GAAS® Association) Finalists Pitching The Role of Microwaves in Contributing to a Sustainable World (Venue: MicroApps)		EuMIC15 Posters	EuMIC/EuMC04 Posters		
Conference Center: Platinum Suite						EuMW Welcome Reception 18:30 – 22:00

■ EuMC   ■ EuMIC   ■ EuRAD   ■ Students   ■ EuMW   ■ Exhibitors

PROGRAMME – CONFERENCE SESSIONS MATRIX

TUESDAY OVERVIEW

Room	09:00 – 10:40	11:20 – 13:00	14:20 – 16:00	16:40 – 18:20	EVENING PROGRAMME
1		**EuMC17** New Design Concepts for Microwave Filters in Planar and Hybrid Technologies		**EuMC22** Advanced Implementations for Substrate-Integrated and Quasi-Planar Filters	
2	**Exhibitor Workshops**		**Exhibitor Workshops**		
3	**Exhibitor Workshops**		**Exhibitor Workshops**		
4	**EuMC13** Non-planar Passive Components	**EuMC18** Frequency Generation, Conversion and Nonlinear Modelling		**EuMW04** Memorial Session for Professor Tatsuo Itoh	
5	**Exhibitor Workshops**		**Exhibitor Workshops**		
6	**SW01** Joint Range-angle Super Resolution MIMO Radar		**SW02** Radar Design from the Ground Up		
7	**WW01** Technologies for 6G Front End Modules				
8					
9	**EuRAD01** Opening Session	Defence, Security and Space (DSS) Forum			
10					
11					
12	Automotive Forum				
13	**EuMC14** Electromagnetic Scattering and Diffraction Effects	**EuMC19** 3D Printing: Processes and Reliability	**EuMC23** 5G Communication and Beyond"	**EuMC26** Novel 3D Printing Approaches for mmWave Applications	
14	**EuMC15** Metamaterial Based Devices and Applications	**EuMC20** Advanced High Efficiency Power Amplifier Techniques	**EuMC24** Advances in Electromagnetic Modeling and Numerical Techniques	**EuMC27** Measurements for 5G and 6G Systems	
15		**IEEE Young Professionals Lunch**	**IEEE Young Professionals Session**		
16					
17	**EuMC16** Integrated Components for Transceivers	**EuRAD02** Radar Applications	**EuRAD03** Emerging Radar Applications	**EuMC28** 5G and mmWave Arrays	
Exhibition Hall		**EuMC21** Posters	**EuMC25** Posters		
North Greenwich Pier (by the O2)					The EuMW Cruise on the River Thames 19:00 – 22:00
Off-site					Automotive Forum Networking Dinner 19:00 – 22:00

EuMC · EuMIC · EuRAD · Students · EuMW · Exhibitors

Load-pull measurement of SiGe:C HBT in BiCMOS 55 nm featuring 11 dBm of output power at 185 GHz

C. Maye[#], S. Lepilliet[#], E. Okada[#], E. Brezza[*], A. Gauthier[*], M. Margalef-Rovira[#], D. Gloria[*], G. Ducournau[#], C. Gaquière[#]

[#]Univ. Lille, CNRS, Central Lille, Univ. Polytechnique Hauts-de-France, UMR 8520 – IEMN, F-59000 Lille, France

[*]STMicroelectronics, 38920 Crolles, France

{caroline.maye.etu, sylvie.lepillet, etienne.okada, guillaume.ducournau, christophe.gacquiere}@univ-lille.fr,

{edoardo.brezza, alexis.gauthier, daniel.gloria}@st.com, marc.margalef-rovira@iemn.fr

Abstract — In this paper, we report high power performances at 185 GHz on single ended heterojunction bipolar transistors in the BiCMOS055 technology. Three device sizes were designed and characterized up to their saturation region. This was possible thanks to a power setup which offers an available input power up to 13.9 dBm at 185 GHz. An innovative integrated tuner was also designed with each transistor to vary the load at the on-wafer output DUT plane. The output power, power gain and power added efficiency are finally extracted under different load conditions. An output power of 11.3 dBm ± 0.5 dB is reached for the larger effective emitter area that is $1.112\ \mu m^2$.

Keywords — Millimeter wave, load-pull measurement, integrated tuner, SiGe bipolar transistor, BiCMOS055.

I. INTRODUCTION

Load-pull measurement is a characterization step used to measure the power behavior of active devices under different sets of load impedances. Nowadays, the recent silicon technologies offer active devices with maximum working frequencies in the millimeter-wave and sub-millimeter-wave range [1]. However, appropriate test benches are still in development. Hence, non-linear device modelling is based on low frequency measurement and a subsequent extrapolation to high frequency.

Using a developed power test bench at 185 GHz that has demonstrated its potentiality [2], we have investigated load-pull measurement of heterojunction bipolar transistor (HBT) of the BiCMOS055 technology manufactured by STMicroelectronics.

This paper is organized as follows: in section I we present the device technology and the devices under test (DUTs). In section II, we describe the power set-up used to perform load-pull measurement. That consists in the power test-bench, the impedance tuner and the accuracy of the measurement. In section III, the power performances of the DUTs are given and compared to their HICUM/L2 model [3]. We finally compare output power densities of the measured HBTs and benchmark them to those reported in the literature at 94 GHz and 200 GHz that are the highest frequency power measurements reported for BiCMOS transistors. Finally, section IV presents the main conclusions of this work.

II. DEVICE TECHNOLOGY

A. Technology

BiCMOS055 technology from STMicroelectronics, also known as B55, is a 55-nm BiCMOS technology developed on a 300-mm production line derived from a 55-nm CMOS technology.

In addition to the multiple options available for NMOS and PMOS related to threshold voltages and power consumptions, three Heterojunction Bipolar Transistor (HBT) flavors with different $f_T \times BV_{CEO}$ trade-off are available: High Speed (HS), Medium Voltage (MV) and High Voltage (HV). The HBTs feature a traditional Double-Polysilicon Self-Aligned architecture with Selective Epitaxial Growth (DPSA-SEG) [4] and have specific collector modules depending on the flavor. Standard devices feature a $0.2 \times 5.56\ \mu m^2$ (drawn dimensions) emitter window in a collector-base-emitter-base-collector (CBEBC) configuration but modifications such as length variation or multiple emitter configuration (C(BE)x-C) are available. The back-End Of Line (BEOL) features 8 copper metal layers with one aluminium capping layer and passive devices specifically designed for mmW applications.

The devices used for the presented high-frequency power measurement are High-Speed with typical maximum frequencies $(f_T, f_{MAX}) = (320, 370)$ GHz and breakdown voltage $BV_{CEO} = 1.45$ V.

B. SiGe HBT devices under test

In the recent technologies, high f_T/f_{max} frequencies are obtained by reducing the device dimensions and optimizing the electrical contacts/transistor topology for parasitics (capacitances, resistances) reduction.

In this paper, we analyzed three CBEBC HS bipolar transistors. Their dimensions are specified in Table 1, based on the schematic of the Fig. 1.

Table 1. Drawn dimensions of the bipolar transistor structures

DUT number	le (μm)	we (μm)	Ne
1	4.5	0.2	1
2	5.56	0.2	1
3	5.56	0.2	2

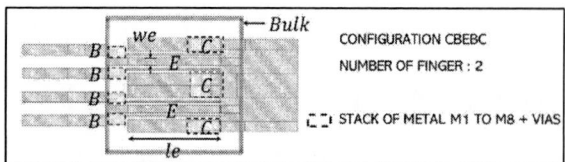

Fig. 1. Device structure of the HS bipolar transistor in configuration N-(CBEBC) with: we, the emitter width, le, the emitter length and Ne, the number of emitter finger.

Note that the effective emitter width is reduced down to 0.1 μm by the use of inside spacers.

III. MEASUREMENT SET-UP

A. Power test bench

The power set-up used to characterize the devices was presented in [2]. It has been modified to increase the available power of the source, P_{inj}, up to 13.9 dBm at the device input plane. The new setup is presented in Fig. 2. which indicates the reference planes when measuring a DUT integrated together with an impedance tuner. The injected power is varied by controlling the mm-wave source driven by microwave synthesizer (PSG). The injected power is then measured with a Schottky diode detector through a 3-ports coupler. Note that the quality of the signal has been validated by performing spectrum analysis.

On the other hand, the output power is measured by using a thermal power detector called PM5, which is our reference detector. The output RF probe and waveguide taper were measured together to calibrate the output power detection at the on-wafer output plane that we note $P_{out\,Tuner}$ on Fig. 2. This notation comes from the use of the impedance tuner that is integrated on-wafer with the DUT. Hence, an offset of power, $Gp_{Tuner}(i)$, is introduced by the tuner for each impedance state, i. It is expressed in equation (1) where S_{21} is the transmission coefficient of the tuner and S_{11} is its input reflection coefficient. $Gp_{Tuner}(i)$ is then considered to calculate the output power of the DUT, $P_{OUT\,DUT}(i)$ with equation (2).

$$Gp_{Tuner}(i) = \frac{|S_{21}(i)|^2}{1-|S_{11}(i)|^2} \qquad (1)$$

$$P_{OUT\,DUT}(i) = \frac{P_{OUT\,Tuner}}{Gp_{Tuner}(i)} \qquad (2)$$

In parallel, the measurement of the hot input reflection coefficient of the DUT and tuner assembly, S'_{11}, versus the injected power at the input plane of the DUT, P_{inj}, is of major importance to calculate the absorbed power P_{abs}, expressed in equation (3).

$$P_{abs}(i) = P_{inj} \cdot (1 - |S'_{11}|^2) \qquad (3)$$

The measurement of S'_{11} requires the detection of the reflected power using an additional coupler with a detector at the input of the DUT. However, this coupler adds losses and reduces the

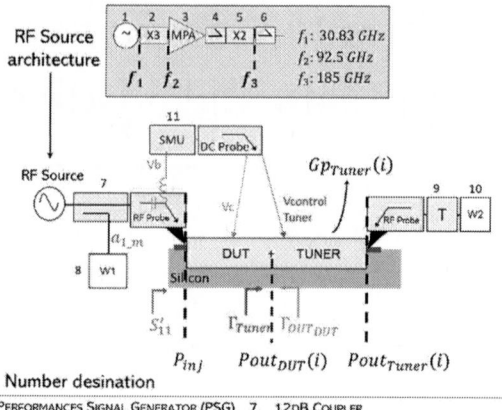

Fig. 2. Load-pull test bench developed at 185 GHz using an external source, an integrated tuner and two power detectors.

available power of the source whereas high injected power is necessary to reach the saturation region. For this reason, we will consider the S'_{11} in small signal condition.

B. Integrated impedance tuner

First, the architecture of the impedance tuner is based on a variable impedance stub terminated by a short-circuit. The transmission line of the stub is then coupled to allow the detection of the output power of the DUT. Then, a bias network is placed at the input to feed the collector access of the DUT. This architecture is illustrated in Fig. 3.

The impedance state is handled with the gate voltage of two n-MOSFET transistors M_1 and M_2. The use of a coupler, with a coupled-line architecture, ensures a good output matching and sets the losses to a relatively constant level for every state.

The input reflection coefficient, $\Gamma_{inTuner}$, output matching S_{22} and transmission coefficient S_{21} are presented in Fig. 4. The tuner features a reflection coefficient up to 0.75 dB at 174 GHz and 0.71 at 185 GHz. At this frequency, the losses, S_{21}, and the output matching S_{22} are at around 14 dB and -20 dB respectively. This output matching involves negligeable reflections between the tuner and the output RF probe.

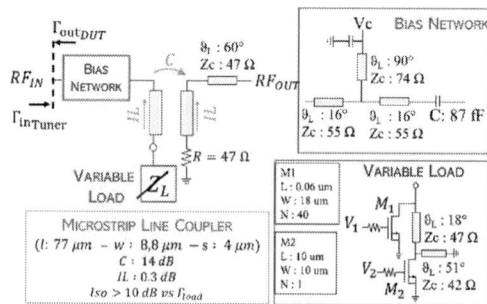

Fig. 3. Architecture of the impedance tuner dedicated to load-pull measurement in the 140 - 220 GHz frequency range.

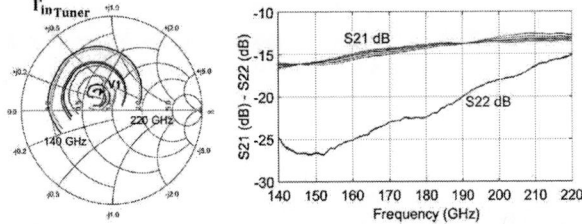

Fig. 4. Left: Reflection coefficient $\Gamma_{IN\,Tuner}$ at the input of the impedance tuner – Right: output matching S_{22} dB and transmission losses S_{21} dB from 140 to 220 GHz.

C. Accuracy of the test bench

The calibration of the test bench was performed using a transmission line from an alumina impedance standard substrate (ISS) from Form Factor. The calibration concerns the injected and output power detections. First, a transmission line, with 2.2 dB of losses, allows to perform correction of the injected power detection up to 8 dBm. Higher power will saturate the PM5 which is set at the 2 mW range. We then use the passive impedance tuner dedicated to characterize our transistor DUTs. This device presents losses in the range of 14 dB and hence we increase the injected power up to 14 dBm without entering in the saturation region of the PM5. Moreover, its measurement aims to validate the calibration on copper metal on a silicon substrate. Fig. 5. shows the δG_T parameter that corresponds to the residual uncertainty on the transducer gain after calibration step. The residual uncertainty indicates that accuracy in the range of ± 0.5 dB is ensured for output power higher than -25 dBm at the output RF probe plane. This condition sets our detectable dynamic of power.

IV. LOAD-PULL CHARACTERIZATION OF THE DUTs

Fig. 6. shows an HBT device and the integrated tuner assembled on the same chip. The voltage of the base-emitter junction, V_b, is handled through the input RF probe whereas the voltage of the collector-base junction, V_c, is supplied with a DC probe through the bias network designed as a part of the tuner.

The analysis of the power behavior of the DUTs consists in the measurement of the output power versus the injected power at different load conditions. First, we compared the power measurements with the power contours provided by simulations. The simulations are based on the parasitic extraction cells of the DUTs. These cells take into account the stacks of metal from level 1 to 8 which are connected with metallic vias and the HICUM L2 model is used. Power contours were simulated by setting V_C to a constant value of 1.2 V. The injected power, P_{inj}, was set to 5 dBm and the output load was varied using an ideal variable load. The power contours from simulation and the output powers obtained from load-pull measurement are presented in Fig. 7. for the three HBTs presented in Table 1. The crosses and circles represent the impedance states offered by the impedance tuner. Circles are the impedance states used for the power measurement. Good agreement between measurement and simulations can be observed, which allowed us to pursue the characterization with a power sweep.

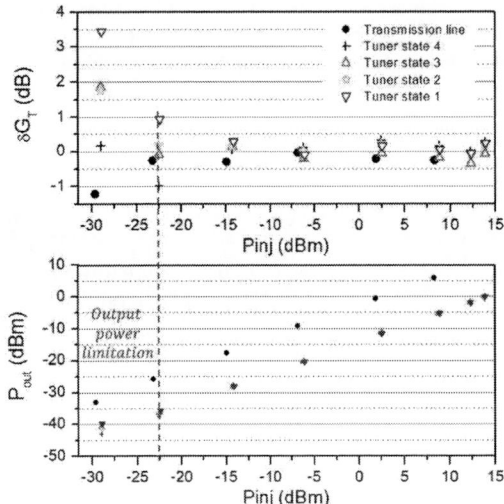

Fig. 5. Residual uncertainty δG_T on the injected power measurement.

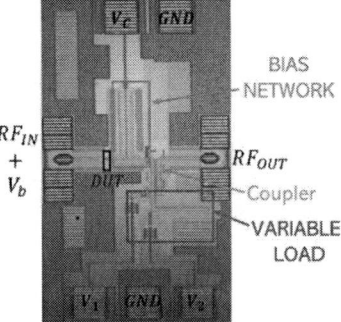

Fig. 6. Chip photography of a DUT and integrated impedance tuner assembly.

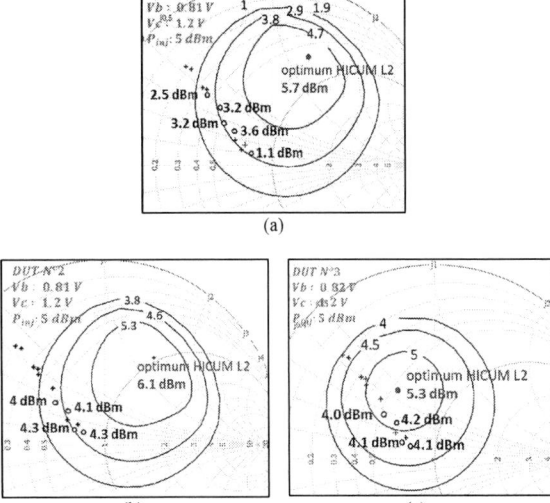

Fig. 7. Power contours based on HICUM/L2 and output power measurement at P_{inj} 5dBm. (Top) DUT N°1 – (Bottom left) DUT N°2 – (Bottom right) DUT N°3.

The maximum output power was achieved by varying the base voltage, the impedance state and the collector voltage. Low V_b involves lower gain but higher compression point. Best performances of the three structures, based on their highest saturation point, are presented in the Fig. 8. with comparison of the simulation. Results of measurement at 185 GHz of the three devices are reported in Table. 2.

Table 2. Reported maximum output powers of DUTs obtained at 185 GHz.

DUT	V_b/V_c (V)	$\Gamma_{in_{Tuner}}$: mag/arg (°)	OPsat (dBm) ± 0.5 dB	Pabs (dBm) at OPsat	PAE max %
DUT N° 1	0.81 /1.6	0.359/130°	6.6	12	4.4
DUT N° 2	0.81 /1.8	0.20/142°	8.6	11	3.4
DUT N° 3	0.82 /1.8	0.285/129°	11.3	11	11.8

Even though uncertainty of the measurement impacts the power results, especially low values of the power gain and PAE, a good agreement is nevertheless obtained between simulation and measurement. High absorbed power was also necessary to let the DUTs work in their saturation region. This justifies the modification of the power setup.

In parallel, the power densities are compared in Table 3 with values reported in the literature above 94 GHz. We note the increase of the power densities between the 130 nm and 55 nm process at 200 and 185 GHz respectively.

Table 3. Reported output power densities on bipolar devices above 94 GHz.

Ref	Process	Freq (GHz)	Emitter Area (um²)	Pc (mW/um²)
[5]	130 nm	94	0.11x4.9	18.5
[6]	130 nm	200	0.11x3	3.4
[7]	130 nm	94	0.12x4.9	22.26
[8]	55 nm	94	0.1x5.56	30.1
DUT N° 1	55 nm	185	0.1x4.5	10.1
DUT N° 2	55 nm	185	0.1x5.56	13
DUT N° 3	55 nm	185	0.2x5.56	12.1

V. CONCLUSION

In this paper we have presented high-power performances at 185 GHz of three bipolar transistors of the B55 technology with different sizes. First, a power source was modified in order to increase the available power up to 13,9 dBm at 185 GHz. Then, an innovative impedance tuner was integrated with the DUTs. By using a coupler in the design, the tuner offers relatively constant losses versus the impedance states and ensures the output matching. The tuner was first measured in stand-alone to determine the accuracy of the measurement on silicon that was ± 0.5 dB over a power dynamic of 30 dB. Saturation regions of the DUTs are observed in the range of the 10 dBm of absorbed power and power densities above 10 mW/μm^2 were measured. Finally, good agreement between measurements and simulations based on HICUM L2 model is observed.

REFERENCES

[1] T. Zimmer *et al.*, "SiGe HBTs and BiCMOS Technology for Present and Future Millimeter-Wave Systems," in *IEEE Journal of Microwaves*, vol. 1, no. 1, pp. 288-298, winter 2021.

[2] C. Maye, S. Lepilliet, E. Okada, D. Gloria, G. Ducournau and C. Gaquière, "Load-Pull Setup Development at 185 GHz for On-Wafer Characterization of SiGe HBT in BiCMOS 55 nm Technology," in *IEEE Transactions on Microwave Theory and Techniques*

[3] A. Pawlak et al., "SiGe HBT modeling for mm-wave circuit design," 2015 IEEE Bipolar/BiCMOS Circuits and Technology Meeting - BCTM, Boston, MA, USA, 2015, pp. 149-156.

[4] P. Chevalier et al., "A 55 nm triple gate oxide 9 metal layers SiGe BiCMOS technology featuring 320 GHz fT / 370 GHz fMAX HBT and high-Q millimeter-wave passives", IEEE International Electron Devices Meeting, pp. 3.9.1-3.9.3, Dec. 2014.

[5] M. Rickelt, H. -. Rein and E. Rose, "Influence of impact-ionization-induced instabilities on the maximum usable output voltage of Si-bipolar transistors," in IEEE Transactions on Electron Devices, vol. 48, no. 4, pp. 774-783, April 2001.

[6] A. Pottrain, T. Lacave, D. Gloria, P. Chevalier and C. Gaquière, "State of the Art 200 GHz power measurements on SiGe:C HBT using an innovative load pull measurement setup," *2012 IEEE/MTT-S International Microwave Symposium Digest*, Montreal, QC, Canada, 2012, pp. 1-3.

[7] I. Hasnaoui, E. Canderle, P. Chevalier, D. Gloria and C. Gaquiere, "94-GHz Load Pull measurements of SiGe HBT by extracting output power density in W-Band," 2013 European Microwave Integrated Circuit Conference, Nuremberg, Germany, 2013, pp. 400-403.

[8] A. Gauthier, W. Aouimeur, E. Okada, N. Guitard, P. Chevalier and C. Gaquière, "A 30.1 mW / $\mu m2$ SiGe:C HBT Featuring an Implanted Collector in a 55-nm CMOS Node," in *IEEE Electron Device Letters*, vol. 41, no. 1, pp. 12-14, Jan. 2020

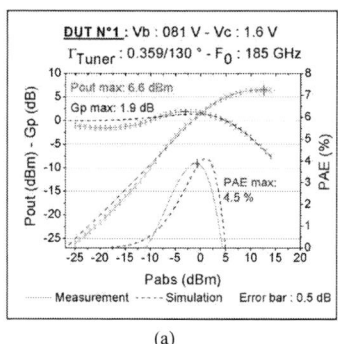 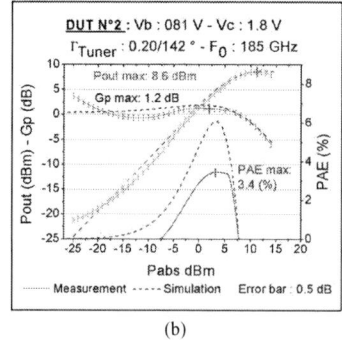 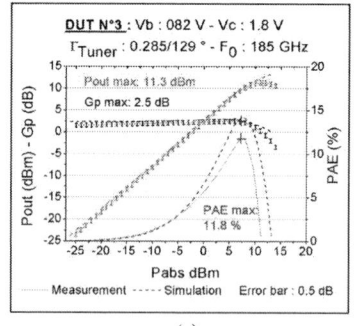

Fig. 8. Power measurement of the transistors DUTs and comparison with simulations based on HICUM L2 model. (a) DUT N°1 – (b) DUT N°2 – (c) DUT N°3.

Proceedings of the 16th European Microwave Integrated Circuits Conference

Nonlinear Characterization of Wideband Power Amplifiers with Frequency-Dependent Match Load

Sanket Chaudhary[*], Nuno Borges Carvalho[*], Marina Jordão[*], Marc Vanden Bossche[#], Adam Cooman[+], Sergio Pires[+]

[*]Instituto de Telecomunicações and Departamento de Eletrónica, Telecomunicações e Informática,
Universidade de Aveiro, Campus Universitário de Santiago, 3810-193 Aveiro, Portugal
[#]NI, Belgium
[+]Ampleon, Nijmegen, The Netherlands
sanket@ua.pt, nbcarvalho@ua.pt

Abstract — This paper investigates key factors in optimum load synthesis method for wideband multi-carrier compatible power amplifiers (PAs) using active modulated load-pull system. The analysis demonstrates that, without compromising the linearity, a 2-6% improvement in power added efficiency (PAE) and 1-2 dB increased gain is achieved near and beyond the 1-dB compression, using a frequency-dependent optimum impedance. The proposed optimum impedance synthesis approach offers more comprehensive information about the device under test (DUT) performance in real-time complex modulated wideband scenario.

Keywords — Wideband modulated load-pull, adjacent channel power ratio (ACPR), PAE, peak-to-average power ratio (PAPR)

I. INTRODUCTION

The revolutionizing wireless communication demands more robust wireless infrastructure development to provide advanced communication capabilities. Over the past decades, mobile telecommunication became one of the largest and most significant platforms for transforming the communication from voice calls to internet services. With more facilitate communications, users and communication devices are increased exponentially, that has led to a massive volume of data traffic demand. To offer services to such increasing data traffic demand, wireless networks are needed to be improved with higher data rates and spectral efficiency while maintaining the 3rd generation partnership project (3GPP) standards.

In today's wireless transceivers, efficiency, bandwidth, and linearity are prime concerns to be considered, specifically in PA stage, to offer highly efficient communication infrastructure. Nevertheless, to provide higher data rates, the emergence of advanced telecommunication standards, such as 4G-LTE (long-term evaluation) and 5G, has demanded wideband multi carrier compatible PAs. PA design is always a trade-off among the linearity, efficiency and gain as a function of bandwidth and time-domain dynamics of the modulated signals [1].

In PA designing, experimental characterization of the active device using the realistic testing scenarios provides more comprehensive information on the performance. As discussed in [1]-[5], Cripps's load-line theory extended with source/ load-pull characterization using actual operating conditions provides significant information about the DUT characteristics.

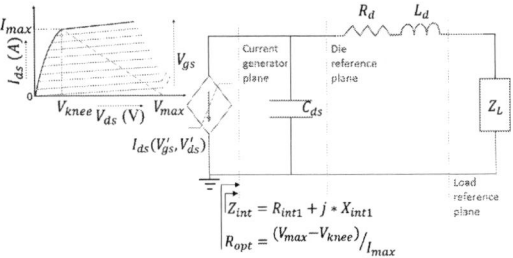

Fig. 1. Simplified unilateral equivalent circuit of the HEMT nonlinear transistor model.

The Cripps' load-line theory in [2] presents information on a purely resistive optimum load at the current generation plane (CGP) based on the static current (I)/ voltage (V) characteristics of the active device. This purely static assumption is then extended in [4] by considering the parasitics at the carrier frequency (F_C) for narrowband/ continuous wave (CW) applications.

The load-pull analysis is carried out with pulsed CW excitation to evaluate the DUT including thermal and trapping effects [6]. In its real application, the DUT is excited by the more complex modulation protocols of 5G, 4G-LTE and other standards. In order to study linearity and the impact of intrinsic/ extrinsic parasitics, an active device is required to be characterized with the wideband modulated signals representing the complexity of protocol signals.

In [7], authors have proposed a new approach to design the wideband modulated compatible PA by providing a platform for impedance synthesis for frequency dependent optimal match. However, the provided study was limited to the gain and PAE analysis of the DUT. Since there is a trade-off between efficiency and linearity for the wideband PA applications, it is also required to study the impact on the linearity for the proposed modulated load-pull synthesis approach.

This paper contributes to the experimental study on modulated load-pull by also including the nonlinearity analysis through a metric such as ACPR. In this paper, it is shown that the proposed approach enhances the PA performance in terms of PAE and gain at and beyond 1-dB compression point

3–4 April 2022, London, UK

without compromising the linearity.

The rest of the paper is organized as follows: in Section II, the analytical study on intrinsic parasitics' impact with respect to frequency for class A and class B PA is provided. Section III includes experimental analysis using a measurement platform synthesizing realistic modulation signal and synthesizing frequency-dependent active loads. Finally, Section IV discusses the experimental results and conclusions are drawn.

II. ANALYTICAL STUDY ON WIDEBAND RESPONSE FROM THE SIMPLIFIED NONLINEAR MODEL

The Cripps' load-line theory in [2] is developed by considering intrinsic parasitics as a part of the extrinsic circuit. With this assumption, optimum impedance (Z_{opt}) is needed to be purely resistive (R_{opt}) at the CGP for the maximum output power. However, in measurements intrinsic and extrinsic parasitics are part of the device under test and therefore, during the characterization of the transistor for wideband applications, the frequency dependency of the device needs to be investigated.

To exemplify the concept, Fig. 1 is considered with a simplified unilateral equivalent circuit of the high-electron-mobility transistor (HEMT) nonlinear transistor model including essential intrinsic parasitics. Assuming the knee voltage as V_{knee}, maximum voltage as V_{max}, and saturation current as I_{max}. R_{opt} is the optimum load resistance obtained from the load-line theory at the CGP. Z_{int} is the impedance that is obtained including the parasitics at the CGP.

As defined in [7], with the assumption from the load-line theory, impedance at the CGP should fulfill the following condition:

$$Z_{int} = R_{int} + jX_{int} = R_{opt} \tag{1}$$

This can be obtained by choosing the appropriate Z_L as,

$$Z_L = R_L + jX_L \tag{2}$$

Where, R_L and X_L can be defined as,

$$R_L = \frac{R_{opt} - R_d C_{ds}^2 R_{opt}^2 \omega^2 - R_d}{C_{ds}^2 R_{opt}^2 \omega^2 + 1} \tag{3}$$

$$X_L = \frac{R_{opt}^2 C_{ds}\omega - L_d C_{ds}^2 R_{opt}^2 \omega^3 - \omega L_d}{C_{ds}^2 R_{opt}^2 \omega^2 + 1} \tag{4}$$

This shows that a frequency-dependent load impedance is required to maintain real R_{opt} at the CGP across a wide bandwidth.

To illustrate the impact of frequency variation on impedance Z_L over the wideband, two classes of operation in PA are considered. Class A and class B with assuming V_{max}=10 V, V_{knee}=2.6 V, I_{max}=1 A, R_{opt}=14.8 Ω, R_d=0.3 Ω, L_d=1.7 nH, C_{ds}=3 pF. The performance parameters, such as output power (P_{out}) and efficiency (η) are evaluated with the following equations,

$$P_{out} = \frac{R_L}{2|Z_L|^2}(V_{max} - V_{knee})^2 \tag{5}$$

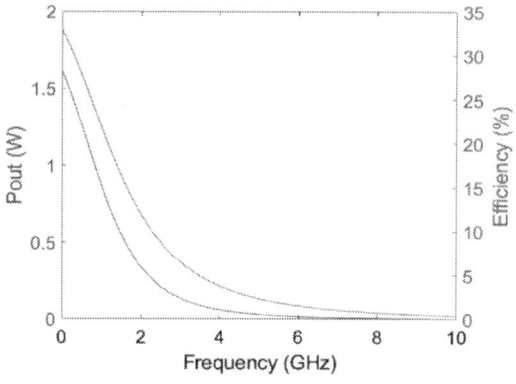

(a)

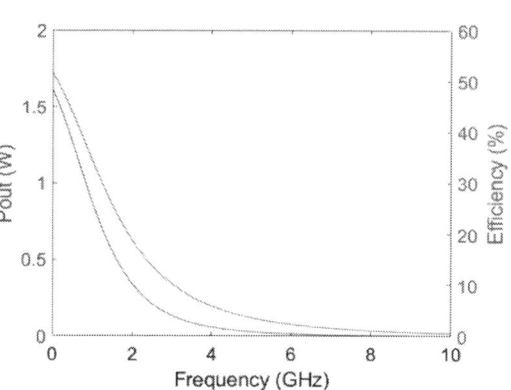

(b)

Fig. 2. Efficiency and output power (P_{out}) of the (a) class A and (b) class B operation mode. Results are obtained from the analytical model (5)-(7) with the load model Z_L considering the intrinsic parasitics effect defined in (2) across the frequency range.

$$\eta_{Class-A} = 100\frac{R_L}{2|Z_L|}\frac{V_{max} - V_{knee}}{V_{max}} \tag{6}$$

$$\eta_{Class-A} = 100\frac{\pi R_L}{4|Z_L|}\frac{V_{max} - V_{knee}}{V_{max}} \tag{7}$$

The results are summarized in Fig. 2 with respect to frequency. Here, Z_L is defined for the model shown in Fig. 1, that is considered from the analysis provided in the [7] to synthesize the real R_{opt} at CGP. From Fig. 2, it can be seen that frequency variation has a significant impact on the DUT performance for wideband frequencies due to parasitics. To optimize the performance over the wide range of the frequencies, optimum load impedance (Z_L) needs to be optimized and varied with the application frequency.

This motivates to synthesize the optimum load that varies with the frequency within the modulation bandwidth to enhance the PA performance in wideband modulated applications situation.

Fig. 3. NI-PXIe5840 based measurement setup for active modulated load-pull characterization

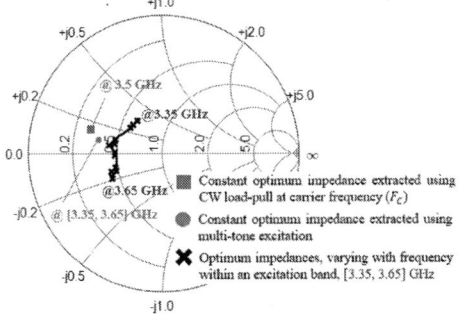

Fig. 4. Optimum impedances synthesized for the analysis using active load-pull setup

III. Experimental Analysis on Nonlinear Characterization Using Active Modulated Load-Pull

As discussed in the previous Section II, constant load Z_L provides a good match at the carrier frequency (F_C). This compensates the intrinsic parasitic impact while maintaining the constant R_{opt} at the CGP to extract optimum output power (P_{out}). However, in wideband applications, to enhance the PA performance, optimum load impedance (Z_L) needs to be varied with the frequency within the application band, to enhance the optimum performance.

In order to study the impact of frequency variation on constant optimum load Z_L and non-constant Z_L over frequency range, the setup presented in Fig. 3 is used. This setup utilizes the NI-PXIe-5840 vector signal transceivers (VSTs) to perform the measurements of in-band and out-band incident and reflected wave. The setup has been implemented for the active load-pull synthesis using modulated excitations.

Presently the study is focused on the principles of impedance tuning and no work was done to extend the maximum generation and impedance tuning capability at higher powers of the PXI system. And hence, an experiment is performed with the low power device (HBT-BFP842ESD). The device is biased in a class B operation mode for the analysis. For the experiment, a multi-tone (Gaussian distribution pdf

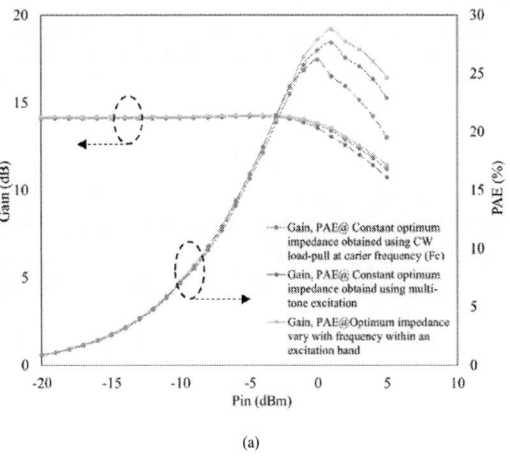

(a)

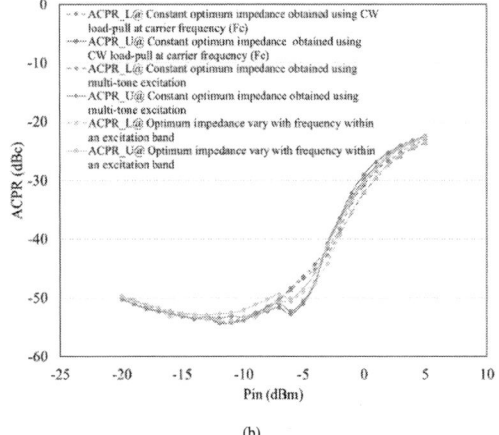

(b)

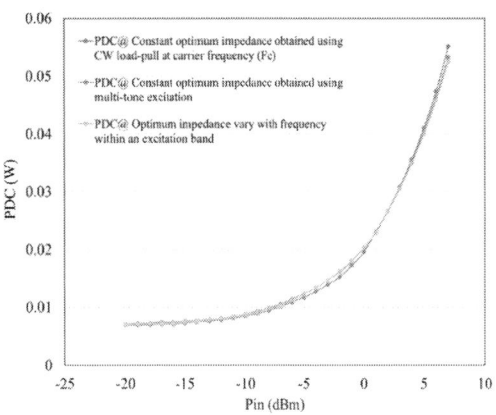

(c)

Fig. 5. a) Gain and PAE comparison, b) ACPR and c) DC power consumption summary obtained from the experimental setup shown in Fig. 3 with considering three different match load approach discussed in Section III.

based 12-tones is considered) with 300 MHz bandwidth has been considered with 7.41 dB PAPR at 3.5 GHz carrier frequency (F_C).

Considering all the above measurement conditions, three types of measurement approaches were performed. First, a multi-tone analysis was performed considering the constant Z_L, which is optimized at the carrier frequency (from CW load-pull setup) for the entire bandwidth. The second approach takes into account the constant Z_L that is optimized under the wideband multi-tone (12-tone) excitation, using multi-tone active load-pull analysis[1]. In the third approach, multiple loads are optimized to be varied with frequency tones within the operational bandwidth. It means that the third approach would perform frequency dependent optimum impedance Z_L for the DUT characterization in the active load-pull measurement setup. In this case each of the tone impedance will be changed individually and this will create a multidimensional problem for the load pull solution, so at the end we will have 12 different load for each tone.

Fig. 4 summarizes the optimum impedances that are extracted from the active load-pull setup for all three approaches. In each measurement, the optimized Z_L is considered for the maximum output power. As shown in Fig. 5 (a), frequency dependent optimum load enhances the PA in terms of the PAE by 2-6% and power gain by 1-2dB at and beyond the 1-dB compression point in the wideband modulated application. Additionally, contrary to what is normal in the trade-off of linearity and efficiency, it can be seen in Fig.5 (b) that linearity of the PA is not compromised while the efficiency and gain are enhanced.

IV. CONCLUSION

In this paper, wideband active modulated load-pull characterization was performed using three different optimal load synthesis approaches. In the first approach, using the traditional CW load-pull, the optimum load is obtained for the analysis. In the second approach, the optimum load is synthesized using multi-tone excitation for the analysis. Finally, in the third approach, multiple optimum load at each frequency tone is obtained using multi-tone load-pull analysis.In each measurement, the optimized Z_L is considered for the maximum output power.

From the comparison of all these different approaches, it can be observed that frequency dependent optimum load obtained from the third approach provides enhanced performance of the PA in terms of the PAE and gain without compromising the linearity of the device at and beyond the 1-dB compression. Moreover, since the main PA is kept in saturation after the back-off level, there is a potential to increase the performance in Doherty configuration.

Furthermore, with the knowledge of the optimum impedance variation across the frequency, we show that designers should consider a frequency varying matching network approach, that will enhance the performance quality of the DUT for the wideband application.

[1] As presented in [8]-[9], multi-sine time-domain statistics could be different for the same PAPR configuration. This fact might change the optimum impedance optimized for the different multi-tone excitation configurations.

ACKNOWLEDGMENT

The authors would like to acknowledge the sponsoring of this work by Ampleon, Nijmegen, The Netherlands, and NI. They would also like to thank Anteverta-mw B.V., Eindhoven, The Netherlands, especially Michele Squillante for providing the impedance synthesis capability and for his technical support on the system. The work of Marina Jordão was supported by the Fundação para a Ciência e Tecnologia (F.C.T.) through Fundo Social Europeu (FSE) and by Programa Operacional Regional do Centro under Ph.D. Grant SFRH/BD/143204/2019. This work was sponsored in part by the Roger Pollard Student Fellowship from the Automatic RF Techniques Group.

REFERENCES

[1] S. Chaudhary, M. Jordão, N. B. Carvalho, M. V. Bossche, and A. Cooman, "PAPR Deviation Impact in the Wideband Power Amplifier Characterization with Realistic Modulated Load-pull System," in *IEEE MTT-S Int. Microw. Symp. Dig.*, Atlanta, GA, USA, 2021, pp. 562–565.

[2] S. C. Cripps, *RF Power Amplifiers for Wireless Communications*, Norwood, MA, USA: Artech House, 1999.

[3] Y. Takayama, "A new load-pull characterization method for microwave power transistors," *IEEE-MTT-S International Microwave Symposium*, Cherry Hill, NJ, USA, 1976, pp. 218-220, doi: 10.1109/MWSYM.1976.1123701.

[4] Ghannouchi, F. M. & Hashmi, Dr. Mohammad., *Load-Pull Techniques with Applications to Power Amplifier* (Springer Series in Advanced Microelectronics), Springer Science & Business Media, Vol. 32. Jan. 2013.

[5] M. Marchetti, M. J. Pelk, K. Buisman, W. C. E. Neo, M. Spirito and L. C. N. de Vreede, "Active harmonic load–pull with realistic wideband communications signals," in *IEEE Trans. Microw. Theory Techn.*, vol. 56, no. 12, pp. 2979-2988, Dec. 2008.

[6] A. Benvegnù et al., "On-Wafer Single-Pulse Thermal Load–Pull RF Characterization of Trapping Phenomena in AlGaN/GaN HEMTs," in *IEEE Trans. Microw. Theory Techn.*, vol. 64, no. 3, pp. 767-775, March 2016, doi: 10.1109/TMTT.2016.2523991.

[7] S. Chaudhary, M. Jordão, N. B. Carvalho, M. V. Bossche, and A. Cooman, "Characterization of the Frequency Dependent Match for Optimal Performance of Wideband Power Amplifiers", in *Proc. ARFTG 97th Conf.*, Atlanta, GA, USA, Jun. 2021.

[8] J. C. Pedro and N. B. Carvalho, "Designing multisine excitations for nonlinear model testing," in *IEEE Trans. Microw. Theory Techn.*, vol. 53, no. 1, pp. 45-54, Jan. 2005.

[9] D. Schreurs, M. Myslinski and K. A. Remley, "RF behavioural modelling from multisine measurements: influence of excitation type," 33rd European Microwave Conference Proceedings (IEEE Cat. No.03EX723C), Munich, Germany, 2003, pp. 1011-1014 Vol.3.

Proceedings of the 16th European Microwave Integrated Circuits Conference

Intermodulation Products of a CMOS SP6T Antenna Switch: Results Comparison Between an Experimental Test-Bench and a Corresponding Simulated Virtual Test-Bench

M. Ben-Sassi[#], H. Saleh[#], O. Sow[#], I. Lahbib[#], Greg D. U'Ren[#], C. Hellepee[*], D. Passerieux[*], G. Neveux[*], D. Barataud[*]

[#]X-FAB, Corbeil-Essonnes, France
[*]XLIM-UMR CNRS n°: 7252, University of Limoges, France
Marwen.Ben-Sassi@xfab.com

Abstract — **This paper presents a comparison between Third Order Intermodulation Distortion (IMD3) measurements and simulations of a RF-SOI SP6T antenna switch. The originality of the paper lies in the development of a simulation tool representing a virtual test-bench corresponding to the experimental set-up used for the measurements of Intermodulation products. All the passive devices (circulators, isolators and filters) used in the experimental test-bench have been characterized in terms of [S] parameters. These [S] matrices have been then implemented into the software environment to simulate the experimental test-bench. A non-linear behavioral model of the SP6T switch from X-Fab foundry has been extracted, validated and implemented in the virtual test-bench. Two-tone Harmonic Balance simulations are performed with this behavioral model. The measured and simulated IMD3 results are finally compared.**

Keywords — **IMD3, Two-tone HB Simulation, SP6T Switch.**

I. INTRODUCTION

Needs for higher data rates in new mobile communication generations are continuously growing. Recently launched 4G TLE-A smartphones can support data rates up to 1 Gbps. These features are driven by the development of revolutionary telecommunication standards and complex front-end architectures such as Multiple Input Multiple Output (MIMO) and Carrier Aggregation (CA). CA allows simultaneous transmission on multiple frequency bands to increases data rates. It is employed in 4G and is expected to grow up in the 5G with the higher number of available frequency bands. A main component in a RF Front-End Module (FEM) of a User Equipment (UE) featuring CA is the antenna switch that selects which transmitter (Tx) or receiver (Rx) path can be connected to the antenna. Therefore, many transmission signals with different frequencies are present at the antenna vicinity and pass through the switch toward their receiver destination. Due to the use of multiple bands simultaneously for uplink and downlink CA, unwanted Inter-Modulation Distortion (IMD) products are generated and could interfere with wanted signals. If the linearity of the antenna switch that is simultaneously processing the Tx, Rx and interferer blocking signals is not high enough, the IMD products of the blocker signal and the Tx signal may fall on the own Rx band. For the IMD products not to cause performance degradation for the receiver, their power levels should not exceed the reference sensitivity power level [3]: the ability of a receiver to receive a wanted signal at its assigned channel frequency in the presence of an unwanted interferer. With the increased number of frequencies in 4G and 5G, interferences and consequently IMD products are more likely to happen. IMD products characterization and modelling of antenna switches is crucial to help designers adopt adequate semiconductor technologies and best linearity architecture for FEM systems.

From practical point of view, RF switches IMD products measurement is challenging due to their low power levels. 3GPP technical specification set receivers' reference sensitivity power level for the different bands between -90 dBm and -100 dBm. Switches' IMD products should not exceed this reference level for receiver performance not to be degraded. Measurement of IMD product levels below -100 dBm requires high linearity test components and very high sensitivity for receiver instruments.

This work aims to numerically reproduce third-order IMD phenomenon in antenna switches. It provides designers with a useful tool that reproduces measurement and can be easily implemented. The proposed approach relies on the development of a behavioral model for the switch linearity and the comparison to measurement. The experimental setup is similar to the one proposed in [1]. The used switch is a SP6T made with a CMOS RF-SOI technology and operates in the cellular band [0.4 GHz – 2.7 GHz]. The IMD simulation and measurement are investigated for Tx signal at 1.74 GHz and a jammer (blocking) signal at 1.65 GHz.

II. BEHAVIORAL RF SWITCH MODEL

A behavioral switch model based on a mathematical equation is defined, as shown in Fig. 1, and used within the virtual test-bench. This model reproduces the transmission behavior between 3 ports of the SP6T but it can easily be expended to the 6 ports.

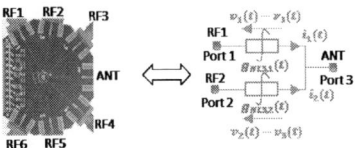

Fig. 1. SP6T switch and its behavioral model

3–4 April 2022, London, UK

It is based on the use of a modified hyperbolic tangent equation to converge with the real operating mode of the switch. The behavioral model is composed of three RF ports which are connected to Antenna (port 3), to RF2 (port 2) and RF1 (port 1). It is based on the use of the mathematical description of two non-linear transconductances $g_{NL3,1}$ and $g_{NL3,2}$, as shown in equations 1 and 2.

$$g_{NL3,1}(t) = \frac{i_1(t)}{v_1(t) - v_3(t)} \quad (1)$$

$$g_{NL3,2}(t) = \frac{i_2(t)}{v_2(t) - v_3(t)} \quad (2)$$

With:

$$i_1(t) = I_1 \text{th}(\beta + \alpha[v_1(t) - v_3(t)])[(1 + V_{1\text{leak}}) - V_4] \quad (3)$$

and $\quad i_2(t) = I_2 \tanh(\beta + \alpha[v_2(t) - v_3(t)])[1 + V_4] \quad (4)$

- $g_{NLi,j}$: Non-Linear Transconductance
- $V_{1leakage}$: Isolation parameter (Real number)
- V_4: Command voltage (Real number)
- I_1, I_2, α, β: Real parameters of non linearities
- $v_i(t)$: voltage at port i

$g_{NL3,1}$ and $g_{NL3,2}$ represent the two transconductances implemented from antenna to RF1 and antenna to RF2 respectively.

Table 1 presents the different parameters used in the behavioral model equations. The parameters have been extracted from [S] parameter measurements and from large signal CW power measurements at frequencies of interest presented in Fig. 2.

Table 1. RF switch model parameters

Parameter	Value	Unit
$V_{1leakage}$	7e-5	V
α	0.03	Real
β	0.05	Real
I_1	-17.5	Real
I_2	2.31	Real

I_1, I_2, α and β are added as real parameters allowing better transmission description for the switch model. The isolation between port (2) and port (3) has been defined within the model using the $V_{1leakage}$ parameter. A DC command (V_4) has been implemented in order to control the transmission configuration of the switch from RF1 to antenna and from RF2 to antenna. In this work, only the signal transmission from antenna to RF1 is taken into account. The DC voltage (V4) is equal to 1 V. For this condition (V4=1 V), $g_{NL3,2}$ tends to 0 while $g_{NL3,1}$ tends to infinity.

Fig. 2 presents the comparison between CW power measurement (RF1 activated) and simulation (V4=1 V) results at fundamental (H0), 2nd (H2) and 3rd (H3) harmonic frequencies at 0.9 GHz, 1.8 GHz and 2.7 GHz respectively. A good agreement between CW power measurements and simulation is obtained.

DC simulation is performed to plot the $g_{NL3,1}$ non-linearity function and to validate the behavioral model of this transconductance.

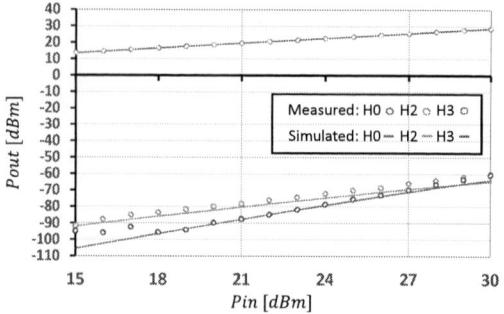

Fig. 2. Comparison of measured and simulated Harmonic distortion results of the model for a 0.9 GHz input fundamental

Fig. 3 presents the output current $i_1(t)$ of the $g_{NL3,1}$ non-linear transconductance versus the voltage difference $v_1(t) - v_3(t)$.

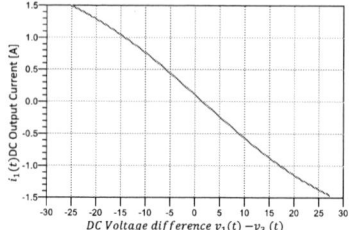

Fig. 3. DC simulation result of the $g_{NL3,1}(t)$ non-linearity

The $g_{NL3,1}(t)$ is a weak non-linear function with an offset from the origin.

Fig. 4 presents a comparison between the measured and the simulated S13 parameters of the switch.

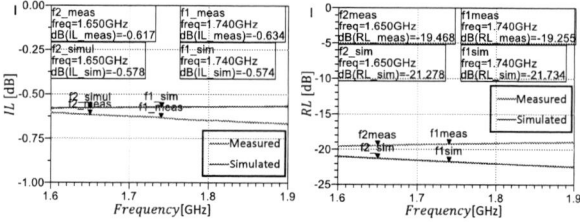

Fig. 4. Comparison of measured and simulated s-parameters of the model

The modelling was focused on S_{13} transmission behavior. S_{11} and S_{33} are matched to 50 Ω. A good agreement between the measured and the simulated insertion loss (S_{13}) is observed.

A little difference between the measured and the simulated return loss is observed because within the behavioral switch model no capacitive elements have been used.

III. PRESENTATION OF THE IMD3 TEST-BENCH

Fig. 5 describes the fundamental architecture of the IMD3 experimental measurement system. Two synchronized RF generators drive the SP6T switch with two RF signals to excite its non-linearities and to measure its potential resulting IMD3. A transmission large signal at the throw port (RF1) and a jammer (blocking) small signal at the antenna port are exciting the SP6T. The RF path to the throw port, at which the transmission is made, is set to ON state to enable the transmission, where the 5 other paths (RF2 to RF6) are in OFF state and 50 Ω loaded.

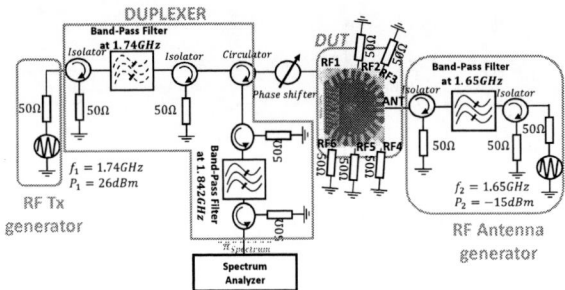

Fig. 5. Block diagram of the IMD3 test-bench with the SP6T configuration

A duplexer is used in this IMD3 test-bench to clearly separate the Rx signal from the Tx signal [1], [2]. It is designed using different band-pass filters associated to 50 Ω matched isolators. The duplexer behavior is described later.

The switch is simultaneously driven with a 1.65 GHz (f_2) jammer signal with a low power (-15 dBm) at the antenna port, and with a 1.74 GHz (f_1) transmission signal with a high power (26 dBm). IMD3 frequency components are measured with a spectrum analyzer which is directly connected to the Rx port of the duplexer.

Table 2 shows the two fundamental frequencies with their associated IMD3 components. The main challenge of the IMD3 characterization is to measure the absolute power levels at the two IMD3 components f_3 and f_4.

Table 2. Presentation of frequencies of interest used in the IMD3 characterization

IMD3 and fundamental frequencies (GHz)		
f_1	f_{Tx}	1.74
f_2	f_{Jammer}	1.65
f_3	$2 \times f_{Tx} - f_{Jammer}$	1.842
f_4	$2 \times f_{Jammer} - f_{Tx}$	1.55

A. Presentation of the duplexer

The main architecture of the duplexer is presented in Fig. 6 (a) [1], [2]. The main challenge for the duplexer is to ensure a very high isolation between port (1) and port (3) to avoid direct leakage from transmission to reception paths.

Isolation between port (1) and port (3) paths is ensured by the use of the band-pass filters at $1.842\ GHz$ and $1.74\ GHz$. The circulator used in the duplexer allows the transmissions of the signal from port (1) to the port (2) of the duplexer and also the one from the port (2) to the port (3) of the duplexer.

Duplexer is composed of three principal devices. The first one is a [1 GHz- 2 GHz] circulator with $20\ dB$ of isolation. The second one is a band-pass filter centered at f_1 with $70\ MHz$ bandwidth. The third one is a band-pass filter

centered at f_3 with $70\ MHz$ bandwidth. It enables the Tx frequency f_1 at port (1) of the duplexer to be driven to the switch. On the other hand, the incoming signal from the DUT port at the jammer frequency (f_2) is driven to the port (3) of the duplexer. That is connected to the receiver.

The signals at f_3 and f_4 are the IMD3 produced in the switch, due to its non-linearity, resulting from the intermodulation between f_1 and f_2 frequencies. The four frequencies f_1, f_2, f_3 and f_4 are driven from the switch to the port (3) through the circulator. The band-pass filter centered at $1.842\ GHz$ selects the f_3 IMD3 frequency component while the f_4 is filtered-out. f_3 is then forwarded to the spectrum analyzer connected to the port (3) of the duplexer. Fig. 6 (b) shows the measured s-parameters of the duplexer.

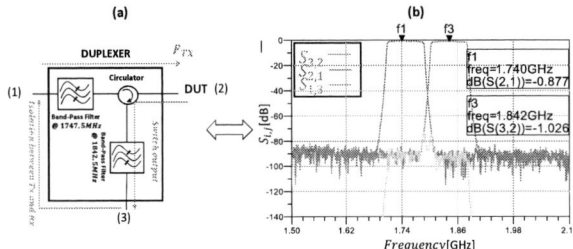

Fig. 6. (a): Block diagram of the duplexer, (b): $S_{3,2}$, $S_{2,1}$ and $S_{1,3}$ measurement results of the duplexer

The band-pass filter centered at $1.842\ GHz$ (S_{32}) allows filtering the f_1 and f_2 components without too much attenuating the IMD3 f_3 frequency component.

The power levels at the two fundamental frequencies f_1 and f_2 are then attenuated to optimize the dynamic range utilization of the spectrum analyzer and avoid compression due to the power levels of f_1 and f_2 signals.

A power calibration between port (2) and port (3) is performed, with reference power sensors at each frequency of interest (f_1, f_2 and f_3).

IV. COMPARISON BETWEEN MEASUREMENTS RESULTS FROM THE EXPERIMENTAL TEST-BENCH AND SIMULATIONS RESULTS OF THE CORRESPONDING VIRTUAL TEST-BENCH

The proposed configuration of the virtual IMD3 test-bench (2-tone HB simulation) is realized with Keysight ADS© software, shown in Fig. 7.

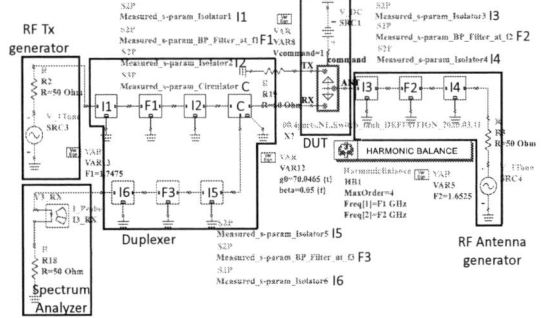

Fig. 7. ADS schematic of the virtual IMD3 test-bench

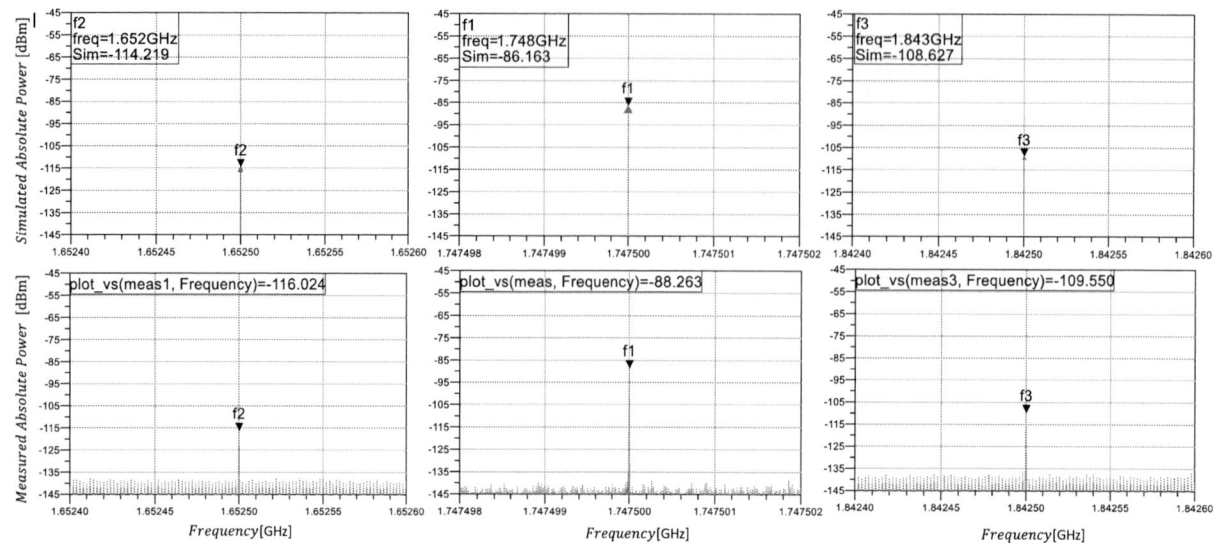

Fig. 8. Comparison between measurements results from the experimental test-bench and simulations results of the corresponding virtual test-bench at f_1, f_2 and f_3.

The virtual test-bench is composed of passive components described by their measured [S] parameters to be representative of the experimental test-bench. The RF transmission and jammer generators are replaced with Keysight ADS© software defined RF sources.

The spectrum analyzer is described with the ideal 50 Ωand the V_{3_Rx} and I_{3_Rx} probes allowing power calculation of all frequency components defined through the HB controller. A 2-tone HB simulation is then performed. Fig. 8 presents the comparison between calibrated measurements results from the experimental test-bench and simulations results of the corresponding virtual test-bench at f_1, f_2 and f_3.

The spectrum analyzer used in the experimental IMD3 test-bench allows performing a narrowband window measurement at each IMD3 frequency (f_1, f_2 and f_3). It allows programming the resolution and video bandwidths, the attenuation and the span separately for each window measurement.

A narrowband window has been used for each frequency component (f_1, f_2 and f_3) for measurements results from experimental IMD3 test-bench and simulations results of the corresponding virtual test-bench.

The absolute power level at the lower IMD3 frequency component ((f_4) filtered-out of the band of interest) is not presented in Fig. 8 because it is beyond the measurement sensitivity of the spectrum analyzer (beyond noise floor).

The absolute power levels of the IMD3 product f_3 from measurement are in a very good agreement with the one simulated with the virtual test-bench. One can see that all parameters from the simulation matched to measurement results. We observed less than 1 dB difference in IMD3 between simulation and measurement.

A good agreement between measurements and simulations is also observed for f_1 and f_2.

To our knowledge, the development of the virtual test-bench using a 2-tone Harmonic Balance simulation associated to a schematic based on measured [S] parameters of every passive devices of the experimental test-bench and applied to a non-linear behavioral frequency defined model of a switch is described for the first time.

V. CONCLUSION

This article presents a simple behavioral RF switch model that can be used within a virtual IMD3 test-bench. The architecture of the experimental IMD3 setup based on the use of a duplexer has been replicated and implemented in an ADS software. A 2-tone Harmonic Balance simulation has been realized. The measured and simulated results are in very good agreement in terms of IMD3 products. For the first time to our knowledge, a corresponding virtual IMD3 test-bench of an experimental IMD3 test-bench has been presented.

REFERENCES

[1] Tero Ranta, Juha Ellä, and Helena Pohjonen, "Antenna Switch Linearity Requirements for GSM/WCDMA Mobile Phone Front-Ends", The European Conference on Wireless Technology, Paris, France, 2005.

[2] "Wideband SP3T RF Switch for RF diversity or RF band selection applications", Available: www.infineon.com, 2016-02-03.

[3] "3rd Generation Partnership Project; Technical Specification Group Services and System Aspects; IP Multimedia Subsystem (IMS) based Packet Switch Streaming (PSS) and Multimedia Broadcast/Multicast (MBMS) User Service; Protocols (Release)", 3GPP TS 26.237 V16.0.0 (2020-07).

Proceedings of the 16th European Microwave Integrated Circuits Conference

A computationally-efficient self-consistent large signal model for GaN HEMTs based on ASM-HEMT

S. Khandelwal[#^], K. Kikuchi[*], H. Yamamoto[*]

[#]University of South Florida, Tampa, USA
[^]Macquarie University, Sydney, Australia
[*]Sumitomo Electric Industries Ltd., Tokyo, Japan
sourabh.khandelwal@mq.edu.au

Abstract— An accurate non-linear large-signal physics-based model for 50 V GaN HEMT technology is presented in this paper. The developed model accounts for the effects of trapping on I-V and C-V characteristics of GaN HEMTs self-consistently. Using the developed model we show the significance of self-consistent modeling of I-V and C-V behavior of the device on its large signal performance in presence of trapping. The computational speed of the developed physics-based model is also presented showing its aptness for use in circuit simulations.

Keywords — GaN HEMTs, Compact models, Large signal, RF PAs, Physics-based compact models.

I. INTRODUCTION

GaN HEMTs are extremely promising devices for 5G communication systems, and for technologies beyond 5G [1]. However, it is well-known that GaN technology suffers from trapping effects which limit its performance [2]. Trapping effects are being intensely studied covering different aspects including mitigation by technology improvement [3], novel characterization techniques [4], design techniques to overcome trapping [5], and modeling techniques [6], [7]. The presence of trapped charges due to the trapping effects change device electrostatics. This leads to change in terminal characteristics including both I-V and C-V behavior of the device. The change in the I-V characteristics as observed by knee-walkout and maximum current reduction has been studied in several works.

The change in electrostatics (due-to trapping) and its impact on the C-V behavior of the device is relatively less studied. The change in I-V and C-V behavior due to trapping originate from the same source, and these effects should be self-consistently modeled. Self-consistent model for I-V and C-V behavior accounting for trapping effects was shown recently in [8]. However, [8] did not include analysis of importance of self-consistent modeling on large signal performance. For GaN HEMTs large-signal models typically require tuning where model accuracy for DC or S-parameters is sacrificed for improved large signal model accuracy. This is attributed to the complex trapping effects. Here we show a physics-based model with good large signal model accuracy without sacrificing DC or S-parameters model accuracy. This is achieved by modeling the effect of trapping on I-V and C-V behavior of the device self-consistently. With

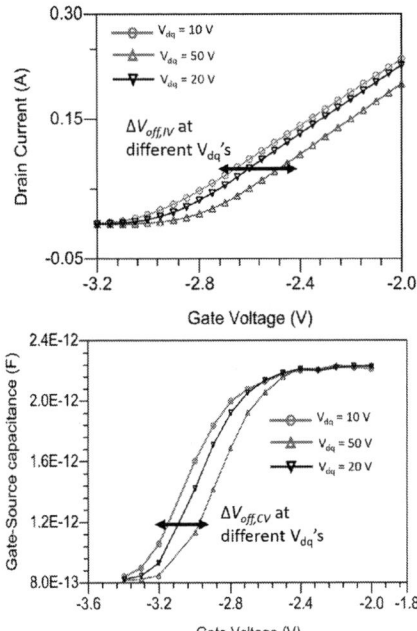

Fig. 1. Change in the cut-off voltage $\Delta V_{off,IV}$ and $\Delta V_{off,CV}$ with V_{dq} observed in the I-V and C-V characteristics of the device respectively due to the trapping effects. Drain-source voltage $V_{ds} = 10$ V.

the self-consistent model, large signal device behavior is only used for model validation and no model tuning is required for a good large signal accuracy.

Self-consistent modeling needs physics-based model formulations. In such models underlying electrostatics as modeled by surface-potential is used in both current and charge/capacitance formulations, giving the needed self-consistency to the model. However, physics-based models are generally believed to be computationally slow. In this paper, we show for the first time computational speed of physics-based self-consistent model for DC, S-parameters, and large signal harmonic balance simulations. The promising model speed shows the aptness of model for circuit simulations

3–4 April 2022, London, UK

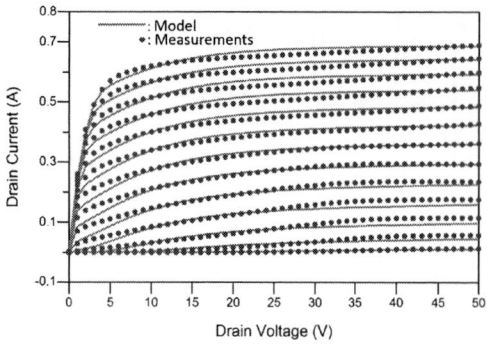

Fig. 2. Accurate model of the complete I-V plane of the device at $V_{dq} = 50$ V. V_{gs} varied from -3 to 0 at 0.25 V step. V_{ds} varied from 0 to 50 V at 1 V step.

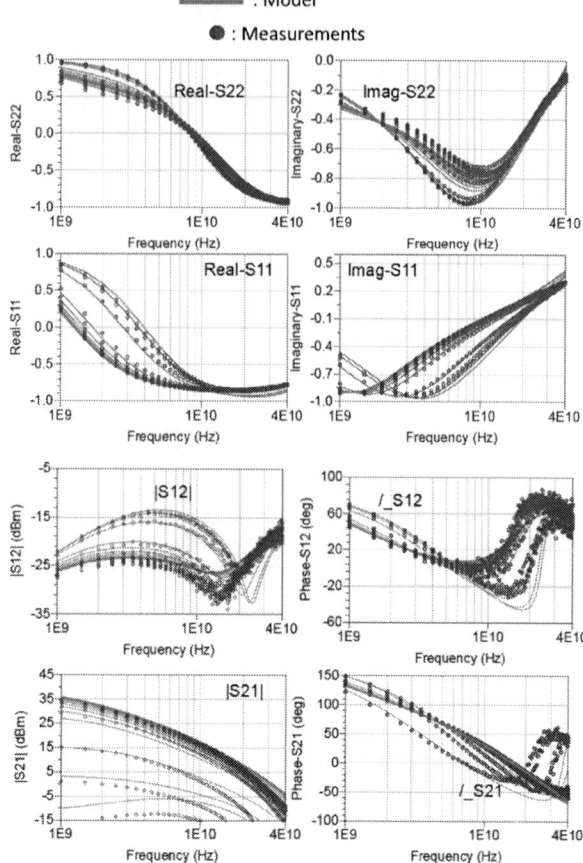

Fig. 3. Accurate model for real and imaginary parts of S11 and S22 for multiple bias conditions from V_{gs} = -3 to 0 at 0.25 V step, and V_{ds} = 50 V. Frequency varied from 0.5 to 40 GHz.

and optimization.

II. MODEL DEVELOPMENT

We use the physics-based model ASM-HEMT for the non-linear modeling performed in this work. ASM-HEMT [9] is a new industry standard which models the I-V and C-V with self-consistent surface-potential-based formulations [10]. GaN HEMT modeled in this work has a total gate width of $W = 1$ mm, and a rated supply voltage $V_{dd} = 50$ V. Our modeling methodology for self-consistent modeling of trapping effects, and developing a computational efficient model is described below.

Pulsed I-V characteristics of the device was measured at multiple quiescent drain voltages (V_{dq}) for a 1 % duty-cycle and 50 μsec pulse-width. As typically seen in GaN HEMTs, the device in this study also showed a change in cut-off (or pinch-off) voltage V_{off} with increasing V_{dq}. This is shown in Fig. 1. We mark the change in cut-off voltage seen in Fig. 1 as $\Delta V_{off,IV}$.

Intrinsic device capacitance C_{gs} at multiple bias points is obtained from pulsed S-parameter measurements for the same V_{dq} conditions. The intrinsic capacitance is extracted after de-embedding the manifolds, and parasitic effects as described in [11]. The change in cut-off voltage of the device can also be observed from C_{gs} versus gate-source voltage (V_{gs}) characteristics as shown in the bottom plot in Fig. 1. The sharp rise in C_{gs} occurs at the cut-off voltage. Different quiescent drain bias measurements are compared as shown in Fig. 1. The change in cut-off voltage seen in Fig. 1 is labelled as $\Delta V_{off,CV}$, and we find that $\Delta V_{off,IV} \approx \Delta V_{off,CV}$. This is expected as the trap-state of the device in these two separate measurements is set by the relatively long quiescent bias condition. Note that we also observed change in other parameters with V_{dq} as reported in [12]. This is accounted for in the developed model. Fig. 1 illustrate the point about consistent change in parameters with V_{dq}. The key observation of consistent change in parameters is used to develop complete non-linear large signal model of the device. The modeling process includes obtaining an I-V model for the

pulsed I-V at the quiescent point equal to the rated voltage. I-V model extraction follows same process as described in [9]. Model results are shown in the next section. Multi-bias pulsed S-parameters at the rated $V_{dq} = 50$ V of the device are used to extract the small-signal behavior of the model. These two are modeled self-consistently with same change in model parameter applied to both I-V and S-parameters. Large signal characteristics including power sweep, and load-pull in class-B, and class-C bias conditions are used for model validation only and model accuracy for I-V and S-parameters is not sacrificed for large signal model performance.

The modeled device in this study had two field-plates. A physical modeling of field-plates introduces internal circuit nodes in the compact model. The circuit nodes are introduced at the electrical terminals of the field-plate model. These internal nodes, and the associated full physical model formulations of field-plate regions increase computation requirements of the model. To improve we adopt the approximation presented in [13]. Quantitative results on model speed are shown in the next section.

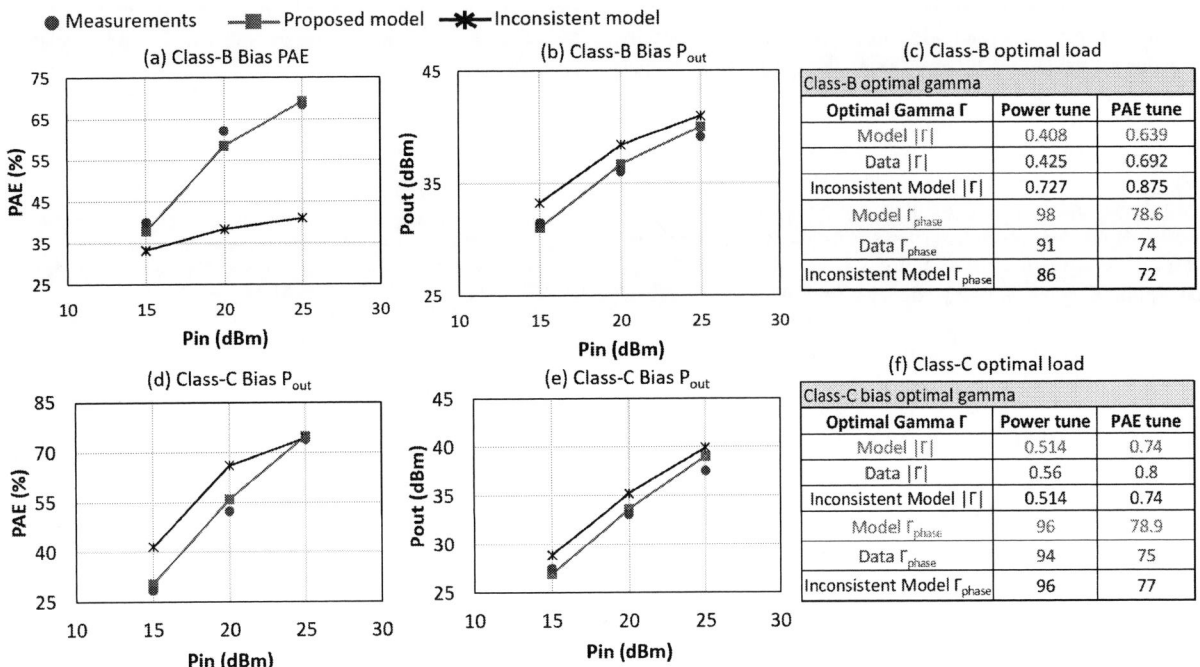

(c) Class-B optimal load

Class-B optimal gamma	Power tune	PAE tune
Model $|\Gamma|$	0.408	0.639
Data $|\Gamma|$	0.425	0.692
Inconsistent Model $|\Gamma|$	0.727	0.875
Model Γ_{phase}	98	78.6
Data Γ_{phase}	91	74
Inconsistent Model Γ_{phase}	86	72

(f) Class-C optimal load

Class-C bias optimal gamma	Power tune	PAE tune
Model $|\Gamma|$	0.514	0.74
Data $|\Gamma|$	0.56	0.8
Inconsistent Model $|\Gamma|$	0.514	0.74
Model Γ_{phase}	96	78.9
Data Γ_{phase}	94	75
Inconsistent Model Γ_{phase}	96	77

Fig. 4. Large signal model results for class-B and class-B biases. Frequency = 4.8 GHz. DC drain bias is set to 50V. Results include both power and PAE tune conditions. Self-consistent model shows excellent agreement with the measurements.

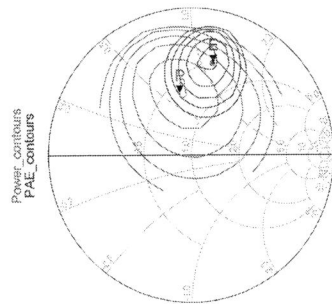

Fig. 5. Typical load-pull contours obtained with the model. Class-C load-pull contours are shown here at 4.8 GHz.

III. RESULTS AND DISCUSSIONS

The I-V model for $V_{dq} = 50$ V is shown in Fig. 2. A good model accuracy can be seen from Fig. 2. For I-V model, the device dimensions as known from the manufacturer are set directly in the model. The parameter extraction flow discussed in [9] is used to extract the other ASM-HEMT model parameters.

Next, the small-signal parameters were extracted. The effects of manifolds have been de-embedded. Additional parasitic effects are modeled following the process described in [9]. The model results for the pulsed S-parameters at $V_{dq} = 50$ V from 0.5 to 40 GHz at multiple bias points from $V_{gs} = -3.0$ V to 0.0 V, and $V_{ds} = 50$ V is Fig.

3 showing good model accuracy. Accurate C-V model is crucial for modeling S-parameters and the developed model self-consistently accounts for trapping effects on both I-V and C-V behavior. The parameters affected by trapping effects such as cut-off voltage V_{off} is shared between the I-V model formulations, and the charge/capacitance model formulations. So the electrostatics of the device as set by the $V_{dq} = 50$ V and its effect of both I-V and C-V behavior of the model is accurately modeled. Next we use the large signal measurements for model validation only.

Large signal measurements used to validate the developed model include load-pull simulations for output power (P_{out}) and power added efficiency (PAE) in tuned conditions. These measurements and model simulations are performed at 4.8 GHz and the rated 50 V drain-bias in class-B and class-C bias classes. We compare the large signal model results of the proposed self-consistent model and an inconsistent I-V and C-V model to clearly highlight the improvements of the self-consistent modeling. These model results are shown in Fig. 4. These results are obtained by multiple load-pull simulations such as one shown in Fig. 5. The top-row in Fig. 4 are model results for class-B, and the bottom row has model results for class-C bias. Fig. 4(a) and Fig. 4(d) show the model simulations and measurements for PAE tuned condition at three different input power levels P_{in} = 15, 20, and 25 dBm. The proposed self-consistent model accurately models the measured PAE in both bias classes for all three input power levels. The inconsistent model where C-V does not account for trapping effects is less accurate.

Figures Fig. 4(b) and Fig. 4(e) are large signal results for power tuned condition. These also show that higher accuracy of self-consistent model. The table in the two rows show the model accuracy for predicting the optimal reflection coefficient Γ_{opt} for the two classes for both power and PAE tuned conditions. Self-consistent model predicts the magnitude and phase of optimal gamma points quite well, while the inconsistent model shows inaccuracy. It is noted that inaccuracy of the inconsistent model is lower for class-C bias condition. Overall, the large signal results with the proposed self-consistent model in Fig. 4 show excellent results for both bias classes, and tuned conditions. Inconsistent model can be tuned for accuracy of large-signal simulations by sacrificing accuracy in DC and S-parameters. Self-consistent model does not need any such sacrifice in accuracy.

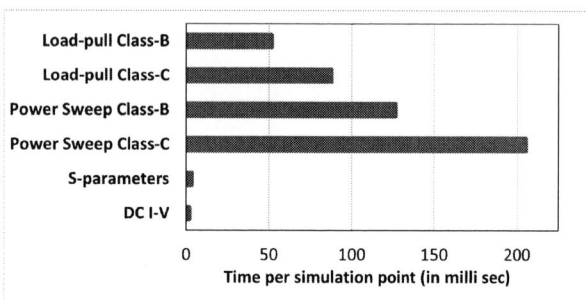

Fig. 6. Simulation time of the developed model for DC I-V, S-parameters, Power sweep and load-pull simulations in class-B and class-C.

The computational performance of the model is evaluated on a typical engineering computer with 16 GB RAM and 2 GHz processor. The model simulation time on this computer for DC I-V, multi-bias S-parameters, input power sweep (for class-B and class-C), and load-pull (for class-B and class-C) is shown in Fig. 6. The simulation time for each type of simulation is obtained by simulating a large number of points, and obtaining the simulation time needed per point. Large signal simulations using the harmonic balance algorithms are found to take the largest amount of time. Nevertheless, the model simulation time for all types of simulations per point is in milli-seconds. This underlines a reasonably fast model performance, for instance a 1000 load impedance points can be computed within a few seconds by the developed model on a typical computer. This excellent speed is quite promising for using this model in circuit simulations.

IV. CONCLUSION

A computationally efficient physics-based large signal model for GaN HEMTs is presented. The proposed model is based on the new industry standard ASM-HEMT model. The impact of trapping effects on I-V and C-V have been self-consistently modeled, and a computationally efficient approach for modeling the field-plates is used in the developed model. Model shows good agreement with measured DC I-V,

multi-bias S-parameters, and large signal load-pull, and tuned load power sweep measurements under multiple bias classes.

REFERENCES

[1] S. Nakajima, "Gan hemts for 5g base station applications," in *2018 IEEE International Electron Devices Meeting (IEDM)*, 2018, pp. 14.2.1–14.2.4.

[2] A. Santarelli, R. Cignani, D. Niessen, G. P. Gibiino, P. A. Traverso, D. Schreurs, and F. Filicori, "Multi-bias nonlinear characterization of gan fet trapping effects through a multiple pulse time domain network analyzer," in *2015 10th European Microwave Integrated Circuits Conference (EuMIC)*, 2015, pp. 81–84.

[3] H. Kim, R. M. Thompson, V. Tilak, T. R. Prunty, J. R. Shealy, and L. F. Eastman, "Effects of sin passivation and high-electric field on algan-gan hfet degradation," *IEEE Electron device letters*, vol. 24, no. 7, pp. 421–423, 2003.

[4] J. L. Gomes, L. C. Nunes, C. F. Gonçalves, and J. C. Pedro, "An accurate characterization of capture time constants in gan hemts," *IEEE Transactions on Microwave Theory and Techniques*, vol. 67, no. 7, pp. 2465–2474, 2019.

[5] T. Cappello, C. Florian, A. Santarelli, and Z. Popovic, "Linearization of a 500-w l-band gan doherty power amplifier by dual-pulse trap characterization," in *2019 IEEE MTT-S International Microwave Symposium (IMS)*, 2019, pp. 905–908.

[6] S. A. Albahrani, D. Mahajan, J. Hodges, Y. S. Chauhan, and S. Khandelwal, "Asm gan: Industry standard model for gan rf and power devices—part-ii: Modeling of charge trapping," *IEEE Transactions on Electron Devices*, vol. 66, no. 1, pp. 87–94, 2019.

[7] X. Du, M. Helaoui, A. Jarndal, T. Liu, B. Hu, X. Hu, and F. M. Ghannouchi, "Ann-based large-signal model of algan/gan hemts with accurate buffer-related trapping effects characterization," *IEEE Transactions on Microwave Theory and Techniques*, 2020.

[8] J. Hodges, D. Schwantuschke, F. van Raay, P. Brückner, R. Quay, and S. Khandelwal, "Consistent modelling of iv and cv behaviour of gan hemts in presence of trapping," in *2019 IEEE MTT-S International Microwave Symposium (IMS)*. IEEE, 2019, pp. 603–606.

[9] S. Khandelwal, Y. S. Chauhan, T. A. Fjeldly, S. Ghosh, A. Pampori, D. Mahajan, R. Dangi, and S. A. Ahsan, "Asm gan: Industry standard model for gan rf and power devices—part 1: Dc, cv, and rf model," *IEEE Transactions on Electron Devices*, vol. 66, no. 1, pp. 80–86, 2019.

[10] S. Khandelwal, Y. S. Chauhan, and T. A. Fjeldly, "Analytical modeling of surface-potential and intrinsic charges in algan/gan hemt devices," *IEEE Transactions on Electron Devices*, vol. 59, no. 10, pp. 2856–2860, 2012.

[11] S. Khandelwal, Y. S. Chauhan, J. Hodges, and S. A. Albahrani, "Non-linear rf modeling of gan hemts with industry standard asm gan model," in *2018 IEEE BiCMOS and Compound Semiconductor Integrated Circuits and Technology Symposium (BCICTS)*. IEEE, 2018, pp. 93–97.

[12] S. Khandelwal, K. Kellogg, C. Hill, H. Morales, L. Dunleavy, G. Drandova, A. Pacheco, and J. Jimenez, "Quiescent drain voltage dependence of pulsed i-v characteristics of gan hemts: Analysis and modeling," in *2019 IEEE BiCMOS and Compound semiconductor Integrated Circuits and Technology Symposium (BCICTS)*, 2019, pp. 1–4.

[13] J. Hodges, S. A. Albahrani, and S. Khandelwal, "A computationally efficient modelling methodology for field-plates in gan hemts," in *2019 IEEE BiCMOS and Compound semiconductor Integrated Circuits and Technology Symposium (BCICTS)*. IEEE, 2019, pp. 1–4.

Proceedings of the 16th European Microwave Integrated Circuits Conference

Large-Signal Modeling for Nonlinear Analysis of Experimental Devices in 22nm FDSOI Technology

Quang Huy Le[†*], Dang Khoa Huynh[†*], Anurag Nayak[†], Steffen Lehmann[#],
Zhixing Zhao[#], Thomas Kämpfe[†], Matthias Rudolph[*]

[†]Center Nanoelectronic Technologies, Fraunhofer IPMS, Dresden, Germany
[*]Ulrich-L.-Rohde Chair of Radio Frequency and Microwave Techniques,
Brandenburg University of Technology, Cottbus, Germany
[#]Fab1, GlobalFoundries, Dresden, Germany

Abstract — The development process of the FDSOI CMOS technology is continuously ongoing to improve the performance for high-frequency applications, especially for power amplification. Meanwhile, customized transistor devices pose difficulties to be integrated fast into the existing foundry's PDK for nonlinear analysis. This paper thereby presents an empirical large-signal modeling approach for experimental devices in the 22nm FDSOI CMOS. In particular, a versatile mathematical expression for the drain current has been developed which can accurately model various device types. Moreover, the gate charge is computed and modeled by using artificial neural network. The modeling approach has been verified by using the latest hardware in the industry's 22nm FDSOI with multi-bias S-parameters up to 110 GHz and non-50-Ω large-signal measurements.

Keywords — 22nm, FDSOI, ANN, charge conservation, large-signal model.

I. INTRODUCTION

Following the recent research activities within the scope of the Important Project of Common European Interest (IPCEI), the industry's fully-depleted silicon-on-insulator (FDSOI) technology, i.e., 22FDX®, has been continuously enhanced for high-frequency performance. Recently optimized devices deliver high potential for radio frequency (RF) front-end applications [1]-[4]. In the development process, the experimental transistor devices undergo various characterization schemes, particularly load-pull measurement for RF large-signal evaluation. Physical constraints from test equipment such as limit of tuning range, frequency band and the ability to control harmonic terminations pose difficulties for a comprehensive large-signal analysis on each device type. In such scenario, an accurate nonlinear model is of high interest to facilitate the hardware development and verification process. However, for cases in which the experimental device's behavior can not be simply adapted based on existing physics-based models provided by the process design kit (PDK), a measurement-based empirical model can be a fast solution. As part of the IPCEI WIN FDSOI project, a suitable empirical large-signal model has been developed and validated well in the mm-wave frequency range [5]. The extraction procedure is simple and allows the model to be quickly constructed from measurement data. This paper accordingly elaborates further the modeling approach

for various experimental devices of the 22FDX® process technology. In the following sections, the modeling is proven to be efficient and versatile and can be simply adapted in different scenarios, e.g., unconventional device layout and asymmetrical source-drain devices. The test structures used in this work consist of the state-of-the-art devices fabricated in the 22nm FDSOI process from GlobalFoundries. The devices under test (DUTs) are embedded in the RF test fixture with 50μm-pitch ground-signal-ground (GSG) pad pattern.

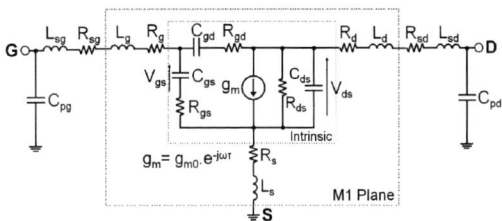

Fig. 1. Small-signal equivalent circuit model of the FDSOI transistor including fixture parasitics from the first metal M1 to probe-tip plane.

II. MODELING APPROACH

A. Model description

The empirical large-signal model for FDSOI transistors presented in [5] is applied as the frame of this work. Fig. 1 illustrates the small-signal equivalent circuit including fixture parasitics. The external parasitics represent the pads and trace from the probe-tip plane to M1 plane. The low-frequency dispersion effect is also addressed within the large-signal model. The model is built from multi-bias S-parameters de-embedded to first metal M1 by using open-short technique. In addition, the modeling of self-heating is chosen to be neglected in the current modeling approach.

B. Charge modeling with artificial neural networks

In this work, terminal charge conservation is considered for robust simulation. The gate capacitances C_{gg} and C_{gd} are derived from an identical gate charge Q_g. This allows the capacitances to fulfill the condition for terminal charge conservation [6] in equation (1).

$$\frac{\partial C_{gg}}{\partial V_{ds}} = -\frac{\partial C_{gd}}{\partial V_{gs}} \qquad (1)$$

3–4 April 2022, London, UK

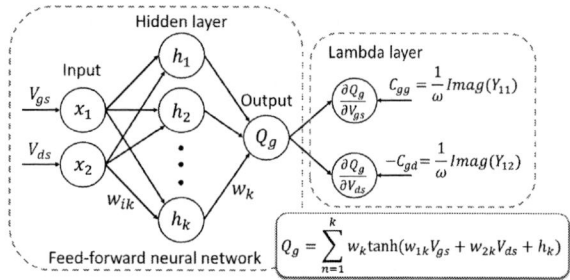

Fig. 2. Feed-forward ANN with one hidden layer utilizing hyperbolic tangent function as activation function for gate-charge computation.

Nevertheless, the gate charge computed from the extracted capacitances often yields inaccuracy in practice. In particular, the line integral of the vector field formed from measured capacitances is path dependent and thus, the model fidelity is compromised when selecting an integration path. Moreover, even after the charge is obtained, fitting of these data remains tedious. To deal with this issue, an artificial neural network (ANN) appears to be a more effective solution. Not restricted to machine learning applications, ANN has been extensively applied in FET device modeling thanks to its usefulness and versatility [8]-[11]. With the advancement of modern computation software, an ANN can be simply deployed and trained to compute the gate charge by using the extracted capacitances. Considering the feed-forward ANN in Fig. 2, the output can also be viewed as a mathematical expression for nonlinear mapping with well-defined functions. This enables the trained ANN to be directly applied in circuit simulators.

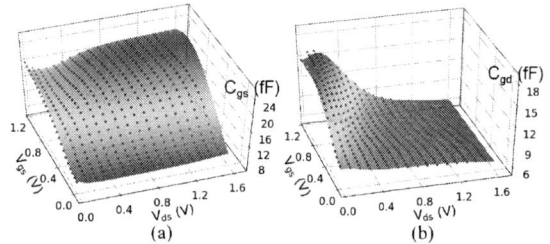

Fig. 3. Measured (blue symbol) and predicted (coloured surface) intrinsic capacitances of an 18nm n-FET: (a) C_{gs} ($C_{gs} = C_{gg} - C_{gd}$) and (b) C_{gd}.

For demonstration, a feed-forward ANN with one hidden layer of 8 perceptrons is defined by using TensorFlow Keras in Python [12]. The hyperbolic tangent function is applied as the activation function, and the Adam optimization algorithm is used to train the ANN [13]. In addition, the Lambda layer is used to set up the training conditions for the ANN. The training data sets, i.e., C_{gg} and C_{gd}, are extracted from multi-bias two-port S-parameters of a multi-finger 0.8V thin-oxide n-channel device (l_g = 18 nm, W_{total} = 32 μm). The ANN is trained without the need of regridding the data sets and applied to predict the gate charge. The gate capacitances are then derived from the computed charge data for comparison with measurement (see Fig. 3). The predicted data is accurate and elegantly extrapolated outside

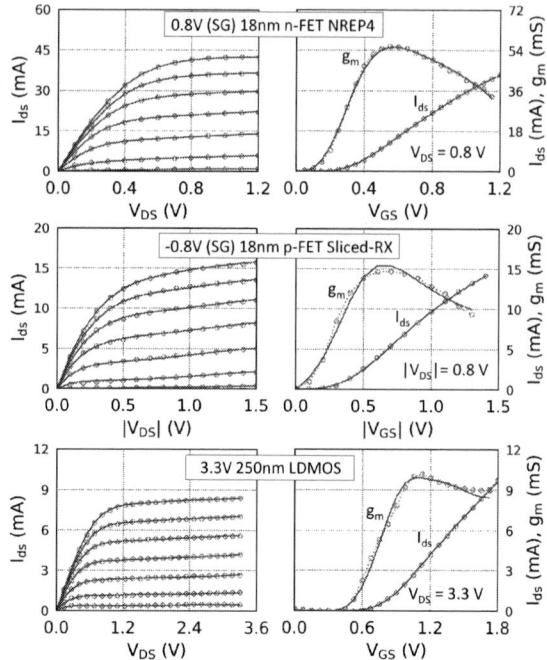

Fig. 4. Modeled (blue solid line) and measured (red symbol) DC-IV characteristics of different 22FDX® transistor types.

the training range. Moreover, the capacitance behaviour in both linear and saturation regions is well described by the model. Accordingly, this approach is versatile to be utilized for different device types to eliminate the need of complicated charge or capacitance models.

C. Nonlinear drain current model

In this work, the drain current characteristics of the transistors are described by using analytical expressions. In particular, the drain current model is taken from the previously developed empirical model for FDSOI in [5]. The formulas are expressed as follows:

$$I_{ds} = I_{pk}(1 + M_{pk}tanh(\psi))tanh(sinh(\alpha V_{ds})) \quad (2)$$

$$I_{pk} = I_0 + I_1V_{ds} + I_2V_{ds}^2 + I_3V_{ds}^3 + I_{br} \quad (3)$$

$$I_{br} = L_{br}e^{E_{br}(V_{ds} - V_{br})} \quad (4)$$

$$M_{pk} = M_{pk0} + M_{pkA}tanh(\psi_M) \quad (5)$$

$$\psi_M = Z_M(V_{gs} - V_{gm0}) \quad (6)$$

$$Z_M = (P_{Z0} + P_{Z1}V_{ds})tanh(\alpha_Z V_{ds}) + P_{Z0}. \quad (7)$$

The drain current model above is modified based on the well-known Angelov model for HEMTs [7]. The parameters ψ and α in (2) are reused from the original Angelov drain current model. Besides, the term I_{br} is to model the break-down effect as well as to guide extrapolation. It has been presented in [5] that the model is able to accurately reproduce the asymmetrical bell-shaped g_m of the FDSOI

18

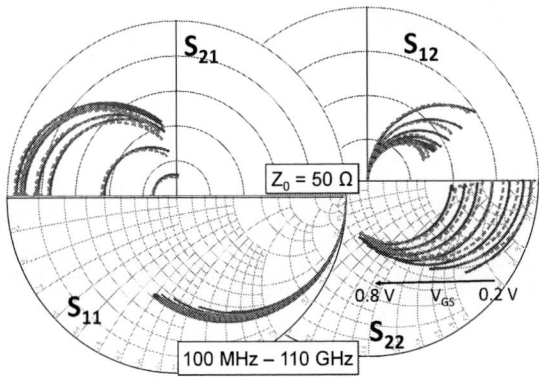

Fig. 5. S-parameters from measurement (red dotted line) and simulation results (blue solid line) at M1 plane of an 18nm n-FET with the bias conditions: $V_{DS} = 0.8$ V and V_{GS} from 0.2 V to 0.8 V.

CMOS with precise derivatives. The drain current model is applied to model the latest experimental devices from the 22FDX®. The results are illustrated in Fig. 4. The devices used for the test are the thin-oxide (SG) n-FET and p-FET with vertical repetition (NREP) configuration and sliced-RX architecture [4], respectively. In addition, the test includes the 3.3V LDMOS, which is an asymmetrical device. The excellent results indicate that the current model is highly flexible to fit different device characteristics.

III. MODEL VERIFICATION AND EVALUATION

An optimized multi-finger 0.8V thin-oxide n-FET with 18nm gate length is selected to verify the proposed modeling approach. The DUT features double-gate contact and 4 vertical repetitions (NREP4) of 16×0.5 μm. The model is implemented in Keysight ADS design environment by using the symbolically defined device (SDD). Harmonic balance simulator is utilized for the large-signal simulation.

In the first step, two-port S-parameters are measured at different gate biases and de-embedded to the first metal M1 for comparison with simulation. The small-signal verification is demonstrated in Fig. 5. The model accordingly delivers a global fit in terms of bias condition. Moreover, the simulation results show good validity in the sub-110 GHz frequency range. The non-50-Ω large-signal measurement is performed by using the nonlinear measurement setup described in [1], [2].

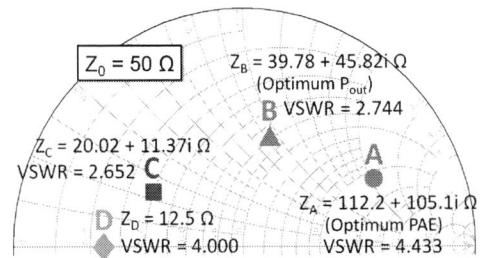

Fig. 6. Load-impedance terminations used for large-signal verification.

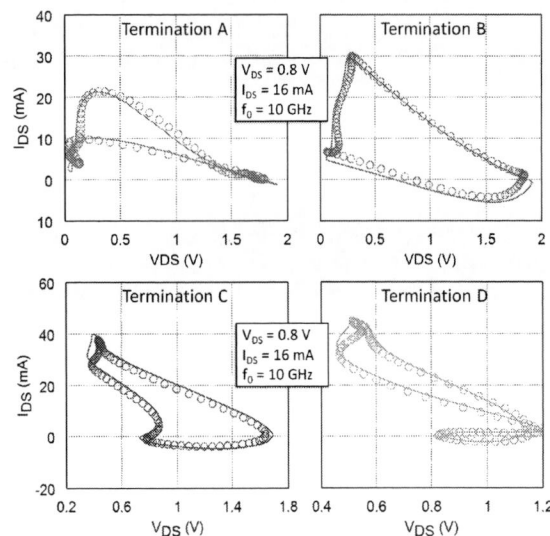

Fig. 7. Dynamic load lines of an 18nm n-FET reconstructed from measurement (symbol) and simulation (solid line) for different impedance terminations at -5 dBm available input power.

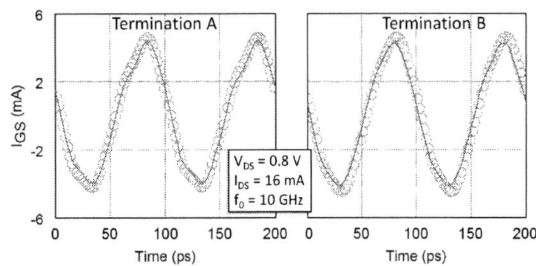

Fig. 8. Gate-source displacement current waveforms of an 18nm n-FET reconstructed from measurement (symbol) and simulation (solid line) for different impedance terminations at -5 dBm available input power.

In addition, the measurement reference plane is at the probe tips. The DUT is presented with a fixed source impedance to enhance the gain, while four distinct load terminations are used consecutively for the validation (see Fig. 6). The associated harmonic impedances up to the fifth order of each termination are also included in the simulation. Termination A is the load impedance for optimum efficiency, by which the voltage swing is maximized. On the other hand, termination B is for maximum output power. Termination C and D are selected for peaking the drain current swings. The combination of these impedance terminations enables a thorough model validation as well as convergence tests in various scenarios. Fig. 7 illustrates the dynamic load lines between measurement and simulation when the DUT is driven into strong compression with a single tone 10 GHz. It can be seen that the simulation results are in good agreement with the experimental data. Furthermore, convergence is well maintained even when the voltage swings are much larger than V_{DD}. Besides, the time-domain gate-source displacement currents at the same

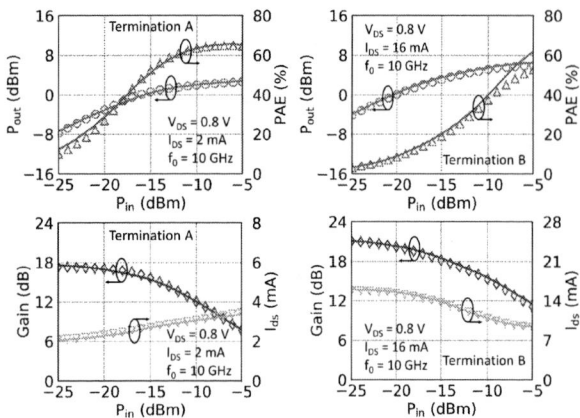

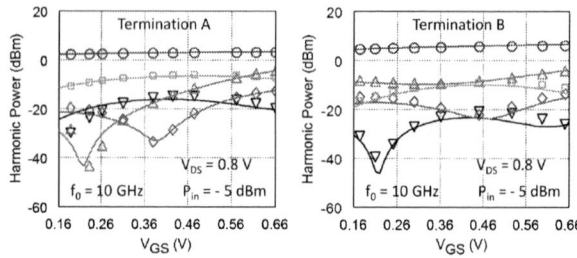

Fig. 9. Power-sweep measurement (symbol) and simulation (solid line) results of an 18nm n-FET for termination A and B at fundamental frequency 10 GHz.

Fig. 10. Harmonic powers versus V_{GS} of an 18nm n-FET from measurement (symbol) and simulation (solid line). Blue circle: fundamental frequency (f_0), red triangle: second component ($2f_0$), orange square: third component ($3f_0$), green diamond: fourth component ($4f_0$), black upside-down triangle: fifth component ($5f_0$).

condition for termination A and B are shown in Fig. 8. The excellent agreement between simulation and measurement indicates that the gate charge has been accurately modeled. For further evaluation, Fig. 9 demonstrates the power-sweep measurement results of the DUT in deep class AB and class A operation with termination A and B, respectively. In addition, the harmonic powers up to the fifth order in terms of V_{GS} are depicted in Fig. 10. The simulation results are accurate and remain stable even in deep compression region.

IV. CONCLUSION

This paper presents a large-signal modeling approach for platform and experimental devices in the 22nm FDSOI CMOS process. The gate charge is computed by using ANN for charge conservative modeling. The accurate drain current model can be well applied to various device types. Those features provide versatility for the development process of mm-wave CMOS devices in the industry's 22nm technology node. The modeling has been applied, and the small-signal simulation accuracy is verified in the sub-110 GHz band. A comprehensive large-signal verification at 10 GHz shows excellent matching to the measurement results. Moreover, as observed from all demonstrated simulation results, the model is consistently ensured for convergence in complex test conditions. In summary, the proposed flexible modeling delivers a high degree of accuracy and robustness for nonlinear analysis of various device types.

ACKNOWLEDGMENT

The authors would like to thank Fab1, GlobalFoundries, for providing the 22FDX® test structures. This work is funded in part by the ECSEL Joint Undertaking and the German ministry of education and Research through the project BEYOND5 and in part by the German ministry of economics and Free State of Saxony through the Important Project of Common European Interest WIN FDSOI.

REFERENCES

[1] Q. H. Le et al., "DC-110 GHz Characterization of 22FDX® FDSOI Transistors for 5G Transmitter Front-End," in Proc. 49th Eur. Solid-State Device Research Conf., Cracow, Poland, 2019.

[2] Q. H. Le et al., "Assessment of a Thick-Oxide Transistor from the 22FDX® Platform for 5G NR sub-6 GHz FEMs," in Proc. IEEE 2nd 5G World Forum, Dresden, Germany, 2019.

[3] Q. H. Le et al., "W-Band Noise Characterization with Back-Gate Effects for Advanced 22nm FDSOI CMOS," in Proc. IEEE Radio Frequency Integrated Circuits Symp., LA, USA, 2020.

[4] Z. Zhao et al., "22FDX® fMAX Optimization through Parasitics Reduction and GM Boost," in Proc. 49th Eur. Solid-State Device Research Conf., Cracow, Poland, 2019.

[5] Q. H. Le et al., "Empirical Large-Signal Modeling of mm-Wave FDSOI CMOS based on Angelov model," in Trans. Electron Devices, 2021.

[6] D. E. Root, 'Nonlinear charge modeling for FET large-signal simulation and its importance for IP3 and ACPR in communication circuits," in Proc. IEEE Midwest Symp. Circuits and Systems, Dayton, USA, 2001.

[7] I. Angelov, H. Zirath, and N. Rorsman, "A new empirical nonlinear model for HEMT-devices," in Proc. IEEE MTT-S Microw. Symp. Digest, Albuquerque, NM, USA, Jun. 1992.

[8] M. Rudolph, C. Fager and D. Root, "The large-signal model: theoretical foundations, practical considerations, and recent trends," in Nonlinear Transistor Model Parameter Extraction Techniques, Cambridge University Press, 2011.

[9] A. Jarndak, "On Neural Networks Based Electrothermal Modeling of GaN Devices," in IEEE Access, vol. 7, pp. 94205 - 94214, 2019.

[10] J. King and C. Wilson, "Charge conservative FET modelling using ANNs," in Proc. European Microw. Integrated Circuits Conf., Nuremberg, Germany, 2017.

[11] Y. Ko et al., "Artificial Neural Network Model of SOS-MOSFETs Based on Dynamic Large-Signal Measurements," in IEEE Trans. Microw. Theory and Techniques, vol. 62, pp. 491 - 501, 2014.

[12] TensorFlow, "Module: tf.keras". Accessed: Oct. 2020. [Online]. Available: https://www.tensorflow.org/api_docs/python/tf/keras

[13] D. P. Kingma and J. Ba, "Adam: a method for stochastic optimization", in Int. Conf. for Learning Representations, San Diego, USA, 2015.

Gap in pagination due to unavailable paper.

Page 21

Proceedings of the 16th European Microwave Integrated Circuits Conference

Analysis of the Relaxed Contacted-Poly-Pitch Effect on the RF Performance of Strained-SiGe-Channel p-FETs in 22nm FDSOI Technology

Quang Huy Le[†*], Dang Khoa Huynh[†*], Steffen Lehmann[#], Zhixing Zhao[#], Thomas Kämpfe[†], Matthias Rudolph[*]

[†]Center Nanoelectronic Technologies, Fraunhofer IPMS, Dresden, Germany
[*]Ulrich-L.-Rohde Chair of Radio Frequency and Microwave Techniques,
Brandenburg University of Technology, Cottbus, Germany
[#]Fab1, GlobalFoundries, Dresden, Germany
quang.huy.le@ipms.fraunhofer.de

Abstract — **Strained-SiGe channel structures have been widely exploited in advanced nanoscale CMOS technology nodes to enhance the performance of p-FETs, especially in the high frequency domain. By relaxing the contacted poly pitch (CPP) to gain beneficial strain effect, a previous study has reported a significant improvement in the transconductance, and thus, the device's speed is increased for radio frequency (RF) operation. This paper accordingly presents a comprehensive analysis of the effect of the gate CPP on the RF performance of the 22nm FDSOI p-FETs. The scope of the study includes the small-signal characteristics, the W-band noise performance and the large-signal performance in non-50-Ohm environment.**

Keywords — **CPP, 22nm FDSOI, Strained SiGe, p-FET.**

I. INTRODUCTION

For high frequency (HF) operation, carrier mobility dictates many important aspects (e.g., saturation current, power gain, speed and noise performance) of a semiconductor device. As the p-type material suffers from poor hole mobility, radio frequency (RF) CMOS n-channel transistors are superior to their p-channel equivalents in RF performance. For nanoscale CMOS, attempts to enhance the carrier mobility of a p-FET often involve strained channel structures. In the industry's 22-nm fully depleted silicon-on-insulator (FDSOI) CMOS process technology, i.e., 22FDX®, compressively strained Silicon-Germanium (SiGe) channel is exploited for p-type transistors [1], [2]. Recent study on the noise characteristics of the 22FDX® devices in W-band has reported the comparable performance between the p-FETs and n-FETs [3]. In addition, Zhao et al. proposed three specific layout-wise approaches for further RF performance enhancement of the p-FETs [2], among which the device transconductance (g_m) can be improved by relaxing the contacted poly pitch (CPP). Increasing the CPP particularly moves the channel away from the shallow trench isolation (STI) (see Fig. 1a), and therefore, introduces beneficial lateral strain effect to the device. However, the scope of the previous work is restricted to a comparative study of small-signal parameters. In order to gain a better insight on this topic, this work presents a detailed study on the RF characteristics of the optimized 22FDX® p-channel devices including also the W-band noise performance and the large-signal performance in non-50-Ω environment.

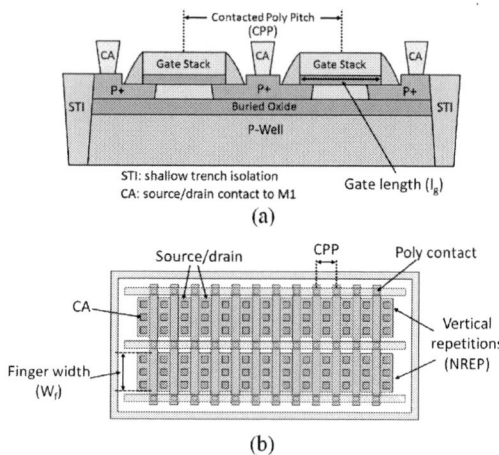

Fig. 1. (a) Simplified cross section of a multi-finger 22FDX® p-FET and (b) top-view layout configuration of an NREP device.

Table 1. Test device description.

Test device	l_g (nm)	W_f (μm)	CPP (nm)	NREP
1×CPP p-FET	18	0.5	104	4
2×CPP p-FET	18	0.5	208	4
3×CPP p-FET	18	0.5	312	4
2×CPP n-FET	18	0.5	208	4

The test structures used in this work consist of common-source 18nm-gate-length thin-oxide (0.8 V) transistors provided by Fab1, GlobalFoundries. Table 1 lists the test structures. The device layout is illustrated in Fig. 1b. The transistors with 0.5-μm finger width feature 4 vertical repetitions (NREP4). The back-gate terminals are grounded. In particular, three p-FETs with different CPPs are available for the study, among which 1×CPP is the standard pitch of 22FDX® devices. A 2×CPP n-FET with similar geometry configuration is used as the reference for the study. In addition, the geometry choices reported in this paper are not for best RF performance, but rather they are for the same dimension comparison between the n-FET and p-FET. The devices under test (DUTs) are embedded in the RF test fixtures with 50-μm ground-signal-ground (GSG) pad pattern.

3–4 April 2022, London, UK

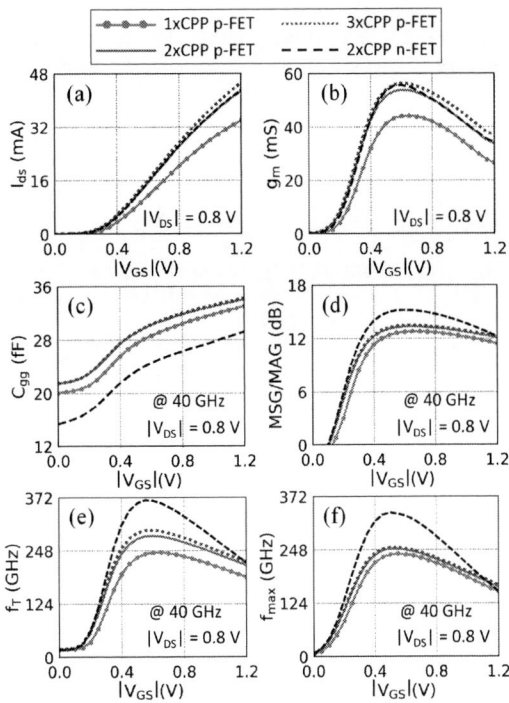

Fig. 2. DC and small-signal parameters versus gate voltage: (a) drain current, (b) dc transconductance, (c) total gate capitance, (d) maximum gain, (e) transit frequency and (f) maximum oscillation frequency.

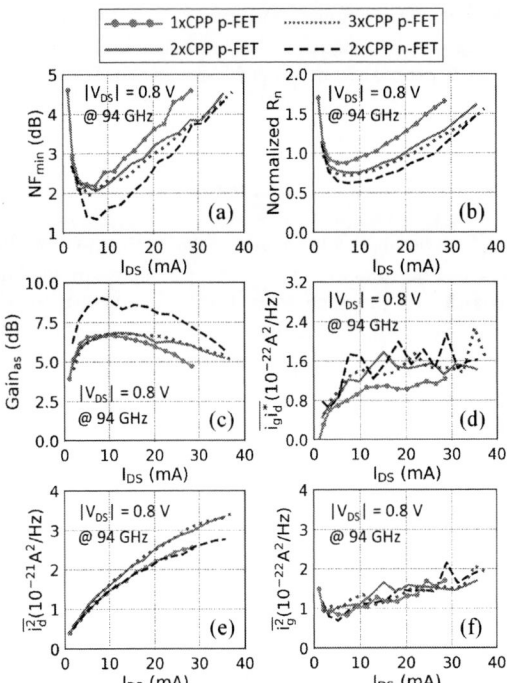

Fig. 3. Noise performance extracted at 94 GHz versus drain current: (a) NF_{min}, (b) normalized R_n, (c) associated gain, (d) noise correlation, (e) channel thermal noise and (f) induced gate noise.

II. DC AND SMALL-SIGNAL CHARACTERISTICS

Two-port S-parameter characterization is performed for the small-signal analysis. The mm-wave measurement setup described in [4] is used for the characterization. The measured S-parameters are de-embedded to the first metal M1 by applying open-short de-embedding technique. In addition, devices from 7 different dies are characterized and the mean values are used for the evaluation. The small-signal parameters considered in this section include the total gate capacitance C_{gg}, the maximum stable gain/maximum available gain (MSG/MAG), the transit frequency f_T and the maximum oscillation frequency f_{max}. Fig. 2 illustrates the mean values of the dc drain current, transconductance g_m and small-signal parameters extracted at 40 GHz. The dc performances of the 2xCPP and 3xCPP p-FETs are clearly improved with respect to the standard 1xCPP p-FET (see Fig. 2a, b). Furthermore, the 3xCPP device exceeds the n-FET in saturation current and the peak g_m. Nevertheless, the 2xCPP and 3xCPP p-FETs have higher C_{gg} than the standard p-FET and the n-FET (see Fig. 2c). Regarding the small-signal performance, the relaxed CPP p-FETs only exhibit clear improvement in f_T while there is minor change in MSG/MAG and f_{Max} (see Fig. 2d-f).

$$f_T \approx \frac{g_m}{2\pi C_{gg}} \quad (1)$$

$$f_{max} \approx \sqrt{\frac{f_T}{8\pi R_g C_{gd}}} \quad (2)$$

Equation (1) and (2) are used to describe f_T and f_{max}, respectively, in terms of the small-signal parameters [2]. It is shown that the gate capacitance has a critical impact on the device's speed. As a consequence, the increase in C_{gg} compromises the RF small-signal performance, particularly the speed. Hence, for the 2xCPP and 3xCPP p-FETs, f_T and f_{max} are still much lower than the 2xCPP n-FET even when g_m has been improved. Moreover, as changing the CPP does not affect the gate resistance R_g [2], the excess gate capacitance is also the reason for an insignificant improvement in f_{max} with respect to the standard 1xCPP p-FET. It can also be seen that the enhancement already reaches saturation at 2xCPP.

III. W-BAND NOISE PERFORMANCE

A tuner-based noise parameter measurement setup [3] is used for the noise analysis in W-band. A calibrated noise source is utilized for the noise receiver calibration. The noise parameters are extracted by applying the cold method. The noise parameters are then de-embedded to the first metal M1 by applying open-short de-embedding technique. Besides, the noise spectral densities are extracted from the Y-representation noise correlation matrix [5]. Fig. 3 demonstrates the mean values of the noise parameters and noise spectral densities extracted at 94 GHz from 9 different dies. As shown in Fig. 3a-c, the noise performance of the relaxed CPP p-FETs is slightly improved with respect to the 1xCPP p-FET. However, the n-FET is still superior in high frequency noise performance. In particular, the CPP relaxation only enables the p-FETs

23

to have comparable noise resistance R_n with the n-FET. As regards the noise spectral densities shown in Fig. 3d-f, the noise correlation and the induced gate noise exhibit nearly identical characteristics for all DUTs. Accordingly, the relaxed CPP effect introduce minor influence to these two aspects. On the other hand, the CPP relaxation causes an increase in the channel thermal noise of the 2xCPP and 3xCPP p-FETs in strong inversion region.

To elaborate the analysis further, the minimun noise factor F_{min} can be related to the small-signal parameters by the Fukui equation (3) [6], in which K_c is a fitting coefficient. Besides, R_n is inversely proportional to the square of g_m according to equation (4) [7], in which k is the boltzmann constant and T is the ambient temperature in Kelvin.

$$F_{min} = 1 + K_c \frac{f}{f_T} \sqrt{g_m(R_g + R_s)}$$
$$= 1 + K_c \frac{2\pi f C_{gg}}{\sqrt{g_m}} \sqrt{R_g + R_s} \tag{3}$$

$$R_n \approx \frac{\overline{i_d^2}}{4kTg_m^2} \tag{4}$$

From equation (3), it can be seen that the gate capacitance C_{gg} not only affects the device's speed mentioned in the previous section but also the minimum noise factor of the 2-port device. It is thereby understandable why NF_{min} of the relaxed CPP devices are merely minimized despite the noticeable improvement in g_m. Nonetheless, as described in equation (4), the enhanced g_m provides a positive impact that helps equalize R_n among the p-FETs and n-FET.

IV. LARGE-SIGNAL PERFORMANCE

The DUTs are characterized in non-50-Ω environment by using the nonlinear large-signal measurement setup with load-pull described in [8], [9]. A source-/load-pull procedure is applied to the DUTs in the first step at the same bias condition (class AB). The optimum source impedances and load impedances determined from source-/load-pull measurement at a fundamental frequency 10 GHz for all 4 DUTs are nearly identical. Accordingly, all devices are also presented with the same associated harmonic impedances. In addition, the source impedance is tuned for maximum gain, while the load impedance is tuned for maximum output power. The power-sweep measurement results are shown in Fig. 4. As for the standard 1xCPP p-FET, the saturation output power and power-added efficiency (PAE) (see Fig. 4a, b) are slightly comparable to its n-FET counterpart despite the poor gain (see Fig. 4c). The 2xCPP p-FET on the other hand shows a significant improvement which is also better than the n-FET. However, it can also be observed that increasing the gate pitch to 3xCPP will not provide further improvement but only an expanse of the footprint area. As regards the linearity, the n-FET exhibits an early gain compression characteristics, which results in the generation of higher harmonic power (see Fig. 4d, e). Fig. 5 illustrates the large-signal performance of the 2xCPP p-FET and 2xCPP n-FET in terms of quiescent drain current. The p-FET clearly outperforms its counterpart

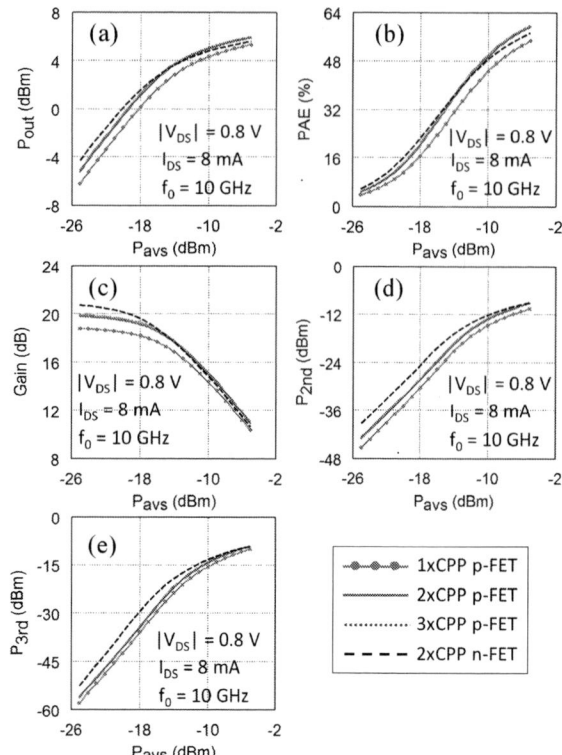

Fig. 4. Single-tone large-signal performance at 10 GHz versus available input power: (a) output power (P_{out}), (b) power-added efficiency (PAE), (c) transducer gain, (d) second-order harmonics and (e) third-order harmonics.

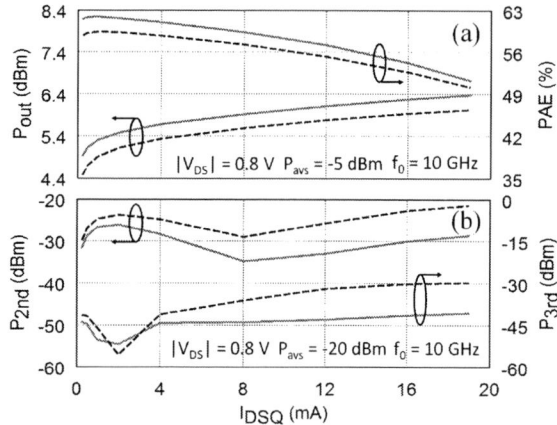

Fig. 5. (a) Large-signal performance versus quiescent drain current and (b) harmonic powers versus quiescent drain current of the 2xCPP n-FET (black dashed line) and 2xCPP p-FET (green solid line).

and features better linearity performance since the premature saturation behaviour of the n-FET becomes pronounced when the conduction angle increases [8]. The optimized p-FET is thereby a better choice for high efficiency and linear power amplifier applications.

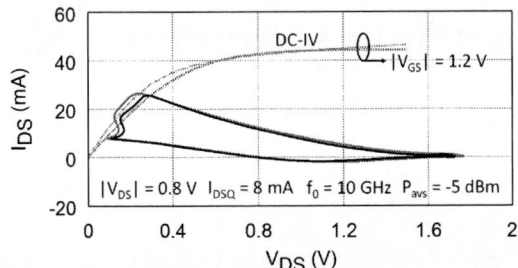

Fig. 6. Experimental dynamic load lines deembedded to M1 plane from the 2xCPP n-FET (black dashed line) and 2xCPP p-FET (green solid line) and dc-IV curves at $|V_{GS}|$ = 1.2 V of the 2xCPP n-FET (blue dotted line) and 2xCPP p-FET (red dotted line).

Moreover, to correctly understand the devices' behaviour under large-signal operation, waveform engineering is applied for the analysis. The instantaneous voltage and current waveforms are reconstructed for visualization of the dynamic load lines. In addition, the fixture parasitics from probe-tip plane to M1 plane are de-embedded by using the approach presented in [10]. Considering the classical load-line theory, the output power and PAE of a transistor device can be estimated by the following equations.

$$P_{out} = \frac{1}{2}(V_{ds} - V_{knee})I_{ds} \qquad (5)$$

$$PAE = \frac{P_{out} - P_{in}}{P_{dc}} \qquad (6)$$

In equation (5), the knee voltage V_{knee} is dictated by the device's on-resistance, and it limits the headroom of the voltage swing. Therefore, the lower V_{knee}, the better the performance. The dynamic load lines of the 2xCPP p-FET and 2xCPP n-FET are depicted in Fig. 6. The dc-IV curves are also included in the figure. Since the two devices are presented with the same load impedance, the dynamic load lines show identical trend. However, the 2xCPP p-FET exhibits lower on-resistance that allows larger current and voltage swings. Combining this feature with equation (5) and (6), the surpassing performance of the p-FET with respect to the n-FET is then clearly explainable.

V. CONCLUSION

In this paper, the effect of the contacted-poly-pitch relaxation on the RF performance of strained-SiGe-channel p-FETs in 22nm FDSOI CMOS technology is comprehensively reported. In particular, the relaxed CPP induces carrier mobility enhancement which can be clearly observed in the dc characteristics. As compared to the device with standard gate pitch, devices with relaxed CPP are improved in high frequency operation. However, due to the large gate capacitance, the p-FETs still deliver slower speed and higher minimum noise figure than the n-FET. On the other hand, it is shown that CPP relaxation allows the p-FETs to outperform their n-FET equivalent in large-signal operation. In short, 2xCPP is found to be the optimal gate pitch for RF applications.

ACKNOWLEDGMENT

The authors would like to thank Fab1, GlobalFoundries, for providing the 22FDX® test structures. This work is funded in part by the ECSEL Joint Undertaking and the German ministry of education and research through the project OCEAN12 and in part by the German ministry of economics and Free State of Saxony through the Important Project of Common European Interest WIN FDSOI.

REFERENCES

[1] R. Carter et al., *"22nm FDSOI Technology for Emerging Mobile, Internet-of-Things, and RF Applications"*, 2016 IEEE Int. Electron Devices Meeting, Dec. 2016.

[2] Z. Zhao et al., "22FDX® fMAX Optimization through Parasitics Reduction and GM Boost," in Proc. 49th Eur. Solid-State Device Research Conf., Cracow, Poland, 2019.

[3] Q. H. Le et al., "W-Band Noise Characterization with Back-Gate Effects for Advanced 22nm FDSOI CMOS," in Proc. IEEE Radio Frequency Integrated Circuits Symp., LA, USA, 2020.

[4] Q. H. Le et al., "DC-110 GHz Characterization of 22FDX® FDSOI Transistors for 5G Transmitter Front-End," in Proc. 49th Eur. Solid-State Device Research Conf., Cracow, Poland, 2019.

[5] H. Hillbrand and P. Russer, "An efficient method for computer aided noise analysis of linear amplifier networks," IEEE Trans. Circuits Syst., vol. 23, no. 4, pp. 235–238, Apr. 1976.

[6] C.-H. Jan et al., "RF CMOS technology scaling in High-k/metal gate era for RF SoC (system-on-chip) applications," in Proc. Int. Electron Devices Meeting, San Francisco, CA, USA, Dec 2010.

[7] S.-C. Wang et al., "Comprehensive Noise Characterization and Modeling for 65-nm MOSFETs for Millimeter-Wave Applications," IEEE Trans. Microw. Theory and Techniques, vol. 58, no. 4, pp. 740-746, Mar. 2010.

[8] Q. H. Le et al., "Assessment of a Thick-Oxide Transistor from the 22FDX® Platform for 5G NR sub-6 GHz FEMs," in Proc. IEEE 2nd 5G World Forum, Dresden, Germany, 2019.

[9] Q. H. Le et al., "Empirical Large-Signal Modeling of mm-Wave FDSOI CMOS based on Angelov model," in Trans. Electron Devices, vol. 68, no. 4, pp. 1446 - 1453, Mar. 2021.

[10] C. Charbonniaud and T. Gasseling, "VNA based load pull harmonic measurement de-embedding dedicated to waveform engineering," in IEEE Int. Conf. Microw., Communications, Antennas and Electronic Systems, Tel Aviv, Israel, Dec. 2015.

Proceedings of the 16th European Microwave Integrated Circuits Conference

Design Methodology of Wide Tuning Range DGS-based VCO for K-band Applications in 0.18-μm CMOS Technology

Baichuan Chen[#], Samundra K. Thapa[#], Nusrat Jahan[*], Adel Barakat[#], Ramesh K. Pokharel[#]

[#]Faculty of Information Science and Electrical Engineering, Kyushu University, Japan

[*]Department of Electrical and Electronic Engineering, Chittagong University of Engineering and Technology, Bangladesh

chen.baichuan.424@s.kyushu-u.ac.jp, pokharel@ed.kyushu-u.ac.jp

Abstract — Recently, defected ground structure (DGS) resonators have been employed to design a low phase noise Voltage-Controlled Oscillator (VCO) for quasi-millimeter wave applications [10]. A few advantages of such VCOs are that a DGS resonator has a higher Q-factor than an on-chip spiral inductor, and it has a higher self-resonance frequency compared to a conventional LC resonator so that a stable inductance value with respect to the frequency can be realized. However, one of its drawbacks is that the tuning range of DGS-based VCOs is limited. In this paper, a method to realize a wide tuning range of DGS-based VCOs has been proposed using two interconnected DGS resonators. As a result, the designed VCO implemented in 0.18-μm 1P6M CMOS technology achieves a 19.8% tuning range with the best phase noise performance of -109.43dBc/Hz at 1-MHz offset frequency in simulations. The experiment result shows that the fabricated VCO has a phase noise of -108.41dBc/Hz at 1-MHz offset frequency with the carrier frequency of 18.51GHz, which exhibits good agreement with the simulation.

Keywords — CMOS integrated circuits, Ka-band, Defected ground structure (DGS), Voltage-controlled oscillator (VCO)..

I. INTRODUCTION

Due to the proliferating demand for high data rate communication, more and more applications in K-band and above have been reported [1]-[10]. Voltage-Controlled Oscillator (VCO), as the key circuit component of a wireless communication system, also attracts a lot of interest from researchers because the system performance is directly affected by the phase noise of a VCO. Recently, many K-band VCOs employing defected ground structure (DGS) as the resonator have been reported [4]-[10], which showed good performance in phase noise and miniaturization. However, because of the trade-off between phase noise and tuning range DGS VCOs usually have a limited tuning range.

In this paper, a new method for designing a wide tuning range DGS-based VCO with good phase noise performance is proposed and analyzed. Also, a new type of DGS resonator is proposed for this concept. By employing the proposed DGS resonator, the varactor can be integrated with this DGS resonator making the chip more compact. Finally, a wide tuning range Ka-band VCO, employing the proposed DGS

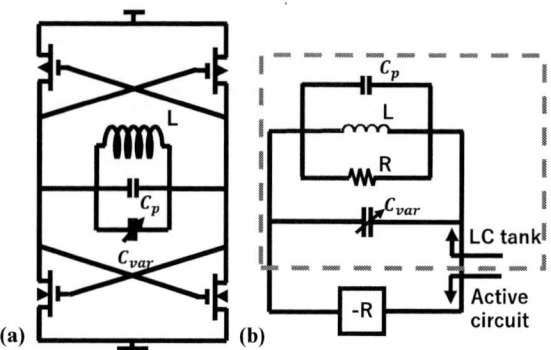

Fig. 1. Conventional push-pull VCO. (a) Its typical schematic. (b) Its equivalent circuit.

resonator, is designed, fabricated, and measured in 0.18-μm 1P6M CMOS technology.

II. DESIGN OF WIDE TUNING RANGE VCO

A. Theory and Concept

Fig. 1 shows the schematic and simplified equivalent circuit of a typical push-pull VCO. This topology shows good performance in controlling power consumption. The resonator, which determines the oscillation frequency, includes an inductor, a parallel capacitor, and the varactor. Its oscillation frequency (f_{osc}) can be calculated by (1).

$$f_{osc} = \frac{1}{2\pi} \cdot \frac{1}{\sqrt{L(C_p + C_{var})}} \tag{1}$$

where C_p is the capacitance of the parallel capacitor, C_{var} is the capacitance of the varactor without any bias and ΔC is the capacitance variation because of the bias. From this equation, we can get the tuning range (the variation of the frequency) in (2).

$$\Delta f = \frac{1}{2\pi} \cdot \left[\frac{1}{\sqrt{L \cdot (C_p + C_{var} + \Delta C)}} - \frac{1}{\sqrt{L \cdot (C_p + C_{var})}} \right] \tag{2}$$

3–4 April 2022, London, UK

Then, with a little transformation, we can get (3). Here C_{total} is used to indicate the sum of the C_p and the C_{var}.

$$\Delta f = \frac{1}{2\pi} \cdot \frac{1}{\sqrt{L \cdot C_{total}}} \left(\frac{1}{\sqrt{1 + \frac{\Delta C}{C_{total}}}} - 1 \right) \quad (3)$$

From (3), it's easy to find that in the normal LC tank VCO's design, there are two ways to improve its frequency tuning range. One is increasing the capacitance variation of the varactor (ΔC) and another is using a parallel capacitor (C_p) as small as possible. Consequently, the inductor should become larger to maintain the oscillation frequency. However, while the tank capacitance decreases, the loaded quality factor of the tank will decrease as well, which may worsen the VCO's phase noise characteristic worse.

B. Implementation of Compact and High-Quality Factor Resonator in DGS Topology

To overcome the problems mentioned above, a new type inductor realized by DGS resonator with compact size and high-quality factor is proposed in this section.

By etching a specific pattern on the ground, the electromagnetic (EM) fields of the transmission line will be interrupted, so that DGS can work as a parallel resonator generating a virtual inductor paralleled with a virtual capacitor [6]. Normally, DGS is used as the compact resonator structure with high quality factors directly. However, if we can decrease its capacitance, an inductor with high quality factor will remain and suit our purpose.

Fig. 2(a) shows the top view of the proposed inductance-specific DGS resonator. According to [6], it's known that the virtual capacitor generated by the DGS resonator is controlled by the width of the gap w_a and the virtual inductor is almost controlled by the length w_b respectively. By lengthen both w_a and w_b, the capacitance part will decrease, and relatively, the inductance will increase. Meanwhile, this also brings enough space for integrating the varactor inside DGS. Besides the above arrangement, the conventional resonator is divided into two small resonators which have the same resonance frequency as shown in Fig. 2(b).

Normally, the resonator occupies most of the chip area, and increasing resonators equals to bigger chip size. However, in this design, using two DGS resonators and setting them as shown in Fig. 3 can decrease the distance between two ports that connects the active circuits. This will help reduce the length of the interconnects to reduce the total layout size and curb the drop of the quality factor [11]. The optimized dimension of the proposed DGS resonator is summarized in Table 1.

The proposed DGS resonator is implemented in 0.18-μm 1P6M CMOS technology and simulated by ANSYS High Frequency Structure Simulator (HFSS). Fig. 4 shows the comparison of the quality factor between the DGS resonator and the conventional spiral inductor. A lumped port is inserted into the excitation gap of DGS to calculate its unloaded Q-factor. The self-inductance (L_{eq}), resistance (R_{EQ}), and Q_u of

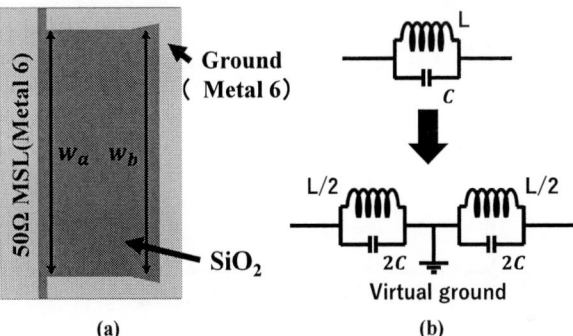

(a)   **(b)**

Fig. 2. (a) Proposed DGS resonator model in CMOS technology. (b) Equivalent circuit of design concept.

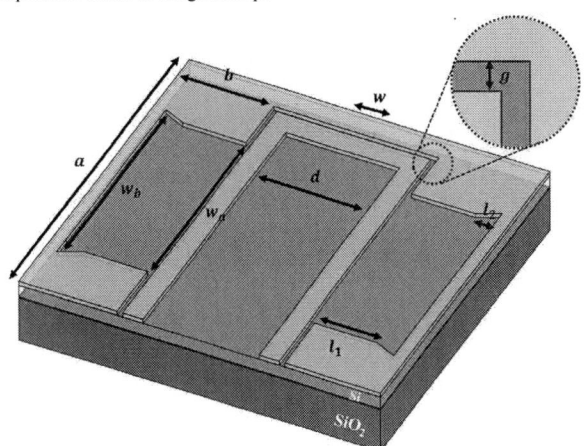

Fig. 3. Overview of the proposed DGS resonator.

Table 1. Design parameters of the DGS resonator

Parameters	a	b	d	g	l_1	l_2	w	w_a	w_b
Dimension(μm)	200	80	52	1	65	15	10	175	180

the DGS can be calculated by (4)

$$L_{eq} = Im(Z_{11})/\omega \qquad R_{EQ} = Re(Z_{11})$$
$$Q_u = Im(Z_{11})/Re(Z_{11}) \qquad (4)$$

From Fig. 4, we can find that the conventional spiral inductor's quality factor will deteriorate rapidly after 15GHz. This is caused by its self-resonance. However, the DGS resonator has over 20 of the quality factors in the Ka-band, which is much higher than the conventional ones at the target frequency band. And Q-factor keeps increasing during the whole band, which means no self-resonance will happen in this band. Therefore, this high-quality factor is expected to remain a good phase noise.

C. Proposed VCO Design

To prove the design concept, the proposed DGS resonator is implemented in a simple Class-B VCO topology and simulated in Agilent ADS as shown in Fig. 7(a). The scattering parameters (S-parameter) and the impedance parameters (Z-parameter) of the proposed DGS resonator are obtained after EM simulation by HFSS.

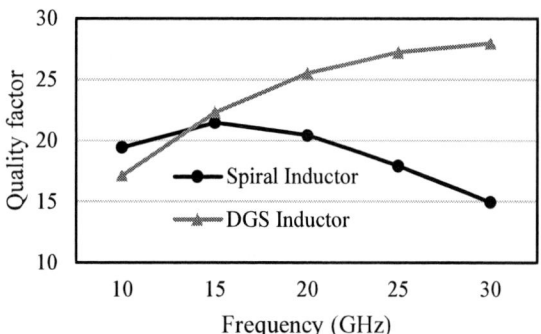

Fig. 4. Comparison of the quality factor between spiral inductor and proposed DGS inductor.

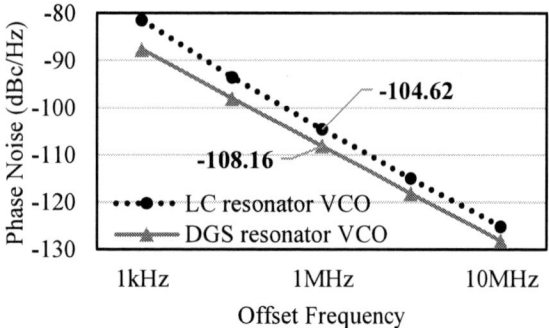

Fig. 5. Comparison of simulated phase noise of the VCO employing an LC resonator and a DGS resonator.

For comparison, a VCO employing conventional spiral inductor working at the same frequency is built and simulated. Both of their simulation result is shown in Fig. 5. From the Fig. 5, it can be found that because of its good quality factor, VCO employing the proposed DGS inductor provides much better phase noise performance than the conventional one. It brings a 4 dBc/Hz phase noise improvement at 1MHz offset.

Fig. 6 shows the simulated phase noise and output frequency of the proposed VCO while changes the tuning voltage. A 19.3 GHz to 22.53 GHz wide tuning range is realized and phase noise is lower than -101 dBc/Hz during the tuning. The designed VCO consumes only 1.38 mA current from a 1.8 V voltage supply, which leads to a -190.42 dBc/Hz of the best figure of merit (FoM) and -194.21 dBc/Hz of the figure of merit considering tuning range (FoM_T). The FoM and FoM_T can be calculated using equations shown as follows:

$$FoM = -PN + 20\log\left(\frac{f_o}{\Delta f}\right) - 10\log(P_{DC})$$

$$FoM_T = FoM_{Peak} + 20\log(\frac{FTR}{0.1})$$

where PN is the phase noise, f_o is the oscillation frequency; Δf is the offset frequency, P_{DC} is the DC power consumption of mW, and FTR is the frequency tuning range.

III. MEASUREMENT RESULT

The designed wide tuning VCO with the integrated DGS resonator was implemented in 0.18-µm 1P6M CMOS

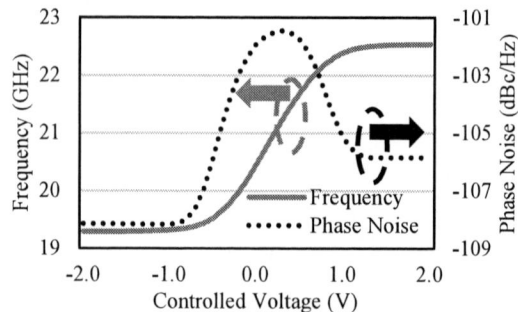

Fig. 6. Simulated phase noise and tuning range during tuning

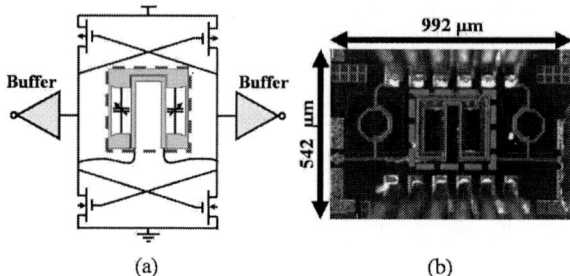

(a)   (b)

Fig. 7. Proposed VCO with DGS resonator. (a)Circuit topology. (b) Fabricated VCO chip.

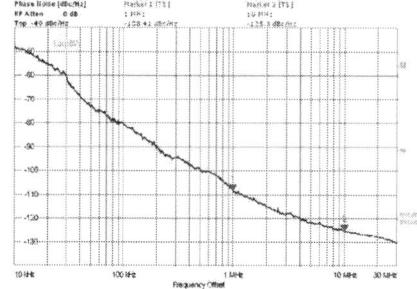

Fig. 8 Measurement result.

technology and its micrograph is shown in Fig. 7. The VCO core is only 0.105 mm². Two DGS resonators are enclosed by red lines. A signal source analyzer (ROHDE&SCHWARZ FSUP.SSA) was used in the measurement. The measured output frequency and phase noise are shown in Fig. 8, respectively. The measurement shows that the VCO has a -108.41 dBc/Hz of phase noise at a 1-MHz offset chip with a carrier frequency of 18.5 GHz. The proposed VCO also showed a good voltage sensitivity and temperature drift.

IV. CONCLUSION

A novel method using DGS as the inductor is proposed to design a low phase noise Ka-band VCO with a wide tuning range in 0.18-µm CMOS technology. The proposed DGS resonator has a compact size and shows much higher Q_u than the conventional spiral inductor. A new connection way of the varactor is proposed which can save the chip area. The proposed VCO employing the novel DGS resonator is demonstrated in 0.18-µm CMOS technology and the measurement result showed good agreement with simulation in phase noise and carrier frequency.

ACKNOWLEDGMENT

This work was supported in part by Casio Science Foundation, in part by The Murata Science Foundation, in part by the Telecommunication Advancement Foundation, in part by the VLSI Design and Education Center (VDEC), The University of Tokyo in collaboration with Cadence and Keysight Corporations.

REFERENCES

[1] Z. Peng, D. Hou, J. Chen, Y. Xiang, and W. Hong, "A 28 GHz Low Phase-Noise Colpitts VCO with Wide Tuning-Range in SiGe Technology," *in Proc. IEEE Int. Symp. on Radio-Freq. Integration Technol.,* Melbourne, VIC, Australia, 2018, pp. 1-3.

[2] A. Masnadi, M. Mahani, H. M. Lavasani, S. Mirabbasi, S. Shekhar, R. Zavari, H. Djahanshahi, "A Compact Dual-Core 26.1-to-29.9GHz Coupled-CMOS LC-VCO with Implicit Common-Mode Resonance and FoM of-191dBc/Hz at 10MHz," *in Proc. IEEE Custom Integr. Circuits Conf.,* Boston, MA, USA, 2020, pp. 1-4.

[3] Y. Shang, H. Yu, D. Cai, J. Ren, and K. S. Yeo, "Design of High-Q Millimeter-Wave Oscillator by Differential Transmission Line Loaded With Metamaterial Resonator in 65-nm CMOS,*" IEEE Trans. Microw.. Theory. Techn.,* vol. 61, no. 5, pp. 1892-1902, May 2013.

[4] N. Jahan, C. Baichuan, R. K. Pokharel, and A. Barakat, "A K-Band VCO Employing High Active Q-factor Defected Ground Structure Resonator in 0.18-μm CMOS Technology," *in Proc. IEEE Int. Symp. on Circuits and Syst.,* Florence, Italy, 2018, pp. 1-5.

[5] N. Jahan, C. Baichuan, A. Barakat, and R. K. Pokharel, "Analysis and Application of Dual Series Resonances for Low Phase Noise K-Band VCO Design in 0.18-μm CMOS Technology," *in Proc. IEEE Int. Symp. on Circuits and Syst.,* Sapporo, Japan 2019, pp. 1-5.

[6] *N. Jahan,* S. A. E. Ab Rahim, A. Barakat, T. Kaho, and R. K. Pokharel, "Design and Application of Virtual Inductance of Square-Shaped Defected Ground Structure in 0.18-μm CMOS Technology," *IEEE J. Electron Devices Soc.,* vol. 5, no. 5, pp. 299–305, Sep. 2017.

[7] N. Jahan, A. Barakat, and R. K. Pokharel, "Ku-band oscillator using integrated defected ground structure resonator in 0.18-μm CMOS technology," *in Proc. 47th Eur. Microw. Conf.,* Nuremberg, Germany, 2017, pp. 763-766.

[8] *N. Jahan,* A. Barakat, and R. K. Pokharel, "Study of phase noise improvement of K-band VCO using additional series resonance realized by DGS resonator on CMOS technology," *in Proc. IEEE Asia Pacific Microw. Conf.,* Kuala Lumpur, Malaysia, 2017, pp. 1014-1017.

[9] N. Jahan, A. Barakat, and R. K. Pokharel, "A −192.7dBc/Hz FoM Ku-Band VCO Using a DGS Resonator With a High-Band Transmission Pole in 0.18-μm CMOS Technology," *IEEE Microw. Wireless Compon. Lett.,* vol. 29, no. 12, pp. 814-817, Dec. 2019.

[10] *N. Jahan,* C. Baichuan, A. Barakat, and R. K. Pokharel, "Utilization of Multi-Resonant Defected Ground Structure Resonators in the Oscillator Feedback for Phase Noise Reduction of K-Band VCOs in 0.18-μm CMOS Technology," *IEEE Trans. Circuits Syst. I, Reg. Papers,* vol. 67, no. 4, pp. 1115-1125, April 2020.

[11] C. Baichuan, N. Jahan, A. Barakat, and R. K. Pokharel, "Experimental Study of the Effect of Interconnects on Phase Noise of K-Band VCO in 0.18μm CMOS Technology," *in Proc. IEEE Asia Pacific Microw. Conf.,* Singapore, 2019, pp. 1670-1672.

Proceedings of the 16th European Microwave Integrated Circuits Conference

Linearity Assessment of GaN HEMTs on Si using Nonlinear Characterisation

Rana ElKashlan[#*1], Ahmad Khaled[#], Raul Rodriguez[#], Vamsi Putcha[#], Uthayasankaran Peralagu[#], AliReza Alian[#], Nadine Collaert[#], Piet Wambacq[*#], and Bertrand Parvais[*#]

[#]imec, Leuven, Belgium

[*]VUB, Brussels, Belgium

[1]rana.y.elkashlan@imec.be

Abstract — We investigate the effect of varying the gate-to-drain spacing and the gate field-plate on the device linearity of GaN HEMTs on Si for 0.11μm, 0.15μm, and 0.19μm gate lengths. The gain compression, phase distortion, and harmonic distortion metrics are measured using a nonlinear characterisation setup calibrated at 6GHz up to the third harmonic. The acquired nonlinearity metrics are correlated with the extrinsic device parasitics extracted from S-parameter measurements. We observe that excessive gate field-plate length scaling down to 0.05μm lowers the total phase distortion at the expense of gain linearity and harmonic distortion in Class AB while minimising the gate-to-drain spacing alleviates the harmonic distortion only for devices of 0.19μm gate length. Further evaluation, under matched conditions, using a passive load-pull measurement setup points to a decline in the peak achievable PAE and P_{SAT} at gate field-plates smaller than 0.12μm.

Keywords — GaN HEMTs, nonlinear distortion, phase distortion, distortion measurement.

I. INTRODUCTION

GaN HEMTs are suitable contenders for RF and high power applications because of their ability to meet the high efficiency, output power density and bandwidth requirements [1]. However, modern wireless communication systems are also required to be spectrally efficient, and this demands utilising high linearity power amplifiers (PAs) in the RF front end modules [2]. Typically, circuit-level linearisation techniques are the solution of choice at the expense of efficiency. GaN HEMTs are notable for their distinct nonlinear behaviour, which results in significant transconductance (g_m) derivatives, thereby not only causing gain compression (AM/AM distortion) but also generating substantial harmonics and intermodulation distortion (IMD) [3], [4].

AM/AM and AM/PM (phase distortion) are also known to contribute to spectral regrowth. Additionally, AM/PM can cause difficulties in detecting the received signal. The combined contribution of gain compression and phase distortion results in out-of-band interference of the transmitted signal and bit errors of the received signal [5]. Hence, device-level linearity improvement is vital to ease the performance trade-off without increasing the circuit complexity.

In [4], the authors perform a device linearity analysis using the MSVG model to ensure its predictive ability, in addition to investigating the physical origins of GaN nonlinearity. The work in [6] considers the influence of the AlGaN barrier thickness of the phase distortion using the ASM-G model. This work intends to give additional insight for device-level linearity enhancement through investigating numerous geometrical configurations comprising different gate field-plate lengths, gate-to-drain spacings, and gate lengths using both nonlinear and small-signal characterisation methods to deduce and correlate linearity metrics, namely AM/AM, AM/PM and harmonic distortion, with extracted extrinsic device data from DC and S-parameters measurements.

II. DEVICES UNDER TEST

The 8-finger AlGaN/GaN HEMT devices under test (DUT) are fabricated in a 3-level Cu BEOL process on 200mm Si wafers (Fig. 1) as described in [7]. During the device processing, an issue occurred that caused an increase in the parasitic capacitance by moving the ohmic metal closer to the gate head than anticipated on both the drain and source sides.

We study the impact of varying the gate length (L_G), gate field-plate (L_{GFP}) and gate-to-drain spacing (L_{GD}) on the device linearity for 8x25μm wide devices. The devices have an off-state lateral breakdown voltage (V_{BD}) of ~105V at a breakdown criterion of 1mA/mm, and a maximum f_T/f_{MAX} of 60/135 GHz at $L_G = 0.11$μm [7]. For the assessed 0.14μm gate length devices, the DC and RF metrics are obtained from the characterisation data at 8V, while the parasitic capacitance and gate resistance (Table 1) are extracted using the same methods described in [8]. It is worth noting that, when considering the influence of the L_{GFP}, to ensure that the observed behaviour is

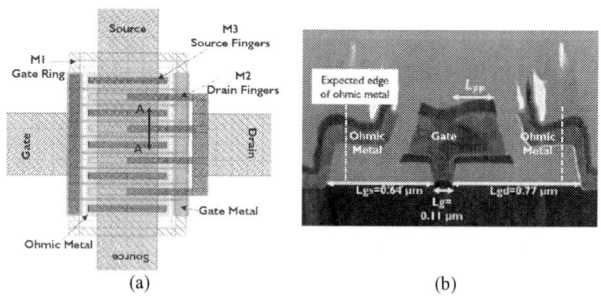

(a)   (b)

Fig. 1. (a) 8-finger GaN HEMT device layout with 3 BEOL metal layers. (b) TEM cross-section of the device L_G=0.11μm, with the definition of the dimensions.

3−4 April 2022, London, UK

Table 1. DC and RF metrics for the studied devices at V_D=8V. Device parasitics are extracted at V_G=-6V, V_D=0V. L_G=0.14µm, W_f=25µm, N_f=8.

L_{FP} (µm)	L_{GD} (µm)	$DC\ g_m$ (mS/mm)	R_G (Ω)	C_{GD} (pF/mm)
0.05	0.22	417	8.5	0.42
0.25	0.42	409	5.6	0.48
0.25	0.67	392	5.8	0.49
0.45	0.62	390	5.1	0.56

a result of solely the L_{GFP}, L_{GD} is adjusted according to the L_{GFP} length to maintain a constant effective gate-to-drain distance. Hence, the L_{GD} values for the 0.05µm, 0.25µm, and 0.45µm examined gate field-plate lengths are 0.22µm, 0.42µm, and 0.62µm, respectively.

III. CHARACTERISATION SETUP

The device nonlinearity is measured using a 50Ω Nonlinear Vector Network Analyser (NVNA) setup calibrated at a fundamental frequency f_0 of 6GHz up to the third harmonic ($3f_0$) at 18GHz (Fig. 2)[9]. The reference plane is at the tip of the GSG probes at the on-wafer DUT. Phase measurement requires highly sensitive phase calibration that allows the NVNA to compare the phase at the DUT input with what it is receiving, thereby finding the phase relation between the A&B waves. Only one phase reference is employed during the measurements considering that the second one is solely used in the setup calibration. The device performance is evaluated under matched conditions using a 6GHz passive load-pull characterisation setup (Fig. 3).

The devices are biased in Class AB at I_D = 320mA/mm. AM/PM data at different gate-bias points (Fig. 4) indicates that this operation class results in the highest phase distortion as expected, which allows for a suitable bias point choice for this study.

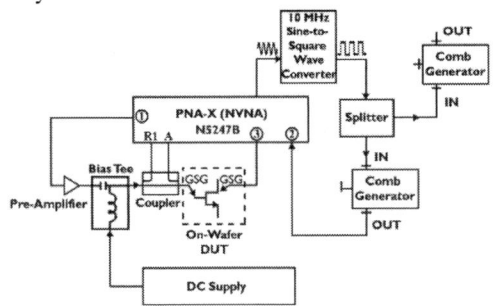

Fig. 2. On-wafer nonlinear VNA characterisation setup.

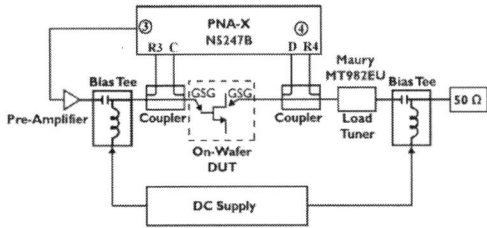

Fig. 3. Passive load-pull characterisation setup.

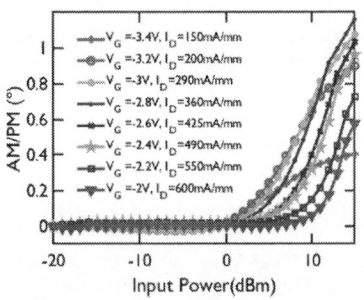

Fig. 4. AM/PM distortion for 8 bias points obtained from the NVNA characterisation at V_D = 8V.

IV. IMPACT OF THE GATE FIELD-PLATE AND GATE-TO-DRAIN SPACING ON LINEARITY

A. Amplitude and Phase Distortion

AM/PM is strongly dependent on the gate-to-source capacitance (C_{GS}) variation with the input signal [6], [10]. Consequently, the phase distortion is affected by the chosen operation class (Fig. 4). In class AB operation, V_G is set slightly higher than V_{TH}; when considering the input signal swing variation, C_{GS} varies between C_{GSmax} and C_{GSmin} values, thus increasing the distortion.

The devices with a longer L_{GFP} show higher P_{1dB} values than devices with shorter ones (Table 2), resulting in improved linear power (Fig. 5(a)). However, phase distortion begins at a lower input power for the 0.25µm and 0.45µm L_{GFP}, resulting in an overall higher AM/PM distortion. Additionally, longer gate field-plates reduce the power gain.

The longest L_{GFP} of 0.45µm length exhibits a 34% higher capacitance variation for a given V_G range ($\partial C_{GS}/\partial V_{GS}$) in contrast with the shortest 0.05µm L_{GFP} (Fig. 6(a)). It is apparent from the collected data that the optimal L_{GFP} value lies at an intermediate point with only a slight decrease in power gain compared to the impact of a long L_{GFP} and a marginally higher ΔC_{GS} compared to that of an extremely short L_{GFP}. This inference is evident in the data for the 0.25µm L_{GFP}, with merely a 0.2dB decrease in gain, a 2.5 dBm reduction in P_{1dB}, and a 2% increase in ΔC_{GS}.

The marginal change, ~10%, in ΔC_{GS} observed for the two examined L_{GD} variants (Fig. 6(b)) clarifies the lack of a profound effect on the AM/AM and AM/PM characteristics (Fig. 5(b)).

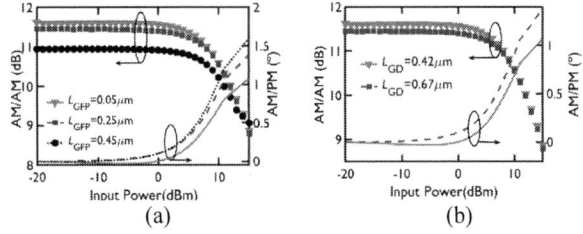

Fig. 5. AM/AM and AM/PM characteristics obtained using the NVNA setup for (a) three different L_{GFP} with an L_G = 0.14µm, and (b) different L_{GD} with an L_{GFP}=0.25µm and L_G = 0.19µm. I_D=320mA/mm, V_D=8V. W_f = 25µm, and N_f=8.

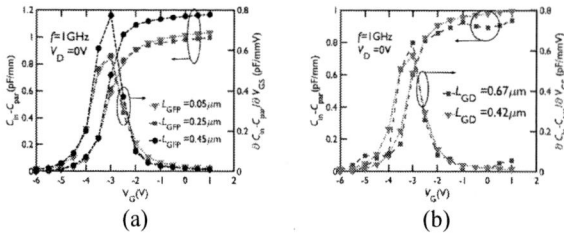

Fig. 6. $C_{GS} - C_{parasitic}$, and $\partial C_{GS}/\partial V_{GS}$ versus V_G for different (a) gate field-plate lengths with an $L_G = 0.14\mu m$, and (b) gate-to-drain spacing lengths with an $L_{GFP}=0.25\mu m$ and $L_G = 0.19\mu m$. $W_f=25\mu m$, $N_f=8$, $V_D = 4V$, and $f = 1GHz$.

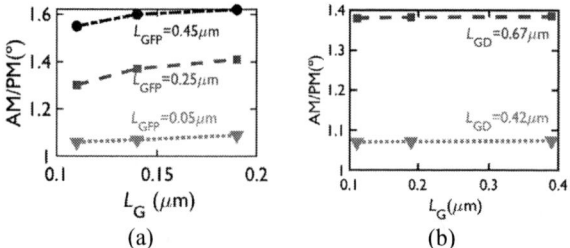

Fig. 7. AM/PM characteristics obtained using the NVNA setup for different gate lengths and (a) gate field-plate lengths (b) gate-to-drain spacings, at $I_D=320mA/mm$, $V_D=8V$, $W_f=25\mu m$, $N_f=8$, and $L_{GFP}=0.25\mu m$.

Nevertheless, the longer L_{GD} (0.67μm) causes the phase to distort slightly starting from the linear region (at a lower P_{in}), consequently leading to a higher AM/PM value.

Further characterisation of the same geometrical alterations for different L_G (Fig. 7) suggests no influence of the gate length on AM/PM, although there is minimal phase distortion improvement for the 0.45μm L_{GFP} at the most scaled L_G.

B. Harmonic Distortion

The nth-order harmonic distortion (HD_n) is the ratio between the output distortion component at the nf_0 frequency and the component at the fundamental frequency f_0. Since the NVNA measures A&B waves in RMS, HD_n is computable from the measured data in dB for the second and third harmonics using the following formula:

$$HD_n = 10\log_{10}\left|\frac{b_2(@\,nf_0)}{b_2(@\,f_0)}\right|^2, n = 2,3 \quad (1)$$

where b_2 is the transmitted wave.

The second (HD2) and third (HD3) harmonic distortions at 7dB back-off power for the same devices across three dies (Fig. 8) show that the influence of HD3 is more pronounced than HD2 at this operation class. There is minimal deviation across the measured dies for HD2 and HD3. Furthermore, harmonic distortion becomes more drastic along with gate length scaling for all L_{GFP}. At a 7dB back-off power, there is a 5dB higher HD3 for the 0.05μm L_{GFP} compared to the 0.45μm L_{GFP} for an L_G of 0.11μm. The deterioration in HD3 with L_{GD} scaling is most apparent at the longest $L_G = 0.19\mu m$ (Fig. 8). However, the increase in HD3 with L_G scaling renders the L_{GD} influence negligible.

HD is dependent on the g_m nonlinearity, which is defined by its first (g_{m2}) and second (g_{m3}) derivatives. The HD dependency on g_m nonlinearity can be approximated by [11]:

$$HD_{nDC} \approx \frac{A^{n-1}}{2}\left|\frac{g_{mn}}{g_m} - A_{vDC}\frac{g_{dn}}{\frac{1}{Z_L}+g_d}\right|, n = 2,3 \quad (2)$$

where A is the amplitude of the AC input signal, $A_{vDC} = g_{m1}/(\frac{1}{Z_L} + g_{d1})$, and Z_L is the load impedance for the device.

The increase in g_{m3} at the current density of operation compared to g_{m2} (Fig. 9) explicates the prevalence of HD3 over HD2. Moreover, the higher g_m, g_{m3} of the 0.05μm L_{GFP} device translates into the seen rise in distortion. From a device perspective, the region below the field-plate acts as a transition between the nonlinear access region and the intrinsic device. The nonlinear access resistance is a contributor to the g_m roll-off, and its impact increases with direct proportionality to the drain current because of the rise of the electric field in the access region. The observed higher g_m roll-off for the short L_{GFP} justifies the faster AM/AM compression seen at high input power levels (Fig. 5(a)), which will affect linearity in the saturated power regime and lower the peak achievable PAE.

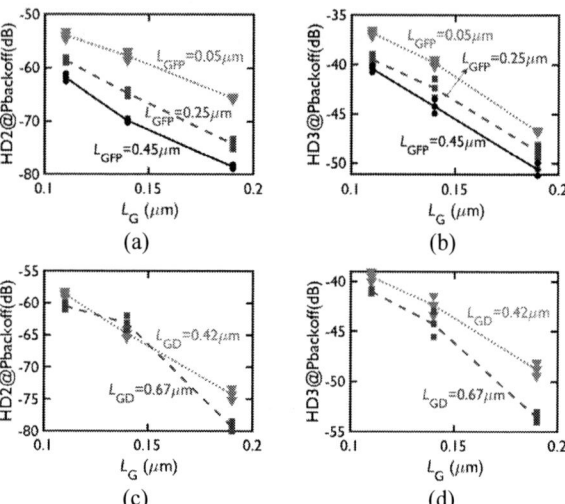

Fig. 8. (a,c) HD2 (b,d)HD3 at a 7 dB back-off for different gate lengths and (a,b) gate field-plate lengths (c,d) gate-drain spacings measured across three dies using the NVNA setup. $W_f=25\mu m$, $N_f=8$, at $I_D=320mA/mm$ and $V_D=8V$.

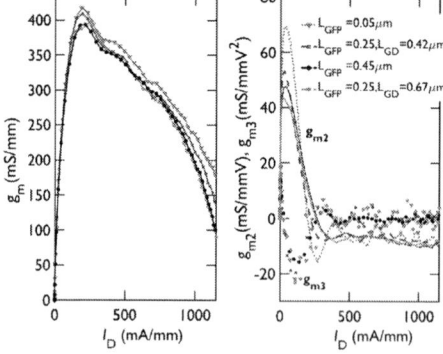

Fig. 9. g_m, g_{m2}, and g_{m3} for 0.05μm, 0.25μm, and 0.45μm gate field-plate lengths, and two gate-to-drain spacings at $V_D=8V$. $L_G = 0.14\mu m$, $W_f= 25\mu m$, and $N_f = 8$.

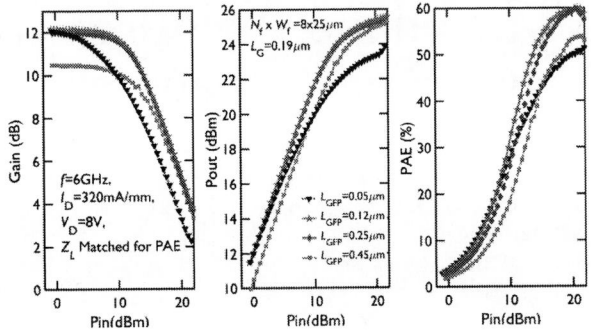

Fig. 10. 6GHz load-pull characterisation results for L_{GFP}=0.05μm, 0.12μm, 0.25μm, and 0.45μm, at V_D=8V, I_D=320mA/mm, W_f= 25μm, and N_f= 8.

Table 2. Summary of the linearity and large-signal metrics at V_D = 8V for various geometry configurations at L_G = 0.19μm, W_f = 25μm, and N_f = 8.

L_{FP} (μm)	L_{GD} (μm)	NVNA			Load Pull (Matched PAE)		
		P_{1dB} (dBm)	AM/PM	HD3 (dB)	P_{1dB} (dBm)	PAE	P_{SAT} (W/mm)
0.05	0.22	9.5	1.08°	-46	7	51%	1.2
0.12	0.44	11	1.2°	-48	12	60%	1.95
0.12	0.54	11	1.29°	-48	12	60%	2
0.25	0.42	10.5	1.39°	-49	11.2	60%	1.86
0.25	0.67	11	1.4°	-49	12.4	60%	1.94
0.45	0.62	12.5	1.6°	-53	13	54%	1.7

V. LOAD-PULL CHARACTERISATION

The load-pull characterisation (Fig. 3) results of the devices complement the observations from the NVNA characterisation (Table 2). The rapid gain compression observed for the shortest L_{GFP} results in a subsequent P_{OUT} saturation, limiting P_{SAT} to 1.2W/mm and the peak PAE to 51% (Fig. 10). Based on the observed trends, an L_{GFP} of 0.12μm is additionally studied for L_{GD} of 0.44μm and 0.54μm (Table 2). The trade-off between AM/AM and HD3 for devices with a L_{GFP} of 0.12μm and 0.25μm results in a PAE of ~60%, which is perceptibly higher than the most and least scaled L_{GFP}. Furthermore, we note a ~2.5% improvement in P_{SAT} for the shorter 0.12μm L_{GFP} alongside the devices with the less scaled L_{GD}. The optimal large-signal performance is exhibited by the 0.12μm L_{GFP} device with a 0.54μm L_{GD}, demonstrating P_{SAT} of 2W/mm. Nonetheless, the overall minor impact of L_{GD} variation is unexpected. We attribute this to the previously mentioned issue in Section II.

VI. CONCLUSION

Nonlinear characterisation of several geometrical configurations of CMOS compatible GaN HEMTs shows the necessity of optimising the L_{GFP} for an appropriate trade-off of the phase and harmonic distortion, where phase distortion is mitigated with L_{GFP} scaling at the expense of the harmonic distortion and gain linearity. At class AB operation, HD3 dominates due to the increase in g_{m3} as compared to g_{m2}. Shortening L_G beyond 0.14μm exacerbates HD3 by roughly 22% when combined with aggressive L_{GFP} reduction below 0.12μm

as a result of a noticeable increase in the g_m roll-off. L_{GD} reduction is shown to have a minimal impact on HD3 for L_G<0.19μm devices and an L_G independent improvement of AM/PM of ~20%. Large-signal characterisation under matched conditions shows performance improvement when optimising the L_{GFP}, L_{GD} combination for an AM/AM, AM/PM and HD3 trade-off, with a peak PAE and P_{SAT} of 60%, 2W/mm at 8V for an 0.19μm L_G device.

REFERENCES

[1] A. S. A. Fletcher and D. Nirmal, "A survey of Gallium Nitride HEMT for RF and high power applications," *Superlattices Microstruct.*, vol. 109, pp. 519–537, Sep. 2017.

[2] P. M. Asbeck *et al.*, "Power Amplifiers for mm-Wave 5G Applications: Technology Comparisons and CMOS-SOI Demonstration Circuits," *IEEE Trans. Microw. Theory Tech.*, vol. 67, no. 7, pp. 3099–3109, Jul. 2019.

[3] J. C. Pedro *et al.*, "Soft compression and the origins of nonlinear behavior of GaN HEMTs," in *2014 9th European Microwave Integrated Circuit Conference*, Rome, Italy, 2014, p. 4.

[4] U. Radhakrishna *et al.*, "Study of RF-circuit linearity performance of GaN HEMT technology using the MVSG compact device model," in *2016 IEEE International Electron Devices Meeting (IEDM)*, San Francisco, CA, USA, Dec. 2016, pp. 3.7.1-3.7.4.

[5] S. Sen *et al.*, "Low-cost AM/AM and AM/PM distortion measurement using distortion-to-amplitude transformations," in *2009 International Test Conference*, Austin, TX, USA, Nov. 2009, pp. 1–10.

[6] S. Khandelwal *et al.*, "Dependence of GaN HEMT AM/AM and AM/PM Non-Linearity on AlGaN Barrier Layer Thickness," 2017, p. 4.

[7] B. Parvais *et al.*, "GaN-on-Si mm-wave RF Devices Integrated in a 200mm CMOS Compatible 3-Level Cu BEOL(Invited)," presented at the IEEE International Electron Devices Meeting, San Francisco, 2020.

[8] R. Y. ElKashlan *et al.*, "Analysis of Gate-Metal Resistance in CMOS-Compatible RF GaN HEMTs," *IEEE Trans. Electron Devices*, vol. 67, no. 11, p. 5, 2020.

[9] L. Betts and D. Root, "Nonlinear Vector Network Analyzer Applications," Agilent Technologies, [Online].

[10] S. K. Dhar *et al.*, "Impact of Input Nonlinearity on Efficiency, Power, and Linearity Performance of GaN RF Power Amplifiers," in *2020 IEEE/MTT-S International Microwave Symposium (IMS)*, Los Angeles, CA, USA, Aug. 2020, pp. 281–284.

[11] B. Parvais and A. Siligaris, "Modeling the SOI MOSFET nonlinearities. An empirical approach," in *Transistor Level Modeling for Analog/RF IC Design*, W. Grabinski, B. Nauwelaers, and D. Schreurs, Eds. Dordrecht: Springer Netherlands, 2006, pp. 157–180.

Proceedings of the 16th European Microwave Integrated Circuits Conference

Reconfigurable and Scalable Monolithic Band Reject Circuit Utilizing Phase-Change Switch Matrices

Tejinder Singh, Raafat R. Mansour

Centre for Integrated RF Engineering, Department of Electrical and Computer Engineering, and
Waterloo Institute for Nanotechnology
University of Waterloo, Waterloo, ON N2L 3G1, Canada
tejinder.singh@uwaterloo.ca

Abstract—This paper reports a reconfigurable band reject circuit utilizing a chalcogenide phase change material (PCM) germanium telluride (GeTe) based switch matrix. A highly miniaturized and scalable 2×2 switch matrix is developed in-house using a custom eight-layer microfabrication process, including a thermally conductive dielectric for efficient thermal energy transportation. As a proof-of-concept, two band-rejection notch filters are monolithically integrated with the 2×2 switch matrix to demonstrate a reconfigurable 2-bit band reject circuit. The fully integrated circuit is designed with overall device core footprints under $0.68 \, \text{mm}^2$. The circuit does not consume any static dc power consumption and offers non-volatile operation. The circuit is reconfigurable to either reject f_1 at 2.6 GHz, f_2 at 5.8 GHz, both $f_1 + f_2$ or bypass the notch filters. The proposed circuit exhibits a measured insertion loss of less than 2.2 dB and a return loss of better than 17 dB. The approach is scalable to a larger number of reconfigurable bands rejection using the 2×2 unit-cell in an upscaled switch matrix.

Keywords—Band rejection circuit, GeTe, Phase change RF switches, PCM, reconfigurable circuits, switch matrix

I. Introduction

With an ever-increasing demand for connected devices [1], [2], there is a need to develop highly miniaturized reconfigurable devices such as tunable filters, phase shifters [3], attenuators [4], switch matrices [5], [6], and tunable impedance matching networks. Microwave switches are widely used as a basic building block in a majority of switchable radio frequency (RF) devices. Electromechanical and semiconductor technology-based switches dominate the market, with electromechanical switches shines at the forefront of RF performance but are not practical on the receiver side and forfeit the miniaturization concept. While the semiconductor switch technologies such as SiGe, GaAs, GaN, BiCMOS are a viable choice for miniaturized devices, they exhibit poor linearity performance and low power handling capability [7], [8]. Moreover, semiconductor-based devices do not offer non-volatile functionality.

Microelectromechanical systems (MEMS) based circuits offer RF performance close to what electromechanical switches offer in a compact package. However, the microscale movable components in the RF MEMS-based devices have reliability concerns and require expensive hermetic packaging, which adds up to the size and cost [9]. RF MEMS-based devices also require higher actuation voltage compared to other commonly available switch technologies. Recently available commercial

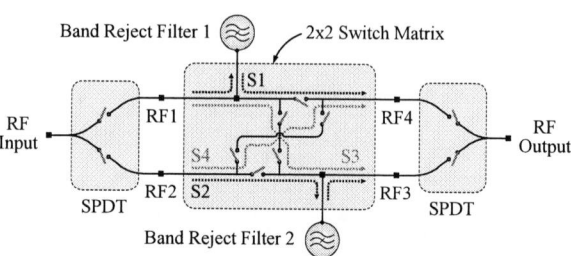

Fig. 1. Schematic of the proposed reconfigurable band rejection circuit utilizing 2×2 switch matrix and single-pole double-throw (SPDT) switches. Four discrete signal routes S1 to S4 are highlighted for routing the signal with either including or bypass the integrated band reject notch filters.

RF MEMS switches demonstrated outstanding reliability and endurance but have a limited operation bandwidth [10], [11].

The emerging RF switches based on phase change material (PCM) technology offer at least a magnitude high figure-of-merit (FoM) [12]–[14] compared to semiconductor counterparts, including GaN, GaAs, InP, and CMOS. High FoM translates to better isolation and lower insertion loss of a switching circuit. PCM technology also offers latching functionality due to the non-volatile property of the germanium telluride (GeTe) material [14], [15].

In this paper, a highly miniaturized and scalable reconfigurable PCM GeTe-based band rejection circuit is presented utilizing a scalable switch matrix, as shown in Fig. 1. A 2-bit reconfiguration scheme is demonstrated, and the scalability approach is discussed in detail in section IV. The presented proof-of-concept band rejection circuit has four discrete signal routes and can be reconfigured to reject f_1, f_2, $f_1 + f_2$, or no rejection. The presented reconfigurability approach is not just limited to filters but can be used for any fixed RF circuits. The measured RF performance of monolithically integrated single-pole single-throw (SPST) and single-pole double-throw (SPDT) switches is presented. The proposed monolithic circuit is highly compact and is developed in-house utilizing a custom eight-layer microfabrication process. The reported fabrication process offers the flexibility to monolithically integrate various planar RF devices [3], [16], [17]. The proposed circuit is the first implementation of a tunable circuit using a scalable matrix approach to the best of our knowledge.

3–4 April 2022, London, UK

II. DEVICE DESIGN AND OPERATION PRINCIPLE

A. Phase-change RF Switching Elements

The monolithically integrated non-volatile phase-change GeTe-based series RF switches are designed with an area under $20\,\mu m \times 20\,\mu m$. The PCM switches exhibit more than five orders of resistance change with the application of a thermal pulse through the refractory micro-heater. The switches can be actuated between two states of the GeTe, namely, the crystalline state (low-resistance state) and the amorphous state (high-resistance state). Applying a 200 ns pulse of 12 V amplitude allows switching from the crystalline state to the amorphous state, while a pulse of $1.2\,\mu s$ width and 7.8 V amplitude switches back the amorphous state to crystalline. Utilizing the reported microfabrication process, we have recently demonstrated high-performance DC–67 GHz RF phase-change switches with cut-off frequency, F_{co}, exceeding 13 THz, measured insertion loss lower than 0.4 dB, CW RF power handling of +35.5 dBm, IIP3 better than +41 dBm, and experimentally measured cycle endurance with more than 1 million reliable switching cycles with no point of failure. Details on the switches are presented in [14], [17].

B. Input/Output SPnT Switches

The phase-change SPDT switches in the circuit, as shown in Fig. 2(c), have a core area under $76\,\mu m \times 76\,\mu m$. The RF cross-over shown in Fig. 2(c) also measures $76\,\mu m \times 76\,\mu m$. The SPDT switch has three RF ports: a common input port, RFC, two output ports, and RF1 and RF2. The series SPST switches are designed close to the junction for minimizing unwanted RF reflections from the CPW T-discontinuity. Although the switches require more than 750 °C heat for an efficient melt-quench switching sequence, it is worth highlighting that the heat generated using a thermal pulse is for a maximum of 200 ns only. With densely packed phase-change switches, no thermal crosstalk has been observed in the nearby switches [5]. Micro-heaters bias lines are replaced with a conductive metal layer to improve the switches' RC time constant.

C. Switch Matrix Unit-Cell

The miniaturized switch matrix cell shown in Fig. 1 is designed with six phase-change series SPST RF switches arranged to route the signal between four ports, namely, RF1, RF2, RF3, and RF4, in two operational states: I and II. In state-I, ports RF1–RF4 (route S1) and ports RF2–RF3 (route S2) are connected, while in state-II, ports RF1–RF3 (route S3) and ports RF2–RF4 (route S4) are connected. Route S3 and S4 in state-II use two series switches instead of a single SPST switch for improving the isolation and RF matching. The control pads C1-C10 provide bias signals to the respective phase-change switches. The switch matrix unit-cell core area is under $300\,\mu m \times 300\,\mu m$. The device layout is optimized to keep the RF performance between two routes of the state-II identical. Further details on a phase-change four-port unit-cell functionality and operational states is discussed in [16].

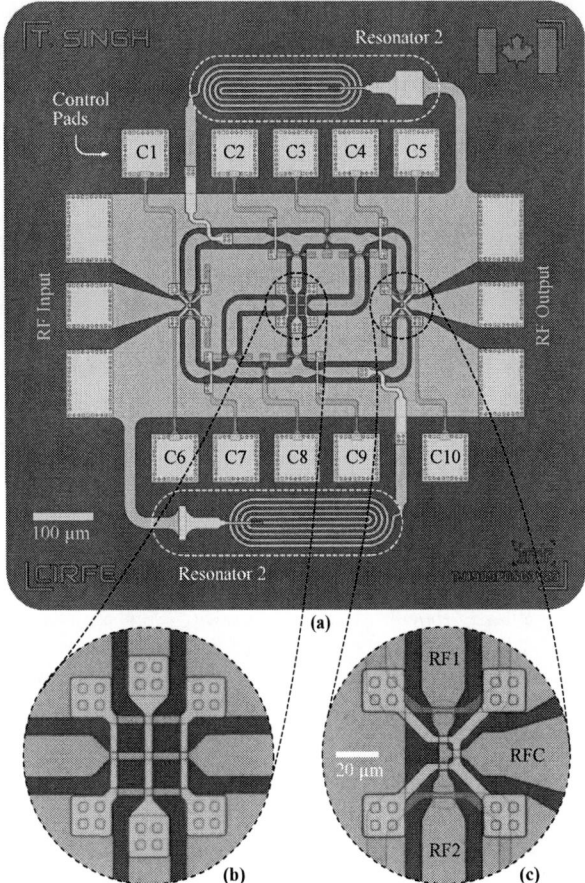

Fig. 2. Optical micrograph: (a) Phase-change GeTe-based reconfigurable band rejection circuit utilizing four-port 2×2 switch matrix highlighting resonators and control pads, (b) RF cross-over junction, (c) phase-change SPDT switch.

D. Integrated Band Rejection Circuit

The switched reconfigurable band rejection circuit is designed by monolithically integrating two notch filters (Resonator 1 and Resonator 2) in shunt with two RF paths in a 2×2 switch matrix, as shown in Fig. 1. Two SPDT switches are integrated at the switch matrix's input and output sides to route between the switch matrix's available paths. The optical micrograph of the fully integrated phase-change band rejection circuit is shown in Fig. 2(a). The RF cross-over junction and SPDT switch core are highlighted in Fig. 2(b) and (c), respectively.

The resonators connected in shunt with the RF signal paths provide band rejection at $f_1 = 2.6$ GHz and $f_2 = 5.8$ GHz. The RF signal can be routed through the available four routes of the circuit (S1, S2, S3, or S4). Route-S1 rejects f_1 band and route-S2 provides signal rejection at band f_2 while keeping an identical performance to route-S1. Route-S3 includes rejection at $f_1 + f_2$ bands, and to achieve this functionality, the RF signal routes through resonator 1 followed by an SPST switch, cross-over junction, and a second SPST switch before adding

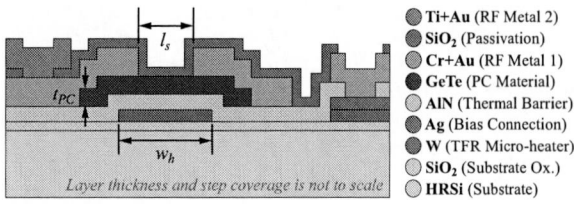

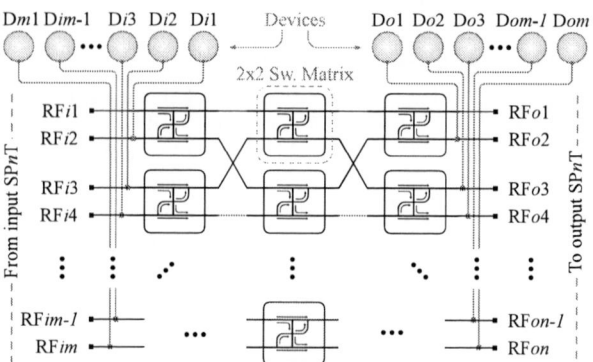

Fig. 3. Fabrication process cross-section and description of various layers for developing PCM RF switches and switch matrices. Micro-heater width (w_h), PCM channel length (l_s), and PCM layer thickness (t_{PC}) is highlighted.

Fig. 4. Scalability approach to higher order $m \times n$ matrices for precise multi-bit tuning of RF devices.

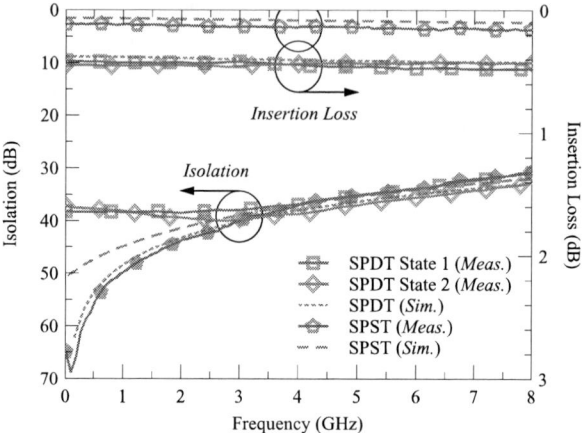

Fig. 5. Measured and simulated RF performance of monolithically integrated phase-change SPST and SPDT switched over DC-8 GHz.

the response of resonator 2 in series to the signal path. Route-S4 bypasses both the resonators and transfer signal from the RF input to the output port without rejecting any frequency band. The inductors' resistance is reduced by adding the M1 (first Au metallization) layer parallel to M2 (second Au metallization). However, the Q can be improved by utilizing a relatively thicker metallization.

III. FABRICATION PROCESS FLOW

A custom eight-layer microfabrication process is developed and optimized in-house for the integrated RF phase-change devices. The cross-section of the process flow is shown in Fig. 3. The devices are fabricated on an intrinsic high resistivity silicon substrate with $\rho = >20\,\mathrm{k\Omega \cdot cm}$. The substrate is passivated with a thin silicon oxide (SiO$_2$) layer, followed by the deposition and reactive ion etching (RIE) of a refractory tungsten (W) layer. Silver (Ag) is sputtered and patterned using RIE for conductive bias circuit for low RC time constant. A thin-film of aluminum nitride (AlN) is RF sputtered and patterned using RIE as a barrier layer. A thin GeTe layer is sputtered and patterned using ion-milling followed by a gold (Au) layer as the first metallization for RF signal flow with chromium (Cr) as a seed layer. The first metallization layer is patterned using the lift-off technique. A passivation layer of SiO$_2$ is further deposited using PECVD and patterned using RIE. A second metallization layer of Au is deposited using titanium (Ti) as a seed layer and patterned using lift-off. Further details on the microfabrication process are given in [3], [17]

IV. SCALABILITY APPROACH

The miniaturized phase-change GeTe-based switch matrix unit-cell offers scalability to higher-order matrices through cascading the cells and adding intermediate connections, as demonstrated in Fig. 4. A switch matrix unit-cell has only six SPST switches for a wideband operation. 4-bit reconfigurability can be achieved by utilizing six 2×2 switch matrix cells to form a 4×4 switch matrix for routing six different devices (Di1, Di2, Di3, Do1, Do2, Do3, and/or a combination of any device) with a total of sixteen states as shown in Fig. 4. Similarly, an 8×8 switch matrix require twenty 2×2 switch matrix unit-cells. The proposed switch matrix offers a better insertion loss in comparison to popular crossbar switch matrix topologies, which exhibit a high insertion loss due to the variation in the lengths of RF paths.

In the higher insertion loss states, the RF signal in a typical 8×8 crossbar switch matrix would go through at least fifteen cascaded SPST switches, while it goes through only ten switches (five switches in case of narrowband operation) using the proposed topology shown in Fig. 4. Thus an 8×8 matrix of this type offers a 33% reduction in the required switching elements than a common crossbar topology. For scaling to an $m \times n$ functionality, i.e., m-inputs and n-outputs can be configures as shown in Fig. 4. It should be mentioned that the SPnT switches need at the input and output, are required only when employing the switch matrix to realize a multi-state reconfigurable circuit with only one RF input and output port.

V. RESULTS AND DISCUSSION

The measured and simulated RF results of the individual SPST switching elements and SPDT switches are shown in Fig. 5. The SPST switch exhibits measured insertion loss less than $0.18\,\mathrm{dB}$ and isolation higher than $30\,\mathrm{dB}$, while the SPDT switches demonstrate insertion loss lower than $0.45\,\mathrm{dB}$ and isolation better than $30\,\mathrm{dB}$ over the $8\,\mathrm{GHz}$ operational bandwidth. The closely matched EM simulations are carried

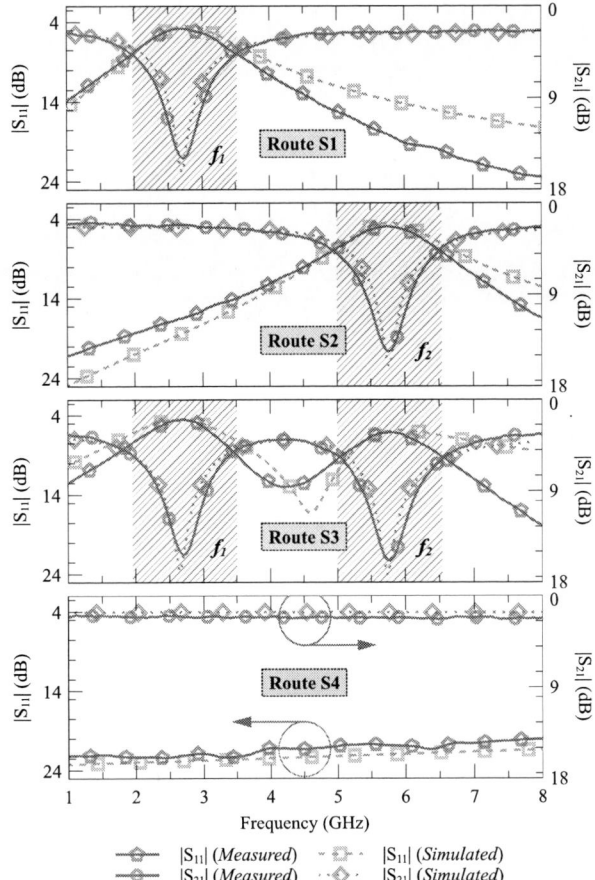

Fig. 6. Measured and simulated RF response of fully integrated 2-bit phase-change reconfigurable band rejection circuit in all possible states.

out in Ansys HFSS. The measured and simulated performance of the reconfigurable band rejection circuit is shown in Fig. 6. Route-S4 shows the measured insertion loss of lower than 2.2 dB and return loss better than 19 dB over 1 GHz to 8 GHz. It should be mentioned that a higher rejection and a better insertion loss can be realized by coupling more than one resonator, i.e., integrating the switch matrix with higher order band-stop filters. Only a single resonator is used for each band in the prototype presented in this paper to demonstrate the concept. The insertion loss can be improved with the use of thick metal layers for both the switch matrix and resonators.

VI. CONCLUSION

A miniaturized reconfigurable band rejection circuit is reported utilizing a scalable miniaturized monolithically integrated phase-change GeTe-based switch matrix. The scalability approach is proposed to reduce the number of switches for higher-order $m \times n$ multi-bit tuning. The integrated circuit is developed in-house using a custom microfabrication process. The band rejection circuit as a proof-of-concept demonstrates the potential of the proposed

routing and scaling approach for reconfiguring various RF circuits by replacing the resonators with higher-order filters, antennas, delay lines, or resistor networks, to name a few. The proposed circuit provides non-volatile operation in a highly miniaturized and is useful in reconfigurable front-end modules.

REFERENCES

[1] S. Rangan, T. S. Rappaport, and E. Erkip, "Millimeter-wave cellular wireless networks: Potentials and challenges," *Proc. IEEE*, vol. 102, no. 3, pp. 366–385, 2014.

[2] L. A. Belov, S. M. Smolskiy, and V. N. Kochemasov, *Handbook of RF, microwave, and millimeter-wave components.* Boston, MA, USA: Artech house, 2012.

[3] T. Singh and R. R. Mansour, "Loss compensated PCM GeTe-based wideband 3-bit switched True-Time-Delay phase shifters for mmWave phased arrays," *IEEE Trans. Microw. Theory Techn.*, vol. 68, no. 9, pp. 3745–3755, Sept. 2020.

[4] T. Singh and R. R. Mansour, "Scalable mmWave non-volatile phase change GeTe-based compact monolithically integrated wideband digital switched attenuator," *IEEE Trans. Electron Devices*, vol. 68, no. 5, pp. 2306–2312, May 2021.

[5] T. Singh and R. R. Mansour, "Miniaturized DC–60 GHz RF PCM GeTe-based monolithically integrated redundancy switch matrix using T-Type switching unit cells," *IEEE Trans. Microw. Theory Techn.*, vol. 67, no. 12, pp. 5181–5190, Dec. 2019.

[6] T. Singh and R. R. Mansour, "Scalable non-volatile chalcogenide phase change GeTe-based monolithically integrated mmWave crossbar switch matrix," in *Proc. IEEE MTT-S Int. Microw. Symp. Dig.*, Atlanta, GA, USA, Jun. 2021, pp. 1–4.

[7] C. F. Campbell and D. C. Dumka, "Wideband high power GaN on SiC SPDT switch MMICs," in *Proc. IEEE MTT-S Int. Microw. Symp. Dig. (IMS)*, Anaheim, CA, USA, May 23–28 2010, pp. 145–148.

[8] P. Sun, P. Liu, P. Upadhyaya, D. Jeong, D. Heo, and E. Mina, "Silicon-based pin spst rf switches for improved linearity," in *Proc. IEEE MTT-S Int. Microw. Symp. Dig. (IMS)*, Anaheim, CA, USA, May 23–28 2010, pp. 948–951.

[9] O. Tabata, T. Tsuchiya, O. Brand, G. K. Fedder, C. Hierold, and J. G. Korvink, *Reliability of MEMS: Testing of Materials and Devices*, O. Tabata and T. Tsuchiya, Eds. Hoboken, NJ, USA: John Wiley & Sons, Mar. 2013.

[10] Menlo Microsystems, Inc., "DC to 26 GHz high power RF switch," MM5130 datasheet, *Rev.* 1.9, Nov. 2020.

[11] Analog Devices, Inc., "0 Hz/dc to 14 GHz, single-pole, four-throw MEMS switch with integrated driver," ADGM1304 datasheet, Jan. 2020.

[12] T. Singh and R. R. Mansour, "Experimental investigation of performance, reliability, and cycle endurance of non-volatile DC–67 GHz phase-change RF switches," *IEEE Trans. Microw. Theory Techn.*, pp. 1–14, 2021. doi: 10.1109/TMTT.2021.3105413 (early access article).

[13] T. Singh and R. R. Mansour, "Non-volatile multiport DC–30 GHz monolithically integrated phase-change transfer switches," *IEEE Electron Device Lett.*, vol. 42, no. 6, pp. 867–870, 2021.

[14] T. Singh and R. R. Mansour, "Characterization, optimization and fabrication of phase change material germanium telluride based miniaturized DC–67 GHz RF switches," *IEEE Trans. Microw. Theory Techn.*, vol. 67, no. 8, pp. 3237–3250, Aug. 2019.

[15] A. Hariri, A. Crunteanu, C. Guines, C. Hallepee, D. Passerieux, and P. Blondy, "Double-Port Double-Throw (DPDT) switch matrix based on phase change material (PCM)," in *Proc. 48th Eur. Microw. Conf. (EuMC)*, Madrid, Spain, Sept. 23–27 2018, pp. 479–482.

[16] T. Singh and R. R. Mansour, "A miniaturized monolithic PCM based scalable four-port RF switch unit-cell," in *Proc. 49th Eur. Microw. Conf. (EuMC)*, Paris, France, Sep. 29 – Oct. 4 2019, pp. 180–183.

[17] T. Singh, "Monolithically integrated phase change material GeTe-based RF components for millimeter wave applications," Ph.D. dissertation, Dept. Elect. Comput. Eng., Univ. of Waterloo, Waterloo, ON, Canada, Apr. 2020. [Online]. Available: http://hdl.handle.net/10012/15906

Gap in pagination due to unavailable paper.

Page 38

Proceedings of the 16th European Microwave Integrated Circuits Conference

A Bidirectional 28 GHz RF Transceiver Front-End with Test and Calibration Interface for 5G Phased Arrays

Katharina Kolb[#1], Julian Potschka[#2], Tim Maiwald[#3], Klaus Aufinger[*4],
Amelie Hagelauer[§5], Marco Dietz[#6], Robert Weigel[#7]

[#]Institute for Electronics Engineering, Friedrich-Alexander-University Erlangen-Nuremberg, Germany
[*]Infineon Technologies AG, Germany
[§]Chair of Micro- and Nanosystems Technology, Technical University of Munich, Germany
{[1]katharina.kolb, [2]julian.potschka, [3]tim.maiwald, [6]marco.dietz, [7]robert.weigel}@fau.de

Abstract — **This paper presents a bidirectional 28 GHz RF transceiver (TRX) front-end for 5G that provides interfaces for calibration and angle of arrival (AoA) estimation by adding area-optimized broadband 10 dB broadside couplers. In addition, possible deployment scenarios and corresponding architectures in large arrays are described. The front-end integration level includes a receive (RX) chain, a transmit (TX) chain and a serial peripheral interface (SPI) for digital reconfigurability. Fabricated in a 130 nm SiGe BiCMOS technology, it delivers up to 22.6 dB and 22.5 dB gain for RX and TX modes, respectively, and offers measured control ranges of 9.1 dB / 9.2 dB for amplitude and 360 ° for phase. An input-referred 1 dB-compression point ($P_{1\,dB}$) of -24.3 dBm and an output-referred $P_{1\,dB}$ of 3.0 dBm is achieved with a DC power consumption of 227.8 mW and 212.2 mW for RX and TX, respectively.**

Keywords — **5G, built-in-self-test, calibration, millimeter-wave integrated circuits, phased arrays**

I. INTRODUCTION

With the rapid growth in data demand and thus the need for a larger bandwidth, 5G introduces millimeter-wave bands located at 28 and 39 GHz for wireless communication. However, the move to higher frequencies for 5G entails higher free space attenuation compared to those of former communication standards [1]. To overcome higher path losses, phased array systems are essential for bridging longer distances between a base station and an end user. Thus, the signal is transmitted and received as a steered beam, allowing both a directed line-of-sight link and hybrid beamforming. RF-path beamsteering TRXs suffer from gain variations over the phase tuning range [2]. Furthermore, beamsteering quality is degraded by phase drift and gain expansion or suppression with supply and temperature changes, non-linearities as well as aging. Therefore, phased array systems must be calibrated in order to achieve optimum performance and high accuracy, especially for narrow beam angles. Many calibration methods have been published, a large number of which are over-the-air [3]. However, to enable a real-time calibration or built-in-self-test (BIST), internal feedback-loops need to be implemented. Recent publications proposed BISTs to calibrate large 5G phased array systems [4], [5]. This work presents the implementation of test and calibration interfaces by adding

area-optimized broadband 10 dB broadside couplers in RX and TX path in order to allow a number of application scenarios, which include not only a use for calibration during operation but also AoA estimation [6] in RX mode or the use of digital predistortion (DPD) algorithms [7] in TX mode. The structure of this work is as follows: we first introduce basic test and calibration architectures and possible application scenarios in Section II. Then, we present design and measurement results of the fabricated bidirectional SiGe BiCMOS RF TRX front-end within Section III and Section IV, respectively.

II. TEST AND CALIBRATION ARCHITECTURES

For calibration, amplitude and phase information of the RF signal are required. Therefore, a portion of the RF signal must be extracted, which allows the implementation of a calibration setup in the TX and the RX paths. Two possible application scenarios are illustrated using the block diagrams from Fig. 1, namely AoA estimation as well as real-time calibration.

First, to align antenna beam pattern, an estimation of the angle of incidence can be determined within RX arrays. For

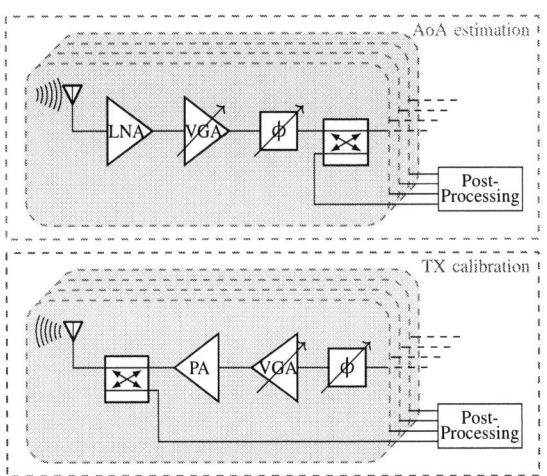

Fig. 1. Possible calibration architectures for RX and TX.

3–4 April 2022, London, UK

this purpose, the phase difference of adjacent RX channels can be detected to re-adjust the phase control as shown in [6]. A possible RX architecture for AoA estimation can be seen in the upper block diagram of Fig. 1. Second, the lower block diagram of Fig. 1 shows a possible TX architecture for TX real-time calibration. In particular, this is important because narrow beam angles in large antenna arrays require higher alignment accuracy. In addition, the DC power of the power amplifier (PA) leads to non-uniform temperature distribution, especially in large arrays. This results in phase and amplitude variations between each TX path, which has a negative influence on the link quality and can be improved by a real-time phase and amplitude calibration. An additional approach regarding TX systems is that nonlinear effects in PAs increase Error-Vector-Magnitude (EVM), which can be compensated by a DPD as shown in [7]. The following section presents an integrated TRX implementation that covers both application scenarios.

III. REALIZATION OF THE RF TRX FRONT-END WITH CALIBRATION INTERFACE

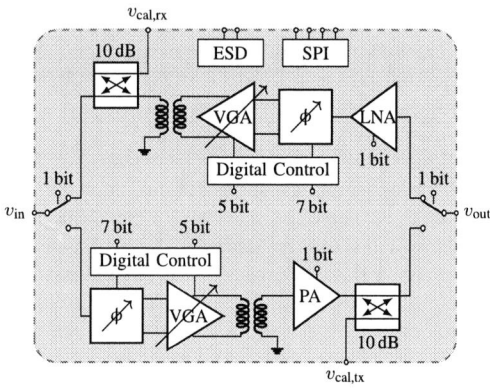

Fig. 2. Block diagram of the bidirectional 28 GHz TRX front-end with calibration interface.

The RF TRX front-end implements interfaces for the application scenarios of the previous section. Fig. 2 shows the block diagram of the integrated bidirectional RF front-end. It includes switches, 7-bit phase shifters, 5-bit variable gain amplifiers (VGA) as well as a low noise amplifier (LNA) and a PA for RX and TX mode of operation, respectively. In addition, a 10 dB broadside coupler is added in each of the two RF paths to provide a coupled signal for post-processing and, thus, for calibration or BIST. Switching between TX and RX mode as well as (calibrated) phase and gain settings are controlled via SPI. A 130 nm SiGe BiCMOS technology with nominal values of 250 GHz and 370 GHz for f_t and f_{max}, respectively, is used for the front-end implementation. Table 1 summarizes the performance of all main building blocks of the integrated 28 GHz TRX channel. A circuit design overview is given in the following.

A. RX/TX Switching

Two single-ended quarter-wave single-pole double-throw switches (SPDT) allow switching between RX and TX mode. In isolation mode, the corresponding transistor is turned on. The smaller the on-resistance, the higher the isolation. The other transistor operates in off-mode, at the same time. To minimize insertion loss, the resistance should be as high as possible with a minimum parasitic off-capacitance [8]. The design reaches a simulated noise figure (NF) of 1.7 dB.

B. RX Mode

The 28 GHz RX chain includes a LNA, a phase shifter with 7-bit phase control and a VGA with 5-bit gain control. According to Friis' formula, the NFs of the SPDT and the LNA dominate the overall noise performance of the whole receiver chain. Therefore, the LNA [9] has to provide low noise and high gain. The LNA is followed by a 7-bit phase shifter [10], which adjusts the phase for antenna beam-steering. It consists of a passive balun, a quadrature-allpass-filter and a vector modulator. The phase shifter is followed by a VGA [10] with 5-bit gain control and a steering range of 9.1 dB. The corresponding architecture is based on a differential current-steering topology. It compensates the power discrepancies of the phase shifter and adjusts the gain of the antenna elements to achieve sidelobe suppression. Further, the PA is switched off in RX mode to reduce the thermal noise leakage of the TRX channel.

C. TX Mode

The 28 GHz TX chain uses the same 7-bit vector modulator phase shifter as the RX chain followed by a VGA, which is based on a differential current-steering topology, similar to the RX-VGA. The 28 GHz PA [11] provides the output power level for the antenna element. In TX mode of operation, the LNA is switched off.

D. Calibration Interface

Each operation link includes a broadside coupler topology for calibration signal extraction. An area efficient design is obtained by applying a meandered line pattern as shown in [12]. The power is transferred between two transmission lines and results in a broadband coupling of -10 dB at the target frequency of 28 GHz. The coupler was placed in front of the SPDT and not directly in front of the antenna in order not to degrade the NF.

Table 1. Component performance of the RF TRX front-end building blocks.

	S_{21} (dB)	P_{dc} (mW)	A (μm^2)
SPDT	-2.7	4.0	155×480
LNA	21.1	43.9	250×330
Phase shifter	-1...1	33.7	595×440
VGA-RX	10.2	55.6	240×320
Balun	-1.6	-	160×330
Coupler	-1.7	-	430×280
VGA-TX	17.1	53.4	240×320
PA	14.1	40.8	220×400

IV. MEASUREMENT RESULTS

A micrograph of the TRX channel is shown in Fig. 3. The chip area excluding the pads has a size of 2660 μm × 1000 μm. The integrated TRX channel consumes a DC power of 227.8 mW and 212.2 mW for RX and TX, respectively. For on-wafer measurement, GSG-Probes with 100 μm pitch are used and digital as well as DC pads are bonded on a test board.

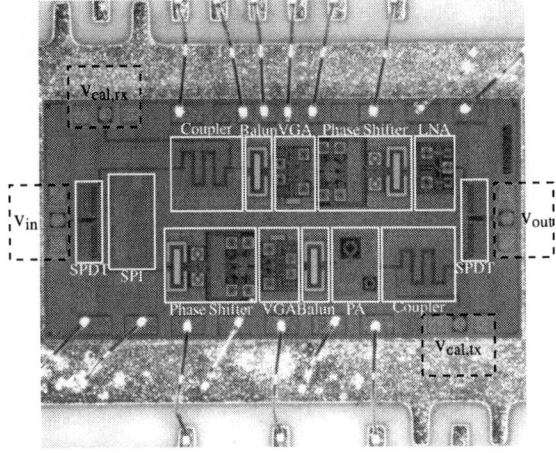

Fig. 3. Chip micrograph (2660 μm × 1000 μm without pads).

Fig. 4 and Fig. 5 show the measured S-parameters of the TRX front-end in RX mode and TX mode, respectively. It achieves up to 22.6 dB and 22.5 dB gain at a frequency of 28 GHz for RX and TX mode. The figures also illustrate the adjustable gain of the investigated TRX front-end. It achieves a controllable amplitude range of 9.1 dB at 28 GHz for RX as well as an adjustable gain of 9.2 dB at 28 GHz for TX. The simulated NF for RX mode is 4.5 dB at the target frequency. The simulation results already consider extracted EM-models of layout and matching networks. The design delivers a 3 dB-bandwidth of 4.4 GHz and 5.5 GHz for RX and TX mode, respectively. The coupled output port for calibration results in a measured coupling $S_{31} - S_{21}$ between −10.7 dB and −11.5 dB for the whole 3 dB-bandwidth and in both modes of operation. The measured input and output reflection coefficients are lower than −12 dB for both RX and TX over the entire bandwidth. Measurement and simulation show a shift below 2 %, which is caused by a slightly overestimation of parasitics during the EM-mode extraction.

Fig. 6 shows the measured and simulated $P_{1\,dB}$ of the RX and the TX chain. The measured 28 GHz input-referred $P_{1\,dB}$ of the whole RX chain achieves −24.3 dBm. The measured output-referred $P_{1\,dB}$ of the whole TX chain is 3.0 dBm. For the measurement, the power of the input and the output of the design is measured and cable losses are considered.

In Fig. 7 the relative phase of every fourth phase step for a fixed gain set can be seen. Additionally, Fig. 8 shows the measurement results for magnitude and phase in polar plots for RX and TX, respectively, at 28 GHz. This shows that via

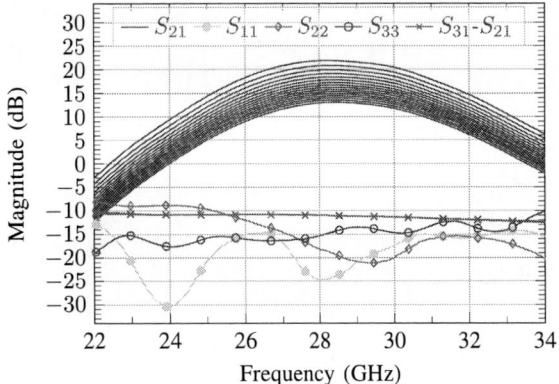

Fig. 4. Measurement results for S-parameters in RX mode.

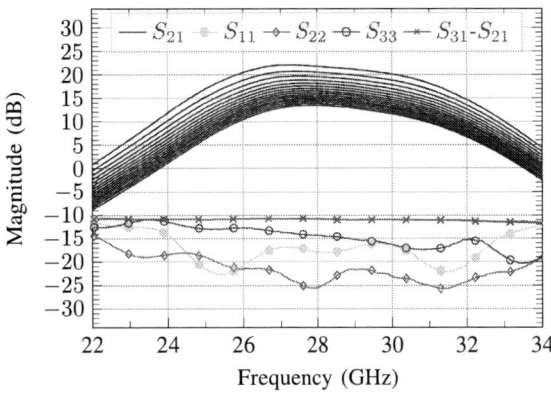

Fig. 5. Measurement results for S-parameters in TX mode.

calibration or BIST even small phase or amplitude errors of the TRX can be compensated by applying an offset to the digital words of both, VGA and phase shifter control.

The measurement results are in good agreement with the simulation results and are summarized in Table 2.

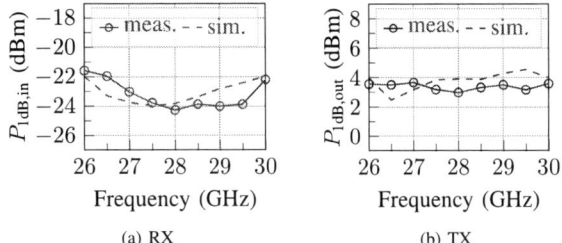

(a) RX   (b) TX

Fig. 6. Measurement and simulation results for input-referred $P_{1\,dB}$ of RX and output-referred $P_{1\,dB}$ of TX.

V. CONCLUSION

This paper presents a bidirectional 28 GHz RF TRX front-end with interfaces for calibration and AoA estimation by adding area-optimized 10 dB broadside couplers. The TRX is fabricated in a 130 nm SiGe BiCMOS technology and

Table 2. Performance summary and comparison to state-of-the-art implementations.

	[13]	[14]	[15]	[16]	This work
Technology	0.13 μm BiCMOS	65 nm CMOS	0.18 μm BiCMOS	0.25 μm BiCMOS	0.13 μm BiCMOS
Frequency (GHz) TX/RX	30-40/30-40	28/28	28/28	29.5-30.8/19.7-21	28/28
No. of channels TX/RX	1/1	8/8	4/4	2/1	1/1
Gain / channel TX/RX (dB)	14.0/17.0	12.0/12.0	20.0/20.0	37.7/33.7	22.5/22.6
Gain control TX/RX (dB)	-	31.0/31.0	14.0/14.0	8.0/20.0	9.2/9.1
Phase control (°)	360	360	360	360	360
Phase resolution (bit)	5	6	6	16	7
$P_{1\,dB,out/in}$ TX/RX (dBm)	3.0/-1.0	-2.5/-	10.5/-22.0	12.9/-38.3	3.0/-24.4
DC power TX/RX (mW)	528.0/1587.0	1450.0/1450.0	800.0/520.0	162.0/15.5	212.2/227.8

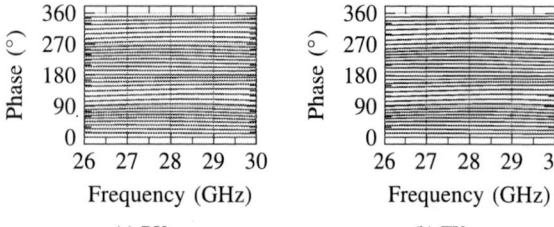

(a) RX (b) TX

Fig. 7. Measurement results for relative phase.

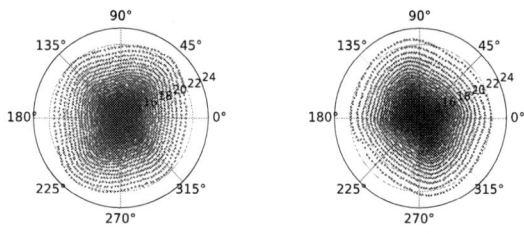

(a) RX (b) RX

Fig. 8. Measurement results for magnitude and phase at 28 GHz.

delivers up to 22.6 dB and 22.5 dB gain for RX and TX mode, respectively, and offers steering ranges of 9.1 dB / 9.2 dB for amplitude and 360° for phase. Compared to state-of-the-art implementations (Table 2), the design achieves comparable performance while providing an additional calibration interface that allows to calibrate very small phase and amplitude errors for optimum performance. Even though the couplers are already area efficent, in exchange for bandwidth and coupling factor one can reduce the size of the coupling structure even further.

ACKNOWLEDGMENT

The authors would like to thank Infineon Technologies AG for fabrication of the integrated circuits. This work was supported by the Federal Ministry of Education and Research (BMBF) and the European Union in frame of the funded project TARANTO under grant number ESECS16 104.

REFERENCES

[1] M. Shafi *et al.*, "5G: A Tutorial Overview of Standards, Trials, Challenges, Deployment, and Practice," *IEEE Journal on Selected Areas in Communications*, vol. 35, no. 6, pp. 1201–1221, 2017.

[2] M. Fakharzadeh, P. Mousavi, S. Safavi-Naeini, and S. H. Jamali, "The Effects of Imbalanced Phase Shifters Loss on Phased Array Gain," *IEEE Antennas and Wireless Propagation Letters*, vol. 7, pp. 192–196, 2008.

[3] M. Jokinen *et al.*, "Over-the-Air Phase Measurement and Calibration Method for 5G mmW Phased Array Radio Transceiver," in *2019 93rd ARFTG Microwave Measurement Conference (ARFTG)*, 2019, pp. 1–4.

[4] Y. Wang *et al.*, "A 39-GHz 64-Element Phased-Array Transceiver With Built-In Phase and Amplitude Calibrations for Large-Array 5G NR in 65-nm CMOS," *IEEE Journal of Solid-State Circuits*, vol. 55, no. 5, pp. 1249–1269, 2020.

[5] Y. Aoki *et al.*, "Inter-Stream Loopback Calibration for 5G Phased-Array Systems," in *2020 IEEE Radio Frequency Integrated Circuits Symposium (RFIC)*, 2020, pp. 359–362.

[6] B. Laemmle *et al.*, "A 77-GHz SiGe Integrated Six-Port Receiver Front-End for Angle-of-Arrival Detection," *IEEE Journal of Solid-State Circuits*, vol. 47, no. 9, pp. 1966–1973, 2012.

[7] T. Ackermann *et al.*, "A Robust Digital Predistortion Algorithm for 5G MIMO: Modeling a MIMO Scenario With Two Nonlinear MIMO Transmitters Including a Cross-Coupling Effect," *IEEE Microwave Magazine*, vol. 21, no. 7, pp. 54–62, 2020.

[8] A. C. Ulusoy *et al.*, "A Low-Loss and High Isolation D-Band SPDT Switch Utilizing Deep-Saturated SiGe HBTs," *IEEE Microwave and Wireless Components Letters*, vol. 24, no. 6, pp. 400–402, 2014.

[9] K. Kolb *et al.*, "A 28 GHz Broadband Low Noise Amplifier in a 130 nm BiCMOS Technology for 5G Applications," in *2020 23rd International Microwave and Radar Conference (MIKON)*, 2020, pp. 192–195.

[10] K. Kolb *et al.*, "A 28 GHz Highly Accurate Phase- and Gain-Steering Transmitter Frontend for 5G Phased-Array Applications," in *2020 IEEE 63rd International Midwest Symposium on Circuits and Systems (MWSCAS)*, 2020, pp. 432–435.

[11] J. Potschka *et al.*, "A Highly Linear and Efficient 28 GHz Stacked Power Amplifier for 5G using Analog Predistortion in a 130 nm BiCMOS Process," in *2019 IEEE Asia-Pacific Microwave Conference (APMC)*, 2019, pp. 920–922.

[12] J. Potschka *et al.*, "A Switchable, Passively Tuneable 28 GHz to 39 GHz Upconversion Link for a 5G Repeater using a Broadside Coupler and Analog Predistortion in a 130 nm BiCMOS Technology," in *2020 18th IEEE International New Circuits and Systems Conference (NEWCAS)*, 2020, pp. 30–33.

[13] C. Liu *et al.*, "A Ka-Band Single-Chip SiGe BiCMOS Phased-Array Transmit/Receive Front-End," *IEEE Transactions on Microwave Theory and Techniques*, vol. 64, no. 11, pp. 3667–3677, 2016.

[14] J. Park, D. Baek, and J.-G. Kim, "A 28 GHz 8-channel Fully Differential Beamforming IC in 65nm CMOS process," in *2019 49th European Microwave Conference (EuMC)*, 2019, pp. 476–479.

[15] K. Kibaroglu, M. Sayginer, and G. M. Rebeiz, "A Low-Cost Scalable 32-Element 28-GHz Phased Array Transceiver for 5G Communication Links Based on a 2 × 2 Beamformer Flip-Chip Unit Cell," *IEEE Journal of Solid-State Circuits*, vol. 53, no. 5, pp. 1260–1274, 2018.

[16] F. Tabarani *et al.*, "Power-efficient full-duplex K/Ka-band phased array front-end," *IET Microwaves, Antennas & Propagation*, vol. 14, no. 4, pp. 268–280, 2020.

Proceedings of the 16th European Microwave Integrated Circuits Conference

A 2-channel TX and 4-channel RX in SiGe BiCMOS for X-band MIMO Radar Applications

M. Kucharski, B. Błaszczuk, M. Klemm, R. Piesiewicz

SIRC, Poland

{m.kucharski, b.blaszczuk, m.klemm, r.piesiewicz}@si-research.eu

Abstract — **This work presents a 2-channel transmitter (TX) and a 4-channel I/Q receiver (RX) for X-band multiple-input multiple-output (MIMO) radar applications. The circuits are fabricated using cost-effective SiGe BiCMOS technology. The TX comprises a common-collector Colpitts VCO providing 15.5% tuning range (TR) and -105.5 dBc/Hz worst-case phase noise at 1 MHz offset from the carrier. The TX provides more than 6 dBm output power at each output. The RX entails an efficient LO distribution network resulting in -15 dBm minimum LO input power for saturated conversion gain (CG). It provides peak CG of 26 dB and more than 23 dB in 7-12.5 GHz range. Despite wideband characteristics the RX yields low gain and phase mismatch between I and Q channels. The TX and RX consume 237 mW and 574 mW power and occupy 1 mm² and 2 mm² silicon area, respectively.**

Keywords — **radar, X-band, VCO, MIMO.**

I. Introduction

Rapid development of silicon technologies enabled high-level of integration allowing for complete systems to be embedded on a single die (SoC). Silicon-Germanium technology proved to be particularly useful for its capability to combine high-performance hetero-junction transistors (HBT) suitable for RF parts with low-power CMOS section optimized for digital processing. Radars are undoubtedly important for their unique sensing features. Recent emergence of unmanned aerial vehicles (UAV), in particular drones, created a danger for safety of sensitive venues, like airports. Counter-drone systems use radars to provide required security against unauthorized UAVs. Among available frequency range X-band proved to be suitable for long-range sensing.

II. MIMO Radar System

The concept of the X-band MIMO radar is presented in Fig. 1. It is composed of a 2-channel transmitter (TX) including an on-chip signal source. The TX drives a high-power amplifier (HPA) typically implemented using a commercially available IC based on III-V semiconductors, such as GaN or GaAs. A small part is coupled to the RX and serves as a LO signal for on-chip quadrature mixers. The RX consists of 4 channels and LO distribution network. Multiple TX and RX channels enable beam-forming that can be used to scan the area in azimuth without mechanical rotation. The on-chip signal source is intended to be used with an external phase-locked loop and can be employed in various radar modes, such as pulsed or FMCW.

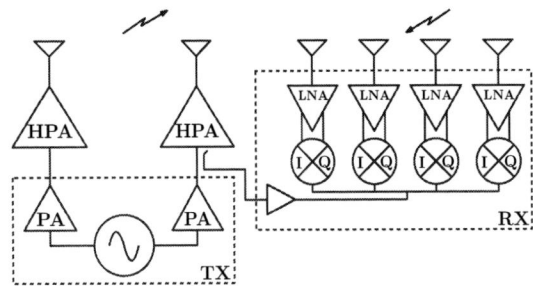

Fig. 1. Conceptual MIMO radar system based on designed TX and RX.

III. Transmitter

Block diagram including selected schematics is presented in Fig. 2. A common-collector Colpitts VCO [1] is used to generate an X-band signal along with two cascaded flip-flop based frequency dividers to meet maximum input frequency requirements of an external PLL. The frequency division is conveniently achieved using a master-slave latch configuration based on emitter-coupled logic (ECL) [2]. In the VCO, instead of shorting collector terminal to ground the signal is taken to a cascode buffer stage using a small 15 Ω. Such arrangement results in an enhanced VCO core isolation with respect to external load. The prescaler input signal is derived directly from emitter inductive divider via small 60 fF capacitor. The VCO signal drives two power amplifiers (PA) using a 1:2 transformer implemented with top thick metal layers for low loss. Single PA is implemented as a common-emitter stage using high-voltage HBTs with BV_{CE0}=7 V. 20 fF neutralization capacitors improve stability and power gain. The PA is loaded with a transformer to convert the differential signal to single-ended form. The transformer and the bondwire provide matching to external 50 Ω resistance for maximum output power. The optimum load impedance at the collectors of the common-emitter stage was obtained from load-pull simulations.

IV. Receiver

Block diagram including schematics is presented in Fig. 3. A single-ended LO input signal is converted to differential and amplified before distribution to RX channels using a 1:4 transformer. Each channel is composed of a two-stage

3–4 April 2022, London, UK

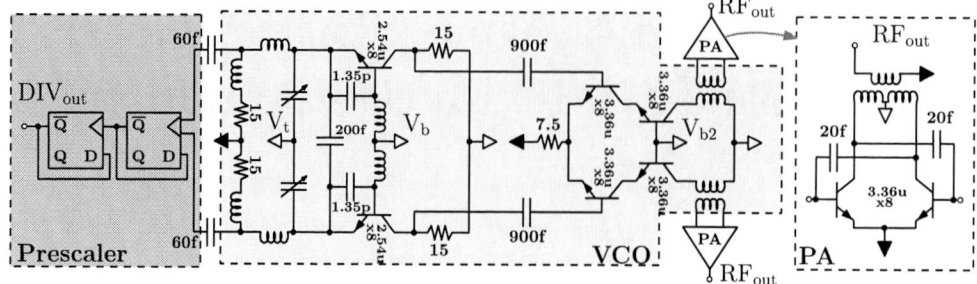

Fig. 2. Schematic of the TX chip.

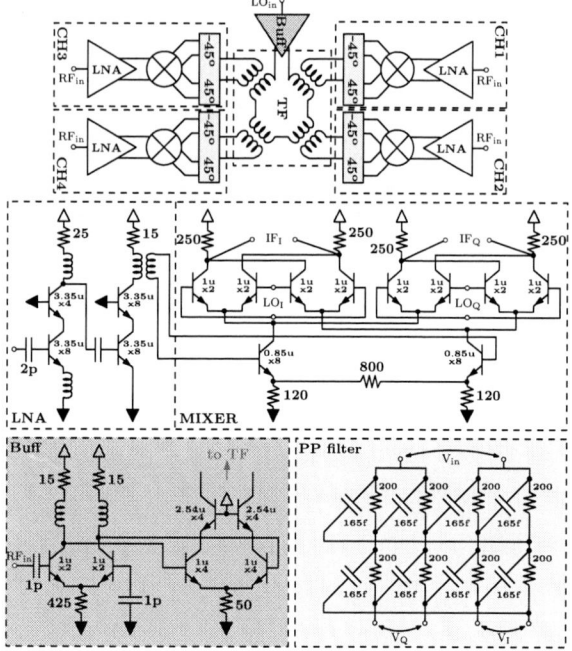

Fig. 3. Schematic of the RX chip.

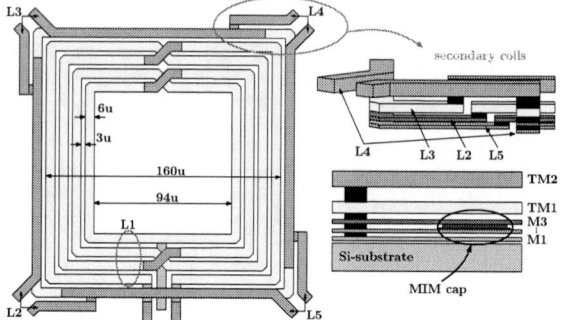

Fig. 4. Layout of the 1:4 LO transformer including BEOL cross-section of SGB25V process.

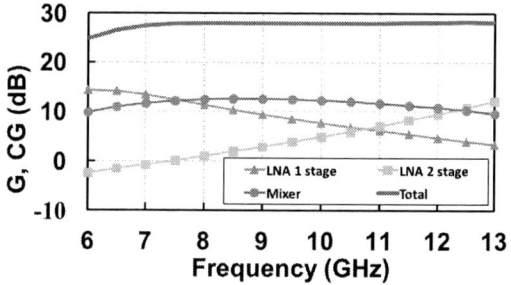

Fig. 5. Simulated gain contributions of each RX component.

low-noise amplifier (LNA) followed by a quadrature mixer. The LNA is implemented as two cascaded cascode circuits loaded by a transformer to convert single-ended input signal to differential. A degeneration inductor at the emitter of the input stage is used to achieve simultaneous matching for noise and gain. Resistors at the inductive collector loads help to improve stability and increase the bandwidth. The mixer follows Gilbert-cell topology with $800\,\Omega$ differential degeneration resistor for enhanced linearity. The input g_{m}-stage amplifies the signal and feeds two switching quads driven by quadrature LO signal. A poly-phase filter (PPF) is used to generate LO signals shifted by 90 degrees. It consists of two cascaded low-pass RC-networks. Due to lack of inductors the PPF has a small footprint traded for increased loss that must be compensated by the input LO buffer. The buffer serves both as balun and amplifier. The signal conversion is achieved by

setting AC-ground at one of the inputs of the differential pair. The signal symmetry is enhanced by the second cascode stage comprising a tail resistor for improved common-mode rejection. The LO buffer provides 25 dB simulated gain with maximum gain and phase deviation of 0.4 dB and 0.6 degree, respectively, between the outputs.

Fig. 4 depicts the layout of the 1:4 transformer used to efficiently distribute LO signal to four RX channels. The primary coil consists of four windings implemented in TM1 layer. Each secondary coil is a single winding inductor with metal line sections stepping from TM2 down to M2 layer at each corner of the transformer. All secondary coils are identical providing the same loading conditions of the PPF. The symmetry of the coupled signal is improved by providing AC-ground at the center tap of each secondary coil by means

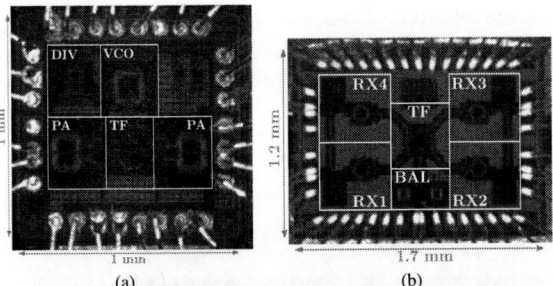

(a)   (b)

Fig. 6. Photograph of mounted and bonded (a) TX and (b) RX IC.

Fig. 7. Breadboard used to characterize the RX chip. A bare-die is mounted, bonded and protected using epoxy resin.

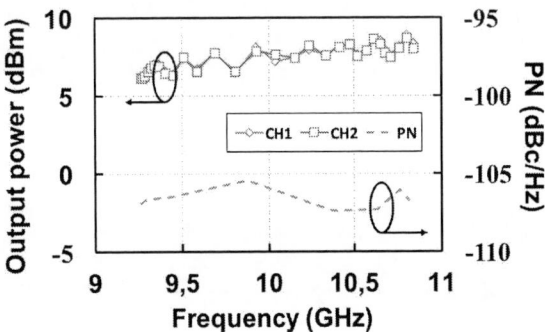

Fig. 8. Measured TX output power and phase noise of the VCO against frequency.

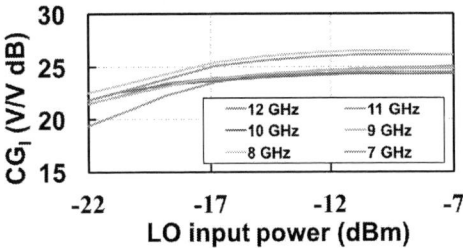

Fig. 9. Measured RX conversion gain for in-phase channel versus LO input power at different LO frequency. The IF frequency is set to 1 MHz.

of shunt MIM capacitor.

Fig. 5 depicts complete CG simulation against frequency with a total gain break-out to see gain contribution of each stage in the RX chain. Notably, wideband characteristics are obtained by shifting the peak gains of the LNA stages to different frequencies. The resulting total gain is flat in 7-12.5 GHz range reaching 28 dB. The simulated input-referred third-order intercept point is -11.6 dBm at 10.5 GHz and larger than -19.1 dBm in 7-12.5 GHz band.

V. EXPERIMENTAL RESULTS

Both chips were fabricated using SGB25V technology from IHP [3]. It is a cost-effective SiGe process offering fast HBTs with f_T/f_{MAX}=75/95 GHz and CMOS section intended for baseband signal processing. The photographs of the ICs is presented in Fig. 6. A dedicated breadboard was prepared for each circuit to measure its characteristics including bondwire inductances resulting from chip-to-PCB mounting. An example of RX breadboard is shown in Fig. 7 where a bare-RX die in mounted and wire-bonded to PCB. The chip is protected using an epoxy resin.

The RF characteristics of the TX were measured using a standard spectrum analyzer. The phase noise of the VCO was measured at the frequency divider output and extrapolated to carrier frequency by adding 20log(4)=12 dB. The RF power at each channel output including VCO phase noise is presented in Fig. 8. The VCO is tunable in 9.25-10.8 GHz range corresponding to 15.5% TR. Worst-case VCO phase noise at 1 MHz offset from the carrier is -105.5 dBc/Hz. The

RF output power at both channels exceeds 6 dBm within the total VCO frequency range.

The RX chip was characterized using two frequency generators to provide RF and LO input signals. The IF outputs were recorded using a digital oscilloscope set in high input impedance mode. The RF and LO frequency was set to keep the IF frequency constant at 1 MHz. CG was calculated by dividing IF voltage amplitude by equivalent RF input voltage amplitude derived from RF input power at 50 Ω. The loss of feeding cables was de-embedded from measurements which corresponds to calibration surface at the SMA connector interface. Thus, the loss due to PCB traces and bondwires is included in presented characteristics. RX CG for in-phase channel 1 against LO input power at different frequencies is presented in Fig. 9. The CG saturates at -15 dBm which beneficial in terms of efficient LO signal distribution in the target radar system. The I/Q conversion gain for all channels against frequency including simulated noise figure (NF) is shown in Fig. 10. The CG reaches 26 dB at 7.5 GHz and more than 23 dB in 7-12.5 GHz range. The gain deviation between the channels is less than 1 dB in 9-12 GHz range. The figure also presents input referred 1-dB compression point (IP_{1dB}) for all channels in 9-11 GHz range. Worst-case IP_{1dB} is -24.4 dBm at 9 GHz input frequency. Gain and phase I/Q balance for each channel is shown in Fig. 11. Gain difference between I and Q channels is kept below 1 dB in 6.5-13 GHz whereas phase deviation from 90 degrees is less than 4 degrees in 7.5-11 GHz range.

Table 1. Summary of selected X-band receivers.

Ref.	Process	Freq. [GHz]	CG_{MAX} [dB]	NF [dB]	IP_{1dB}	P_{DC} [mW]	Area [mm²]
[4]	SiGe 130 nm	8-18	53	6.7	-63	180	1.81
[5]	SiGe 130 nm	8-9	44.7	2.4[d]	-35	326[b]	2.6[c]
[6]	CMOS 90 nm	8-12	65.3	3.3	-64.1	24	0.96
[7]	CMOS 65 nm	9.1-10.6	24.1	4.8	-33.6	66[a]	10.6[e]
This Work	**SiGe 250 nm**	**7-12.5**	**26**	**4**[d]	**-24.4**	**188**[a]	**2**

[a] single channel, [b] incl. BB and ADC, [c] est. graphically, [d] sim., [e] incl. 4TX and PLL,

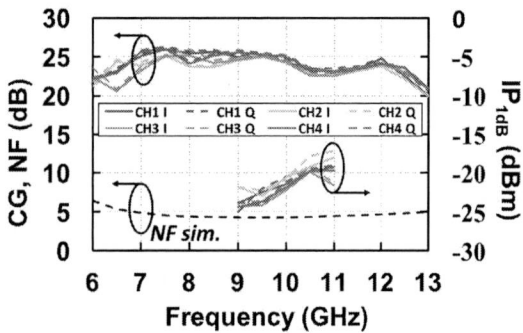

Fig. 10. Measured RX I/Q conversion gains, input referred compression point and simulated noise figure versus frequency.

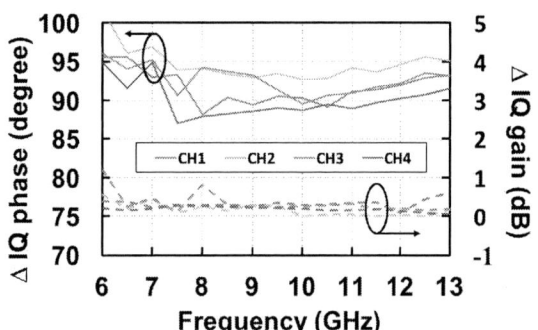

Fig. 11. Measured RX I/Q phase and gain balance.

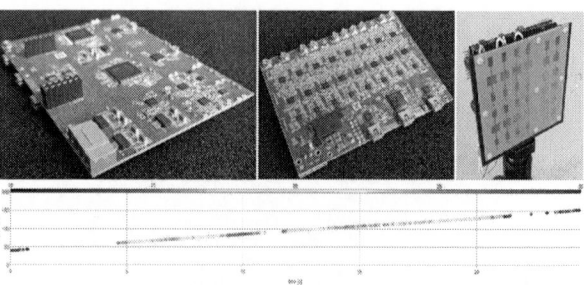

Fig. 12. X-band radar prototype with packaged TX and RX chips including patch antennas and baseband processing. The radar achieves 150 m maximum range.

ACKNOWLEDGMENT

We acknowledge NCBiR national funding (Contract number POIR.01.01.01-00-0393/18-00) that led to creation of the results presented in this paper.

REFERENCES

[1] H. Li and H. M. Rein, "Millimeter-wave VCOs with wide tuning range and low phase noise, fully integrated in a SiGe bipolar production technology," *IEEE J. Solid-State Circuits*, vol. 38, no. 2, pp. 184–191, Feb. 2003.

[2] M. Kucharski, A. Ergintav, W. A. Ahmad, M. Krstić, H. J. Ng, and D. Kissinger, "A scalable 79-GHz radar platform based on single-channel transceivers," *IEEE Transactions on Microwave Theory and Techniques*, vol. 67, no. 9, pp. 3882–3896, 2019.

[3] H. Rücker *et al.*, "A 0.13 SiGe BiCMOS technology featuring f_T/f_{max} of 240/330 GHz and gate delays below 3 ps," *IEEE J. Solid-State Circuits*, vol. 45, no. 9, pp. 1678–1686, Sep. 2010.

[4] D. Ma, F. F. Dai, R. C. Jaeger, and J. D. Irwin, "An X- and Ku-band wideband recursive receiver MMIC with gain-reuse," *IEEE Journal of Solid-State Circuits*, vol. 46, no. 3, pp. 562–571, 2011.

[5] J. Yu, F. Zhao, J. Cali, F. F. Dai, D. Ma, X. Geng, Y. Jin, Y. Yao, X. Jin, J. D. Irwin, and R. C. Jaeger, "An X-band radar transceiver MMIC with bandwidth reduction in 0.13 um SiGe technology," *IEEE Journal of Solid-State Circuits*, vol. 49, no. 9, pp. 1905–1915, 2014.

[6] P. Wang, Y. Shen, T. Wu, M. Chen, Y. Chang, D. Chang, and S. S. H. Hsu, "A low-power and low-noise X-band receiver MMIC in 90nm CMOS," in *2015 European Microwave Conference (EuMC)*, 2015, pp. 905–908.

[7] K. Tang, L. Lou, T. Guo, B. Chen, Y. Wang, Z. Fang, C. Yang, W. Wang, and Y. Zheng, "A 4TX/4RX pulsed chirping phased-array radar transceiver in 65-nm CMOS for X-band synthetic aperture radar application," *IEEE Journal of Solid-State Circuits*, vol. 55, no. 11, pp. 2970–2983, 2020.

The chips were used in an X-band radar prototype presented in Fig. 12. Phase and amplitude calibration is done during field testing by placing the radar in front of the corner reflector. The measurements results serve as a reference used to compensate for any phase and amplitude mismatch in the RX.

VI. CONCLUSION

A SiGe BiCMOS 2-channel TX and 4-channel I/Q RX were designed, fabricated and measured. The TX provides 9.25-10.8 GHz tuning range with worst-case phase noise of 105.5 dBc/Hz at 1 MHz offset and more than 6 dBm output power. The RX provides peak 26 dB conversion gain for each channel and more than 23 dB in 7-12.5 GHz range at minimum of -15 dBm LO input power. The RX performance is summarized in Tab. 1.

Two-Element 81-86 GHz SiGe Transmitter Beamformer for Backhaul Applications

Roee Ben Yishay, Oded Katz, Danny Elad

ON Semiconductor, Image Sensor Group, Haifa, Israel
roee.ben-yishay@onsemi.com

Abstract — **This paper presents the design and characterization of a two-element 81-86 GHz phased-array beamformer in SiGe BiCMOS for backhaul communication applications. Each transmitter path integrates power-amplifier (PA), power detector, variable attenuators, driver amplifier and a 6-bit phase shifter. The TX front-end achieves 24 dB small signal gain, gain bandwidth of 12 GHz, 17 dBm saturated output power and 14 dBm output-referred 1 dB compression point. The phase shifter achieve full 360° phase span with 6-bit phase resolution, 1.9° rms phase error and 0.11 dB rms gain error at 83 GHz. The IC occupies area of 5.1 mm² (including pads) and consumes 312 mW per element at P1dB.**

Keywords — **phased-array, transmitter, phase shifter, SiGe BiCMOS, millimeter wave integrated circuits, E-Band**

I. INTRODUCTION

The E-band is a licensed band with total of 10 GHz bandwidth (71-76 GHz, 81-86 GHz) for high capacity wireless links. Recently, it has been proposed for ground-to-air backhaul links to provide internet access to areas where infrastructure might be economically unfeasible [1]. A scalable phased array is required for backhaul base stations to obtain the desired SNR and EIRP through spatial power combining and interference mitigation. SiGe BiCMOS technology is an optimal candidate for such application, as it offers high-speed and high output power HBTs while enabling large-scale integration level. So far, Si-based beamformers at W/E-Band [2]-[5] demonstrated only low-power (P1dB~4 dBm) transmitters, which lead to large number of elements, multiple IC's and in turn, higher assembly complexity. Moreover, recent research using 64-256 elements phased-arrays [4] showed that a fully-integrated solution is very challenging in terms of heat dissipation and IC yield. An alternative more flexible approach is utilizing 2-4 element beamformer sub-arrays [6]-[7]. Besides higher yield and more uniform heat distribution over the PCB, this approach also greatly simplifies testing prior assembly.

This work demonstrates an 81-86 GHz two-element transmitter (TX) front-end in SiGe BiCMOS, featuring high output power (P1dB~14 dBm), 30 dB amplitude control and high precision 6-bit phase resolution. The high output power enables significant reduction in the number of elements used, thereby reducing cost and complexity.

The presented transmitter is a part of a 32-element transceiver designed to support 50 dBm EIRP (at 1 dB compression) for +2 km links.

II. TRANSMITTER CIRCUITS

The IC is designed and fabricated in Global-Foundries 120 nm SiGe8XP technology that provides high speed HBTs with f_T/f_{max}=260/340 GHz. The process offers 7-layer metallization with a 4 µm thick Al upper metal.

The beamformer includes two TX paths, connected by a Wilkinson power divider to a single RF input (Fig. 1). Each TX path includes power amplifier (PA) with power detector, driver amplifier, variable attenuator, 6-bit fine-tuned phase shifter (0-180°) and phase inverter (1-bit phase shifter). All sub-blocks make use of high-Q MIM capacitors and G-CPW transmission lines for matching, realized between the topmost Al layer (signal) and the 5th Cu layer (ground).

In the following sections detailed description is given for the main RF circuits, as well as measurement results.

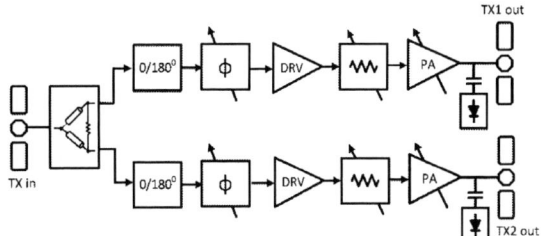

Fig. 1. Block diagram of the two-element transmitter

A. Power Amplifier and Attenuator

The PA (Fig. 2a) consists of five common-emitter (CE) stages, operating with 1.6 V and PTAT biased using λ/4 t-lines. The PA, based on [8], provides 24 dB small signal gain (Fig. 3a), 14 dBm OP1dB, 17 dBm Psat and 14% peak PAE (Fig. 3b) at 84 GHz. To maximize the output power, an in-phase power-combining is applied in the output stage by splitting the power into two transistors and combining back by a power combiner with a parallel 100 Ω resistor for common-mode stability.

B. Attenuator

The PA includes an attenuator to control the output power back-off level. It is inserted between the 2nd and 3rd PA stages by using the matching network series transmission line both for resonating out the 'off' capacitance of the attenuator's FET

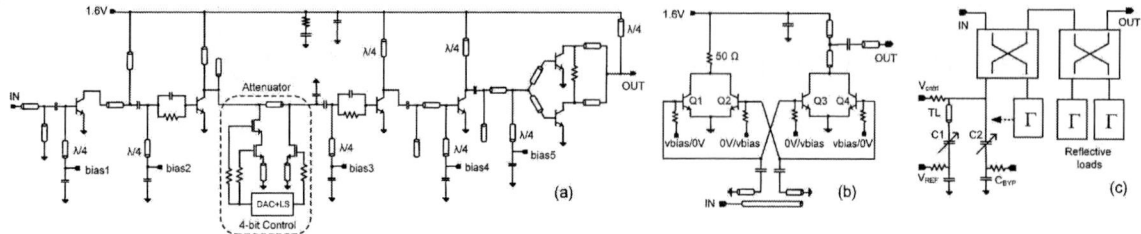

Fig. 2. Schematic of the (a) PA with embedded attenuator (dashed), (b) phase inverter and (c) phase shifter

devices and performing the required impedance transformation. The attenuator utilizes two stages with digital control of the attenuation level. The first stage comprises a stacked thin oxide nMOS while the second utilizes thick oxide nMOS (Fig. 2a). Both stages use a small inductive degeneration for phase compensation and controlled by a 4-bit DAC followed by a level shifter (LS) circuit that triggers the devices with an offset, which enhances the attenuator linearity. Measurements on the PA breakout shows 15 dB gain tuning range with negligible effect on matching. A similar attenuator (with modified matching) is added between the PA and the driver for amplitude tapering and gain compensation (if needed).

C. Phase Inverter

The phase inverter (PI) is placed first in the chain to buffer the Wilkinson divider from variation of the phase shifter input impedance over different phase states The PI (Fig. 2b) comprises a sub-quarter wave balun and two switching CE amplifier pairs [9]. In each state (0°/180°) only two stages are active - one CE is serving as the RF path, either from the 0° or 180° outputs of the balun, while another CE stage is used as a load replica to balance the impedance seen by the balun's outputs. This arrangement allows maintaining a wideband and near-perfect amplitude and phase balance, limited by the balun performance. Fig. 4 shows the measured gain and phase in 0°/180° modes. The gain imbalance at 81-86 GHz is <0.25 dB while the phase imbalance is <1.5°.

D. Phase Shifter

The reflection-type topology is chosen for the 0-180° phase shifter (Fig. 2c) for its high linearity and fine phase resolution. The design is described in [10]. It uses two cascaded Lange couplers with reactive L-C-L loads, optimized for low IL variation. The accumulation-mode MOS varactors are controlled by 6-bit voltage DAC (LSB can be used for phase trimming). The IL vs. phase shift in different frequencies is plotted in Fig. 5. The IL is 7.2-8.3 dB at 81-86 GHz across all phase states. The IL variation (up to 180° phase shift) over different phase states is up to ±0.4 dB. This low amplitude variation is achieved without any means of calibration and was observed in multiple breakout measurements.

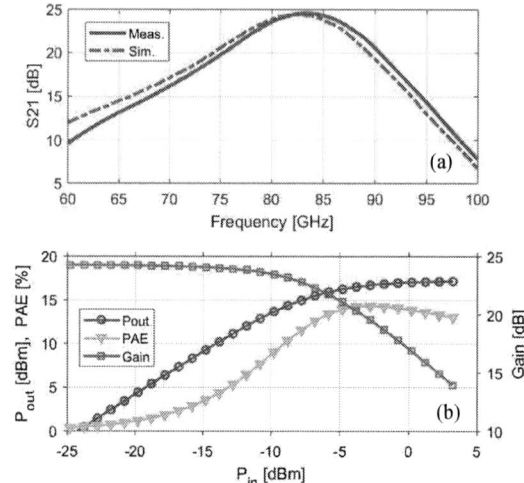

Fig. 3. PA measured small signal gain (a) and power sweep at 84 GHz (b)

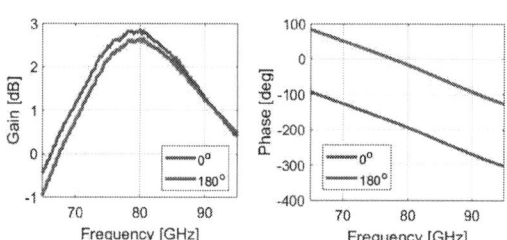

Fig. 4. Phase inverter measured gain and phase

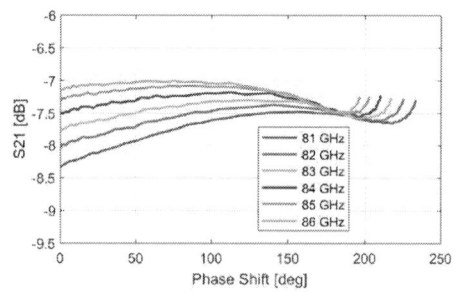

Fig. 5. Phase shifter measured IL vs. phase shift

III. TRANSMITTER MEASUREMENTS

The IC, shown in Fig. 6 occupies area of 5.1 mm² and characterized using on-wafer probing. S-Parameters measurements are performed using Keysight PNA-X N9030A. Large signal measurements are also performed with the same setup, using the calibrated power sweep capabilities. Each TX channel consumes 140 mA from 1.6 V in small signal and 195 mA at 1dB compression point. The chain's insertion phase and small signal gain (including the Wilkinson divider loss) at all 128 phase states are plotted in Fig. 7 and Fig. 8, respectively. A peak gain of 23-24 dB, is achieved at 81-83 GHz (depending on phase state) and drops by up to 2 dB at 86 GHz. The input and output matching of the TX are below -15 dB and -11 dB (Fig. 9), respectively, and show minor dependency on phase state due to the reverse isolation provided by the PI and PA. The maximum amplitude variation over phase states at 81-86 GHz is ±0.55 dB, which corresponds to 0.23 dB maximum RMS error (0.15 dB averaged over frequency), as plotted in Fig. 10. Using the LSB as trim bit, the phase settings are optimized for 5.6° resolution in 1 GHz steps and are loaded from the on-chip memory according to the central PLL frequency setting (not included in this work). The RMS phase error is on average 1.9° and plotted in Fig. 10. The TX amplitude invariance with respect to phase is illustrated by the 6-bit complex S21 constellation at different frequencies, shown in Fig. 11.

The amplitude tuning of the front-end is demonstrated with the attenuator embedded between the PA stages. Fig. 12a shows the gain profile for several attenuation states. The attenuator enables the gain to be varied by 15 dB and measurement show no variation in peak gain frequency for different settings. The resulting RMS phase error associated with the gain control is lower than 2.4° (Fig. 12b). With the second attenuator the front-end has in total 30 dB gain control range.

Both TX channels were characterized under large signal to verify performance uniformity. Output P1dB is in the range of 13.3-14.4 dBm at 81-86 GHz with up to 0.3 dB variation between channels (Fig. 13). The saturated power is measured to be 16-17.2 dBm with up to 0.2 dB variation.

The coupling (leakage) between the two channels is evaluated in terms of gain and phase error in channel 1 (kept in constant phase and measured for 2-port S-parameters) in the presence of 0-360° 6-bit phase sweep performed in channel 2 (left unterminated). The measured maximum gain and phase deviations (from 0° reference, over all phase states) are plotted

in Fig. 14 with peak values of 0.32 dB and 1.2°, respectively, at 81–86 GHz. An isolation of 38-40 dB is measured between the channels (not shown for brevity). The high isolation is obtained by using shielded transmission lines, ground planes separation and oxide deep-tranches patterns between the channels, but assumed to be degraded somewhat once the IC is packaged (which is not within this paper scope).

Table 1 summarizes the measured results with comparison to previously reported state of the art SiGe W/E-Band beamformer transmitters.

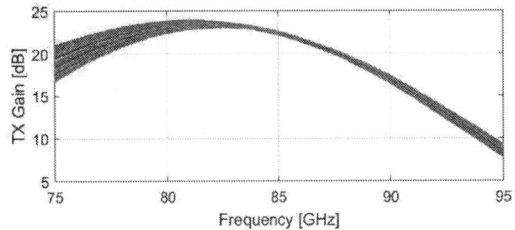

Fig. 7. Measured TX gain at all 128 phase states

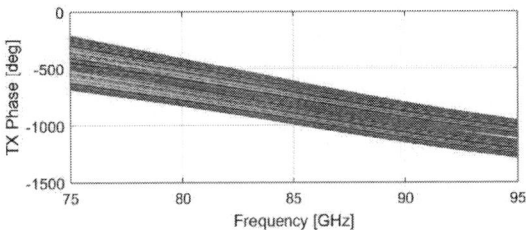

Fig. 8. Measured TX phase at all 128 phase states

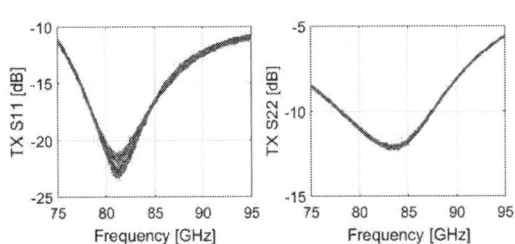

Fig. 9. Measured TX input/output matching at all 128 phase states

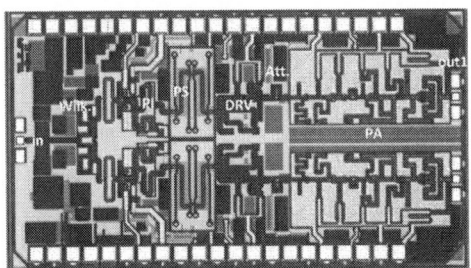

Fig. 6. Two-element transmitter micrograph (3 mm × 1.7 mm)

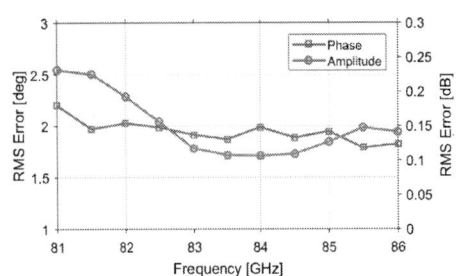

Fig. 10. Measured TX RMS phase/amplitude error

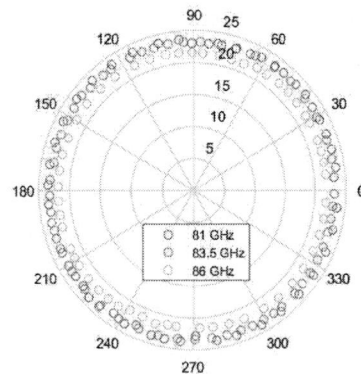

Fig. 11. Measured TX 6-bit S21 constellation

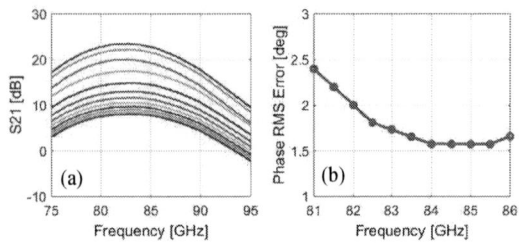

(a)   (b)

Fig. 12. Measured TX gain in different attenuator states (a) and corresponding RMS phase error (b)

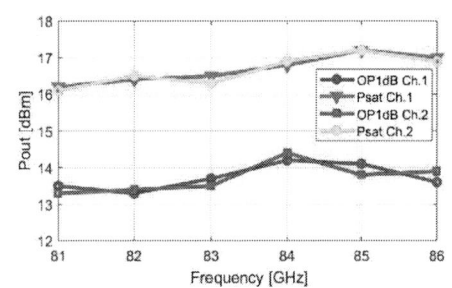

Fig. 13. Measured TX1 and TX2 output P1dB and Psat

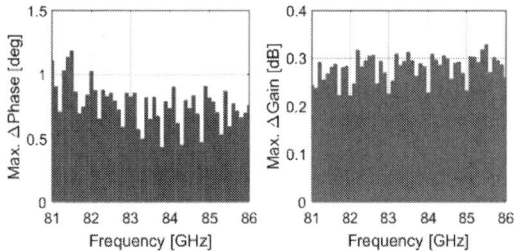

Fig. 14. Measured TX1 phase and amplitude variation (maximum over all phase states) due to TX2 phase sweep

IV. CONCLUSION

An integrated 81-86 GHz two-element beamformer for backhaul base stations applications, is demonstrated in 120 nm SiGe BiCMOS technology. To the best of our knowledge, it achieves simultaneously the highest output power and dynamic

range reported to date, with low gain and phase errors. It is a part of an E-Band scalable phased array that allows broadband wireless links for communication at data-rates of multi Gb/s.

Table 1. Results Summary and comparison

Single element performance	This work	[2]	[3]	[6]
Process	SiGe 0.12 μm	SiGe 0.18 μm	SiGe 0.12 μm	SiGe 90 nm
Freq. (GHz)	81-86	70-100	93-96	71-86
Gain (dB)	24	25	35	20
OP1dB (dBm)	14	-	4	7.5
Psat (dBm)	17	8	7.8	10
Gain cntrl. (dB)	30	-	15	-
Phase resolution (°)	5.6	22.5	1.6	5
Phase RMS error (°)	1.9	2.7	0.77	2
Amp. RMS error (dB)	0.15	0.57	0.36	0.4
P_{DC} (mW)	312$^{(*)}$	250	154	386
size (mm²)	2.55	-	1.5$^{(**)}$	1.78

(*) at P1dB (**) Evaluated from chip photograph

REFERENCES

[1] I. Thomson, (Jan. 2016), Loons in balloons: Google asks FCC to Approve Net plan. [Online]. Available: https://www.theregister.com

[2] S. Shahramian, M. J. Holyoak and Y. Baeyens, "A 16-Element W-Band Phased-Array Transceiver Chipset With Flip-Chip PCB Integrated Antennas for Multi-Gigabit Wireless Data Links," in *IEEE Transactions on Microwave Theory and Techniques*, vol. 66, no. 7, pp. 3389-3402, July 2018

[3] W. Lee et al., "Fully Integrated 94-GHz Dual-Polarized TX and RX Phased Array Chipset in SiGe BiCMOS Operating up to 105 °C," in *IEEE Journal of Solid-State Circuits*, vol. 53, no. 9, pp. 2512-2531, Sept. 2018

[4] S. Zihir, O. D. Gurbuz, A. Kar-Roy, S. Raman and G. M. Rebeiz, "60-GHz 64- and 256-Elements Wafer-Scale Phased-Array Transmitters Using Full-Reticle and Subreticle Stitching Techniques," in *IEEE Transactions on Microwave Theory and Techniques*, vol. 64, no. 12, pp. 4701-4719, Dec. 2016

[5] S. Shahramian, M. J. Holyoak, A. Singh and Y. Baeyens, "A Fully Integrated 384-Element, 16-Tile, W-Band Phased Array With Self-Alignment and Self-Test," in *IEEE Journal of Solid-State Circuits*, vol. 54, no. 9, pp. 2419-2434, Sept. 2019

[6] N. Ebrahimi, P. Wu, M. Bagheri and J. F. Buckwalter, "A 71–86-GHz Phased Array Transceiver Using Wideband Injection-Locked Oscillator Phase Shifters," in *EEE Transactions on Microwave Theory and Techniques*, vol. 65, no. 2, pp. 346-361, Feb. 2017

[7] S. Afroz and K. Koh, "A D-Band Two-Element Phased-Array Receiver Front End With Quadrature-Hybrid-Based Vector Modulator," in *IEEE Microwave and Wireless Components Letters*, vol. 28, no. 2, pp. 180-182, Feb. 2018

[8] R. B. Yishay and D. Elad, "A Variable Gain E-Band Power Amplifier using Highly Linear Embedded Attenuator," *2019 IEEE 19th Topical Meeting on Silicon Monolithic Integrated Circuits in RF Systems (SiRF)*, Orlando, FL, USA, 2019, pp. 1-3

[9] R. Ben Yishay and D. Elad, "D-Band 360° Phase Shifter with Uniform Insertion Loss," *2018 IEEE/MTT-S International Microwave Symposium - IMS*, Philadelphia, PA, 2018, pp. 868-870.

[10] R. Ben Yishay and D. Elad, "E-band reflection-type phase shifter with uniform insertion loss," *2018 IEEE 18th Topical Meeting on Silicon Monolithic Integrated Circuits in RF Systems (SiRF)*, Anaheim, CA, USA, 2018, pp. 75-78

Proceedings of the 16th European Microwave Integrated Circuits Conference

A W-Band Single-Chip Receiver in a 60 nm GaN-on-Silicon Foundry Process

Robert Malmqvist[#1], Rolf Jonsson[#2], Mingquan Bao[*3], Rémy LeBlanc[£4], Koen Buisman[€5], Christian Fager[€6], Kristoffer Andersson[*7]

[#]Swedish Defence Research Agency (FOI), Sweden

[*]Ericsson Research, Ericsson AB, Sweden

[£]OMMIC S.A.S., France

[€]Chalmers University of Technology, Sweden

{[1]rma, [2]roljon}@foi.se, {[3]mingquan.bao, [7]kristoffer.andersson}@ericsson.com, [4]r.leblanc@ommic.com, {[5]koen.buisman, [6]christian.fager}@chalmers.se

Abstract — **This paper presents a compact W-band heterodyne receiver MMIC realised in a 60 nm GaN-on-Si foundry process. The single-chip receiver consists of an LNA with a measured NF of 4.4-5.5 dB at 90-95 GHz and a resistive down-conversion mixer with a frequency doubler for the multiplication of the LO signal. The measured receiver conversion gain is 0-7.4 dB/0-6.4 dB at 75-91 GHz when the LO/2 power is 20.5 dBm (IF=5 GHz/2 GHz). The measured receiver 1 dB compression point is found to occur at an input power level of -12 dBm and -9 dBm at 80 GHz and 91 GHz, respectively (IF=2GHz). To the best of our knowledge, this single-chip receiver achieves the smallest size (4 mm^2), widest RF and IF bandwidths and highest linearity among reported GaN single-chip receivers in this frequency range. The receiver noise figure can be reduced by using an alternative GaN-on-Si LNA design with a measured average NF of 3.6-4.0 dB at 75-95 GHz and a measured maximum input 1 dB compression point of -3 dBm at 95 GHz.**
Keywords — **MMIC, low-noise amplifier, frequency conversion, linearity, Gallium Nitride, millimeter-wave.**

I. INTRODUCTION

Gallium Nitride-on-Silicon (GaN-on-Si) semiconductor processes offer several attractive features for Monolithically Microwave Integrated Circuits (MMICs) addressing wireless communications and radar applications up to the millimetre (mm)-wave range incl. 5G NR solutions [1]-[6]. The maximum power handling of GaN-on-Si high electron mobility transistors (HEMTs) is typically below that of GaN-on-Silicon Carbide (GaN-on-SiC) devices while silicon wafers are less expensive and more accessible on a worldwide market. GaN-on-Si processes can scale the substrate dimensions up to 6-8" and higher in order to reduce the chip fabrication cost [2]-[3]. Furthermore, it can be more cost- and power-efficient to use the same technology for transmit and receive modules, especially in the mm-wave range where applications such as 5G may benefit from using electronically steerable active antenna arrays. Several of the reported GaN-on-Si MMICs and processes are targeting applications up to 40 GHz while only a few studies investigated possible extensions of such technology up to W-band (75-110 GHz). W-band low noise amplifier (LNA), power amplifier and up-converter circuits were recently realised in OMMIC's 60 nm gate length GaN-on-Si process [3]-[5].

Compared with other semiconductor technologies, GaN LNAs are more robust since they can sustain higher input power levels and may be used to avoid a protective power limiter thereby reducing the receiver noise figure (NF) and complexity [6]. Single-chip GaN front-ends can enable more highly integrated active antenna arrays at these frequencies where the package loss and cost are higher [6]-[7]. A 100 nm GaN-on-Si LNA obtained an NF of 4.5-5.2 dB at 78-90 GHz and a 40 nm GaN-on-SiC LNA reported an NF < 4 dB at W-band [6], [8]. A single-chip receiver fabricated in a 100 nm GaN-on-SiC technology provided 0-11 dB of conversion gain at 75-80 GHz and the input 1dB compression point (P$_{1dBin}$) was -14 dBm at 77 GHz [9]. A W-band GaAs single-chip receiver has a simulated conversion gain and average NF of 6 dB and 5 dB at 77-92 GHz (P$_{1dBin}$=-15 dBm) [10]. In this work, we present measured and simulated results of a W-band single-chip receiver (LNA, mixer and frequency doubler) realised in a 60 nm GaN-on-Si process. To the best of our knowledge, this is the first time experimental results of such a compact GaN-on-Si single-chip receiver (4 mm^2) are reported up to 100 GHz. The LNA used in this single-chip receiver has a measured NF of 4.4-5.5 dB at 90-95 GHz whereas an alternative 60 nm GaN-on-Si LNA design has a lowest measured average NF of 3.6-4.0 dB at 75-95 GHz and P$_{1dBin}$ ≤-3 dBm showing that further enhancements are feasible. Sections II, III and IV present the results of the receiver parts (LNA and down-converter) and the combined receiver MMIC. Finally, our conclusions of the results are given in section V.

II. 60 NM GAN-ON-SI W-BAND LOW NOISE AMPLIFIER

The fabricated LNAs and single-chip receiver MMICs were realised using OMMIC's 60 nm GaN-on-Si process (D006GH). The current gain cut-off frequency (f$_T$) of 60 nm gate length HEMTs within this process is 130 GHz and the maximum drain current is 1.6 A/mm. The 100 μm thick silicon substrate allows cost-effective processing compatible with 6-8 inch wafers. Grounded co-planar waveguides (GCPWs) are used in the LNA and RF part of the down-converter to supress undesired higher-order transmission modes. CPWG lines have up to 0.6 dB/mm of measured losses below 110 GHz [4].

3–4 April 2022, London, UK

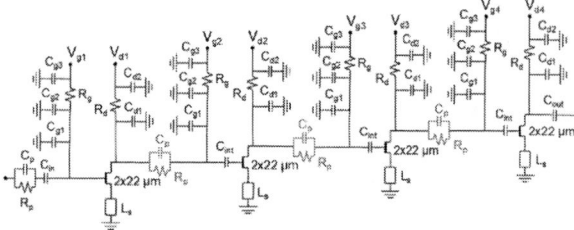

Fig. 1. Simplified circuit schematic used for two W-band LNA designs implemented in a 60 nm GaN-on-Si MMIC process (LNA1 and LNA2).

Figure 1 depicts a simplified schematic used for two W-band LNA designs fabricated in this 60 nm GaN-on-Si process (denoted LNA1 and LNA2). The LNA1 circuit has also been combined with a frequency down-converter circuit in a W-band single-chip heterodyne receiver MMIC (see Fig. 6). The LNA topology consists of four cascaded common-source stages with a transistor gate width of 2×22 μm in all the stages and was selected to provide a high gain and reasonably low NF over a relatively wide band [4], [11]. The gate and drain voltages (V_{gi} and V_{di}) are applied using separate DC bias lines that each are connected to a tee-junction close to the gate and drain terminals. The matching networks are realised using the capacitors C_p, C_{in}, C_{g1}, C_{g2}, C_{g3}, C_{d1}, C_{d2}, C_{int}, C_{out} and transmission lines (not shown in the schematic). The on-chip resistors $R_p=R_g=50$ Ω and $R_d=20$ Ω are used to ensure the amplifier is unconditionally stable. The resistor R_p and capacitor C_p are only used in the first stage of LNA1 while they are used in all the stages of another LNA circuit with modified component values in the matching networks (LNA2). Two lines (L_s) each with a length of 21 μm are connected between the source terminals and ground plane to minimise the noise figure using inductive source peaking.

The RF measurements of the two fabricated 4-stage LNAs (and the single-chip receiver shown in Fig. 6) were done after assembly in test fixtures with off-chip decoupling capacitors and the DC connections to the chip provided using bond-wires. The LNA s-parameter characterisation was done with RF probes connected to an HP 8510C/8510XF network analyzer and an on-wafer TRL cal-kit. The active circuits were simulated in ADS using the EM-simulated data of the passive parts and a transistor model in the foundry process design kit. Figure 2 shows measured s-parameters of the LNA1 and LNA designs when $V_d=10$ V and I_d is equal to 70 mA and 30 mA, respectively. For LNA1 the measured $|s_{21}|$ is then equal to 23.2 dB at 74.4 GHz and 15.0 dB at 94.9 GHz (25.2 dB and 10.2 dB simulated). LNA2 has on the other hand slightly lower measured gain at 70-90 GHz but some dBs higher measured gain above 90 GHz (18.3 dB and 18.7 dB at 74.4 GHz and 94.9 GHz, respectively). The measured $|s_{11}|$ and $|s_{22}|$ of LNA1 when $V_d=10$ V and $I_d=70$ mA are between -7.2 dB to -0.6 dB and -32.9 dB to -14.6 dB at 75-95 GHz, respectively. The measured $|s_{11}|$ and $|s_{22}|$ of LNA2 when $V_d=10$ V and $I_d=30$ mA are between -10.3 dB to -2.6 dB and -8.8 dB to -14.2 dB at 75-95 GHz, respectively. Compared with LNA1, LNA2 has a higher gain above 90 GHz and its input matching is better at 75-88 GHz when I_d and the DC power consumption (P_{DC}) are reduced with more than a factor of two.

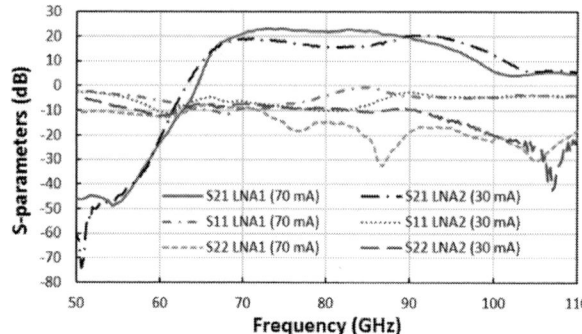

Fig. 2. Measured s-parameters of two W-band LNAs (LNA1 and LNA2) realised in a 60 nm GaN-on-Si MMIC process ($V_d=10$ V, $I_d=30$ mA and 70 mA).

The LNA Noise Figure measurements were done using a Keysight N9041B spectrum analyser (SA) and similar setups as in [4]. The LNA1 and LNA 2 circuits were measured using a 90-100 GHz noise source (Millitech NSS-10-R1520) and a 75-110 GHz noise source (ELVA-1 ISSN-10), respectively. Figure 3 shows measured and simulated NF of LNA1 and LNA2 when $V_d=10$ V and $I_d=70/110$ mA and 30mA/60 mA, respectively. The measured NF of LNA2 is equal to 2.5-4.7 dB/3.0-5.0 dB (3.6 dB and 4.0 dB of average values) at 75-95 GHz with this DC bias. LNA1 has a measured NF of 4.4-5.5 dB at 90-95 GHz when I_d is 70 mA and it is 2-3 dB higher when I_d is 110 mA.

The linearity measurements of LNA1 were done using an RF source connected to an up-converter and the calibrated output power level was measured with a W-band power sensor using the same setup as in [4]. This resulted in a measured P_{1dBout} of 9 dBm and 13 dBm at 90-95 GHz when $V_d=10$ V and I_d was 55 mA and 110 mA, respectively (the measured P_{1dBin} is between -10 dBm and 0 dBm at 90-95 GHz). The LNA2 linearity measurements were done with a network analyser and frequency extenders. The calibrated output power level was measured with the same SA as in the NF measurement setup. The measured P_{1dBout} of LNA2 at 75/80/85/90/95 GHz occurred at 3-10 dBm and 7-12 dBm when $V_d=10$ V and I_d was 30 mA and 60 mA, respectively (measured P_{1dBin} is between -14 dBm and -3 dBm). The simulated LNA2 P_{1dBout} is equal to 4-12 dBm and 11-16 dBm at 75-95 GHz, respectively, with this DC bias.

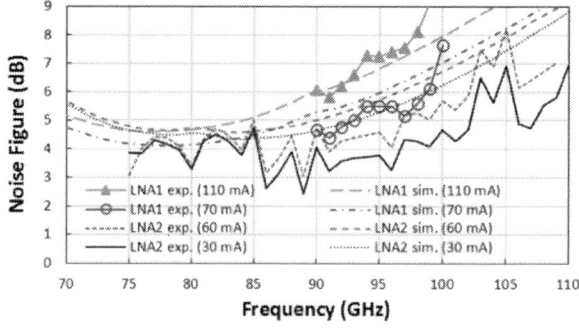

Fig. 3. Measured and simulated NF of two W-band LNAs (LNA1 and LNA2) realised in a 60 nm GaN-on-Si MMIC process ($V_d=10$ V, $I_d=30$ mA, 60mA, 70 mA and 110 mA, respectively).

III. 60 nm GaN-on-Si W-Band Down-Converter

Figure 4 shows a circuit schematic of the frequency down-converter used in the fabricated 60 nm GaN-on-Si W-band single-chip receiver (see Fig. 6). The down-converter is based on a balanced resistive mixer where the two transistors Q1 are fed by the LO signal. The LO is generated with a balanced cascode frequency doubler using two pairs of transistors Q2 and Q3 that are connected 90° out of phase using a Lange-coupler. Due to the frequency doubling (from LO/2) the two branches will be 180° out of phase. The two transistors Q1 have their drain terminals connected to the RF signal via a DC block capacitor and a quarter wave-length impedance transformer (L4). The differential intermediate frequency signals (IF+ and IF-) are extracted through high RF impedance connections at the source terminals. The doubler and mixer drain currents (I_{d1}, I_{d2}) are controlled by the DC bias voltages V_1, V_2, V_3 and V_4.

Figure 5 shows simulated down-converter conversion gain at W-band with different power levels at the RF and LO/2 ports (P_{RF} and $P_{LO/2}$) when V_1=1 V, V_2=-2.8 V, V_3=3.9 V, V_4=6.0 V, and the total drain current (I_d=I_{d1}+ I_{d2}) is not exceeding 81 mA. The simulated receiver conversion gain using the measured LNA1 s-parameters (shown in Fig. 2) is also shown in the same graph. The simulated down-converter conversion loss is in the range of 12-20 dB at 75-110 GHz and the simulated receiver conversion gain lies between 2 dB and 10 dB at 75-95 GHz (P_{RF}=-30 dBm, $P_{LO/2}$=14.5-20.5 dBm, IF=2 GHz).

IV. 60 nm GaN-on-Si W-Band Single-Chip Receiver

Figure 6 shows a micrograph of a W-band single-chip receiver (LNA and down-converter) realised in a 60 nm GaN-on-Si process. The receiver conversion gain was measured with RF probes using a 4-port Rohde&Schwarz ZVA 67 network analyser. The RF input signal to the DUT was supplied via a ZVA-Z110E up-converter connected to two ports of the VNA (for the RF and LO signals). The down-converter LO/2 signal was supplied by the VNA (via a power amplifier) at a frequency of (RF-IF)/2 (the division by 2 is since the down-converter has an on-chip frequency doubler). Only one part of the differential output signal was measured at IF using a differential probe while the other part of the IF pad was terminated by a 50 Ω load (3 dB was added to measured IF power to compensate for this). Estimated values of RF losses in cables, transitions and probes were used to compensate the measured power levels of the DUT.

Figure 7 shows the measured conversion gain of the W-band single-chip receiver MMIC. This measurement was done with $P_{LO/2}$=19.5 dBm and P_{RF} increased from -30 dBm to -9 dBm (IF=2 GHz). The following DC biases were used for the LNA and down-converter: V_d=10 V, I_d=70 mA, V_1=1.2 V, V_2=-2.1 V, V_3=4.6 V, V_4=4.6 V, I_{d1}=78 mA and I_{d2}=4 mA. The measured receiver conversion gain is in the range of 0-6 dB at 75-90 GHz when P_{RF} does not exceed -20 dBm. The gain compression is seen up to 91 GHz when a higher P_{RF} is used. The measured results imply the down-converter losses are 1-7 dB higher than anticipated at 75-95 GHz. Figure 8 shows measured receiver conversion gain when P_{RF}=-12 dBm and $P_{LO/2}$ = 10.5-20.5 dBm (IF=2 GHz). The measured conversion gain is saturated (between 0-6 dB at 75-90 GHz) when $P_{LO/2}$=16-18 dBm or higher i.e. similar as simulated (see Fig. 5).

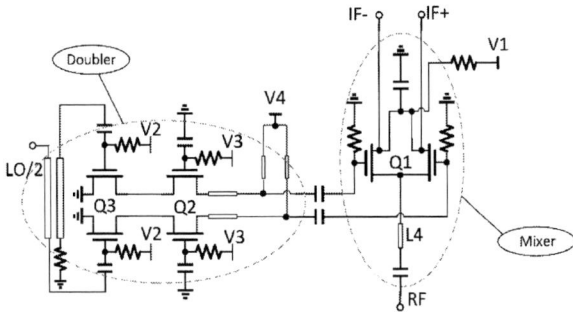

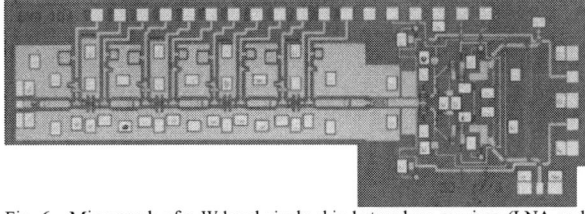

Fig. 6. Micrograph of a W-band single-chip heterodyne receiver (LNA and down-converter) realised in a 60 nm GaN-on-Si process (3.68 mm × 1.19 mm).

Fig. 4. Simplified circuit schematic of a W-band frequency down-converter (mixer and LO doubler) implemented in a 60 nm GaN-on-Si MMIC process.

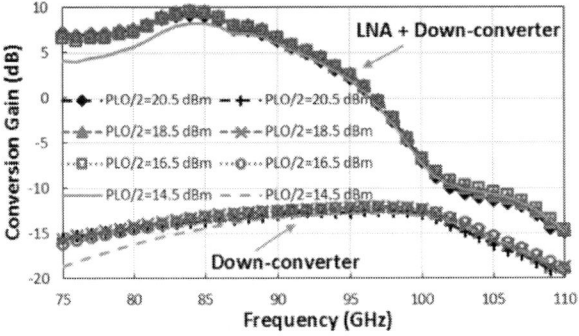

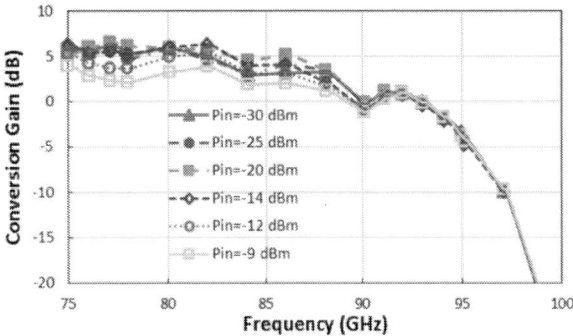

Figure 5. Simulated conversion gain of a W-band down-converter and a W-band single-chip receiver designed in a 60 nm GaN-on-Si process (P_{RF}=-30 dBm, $P_{LO/2}$=14.5-20.5 dBm, IF=2 GHz, LNA Vd=10 V, Id=70 mA, down-converter V_1=1.0 V, V_2=-2.8 V, V_3=3.9 V, V_4=6.0 V, $I_d \leq$81 mA.

Figure 7. Measured conversion gain of a 60 nm GaN-on-Si W-band receiver (LO/2=45 GHz, $P_{LO/2}$=19.5 dBm, IF=2 GHz, LNA V_d=10 V, I_d=70 mA, down-converter V_1=1.2 V, V_2=-2.1 V, V_3=4.6 V, V_4=4.6 V, I_{d1}=78 mA and I_{d2}=4 mA).

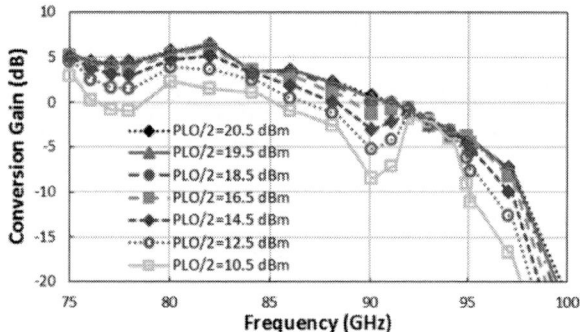

Figure 8. Measured conversion gain of a 60 nm GaN-on-Si W-band receiver (LO/2=45 GHz, P_{RF}=-12 dBm, IF=2 GHz, LNA V_d=10 V, I_d=70 mA, down-converter V_1=1.0 V, V_2=-2.8 V, V_3=3.9 V, V_4=6.0 V, Id_1=70-84 mA, I_{d2}=5 mA).

Figure 9 shows the measured receiver conversion gain at different values of IF (1-6 GHz) when P_{RF}=-12 dBm and $P_{LO/2}$=20.5 dBm. The measured receiver conversion gain is highest when IF is equal to 5 GHz and lowest when IF is 1GHz. The conversion gain is 7.4 dB at 80 GHz (IF=5 GHz) and between 0-7 dB at 75-91 GHz when an IF of 2-6 GHz is used. Table 1 compares the results of the 60 nm GaN-on-Si W-band single-chip receiver with an earlier reported 77 GHz GaN-on-SiC receiver MMIC [9]. The GaN-on-Si receiver has a wider RF bandwidth (75-91GHz) and can be used with different values of IF (1-6 GHz). The measured receiver $P_{1 dB}$ (IF=2 GHz) occurs at 2-5 dB higher input power levels (-12 dBm and -9 dBm at 80 GHz and 91 GHz, respectively). The area is 4 mm².

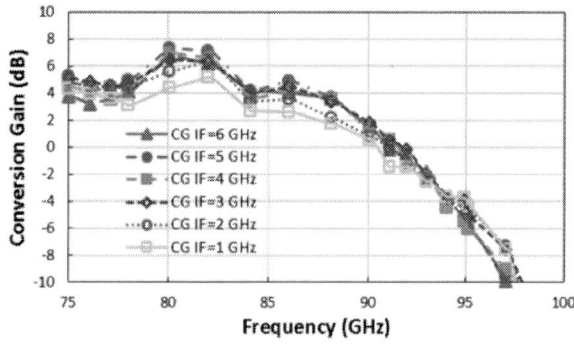

Figure 9. Measured conversion gain of a 60 nm GaN-on-Si W-band receiver (LO/2=45 GHz, P_{RF}=-12 dBm, P_{LO2}=20.5 dBm, LNA V_d=10 V, I_d=70 mA, down-converter V_1=1.0 V, V_2=-2.8 V, V_3=3.9 V, V_4=6.0 V, Id_1=83-89 mA, I_{d2}=5 mA).

Table 1. RF performance of W-band single-chip heterodyne receivers realised in 100 nm GaN-SiC and 60 nm GaN-Si based MMIC processes, respectively.

Reference	[9]	This work
Technology	100 nm GaN-on-SiC	60 nm GaN-on-Si
RF	75-80 GHz	75-91 GHz
IF	0.1 GHz	1-6 GHz
Conversion Gain	0-11 dB	0-7.4 dB (IF=5 GHz) 0-6.4 dB (IF=2 GHz)
Input P₁dB	-14 dBm @77 GHz	-12 dBm @80 GHz -9 dBm @91 GHz
Area	5 mm²	4 mm²

V. CONCLUSION

The performance of a W-band single-chip receiver in a 60 nm GaN-on-Si process is demonstrated for the first time. This receiver is composed of a wideband LNA and a down-converter with an integrated frequency doubler for the LO generation. The LNA has a measured gain of 10-23 dB at 66-98 GHz and a measured noise figure of 4.4-5.5 dB at 90-95 GHz when consuming 700 mW of DC power. The measured receiver conversion gain is between 0-7.4 dB (0-6.4 dB) at 75-91 GHz when the LO/2 power is 20.5 dBm and the IF is 5 GHz (2GHz). The measured receiver 1 dB compression point occurred at -12 dBm and -9 dBm of input power at 80 GHz and 91 GHz, respectively (IF=2 GHz). This single-chip receiver has benefits such as a relatively high conversion gain/linearity, wide RF/IF bandwidths and small circuit area. The receiver noise figure can be improved using a modified 60 nm GaN-on-Si LNA circuit with a measured average NF of 3.6-4.0 dB at 75-95 GHz and a measured maximum input 1 dB compression point of -3 dBm at 95 GHz while dissipating 300-600 mW of DC power.

ACKNOWLEDGEMENTS

Stig Leijon is acknowledged for assembly of MMICs in test fixtures. The European Union is acknowledged for funding of the H2020 ICT project SERENA (Grant Agreement no.779305)

REFERENCES

[1] T. Boles "GaN-on-Silicon present challenges and future opportunities", in *Proc. EuMIC'2017*, 2017, pp. 21-24.

[2] H.-F. Huang, X.-Y. Liu, J.-S. Shi, L.-Q. Zhang, S.-X. Zhao, M.-Z Lin, B. Wu, P.-F. Wang, "Investigation of a GaN-on-Si HEMT optimized for the 5th-generation wireless communication," *Proc. ASICON'2015*, 2015, pp. 1-4.

[3] R. Leblanc, M. El Kaamouchi, A. Gasmi, J. Moron, J. Poulain, O. Rehioui, F. Lecourt, P. Frijlink, F. Gamand, O. Leveille, "An industrial foundry offer for a 100nm GaN/Si process for applications up to V band," *9th wide band gap semiconductor and components workshop*, 2018.

[4] R. Malmqvist, R. Jonsson, A. Bernland, M. Bao, R. LeBlanc, K. Buisman, C. Fager, K. Andersson, "E/W-band CPW-based amplifier MMICs fabricated in a 60 nm GaN-on-Silicon foundry process," in *Proc. EuMIC'2020*, 2020, pp. 1-4.

[5] M. Bao, R. Malmqvist, R. Jonsson, J. Hansryd, K. Andersson, "A W-band up-conversion mixer with integrated LO frequency doubler in a 60 nm GaN technology," submitted *EuMIC'2021*.

[6] X. Tong, P. Zheng, L. Zhang, "Low-noise amplifiers using 100-nm gate length GaN-on-Silicon in W-band," *IEEE Microwave and Wireless Comp. Lett.*, vol. 30, pp. 957-960, Oct. 2020.

[7] A. Fung, L. Samoska, P. Kangaslahti, G. Sadowy, A. Brown, S. O'Connor, D. Gritters, "W-band Gallium Nitride MMIC amplifiers for cloud doppler radar arrays" in *Proc. IRMM-THz'2015*, 2015, pp. 1-1.

[8] A. Kurdoghlian, H. Moyer, H. Sharifi, D. F. Brown, R. Nagele, J. Tai, R. Bowen, M. Wetzel, R. Grabar, D. Santos, M. Micovic, "First demonstration of broadband W-band and D-band GaN MMICs for next generation communication systems", *Proc. IEEE MTT-S IMS' 2017*, 2017, pp. 1126-1128.

[9] I. Kallfass, R. Quay, H. Massler, S. Wagner, D. Schwantuschke, C. Haupt, R. Kiefer, O. Ambacher,, "A single-chip 77 GHz heterodyne receiver MMIC in 100 nm AlGaN/GaN HEMT technology," in *Proc. IEEE MTT-S IMS'2011'*, 2011, pp. 1-4.

[10] L. Li, Y. Gao, R. Qian, X. Sun, "W-band four-channel receiver MMIC in 0.1um GaAs PHEMT technology," in *Proc. APMC'2019*, 2019, pp. 1512-1514.

[11] I. Kallfass, H. Massler, S. Wagner, D. Schwantuschke, P. Brückner, C. Haupt, R. Kiefer, R. Quay, O. Ambacher, "A highly linear 84 GHz low noise amplifier MMIC in AlGaN/GaN HEMT technology," *Int. Microw. Workshop Series on Millimeter Wave Integr. Tech.*, 2011, pp. 144-147.

Proceedings of the 16th European Microwave Integrated Circuits Conference

A 30-to-38 GHz Active and Passive Combined Down-conversion Variable Gain Mixer with Low OP_{1dB} Variation in 65-nm CMOS

Mu-Heng Li, Chun-Nien Chen, Yunshan Wang, Huei Wang

Graduate Institute of Communication Engineering, National Taiwan University, Taipei 106, Taiwan

Abstract — **In this paper, a 30 to 38 GHz variable gain mixer with low OP_{1dB} variation in 65-nm CMOS technology for the 5G application is presented. An active and passive combined mixing cores (APC) technique is used to control the gain state and noise performance of the mixer. By using the APC structure, the proposed circuit can operate in variable gain states with low OP_{1dB} variation. The measured peak conversion gain is 4.3 dB with 3-dB bandwidth that covers from 30 to 38 GHz, and the measured effective gain control range is 5.8 dB with 1.2 dB OP_{1dB} variation at 38 GHz. The mixer consumes only 10.4-mW dc power, including the IF buffers, with a compact chip area of 0.62 × 0.74 mm².**

Keywords — **CMOS, variable-gain mixer (VGM), 5G mobile communication, mm-wave.**

I. INTRODUCTION

Recently, the milimeter-wave (MMW) spectrums of 28-GHz and 38-GHz bands draws extensive attention to future network with the need for fifth-generation (5G) cellular systems. Variable gain amplifiers (VGA) and variable-gain low noise amplifiers (VGLNA) play vital roles in receivers because of their gain control functions. However, although the VGA can achieve wide gain control range (GCR), the OP_{1dB} may drop when changing the gain state which lead the overall GCR of the beamformer is limited [1]. In addition, the dc-offset and power consumption is a critical problem when the VGA is used after the mixer in a receiver [2].

Typically, although the modified current steering technique can be used to change the gain and obtain a constant OP_{1dB}, the GCR is limited to a small range [3]. To enhance the GCR of the receiver, the star variable gain mixer was published in pHEMT process for Ka-band application [4]. Nevertheless, this structure can only achieve low GCR in narrow bandwidth. Another approach is to use switch in transconductance stage design which can reconfigure the gate width of the transistors to obtain three gain-states [5]. For wider GCR and more gain states, the dc free variable resistance was implemented in [2]. However, this structure may lead to high noise figure (NF) variation and critically OP_{1dB} variation when it operates in wide GCR. Thus, it is necessary to define an effective GCR to avoid affecting the overall receiver performance.

In this paper, we propose the dc free variable gain mixer with active and passive combined mixing cores (APC) technique. The APC technique does not provide gain control function only, but also obtains constant OP_{1dB}. The overall

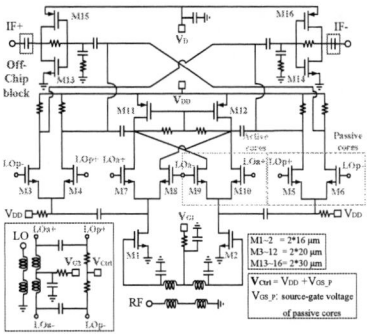

Fig. 1. Circuit Schematic of the proposed down-conversion mixer.

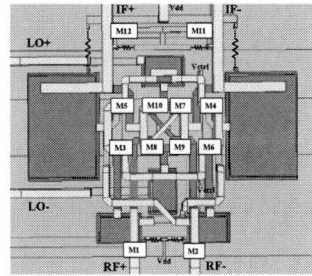

Fig. 2. Layout of proposed active and passive combined mixing.

benefits of this work are achieved by using APC technique and a competitive performance is achieved among published CMOS mixer and variable gain mixer. The details of circuit design and analysis are discussed in Section II with measurement results at Section III.

II. CIRCUIT DESIGN

Fig. 1 presents the schematic of the proposed variable gain mixer implemented in TSMC 65nm CMOS process. The variable gain mixer is composed of a double balanced mixer with IF buffers

The active and passive combined mixer is presented to achieve low flicker noise in [6]. Nevertheless, this structure may generate complicated harmonic product because the series active and passive mixing cores down-convert the frequency twice. The proposed structure is composed of active and passive core parallel-combined, as shown in Fig. 1. In addition, the layout of the proposed active and passive

3–4 April 2022, London, UK

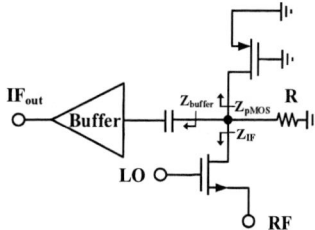

Fig. 3. Half circuit of the active mixing cores.

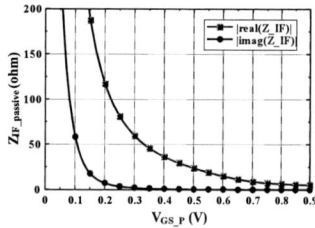

Fig. 4. Simulated Z_{IF} of passive mixing cores versus V_{GS_P}.

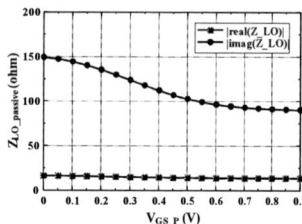

Fig. 5. Simulated Z_{LO} of passive mixing cores versus V_{GS_P}.

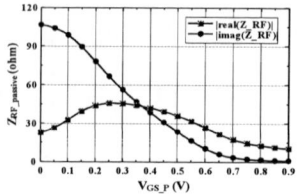

Fig. 6. Simulated Z_{RF} of passive mixing cores versus V_{GS_P}.

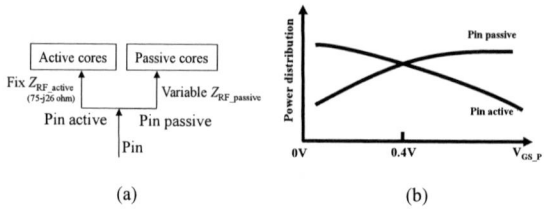

Fig. 7. (a) Schematic diagram of operation of the proposed architecture (b) input power delivered to active and passive cores.

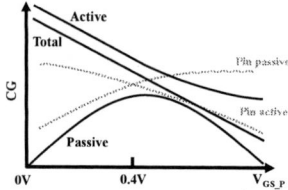

Fig. 8. CG and input power versus V_{GS_P}.

combined mixing cores is shown in Fig. 2. The source-drain voltages (V_{DS}) of passive mixing cores which are composed of the transistors M_3 to M_6 is 0 V and operate in the deep triode region. The active mixing cores which are composed of the transistors M_7 to M_{10} are biased with pMOS active load. Additionally, the dc-biasing of active and passive mixing cores does not affect each other by on-chip block.

To give a clear description, the operation of active mixing cores which provide positive gain can be analyzed by the half circuit as shown in Fig. 3. In general, the pMOS active load is self-biased and the gain of active mixing cores can be expressed as

$$\mathrm{Av}_{active}(j\omega) = \frac{2}{\pi} \times g_{m\_active} \times Z_{OUT}, \tag{1}$$

where

$$Z_{OUT} = Z_{IF} \| Z_{buffer} \| Z_{pMOS} \| R, \tag{2}$$

and Z_{OUT} is total output impedance which compose of the output impedance of mixing cores Z_{IF}, input impedance of buffer Z_{buffer}, pMOS active load Z_{pMOS}, and load R. If R is properly chosen, the Av_{active} is almost not affected by R. In addition, Z_{IF} can be expressed as

$$Z_{IF} = Z_{IF\_active} \| Z_{IF\_passive} \tag{3}$$

by paralleling the passive cores with active cores. To achieve the variable gain and dc free function, we can control the $Z_{IF\_passive}$ by changing the source-gate voltages (V_{GS}) of passive

mixing cores (V_{GS_P}). Fig. 4 shows the simulated $Z_{IF\_passive}$ versus V_{GS_P}, and the $Z_{IF\_passive}$ decrease significantly as V_{GS_P} rises until V_{GS_P} is greater than 0.5 V which means Av_{active} will also decrease as V_{GS_P} rises. Fig. 5 shows the simulated LO input impedance of passive mixing cores $Z_{LO\_passive}$ versus V_{GS_P}. The $Z_{LO\_passive}$ does not change much as V_{GS_P} rises which means the LO input power (P_{LO}) can be fixedly divided to both active and passive mixing cores.

Generally, passive mixing cores are high linearity [7] and active mixing cores have good conversion gain (CG). If we combine active and passive mixing cores, the combined mixing cores can be adjusted between CG and linearity. Fig. 6 shows the simulated RF input impedance of passive mixing cores $Z_{RF\_passive}$ versus V_{GS_P}, and the imaginary part of $Z_{RF\_passive}$ decrease significantly as V_{GS_P} rises until V_{GS_P} is greater than 0.5 V and real part of $Z_{RF\_passive}$ is not change much. In the proposed parallel-combined mixing cores, the RF input power is divided to the active and passive mixing cores and controlled by variable $Z_{RF\_passive}$ as shown in Fig. 7 (a). Fig. 7 (b) shows that the RF input power of passive cores increases gradually as V_{GS_P} rises since the decreasing of $Z_{RF\_passive}$ causes more RF current mixed by paralleled passive cores. As stated above, when V_{GS_P} rises, the combined mixing cores operates in passive cores, which means the higher IP_{1dB} and lower CG.

Fig. 8 shows CG and RF input power versus V_{GS_P}. Based on (1), the CG of active mixing cores changes with Z_{OUT} until V_{GS_P} is greater than 0.5 V, but the total CG still decreases by

56

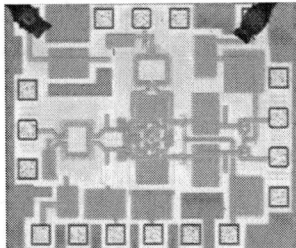

Fig. 9. Micrograph of the proposed mixer with a chip size 0.62 × 0.74 mm².

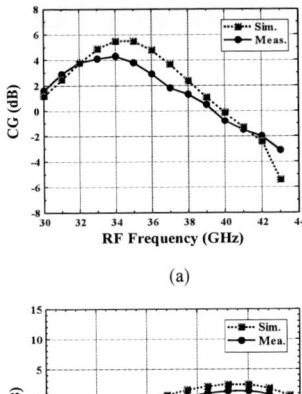

(a)

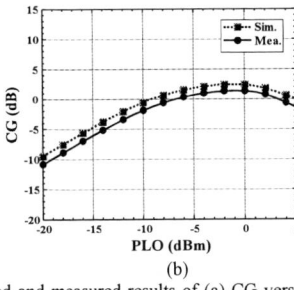

(b)

Fig. 10. Simulated and measured results of (a) CG versus RF frequency, (b) CG versus P_{LO} under -20 dBm P_{RF} as V_{GS_P} is 0 V.

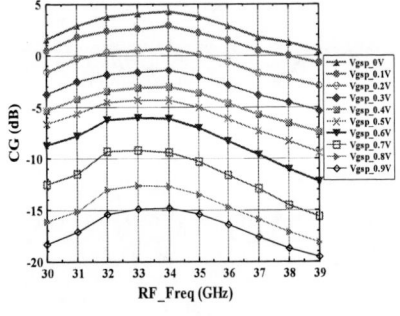

(a)

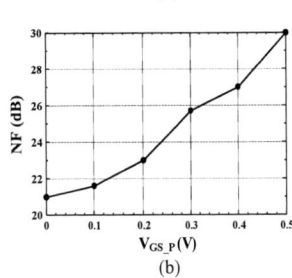

(b)

Fig. 11. (a) Measured CG versus RF frequency under different gain state. (b) Measured NF under different gain state at RF 38GHz.

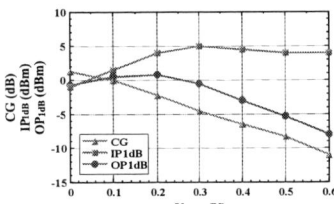

Fig. 12. Measured CG, IP_{1dB}, OP_{1dB} versus V_{GS_P}.

passive mixing cores. As V_{GS_P} is greater than 0.5 V, more RF input power is delivered to passive mixing cores and the total CG is dominated by the passive mixing cores.

In summary, by controlling the V_{GS_P}, the APC can trade-off CG and IP_{1dB}. Although, the APC structure can achieve high GCR, the IP_{1dB} improvement is limited by $Z_{RF\_passive}$ variation and transistors saturation in a specific GCR. Thus, the proposed APC structure can achieve low OP_{1dB} variation in an effectively GCR.

III. MEASUREMENT RESULTS

Fig. 9 shows the photograph of the proposed variable gain mixer implemented in 65-nm CMOS process with the overall chip area of 0.62 × 0.74 mm². The measurement results of the proposed mixer is obtained through on-wafer probing under 1-V and 1.69-V drain bias and consumes 10.4-mW dc power. The Keysight signal generators E8257D are used as RF and LO source, respectively. The N9041B UXA signal analyzer is used to measure the output power and noise figure (NF). Fig. 10(a) shows simulated and measured CG versus RF frequency, with fixed IF of 1 GHz and P_{LO} of 0 dBm. The peak CG is 4.3 dB at RF 34 GHz and the 3-dB bandwidth covering 30 to 38

GHz as V_{GS_P} is 0 V. Fig. 10(b) shows simulated and measured CG versus P_{LO}, with fixed RF of 38 GHz, LO of 37 GHz and RF input power (P_{RF}) of -20 dBm as V_{GS_P} is 0 V. Compare with simulated results, the measured CG agrees reasonably with simulation. In addition, when the P_{LO} is higher, the drain voltage of mixing cores will drive the IF buffer into low gain region which leads to voltage drop under large P_{LO} [8].

Fig. 11(a) shows measured CG versus RF frequency as V_{GS_P} increases from 0 to 0.9 V, the CG drops from 1.3 to -18.7 dB at 38 GHz and the CG flatness keeps well. Additionally, the CG curves of each gain state are similar because of the APC structure. Fig. 11(b) shows measured NF versus V_{GS_P}, at RF 38 GHz. The measured NF increases significantly when V_{GS_P} is higher than 0.5V, which is caused by variable Z_{OUT}. And a specific operating range should be defined for the variable gain mixer to avoid affecting the overall receiver performance. Fig. 12. shows measured CG, IP_{1dB}, OP_{1dB} versus V_{GS_P}, the IP_{1dB} increases as CG drops until V_{GS_P} is higher than 0.3 V, which makes the proposed APC structure exhibits a low OP_{1dB} variation in specific GCR. Therefore, we define the 5.8 dB effectively GCR which V_{GS_P}

Table 1. Performance comparison of recently published MMW CMOS mixers.

Ref.	Process (CMOS)	Design feature	RF freq. (GHz)	CG (dB)	GCR (dB)	NF(ΔNF) (dB)	IP_{1dB} (dBm)	OP_{1dB} (dBm)	ΔOP_{1dB} (dB)	P_{LO} (dBm)	P_{dc} (mW)
[9] MWCL'18	180-nm	Distributed derivative superposition	23~25	-4.5 ±0.6	N/A	N/A	-5*	-11	N/A	5	16
[10] APMC'18	180-nm	Current-bleeding neutralization	21.7~24.2	10.7	N/A	N/A	-13.6*	-3.9	N/A	3	23
[2] MWCL'14	65-nm	Variable resistance load	10~67	18.6 ±2.1	40	N/A	-28.5*	-8.8	N/A	-2.5	1.44*#
[11] EL'15	65-nm	Cross-coupled pair	60	7.5/3	4.5	N/A	-7.2	-0.7*	N/A	0	6.2
This work	65-nm	Active and passive mixing cores combined	30~38	4.3	5.8	21(4.7)	-1	-0.7	1.2	0	10.4

* estimated from figure, high CG mode/low CG mode, Δ: variation, #: power consumption of mixing cores

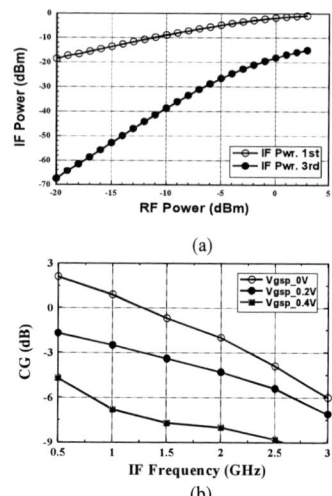

Fig. 13. (a) The measured IF power versus RF power at 38GHz RF frequency under high gain state. (b) The measured CG versus IF frequency under -20 dBm RF input power.

should be controlled between 0 to 0.3 V to achieve both 1.2 dB low OP_{1dB} and 4.6 dB NF variation.

Fig. 13(a) shows the two-tone measurements at RF frequency of 38 GHz ± 20 MHz with fixed LO 37 GHz, with P_{LO} of 0 dBm. The measured IIP_3 is 4 dBm at 38 GHz as V_{GS_P} is 0 V with power sweep measurement. Fig. 13(b) shows the measured CG versus IF frequency, with fixed LO 37 GHz and P_{LO} of 0 dBm.

Table I summarizes the performances of the proposed variable gain mixer and the recently published MMW CMOS mixers with and without variable gain function. Compared with other published CMOS mixers in recent years, the proposed mixer shows good CG, P_{1dB} and bandwidth with low power consumption. Compared with other published CMOS variable gain mixers in recent years, the proposed mixer shows low OP_{1dB} and NF variation, which play an important role in receiver.

IV. CONCLUSION

A 38 GHz variable gain mixer fabricated in 65-nm CMOS process is presented. By using APC structure in variable gain mixer design, it achieves the reasonable GCR and avoids down-converting the signal twice. In addition, based on the definition of effectively GCR, the OP_{1dB} and NF variation is considered for better overall receiver performance. The proposed mixer shows competitive performances with low OP_{1dB} and NF variation among the published CMOS mixers.

ACKNOWLEDGMENT

The chip of proposed mixer is fabricated by Taiwan Semiconductor Manufacturing Company (TSMC), Hsinchu, Taiwan through Taiwan Semiconductor Research Institute (TSRI) in Taiwan.

REFERENCES

[1] C. Chen *et al.*, "36–40 GHz Tx/Rx Beamformers for 5G mm-Wave Phased-Array," *2018 Asia-Pacific Microwave Conference (APMC)*, Kyoto, Japan, 2018.

[2] C. C. Boon and X. Yi, "A 10–67 GHz 1.44 mW 20.7 dB Gain VGA-Embedded Downconversion Mixer With 40 dB Variable Gain Range," in *IEEE Microwave and Wireless Components Letters*, vol. 24, no. 7, pp. 466-468, July 2014.

[3] Y. Chang, Y. Chen and H. Lu, "A 38 GHz Low Power Variable Gain LNA Using PMOS Current-Steering Device and Gm-Boost Technique," *2018 Asia-Pacific Microwave Conference (APMC)*, Kyoto, 2018.

[4] C. Lin, J. Chiu, C. Lin, Y. Lai and Y. Wang, "A Variable Conversion Gain Star Mixer for Ka-Band Applications," in *IEEE Microwave and Wireless Components Letters*, vol. 17, no. 11, pp. 802-804, Nov. 2007.

[5] M. Wang and C. E. Saavedra, "Reconfigurable broadband mixer with variable conversion gain," *2011 IEEE MTT-S International Microwave Symposium*, Baltimore, MD, 2011.

[6] D. Seo, J. Lee and T. Yun, "Active and Passive Combined Mixer for Low Flicker Noise and Low dc Offset," in *IEEE Microwave and Wireless Components Letters*, vol. 25, no. 7, pp. 463-465, July 2015.

[7] C. Chen, Y. Chen, T. Kuo and H. Wang, "A 35-39 GHz CMOS Linearized Receiver with 2 dBm IIP3 and 16.8 dBm OIP3 for the 5G Systems," *2019 14th European Microwave Integrated Circuits Conference (EuMIC)*, Paris, France, 2019.

[8] J. Kuo *et al.*, "Design and Analysis of Down-Conversion Gate/Base-Pumped Harmonic Mixers Using Novel Reduced-Size 180° Hybrid with Different Input Frequencies," in *IEEE Transactions on Microwave Theory and Techniques*, vol. 60, no. 8, pp. 2473-2485, Aug. 2012.

[9] H. Lin, Y. Lin and H. Wang, "A High Linearity 24-GHz Down-Conversion Mixer Using Distributed Derivative Superposition Technique in 0.18-CMOS Process," in *IEEE Microwave and Wireless Components Letters*, vol. 28, no. 1, pp. 49-51, Jan. 2018.

[10] D. Lin, K. Kao and K. Lin, "A K-Band High-Gain Linear CMOS Mixer with Current-Bleeding Neutralization Technique," *2018 Asia-Pacific Microwave Conference (APMC)*, Kyoto, 2018.

[11] C. Wang, Z. Li, Q. Li and Z. Wang, "60 GHz broadband variable gain mixer using positive feedback in 65 nm CMOS", *Electronics Letters*, vol. 51, 2015.

Proceedings of the 16th European Microwave Integrated Circuits Conference

Analog Linearization of a 10-W GaN Power Amplifier by Baseband Feedback

Mathani Eltayeb[#], Morten Olavsbråten[*], Gian Piero Gibiino[#], Alberto Santarelli[#]

[#]Dept. Electrical, Electronic, and Information Engineering, University of Bologna, Italy

[*]Norwegian University of Science and Technology (NTNU), Norway

mathani.eltayeb@studio.unibo.it, morten.olavsbraten@ntnu.no, {gianpiero.gibiino, alberto.santarelli}@unibo.it

Abstract — **This work presents an experimental investigation on power amplifier (PA) linearization using baseband feedback (BBF). The technique takes advantage of the even-order intermodulation tones produced at the baseband frequencies, which are proportional to the envelope amplitude of the radio-frequency (RF) input signal. These components are fed from the drain terminal to the gate terminal of the transistor of the considered single-stage PA. A suitable gain and phase shift are applied to minimize the third-order intermodulation products (IMD3) across the entire input power sweep range up to compression. The experimental results, carried out on a 10-W Gallium Nitride (GaN) High-Electron-Mobility Transistor (HEMT) proof-of-concept prototype, demonstrate improvements of up to 18 dBc in the in-band IMD3 for various tone spacings from 5 MHz to 20 MHz. In addition, the method shows a 3-dBc improvement in the Adjacent-Channel Power Ratio (ACPR) and 1% reduction in Error-Vector Magnitude (EVM) for a 1-MHz 16-QAM signal.**

I. INTRODUCTION

Radio-Frequency (RF) power amplifier (PA) designers often have to trade off the conflicting requirements of high power efficiency and high linearity. High power efficiency is important to minimize power consumption, preserving battery lifetime in handheld devices and limiting cooling in base stations. On the one hand, PAs should necessarily be driven into nonlinear operation to achieve high power efficiency. On the other hand, a demanding level of linearity, typically expressed in terms of maximum adjacent-channel power ratio (ACPR) or error-vector magnitude (EVM), is requested for communications applications.

The intermodulation distortion (IMD) products that fall closest to the fundamental RF frequency are the most detrimental to the integrity of the signal, as they cannot be easily filtered out. The most significant of these products are the third-order IMD products (IMD3s). The asymmetry between the upper and the lower IMD3s as well as the variation of the IMD3 across the bandwidth are often observed in RF PAs [1], [2]. This behavior hints at the presence of memory effects, which can be caused by different mechanisms like spurious gate/drain voltage modulation due to the bias-tees, the nonlinear capacitance effect due to the semiconductor junctions within the transistor, charge-trapping effects at microelectronic level, dynamic self-heating, or the linear matching networks.

These effects are of particular concern for broadband modulated signals [3]. Indeed, the sideband asymmetry degrades the performance of digital predistortion (DPD) techniques in meeting the required linearity level. Therefore, analog techniques that could minimize the presence of nonlinear memory are nowadays investigated [4].

The behavior of the IMD3 at RF is directly correlated to the IMD products generated at the drain baseband node [5]. These relationships could be modeled at behavioral level [6], [7], and exploited to manipulate linearity and efficiency by modifying the baseband termination at the drain terminal [8].

An alternative technique deals with the modification of the baseband termination at the gate port. Indeed, the distortion produced by the nonlinear behaviour of the PA greatly depends on the nonlinear gate-source capacitance controlled by the gate voltage [5]. Hence, presenting a short circuit at the baseband frequency (or even a negative impedance) has been proven to reduce IMD3 [5], [9]. The one of the second order is the most important among the even-order products falling at baseband, and its manipulation can be effectively used to reduce the RF in-band distortion by modulating the voltage at the gate to synthesize the desired baseband impedance effect [10], [11], [12].

In this work, the effect of gate baseband termination is experimentally tested by a prototype circuit implementing a baseband feedback (BBF) analog module. Thanks to this solution, external signal injection at the gate is avoided. The study involves both simulation and measurements, highlighting the advantages and disadvantages of the technique and showing that the asymmetry between the upper and lower IMD3 sidebands is directly affected by the phase of the feedback which, in turn, varies depending on the tone spacing.

The article is organized as follows. Section II describes the design of the BBF module, tailored to a Gallium Nitride (GaN) High-Electron-Mobility Transistor (HEMT). Such a module can be adopted in place of a standard DC bias feed module. In Section III, the experimental results show that the proposed BBF prototype allows for improvements in both IMD3 sideband amplitude and asymmetry for two-tone signals at different frequency spacings, and for a 16-QAM signal test. Conclusions are drawn in Section IV.

II. DESIGN OF THE BBF MODULE

The transistor under test targeted in this study is the CG2H40010F, a 10-W GaN HEMT from Wolfspeed. Using this device, an RF PA is designed by implementing RF input and output matching networks for operation at 2-GHz center

3−4 April 2022, London, UK

frequency. The PA is operated in class-AB at -2.77 V gate bias voltage and 28 V drain supply voltage. The saturated RF output power is 40 dBm.

The investigation on the feedback design was initially performed in Keysight Advanced Design System (ADS) using two-tone harmonic-balance simulations of the PA schematic shown in Fig. 1. In particular, the phase and gain of the BBF were swept to evaluate the IMD3 asymmetry at various tone spacings. Also, the IMD3 magnitude at both sidebands across the whole RF input power sweep range up to compression was characterized. As shown in Fig. 2, the IMD3 asymmetry and magnitude heavily depend on the phase shift applied by BBF at different frequencies.

Once the signal excitation for the target application is defined, the amplitude and phase of the BBF signal can be tuned accordingly, allowing to operate in regions of combined minimum distortion and reduced memory effects. The actual

tuning can be implemented by suitable signal conditioning in the BBF module using low-cost electronics.

As a case study, consider a two-tone excitation with 5-MHz frequency spacing. As shown in Fig. 3, at P_{OUT} = 30 dBm the IMD3 sidebands feature a minimum for a BBF phase of 180°. This is an interesting case, which was used for the implementation of the BBF demonstrator. Indeed, 180° phase shift is easily implementable by inverting the sign of the feedback signal (a multiplication by -1 is shown in Fig. 1). With this BBF configuration, the simulation results in Fig. 4 show that most of the reduction in IMD3 occurs between 25 to 35 dBm of P_{OUT}, while only a small improvement is notable at compression.

The actual circuit implementation (Fig. 1) uses two pull-up resistors at the gate and drain sides to prevent the baseband

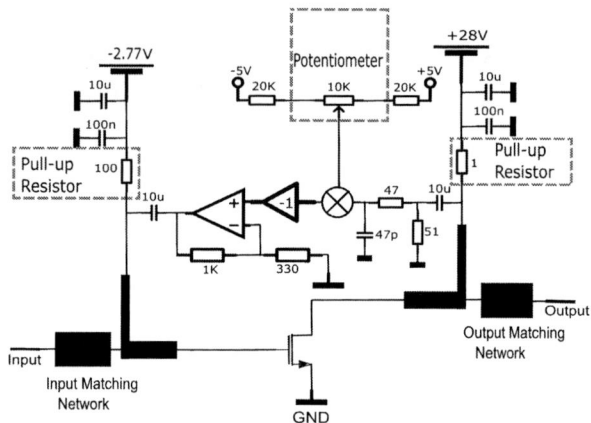

Fig. 1. Circuit diagram of the PA including the proposed baseband feedback (BBF) solution.

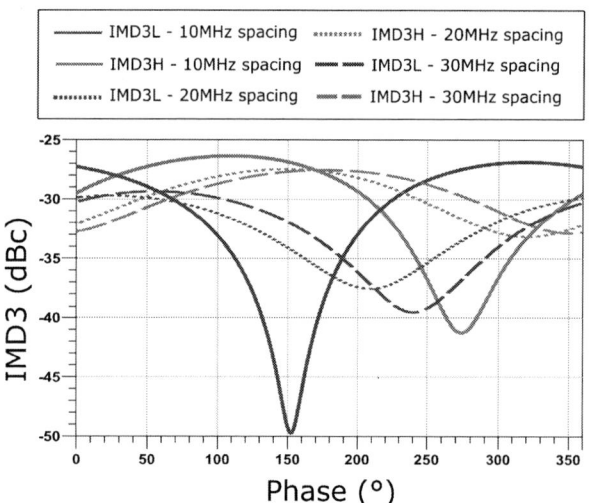

Fig. 2. Effect of feedback phase variation on the IMD3 sidebands for two-tone frequency spacings of 10 MHz, 20 MHz and 30 MHz (P_{OUT} = 30 dBm).

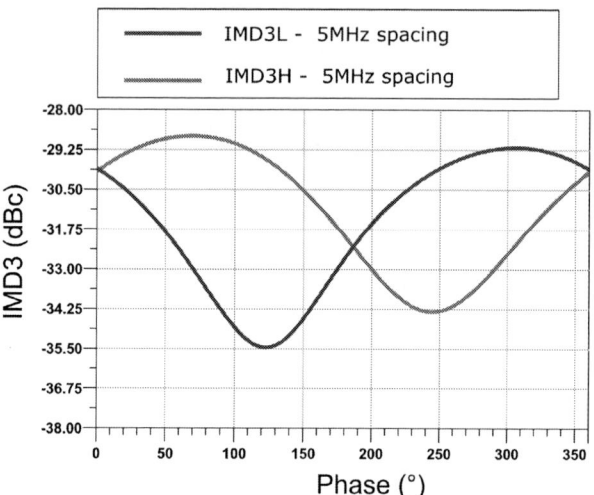

Fig. 3. Effect of feedback phase variation on the IMD3 sidebands asymmetry for 5-MHz two-tone spacing (the implemented case) at P_{OUT} = 30 dBm.

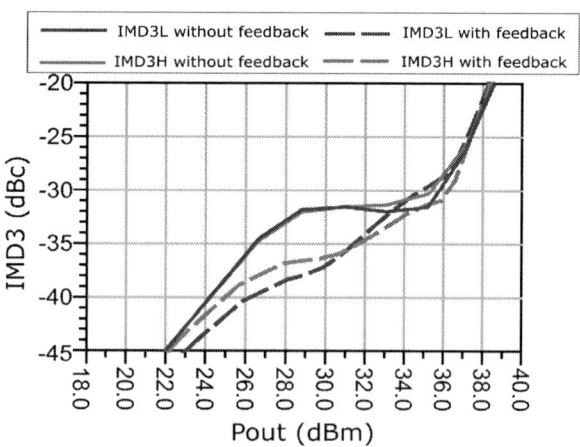

Fig. 4. IMD3 simulation results for 5-MHz spacing with 180° baseband feedback compared to the case without baseband feedback.

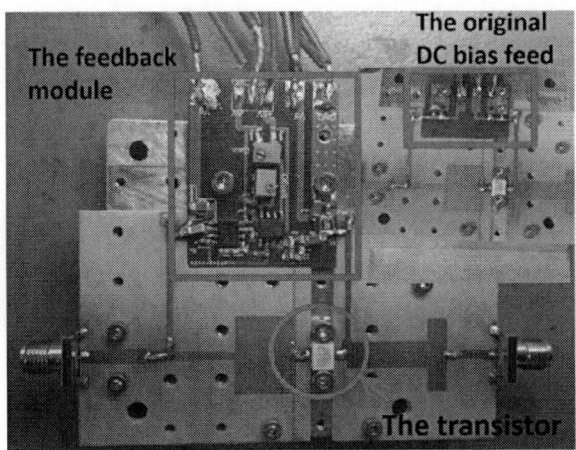

Fig. 5. The PA prototype with the baseband feedback (BBF) module board mounted on an aluminum heatsink.

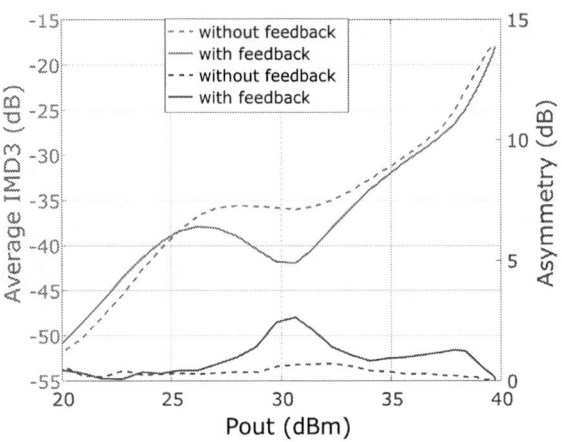

Fig. 6. IMD3 measured for the prototype for two-tone test with 5-MHz frequency spacing.

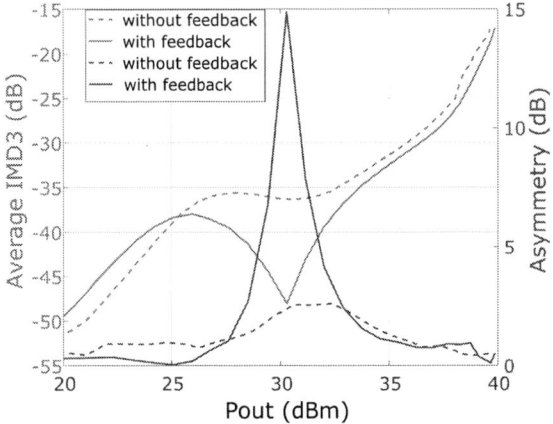

Fig. 7. IMD3 measured for the prototype for two-tone test with 20-MHz frequency spacing.

tones from being shorted. In particular, a small resistor of $1\ \Omega$ at the drain side minimizes power consumption. Also, a potentiometer was incorporated in the design to allow for tuning the gain manually, as needed to match the amplitude levels necessary for IMD compensation. In this design, the potentiometer provides a voltage output range between -1 V to 1 V, which is then multiplied by the baseband tones fed back from the drain. This mixer product in combination with the non-inverting amplifier ($\times 4$ voltage gain) allows for suitable voltage levels at the gate bias node.

The BBF module was designed using standard PCB technology and SMD components, then connected to the PA board by replacing the original bias feed. In this way, the main RF path remains unaltered (see Fig. 5).

III. EXPERIMENTAL RESULTS

The BBF prototype was tested by applying two-tone signals at different frequency spacings from 5 MHz to 20 MHz. In particular, Figs. 6 and 7 report the third-order sidebands with and without BBF for 5 MHz and 20 MHz tone spacings. At 5 MHz, the BBF effectively reduces the amplitude of both sidebands, while maintaining negligible sideband asymmetry.

While the design of the BBF was tuned at 5-MHz spacing, the addition of BBF reduces both sidebands also for 20-MHz spacing at P_{OUT} = 30 dBm, although it also displays an increase in IMD3 asymmetry at the same level of P_{OUT}. In fact, at 20-MHz spacing the BBF configuration favours the nonlinearity reduction while increasing the memory effects. Nevertheless, BBF phase shift and amplitude could be easily adapted to the specifications of the application for the 20-MHz case (or any other test signal).

Figure 8 shows that the introduction of the BBF accounts for a minor gain decrease at the fundamental tones. This result is also confirmed by Fig. 9, which displays the AM-AM characteristic at the fundamental frequency.

Finally, as a general-purpose test, the prototype was tested with a 1-MHz 16-QAM signal, showing 3-dBc improvement in the Adjacent-Channel Power Ratio (ACPR), and 1% in the Error-Vector Magnitude (EVM), as reported in Table 1.

IV. CONCLUSION

The implemented baseband feedback (BBF) method has shown the capability to effectively manipulate IMD3s in RF PAs, thus allowing for distortion reduction. Such a reduction mostly occurred in the back-off region, between 25 dBm and 35 dBm of P_{OUT}. The presented results also display the effect of the BBF phase on the asymmetry of the sidebands, hence control on the displayed memory effects, depending on the excitation signal.

This study also highlights a compromise, namely, that the IMD3 can be reduced significantly on one sideband at the expense of the symmetry between the sidebands. Indeed, the

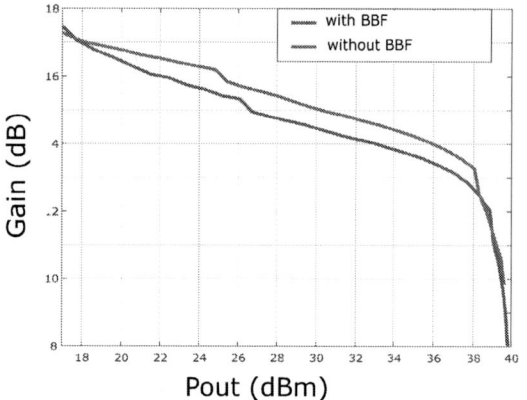

Fig. 8. Gain characteristic at the fundamental frequency (2 GHz) with and without BBF.

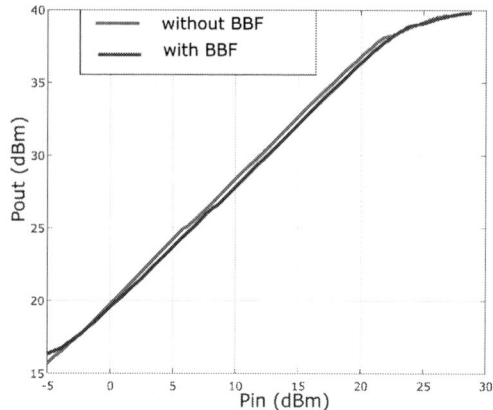

Fig. 9. AM-AM characteristic at the fundamental frequency (2 GHz) with and without BBF.

Table 1. Measurement results for 1-MHz 16-QAM signal.

$P_{in\,avg}$ (dBm)	w/o BBF		w/ BBF	
	ACPR (dBc)	EVM (%)	ACPR (dBc)	EVM (%)
17	[-40.5,-40.6]	3.0	[-43.9,-43.2]	1.0
19	[-37.1,-37.1]	3.2	[-40.1,-39.9]	2.5
21	[-34.0,-33.1]	4.5	[-36.7,-36.5]	3.5

BBF phase shift value must be accurately matched with the characteristics of the input signal, i.e., the signal's bandwidth and statistics. While the linearity improvements obtained with the actual preliminary implementation can be considered as rather local ones, more sophisticated BBF control strategies could be envisioned for obtaining better linearity performances and more general operating conditions.

The implemented BBF module features a straightforward design and little complexity, hence it can be easily employed as an analog method to adjust PA linearity and memory effects, possibly in combination with DPD techniques.

ACKNOWLEDGMENT

The authors would like to thank the NTNU, Norway for the help in the BBF prototype realization. Mathani Eltayeb was supported by a scholarship for thesis abroad by the Dept. Electrical, Electronic, and Information Engineering 'G. Marconi', University of Bologna, Italy.

REFERENCES

[1] N. Borges de Carvalho and J. C. Pedro, "A comprehensive explanation of distortion sideband asymmetries," in *IEEE Trans. Microw. Theory Techn.*, vol. 50, no. 9, pp. 2090-2101, Sept. 2002

[2] J. S. Kenney and P. Fedorenko, "Identification of RF Power Amplifier Memory Effect Origins using Third-Order Intermodulation Distortion Amplitude and Phase Asymmetry," in *IEEE MTT-S Int. Microw. Symp. Dig.*, San Francisco, CA, 2006, pp. 1121-1124

[3] A. M. Angelotti, G. P. Gibiino, C. Florian and A. Santarelli, "Broadband Error Vector Magnitude Characterization of a GaN Power Amplifier using a Vector Network Analyzer," *IEEE/MTT-S Int. Microw. Symp. Dig.*, pp. 747-750, 2020.

[4] P. M. Tomé, F. M. Barradas, T. R. Cunha and J. C. Pedro, "Hybrid Analog/Digital Linearization of GaN HEMT-Based Power Amplifiers," *IEEE Trans. Microw. Theory Techn.*, vol. 67, no. 1, pp. 288-294, Jan. 2019.

[5] J. Brinkhoff and A. E. Parker, "Effect of baseband impedance on FET intermodulation," in *IEEE Trans. Microw. Theory Techn.*, vol. 51, no. 3, pp. 1045-1051, Mar. 2003

[6] G. P. Gibiino, G. Avolio, D. Schreurs, A. Santarelli and F. Filicori, "A Three-Port Nonlinear Dynamic Behavioral Model for Supply-Modulated RF PAs," in *IEEE Trans. Microw. Theory Techn.*, vol. 64, no. 1, pp. 133-147, Jan. 2016

[7] G. P. Gibiino, J. Couvidat, G. Avolio, D. Schreurs, A. Santarelli, "Supply-terminal 40 MHz BW characterization of impedance-like nonlinear functions for envelope tracking PAs," in *ARFTG Microwave Measurement Conf.*, 2016, pp. 1-4

[8] G. P. Gibiino, G. Avolio, S. Schafer, D. Schreurs, Z. Popović, A. Santarelli, F. Filicori, "Active baseband drain-supply terminal load-pull of an X-band GaN MMIC PA," in *Proc. European Microwave Conf. (EuMC)*, Paris, 2015, pp. 1204-1207

[9] M. Akmal et al., "The effect of baseband impedance termination on the linearity of GaN HEMTs," in *Proc. European Microwave Conf. (EuMC)*, Paris, 2010, pp. 1046-1049

[10] W. Sear and T. W. Barton, "A Baseband Feedback Approach to Linearization of a UHF Power Amplifier," in *IEEE MTT-S Int. Microw. Symp. Dig.*, 2019, pp. 75-78

[11] E. Cijvat, K. Tom, M. Faulkner and H. Sjoland, "A GaN HEMT power amplifier with variable gate bias for envelope and phase signals," in *Proc. Norchip*, 2007, pp. 1-4

[12] P. Medrel, A. Martin, T. Reveyrand, G. Neveux, D. Barataud, P. Bouysse, J.-M. Nébus, L. Lapierre, and J.-F. Villemazet, "A 10-W S-band class-B GaN amplifier with a dynamic gate bias circuit for linearity enhancement," in *Int. J. Microw. Wirel. Tech.*, vol. 6, no. 1, pp. 3–11, 2014.

Statistical Modeling of GaN HEMTs by Direct Transfer of Variations to Model Parameters

Petros Beleniotis[1], Serguei Chevtchenko[2], Matthias Rudolph[1,2]

[1]Brandenburg University of Technology Cottbus-Senftenberg (BTU),
Ulrich L. Rohde Chair of RF and Microwave Techniques, Germany

[2]Ferdinand-Braun-Institut gGmbH, Leibniz-Institut für Höchstfrequenztechnik (FBH), Germany

petros.beleniotis@b-tu.de

Abstract — A statistical physics-based model for GaN HEMTs with a direct transfer of variations to model parameters is proposed in this paper. This statistical approach includes the collection of available data from process control monitor (PCM) on-wafer measurements and their transfer to the respective physical model parameters. Three main variables are analyzed and used in the simulations. Those are the semiconductor sheet resistance R_{sh}, the ohmic contact resistance R_c and the threshold voltage V_{th}. The model's performance is validated by comparing Monte Carlo large-signal simulations and measurements of several devices on the same wafer.

Keywords — Compact model, GaN HEMT, large signal, Monte Carlo, statistical modeling.

I. Introduction

GaN high electron mobility transistors (HEMTs) remarkable performance at high-power and high-frequency applications requires highly accurate device modeling. Power amplifier design can be a demanding task due to fluctuations in device performance. Such fluctuations usually come from variations in device processing or semiconductor growth. Statistical models are therefore needed for an accurate representation of variations and circuit yield estimation.

Statistical modeling of electronic devices has been rendered an important field of study for a long time. Various statistical models have been created for GaN HEMTs with an empirical basis [1], [2], [3]. However, the empirical character cannot directly connect process or semiconductor variations with the model parameters. Quasi-physical statistical models [4] can correct this disadvantage but still demands the extension of the model to absorb a new variation. On the other hand, most of the statistical models follow a typical procedure for the extraction of the varying model parameters. This procedure at first contains the individual parameter extraction of several devices on a wafer and then the analysis of statistical variations of the model parameters. The model parameter extraction by the direct fit to the variations has been previously proposed for MOSFETs [5], but to the best of the authors' knowledge, it has not been applied on GaN-based devices and a physics-based model.

In this paper, we demonstrate a new approach for statistical modeling of GaN HEMTs. We use the surface-potential model ASM-HEMT, equipped with our drain-lag description previously proposed in [6]. Statistical simulation results with ASM-HEMT have been already published but with the typical procedure described before [7]. In our study, we take advantage of available process control monitoring (PCM) on-wafer measurements. We analyze their statistical behavior and directly transfer the measured variations to the respective model parameters. For the validation of the proposed statistical model with Monte Carlo large-signal simulations, we use $0.25\,\mu$m HEMT devices from the GaN-on-SiC process of the Ferdinand-Braun-Institut, Leibniz-Institut für Höchstfrequenztechnik [8].

II. Statistical Model Description

A. Measurements and Analysis

The extraction procedure of compact model parameters usually requires accurate yet rigorous measurements due to the significant trapping effects in GaN HEMTs. The main scope of this study is to avoid such kinds of measurements for numerous individual devices. Instead, we extract the ASM-HEMT model parameters employing pulsed I-V and S-parameter measurements for only one device, as described in [6]. Then, we can take advantage of available PCM data. Those measurements usually contain important information for the whole wafer that can be transferred to a compact model and randomly vary specific model parameters. In this paper, we consider the statistical variation of the threshold voltage V_{th}, the semiconductor sheet resistance R_{sh}, and the resistance of the ohmic contacts R_c. The characterization of R_{sh} is obtained using the Van der Pauw method [9], and the contact resistances according to the Transfer-Length-Method (TLM) [10]. V_{th} was extracted from the transfer characteristic with $V_{ds} = 10$ V.

Fig.1 presents the probability density for the three variables, and the data fit by a normal distribution. V_{th} and R_c follow a similar to the normal distribution's behavior. On the other hand, the mean value of R_{sh} exhibits a very low standard deviation making its representation with a normal distribution inaccurate. At this point, we could assume that the variation of the sheet resistance could be neglected due to the very low fluctuation of its values. However, we include R_{sh} variations for this paper, where data of R_{sh} could be implemented in a physics-based statistical model.

PCM data presented in Fig.1 can be further analyzed, providing the mean values and the respective standard

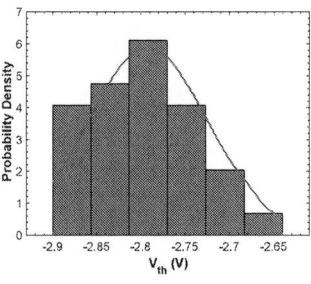 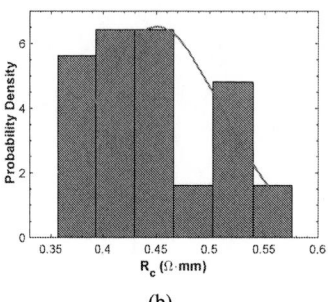 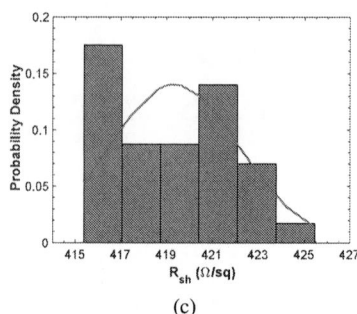

(a)   (b)   (c)

Fig. 1. Probability density for the three variables that analyzed for the creation of the proposed statistical model. (a) presents threshold voltage V_{th}, (b) the contact resistance R_c, and (c) the sheet resistance R_{sh}.

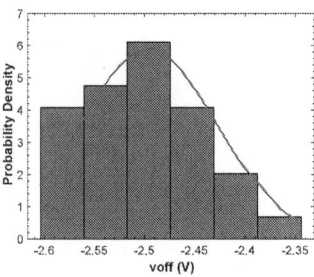 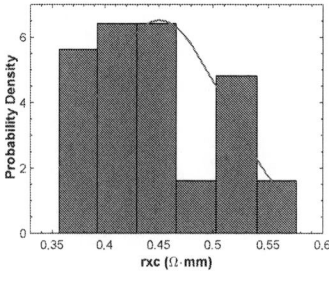 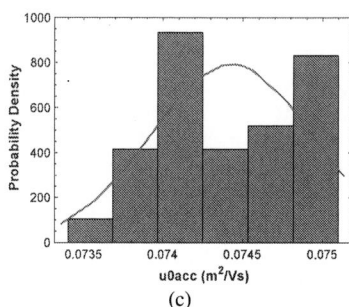

(a)   (b)   (c)

Fig. 2. Probability density for three model parameters. Parameter values are transferred directly from PCM data. (a) presents voff, (b) the contact resistance rxc, where x can be s for source or d for drain, and (c) the access region mobility u0acc.

deviation of the variables. However, except R_c, none of the other two can be used directly in ASM-HEMT as model parameters. Thus, we have to observe any relations between them and transfer the experimental data to parameters' values.

B. Transferring Variations to Model Parameters

Values of PCM data variables cannot always be directly transferred to a compact model. There are usually different extraction procedures or an extended set of parameters that simulates one of the statistically varied variables through specific equations. In this part, we use the suitable equations of ASM-HEMT to transfer the variations obtained from our PCM measurements to our statistical model.

In ASM-HEMT, the parameter $voff$ represents the threshold voltage V_{th}, and requires a specific procedure for its extraction. The user always extracts $voff$ from transfer characteristics with the lowest possible V_{ds} conditions. However, in PCM measurements, V_{th} could be extracted at various V_{ds}, as in our case at 10 V. In devices with short channel effects, such as Drain-Induced Barrier Lowering (DIBL), a mismatch between $voff$ and V_{th} could appear. Thus, we use ASM-HEMT's formulation for the DIBL effect to calculate variations of $voff$ from the V_{th} measurements. The DIBL-affected $voff$ [11] is calculated as follows:

$$V_{off,DIBL} = voff - eta0 \cdot \left(\frac{V_{ds} \cdot vdscale}{\sqrt{V_{ds}^2 + vdscale^2}} \right) \quad (1)$$

where $voff$ is the model parameter for the threshold voltage, and $eta0$ and $vdscale$ are model parameters for simulating the DIBL effect. We can change the above equation to obtain $voff$ as follows

$$voff = V_{th} + eta0 \cdot \left(\frac{V_{ds,th} \cdot vdscale}{\sqrt{V_{ds,th}^2 + vdscale^2}} \right) \quad (2)$$

where V_{th} is the measured threshold voltage presented in Fig. 1a and $V_{ds,th}$ is the V_{ds} at which V_{th} has been extracted through measurements, 10 V in our case. Fig. 2a presents the calculated $voff$ using (2) and the V_{th} values from Fig. 1a. The model parameters $eta0$ and $vdscale$ have been extracted during the parameter extraction as stated at the beginning of the previous sub-section. Their values used in our calculations are 0.0593 (unitless) and 20.11 V, respectively.

The next step is to absorb the experimental results for the contact resistance R_c. Here, model allows us to use them directly. The two parameters rdc and rsc represent the contact resistance for the drain and the source side, respectively. We can transfer the R_c variations directly to both parameters.

Finally, we have experimental results for the sheet resistance R_{sh}. In ASM-HEMT, the access region resistance model simulates the respective region of the device. To transfer the R_{sh} variations, we have to analyze the access region model. Below, we can find the equation that ASM-HEMT employs for the access region resistances, as described in [12].

$$R_{acc} = \frac{R_{acc0}}{\left[1 - \left(\frac{I_d}{I_{max}}\right)^{\beta}\right]^{\frac{1}{\beta}}} \quad (3)$$

where $R_{acc0} = L_{acc}/(W \cdot NF \cdot q \cdot ns0acc \cdot u0acc)$ with L_{acc} the length of the access region, W the gate width, NF the number of fingers, $ns0acc$ the 2DEG density at the access regions, and $u0acc$ the carrier mobility. Also, $I_{max} = W \cdot NF \cdot q \cdot ns0acc \cdot vsataccs$, where $vsataccs$ is the saturation velocity at the source side access region. During the Van der Pauw method, the current between the two ohmic contacts is much lower than the I_{max} of the ASM-HEMT. Thus, R_{acc} can be simplified and be equal to R_{acc0}

$$R_{acc} = \frac{L_{acc}}{W \cdot NF \cdot q \cdot ns0acc \cdot u0acc} \quad (4)$$

Knowing from [10] that the total resistance of a semiconductor sample with length L and width W is given by $R = R_{sh} \cdot L/W$, we can connect R_{sh} with the model parameters as follows

$$R_{sh} = \frac{1}{q \cdot ns0acc \cdot u0acc} \quad (5)$$

We can now identify two model parameters connected with the experimental R_{sh}. The model parameter $ns0acc$ is included in the I_{max} of (3) and affects the non-linear part of the access region resistances. Thus, to keep our model simple, we suppose that $ns0acc$ is remaining constant, and the observed variations of R_{sh} would be caused only by fluctuations of the mobility at the access region. According to (5) and the R_{sh} values from Fig. 1c, we calculate the values of the parameter $u0acc$ as shown in Fig. 2c. The extracted value of the $ns0acc$ that was used for the calculation was $2 \cdot 10^{17} m^{-2}$.

III. LARGE-SIGNAL VALIDATION

The validation of the statistical model is discussed in this section. Monte Carlo simulations in Keysight ADS EDA software were used for the statistical analysis of large-signal performance. In this section, we compare the Monte Carlo results with RF power sweep measurements at 2 GHz. Load impedance was tuned for optimum output power. Our validation uses again on-wafer measurements obtained by the foundry during the performance verification. Five devices, which are located in different sites on the wafer for reflecting the possible fluctuations, were measured. Fig. 3 shows the five selected devices and their location on the wafer.

Monte Carlo simulations consisted of 250 trials. Five model parameters were allowed to follow a Gaussian distribution with mean values and standard deviations extracted from the previous section's results presented in Fig. 2. All the other model parameters and the extrinsic elements of the device were kept constant. It is important to emphasize that the main model parameters (not the statistically varied ones) were extracted using pulsed I-V and pulsed S-parameter measurements of a device not contained in the set of the five ones that take part in our validation.

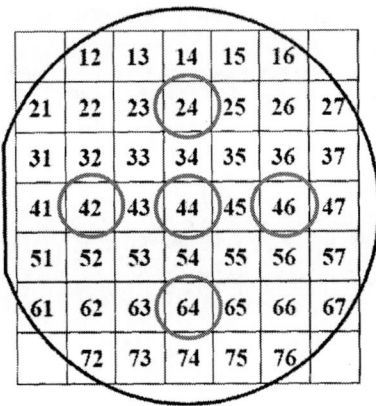

Fig. 3. Location of the 5 devices used for the validation of the model.

Table 1 summarizes the extracted mean values and the respective standard deviation for the five model parameters. We transferred the variations from the three PCM variables but we have to adjust five model parameters. That happens because there are two sides for each resistance in our model, one for the source and another one for the drain side. Thus, we have two parameters for the contact resistance, rdc and rsc, as well as two parameters for the access region mobility, $u0accd$ and $u0accs$.

Table 1. Statistical model parameters

Parameter	μ	$\sigma\%$
voff	-2.496 V	2.72
rdc, rsc	0.45 $\Omega \cdot mm$	13.58
u0accd, u0accs	0.0744 $m^2/V \cdot s$	0.6770

Fig. 4 shows the comparison between the Monte Carlo simulation results (Fig. 4a & 4c) and the load-pull measurements (Fig. 4b & 4d). We are comparing output power (P_{out}), gain, power-added efficiency (PAE), and mean I_{ds} for $V_{ds} = 28 V$ and $V_{gs} = -1.7 V$. We observe the accurate prediction of the measurements for the five devices. In all cases, the fluctuation of device performance lies within the predicted range of our statistical model's results. A slight discrepancy is observed right above 4 dBm of input power but it is caused by the exclusion of higher harmonics' impedances in our simulations and not the statistical model. In our harmonic balance simulations, we considered the source and load impedance only for the fundamental frequency due to the absence of information on higher harmonic terminations from measurement data. More validation with the consideration of the higher harmonics may assist in further enhancement of our statistical model.

IV. CONCLUSION

The proposed physics-based statistical model allows for accurate simulation of fluctuations observed in GaN HEMTs'

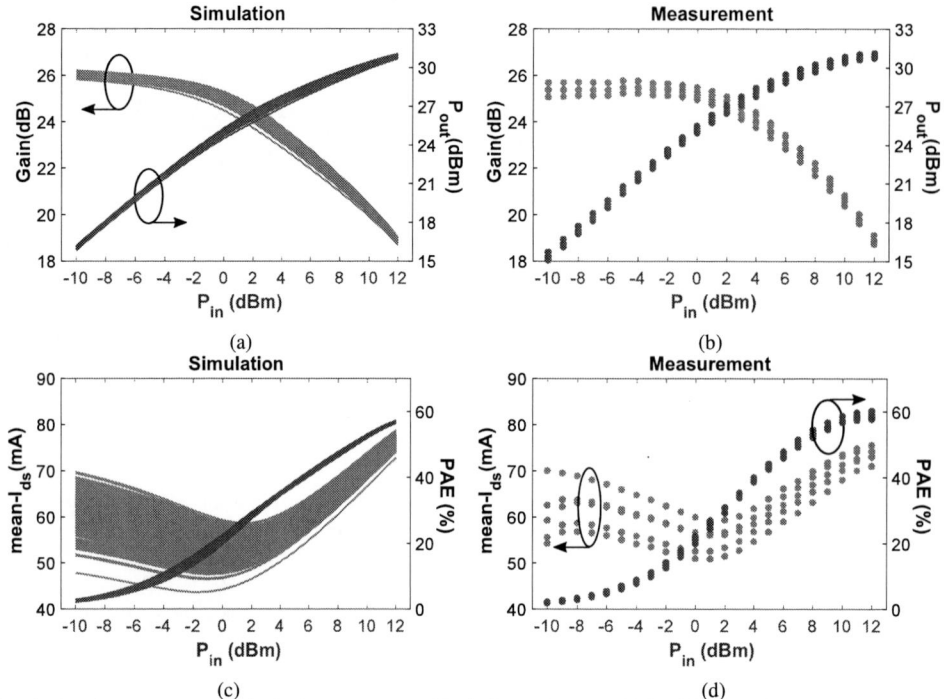

Fig. 4. Simulations (a, c) and measurements (b, d) of gain, output power (P_{out}), I_{ds}, and power-added efficiency (PAE) at 2 GHz. Measurements contain results from five different devices, whereas simulations from 250 Monte Carlo trials. V_{ds} is at 28 V and V_{gs} at -1.7 V.

large-signal performance. We made use of data available from process control monitoring and transferred variations directly to model parameters. For our approach, we used the physics-based ASM-HEMT model, but similar steps can be applied to any physics-based compact model. We compared Monte Carlo large-signal simulations with load-pull measurements and found a good agreement.

ACKNOWLEDGMENT

Financial support by Deutsche Forschungsgemeinschaft (DFG) under grand no. 440549658, and by the German Ministry of Education and Research (BMBF) under the project reference 16FMD02 (Forschungsfabrik Mikroelektronik Deutschland), is gratefully acknowledged.

REFERENCES

[1] Z. Chen, Y. Xu, C. Wang, Z. Wen, Y. Wu, and R. Xu, "A Large-Signal Statistical Model and Yield Estimation of GaN HEMTs Based on Response Surface Methodology," *IEEE Microwave and Wireless Components Letters*, vol. 26, no. 9, pp. 690–692, 2016.

[2] W. Stiebler, P. Kolias, and J. Sanctuary, "Statistical large-signal model enabling yield optimization in high-power amplifier design," in *2007 IEEE Compound Semiconductor Integrated Circuits Symposium*, 2007, pp. 1–4.

[3] Z. Chen, Y. Xu, B. Zhang, T. Chen, T. Gao, and R. Xu, "A GaN HEMTs Nonlinear Large-Signal Statistical Model and Its Application in S-Band Power Amplifier Design," *IEEE Microwave and Wireless Components Letters*, vol. 26, no. 2, pp. 128–130, 2016.

[4] Z. Wen, S. Mao, Y. Wu, R. Xu, B. Yan, and Y. Xu, "A Quasi-Physical Large-Signal Statistical Model for 0.15 AlGaN/GaN HEMTs Process," in *IEEE MTT-S International Microwave Symposium Digest*, vol. 2019-June, 2019, pp. 208–211.

[5] K. Takeuchi and M. Hane, "Statistical Compact Model Parameter Extraction by Direct Fitting to Variations," *IEEE Transactions on Electron Devices*, vol. 55, no. 6, pp. 1487–1493, 2008.

[6] P. Beleniotis, F. Schnieder, and M. Rudolph, "Simulating Drain Lag of GaN HEMTs with physics-based ASM model," in *2020 15th European Microwave Integrated Circuits Conference (EuMIC)*, 2021, pp. 165–168.

[7] S. A. Ahsan, S. Ghosh, S. Khandelwal, and Y. S. Chauhan, "Statistical Simulation for GaN HEMT Large Signal RF performance using a Physics-based Model," no. December 2016, 2017.

[8] S. A. Chevtchenko, P. Kurpas, N. Chaturvedi, R. Lossy, and J. Würfl, "Investigation and Reduction of Leakage Current Associated with Gate Encapsulation by SiNx in AlGaN/GaN HFETs," *2011 International Conference on Compound Semiconductor Manufacturing Technology, CS MANTECH 2011*, pp. 237–240, 2011.

[9] L. J. van der PAUW, "A METHOD OF MEASURING SPECIFIC RESISTIVITY AND HALL EFFECT OF DISCS OF ARBITRARY SHAPE," in *Semiconductor Devices: Pioneering Papers*. WORLD SCIENTIFIC, mar 1991, pp. 174–182.

[10] D. K. Schroder, *Semiconductor Material and Device Characterization, 3rd edition*. John Wiley & Sons, Ltd, 2006.

[11] S. Khandelwal, S. Ghosh, S. A. Ahsan, A. Dasgupta, and Y. S. Chauhan, *Advanced SPICE Model for HEMTs - Technical Manual*, 2018.

[12] S. Khandelwal, Y. S. Chauhan, T. A. Fjeldly, S. Ghosh, A. Pampori, D. Mahajan, R. Dangi, and S. A. Ahsan, "ASM GaN: Industry Standard Model for GaN RF and Power Devices—Part 1: DC, CV, and RF Model," *IEEE Transactions on Electron Devices*, vol. 66, no. 1, pp. 80–86, 2019.

Proceedings of the 16th European Microwave Integrated Circuits Conference

Design of Terahertz InP pHEMT Using Machine Learning Assisted Global Optimization Techniques

Jing Wang, Li-Yuan Xue, Bo Liu, Chong Li

James Watt School of Engineering, University of Glasgow, UK

{j.wang.6, l.xue.1}@research.gla.ac.uk, {bo.liu, chong.li}@glasgow.ac.uk

Abstract—**This paper presents an optimal design of terahertz InP-based pseudomorphic high electron mobility transistors (pHEMT) powered by an artificial intelligence (AI) technique. Unlike the traditional physics-based design optimization methods, the new technique employs a machine learning-assisted global optimization algorithm. A state-of-the-art commercial pHEMT operating at millimeter-wave frequencies was used to calibrate the physics-based model. Based on the pHEMT, the proposed machine learning-assisted optimization method was implemented with the constraint of gate length, i.e., 100 nm. The simulation results show significant improvement in terms of cut-off frequency, i.e., 57%, and maximum oscillation frequency, i.e., 30%, compared to the commercial design. To the best of our knowledge, this is the first time to employ machine learning-assisted global optimization techniques to pHEMT design, showing high potential in terms of numerical simulation and device design for ultrafast semiconductor devices.**

Keywords — **pHEMT, terahertz, machine learning, global optimization.**

I. Introduction

Terahertz (THz) frequency, covering 100 GHz to 3 THz, has attracted attention in many applications, including security screening [1], next-generation autonomous radars [2], and high data rate mobile communications beyond 5G [3]. Transistors are the core devices in the front-ends of RF systems, and indium phosphide (InP)-based pseudomorphic high electron mobility transistors (pHEMTs) have a cut-off frequency f_T over 700 GHz and a maximum oscillation frequency f_{max} close to 1.5 THz [4]. This makes them ideal enabling technologies for terahertz monolithic integrated circuits, which have the advantages of low cost, ease of integration, and small size over other technologies.

However, the design of pHEMTs is not trivial, which involves both material (epitaxial layers) and structure (mainly electrodes) optimization. The total number of parameters to be optimized could be more than 20. For example, to achieve the highest operating frequency, one should first optimize materials including material composition, layer thickness, and doping method and level for as high electron mobility as possible and then optimize transistor structure, including gate length, gate shape, source-drain separation, the gate to drain/source distance, etc. to reduce parasitic capacitance. In addition, metal alloys/stacks for gate and source/drain contacts should also be optimized. In the past, the above optimization process mainly relied on trial-and-error with fabricated devices. Clearly, it is a time-consuming and costly method and may also obtain suboptimal results. Recently, numerical simulation models have become available. However, the drawbacks on efficiency and optimization quality of the trial-and-error method still exist. This calls for the need to employ modern optimization techniques.

This paper introduces a state-of-the-art machine learning-assisted global optimization method to design pHEMTs for the first time. The employed algorithm is called surrogate model assisted differential evolution for antenna synthesis (SADEA), initially developed for antenna design [5]. We employ SADEA to a commercial pHEMT structure and obtain the optimized new design, showing 57% and 37% improvements on f_T=336 GHz and f_{max}=770 GHz, respectively, without changing the gate length. The optimization time using a normal desktop computer is 16 hours, significantly shorter than the conventional methods, which may take several days. The new method shows great potential in terms of efficiency and optimization quality for ultrafast transistor design. It could also be extended to other advanced semiconductor devices and circuits.

The remainder of this paper is organized as follows. In Section II, the physical model of pHEMTs and the SADEA algorithm are first introduced. In Section III, a case study of optimizing commercial InP pHEMTs using the SADEA is demonstrated. Conclusions are finally provided in Section IV.

II. SADEA-based pHEMT Design Optimization Method

Depending on applications, the design focus of HEMTs varies. For ultrafast applications, f_T and f_{max} are the two main figures of merit and should be maximized wherever possible. This section will introduce pHEMTs, the conventional design method, and the SADEA-based design method.

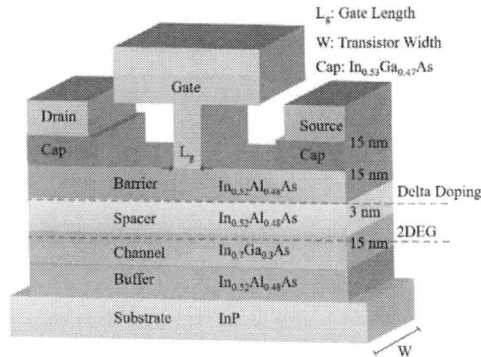

Fig. 1. Illustration of epilayers and electrodes of a pHEMT.

3–4 April 2022, London, UK

67

A. Introduction to pHEMTs

Fig. 1 shows a typical InP pHEMT. The epilayers are grown on a semi-insulating InP substrate using either Metal-Organic Chemical Vapor Phase Deposition (MOCVD) or Molecular Beam Epitaxy (MBE). The channel layer, made of $In_yGa_{1-y}As$ where y indicates indium mole fraction, is sandwiched by two wider bandgap $In_{0.52}Al_{0.48}As$ layers. A quantum well is formed in the channel, making electrons well confined. The top $In_{0.52}Al_{0.48}As$ is further divided into halves by a silicon delta doping layer that provides electrons to the channel. The lower $In_{0.52}Al_{0.48}As$ adjacent to the channel is a spacer that separates electrons from the dopants (i.e., silicon), reduces the scattering of impurities, and improves electron mobility in the channel. The upper $In_{0.52}Al_{0.48}As$ is a barrier where the gate sits, controlling the channel's electron flow. A heavily doped n-type $In_{0.53}Ga_{0.47}As$ layer above the barrier forms the ohmic contacts for low contact resistance. The cap layer under the gate is removed to allow the formation of a Schottky barrier.

The epilayers play a significant role in determining the electron mobility and the sheet resistance of the channel, contact resistance, parasitic capacitance, f_T, and f_{max}. Researchers have attempted to improve carrier mobility by optimizing the thickness, mole fractions, and material compositions in the barrier, spacer, and channel. For example, lattice-matched InAlAs/$In_yGa_{1-y}As$/InAlAs ($y>0.53$) composite channel with high indium mole fraction was investigated in [6]. InGaAs/InP composite-channel HEMT was found to have mobility as high as 11,000 cm²/Vs⁻¹ [7]. Recently, InAs/InGaAs composite-channel HEMT was reported with electron mobility approaching 15,000 cm²/Vs⁻¹ [8].

Apart from materials, electrodes also play a significant role on f_T and f_{max}. The smaller the gate length L_g, the lower the gate resistance, leading to higher f_T and f_{max}. As shown in Fig.1, a T-shaped Schottky gate is often used to balance current capacity, narrow gate length, and parasitic capacitance. However, the gate foot shorter than 50 nm becomes challenging to make. A 10 nm T-gate has been demonstrated [9], but its mechanical support provided by the foot is not sufficiently strong, and therefore the yield becomes comparatively low. In this work, the foot length of the T gate is kept at 100 nm.

B. State-of-the-art pHEMT design optimization

A widely used method for simulation-based pHEMT design optimization is based on equivalent circuits, as shown in Fig. 2. Equations (1) and (2) give the main parameters from the small-signal equivalent circuit that affects f_T and f_{max}, and Table 1 explains those parameters. Deriving such a circuit relies on accurate measurement, which is known notoriously time-consuming. Hence, most foundries can only provide equivalent circuit models for a few biasing conditions. This restricts the flexibility of optimization and often obtains sub-optimal results.

An alternative method is modeling devices using physics-based CAD tools (e.g., Sentaurus). The advantages over circuit models include high accuracy and flexibility. However, the

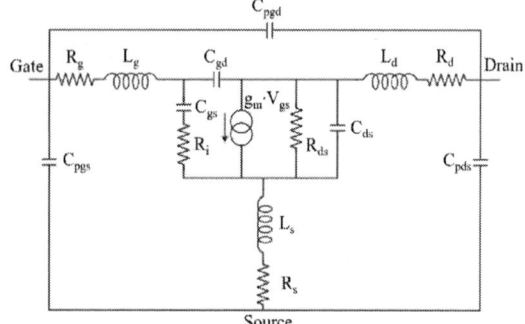

Fig. 2. The complete small-signal equivalent circuit of a pHEMT.

$$f_T = \frac{g_m}{2\pi(C_{gs}+C_{gd})(1+\frac{(R_s+R_d)}{R_{ds}})+C_{gd}g_m(R_s+R_d)} \quad (1)$$

$$f_{max} = \frac{f_T}{2\sqrt{\frac{R_g+R_i+R_s}{R_{ds}}+2\pi f_T R_g C_{gd}}} \quad (2)$$

Table 1. The main parameters from the small-signal equivalent circuit affect the operating frequencies.

Parameters	Description
C_{gs}	Source-gate capacitance
C_{gd}	Drain-gate capacitance
R_s	Parasitic source resistance
R_d	Parasitic drain resistance
R_{ds}	Output resistance
R_i	Intrinsic channel resistance
g_m	Transconductance

drawbacks located in the optimization methods. Most available methods neither have sufficient optimization ability (e.g., quasi-Newton method) nor efficient (e.g., genetic algorithm). Note that the simulation of a physics-based device model needs numerical techniques (e.g., finite element analysis), which is computationally expensive, and standard global optimization methods such as genetic algorithms often need thousands to tens of thousands of such simulations.

C. pHEMT design optimization by SAEDA

To reduce the optimization time to a practical level while maintaining the optimization quality as standard global optimization algorithms, surrogate model-assisted evolutionary algorithms (SAEA) are introduced into pHEMT design in this work. In SAEA, the surrogate model mapping the inputs (i.e., design variables) and outputs (i.e., performances) are often constructed by machine learning techniques. By replacing the computationally expensive Sentaurus simulations with computationally cheap surrogate model predictions, the optimization time can be considerably reduced. In this paper, the selected algorithm is SADEA. SADEA is initially designed for microwave antenna design exploration; however, good performances are also found in other applications. This is the first attempt to implement the algorithm on HEMTs. Fig. 3 illustrates how MATLAB, which runs the SADEA code, collaborates with Sentaurus, which

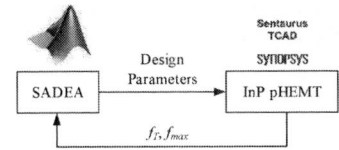

Fig. 3. The simulation-driven optimization flow.

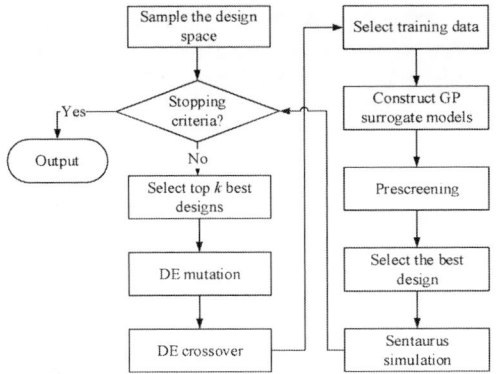

Fig. 4. The flow diagram of SADEA.

runs HEMTs' simulation.

In SADEA, the machine learning method is Gaussian process (GP), the evolutionary algorithm is differential evolution (DE), and the model management method is surrogate model-aware evolutionary search framework [10]. The flow diagram of SADEA is shown in Fig. 4. Here is how it works. A Latin hypercube sampling is first implemented to initialize the design space. Then, in each iteration, the k top-ranked candidate designs are selected to form the parent population. From the parent population, the new population is generated by applying DE mutation and crossover operators. The used operator here is DE/best/1 [11], showing fast convergence speed. Then, GP surrogate models are built for each candidate design in the generated new population. The training data are the nearest designs (based on Euclidean distance) from the simulated designs. Then, a prescreening method, lower confidence bound [12], is used to consider both prediction uncertainty and performance to find the expected most promising design in the current iteration, which will then be simulated by Sentaurus. This process continues until the stopping criterion is met. More details about SADEA can be found in [5].

III. CASE STUDY

In this case study, our optimal design is compared with a commercial pHEMT by Diramics [13]. We will first calibrate the device model using the data provided by Diramics, then SADEA is employed to optimize the design parameters. Note gate length is not optimized in this work.

A. Model Calibration

Commercial pHEMT's datasheet is available from [13]. The models used in Sentaurus include Hydrodynamic transport models for electrons, High-field mobility, and the

Recombination model includes Shockley–Read–Hall (SRH), Auger, and Radiative. The ohmic contact or Schottky contact are specified. Table 2 lists material properties used in the simulation. All the parameters are referred to room temperature conditions. To guarantee the accuracy of the simulation model, the mesh density below the gate region is set high.

Table 2. List of main semiconductor parameters used in the modeling.

Parameter	InP	In$_{0.53}$Ga$_{0.47}$As	In$_{0.52}$Al$_{0.48}$As
Lattice constant (Å)	5.86	5.86	5.86
Band gap (eV)	1.34	0.72	1.48
Dielectric constant (static)	12.4	14.3	12.4
Electron affinity (eV)	4.44	4.55	4.27
Effective mass m$_e^*$ /m$_0$ at central valley	0.079	0.047	0.081

The simulated DC (Fig. 5) and RF (Table 3) show good agreement with the experimental results [13]. The final thickness of the epilayers is shown in Fig. 1.

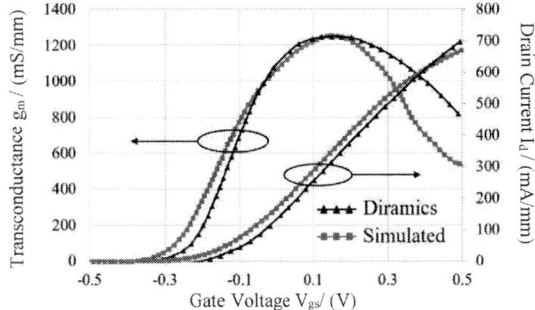

Fig. 5. IV characteristics and transconductance of the experimental (Diramics) and simulated 100 nm pHEMT at ambient temperature, Vds=0.9 V.

B. Device Optimization

In this study, 16 design parameters are selected for optimization. Table 4 shows the search range of the parameters. To ensure the optimized parameters are physically realizable, geometric constraints are set as follows: let x denote the design parameters, and the pHEMT design optimization problems can be described as follows.

$$\arg \max_{x} \ (f_T, f_{\max})$$

$$\text{subeject to:} \quad f_T^* \geq 220 \text{ GHz}$$
$$f_{\max}^* \geq 550 \text{ GHz}$$
$$x_4 + x_5 - x_9 \geq 3 \text{ (nm)}$$
$$x_9 - x_{11} \geq 2 \text{ (nm)} \quad\quad (3)$$
$$x_{12} - x_{13} \geq 0.35 \text{ (um)}$$
$$x_{12} + x_{14} \leq 1.15 \text{ (um)}$$

C. Results

The optimized 100 nm pHEMT design is shown in Table 4, the optimization results show that the Diramics model is within a reasonable boundary value range. The values of the epitaxial layer and structure of the device tend to the boundary

Table 3. Key performances of 100 nm pHEMT between the Diramics, simulated and optimized.

Performance	Diramics	Simulated	Optimized
Transconductance (mS/mm)	1250	1255	1672
Drain Current (mA/mm)	700	675	775
f_T (GHz)	220	215	336
f_{max} (GHz)	550	542	770
C_{gs} (fF)	38	10-22	30-53
C_{gd} (fF)	11	3-24	3-17

Table 4. The search ranges of the design parameters and the optimized value.

x	Parameter name	Search range	Optimized value
1	Substrate thickness (um)	50-100	61
2	Buffer thickness (um)	0.2-0.8	0.2
3	Channel thickness (nm)	4-20	4
4	Spacer thickness (nm)	2-10	2
5	Barrier thickness (nm)	8-15	8
6	Cap thickness (nm)	8-25	8.5
7	Delta-doping concentration (cm^2 /Vs^{-1})	1×10^{12}-1×10^{13}	9.4×10^{12}
8	Indium fraction of channel layer	0.6-0.85	0.85
9	Location of delta doping (nm)	4-21	7
10	Passivation thickness (nm)	40-100	40
11	Recessed thickness (nm)	0-7	5
12	Gate foot location (um)	0.45-1.05	0.47
13	Gate-source separation (um)	0.1-0.6	0.1
14	Gate-drain separation (um)	0.1-0.6	0.1
15	Schottky barrier (eV)	0.6-1.5	0.6
16	Contact resistance (ohm·um)	30-120	30

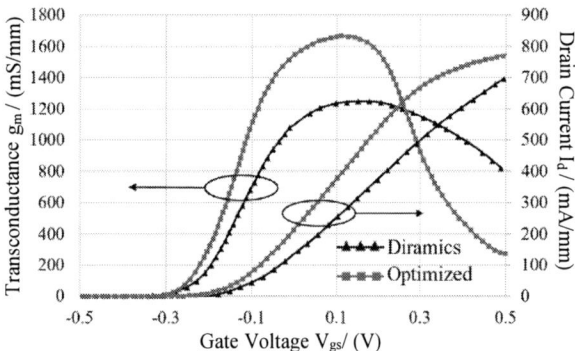

Fig. 6. IV characteristics and transconductance of the experimental (Diramics) and the optimized 100 nm pHEMT at ambient temperature, V_{ds}=0.9 V.

conditions and good results can be obtained. The f_T and f_{max} are improved to 336 GHz and 770 GHz compared to 220 GHz and 550 GHz that are obtained by Diramics, respectively. In addition, Fig. 6 shows the transfer characteristic of the transistor of the optimized HEMT is higher than the Diramics and demonstrates its lower at a higher frequency. It can be seen that after optimization: (1) the indium content in the channel layer is increased to 85%; (2) the mobility approaches to 14000 cm^2/Vs^{-1}; (3) the parasitic capacitance is reduced by the asymmetric structure. Therefore, this optimized pHEMT

has a higher operating frequency in f_T and f_{max}. The results are summarized in Table 3.

IV. CONCLUSION

In this paper, machine learning-assisted global optimization techniques are introduced to design pHEMT. Remarkably, the Diramics' 100 nm pHEMT is optimized from its epitaxy layer and device structure. After optimization, the cut-off frequency and maximum oscillation frequency are improved from 215 GHz to 336 GHz and 542 GHz to 770 GHz, respectively. Moreover, the transconductance and current are improved to 1.6 S/m and approach 0.8 A/mm, respectively. The total number of Sentaurus simulation iterations is 240, costing 16 hours in a normal desktop computer. This new method shows high potential in terms of efficiency and optimization quality for transistor design.

REFERENCES

[1] R. Appleby and H. B. Wallace, "Standoff detection of weapons and contraband in the 100 GHz to 1 THz region," *IEEE Transactions on Antennas and Propagation.*, vol. 55, pp. 2944-2956, Nov. 2007.

[2] D. Jasteh, M. Gashinova, E. G. Hoare, T. Tran, N. Clarke, and M. Cherniakov, "Low-THz imaging radar for outdoor applications," *in 2015 16th International Radar Symposium (IRS)*, 2015, pp. 203-208.

[3] *World Radiocommunication conference 2019 (WRC-19) Provisional Final Acts*, ITU, 2019, p. 60.

[4] X. Mei et al., "First demonstration of amplification at 1 THz using 25-nm InP high electron mobility transistor process," *IEEE Electron Device Letters.*, vol. 36, pp. 327-329, Apr. 2015.

[5] B. Liu, H. Aliakbarian, Z. Ma, G. A. E. Vandenbosch, G. Gielen, and P. Excell, "An efficient method for antenna design optimization based on evolutionary computation and machine learning techniques," *IEEE Transactions on Antennas and Propagation.*, vol. 62, pp. 7-18, Jan. 2014.

[6] H. Sugiyama, H. Matsuzaki, H. Yokoyama, and T. Enoki, "High-electron-mobility $In_{0.53}Ga_{0.47}As/In_{0.8}Ga_{0.2}As$ composite-channel modulation-doped structures grown by metal-organic vapor-phase epitaxy," *in 2010 22nd International Conference on Indium Phosphide and Related Materials (IPRM)*, 2010, pp. 1-4.

[7] Y. C. Chen et al., "Composite-channel InP HEMT for W-band power amplifiers," *in Eleventh International Conference on Indium Phosphide and Related Materials (IPRM'99)*, 1999, pp. 305-306.

[8] M. D. Lange et al., "InAs/InGaAs composite-channel HEMT on InP: Tailoring InGaAs thickness for performance," *in 2008 20th International Conference on Indium Phosphide and Related Materials*, Versailles, 2008, pp. 1-4.

[9] J. Hao, J. Wan, B. Lu, and Y. Chen, "A theoretical study of gating effect on InP/InGaAs HEMTs by tri-layer T–shape gate," *Microelectronic Engineering*, 2019, pp. 54-59.

[10] B. Liu, Q. Zhang, and G. G. E. Gielen, "A gaussian process surrogate model assisted evolutionary algorithm for medium scale expensive optimization problems," *IEEE Transactions on Evolutionary Computation*, vol. 18, pp. 180-192, Apr. 2014.

[11] R. Storn and K. Price, "Differential evolution – A simple and efficient heuristic for global optimization over continuous spaces," *Journal of Global Optimization*, 1997, vol. 11, pp. 341–359.

[12] Z. Wang, B. Shakibi, L. Jin and N. Freitas, "Bayesian multi-scale optimistic optimization," *Artificial Intelligence and Statistics*, vol. 33, pp. 1005-1014, Feb. 2014.

[13] The Diramics website. [online] Available: https://diramics.com/wp-content/uploads/downloads/DIRAMICS-pH-100-4F20.pdf

Proceedings of the 16th European Microwave Integrated Circuits Conference

Low-Power *Ka*- and *V*-Band Miller Compensated Amplifiers in 130-nm SiGe BiCMOS Technology

Batuhan Sutbas[*#1], Herman Jalli Ng[†], Jan Wessel[*], Alexander Koelpin[‡], Gerhard Kahmen[*#]

[*]IHP – Leibniz-Institut für innovative Mikroelektronik, Frankfurt (Oder), Germany
[#]Brandenburg University of Technology, Cottbus, Germany
[†]Karlsruhe University of Applied Sciences, Karlsruhe, Germany
[‡]Hamburg University of Technology, Hamburg, Germany
[1]suetbas@ihp-microelectronics.com

Abstract — **This paper presents the design of two low-power differential amplifiers at 30 GHz and 60 GHz in a 130-nm SiGe BiCMOS technology. The amplifiers use cross-connected compensation transistors in a common–emitter configuration to improve the gain-bandwidth product. This technique also provides unconditional stability which is ensured by analyzing the core circuit at the process corners for various operation temperatures. The design procedure is verified by experimental results which show that the 30 GHz amplifier consumes 5.3 mW and has a peak small-signal gain of 14.9 dB, while the 60 GHz amplifier consumes 12 mW with a peak small-signal gain of 12.5 dB. In addition, 6 dBm output power is measured from the 30 GHz amplifier at 1 dB compression and more than 10 dBm output power is expected from the 60 GHz amplifier. The low power consumption, high gain, and high linearity of the proposed amplifier blocks show that the described approach is promising for low-power and high-efficiency millimeter-wave RF frontends.**

Keywords — **integrated circuit, low-power, Miller effect, mm-wave, silicon-germanium.**

I. Introduction

Silicon-germanium (SiGe) BiCMOS technologies offer high level integration with moderate mask costs. Their main advantage of high transit frequency (f_T) and high linearity transistors over advanced CMOS nodes make them attractive for millimeter-wave (mm-wave) RF frontends aiming low-power applications. Recent works have already demonstrated highly miniaturized and high performance radar sensors which are promising for various commercial applications [1]–[3]. System- and circuit-level optimization toward lower power dissipation will undeniably extend their use into mobile applications such as vital-sign detection and gesture tracking. The high f_T headroom of the SiGe BiCMOS technologies can enable a reduction of the current consumption without compromising the high performance.

A typical approach in mm-wave radar sensors is to use a subharmonic frequency synthesizer along with a multiplier to generate the LO signal at the desired frequency. As a result, several amplifier blocks are required at the subharmonic and fundamental frequencies. In this paper, the design of two amplifiers for a low-power and high-efficiency 60 GHz radar sensor is described. The circuits are fabricated using a 130-nm SiGe BiCMOS technology and verified by on-wafer measurements. The transistors are operated in the low-current

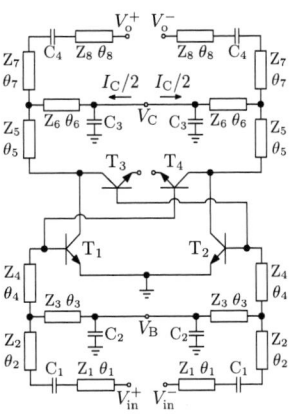

Fig. 1. Circuit schematic of the differential amplifiers.

regime where a sufficiently high gain is still attainable thanks to the high f_T headroom of the technology. The available gain is further improved by employing a Miller capacitance cancellation technique. The presented results show that low power consumption and high gain-bandwidth product can be obtained at the same time when the single-stage amplifier is carefully designed according to this approach.

II. Design Considerations

A. Collector–Base Capacitance Cancellation

The collector–base capacitance of a BJT is one of the limiting factors on the gain-bandwidth product. Although the cascode configuration mitigates this effect, it is not suitable for low-power circuits since a higher voltage supply is required. Instead, effectively negative capacitances can be used in a differential common–emitter configuration to cancel the collector–base capacitances [4], [5]. When the cancellation is designed carefully the circuit is unconditionally stabilized and the maximum available gain of the transistor is increased.

In this work, the proposed Miller effect cancellation technique is employed to design two low-power differential amplifiers at 30 GHz (DA30) and 60 GHz (DA60). A common circuit schematic for both amplifiers is shown in Fig. 1. The active part is composed of the amplifying common–emitter transistors T_1 and T_2, and the cross-connected compensation

3–4 April 2022, London, UK

transistors T_3 and T_4. The compensating pair of transistors have their emitters left open however they have the same collector–base voltage with the amplifying transistors. Therefore, when the collectors of the compensating transistors are connected to the inverting collectors of the amplifying transistors, these collector–base capacitors effectively behave as negative capacitances which cancel the Miller capacitances. Both amplifier circuits are biased with a collector current less than 0.4 mA per emitter finger which drops the effective f_T by about 40%. Nevertheless, high gain can be still obtained thanks to the compensation technique and high f_T of the technology.

B. Stability

Although the best improvement in the maximum available gain of the transistor is attained when the collector–base capacitance is completely eliminated, a deviation from the design expectations might cause a positive feedback which must be avoided at all cost. Therefore, a more practical design approach is to guarantee that the collector–base capacitance is not fully canceled across process and temperature variations and the compensating capacitances must be designed smaller than the optimum capacitance values as a safety measure.

Fig. 2 shows the optimum capacitance value required in the Miller effect cancellation loop to stabilize and maximize the available gain from a unit single-emitter transistor as a function of the collector–base voltage at various temperatures and at the process corners. The typical value is 1.75 fF and might vary from 1.6 fF to 1.95 fF where the presented amplifiers operate. Compensating pair of transistors instead of fixed MIM or MOM capacitors are used in the design to have a better matching to the collector–base capacitance across process and temperature variations. The metal stack on the amplifying transistors are raised higher in DA30 and the compensating transistors are made smaller in DA60 to ensure that the collector–base capacitance is not fully canceled.

C. Biasing and Matching Networks

A single supply voltage of 1.5 V is used in both DA30 and DA60. The base voltages are generated internally using single-emitter mirror transistors (not shown in Fig. 1). The bases and collectors of the transistors are biased via

Table 1. Design parameters of the 30 GHz and 60 GHz amplifiers.

Prm.	DA30	DA60	Prm.	DA30	DA60	Prm.	DA30	DA60
Z_1	75 Ω	80 Ω	θ_1	2°	15°	C_1	0.06 pF	0.3 pF
Z_2	85 Ω	85 Ω	θ_2	2°	9°	C_2	1.4 pF	1.4 pF
Z_3	85 Ω	85 Ω	θ_3	45°	15°	C_3	1.4 pF	1.4 pF
Z_4	75 Ω	90 Ω	θ_4	6°	8°	C_4	0.06 pF	0.75 pF
Z_5	95 Ω	80 Ω	θ_5	3°	7°	T_1	4×0.9u	10×0.9u
Z_6	90 Ω	70 Ω	θ_6	45°	27°	T_2	4×0.9u	10×0.9u
Z_7	80 Ω	85 Ω	θ_7	1°	3°	T_3	4×0.9u	8×0.9u
Z_8	95 Ω	75 Ω	θ_8	3°	8°	T_4	4×0.9u	8×0.9u
V_C	1.5 V	1.5 V	V_B	0.84 V	0.84 V	I_C	3.5 mA	8 mA

conventional bias tee circuits where the transmission lines are incorporated into the matching by forming T-type networks.

The amplifying transistors in DA60 are made ×2.5 larger compared to DA30 to reach the higher output power goal required at 60 GHz. DA30 is designed for better input and output return losses aiming higher gain, while DA60 is optimized for higher output power and linearity. All transmission lines are meandered to save chip area. Both the input and output matching networks utilize series capacitors to serve as dc blockers when they are integrated with the other building blocks. The selected component parameters of the amplifiers are listed in Table 1.

III. EXPERIMENTAL RESULTS

A. Technology and Experimental Setup

The presented amplifier designs are realized using the high performance IHP SG13G2 process. The 130 nm SiGe BiCMOS technology features high f_T and f_{max} of 300 GHz and 500 GHz [6]. The topmost 3 μm-thick aluminum layer is used for the microstrip lines. Micrographs of the fabricated amplifiers are shown in Fig. 3. Here, the functional parts of the amplifiers, which exclude the compensated on-wafer measurement pads and the additional 100 Ω differential feed lines, are indicated by the dotted frames. DA30 and DA60 effectively occupy 0.20 mm × 0.56 mm and 0.16 mm × 0.36 mm of chip area, respectively.

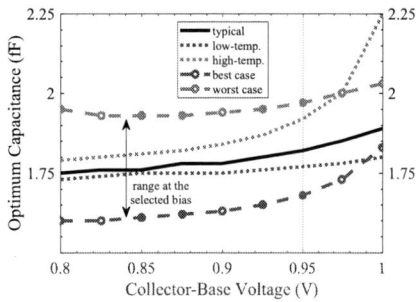

Fig. 2. Optimum capacitance value to stabilize and maximize the available gain from a unit single-emitter npn-type BJT as a function of the collector–base voltage at typical (25°C), low-temperature (−40°C), and high-temperature conditions (85°C) as well as at best case and worst case process corners.

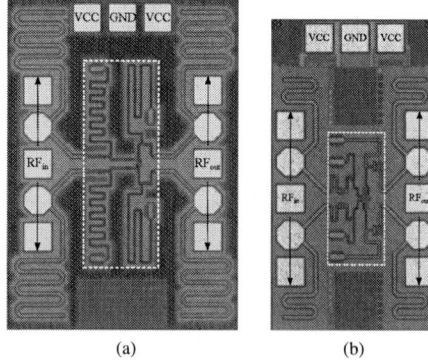

(a)   (b)

Fig. 3. Micrographs of the fabricated (a) 30 GHz and (b) 60 GHz amplifiers.

On-wafer small-signal measurements are performed using differential RF[1] and dc[2] probes together with a 4-port vector network analyzer (ZVA67[3]). In addition, a power sensor (NRP-Z57[3]), a spectrum analyzer (FSW67[3]), and a programmable power supply (E36312A[4]) are utilized for the large-signal measurements. Contribution of the RF pads to the measurement results are de-embedded with the help of a back-to-back pad test structure. The de-embedded results are in good agreement with the simulation expectations supported by the transistor models and Momentum[4].

B. Small-Signal Measurement Results

Fig. 4 shows the simulated and measured S-parameters of DA30. The peak gain of 14.9 dB is measured at 29 GHz and the measured 3 dB bandwidth covers 26 GHz to 32.5 GHz with a fractional bandwidth of 22%. The measured reverse isolation is around 35 dB thanks to the Miller effect cancellation technique, which is otherwise difficult to achieve in a conventional common–emitter topology.

Fig. 5 shows the simulated and measured S-parameters of DA60. The peak gain of 12.5 dB is measured at 61.5 GHz. The 3 dB bandwidth starts from 51 GHz and extends beyond the maximum measurement frequency of 67 GHz. The extrapolated upper frequency limit is 74.5 GHz which yields an estimated fractional bandwidth of 37%.

C. Large-Signal Measurement Results

Fig. 6a shows the simulated and measured power gain, output power and power-added efficiency (PAE) of DA30 at 30 GHz for an available input power sweep from −12.5 dBm to 2.5 dBm. When the input drive is about −8.5 dBm, the power

[1]Cascade Microtech, Inc., Beaverton, OR, USA
[2]GGB Industries, Inc., Naples, FL, USA
[3]Rohde & Schwarz GmbH & Company KG, Munich, Germany
[4]Keysight Technologies, Inc., Santa Rosa, CA, USA

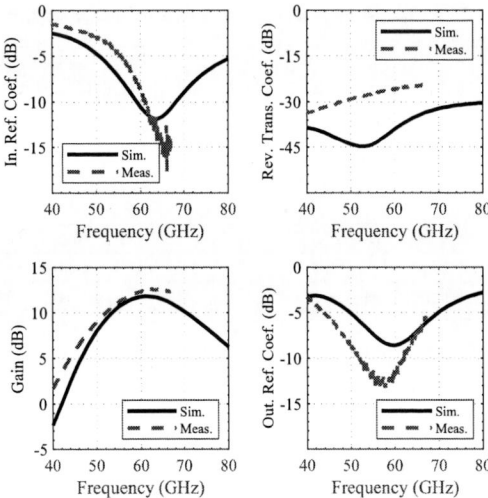

Fig. 5. Simulated and measured (from top left to bottom right): $|S_{11}|$, $|S_{12}|$, $|S_{21}|$, and $|S_{22}|$ of DA60 as a function of frequency.

gain compresses by 1 dB to 13.3 dB. Here, the output power and PAE are measured as 4.8 dBm and 16.1%, respectively. The experimental large-signal performance of DA30 is slightly deteriorated compared to the simulation expectations as the output matching is shifted to lower frequencies. Fig. 6b shows the simulated power gain, output power and PAE of DA60 at 60 GHz for an available input power sweep from −15 dBm to 5 dBm. The output power and PAE at 1 dB compression are simulated as 9.8 dBm and 28.8%, respectively.

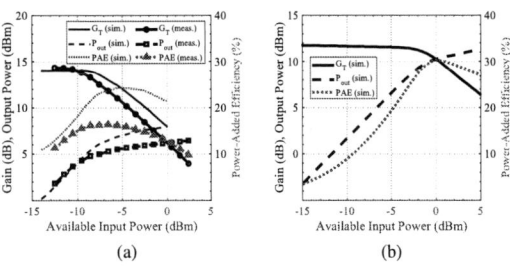

Fig. 6. Power gain, output power, and power-added efficiency of (a) DA30 (simulated and measured) and (b) DA60 (simulated) as a function of available input power.

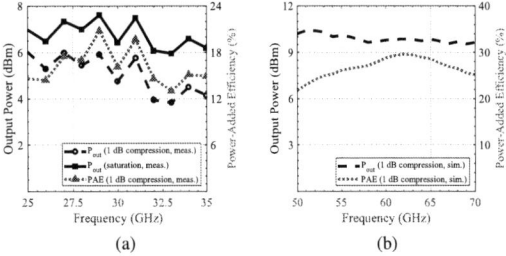

Fig. 7. Output power and power-added efficiency of (a) DA30 (measured) and (b) DA60 (simulated) as a function of input frequency.

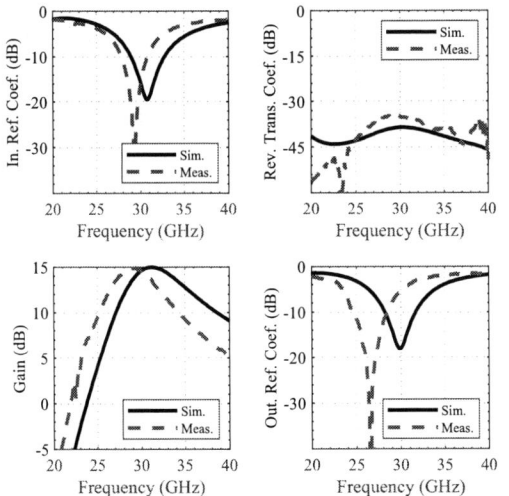

Fig. 4. Simulated and measured (from top left to bottom right): $|S_{11}|$, $|S_{12}|$, $|S_{21}|$, and $|S_{22}|$ of DA30 as a function of frequency.

Table 2. Performance of recent SiGe BiCMOS low-power amplifiers.

Ref.	f_T/f_{max} (GHz)	f_c (GHz)	BW_{3dB} (GHz)	G_T (dB)	V_{cc} (V)	P_{dc} (mW)	$P_{out,1dB}$ (dBm)	Size (mm^2)
[7]	240/270	61	43–67	32.5	1.8	11.7	-6.5	0.39
[8]	200/200	27	22–32	18.6	1.8	9.0	3.0	0.46
[9]	240/270	60	52–66	15.0	2.5	11.8	0.3	0.11
[10]	270/212	32.5	29–36	18.0	0.85	5.9	-16	0.23
[11]	250/500	24	22–25	14.4	1.5	10.5	1.6	0.93^p
[12]	300/450	70	54–77	23.8	2.0	8.0	-3.6	0.41^p
[13]	300/500	61	44–67^m	13.3	1.5	5.1	7.1	0.06
DA30	300/500	30	26–33	14.5	1.5	5.3	4.8	0.11
DA60	300/500	61	51–67^m	12.5	1.5	12.0	9.8^s	0.06

^{m}measurement limitation, ^{s}simulated, ^{p}including pads

Fig. 7 shows the output power and PAE of the amplifiers at 1 dB compression as a function of frequency. The measured $P_{out,1dB}$ and PAE_{1dB} of DA30 vary between 3.8 dBm and 6 dBm, and between 13% and 21%, respectively, from 25 GHz to 35 GHz. The simulated $P_{out,1dB}$ and PAE_{1dB} of DA60 is at least 9.5 dBm and 22%, respectively, from 50 GHz to 70 GHz.

Performance parameters of recent low-power amplifiers based on SiGe BiCMOS technologies are summarized in Table 2. Here, the values are reported at the center frequencies f_c of the amplifiers. The compact amplifiers presented in this paper have higher gain per stage and higher output power at 1 dB compression while having similar power consumption which make them favorable in the design of low-power and highly efficient radar sensors.

IV. CONCLUSION

Two low-power differential amplifiers at 30 GHz and 60 GHz are designed using a Miller capacitance cancellation technique with an optimized core circuit to be stable in case of process and temperature variations. Instead of using passive MIM or MOM capacitors, bipolar transistors are employed to realize effectively negative capacitances. The circuits implemented in a SiGe BiCMOS technology demonstrate high gain and high linearity while consuming low dc current and occupying small area. Therefore, the presented design approach is promising for the amplifiers required by low-power and high-efficiency mm-wave RF frontends.

ACKNOWLEDGMENT

This work has been funded by the Federal Ministry of Education and Research of Germany (BMBF) within the iCampus Cottbus Project under Grant 16ES1128K.

REFERENCES

[1] I. Nasr, R. Jungmaier, A. Baheti, D. Noppeney, J. S. Bal, M. Wojnowski, E. Karagozler, H. Raja, J. Lien, I. Poupyrev, and S. Trotta, "A highly integrated 60 GHz 6-channel transceiver with antenna in package for smart sensing and short-range communications," *IEEE J. Solid-State Circuits*, vol. 51, no. 9, pp. 2066–2076, Sep. 2016.

[2] H. J. Ng, M. Kucharski, W. Ahmad, and D. Kissinger, "Multi-purpose fully differential 61- and 122-GHz radar transceivers for scalable MIMO sensor platforms," *IEEE J. Solid-State Circuits*, vol. 52, no. 9, pp. 2242–2255, Sep. 2017.

[3] H. J. Ng and D. Kissinger, "Highly miniaturized 120-GHz SIMO and MIMO radar sensor with on-chip folded dipole antennas for range and angular measurements," *IEEE Trans. Microw. Theory Techn.*, vol. 66, no. 6, pp. 2592–2603, Jun. 2018.

[4] C. C. Cheng, "Neutralization and unilateralization," *IRE Trans. Circuit Theory*, vol. CT-2, no. 2, pp. 138–145, Jun. 1955.

[5] J. A. Mataya, G. W. Haines, and S. B. Marshall, "IF amplifier using C_c compensated transistors," *IEEE J. Solid-State Circuits*, vol. SSC-3, no. 4, pp. 401–407, Dec. 1968.

[6] H. Rucker, B. Heinemann, and A. Fox, "Half-terahertz SiGe BiCMOS technology," in *Proc. IEEE 12th Top. Meeting Silicon Monolithic Integr. Circuits RF Syst.*, Santa Clara, CA, USA, Jan. 2012, pp. 133–136.

[7] S. Jang and C. Nguyen, "A high-gain power-efficient wideband V-band LNA in 0.18-μm SiGe BiCMOS," *IEEE Microw. Wireless Compon. Lett.*, vol. 26, no. 4, pp. 276–278, Apr. 2016.

[8] C. Geha, C. Nguyen, and J. Silva-Martinez, "A wideband low-power-consumption 22-32.5-GHz 0.18-μm BiCMOS active balun-LNA with IM2 cancellation using a transformer-coupled cascode-cascade topology," *IEEE Trans. Microw. Theory Techn.*, vol. 65, no. 2, pp. 536–547, Feb. 2017.

[9] S. Zihir and G. M. Rebeiz, "A wideband 60 GHz LNA with 3.3 dB minimum noise figure," in *IEEE MTT-S Int. Microw. Symp. Dig.*, Honolulu, HI, USA, Jun. 2017, pp. 1969–1971.

[10] A. A. Alhamed and G. M. Rebeiz, "A 28-37 GHz triple-stage transformer-coupled SiGe LNA with 2.5 dB minimum NF for low power wideband phased array receivers," in *Proc. IEEE BiCMOS Compound Semiconductor Integr. Circuits Technol. Symp.*, Monterey, CA, USA, Nov. 2020.

[11] V. Issakov and A. Werthof, "A 10 mW LNA with temperature compensation for 24 GHz radar applications in SiGe BiCMOS," in *Proc. IEEE BiCMOS Compound Semiconductor Integr. Circuits Technol. Symp.*, Monterey, CA, USA, Nov. 2020.

[12] A. Ferschischi, H. Ghaleb, S. U. Rehman, C. Carta, and F. Ellinger, "A broadband 60-GHz low noise amplifier with 3.2 dB noise figure and 24 dB gain," in *Proc. 15th Eur. Microw. Integr. Circuit Conf.*, Utrecht, Netherlands, Jan. 2021, pp. 233–236.

[13] B. Sutbas, H. J. Ng, J. Wessel, A. Koelpin, and G. Kahmen, "A V-band low-power compact LNA in 130-nm SiGe BiCMOS technology," *IEEE Microw. Wireless Compon. Lett.*, vol. 31, no. 5, pp. 497–500, May 2021.

Gap in pagination due to unavailable paper.

Page 75

A High GBW High Power Wideband Power Amplifier for Automotive Radar Application

Kambiz Hadipour, Dominik Amschl, Daniel Knauder, Stefano Di Martino

Infineon Technologies Austria, Austria

{kambiz.hadipourabkenar, dominik.amschl, daniel.knauder, stefano.dimartino}@infineon.com

Abstract —Availability of large fractional bandwidth has made mm-wave frequencies the frequency range of choice for several well-established applications. Design of high gain, wideband power amplifier has always been a challenge since efficiency as one key performance parameter trades with the gain-bandwidth product. The gain-bandwidth is mainly limited by the parasitics due to the large form factor of active stages needed to deliver the required power. In this paper wideband, inductively-coupled transformer-based inter-stage/output matching networks together with improved layout techniques for transformers are employed to implement a four stage power amplifier. The E-band PA is realized in standard 28nm CMOS technology and delivers 18.0dBm of saturated output power with a peak power added efficiency of 17.9% and a gain-bandwidth product of 567.7GHz. The output power of PA is adjustable via some digital controls and its performance is robust versus Process-Voltage-Temperature (PVT) variations which makes it suitable for automotive radar applications.

Keywords — Power amplifier, inductively coupled resonator, mm-wave, gain-bandwidth, CMOS integrated circuits.

I. Introduction

Mm-waves frequencies have emerged as natural host for the development of wideband wireless transceivers suitable for high data rate communication, imaging and automotive applications. Most of these applications demand wide fractional bandwidths to achieve high spatial resolution and minimize acquisition time. Multi-band mm-wave amplifiers are attractive solutions. Nevertheless it is hard to maintain high performance over the full range of operation.

To increase the safety and comfort of passengers in automotive industry, long- and short-range radars are exploited for adaptive cruise control (ACC), collision warning, blind spot detection, and many other features [1]. Due to its lower cost and higher integration capabilities, CMOS is the technology of choice for the next generation of automotive radars. However, this comes with new challenges as advanced CMOS nodes lend themselves into narrowband operation requiring leverage of high order matching networks to maintain high gain and bandwidth demands.

In this paper 4th order inter-stage/output matching networks are employed to achieve a high gain within a wide bandwidth for the power amplifier while keeping the efficiency high. The 4th order matching networks are then replaced by their equivalent transformer-based structures to minimize the number of components, realize a compact design and offer the possibility for DC decoupling of different amplifier stages. To

increase output power without compromising efficiency, power combining is exploited at the output of PA. To increase the power transfer efficiency of the transformer-based combiner, some layout techniques are proposed which increase the mutual coupling of the coils and reduce the insertion loss of the combiner, improving the power added efficiency (PAE).

This paper is organized as follows: in Section II realization of the wideband matching network is described. Since layout is of extreme importance at mm-wave, section III discusses the ideas to optimize the layout of the transformers. Design of the power amplifier is described in section IV with the measurement realized presented in section V.

II. Wideband Matching Network

In addition to the conventional power amplifiers' performance parameters such as output power and efficiency, several emerging applications demand higher operating bandwidth. This not only helps for having a robust flat output power versus PVT variations but also allows usage of simpler modulation schemes for low complexity communication systems or higher resolution for imaging and radar applications.

Achieving good performance over a large bandwidth is very challenging for mm-wave PAs where the large device sizes cause considerable parasitics which not only limits the power gain of transistors but also impacts the bandwidth. Since for a power amplifier the focus is to increase the PAE, the last stages are usually biased at a low level and the driver stages are sized to have a small form factor both of which will reduce the transconductance and the gain-bandwidth (GBW). To realize the required impedance transformation and to increase the GBW, inductively coupled inter-stage/output matching networks similar to [2, 3] are employed in this design.

Fig. 1(a) shows the inductively-coupled resonator which is able to accommodate the large parasitic capacitances of the devices. In this figure C_1 and R_1 denote the output impedance of the first stage and C_2 and R_2 resemble the input impedance of the second stage (intrinsic device parasitics). L_1 and L_2 resonate at the centre frequency of the band with C_1 and C_2, respectively, while L_3 realizes the inductive coupling between the two sides. This network has simple topology and thus the insertion loss can be small.

The transimpedance (V_{out}/I_{in}) frequency response of this network shows two peaking frequencies:

$$\omega_L = \omega_0$$

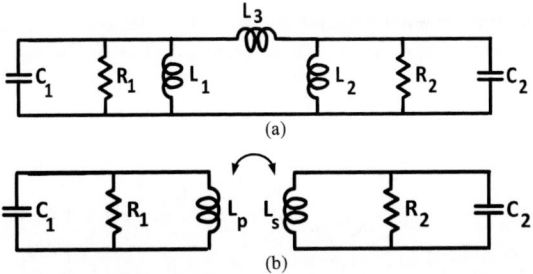

Fig. 1. (a) Inductively coupled resonator (b) transformer based equivalent of inductively coupled resonator

$$\omega_H = \omega_0 \sqrt{1 + \frac{L_1}{L_3} + \frac{L_2}{L_3}}.$$

Where ω_0 is the resonance frequency of C_1-L_1 and C_2-L_2. The bandwidth and the ripple of the matching network are defined by the coupling inductor L_3 and the quality factor (Q). For a fixed Q, the bandwidth can be increased by decreasing the size of L_3 at the cost of more in-band ripple. This in-band ripple can then be compensated by stagger tuning different amplifier stages.

The input impedance of the coupled resonator remains constant across large bandwidth. This is particularly important for the design of the output matching network of the PA where presenting the optimum load impedance is of utmost importance.

To reduce the number of elements and enable a compact layout, the three inductors in Fig. 1(a) are replaced by their transformer equivalents after Norton transformation, inductor combining and splitting as shown in Fig. 1(b). The transformers also make DC decoupling of the stages possible to separately provide bias and supply.

III. TRANSFORMER LAYOUT OPTIMIZATION

Transformers and transformer-based baluns and power combiners with low insertion loss are extensively used in modern power amplifier design. Two important parameters for the insertion loss of such transformers are the quality factors of the coils and the coupling between the primary and secondary windings. As the frequency of operation and the form factor of active stages increase, the size of transformers become smaller which makes achieving a high coupling coefficient more difficult.

Different approaches have been employed to reduce the insertion loss of transformers/combiners. These include using stacked transformers, interleaved structures with minimum spacing between transformer turns or even new layout implementation such as Quadrafilar combiners or combiners based on differently sized transformers.

A metal stack with two thick metals usually is the preferred option for transformer design especially at mm-wave; however, where other constraints dictate a single thick metal, the approach used in this paper offers its substantial merits [4]. In such case, interleaved structures on the top thick metal become the first option for realizing the transformers. Current handling

capabilities - critical for Electromigration - as well as quality factor benefit from this choice. Nevertheless, due to the minimum spacing requirement imposed by DRC rules such coils cannot achieve very low insertion loss as this spacing limits the achievable coupling between the primary and secondary sides.

In this work we employ a combination of interleaved and stacked structures to use the advantages of both the approaches [4]. The transformers are initially realized as interleaved coils by placing both primary and secondary turns on the uppermost/thickest metal layer. Furthermore, an additional winding geometry is connected in parallel to either primary or secondary of the coil on a lower metal layer which goes below the other turn (Fig. 2). Consequently, the coupling between the two coils increases.

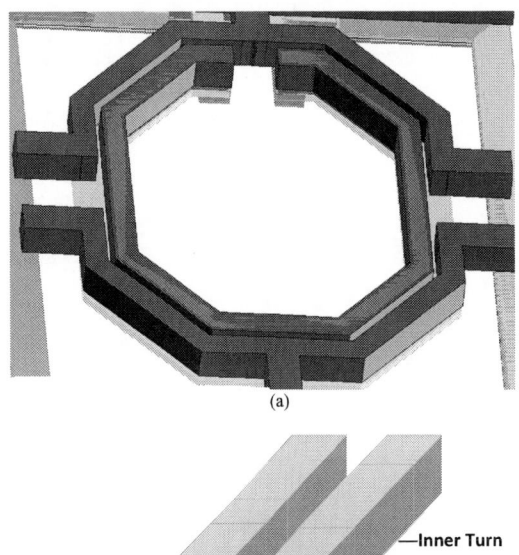

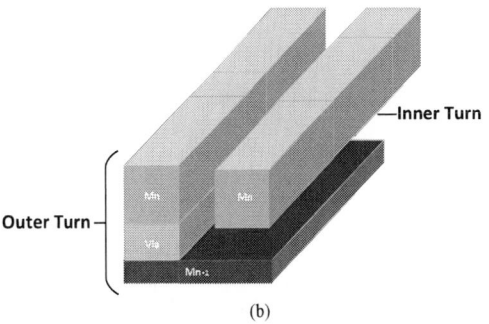

Fig. 2 Interleaved-stacked transformer (a) 3-D view (b) cross-side view

The addition of the stacked turn can slightly change the quality factor of the coil and add more parasitic capacitance to it. However, in many applications the quality factor of the matching network is dominated by the quality of impedances seen at the input/output of the coil. As a result the potential reduction in the Q of the primary or secondary turn will not be an issue. The increased parasitic capacitance shifts the self-resonance frequency (SRF) of the coil by some GHz. Since the self-resonance frequency is usually far beyond the frequency of operation of the coil, a shift in the range of few GHz in the SRF cannot have big impact on the coil's operation. Fig. 3 illustrates the improvement in K and insertion loss of a 1:1 transformer after employing the stacked-interleaved approach.

Non-isolated combiners are extensively used due to the compact dimension and low losses. To further reduce the size of the power combiner and minimize the insertion loss of the series combiner at the output of the PA – which is critical for the overall output power and efficiency of the amplifier- it is realized as shown in Fig. 2(a) which alleviates the need for any additional connection tapers on the secondary side.

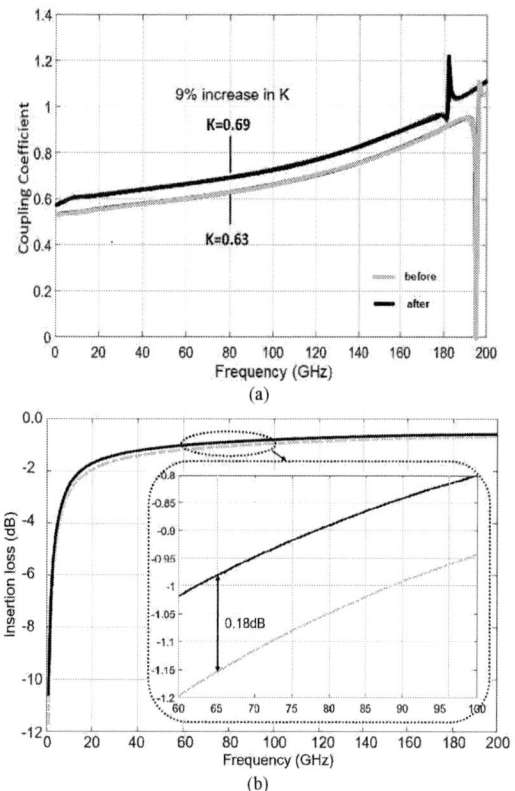

Fig. 3. Improvement in (a) K and (b) insertion loss of a sample transformer after employing the interleaved-stacked approach

IV. DESIGN OF THE POWER AMPLIFIER

Fig. 4(a) shows the architecture of the PA. Since the PA needs to deliver a saturated output power of more than 15 dBm for an input power lower than 0 dBm, it is composed of four amplifying stages to satisfy the need for high power gain. All the stages employ the conventional (Pseudo)-differential common source architecture with cross-coupled neutralization. The first two stages are based on differential stages with bottom current source while the last two stages employ Pseudo-differential architecture to save the voltage headroom required for high output power (Fig. 4(b)). The Neutralizing cross coupled capacitor C_1 and C_2 are realized by MOS devices having half the width of main amplifying transistors M_1 and M_2 to make the neutralization robust with respect to PVT variations.

A 2-way transformer-based series combiner and several 1:1 inter-stage transformers derived from the broadband 4th order matching network described in section II are employed to enhance the GBW product without inserting additional lossy components. The pad capacitance is considered as part of the input/output matching network so that a shunt stub is not required to resonate it out for measurement purposes. To further extend the GBW and compensate the in-band ripple in the frequency response of different stages, the amplifying stages are stagger tuned.

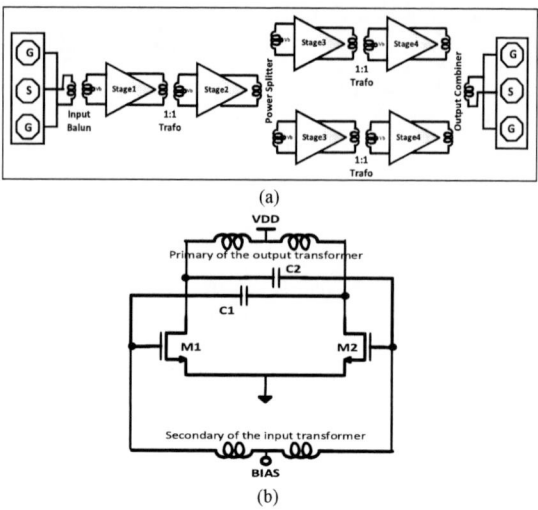

(a)

(b)

Fig. 4. (a) Architecture of PA (b) schematic of the amplifying stages

The bias of all the PA stages are controlled through DACs which makes the change of output power with fine steps possible as required by automotive applications.

V. MEASUREMENT RESULTS

The power amplifier is realized in standard 28nm CMOS technology. The chip micrograph is shown in Fig. 5. Thanks to the simple realization of the inter-stage/output matching networks and the compact floor plan the PA has a relatively small area.

Fig. 6 shows the simulated vs. measured small signal parameters. The PA achieves a peak gain of 31.0 dB in a 3-dB bandwidth of 16 GHz extending from 70.0 GHz to 86.0 GHz translating to a GBW product of 567.7 GHz at E-band.

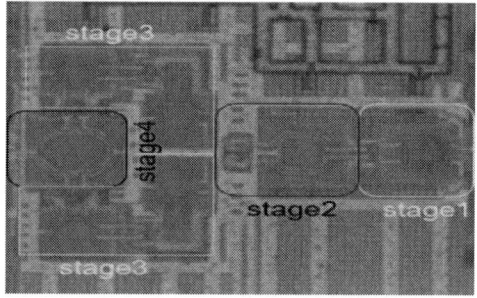

Fig. 5. Chip micrograph (supply pads not shown)

Simulated vs. measured large-signal continuous-wave (CW) performance against P_{in} are shown in Fig. 7. The PA achieves a saturated output power of 18.0 dBm from a 1.0 V supply with a peak power added efficiency of 17.9% at 78.0 GHz.

Table 1. Comparison with state of the art

Ref.	Tech.	Center freq (GHz)	S21 (dB)	BW (GHz)	PSAT (dBm)	PAE (%)	FOM ITRS*	Core Area (um x um)
[1]	28nm	73.0	43.0	18.0	19.5	15.0	111.5	760 x 340
[5]	28nm	79.0	17.0	24.5	12.3	13.8	78.6	360 x 360
[6]	32nm SOI	78.0	12.0	27.0	18.7	24.0	82.3	280 x 265
[7]	65nm	79.0	25.1	15.0	11.5	13.5	85.9	1240 x 400
This work	**28nm**	**78.0**	**31.0**	**16.0**	**18.0**	**17.9**	**99.0**	**520 x 315**

$* \ FOM \ ITRS = P_{SAT}[dBm] + Gain\ [dB] + 20\log(f[GHz]) + 10(PAE[\%])$

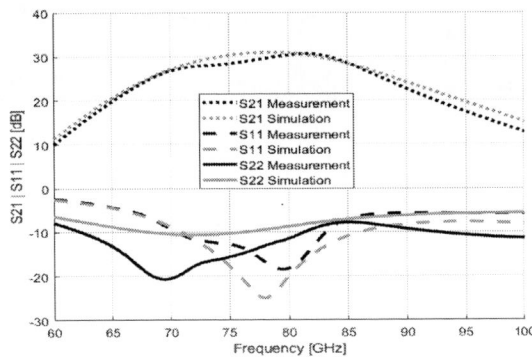

Fig. 6. Small signal measurement vs. simulation results

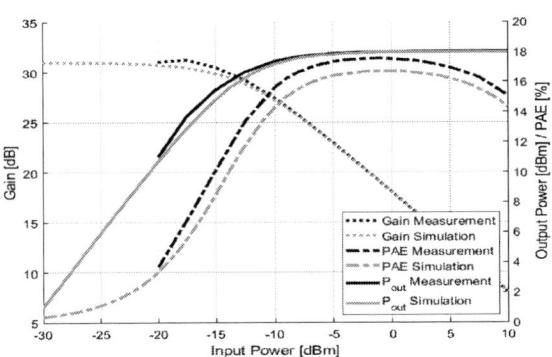

Fig. 7. Measured gain, output power and PAE at 78GHz

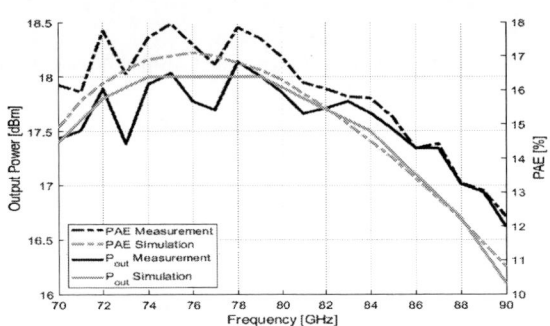

Fig. 8. Output power vs. frequency

VI. CONCLUSION

Design of a high gain-bandwidth E-band CMOS power amplifier is presented in this paper. By optimizing the layout of active stages for low parasitic, employing 4th order matching network and stagger tuning of amplifier stages a high gain over a wide operating range has been achieved. Furthermore, optimization techniques have been employed to enhance the layout of transformers/combiners used in the inter-stage/output matching networks to increase the coupling coefficient and reduce the insertion loss crucial for output power and efficiency of the power amplifier. The PA achieves a saturated output power of 18.0dBm with 17.9% PAE and a GBW of 602.9 GHz.

ACKNOWLEDGMENT

This work is partially supported by the European Commission in the framework of Prystine project (PID:783190).

REFERENCES

[1] D. Reiter, H. Li, H. Knapp, J. Kammerer, J. Fritzin, S. Majied, B. Sene, N. Pohl, „A 19.5dBm Power Amplfiier with Highly Accurate 8-bit Power controlling for Automotive Radar Applications in a 28nm CMOS Technology", *IEEE BCICTS*, Jan 2020

[2] K. Hadipour, A. Ghilioni, A. Mazzanti, M. Bassi, F. Svelto, "A 40GHz to 67GHz Bandwidth 23dB Gain 5.8dB Maximum NF mm-wave LNA in 28nm CMOS", *IEEE RFIC Symp.*, pp. 327-330, May 2015.

[3] M. Bassi, J. Zhao, A. Bevilacqua, A. Ghilioni, A. Mazzanti, F. Svelto, "A 40-67 GHz Power Amplifier With 13 dBm PSAT and 16% PAE in 28nm CMOS LP", *IEEE JSSC*, vol 50, no 7, pp. 1618-1628, Jul 2015.

[4] K. Hadipour, M. Dechant, "Stacked and Interleaved transformer layout", U.S. Patent 2020E00223AT, Jul 2020.

[5] A. Medra, V. Giannini, D. Guermandi, P. Wambacq, " A 79GHz Variable GainLow_Noise Amplifier and Power Amplifier in 28nm CMOS Operating up to 125C", *40th ESSCIRC*, pp. 183-186, May 2016.

[6] J. A. Jayamon, J. F. Buckwalter, P. M. Asbeck, "A PMOS mm-wave power amplifier at 77 GHz with 90 mW output power and 24% efficiency", *IEEE RFIC Symposium*, pp. 262–265, May 2016.

[7] I. Soga, Y. Yagishita, H. Matsumura, Y. Kawano, T. Suzuki, "A 76-81 High Efficiency Power Amplifier for Phased Array Automotive Radar Applications", *IEEE RFIT*, Aug 2015

Fig. 8 shows the output power and efficiency of the PA vs. frequency. The saturated output power is very flat with less than 1 dB variation for a wide frequency range of 67.0 to 87.0 GHz and shows very good agreement with simulations. One important parameter for power amplifier used in radar application is robust behavior vs. PVT variations. In order to investigate PA robustness, on-wafer temperature and statistical measurements have been performed for multiple chips measured across multiple wafers. The output power can be controlled to have minimum variation by changing the PA DAC setting with fine resolution. Phase and amplitude noise of the PA were also tested at 1MHz offset which show a value of -157.7 dBc/Hz and -166.7 dBc/Hz at 10MHz offset from carrier, respectively.

Table 1 summarizes the measured results and provides a comparison with state of the art E-band CMOS PAs. Only ref [1] shows a higher GBW product than the class AB PA in this design. However, it is using a non-standard supply for 28nm bulk CMOS and its gain shows more than 3dB variation within the specified bandwidth.

Proceedings of the 16th European Microwave Integrated Circuits Conference

A highly rugged 39 GHz 19.3 dBm Power Amplifier for 5G Applications in 45nm SOI Technology

Alice Bossuet[#], Baudouin Martineau[#], Cédric Dehos[#], Benjamin Blampey[#], Alexis Divay[#], Yvan Morandini[*]

[#]CEA-Leti, France

[*]Soitec, France

{alice.bossuet, baudouin.martineau}@cea.fr

Abstract —This paper describes a highly rugged power-amplifier for the fifth generation (5G) FR2 new radio (NR) application implemented in a 45nm SOI process (45RFSOI). By using device stacking technique together with an optimized supply voltage reduction, the power amplifier achieves 20 dBm P_{sat} and 23 % PAE_{max}. A P_{avg} of 10 dBm and a PAE_{avg} of 8% is achieved in 64-QAM 200MHz bandwidth OFDM at EVM_{avg} of 6.2% (-24.1dB) without the use of digital predistortion. The 4:1 VSWR measurement shows an excellent PA reliability even under the worst mismatch condition. These results enable an efficient, high power and highly reliable power amplifier for 5G applications.

Keywords — power amplifier, 5G OFDM, 45RFSOI, stacking, realiability

I. INTRODUCTION

While early 5G systems use the sub-6 GHz frequency band (FR1), with one PA operating at high power to amplify a single-beam omnidirectional signal, the 5G FR2 working at millimeter-wave frequencies uses a different approach. Indeed, most 5G mm-wave applications use phased array antennas to focus and steer multiple beams. This ability to divide the transmission among multiple beams implies that each PA operates at lower power making possible to use silicon technologies instead of GaAs. Among these technologies, the partially depleted CMOS SOI (or RFSOI) has already demonstrated its ability to address handset switches, LNA and antenna tuners. Indeed, thanks to SOI substrate with trap rich layer, sub-6 GHz functions have been drastically improved over years. A better performance for switches linearity, RF-Digital isolation or passive component quality factor has been proved [2]. The remain question is does such type of technology can be used for integrated front-end devices needed for 5G mm-wave handsets, access points or base stations ? This work proposes to demonstrate that a 20 dBm output power at the highest 5G frequency band of 40 GHz can be carry out with a high level of robustness with the 45RFSOI technology.

This paper first discuss on circuit design and specificities of this work. Then, the measured performances in nominal and under stress condition are presented. Finally, a comparison with the state of the art is presented.

II. POWER AMPLIFIER DESIGN

Implementing a >20 dBm level efficient PA in deep sub-micron CMOS at 40 GHz is a challenging task due to the low breakdown voltage of these technologies 0,9 − 1 V. One strategy to generate high output power is to combine or stack unit power cells to achieve the maximum output power.

In this work, we present a two stages PA based on the stack topology using the recently added PAFET drain extended devices that present improved hot carrier injection (HCI) performances and an improved off-state junction breakdown [1]. In any power amplifier in class A, AB or B, the drain to source RF voltage swing of the output stage is $2xV_{supply}$. Therefore, any devices biased at nominal technology voltage is exposed to reliability issue in large signal operation, making them sensitive to HCI and time dependent dielectric breakdown TDDB. It appears that most designs are biased at upper limit, or even boosted sometimes [5][6][7][8]. To design a PA conform for manufacturing, this aspect must be take into account when RF performances are benchmarked. In that work, to solve that issue, power supply is derated by 20 % to secure aging and reliability.

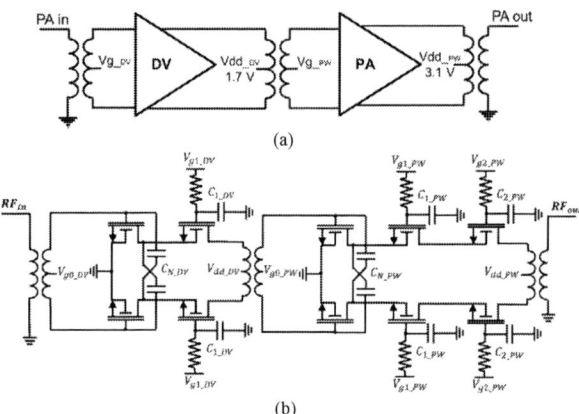

Fig. 1. Power amplifier synoptic (a) and detailed schematic with the DC voltages names (b).

The power amplifier Fig. 1 is composed of a driver stage followed by a power stage to meet both gain and high output power performances. The cascode driver stage is composed of two stacked thin oxide adnfet. The power stage is based on three stacks transistors, two thin oxide L=32 nm PAFET and one thick oxide L=112 nm tonfet. A stack of three transistors alleviates breakdown voltage limits by dividing the voltage swing across several devices. Thanks to the high resistivity substrate, phase variation and losses from one cell of the stack

3−4 April 2022, London, UK

to another are reduced offering a better power efficiency for stacked power amplifier design [2]. The voltage distribution is obtain with a proper biasing and loading, performed with the gate capacitors on the common gate stages. C_{1_pw} and C_{2_pw} with C_{gs1_pw} and C_{gs2_pw} form a voltage divider that distribute the voltage and respect the breakdown V_{ds} limits and the recommended maximum oxide voltage to secure TDDB degradation.

Common source transistor are stabilized thanks to the neutralization capacitance C_N. The 45RFSOI technology advocate a maximum Vdd voltage of 1.1 V for a thin-oxide transistor and 1.65 V for a thick oxide transistor. De-rating of 20 % is applied on these values giving $V_{dd\_nfet\_max}$=0.88 V and $V_{dd\_tonfet\_max}$=1.32 V. Optimization sizing of transistors size lead to a total width of 200 um for the low voltage FETs and a total width of 300 um for the high voltage FET. The custom-made optimum transformers for matching networks are shown Fig. 2.

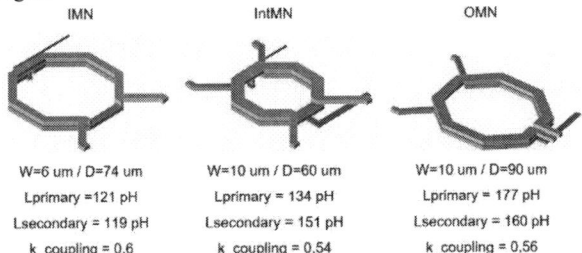

Fig. 2. Matching Network transformers of the Power Amplifier.

III. MEASUREMENT RESULTS

The PA is fabricated in 45RFSOI process and occupies a 0.5x0.28 mm² core area (Fig. 3).

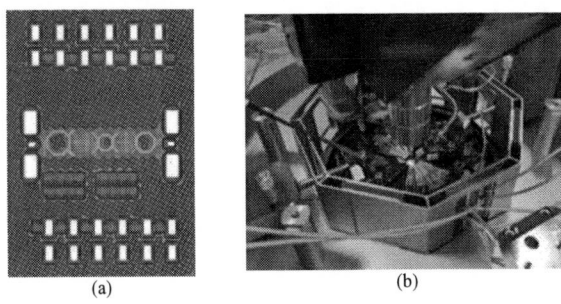

Fig. 3. Die micrograph of the fabricated PA chip in 45-nm CMOS SOI process (a) and on wafer probing setup (b).

A. Small and Large Signal performance

Small signal measurement are performed with a vector network analyzer (R&S ZVA67). Dedicated coplanar probes are used for 39 GHz measurement. The DC measured current of the driver and power stage are 12 mA and 42 mA respectively. The simulated and measured S-parameters are presented Fig. 4.

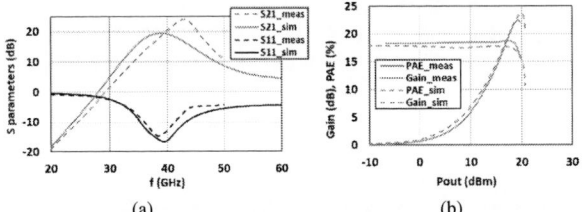

Fig. 4. S parameters (dB) vs. frequency (GHz) (a) and Gain and PAE vs. Pout of the PA biased at V_{g0_DV}=0,2V, V_{g1_DV}=1.2V, V_{dd_DV}=1.7V, V_{g0_PW}=0.32V, V_{g1_PW}=1.22V, V_{g2_PW}=2.45V, V_{dd_DV}=3.1 V (b).

The power amplifier achieved 19.3 dBm output compression point OCP1dB and saturated power of 20 dBm at 39 GHz (Fig. 4.b). The measured gain is 18.5/23.4 dB at 39/42 GHz and the input return loss S_{11} is <−10 dB over the 36 - 41.3 GHz bandwidth (Fig. 4.a). The maximum PAE is 23 %.

Optimum Z_{opt} is verified by a large-signal load-pull measurement using tuner (Fig. 5) where it can be observed that the PA is well adapted to max output power and efficiency.

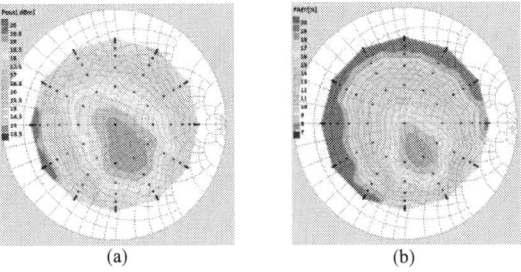

Fig. 5. Load pull measurement at 39 GHz. Power contour level (a) and PAE contour level (b).

On Fig. 6.a AM-AM and AM-PM variations are represented and show less than 0.5 dB and 5° up to 14 dBm. For the IMD measurement, two continuous wave (CW) tones with equal tone power and spaced by 100 MHz are applied to the PA input. The IMD obtain is -30.8 dBc at 10 dBm output (Fig. 6.b).

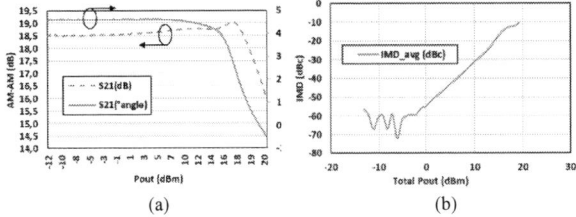

Fig. 6. Measured AM-AM (dB) and AM-PM (deg) vs. Pout (a) and IMD3 vs. Pout (b)

B. Dynamic Performance

For modulation measurements, an arbitrary waveform generator (AWG Tektronix 7122C) is used to generate baseband I/Q signals (Fig. 7). The signals are up-converted to 39 GHz using an external mixer with a signal generator to feed the Power Amplifier under Test. The chip output is down converted and analyzed using a real-time oscilloscope

(Tektronix 70 GHz) and Matlab post-processing for OFDM demodulation. The PA is tested using FR2 5G NR OFDM signals with no Digital Pre Distortion and no equalization is applied before demodulation. The PA non linearity contribution to EVM is extracted from the total EVM, as the setup generated unwanted impairments.

Fig. 8 shows modulation tests of 5G NR FR5 200 MHz frame at 39 GHz, modulated with OFDM 64QAM signal. For a power back-off of 9 dB on ICP1dB, the measured Poutavg is 10 dBm and PAEavg is 8 %, while the nonlinear contribution of the PA to EVM is less than 3 % at 39 GHz.

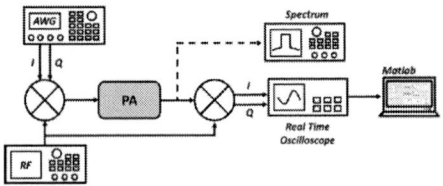

Fig. 7. setup of OFDM modulated measurement.

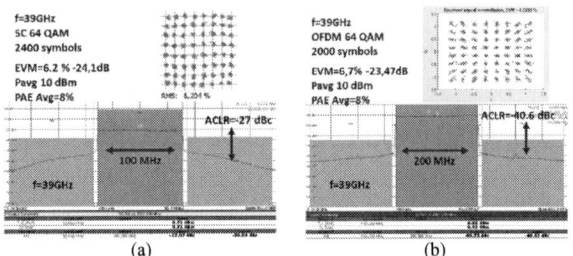

Fig. 8. Modulation PA measurement results with SC 64 QAM signal BW=100 MHz (a) and with OFDM 64 QAM BW=200 MHz modulation measurement results at 39 GHz (b).

The Fig 9 shows the EVM degradation induced by the PA function of its power input back off. As a comparison in ETSI TS 138 104, the EVM requirement for Base Station is 8% for 64QAM, measured on 50 MHz bandwidth.

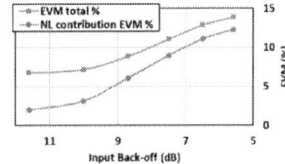

Fig. 9. Measured PA contribution on EVM (%) vs. power input Back Off for OFDM 64QAM modulation, 200MHz Bandwidth.

C. Reliability Measurements

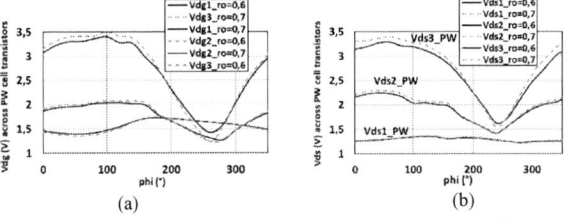

Fig. 10. Simulation of expected Vds max voltage across power cell stage for different impedances at VSWR 4:1 (rho=0,6) and VSWR 6:1 (rho=0,7) for Vds_PW (a) and Vdg_PW (b).

In a use-case environment, the power amplifier is the latest stage before antenna, making this bloc highly sensitive to load variation. Independently of its performance, its robustness to impedance variation VSWR is of most concern to ensure its functionality. Indeed, the voltage variation expected across transistors versus load impedance can far exceeds the recommended V_{ds} operating voltage.

The simulation Fig. 10 is performed at $P_{out}=P_{1dB}$ for rho=0.6 and 0.7 and covering all phases. The upper thick oxide transistor sees a max $V_{ds3max}=3.5$ V / $V_{dg3max}=3.3$ V, and the two cascoded thin oxide transistor a max $V_{ds2max}=2$ V / $V_{dg2max}=2.4$ V and $V_{ds1}=1.6$ V / $V_{dg1}=0.8$ V. The V_{dsmax} of thick and thin oxide transistors are 2.35 V and 1.8 V respectively. Considering the CCDF of the 5G FR2 modulation scheme, a 0.1-year lifetime at P1dB is evaluated. Therefore, the max voltage should be V_{dgmax} 2.79 V and 1.73 V. Two tests have been performed. A soft stress of 27 min loading presented in Fig. 11, and a hard test corresponding to a longer 15 hours stress in Fig. 12. Blue curves show results before stress and orange curves the results after stress. In Fig. 11, the input power is first fixed at IP1dB, the gamma_mag is set to VSWR 4:1 and the gamma phase is swept 3 times from 0 to 350 degree by 10 deg. A 15 sec waiting time is set at every point leading to a total stress of 27 min.

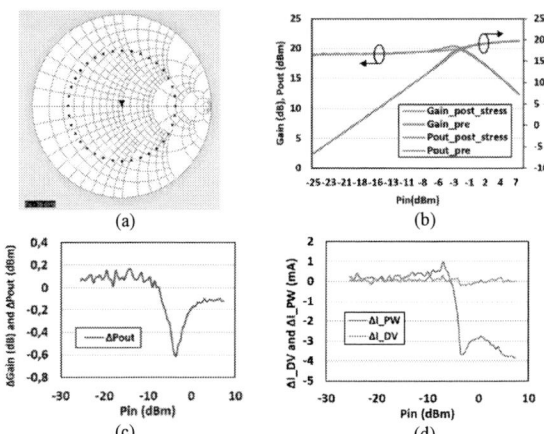

Fig. 11. pre and post 27 min VSWR 4:1 stress test (a): Gain (dB) and Pout (dBm) vs. Pin (b), ΔGain (dB) and ΔPout (dB) vs. Pin (c), ΔI_PW (mA) and ΔI_PW (mA) vs. Pin (d).

Same operation in Fig. 12 is performed for hard test with gamma phase swept 100 times leading to a total stress of 15 h. No breakdown is observed even after 15h of 4:1 VSWR, which show a very good robustness against load impedance variation. This measurement was also performed up to 7:1 for 30 min with no breakdown observed. At OCP1dB, after 15h of stress, a degradation of 0.6 dB for the gain and 1.2 % for the PAE is measured and almost no degradation at P_{avg}. However, a more significant degradation is measured before compression. The decrease of this current linearize the PA changing its class of operation. The measurements were made on five PA chips and are perfectly reproducible. Almost no current degradation is observed on the driver current, as this stage is well isolated from output load variation because of the

82

power call stage between. Degradation can be expected if upper limit supply voltage is applied. Consequently, to insure a given lifetime for a specific power amplifier, the DC and RF voltage must be examined carefully and not over-stressed.

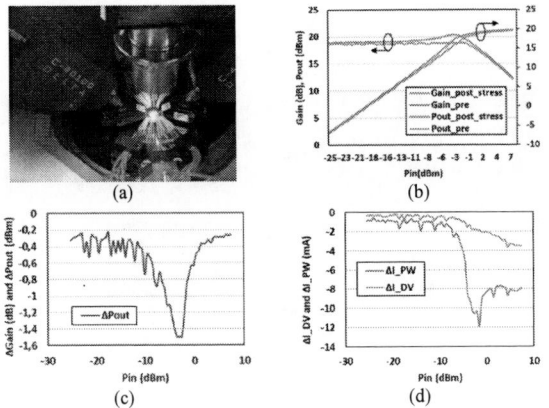

Fig. 12. loadpull setup (a), pre and post 15h VSWR 4:1 stress test: Gain (dB) and Pout (dBm) vs. Pin (b), ΔGain (dB) and ΔPout (dB) vs. Pin (c), ΔI_PW (mA) and ΔI_PW (mA) vs. Pin (d).

Table 1. Summary and comparison of measured performances.

	[4]	[5]	[6]	[7]	[8]	This work
freq. (GHz)	39	38.5	39	39	39	39
Tech.	65nm CMOS	28nm CMOS	22nm SOI CMOS	28nm CMOS	45nm SOI CMOS	45nm SOI CMOS
Process nom. Voltage (V)	1.2	0.9+0.9*	0.8+0.8*	0.9+0.9*	1+1*	1+1.5+1.5*
Supply (V)	-	1.8	1.6	2.2	2	3.1
Gain (dB)	20.3	25.8	15	38	20.5	18.5
P_{sat} (dBm)	16.7	16.8	20.4	26	19.1	19.7
OCP1 (dBm)	13.5	14.9	17.8	21.5	18	19.3
PAE_{max} (%)	22.2	32.9	25.6	26.6	38.6	23
PAE_{P1dB} (%)	14.9	28.8	20.9	13.5	37.3	22.5
Mod. scheme	64QAM	64-QAM 200MHz OFDM	64QAM	5G NR FR2 64-QAM 50MHz OFDM	5G NR FR2 64-QAM 800MHz OFDM	5G NR FR2 64-QAM 200MHz OFDM
EVM (dB)	-26.3**	-25.3	-25	-26	-25.1	-24.1
ACPR (dBc)	-	-30	-24.5	-33	-26.1	-27
P_{avg} (dBm)	7.1**	10.3	13.5	14.7	10.2	10
PAE_{avg} (%)	-	16.9	10.6	-	13.4	8
Core Area (mm²)	-	0.07	-	-	0.21	0.14

*Added values when devices are stacked,　　**At 28GHz

IV. CONCLUSION

Table 1 summarizes the performances of this work and the recently reported 5G SOI/Bulk CMOS PA power amplifiers in the mmW 39 GHz band. The presented PA has been characterized with true 5G NR FR2 modulation scheme and verified under VSWR 4:1 to accurately evaluate the robustness and the capability of the technology.

Maximum output power requirement for beamforming system reduces as number of elements increases, making SOI technologies a good candidate for 5G applications. Aging and reliability are most concern: power amplifier must be robust from environment variation while keeping good performances over time. In this paper, a highly rugged 39 GHz power amplifier with an output power at P_{1dB} of 19.3 dBm, 18.5 dB gain and a PAE_{P1dB} of 22.5 % is presented. These good results were achieved thanks to stacking device technique, offering high output power while securing FET from breakdown. As the author's knowledge, this is the only PA published at 39 GHz meeting 5G required performances with derating supply voltage and where reliability has been verified.

REFERENCES

[1] Shafiullah Syed et al., "A highly rugged 19 dBm 28GHz PA using novel PAFET device in 45RFSOI technology achieving peak efficiency above 48%", in IEEE/MTT-S International Microwave Symposium (IMS), 4-6 Aug. 2020.

[2] B. Martineau and D. Belot, "Si and SOI CMOS technologies for millimeter wave wireless applications", in IEEE International Electron Devices Meeting (IEDM), 2020.

[3] T. Chen et al., "Excellent 22FDX Hot-Carrier Reliability for PA Applications," in IEEE Radio Frequency Integrated Circuits Symposium (RFIC), 2019.

[4] S. Mondal, R. Singh and J. Paramesh, "A Reconfigurable Bidirectional 28/37/39GHz Front-End Supporting MIMO-TDD, Carrier Aggregation TDD and FDD/Full-Duplex with Self-Interference Cancellation in Digital and Fully Connected Hybrid Beamformers," in IEEE International Solid- State Circuits Conference - (ISSCC), 2019.

[5] H. Park et al., "A High Efficiency 39GHz CMOS Cascode Power Amplifier for 5G Applications," in IEEE Radio Frequency Integrated Circuits Symposium (RFIC), 2019.

[6] Z. Zong, et al. , "A 39GHz T/R front-end module achieving 25.6% PAEmax, 20dBm Psat, 5.7dB NF, and -13dBm IIP3 in 22nm FD-SOI for 5G communications," in IEEE Radio Frequency integrated Circuits (RFIC) Symposium - Digest of Papers, June 2021.

[7] K. Dasgupta, S. Daneshgar, C. Thakkar, J. Jaussi and B. Casper, "A 26 dBm 39 GHz Power Amplifier with 26.6% PAE for 5G Applications in 28nm bulk CMOS," in IEEE Radio Frequency Integrated Circuits Symposium (RFIC), 2019.

[8] F. Wang and H. Wang, "24.6 An Instantaneously Broadband Ultra-Compact Highly Linear PA with Compensated Distributed-Balun Output Network Achieving >17.8dBm P1dB and >36.6% PAEP1dB over 24 to 40GHz and Continuously Supporting 64-/256-QAM 5G NR Signals over 24 to 42GHz," in IEEE International Solid- State Circuits Conference - (ISSCC), 2020.

44 dBm Output Power and High Gain K-band GaN Power Amplifier for Satellite Communication

Takuma Torii, Yoshifumi Kawamura, Eigo Kuwata, Masaomi Tsuru

Information Technology R&D Center, Mitsubishi Electric Corporation
Ofuna 5-1-1, 247-8501 Kamakura, Japan
Torii.Takuma@dr.MitsubishiElectric.co.jp

Abstract — **This paper presents a high gain K-band GaN power amplifier. The fabricated GaN power amplifier is utilized 0.15 µm gate-length and SFP structure to enhance gain. The GaN power amplifier demonstrates output power of over 42.8 dBm, PAE of over 26 % with power gain over 25dB over 17 to 21.4 GHz under CW condition.**

Keywords — **Power Amplifiers, GaN HEMT, MMIC, K-band, Satellite Communications.**

I. INTRODUCTION

In recent years, a large communication capacity is required to enhance data traffic of satellite communication system. In such system, an active phased array system will be adopted to achieve high throughput. TWTAs (Traveling Wave Tube Amplifier) were conventionally best candidate for such satellite communication systems due to its high output power and efficiency. Although TWTAs present high performance, it is not suitable for phased array system because it enlarges total mass of transmitter. A 5 to 20W class small size power amplifiers will be preferred as an antenna element, then TWTAs can be replaced to GaN power amplifiers due to its high power density. Since GaN power amplifiers meet the demand of small size and output power, So far, output power of over 10W K-band power amplifiers were reported [1, 4].

Efficiency, gain and output power of GaN power amplifier needs to be maximized in order to meet requirements for transmitter module. Although Gain becomes higher as the number of stage increase, it decreases the efficiency due to power consumption of driver stage. Additionally, it enlarges chip size and pushes up costs. Transistor gain at higher frequency is deteriorated by negative feedback through parasitic capacitance between gate and drain node(C_{gd}). Source field plate (SFP) is known as a technique to reduce C_{gd}, and it results in gain improvement.

In this paper, high gain K-band power amplifier is presented. The performance of GaN utilizing SFP is shown at section II. The design of power amplifier will be discussed at section III including circuit design and thermal simulation to achieve output power of 20W. The GaN power amplifier was fabricated with MELCO's process. Fig.1 shows chip photograph of the realized high gain power amplifier. The measurement results of GaN power amplifier is demonstrated at section IV.

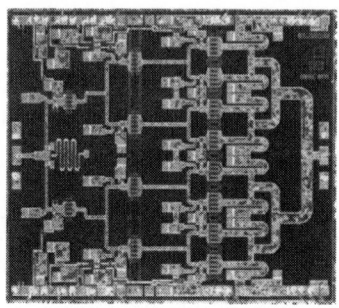

Fig. 1. Chip photograph of the realized high gain power amplifier. Chip Size is 3.9mm × 3.4mm

II. PERFORMANCE OF GAN HEMT

The performance of GaN HEMT is described at this section. Gain is one of the key characteristic to enhance the overall transmitter performance. However, gain of transistor becomes lower at higher frequency because of negative feedback through C_{gd}. Since the reduction of C_{gd} contributes higher gain, SFP structure is attractive. It can attribute enhancement of power amplifier gain and efficiency. Although SFP is known as the technique to cancel C_{gd}, it causes the increasement of gate-source capacitance (C_{gs}) and drain-source capacitance (C_{ds}). Large SFP resulted in higher gain but low operating frequency, on the other hand, small SFP resulted in higher operating frequency but small gain. The length of SFP is chosen to satisfy the frequency requirement. Additionally, unit gate width of transistor cell is one of the parameters because gain becomes smaller as it is enlarged to achieve higher output power.

The GaN HEMT was fabricated using MELCO's GaN-HEMT process which is available for the gate length of 0.15µm and 50 µm thickness of the SiC substrate. Each source of transistor has individual source via which reduces source inductances and provides higher gain. First, the benefit by utilizing SFP is confirmed from the small signal result. Fig. 2 shows measured maximum stable gain (MSG) and maximum available gain (MAG) of the transistor utilizing SFP and without SFP. The measured transistor is 10finger×80µm. It shows that MSG of the transistor utilizing SFP was enhanced by 3dB compared to that of the transistor without SFP. And also, the conversion frequency of MSG and MAG is about 23GHz satisfying the operating frequency of between 17GHz to 21.4GHz. Secondly, large signal measurement was conducted to make sure that performance is improved. Fig. 3 shows the measured input output characteristic of the transistor utilizing SFP and without SFP. Measured transistors are 10finger×80μm with pre-matched circuits. The loss of the pre-matched circuits is de-embedded to calculate the performance at transistor nodes. The large-signal measurement was conducted under conditions of continuous wave(CW), 28V of Vd and optimum load impedance for DE at 18GHz. As shown in Fig.3, the measured performance of the transistor utilizing SFP shows power density of 3.7W/mm, maximum DE of 62%, and power gain of 13dB which is 3dB higher than gain of transistor without SFP. This power density is high enough to achieve output power of over 20W by combining 8 cells.

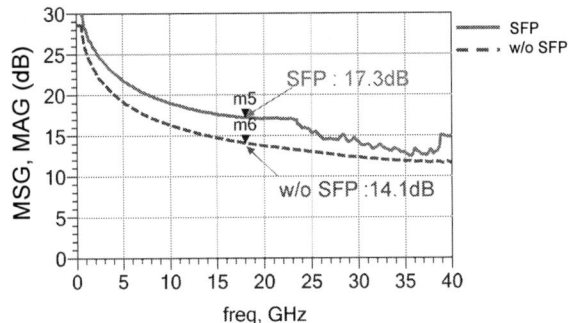

Fig. 2. The measured MSG and MAG of the transistor utilizing SFP and without SFP.

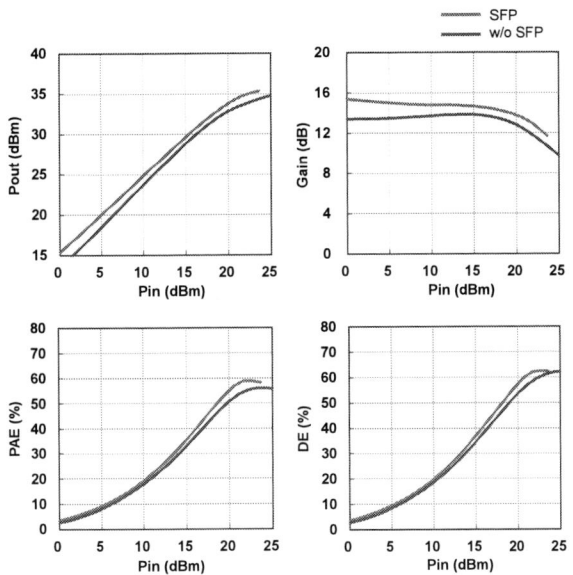

Fig. 3. Measured input output characteristic of transistor utilizing SFP and without SFP.

III. POWER AMPLIFIER DESIGN

A. Output Matching Network Design

The design of the power amplifier is discussed at this section. As mentioned before, transistor cell showed the power density of 3.7W/mm. The output power of over 20W is estimated with considering 8 cells combined and the loss of output matching network (OMN) into accounts. OMN is one of the important components on design procedure because the combiner loss affects efficiency of power amplifier. Fig.4 shows equivalent circuit of the output matching circuit. In order to maximize the output power and efficiency in full-band(17 to 21.4GHz), each parameter should be optimized. Intermediate impedance Z_{mid} given by transmission line of λ/4 affects bandwidth and shunt L and series L are dominant to loss. As shunt L becomes smaller, the mismatch is improving. However it causes reduction of output power and efficiency because OMN loss is increasing. Parameters of OMN is decided by following procedures. Z_{mid} of 4 to 10 Ω are given by varying the characteristic impedance of the transmission line. Additionally, combination of series L and shunt L which maximizing bandwidth, output power and efficiency are selected for each Z_{mid}. Finally, from the most performed combination of Z_{mid}, series L and shunt L is extracted. Fig. 5 shows mismatch and loss of OMN. Mismatch of less than -12dB and OMN loss of 0.6 to 0.4dB are realized. This result is good enough to perform output power of 20W at desired frequency.

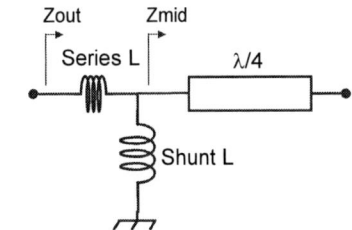

Fig. 4. Equivalent circuit of the output matching network.

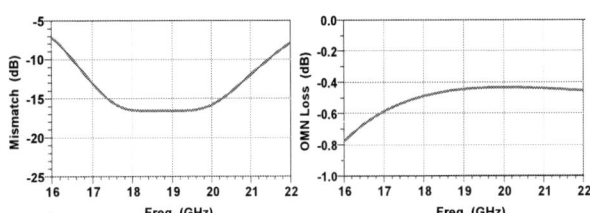

Fig. 5. Mismatch and Loss of OMN.

B. Thermal Simulation

Considering thermal effect at saturated operation is necessary to achieve output power of 20W and durability. Thermal simulation was conducted to calculate base temperature and thermal resistance of the power amplifier. To meet the junction temperature requirement, the distance between each transistor cell should be considered since thermal effect relates each other. Smaller transistor pitch brings a benefit of small ship size, on the other hand, thermal junction temperature becomes high. Package base material is chosen to minimize thermal affect to GaN chip.

Fig 6 shows thermal simulation result. During the simulation, the dissipated power are calculated based on power amplifier design result. Heat sources considered as dissipated power are assigned between each drain and source finger. Thermal flow is simulated assuming material definition mentioned above. It is confirmed that junction temperature is about 160 deg.C.

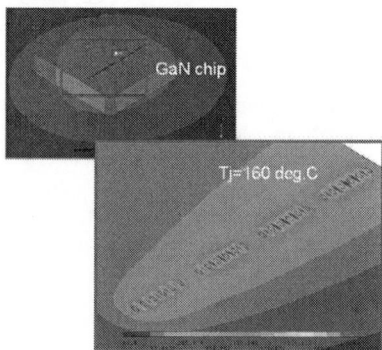

Fig. 6. Thermal simulation result

IV. MEASUREMENT RESULT

Measurement result of the power amplifier is presented at this section. The fabricated power amplifier has 3stage. The measurement was conducted with drain bias voltage of 28V and quiescent current of 25mA/mm. Fig. 7 shows on-wafer S-parameter measurement result of multiple samples. As seen from fig.7, S11, S21, S22 very suite to simulation result, and Variation of S21 is quite low.

Large-Signal characteristic was performed with both pulse and CW input signal. The GaN power amplifier is mounted on carrier and input-output port is connect to alumina microstrip lines by bonding wire. 1000 pF and 6600 pF bypass capacitors are implemented and 0.1uF is used on the outside bias board. Output power of 30W, maximum PAE of 43% and power gain of 28 dB were achieved under pulsed(10%) input signal. Fig.8 shows on-carrier input output characteristic measurement and simulated result. As seen from Fig.8 gain shows relatively flat characteristic and gain of over 25dB is performed. Measurement result well matches with the simulated result. Fig.9 shows frequency response of output power, PAE and junction temperature. The junction temperature was estimated based on calculated thermal resistance, and dissipated power according to measurement result of output power and drain current of each stage. The junction temperature is around 160 deg.C, it well corresponds to the simulated result. Maximum output power of 44.2dBm(26W) and maximum PAE of 32% was achieved with power gain of over 25dB at best performed frequency. As speaking of wide band operation, the power amplifier realized output power of 42.8 dBm (19W) and PAE of over 26%, and power gain of over 25dB at 17 to 21.4GHz.

Performance with the modulated input signal was also conducted. 8PSK modulation (roff = 0.15) was applied as test signal which has 10MHz of bandwidth, peak to average ratio of about 8.4dB. Fig.10 shows measurement result with 8PSK modulated signal. Average output power of 42.4dBm, PAE of 28.9 %, Gain of 26.5 dB, EVM of 5.4% ACPR of −30dBc are performed.

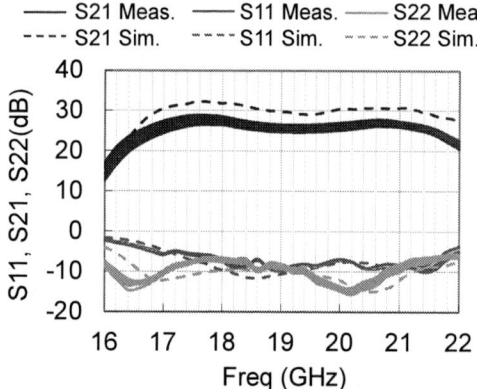

Fig. 7. On-wafer S-parameter measurement result of multiple samples.

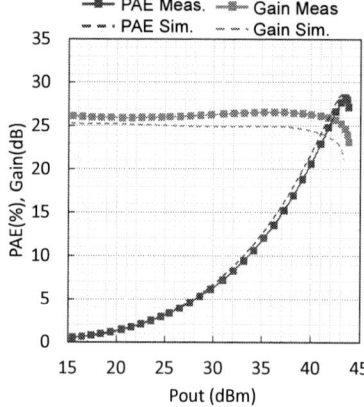

Fig. 8. On-carrier input output characteristic measurement and simulated result.

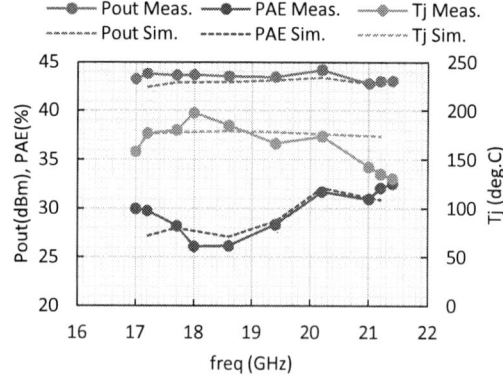

Fig. 9. Frequency response of output power, PAE and junction temperature.

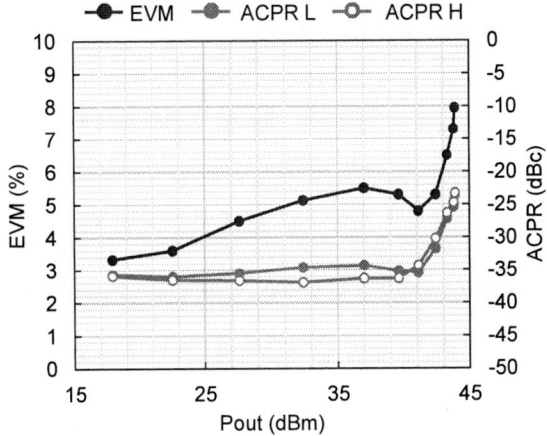

Fig. 10. Measurement result with 8PSK modulated signal.

Table 1 shows state of the art of a K-band GaN power amplifiers. The power gain of the fabricated power amplifier is comparatively higher than other reported K-Band GaN power amplifiers. For author's best knowledge, this is first time to report that over 20W output power was realized by a GaN chip power amplifier.

Table 1. State of the art of a K-band GaN power amplifiers.

	Ref	CW/Pulse	Freq (GHz)	BW (%)	Num. stage	Pout (dBm)	PAE (%)	Gain (dB)
[1]	APN269	CW	17.2 – 20.2	16.0	2	41 – 42	36 – 43	17
[1]	APN272	CW	17.2 – 20.2	16.0	1	39.0	38 – 48	7
[2]	TGA4548	CW	17 – 20	16.2	3	40 – 42	30 – 36	25
[3]	(Doherty PA)	CW	20 – 21	4.9	3	37	35	20
[4]		CW	17 – 21	21.1	2	40	36 – 40	22
[5]	(Doherty PA)	CW	17.3 – 20.3	16.0	2	34	25 – 30	10
[6]		Pulse	18.5 – 23.5	23.8	3	35 – 37	34 – 40	25
[7]		Pulse	17 – 20.2	17.2	3	40 – 41	30 – 36	22.5
[8]	APN279	Pulse	16 – 20.2	23.2	2	41.5	30	15
	this work	Pulse	17.2 – 21.2	20.8	3	43.2 – 44.8	33 – 43	28
	this work	CW	17.2 – 21.2	20.8	3	42.8 – 44.2	26 – 32	25

V. CONCLUSION

In this paper, the design and measurement result of a K-band GaN power amplifier was presented. The fabricated GaN power amplifier was utilized 0.15 gate-length and SFP structure to enhance gain. Thermal effect from each transistor cell was simulated to make sure the durable operation under output power of 20W. The performance of the fabricated power amplifier have been demonstrated with CW condition. The experimental result shows output power of over 42.8 dBm, PAE of over 26 % with power gain over 25dB over 17 to 21.4 GHz. For author's best knowledge, this is first time to report that over 20W output power was realized by a GaN chip power amplifier.

REFERENCES

[1] S. Din, A. M. Morishita, N. Yamamoto, C. Brown, M. Wojtowicz, and M. Siddiqui, "High-power K-band GaN PA MMICs and module for NPR and PAE," *IEEE MTT-S Int. Microw. Symp.* Dig., pp. 1838–1841, 2017.

[2] https://www.qorvo.com/products/p/TGA4548

[3] V. Valenta, I. Davies, N. Ayllon, S. Seyfarth, and P. Angeletti, "High-gain GaN doherty power amplifier for Ka-band satellite communications," *PAWR 2018 - Proc. 2018 IEEE Top. Conf. Power Amplifiers Wirel. Radio Appl.*, vol. 2018–January, pp. 29–31, 2018.

[4] L. Marechal et al., "10W K band GaN MMIC amplifier embedded in Waveguide-based Metal Ceramic Package," *2019 14th European Microwave Integrated Circuits Conference (EuMIC)*, pp. 184–187, 2019

[5] A. Piacibello, R. Giofré, R. Quaglia and V. Camarchia, "34 dBm GaN Doherty Power Amplifier for Ka-band satellite downlink," *2020 15th European Microwave Integrated Circuits Conference (EuMIC)*, Utrecht, Netherlands, pp. 25-28, 2021.

[6] M. R. Duffy, G. Lasser, M. Roberg and Z. Popović, "A 4-W K-Band 40% PAE Three-Stage MMIC Power Amplifier," *2018 IEEE BiCMOS and Compound Semiconductor Integrated Circuits and Technology Symposium (BCICTS)*, 2018, pp. 144-147, 2018

[7] P. Colantonio and R. Giofré, "A GaN-on-Si MMIC Power Amplifier with 10W Output Power and 35% Efficiency for Ka-Band Satellite Downlink," *2020 15th European Microwave Integrated Circuits Conference (EuMIC)*, pp. 29-32, Utrecht, Netherlands, 2021

[8] https://www.northropgrumman.com/wp-content/uploads/Microelectronics-APN279.pdf

125 W Solid State Power Amplifier for 17.3-20.2 GHz SatCom Applications

R. Giofrè[1], P. Colantonio[1], L. Cabria[2], M. Lopez[2]

[1]Department of Electronic Engineering, University of Roma Tor Vergata, Italy

[2]TTI Norte, Spain

Abstract — This paper reports the design, realization and tests of the first prototype of a 125 W Solid State Power Amplifier (SSPA) based on spatial power combining technique for 17.3 GHz to 20.2 GHz SatCom applications. Sixteen 10 W high efficiency Gallium Nitride (GaN) Monolithic Microwave Integrated Circuits (MMICs) Power Amplifiers are combined through a low-lossy Radial structure developed in waveguide (WR-42). The MMICs are individually packaged and equipped with input/output microwave-to-waveguide hermetic transitions showing negligible losses. The Radio Frequency Tray (RFT) of the SSPA includes also a gain control unit, an analogue linearizer and a driver, while a waveguide coupler and an isolator are placed in sequence at the output. From 17.3 GHz to 20.2 GHz, the developed SSPA supplies more than 125 W of saturated output power with a gain and an overall efficiency better than 70 dB and 20%, while satisfying space constraints in terms of de-rating and reliability.

Keywords — Solid State Power Amplifiers, Spatial Combining Techniques, GaN, Ka-Band, PA.

I. INTRODUCTION

Broadband access services are bound to experience an unprecedented growth in the next decade. Already, with the under development generation of mobile communication systems (i.e., 5G), it is expected to achieve data-rate up to 100 Gbit/s with a simultaneous connections density higher than 1 M/km^2. Moreover, the internet of things (IoT) is quickly coming up and thus, together with the traditional services, also billions of daily used objects like washing machines, fridges, etc. will be connected to internet. In order to face this scenario, the underlying network has to be redesigned introducing innovations like higher cell density and thus lower peak power, carrier frequency in the K-bands, extended bandwidth and the use of more efficient spectrum aggregation signals [1]. At the same time, although equipped with these innovations, a network relying upon terrestrial infrastructures only could not be sufficient to satisfy all the requirements. Therefore, its integration with a satellite network composed by Geostationary Earth Orbit (GEO) Very High Throughput Satellites (vHTSs) it is envisaged [2]. Indeed, vHTSs can achieve unbelievable data rate, even larger than 1 Tbps, while increasing the flexibility of the overall network, since capacity can be allocated where it is needed. Their architecture foresees the feeder link in Q/V bands (e.g., $f_c = 40$ GHz), whereas the data link resides in the 17.3-20.2 GHz band [1], [2], also known as satellite Ka-band (IEEE K-band designation). Consequently, the development of high performing space-borne components at such frequencies becomes crucial, also accounting for the higher number of used beams, which implies an increasing number of equipment embarked, and therefore a subsequent need of miniaturization, i.e., lower mass and volume, better thermal management and reliability. Such features at spacecraft level are strongly related to the performance of the adopted power amplifier (PA). Indeed, this sub-unit consumes even more than 75% of the payload overall dc power (thus affecting efficiency and thermal management), determines the output power level and the spectral regrowth of the transmitted signal (linearity and power issues), and it is made redundant to overcome possible failures (reliability, mass and volume issues). Nowadays, thanks to the availability of reliable and powerful semiconductor technologies such as Gallium Nitride (GaN) and the adoption of innovative power combining techniques, Solid State PAs (SSPAs) are becoming a valid and, in some cases, preferable alternative to travelling wave tube amplifiers (TWTAs) to implement the power stage of space systems.

This paper discusses the design, realization and tests of the first prototype of a Space-borne GaN SSPA for vHTS applications from 17.3 GHz to 20.2 GHz. It supplies more than 125 W of saturated output power with a gain and an overall efficiency better than 70 dB and 20%, respectively, while satisfying all the space de-rating rules. The results reported hereafter have been carried out in the framework of the European project named FLEXGAN [3].

II. SSPA DESIGN

Fig. 1 reports the selected SSPA architecture. It includes the radio frequency tray (RFT), the power supply unit (PSU) and the electronic power conditioner unit (EPC). The last two subsystems are conceived, respectively, to convert the satellite primary bus voltage (i.e., 50 V) into the required secondary voltages inside the SSPA, and to implement the required functionalities for controlling and monitoring its behaviour.

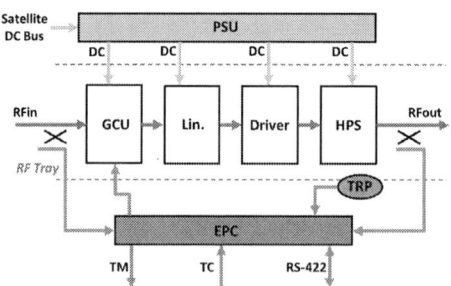

Fig. 1. SSPA Architecture.

On the other hand the RFT, which is the main object of this paper, has to amplify the modulated RF signal from a minimum value of -19 dBm up to the peak of 51 dBm in saturation (i.e., a gain up to 70 dB), while maximizing as much as possible its overall efficiency. Such performance should be guaranteed from -5 to +85°C of back side temperature (T_{BS}), limiting the overall weight to 2 Kg maximum.

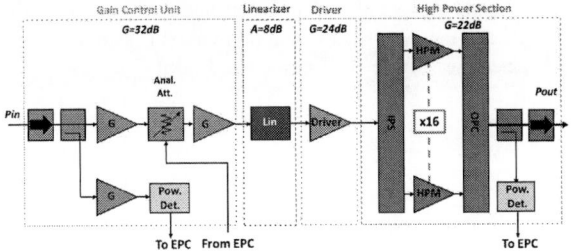

Fig. 2. Power budget of the RFT of the SSPA.

The power budget of the RFT is detailed in Fig. 2 together with its composition. In order to attain 125 W at SSPA level, sixteen high efficiency 10 W MMICs PAs, identified as high power modules (HPMs), are combined by using the low losses radial power combiner structure described in [4]. This subunit, named high power section (HPS), is driven by a single driver based on the same MMIC, referred as to medium power module (MPM). An analogue linearizer is also designed and placed in front of the driver to improve the linearity of the overall chain. Finally, the RFT is completed with a gain control unit (GCU) providing a gain of at least 30 dB, and allowing for the setting of different working modes as well as the thermal/aging compensation.

The TGA4548 from Qorvo® [5] is adopted as a building block for this SSPA. It is a commercial 10 W MMIC developed on 0.15 μm GaN on Silicon Carbide (SiC) process. At the same time, a second version based on the MMICs presented in [6], [7], which were developed on a 0.1 μm GaN on Silicon (Si) from OMMIC, is under development and the experimental results will be available soon. Therefore, a fair evaluation of pros and cons of using either GaN-SiC or GaN-Si components in space-borne SSPAs will be also proposed in the final paper.

A. HPS Design

The TGA4548 delivers more than 10 W of output power with about 30 % of power added efficiency (PAE) and 22 dB of large-signal gain. These features are referred to the MMIC measured on-wafer and at ambient temperature. Since the manufacture declares a maximum operating drain voltage of 29.5 V, in this design it was reduced to $V_{DD} = 22.125 V$ (i.e., 75% of the maximum one, as foreseen by de-rating rules). Moreover, being a space application, the MMIC has to be hermetically shielded and its junction temperature must be kept lower than 160°C in every working condition. Therefore, the design of the package, the related thermal management as well as the mechanical and electromagnetic compatibilities become of primary importance. Hence, several analysis have

been conducted in parallel in order to identify the most suitable solution to translate the RFT architecture in Fig. 2 in a feasible and reliable structure, while maximizing the achievable performance.

For the MMICs, the package has been realized in copper using hermetic feedthroughs to bring the bias, while hermetic waveguide to coaxial transition (probe) at both input and output RF ports have been designed to make it compatible with the radial combiner structure. Fig. 3 shows a picture of one sample of the HPMs with related performances at the nominal bias conditions i.e., $V_{DD} = 22.125 V$ and at ambient temperature (T_{BS}=20 °C). Notably, the assembled MMIC shows almost the same feature as when tested on-wafer, which in turn highlights the negligible losses introduced by the transition.

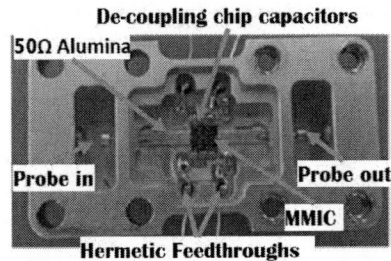

(a)

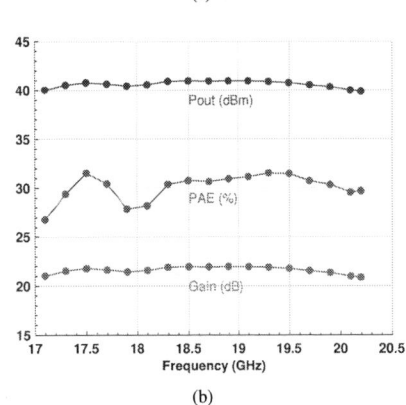

(b)

Fig. 3. Picture of an assembled HPM (a) and measured performance (b).

At SSPA level, clearly the most critical part is the HPS, where most of the dissipated power is concentrated. It is formed by three individual subsystem: an input power splitter, 16 HPMs and a power combiner. Two different configurations were analysed depending on the placement of the HPMs, i.e., horizontal or vertical (with respect to the radial splitter/combiner structure). Fig. 4 shows the sketch of both configurations. The design has been focused in assessing the feasibility to fulfil the thermal and electrical requirements imposed. In the first configuration, identified as horizontal one, the HPMs are rest on the base of the SSPA mechanical housing to achieve direct heat transfer from the MMICs to the satellite panel. The main drawback of this solution is the use of longer waveguide sections in the splitter to connect each access to the corresponding HPM input, thus incrementing the occupied volume and weight of the overall structure. On the contrary, the vertical solution allows straight connection between splitter/combiner and HPMs, saving a significant amount of space and some losses. The main drawback of this solution is related to the thermal dissipation of the HPMs. Indeed, it does not allow a direct heat transfer to the cooling surface (i.e., the satellite panel), which would be the optimal. Taking into account the estimated dissipated power of each HPM (around 23 W), solving this thermal issue ensuring the modules reliability was a really challenging task. To face such an issue, different heat transfer technologies such as heat pipes, thermal straps, vapour chambers, encapsulated graphite, etc. have also been evaluated. However, in any case numerical results showed that the design developed might be suitable from a thermal point of view, but considerably heavier than the horizontal solution, thus making impossible the fulfilment of the weight requirement at SSPA level, i.e., less than 2 Kg.

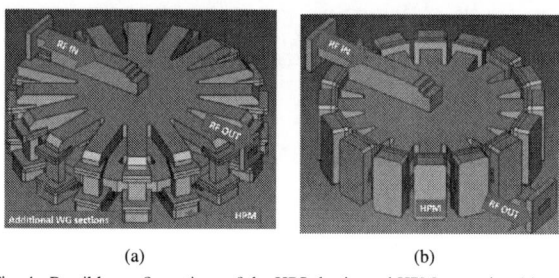

(a)   (b)

Fig. 4. Possible configurations of the HPS: horizontal HPM mounting (a) and vertical HPM mounting (b).

Fig. 5 reports the tridimensional view of the selected solution. In the same figure are also visible the driver (MPM), the output coupler and the isolator, both implemented in waveguide WR42.

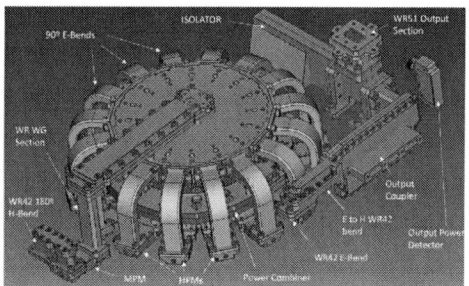

Fig. 5. Tridimensional view of the HPS, driver, output coupler and isolator.

B. SSPA Assembly

Once the configuration of the radial combiner was selected, a structural box to accommodate all the electronic components of the SSPA in the smallest volume and footprint was designed. Fig. 6 shows the conceptual design that has been envisaged for the housing. It consists of the following elements:

- Baseplate. It is the most critical part, providing interface to the satellite panel, the mounting points of most of

the electronic components and the thermal control. It was realized by using a sandwich (5 mm thickness) of materials with high thermal conductivity.
- Fixed walls. They form the chassis of the housing also guesting some electronic components such as the coupler and part of the EPC.
- Removable parts. They are the Upper Lid and the Lateral Door, that have to be dis-mountable for the assembly/inspection of the electronic components. In the case of the Lateral Lid, heat dissipation is also important since part of the PSU is mounted on it.

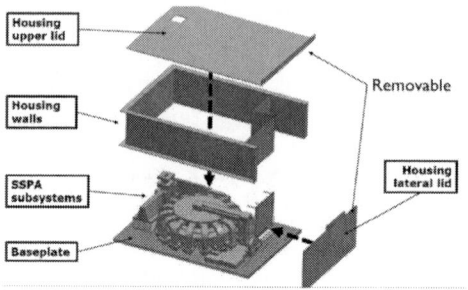

Fig. 6. Conceptual design of the SSPA housing.

Before starting with the SSPA integration, all passive and active sub-units have been individually measured in order to verify their functionality and coherency with the simulations. Fig. 7 shows some pictures of the SSPA during the assembly phases.

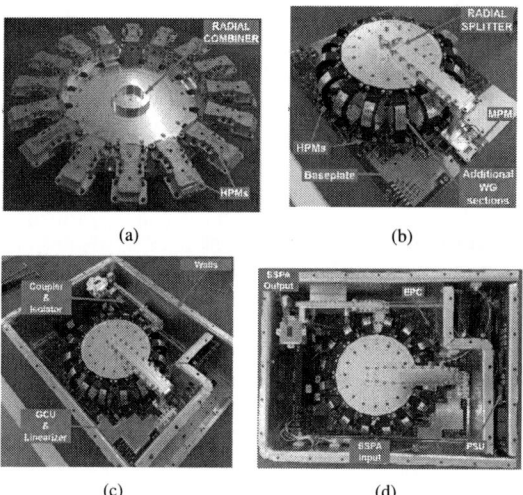

(c)   (d)

Fig. 7. SSPA assembly phases: (a) HPM and radial combiner, (b) HPS onto the baseplate, (c) introduction of the GCU, linearizer, output coupler and isolator, (d) PSU and EPC placement.

C. SSPA Measurement Results

The full validation of the SSPA design foresees a wide and time consuming measuring campaign spanning from mechanical to electrical and electromagnetic compatibility tests, also over a large temperature range and ambient conditions e.g., vacuum. Most of them are very time consuming and are still running. Anyway, some preliminary electrical characterization carried out at room temperature are already available and thus discussed in the following. The small signal characterization of the complete SSPA is shown in Fig. 8, resulting in an overall chain gain larger than 70 dB, with input and output return loss around 20 dB.

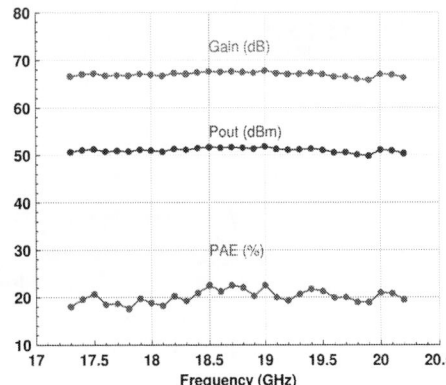

Fig. 10. Output power, PAE and gain as functions of frequency for an input power of -19 dBm.

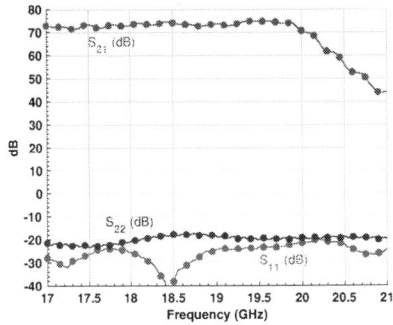

Fig. 8. Measured Scattering parameters of the SSPA at ambient temperature and nominal bias conditions.

A power sweep characterisation has been performed from 17.3 to 20.2 GHz, obtaining the results reported in Fig. 9. An output power larger than 51 dBm (>125 W) has been registered with a PAE close to 20 % in such a frequency band. These features are evaluated including the losses of the output coupler and isolator as well as the PSU efficiency and EPC consumption.

17.3 GHz to 20.2 GHz SatCom applications. In the current version, sixteen 10 W commercially available high GaN/SiC MMICs PAs from Qorvo have been individually packaged and combined through a low-lossy Radial structure realized in WR-42. The reported results demonstrate that the developed SSPA supplies more than 125 W of saturated output power with a gain and an overall efficiency better than 70 dB and 20%, respectively, in the frequency range from 17.3 GHz to 20.2 GHz. Activities are in progress for the replacement of the current MMIC version with fully EU based GaN/Si MMICs developed on 100 nm OMMIC D01GH process.

ACKNOWLEDGEMENT

This project has received funding from the European Union's Horizon 2020 research and innovation programme under grant agreement No. 821830.

REFERENCES

[1] R. Emrick, P. Cruz, N. B. Carvalho, S. Gao, R. Quay, and P. Waltereit, "The sky's the limit: Key technology and market trends in satellite communications," *IEEE Microwave Magazine*, vol. 15, no. 2, pp. 65–78, March 2014.

[2] H. Fenech, S. Amos, A. Tomatis, and V. Soumpholphakdy, "High throughput satellite systems: An analytical approach," *IEEE Transactions on Aerospace and Electronic Systems*, vol. 51, no. 1, pp. 192–202, January 2015.

[3] Flexgan. Project website. [Online]. Available: http://www.h2020-flexgan.eu/

[4] R. Giofrè, P. Colantonio, F. Di Paolo, L. Cabria, and M. Lopez, "Power combining techniques for space-borne gan sspa in ka-band," in *2020 International Workshop on Integrated Nonlinear Microwave and Millimetre-Wave Circuits (INMMiC)*, 2020, pp. 1–3.

[5] Qorvo. Tga4548 datasheet. [Online]. Available: https://www.qorvo.com/products/p/TGA454

[6] P. Colantonio and R. Giofré, "A gan-on-si mmic power amplifier with 10w output power and 35 2020 15th European Microwave Integrated Circuits Conference (EuMIC), 2021, pp. 29–32.

[7] R. Giofrè, P. Colantonio, F. Costanzo, F. Vitobello, M. Lopez, and L. Cabria, "A 17.3–20.2-ghz gan-si mmic balanced hpa for very high throughput satellites," *IEEE Microwave and Wireless Components Letters*, vol. 31, no. 3, pp. 296–299, 2021.

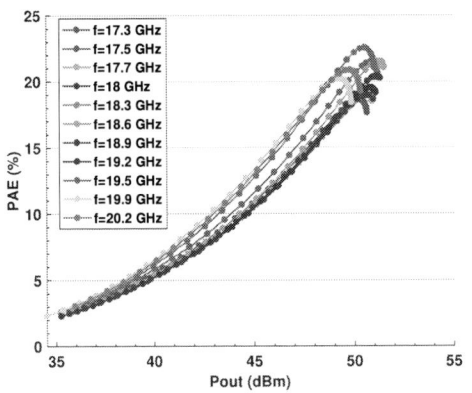

Fig. 9. PAE as a function of the output power from 17.3 GHz to 20.2 GHz of the complete SSPA

For a fixed input power level of -19 dBm, the measured performance in frequency are reported in Fig. 10, where the contribution of all subsystems are accounted for.

III. CONCLUSION

This contribution discussed the design, realization and preliminary tests of a 125 W SSPA prototype, conceived for

Proceedings of the 16th European Microwave Integrated Circuits Conference

A Ka-Band MMIC Single-Chip Frequency Converter for Telecom Satellite Applications

Francesco Scappaviva[#], Davide Resca[#], Luca Cariani[#], Andrea Biondi[#], Francesco Vitulli[*], François Deborgies[°]

[#]MEC – Microwave Electronics for Communications, Bologna, Italy

[*]Thales Alenia Space, Roma, Italy

[°]ESA ESTEC, The Netherlands

{francesco.scappaviva, davide.resca, luca.cariani, andrea.biondi}@mec-mmic.com,
francesco.vitulli@thalesaleniaspace.com, francois.deborgies@esa.int

Abstract — **In this paper, the design and experimental validation of a Ka-band single-chip frequency converter to be exploited in future generation telecom satellite converters is presented. The chip is manufactured exploiting a European 0.15-μm GaAs pHEMT MMIC technology occupying a total die area of 5x5 mm². A conversion gain of about 43 dB is achieved in the broad IF frequency band from 17.2 GHz to 20.2 GHz synthesized by the LO frequency from 4.65 GHz to 6.1 GHz and the RF frequency in the band from 27.5 to 30.0 GHz. An output P1dB of 23 dBm with an output TOI of 33 dBm have been measured. The chip also integrates 20 dB programmable gain attenuation, with 1 dB steps and a continuous 7 dB fine tuning of the gain to compensate for temperature variations. The chip is fully compliant to space derating conditions.**

Keywords — **MMIC, single-chip, doubly balanced mixer, gain variation, space technology.**

I. INTRODUCTION

Nowadays telecommunication satellite constellations need a drastic increase of the integration density [1], implying that the transparent transponder design needs a radically different approach. As far as the frequency translation function is concerned, an additional level of integration can be achieved by integrating the whole downconversion circuitry into a broad-band single-chip but the filter. This latter is anyway needed to select the useful band (that may change for different satellites) and reject the undesired intermodulation products. A variable gain feature is also envisaged to make the chip more versatile and suitable for reconfigurable, flexible payloads.

To this purpose, as already demonstrated at Ku band [2], the single-chip frequency converter (SCFC) architecture has been here introduced and adopted for a Ka to Ka-band frequency conversion scheme, translating the [27.5, 30.0] GHz RF signal down to the [17.2, 20.2] GHz Ka-band IF signal, using an LO signal in the [4.65, 6.10] GHz band, thanks to a sub-harmonic configuration of the mixer core.

The Ka-band SCFC has been designed and manufactured exploiting a 0.15-μm GaAs power pHEMT MMIC technology, integrating into a single MMIC a broadband doubly balanced sub-harmonically pumped diode ring mixer (DBSHPDM), cascaded after an RF variable gain amplification stage (RF VGA) and a single stage RF buffer (RF Buffer) and before the IF amplification stages (a gain stage followed by a Medium

Power Amplifier, MPA). Finally, two different digital step attenuators (DSAs) have been exploited to program the gain. Its RF/IF chain provides 43 dB of maximum conversion gain, including 20 dB of programmable gain control in 1 dB steps and the compensation of the gain variation over temperature.

In the following sections the SCFC architecture, its design strategies, and its experimental characterization are presented.

II. SCFC ARCHITECTURE

The architecture of the SCFC is derived from the requirements of the downconverter system [1] adjusted at the MMIC reference planes, as summarized in Table 1.

Table 1. Main requirements of the Ka-band SCFC

Parameter	Goal	Unit
RF input frequency band	27.5 – 30.0	GHz
LO input frequency band	4.65 – 6.10	GHz
IF output frequency band	17.2 – 20.2	GHz
C/I3 (at maximum gain and Pout = 7 dBm)	51	dBc
Linear Gain	43	dB
Gain Variation	20	dB
Gain Variation Step	1	dB
Continuous Gain control over Temperature	5	dB
Noise Figure	20	dB
LO spurious level (at maximum gain and Pout = 7 dBm)	-20	dBm

Looking at the main performance of Table 1, the 43 dB of overall gain and the required 20 dB of programmable gain variation with 1 dB step, dictate the total number of gain and attenuation stages. The linearity requirement constraints both the power performance of the IF output stage, as well as the positioning of the gain, attenuation and mixing stages in the RF/IF chain.

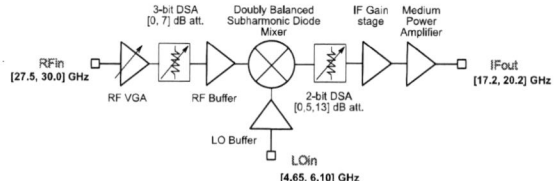

Fig. 1. Ka-band SCFC converter architecture.

Besides, the proper position of the mixer core is also checked against the LO harmonic generation: given the level of those spurious components generated by the mixer, it is possible to increase their attenuation, with respect to the IF spectral components, by moving the mixer core in the right position in the RF/IF gain/attenuation chain. Budget simulations are performed to check the optimal positioning and topology of the subcircuits, based on their stand-alone simulated performance. The best architecture to be exploited for the SCFC is found to be as shown in Fig. 1.

The optimal architecture allows to obtain the best trade-off between gain, linearity, power performance and LO harmonic level, with the least possible DC power consumption. The noise figure of the converter has been taken into account in the budget, but its minimization is not the priority in the trade-off with the linearity (C/I3). This is reasonable knowing that the complete downconversion sub-system, exploiting the SCFC, includes in the front-end a low noise amplifier having 45 dB of gain typically. This makes the high SCFC noise figure of negligible impact at sub-system level.

III. SCFC Design

The SCFC is designed using a 0.15-μm GaAs pHEMT MMIC process, namely the PPH15X-20 process of the European foundry UMS. The main reason of the choice is the highest power density of 800 mW/mm, among their space-qualified GaAs processes, which allows to meet the demanding linearity and power performance of the SCFC in a compact layout. In the following paragraphs, a few details on the strategies adopted for the design of each subcircuit of the SCFC are presented. Eventually, the subcircuits are arranged in the MMIC die as shown in Fig. 2, highlighting in the frames the subcircuit sections. Extensive EM optimization has been carried out to obtain the final layout drawing. The final die size is 5.0 x 5.0 mm^2.

A. Doubly Balanced Sub-Harmonic Diode Mixer Core

The core of the converter is the doubly balanced sub-harmonically pumped diode ring mixer [3] performing the frequency conversion from Ka to Ka band according to the frequency synthesis of Table 1. The double balancing is chosen to keep in check the higher-order LO harmonics, inevitably falling very close to the IF output band, to maximize the port to port isolations and also to avoid using ultrabroadband 180-deg hybrids, which are quite difficult to exploit with MMIC technology, due to the frequency conversion required for the application. Besides, the sub-harmonic configuration for the diode mixer helps in achieving the high linearity and spurious suppression level required for SCFC. Unbiased diode mixer is practically the only possible choice to implement a compact double-balanced structure with the necessary symmetry [3]. Indeed, other possible balanced active solutions (e.g., Cold FET, transconductance mixer, Gilbert cell) would require complex topologies (also for the distribution of bias lines), which are ineffective in terms of performance (due to parasitic effects and spurious couplings) and space occupation. The core of the mixer is composed of four antipodal diode pairs (APDP) in a ring structure used as the nonlinear mixing elements. The APDP

connection makes possible the frequency synthesis IF = RF – 2LO, by halving the LO frequency. The LO pump signal is applied with 180 deg phase displacement at the corresponding nodes of the APDP ring. The RF and IF signals are injected and extracted with 180 deg phase displacement at the APDP ring nodes opposite to those where the LO is injected. An LO balun, an RF balun and an IF balun are exploited to make the circuit work properly. Matching networks between both the balanced output ports of the three baluns and the corresponding ring nodes are required to properly match the diode impedance to the balun impedance, as well as to provide the proper embedding impedances to the diodes at the RF/IF frequency bands. The LO balun takes the form of a Coupled Resonator Transformer Balun, while the RF / IF baluns are 3-conductors Marchand balun, where magnetic/electric coupled T-lines realize the conventional Marchand balun. An LO buffer was included to minimize the LO power requirements.

B. RF Variable Gain Amplifier

The RF side VGA is designed as the first gain stage in the chain with gain ranging from 10 to 17 dB at the input Ka-band from 27.5 GHz to 30.0 GHz. It is a two-stage balanced amplifier, where the first stage is designed for optimal noise figure and the second stage to improve the linearity. A 7 dB continuous gain control is implemented here to compensate the SCFC gain variation over temperature. The gain control is obtained using shunt switch devices as variable matching elements in the interstage and output networks. The supply voltage is 4 V, compliant with space derating, with a total drain current of 82 mA and an overall power consumption of 328 mW. The control voltage for the gain variation ranges from -4 V to 0 V.

C. RF 3-bit Digital Step Attenuator and RF buffer

The RF 3-bit DSA is designed to perform the first steps of the SCFC gain attenuation from 0 dB to 7 dB in the same input Ka-band. It is composed of three stages with attenuations of 1 dB, 2 dB and 4 dB respectively. Each stage exploits a conventional tee-type attenuator with series and shunt switch devices [4]. The insertion loss in the "0" state is 4 dB. A total of two complementary and one single control voltages, with value either -1.5 V or 0 V, are required to properly select the gain configuration. A one-stage buffer is placed at the Mixer input to improve the LO harmonics attenuation and to reach the total required gain.

D. IF Gain stage and 2-bit Digital Step Attenuator

The first IF gain stage after the mixer core is designed to provide 13 dB of maximum gain and to perform the remaining 13 dB of SCFC gain variation. To this purpose, the IF gain stage exploits a two-stage amplifier centred in the IF Ka-band from 17.2 to 20.2 GHz. Each stage is designed exploiting a 4×50 μm pHEMT in common source configuration, with a drain to gate resistor-capacitor (RC) feedback network for stabilization and bandwidth extension. The supply voltage is 4 V, compliant with space derating, with a total drain current of 53 mA and an overall power consumption of 212 mW. It is preceded by two different 1-bit DSA, with 5 dB and 8 dB attenuation selected by two complementary control voltages, with value either -1.5 V or 0 V. This renders possible to correctly program the last 13 dB of the 1 dB step gain variation in the SCFC.

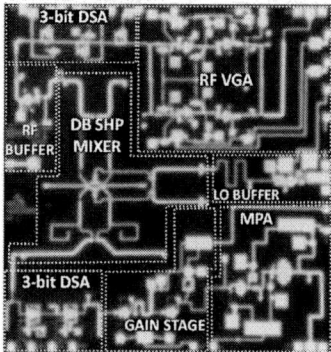

Fig. 2. Ka-band SCFC downconverter MMIC photograph. The main subcircuits are highlighted in the frames. The total die area is 5.0 x 5.0 mm²

E. IF Medium Power Amplifier

The IF output MPA is a two-stage power amplifier designed with main focus on its output power of 23 dBm at 1 dB of gain compression and an output TOI of 35 dBm to get enough margin to the power requirements of the SCFC in Table 1. Thanks to the high output power density of the chosen technology, the required P1dB and linearity are obtained with just a single device on the output stage. The MPA provides also 19 dB of gain to contribute to the overall SCFC gain. The supply voltage is 4 V, compliant with space derating, with a total drain current of 143 mA and an overall power consumption of 572 mW.

Fig. 3. Ka-band SCFC downconverter test board photograph.

IV. SCFC EXPERIMENTAL CHARACTERIZATION

In order to perform the full experimental characterization of the SCFC, the MMIC is mounted on a custom test board shown in Fig. 3. The test board includes a properly designed dc and control conditioning of the MMIC dc interfaces aimed at minimizing the required test instrumentation. Single drain supply voltage VD of 4 V and negative gate voltage VG of -0.65 V are used. The total drain current of 290 mA, resulted in a total dc power consumption of 1.1 W. Five different complementary voltages are also required to program the gain attenuation steps. The gain attenuation is programmed using an external microcontroller. The gain over temperature

compensation is controlled by using a further control voltage VC varying from -4 V to 0 V. The nominal input LO power is 3 dBm, while the RF input power, in order to mimic the actual behaviour of the complete downconverter system, follows the following variation with respect to the gain attenuation state:

$$P_{RF}(A) = \begin{cases} -36 + A \ dBm & A = 0, \dots, 10 \ dB \\ -26 \ dBm & A = 11, \dots, 20 dB \end{cases} \quad (1)$$

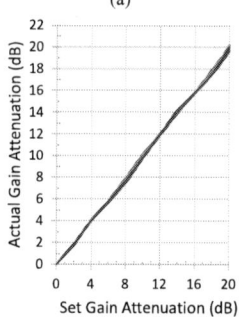 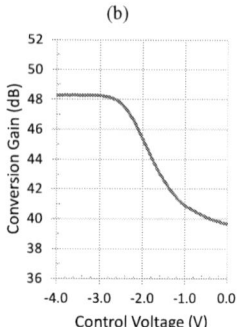

Fig. 4. Ka-band SCFC measured actual gain attenuation as a function of the set attenuation, with LO frequency of 4.975 GHz in the whole RF range from 27.5 to 30 GHz (a). Maximum conversion gain variation as a function of the control voltage VC, from -4.0 V to 0 V for the IF frequency of 19.05 GHz, with input RF frequency and LO frequency set to 29.0 GHz and 4.975 GHz, respectively (b). VD = 4 V, Id$_{tot}$ = 290 mA. The DSAs bits are set to the right value according to the required attenuation level.

To validate the gain attenuation and the gain variation as a function of the control voltage, we report in Fig. 4-(a) the measured actual gain attenuation at 4.975 GHz of LO frequency with respect to the set attenuation. The actual gain attenuation is very good and fairly linear as expected. Fig. 4-(b) shows that nearly 8 dB of conversion gain variation can be achieved spanning the control voltage VC from -4 V to 0 V.

In Fig. 5 we report the measured conversion gain and IF output power in nominal operation, considering the whole frequency conversion matrix with LO frequency swept in the band [4.575, 6.175] GHz, step 400 MHz, and RF frequency swept in the band [27.5, 30.0] GHz with 250 MHz step. The programmable gain was also swept from the maximum value to the minimum, by properly setting the attenuation, and the RF input power according to (1). The measured conversion gain is very flat over the entire IF band from 17.2 GHz to 20.2 GHz. The required 20 dB of gain adjustment were achieved with a variation on the range 25 dB to 45 dB. The measured output power lies in the range 7 to 8 dBm at 0 dB and 10 dB of attenuation and -3 to -2 dBm at 20 dB attenuation. According to the required variation of the RF input power (1) the output power at the minimum gain is 10 dB lower than that at maximum gain.

We report in Fig. 6 the conversion gain compression curves for LO frequency of 4.575 GHz and RF frequency of 29.5 GHz at the maximum, mid and minimum gain. The measured IF output power at 1 dB of gain compression is higher than 23 dBm, for the maximum and mid gain, and 19.5 dBm at minimum gain. Those figures are confirmed also for the other frequency synthesis.

94

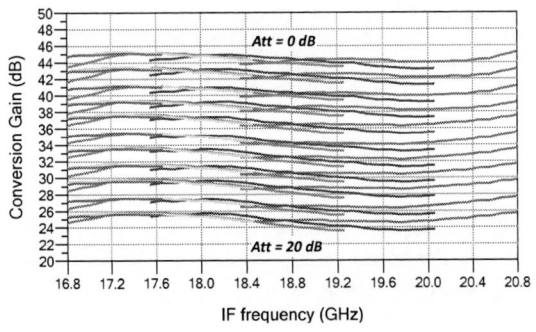

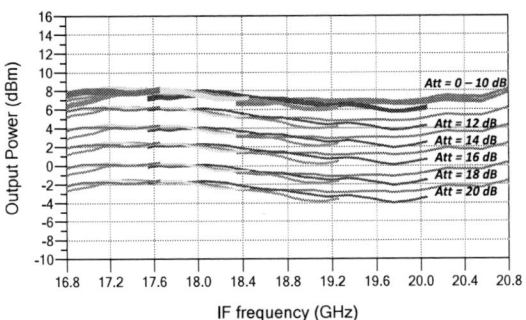

Fig. 5. Ka-band SCFC conversion gain and output power measurements. The Ka-band RF frequency in the band 27.5 GHz to 30.8 GHz, with step 100 MHz is converted down to IF, setting the LO frequency to 4.575 GHz (red), 4.975 GHz (blue), 5.375 GHz (pink), 5.775 GHz (cyan) and 6.175 GHz (purple). VD = 4 V, Id_{tot} = 290 mA, VC = -1.8 V. The DSAs bits are set to the right value to obtain 0 dB up to 20 dB of conversion gain attenuation with step 2 dB.

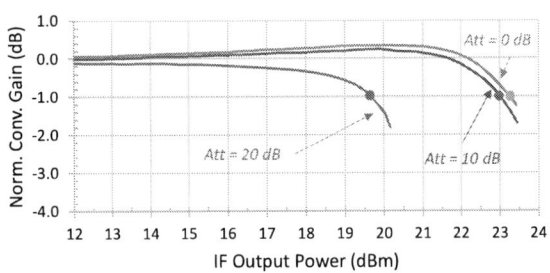

Fig. 6. Ka-band SCFC measured conversion gain compression at LO frequency of 4.575 GHz and IF frequency 20.35 GHz for 0 dB, 10 dB and 20 dB of attenuation. VD = 4 V, Id_{tot} = 280 mA, VC = -1.8 V. The DSAs bits are set to the right value according to the required attenuation level.

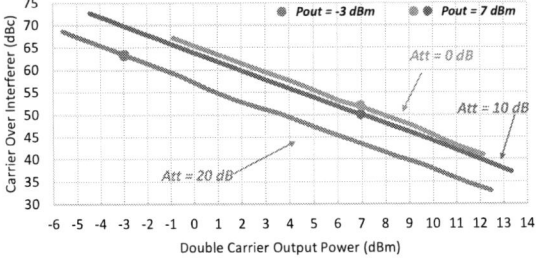

Fig. 7. Ka-band SCFC measured C/I3 at the worst-case IF frequency of 17.25 GHz with LO frequency of 5.375 GHz for 0 dB, 10 dB and 20 dB of attenuation. VD = 4 V, Id_{tot} = 290 mA, VC = -1.8 V. The DSAs bits are set to the right value according to the required attenuation level.

The output P1dB at the maximum attenuation is a little lower than expected from simulations but, thanks to the lower input power (1), this not affects the SCFC linearity, shown in Fig. 7 in terms of output C/I3 for the 3 attenuation states of 0, 10 and 20 dB. The worst case of 49.5 dBc is measured at 10 dB of attenuation at the lower end of the IF frequency, at 7 dBm of output power.

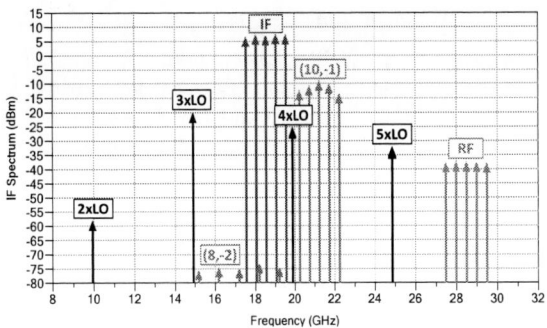

Fig. 8. Ka-band SCFC measured output spectrum at LO frequency of 4.975 GHz and RF frequency in the band [27.5, 29.5] GHz for 0 dB of attenuation. VD = 4 V, Id_{tot} = 290 mA, VC = -1.8 V. The DSAs bits are set to the right value according to the required attenuation level.

In Fig. 8 we report the SCFC measured output spectrum at 4.975 GHz of LO frequency and 0 dB of gain attenuation, in the [27.5, 29.5] GHz RF frequency range. The frequency conversion is well suited to be free of in-band spurious generation apart the (-8,2) component which is at least 80 dB lower than the IF tones. Concerning the LO harmonic, the highest 3xLO spectral component is easily attenuated by the filter because about 3.625 GHz far from the IF band. The most critical component is the 4th LO harmonic, which is 350 MHz above the IF band, but it is attenuated of about 33 dB at the output of the SCFC.

V. CONCLUSION

A MMIC Ka-band frequency converter integrating a broadband mixer, cascaded to many gain and programmable attenuation stages, including also a fine continuous gain tuning to recover the overall gain variation versus temperature has been designed, manufactured and experimentally validated. The space qualified process and the applied space derating make it very suitable for future generation telecom satellite frequency converters. More than 44 dB of maximum conversion gain with 49.5 dBc of C/I3 have been measured when converting the [27, 30] GHz band down to the broad [17.2, 20.2] GHz band using a low frequency broadband LO signal in the band [4.65, 6.10] GHz.

REFERENCES

[1] ESA, "Single-Chip Frequency Converter", ARTES AT 5E.012, [Online]. Available: https://artes.esa.int/funding/single-chip-frequency-converter-artes-5e012-0 .

[2] D. Resca, F. Scappaviva, A. Biondi, L. Cariani and F. Vitulli, "A Ku Band MMIC Single Chip Frequency Converter for Telecom Satellite Applications," IEEE EuMIC, 2021, pp. 57-60.

[3] S.A. Maas, "Microwave mixers," Norwood, MA, Artech House, Inc., 1993.

[4] I. J. Bahl, "Control Components Using Si, GaAs, and GaN Technologies," Norwood, MA, Artech House, Inc., 2014.

Proceedings of the 16th European Microwave Integrated Circuits Conference

A *V*-band Low-Power and Compact Down-Conversion Mixer with Low LO Power in 130-nm SiGe BiCMOS Technology

Batuhan Sutbas[*#1], Herman Jalli Ng[†], Jan Wessel[*], Alexander Koelpin[‡], Gerhard Kahmen[*#]

[*]IHP – Leibniz-Institut für innovative Mikroelektronik, Frankfurt (Oder), Germany
[#]Brandenburg University of Technology, Cottbus, Germany
[†]Karlsruhe University of Applied Sciences, Karlsruhe, Germany
[‡]Hamburg University of Technology, Hamburg, Germany
[1]suetbas@ihp-microelectronics.com

Abstract—In this paper, the design of a *V*-band low-power compact down-conversion mixer in a 130 nm SiGe BiCMOS technology is presented. The differential mixer design replaces the conventional transconductance stage of the Gilbert-cell with a pure passive VI-converter. This approach using only the switching quad as active circuit reduces the dc power consumption and improves linearity. In addition, the LO power requirement is decreased down to −10 dBm to eliminate the need of an additional LO buffer amplifier saving power from an RF frontend system architecture point of view. Measurement results show that the fabricated mixer achieves a peak voltage conversion gain of 16.2 dB and it consumes only 2.1 mW of dc power when it is compressed by 1 dB. The 3 dB RF bandwidth covers 39 GHz to 80 GHz with a 69% fractional bandwidth. Compared to other state of the art high frequency low-power mixers, the proposed circuit achieves higher voltage conversion gain in a wider RF bandwidth while it has a lower LO power requirement.

Keywords—down-conversion mixer, low-power, linearity, wideband, mm-wave, silicon-germanium, integrated circuit.

I. INTRODUCTION

The ISM band at the *V*-band is appealing to many short-range applications. Typical applications are radar systems for vital-sign detection and gesture tracking which have low-power requirements as they are intended for mobile use. Silicon-germanium (SiGe) BiCMOS technologies are especially suitable for such low-power applications thanks to their high performance at millimeter-wave frequencies even when operated at low current allowing more design freedom.

A typical radar system utilizes two down-conversion mixers in an IQ receiver circuit to convert the input RF signal into in-phase and quadrature IF signals at the output. The same radar circuit must also provide the LO signals required for the mixing operations. However, high LO power requirement from the mixers can be a heavy burden on the overall radar power consumption budget. Additionally, a large LO amplitude limits the voltage headroom at the IF output which is quite valuable in a low-voltage and low-power mixer design [1].

Stacked transistors at the transconductance stage of a double-balanced active mixer also reduce the available voltage headroom at the IF output. In addition, they contribute significantly to the overall power consumption and they initiate the gain compression mechanism which impairs the

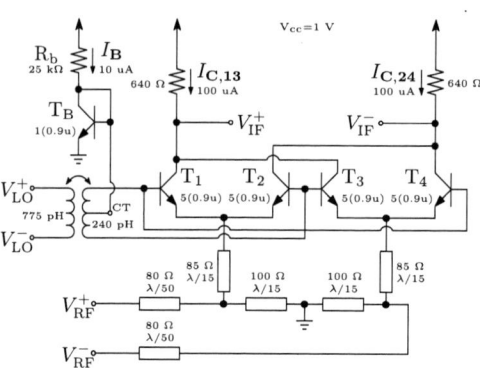

Fig. 1. Circuit schematic of the proposed down-conversion mixer using a pure passive VI-converter at the RF input.

overall linearity. Therefore, the transconductance stage has been eliminated [2] or replaced by Marchand baluns [3] and transmission line based matching networks [4], [5] in recent studies to achieve the low-voltage, low-power, and high-linearity requirements. However, the reported mixers lack high conversion gain and still require relatively high LO power levels. In this paper, we present the design of a compact double-balanced low-power down-conversion mixer at *V*-band which operates with an LO power of −10 dBm and uses only the switching quad as active circuit with a pure passive VI-converter based on transmission lines. The low-power mixer is fabricated and verified by measurements demonstrating high voltage conversion gain and linearity in a wide RF bandwidth.

II. DOWN-CONVERSION MIXER DESIGN

The circuit schematic of the proposed down-conversion mixer design is shown in Fig. 1. The transconductance stage of the double-balanced active mixer is replaced by a pure passive VI-converter transforming the $100\,\Omega$ differential input signal into differential currents at the quad emitters. The switching quad transistors have five emitter fingers each with lengths of $0.9\,\mu m$. Their base voltages are generated internally from the single supply voltage of 1 V.

The transistors can be operated in the low-current regime with collector currents of $10\,\mu A$ per emitter finger thanks to the high transit frequency (f_T) of the process. The reduced current consumption together with the lower supply voltage enables a low-power circuit. The proposed topology and the biasing of the transistors also allow the mixer to operate with a low LO power of $-10\,dBm$ decreasing the total power required by the mixer. The choice of $640\,\Omega$ IF load resistors considers a balance between voltage conversion gain and linearity.

The switching quad draws more current with increasing input RF drive. A larger voltage drop occurs across the IF load resistors and decreases the collector-base voltage of the transistors driving them into saturation leading to compression in conversion gain. In a conventional Gilbert-cell, the transconductance transistors limit the headroom available for this voltage drop on the one hand and also introduce another compression mechanism. By replacing the transconductance stage with a pure passive VI-converter linearity is improved.

A matching network based on transmission lines instead of lumped elements is used at the RF side to have lower loss. The microstrip lines are meandered for a compact size. The center of symmetry of the RF matching network is connected to the ground plane to improve common-mode rejection and suppress potential instabilities. The same network also provides the dc ground for the emitters of the transistors. For the LO matching network, an octagonal transformer is utilized. All metal layers and the p-well formation below the transformer are removed to increase the quality factor of the inductors. The bases of the transistors are biased from the center tap of the transformer. The absence of additional bias networks simplifies the design and facilitates easier matching. The layout of the matching networks is designed to preserve the symmetry.

The simulated RF and LO matching network losses at the center frequency are $0.8\,dB$ and $0.7\,dB$, respectively. The low-loss matching networks allow higher conversion gain and lower required LO power. The measured RF and simulated LO return losses at the center frequency are better than $15\,dB$ and $20\,dB$, respectively, as shown in Fig. 2.

The high-frequency RF and LO paths use the two top metals while the low-frequency IF signals are always on the

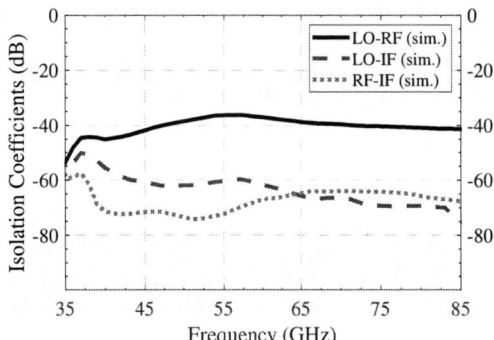

Fig. 3. Simulated LO to RF, LO to IF, and RF to IF isolation coefficients of the down-conversion mixer as a function of frequency.

lower metal layers separated by the ground plane. Thanks to this physical separation, higher LO to IF and RF to IF electrical isolation levels are easier to achieve. Fig. 3 shows that the simulated LO to RF, LO to IF, and RF to IF isolation levels are better than $35\,dB$, $55\,dB$, and $65\,dB$ in the entire V-band.

The IF-bandwidth is not considered in the mixer design because a narrow band signal is sufficient for the targeted frequency-modulated continuous wave radar applications. As the IF output will drive a low-pass filter, IF loads higher than $10\,k\Omega$ are used in the circuit simulations to represent the high-impedance load seen looking into this filter.

III. Experimental Results

A. Technology and Experimental Setup

A high performance 130 nm SiGe BiCMOS technology [6] which features f_T and f_{max} of 300 GHz and 500 GHz is utilized to realize the down-conversion mixer design. The backend process offers a total of seven conductor layers. The two top metals used for the RF networks are $2\,\mu m$- and $3\,\mu m$-thick, respectively. Fig. 4 shows a micrograph of the fabricated circuit. The effective mixer size indicated by the dotted area is $0.2\,mm \times 0.3\,mm$, excluding the compensated on-wafer measurement pads and the additional $100\,\Omega$ feed lines.

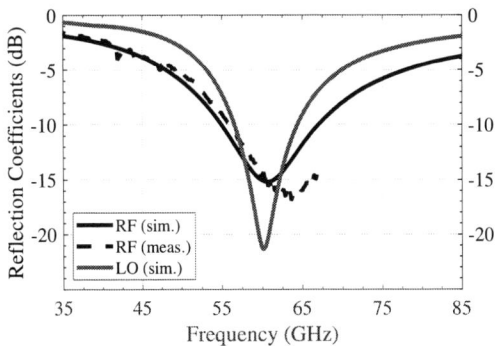

Fig. 2. Simulated and measured RF and LO reflection coefficients of the down-conversion mixer as a function of frequency.

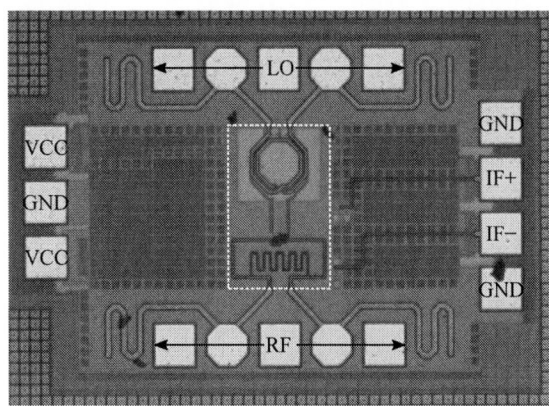

Fig. 4. Micrograph of the fabricated down-conversion mixer.

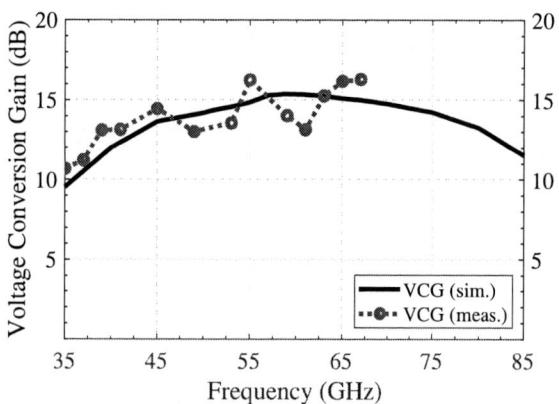

Fig. 5. Simulated and measured voltage conversion gain of the down-conversion mixer as a function of frequency.

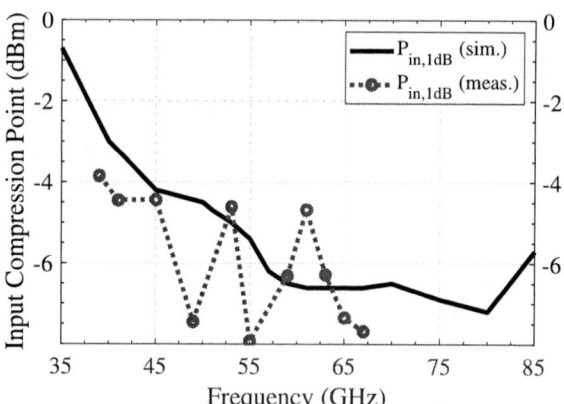

Fig. 7. Simulated and measured input referred 1 dB compression point of the down-conversion mixer as a function of frequency.

The differential RF and LO input signals are provided using a vector network analyzer (ZVA67[1]) and differential RF[2] probes. The differential IF output signal is captured with an oscilloscope (WavePro960[3]) using a dc[4] probe. A multimeter (34411A[5]) is used to record the current consumption. The cutoff frequency of the dc probe is high enough to pass the IF signal at 1 MHz. The RF and LO frequencies are swept simultaneously to keep the IF frequency fixed at 1 MHz and the LO power is set to −10 dBm in all of the measurements. The measured results agree well with the simulations supported by EM modeling[6] of the passive matching structures.

B. Measurement Results

Fig. 5 shows that the simulated and measured voltage conversion gain of the mixer is more than 13.2 dB within a

[1]Rohde & Schwarz GmbH & Company KG, Munich, Germany
[2]Cascade Microtech, Inc., Beaverton, OR, USA
[3]Teledyne Technologies, Inc., Thousand Oaks, CA, USA
[4]GGB Industries, Inc., Naples, FL, USA
[5]Agilent Technologies, Inc., Santa Clara, CA, USA
[6]ADS 2019.01, Keysight Technologies, Inc., Santa Rosa, CA, USA

3 dB RF bandwidth range from 39 GHz to 80 GHz covering the entire V-band. The wideband performance is attainable thanks to the good RF matching as well as the low LO power requirement of the mixer as higher reflected power at the LO port can be tolerated. The simulated double side band noise figure in the 3 dB bandwidth is less than 12.1 dB and reaches its minimum of 6.1 dB at 60 GHz as shown in Fig. 6.

Fig. 7 and Fig. 8 show the input referred 1 dB compression point and the dc power dissipation at this point versus frequency. At the center frequency of 61 GHz, the 1 dB compression occurs at an RF drive of −4.7 dBm where the differential IF output reaches a 1.48 V peak-to-peak amplitude and the circuit draws a total current of 2.1 mA.

A comparison with recent high frequency low-power mixers in silicon technologies is listed in Table 1. Here, the performance parameters are reported at the respective center frequencies. Although the mixers based on Gilbert-cells achieve higher gain, removing the transconductance stage or using a fully passive structure reduces the dc power

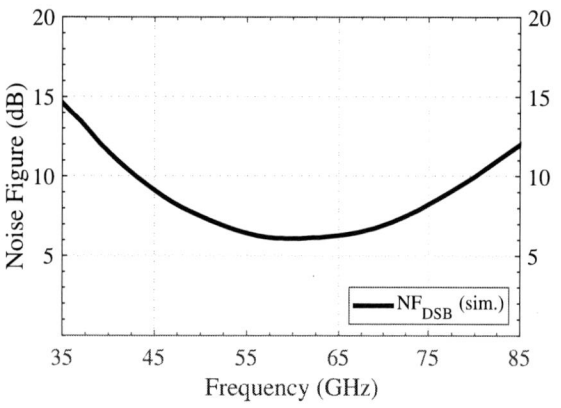

Fig. 6. Simulated double side band noise figure of the down-conversion mixer as a function of frequency.

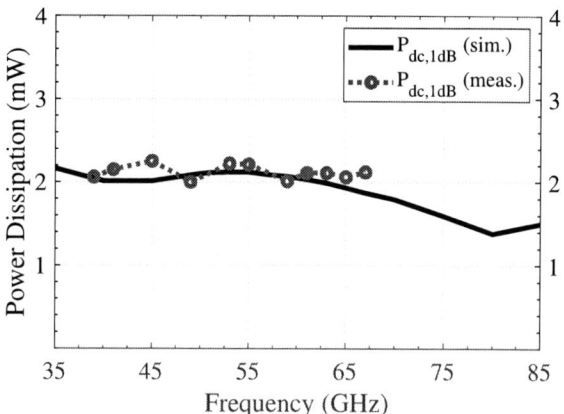

Fig. 8. Simulated and measured dc power dissipation of the down-conversion mixer at the 1 dB compression point as a function of frequency.

Table 1. Performance of recent high frequency low-power down-conversion mixers based on silicon technologies.

Reference	Technology	Topology	Freq. (GHz)	$RF_{BW,3dB}$ (GHz)	P_{LO} (dBm)	VCG (dB)	NF_{DSB} (dB)	V_{cc} (V)	P_{dc} (mW)	$P_{in,1dB}$ (dBm)	Size (mm^2)
This work	130 nm SiGe	no g_m stage	61	39–67^m	-10	13.2	6.1^s	1	2.1	-4.7	0.06
[2] 2020	40 nm CMOS	active loads	78.5	75–82^m	0	4.8	n/a	1	1.5	0.8	0.07
[4] 2019	28 nm CMOS	no g_m stage	60	n/a	-2	2	13.6	0.9	11.7	-4	0.47^p
[5] 2018	28 nm CMOS	no g_m stage	60	57–64^m	-2	4.7	15.3	0.9	1.8	-3	0.24^p
[7] 2016	180 nm CMOS	Gilbert-cell	22	20^m–26	1	24	n/a	1.8	17	-20	0.36^p
[8] 2013	130 nm SiGe	Gilbert-cell	77	n/a	0	13.7	14.4	2	15	2	0.23
[9] 2013	pure SiGe	passive	77	n/a	0	-0.5	13.8^s	3.3	0	-3	0.13

^{m}measurement limitation, ^{s}simulated, ^{p}including pads

consumption significantly. Despite using a much lower LO power compared to other low-power mixers, the mixer presented in this paper has a higher voltage conversion gain while maintaining a similar linearity performance. Also, the widest 3 dB RF bandwidth is reported.

IV. CONCLUSION

A double-balanced active mixer with a pure passive VI-converter is successfully designed to obtain low-voltage, low-power, and high linearity at the same time. The transistors are biased in low-current regime with very low collector currents to further reduce the power consumption. The fabricated and measured circuit achieves high voltage conversion gain and high input compression point. The presented mixer requires the lowest LO power and has the widest RF bandwidth compared to other state of the art low-power high frequency mixers. The low LO power requirement of the mixer can also be quite useful to decrease the number of circuit blocks needed in the LO path of highly efficient RF frontends saving additional power and chip area.

ACKNOWLEDGMENT

The authors would like to thank Wael A. Ahmad, Aniello Franzese, Ahmed Gadallah, and Thomas Mausolf of IHP for providing support in the measurements. This work has been funded by the Federal Ministry of Education and Research of Germany (BMBF) within the iCampus Cottbus Project under Grant 16ES1128K.

REFERENCES

[1] K. L. Fong and R. G. Meyer, "Monolithic RF active mixer design," *IEEE Trans. Circuits Syst. II*, vol. 46, no. 3, pp. 231–239, Mar. 1999.

[2] C. Xu, J. Chen, and D. Zhao, "A 77-GHz 1.5-mW down-conversion mixer using active loads technique in 40-nm CMOS for automotive radar application," in *Proc. IEEE Int. Conf. Integr. Circuits Technol. Appl.*, Nanjing, China, Nov. 2020, pp. 7–8.

[3] J.-H. Tsai, "Design of 1.2-V broadband high data-rate MMW CMOS I/Q modulator and demodulator using modified Gilbert-cell mixer," *IEEE Trans. Microw. Theory Techn.*, vol. 59, no. 5, pp. 1350–1360, May 2011.

[4] R. Ciocoveanu, R. Weigel, A. Hagelauer, and V. Issakov, "Modified Gilbert-cell mixer with an LO waveform shaper and switched gate-biasing for 1/f noise reduction in 28-nm CMOS," *IEEE Trans. Circuits Syst. II*, vol. 66, no. 10, pp. 1688–1692, Oct. 2019.

[5] R. Ciocoveanu, J. Rimmelspacher, R. Weigel, A. Hagelauer, and V. Issakov, "A 1.8-mW low power, PVT-resilient, high linearity, modified Gilbert-cell down-conversion mixer in 28-nm CMOS," in *Proc. IEEE 18th Top. Meeting Silicon Monolithic Integr. Circuits RF Syst.*, Anaheim, CA, USA, Jan. 2018, pp. 19–22.

[6] H. Rucker, B. Heinemann, and A. Fox, "Half-terahertz SiGe BiCMOS technology," in *Proc. IEEE 12th Top. Meeting Silicon Monolithic Integr. Circuits RF Syst.*, Santa Clara, CA, USA, Jan. 2012, pp. 133–136.

[7] Y.-T. Chang, H.-Y. Wu, and H.-C. Lu, "A K-band high-gain down-converter mixer using cross couple pair active load," in *Proc. 11th Eur. Microw. Integr. Circuit Conf.*, London, UK, Oct. 2016, pp. 377–380.

[8] J.-S. Jang, L. Moquillon, P. Garcia, E. Lauga-Larroze, and J.-M. Fournier, "Low power and high linearity millimeter wave differential mixer using passive transformer," in *Proc. Asia-Pacific Microw. Conf.*, Seoul, South Korea, Nov. 2013, pp. 700–702.

[9] H. J. Ng, M. Jahn, R. Feger, C. Wagner, and A. Stelzer, "An efficient SiGe double-balanced mixer with a differential rat-race coupler," in *Proc. 43rd Eur. Microw. Conf.*, Nuremberg, Germany, Oct. 2013, pp. 1551–1554.

A 60 GHz Frequency Doubler with 3.4-dBm Output Power and 4.4% DC-to-RF-Efficiency in 130-nm SiGe BiCMOS

Yu Zhu, Hatem Ghaleb, Vincent Rieß, Niko Joram, Frank Ellinger

Chair of Circuit Design and Network Theory, Technische Universität Dresden, Germany
yu.zhu2@tu-dresden.de

Abstract — This paper presents a power efficient 60 GHz frequency doubler fabricated in a 130-nm SiGe-BiCMOS technology for conversion of an input signal at 30 GHz to an output signal at 60 GHz. In order to improve the conversion gain (CG) and bandwidth, the proposed frequency doubler is implemented by a Gilbert-cell topology utilizing a bootstrapping technique. To realize single-ended input and output, an active balun and a Marchand balun are designed at the input and output of the frequency doubler, respectively. This proposed design achieves up to 3.4 dBm output power at 59 GHz, with a maximum CG of 22.5 dB, and a 4.4% peak DC-to-RF efficiency. Moreover, measurement results show a 3-dB bandwidth of 15 GHz from 51 to 66 GHz and a maximum fundamental suppression of 45 dB. To the best knowledge of the authors, the proposed frequency doubler provides the highest CG at the operating frequencies.

Keywords — Frequency doubler, Gilbert-cell, 60 GHz, microwave integrated circuits, BiCMOS, SiGe

I. Introduction

For mm-wave wireless communication and radar systems, high power, purity and stability signal generation is becoming more crucial as operating frequencies keep increasing into the millimeter-wave band. At lower frequencies a high-performance signal can be generated by a voltage-controlled oscillator (VCO) with a phase-locked loop (PLL) in modern transmitters (TX) and receivers (RX). In fact, the design of low phase noise VCOs with wide tuning range becomes more challenging at higher frequencies, due to low quality factor of the integrated LC tank and limited tuning range of the varactors. To overcome the above shortcomings of millimeter-wave VCOs, a frequency multiplication approach is often used to convert the signal into high frequency requested by many applications. Compared with high frequency fundamental tone VCOs, frequency multipliers can deliver better phase noise and larger tuning range. However, the overall phase noise is theoretically increased by $20 \log_{10} N$ dB adopting the frequency multiplication by an integer factor of N [1]. Furthermore, a good compromise between phase noise, power consumption, and circuit complexity regarding system requirements has to be chosen when using the frequency multiplication approach.

Typically, the frequency doubler is commonly used to generate the second harmonic of the input signal. To achieve frequency doubling three circuit topologies are widely employed: the single stage amplifier [2], the push-push amplifier [3], [4] and the Gilbert-cell [5], [6]. The single stage amplifier

topology is more compact and consumes low power, but an additional bandpass filter is required to suppress undesired harmonics. The push-push amplifier can achieve even-order harmonics superposition at a common node and odd-order harmonics cancellation without a bandpass filter. However, these two topologies have inherently a single-ended output and thus are not preferred for fully differential systems. In addition, the Gilbert-cell topology is widely used by mixing the input signal with itself to produce a signal at the doubled frequency. This approach offers differential output and high output power at the cost of high dc power consumption.

In this work, a double-balanced frequency doubler based on a Gilbert-cell design was chosen for its fully differential signaling, high CG and high output power. In Section II, an improved Gilbert-cell frequency doubler using a bootstrapping technique is described, in order to maximize the CG and the DC-to-RF efficiency. Section III presents the measurement results and a conclusion with a comparison to other doubler designs is described in Section IV.

II. Circuit Design and Implementation

A circuit block diagram representing the proposed frequency doubler is given in Fig. 1(a). An active balun at the input stage transforms the 30 GHz differential signals from a single-ended input. This signal is converted to a frequency of around 60 GHz by the following fully differential bootstrapped Gilbert-cell based frequency doubler. At the output of the frequency doubler, a Marchand balun is used to enable the single-ended signal format around 60 GHz frequency. The whole circuit needs three voltage supplies: $V_{ref} = 2.3$ V and $V_{CCA} = 3.3$ V for the input active balun, and $V_{CCB} = 3.3$ V for the frequency doubler core.

A. Active Balun

As a differential input signal is required by the Gilbert-cell topology doubler, an active balun is integrated at the input [7], due to its smaller size and wider bandwidth. Moreover, common-mode rejection is introduced due to the symmetry of the circuit and of the differential signal at the input [4]. Therefore, the fundamental suppression of the following doubler can be further increased.

The circuit schematic of the active balun is shown in Fig. 2. At the input, a common-base amplifier stage (T_3) is used

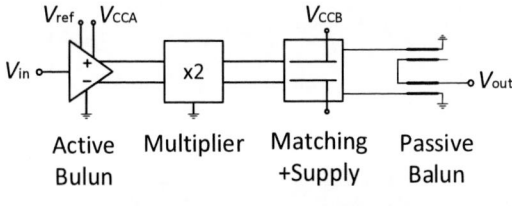

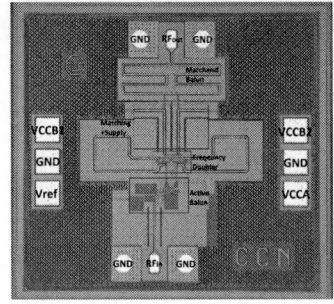

Fig. 1. Designed 60 GHz frequency doubler: (a) Block diagram; (b) Micrograph of the frequency doubler. The chip area (including pads) is $1000\,\mu\text{m} \times 920\,\mu\text{m}$.

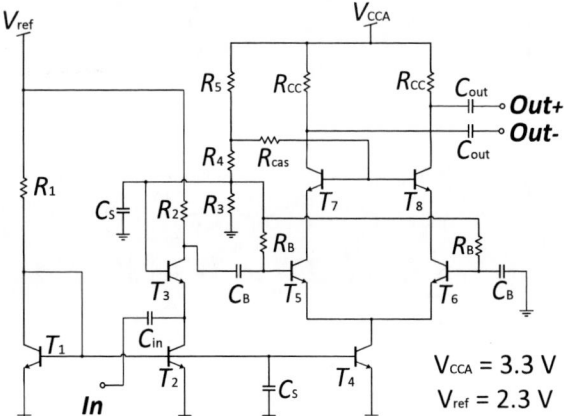

Fig. 2. Schematic of the designed active balun at the input.

to achieve a good broadband matching to 50 Ω input port and simultaneously provides signal pre-amplification. Then by connecting the common-base amplifier output to one input (T_5) of the cascode differential pair (T_5-T_8), while shorting the other input (T_6) of the pair to ac ground, a differential output signal is generated. This way, due to high common-mode suppression of the differential pair, the differential-mode is further amplified. The tail transistors (T_1, T_2 and T_4) are used to realize the current mirrors acting as current sources for the active balun, the reference current path is biased to 1.6 mA.

According to post-layout simulation, this active balun achieves a peak gain of 10.7 dB with a 3-dB bandwidth from 2 GHz to 32 GHz. The simulated input-referred 1-dB compression point is at -18 dBm. The total dc power consumption is about 18 mW. Besides, this balun also offers amplitude imbalance of less than 0.2 dB, and the phase difference of differential output signals is quite close to the ideal 180° with less than 0.7° variation over the whole bandwidth.

B. Bootstrapped Frequency Doubler

Although a push-push topology allows for higher power efficiency, the Gilbert-cell based topology is implemented here since it offers a differential output, which is more favorable for a fully differential system. Moreover, a bootstrapping technique is employed by feeding the input of upper switching quad transistors directly from the collectors of the lower differential pair, which results in higher CG compared with the conventional Gilbert-cell based topology, since the LO signal is amplified before driving the switching transistors [8].

The schematic of the proposed frequency doubler is shown in Fig. 3. The circuit core is realized by one differential

transconductance (g_m) pair (T_5 and T_6) at the input and the switching quad transistors ($T_1 - T_4$). At the RF input of the Gilbert-cell, two extra transmission lines (TL_E) are used with respect to the conventional topology, introducing a phase shift for higher voltage swing at the switching quad. According to [9], higher output power and efficiency can be achieved by optimizing the electrical length of TL_E. At the doubler's output, transmission lines TL_C1, TL_C2 and an output dc blocking capacitor are implemented as impedance matching network to 100 Ω at the second harmonic. A trade-off between bandwidth, output power and power consumption of the circuit can be mainly determined by the choice of the transistor size. Thus, a transistor ($N_\text{X} = 5$) with emitter size of 0.315 μm^2 is chosen here and biased at 5 mA collector current for the highest f_T/f_max. To further improve the suppression of unwanted higher-order even harmonics, which can be caused by leakage or the self-mixing, a tail resistor R_t is added in the g_m-stage. According to the simulation, it improves the rejection of the fourth harmonic by more than 10 dB but at the expense of higher power consumption. In addition, two capacitors (C_E) are used to enhance the harmonic spur rejection by shunting the higher harmonics to ground.

C. Marchand Balun

To enable the conversion of the differential signals to single-ended format at the output, a V-band Marchand balun has been designed, which simplifies the measurement setup. Fig. 4 presents the 3D view of the Marchand balun. In this balun, two lines are implemented in the top two metals, TM2 and TM1 have a thickness of 3 μm and 2 μm, respectively. Therefore, metal losses were significantly minimized, dc current handling capacity has been maximized and the coupling to the lossy substrate was reduced. To prevent the crosstalk in the vicinity of devices and ease modeling, a ground sidewall is implemented surrounding the balun.

The width of conductor lines is chosen to match the single-ended input impedance to 50 Ω and differential output impedance to 100 Ω. In order to achieve better inser-

101

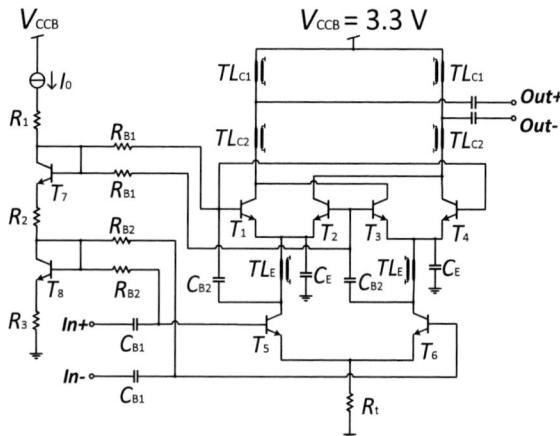

Fig. 3. Schematic of the proposed bootstrapped Gilbert-cell doubler.

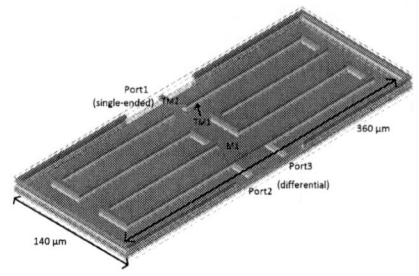

Fig. 4. 3D view of the designed output Marchand balun.

tion loss, bandwidth, the phase and amplitude imbalance, the physical dimension of the balun was tuned and optimized by EM-simulation in ADS momentum. With a size of 360 μm × 140 μm, the simulated insertion loss is about 1.6 dB at 60 GHz. From 50 GHz to 70 GHz, the amplitude imbalance is less than 1.0 dB and the phase imbalance is less than ±3° in the same band.

III. EXPERIMENTAL RESULTS

The frequency doubler has been manufactured in a SiGe 130nm BiCMOS technology. The SiGe HBT achieves a maximum f_T/f_{max} of 300 GHz/500 GHz and has an emitter area of $A_E = N_X \cdot 70$ nm $\cdot$ 900 nm, where N_X is a selectable integer multiplication factor. The chip micrograph with labels is presented in Fig. 1(b). The circuit occupies 0.92 mm² including all bondpads and baluns. The size of the frequency doubler core is only 0.125 mm². Zero-Ohm transmission lines were used for the dc supply connection between all circuit blocks. The total power consumption of all blocks is 50 mW.

To measure the input and output return losses of the frequency doubler from 10 GHz to 67 GHz, a four-port network analyzer Rohde&Schwarz ZVA-67 and GSG probes are used. The measured and simulated $|S_{11}|$ and $|S_{22}|$ are shown in Fig. 5. It can be seen that there is a good agreement between simulated and measured values, although there is

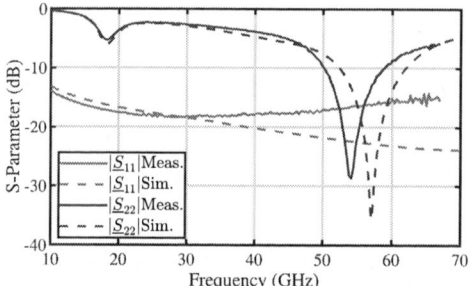

Fig. 5. Measured and simulated input and output return losses.

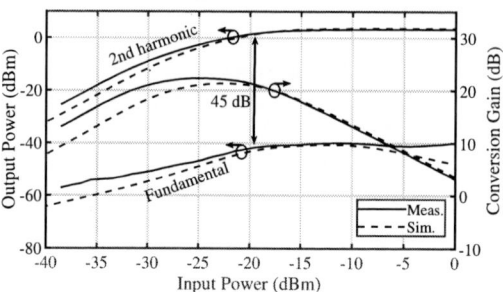

Fig. 6. Measured and simulated output power at the fundamental frequency and at the second harmonic and CG versus input power for $f_{in} =$ 29.5 GHz.

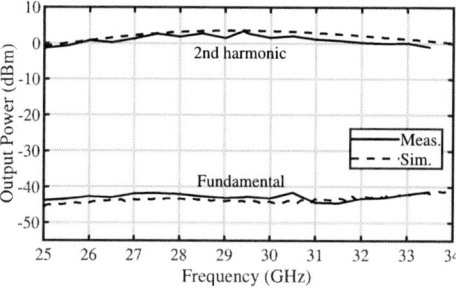

Fig. 7. Measured and simulate output power at the fundamental frequency and at the second harmonic versus input frequency for $P_{in} =$ -7.5 dBm.

a slight variation and shift towards lower frequencies. The measured $|S_{11}|$ is below -15 dB between 15 GHz and 50 GHz, and $|S_{22}|$ is below -8 dB from 48 GHz to 62 GHz, which allows sufficient power transfer to the output at the second harmonic. Thus, the output matching network also serves as bandpass filter to further suppress undesired harmonics.

Fig. 6 presents the measured and simulated output power and CG versus input power at 29.5 GHz input frequency, which are measured by a signal generator (Keysight E8257D) and a spectrum analyzer (Rohde&Schwarz FSW-67). Both the output power at fundamental frequency and at the second harmonic are measured and depicted here. The circuit saturates at 3.4 dBm output power and provides a peak CG of 22.5 dB at the input power level of -25 dBm. It can be observed that measured output power at the second harmonic brings

Table 1. Comparison with state-of-the-art of frequency doublers

Ref.	This work	[5]	[6]	[10]	[11]	[7]	[12]
Technology	130nm SiGe BiCMOS	400nm SiGe Bipolar	180nm SiGe BiCMOS	130nm SiGe BiCMOS	130nm SiGe BiCMOS	130nm SiGe BiCMOS	22nm FDSOI
Topology	Gilbert cell	Gilbert cell	Gilbert cell	Gilbert cell	Push−Push	Push−Push	Push−Push
Input-Output ^{a}	SE−SE	Diff.−Diff.	Diff.−Diff.	SE−Diff.	Diff.−Diff.	SE−Diff.	Diff.−Diff.
Out. freq. (GHz) ^{b}	51−66	22−39	57−71	55−75	55.6−65.6	57−77^c	50−66
Max. P_{out} (dBm)	3.4	0	1.7	0.3	-4	-13	-8.2
Max. Conv. Gain (dB)	22.5	8.6	10.2	2^d	-15	0	-5.7
Max. fund. suppr. (dB)	45	30	36	30	42	35	40
DC power (mW)	50	185	137.3	50	23.5	31	16
Peak efficiency (%)	4.4	0.5	1.1	2.1	1.7	0.16	0.95
Chip size (mm^2)	0.92	0.25	0.27	0.25	0.55	0.48	0.135

^{a} Single-ended (SE) Differential (Diff.) ^{b} −3dB bandwidth ^{c} simulated ^{d} graphically estimated

a strong non-linearity, while it demonstrates a rising slope of 20 dB per decade over the input power before compression. The highest fundamental suppression reaches 45 dB at around -20 dBm input power. In fact, the fundamental frequency is not perfectly canceled out at the output, due to parasitic coupling, mismatch of the amplifiers, amplitude and phase imbalances in the baluns.

With the same measurement setup, the simulated and measured saturated output power at the fundamental frequency and at the second harmonic for an input power of -7.5 dBm versus frequency are shown in Fig. 7. The circuit achieves a 3-dB bandwidth of 15 GHz from 51 GHz to 66 GHz with a maximum output power of 3.4 dBm at 59 GHz. Moreover, the fundamental suppression is larger than 42 dB over the whole 3-dB bandwidth.

IV. CONCLUSION

A 60 GHz frequency doubler has been fabricated and measured in this work. To convert the single-ended input signal to the required differential input signal, an active balun is designed to provide broadband matching and also pre-amplification. At the output, a passive Marchand balun is implemented to enable the single-ended measurements, introducing a bandpass filter to improve the undesired harmonics suppression. Due to the bootstrapping technique, this frequency doubler achieves higher CG and output power compared with standard topology. The results are summarized in Table 1, where different state-of-the-art frequency doublers are compared. The presented circuit delivers a maximum output power of 3.4 dBm with 22.5 dB peak gain and highest efficiency of 4.4% at 59 GHz. Additionally, a high fundamental suppression of 45 dB is reached. In terms of CG, saturated output power and efficiency, the proposed frequency doubler achieved better performance than the conventional Gilbert-cell based and also push-push based frequency doublers.

ACKNOWLEDGMENT

This research was supported by the Federal Ministry of Transport and Digital Infrastructure BMVI, in the funding program Automated and Connected Driving, ATLAS60G.

REFERENCES

[1] S. A. Maas, *Nonlinear Microwave and RF Circuits*, 2nd ed.   Artech house, 2003.

[2] F. Ellinger and H. Jackel, "Ultracompact SOI CMOS frequency doubler for low power applications at 26.5-28.5 GHz," *IEEE Microwave and Wireless Components Letters*, vol. 14, no. 2, pp. 53–55, 2004.

[3] G. Liu, A. Ç. Ulusoy, A. Trasser, and H. Schumacher, "60–80 GHz frequency doubler operating close to f_{max}," in *2010 Asia-Pacific Microwave Conference*.   IEEE, 2010, pp. 770–773.

[4] S. Chakraborty, L. E. Milner, X. Zhu, L. T. Hall, O. Sevimli, and M. C. Heimlich, "A K-Band Frequency Doubler With 35-dB Fundamental Rejection Based on Novel Transformer Balun in 0.13-μm SiGe Technology," *IEEE Electron Device Letters*, vol. 37, no. 11, pp. 1375–1378, 2016.

[5] S. Hackl and J. Bock, "42 GHz active frequency doubler in SiGe bipolar technology," in *2002 3rd International Conference on Microwave and Millimeter Wave Technology, 2002. Proceedings. ICMMT 2002*.   IEEE, 2002, pp. 54–57.

[6] A. Y.-K. Chen, Y. Baeyens, Y.-K. Chen, and J. Lin, "A 36–80 GHz High Gain Millimeter-Wave Double-Balanced Active Frequency Doubler in SiGe BiCMOS," *IEEE Microwave and Wireless Components Letters*, vol. 19, no. 9, pp. 572–574, 2009.

[7] V. Rieß, C. Carta, and F. Ellinger, "A 60 GHz Frequency Doubler with Differential Output in 130 nm SiGe BiCMOS Technology," in *2018 Asia-Pacific Microwave Conference (APMC)*.   IEEE, 2018, pp. 279–281.

[8] S. Yuan and H. Schumacher, "90–140 GHz frequency octupler in Si/SiGe BiCMOS using a novel bootstrapped doubler topology," in *2014 9th European Microwave Integrated Circuit Conference*.   IEEE, 2014, pp. 158–161.

[9] M. Kucharski, M. H. Eissa, A. Malignaggi, D. Wang, H. J. Ng, and D. Kissinger, "D-Band Frequency Quadruplers in BiCMOS Technology," *IEEE J. Solid-State Circuits*, vol. 53, no. 9, pp. 2465–2478, 2018.

[10] S. Yuan and H. Schumacher, "Compact V band frequency doubler with true balanced differential output," in *2013 IEEE Bipolar/BiCMOS Circuits and Technology Meeting (BCTM)*.   IEEE, 2013, pp. 191–194.

[11] V. Rieß, P. V. Testa, C. Carta, and F. Ellinger, "Analysis and Design of a 60 GHz Fully-Differential Frequency Doubler in 130 nm SiGe BiCMOS," in *2018 IEEE International Symposium on Circuits and Systems (ISCAS)*.   IEEE, 2018, pp. 1–5.

[12] M. Cui, C. Carta, and F. Ellinger, "Design of an Ultra Compact Low Power 60 GHz Frequency Doubler in 22 nm FD-SOI," in *2020 IEEE International Symposium on Radio-Frequency Integration Technology (RFIT)*.   IEEE, 2020, pp. 40–42.

Proceedings of the 16th European Microwave Integrated Circuits Conference

A 14.6 GHz – 19.2 GHz Digitally Controlled Injection Locked Frequency Doubler in 45 nm SOI CMOS

Olli Kursu, Timo Rahkonen, Aarno Pärssinen

University of Oulu, Faculty of Information Technology and Electrical Engineering (ITEE), Finland
firstname.lastname@oulu.fi

Abstract— In this paper we present a wide locking range (14.6 GHz – 19.2 GHz and 12.65 GHz – 20.6 GHz, -3 db and -6 dB, respectively) injection locked frequency doubler implemented with 45 nm CMOS SOI technology. The doubler is designed and optimized for a 5G sliding-IF transceiver architecture. It exploits a digitally tunable LC tank to enhance the frequency range. Measured results show 36 – 55 dBc fundamental and 40 – 54 dBc 3rd harmonic suppression as well as 10 dB peak conversion gain. Phase noise performance of the doubler has also been measured. The power consumption varies from 5 mW to 11 mW. The core size is 270 μm x 450 μm.

Keywords— 5G, frequency doubler, injection locked, LC oscillator

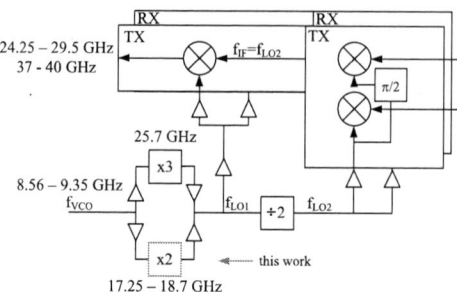

Fig. 1. Block diagram of the LO scheme in the sliding-IF transceiver.

I. INTRODUCTION

Frequency multipliers are widely used in transceiver circuits and radar systems. Several frequency doublers operating on the Ku- and K-band frequency range have been presented in the literature. Frequency multipliers can be divided into three categories: mixer-based, device nonlinearity based and injection locked oscillator based. The first one suffers from complex mixing of harmonics and relatively poor power consumption. Doublers based on amplifying the harmonic distortion resulting from device nonlinearity [1]–[4] often employ transformer baluns to achieve wide operating range and suppression of the fundamental and unwanted harmonic components. This suppression is, however, difficult to achieve due to the imbalance in the transformer resulting from parasitic capacitances between the coils and the fundamental rejection is typically only around 35 dB [1]. One design uses a hybrid coupler to achieve fully balanced operation that achieves 44 dB fundamental rejection but has a narrow tuning range [5].

The injection locked frequency doubler (ILFD) principle uses an active doubler consisting of a differential pair with shorted drain nodes. This inherently cancels the fundamental, injects even harmonics into the loop and locks the oscillation frequency at twice the input frequency [6]. With this topology, balanced operation is achieved without transformer components. Injection-locked oscillators are particularly suitable for phased array transceivers due to their excellent phase noise performance [7]. The phase noise of the output of the multiplier is ideally $20log_{10}(n)$ dB higher than the phase noise of the input source, where n is the locked

harmonic [8]. In the doubler case this results in a 6 dB phase noise degradation.

Distributing the LO signal consumes a lot of power in beam-forming transceivers. One way of minimising this, and distributing gain to different frequencies, is to use a sliding-IF architecture, where LO side injection can be used in the mixers. The presented ILFD is intended to be part of LO generation block shown in Fig. 1. In the frequency planning one of the main constraints was to limit the frequency range of the external phased locked loop (PLL) to less than 1 GHz i.e. 8.56 – 9.35 GHz. This results only to ∼9% tuning range requirement of the VCO. The ILFD presented in this paper is designed for a dual-band sliding-IF phased array transceiver, where the output RF frequency range will be from 24.25 GHz to 29.5 GHz (bands n257 and n258) as well as 39 GHz (band n260). The doubler is designed to cover the lower bands where input VCO frequency and the IF frequency are on the same range and a separate tripler will cover band n260. At the moment we have measured results only from the doubler part, and rest of the paper concentrates on it.

II. INJECTION LOCKED DOUBLER DESIGN

Injection locked frequency multipliers are designed to oscillate without input signal at a so-called self-oscillation or "free-running" frequency, on which the LC tank is tuned. The circuit topology of the doubler is shown in Fig. 2. Input biasing, bias circuits for individual N- and P-side biasing (points V_a and V_b, respectively) and output buffers are not shown in order to improve readability.

3–4 April 2022, London, UK

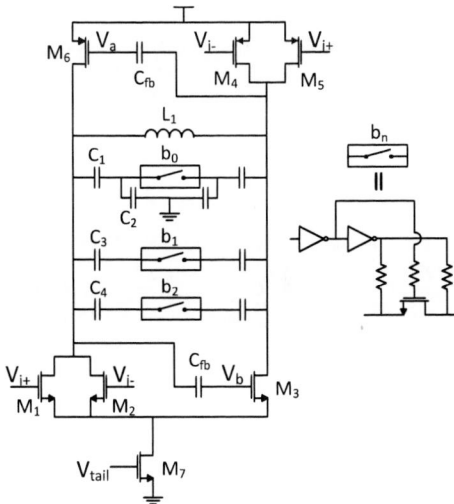

Fig. 2. Doubler schematic.

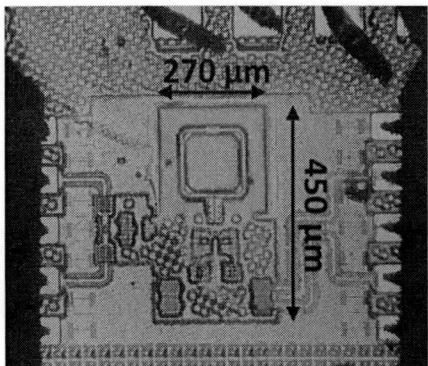

Fig. 3. Doubler chip micrograph.

In injection locked oscillators, locking range can be extended by several means: reducing the oscillator quality factor (Q), increasing the injection current I_{inj} and reducing I_{osc} by reducing the the gm of the amplifier stage. The first technique results in less efficient operation and the last one in reduced oscillation amplitude or even loss of oscillation [7]. Achieving high Q for the LC tank is challenging due to poor Q-value of capacitors and varactors at high frequencies and therefore further Q degradation is often not necessary.

From power consumption perspective, tuning the free-running oscillation frequency is the best choice. The doubler design follows the basic topology as shown for example in [6]. In our circuit, no varactors are used, but a capacitor bank with switchable MIM capacitors with 3-bit control is implemented. The tank Q can be maintained high while still having a large locking range by having a capacitor bank with digital control. With this scheme, the continuous locking range of the doubler can be greatly extended if the individual locking ranges are overlapping.

The performance of the doubler can be optimized for certain frequencies since the conversion gain is peaking in the center point of the locking range which coincides with the free-running frequency. When all of the switches (W/L = 89.6 μm / 40 nm) are off, the oscillation frequency is determined by the series connection of the capacitances C1 and C2 of the b0 switch, which provides the highest oscillation frequency. This way, the "all switches off" oscillation frequency is controlled rather than determined by parasitic capacitances alone. Like in [6], locking range is further increased by attenuating the feedback signal with C_{fb}.

As the doubler is going to be the starting point of the LO path in the transceiver, full voltage swing at the output node is not necessary. On the contrary, the swing should be limited so that the following LO buffer amplifier stages after the doubler are not causing unnecessary distortion. The LO needs to be

divided to several paths, i.e. mmW to IF and IF to BB mixers on both TX and RX paths. A capacitive attenuator consisting of a 75 fF MIM capacitor and the approx. 10 fF input capacitance of the first buffer stage provide a small attenuation for the signal. A second buffer stage is designed to drive a differential 100 ohm probe load.

III. MEASUREMENT RESULTS

A micrograph of the doubler is shown in Fig. 3. The power consumption of the doubler core varies from 5 mW to 11 mW depending on the tunable tail bias setting. The first buffer consumes 4 mW and the second one 11 mW.

The measurement setup is shown in Fig. 4. The setup contains Cascade Microtech differential 67 GHz GSGSG probes, Marki Microwave BAL-0067 baluns and coaxial cables with 1.85 mm connectors. A Keysight PSG E8257C signal generator is used as the signal source and a Keysight N9040B UXA signal analyzer has been used in the spectrum, locking range and input sensitivity measurements. Keysight E4446a spectrum analyzer is used to measure the phase noise. Cable, connector and probes losses have been measured separately without the DUT and are removed from the presented results over the entire frequency range at input and at output separately. Due to probe setup, measurements were conducted in the room temperature only.

The doubler self-oscillation measured with all eight frequency settings is shown in Fig. 5. A typical spectrum result is shown in Fig. 6. The measurement has been conducted with -14 dBm input power and 8 GHz input signal. In the LO path, as the LO signal can be amplified to mixers with tuned amplifiers, fundamental leakage and higher order harmonics are further attenuated. Fundamental leakage varied from -30 dBc to -55 dBc in the measurements. The rejection of the odd harmonics depends on the amplitude and phase balance of the differential inputs, and is dominated by the response of the external input balun.

The -3 dB locking range has been measured by sweeping the spectrum with constant input power of 0 dBm between 7 GHz and 9.8 GHz. Fig. 7 shows the multiplied output with four out of eight different oscillation frequency control values.

105

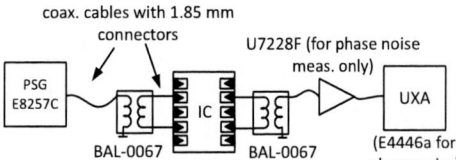

Fig. 4. Doubler measurement setup.

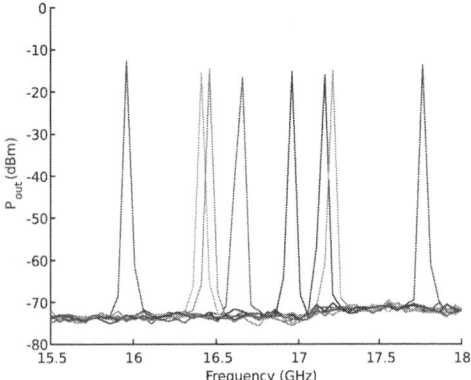

Fig. 5. Doubler self-oscillation frequencies for the eight center frequency settings.

-3 dB locking range is achieved from 14.6 GHz to 19.2 GHz. If the input power is increased to 2 dBm, the -3 dB locking range extends from 14 GHz to 19.6 GHz and -6 dB locking range from 12.65 GHz to 20.6 GHz. Thus, the 3-bit digital control expands the locking range 48% with good margin (overlap) between the controls.

The input sensitivity plot is shown in Fig. 8. The input power was reduced until the lock was lost and the last power level with lock was recorded. The measurement is shown on four of the frequency control values. As expected, conversion gain is at its highest near the free-running frequencies, where the peak conversion gain is approximately 10 dB. From Fig. 8 it can be also seen that the doubler can be operated over a wide

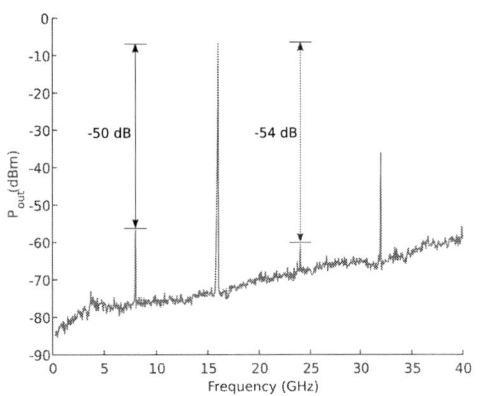

Fig. 6. Doubler spectrum measured at 16 GHz.

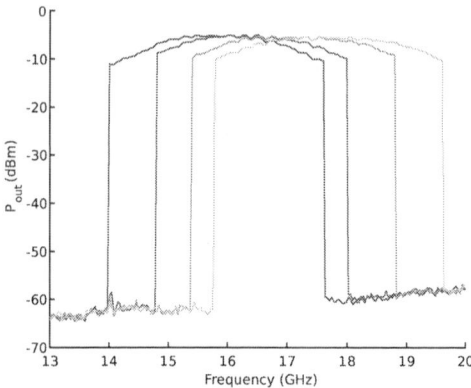

Fig. 7. Doubler locking range for oscillation frequency settings 0 (blue), 2, 5 and 7 (green).

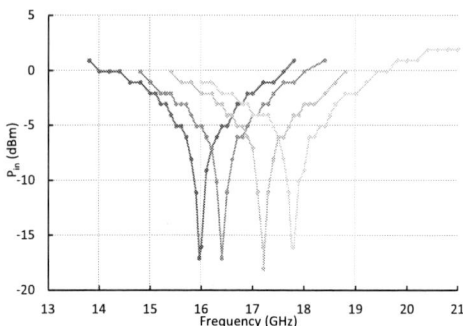

Fig. 8. Doubler input sensitivity measured with oscillation frequency settings 0 (blue), 2, 5 and 7 (green).

frequency range with lower input power level (-5 dBm) while still maintaining lock. This reduces the fundamental and 3rd harmonic level while still maintaining the frequency control range requirement of the multi-band system with good margin.

The measured phase noise at 18 GHz is shown in Fig. 9. The phase noise measurement is impacted by the purity of the signal source, the signal level at the doubler input and the signal level at the receiving phase noise measurement instrument input. For this reason, a buffer amplifier (Keysight U7228F) has been used to boost the signal level for the spectrum analyzer. The same amplifier is present in the signal generator phase noise measurement (lower plot) and the phase noise has been measured with the same 6 dBm power in both cases. The measured doubler phase noise is mostly approximately 6 dB higher than the signal generator phase noise that has been measured at 9 GHz.

The implemented doubler is compared against recent works in Table 1. It achieves state-of-the art performance with wide tuning range, low area, supply and power consumption demonstrating the benefits of digital control for also above 10 GHz solutions. The wide tuning range in [2] has been measured with high input power of 5 dBm and with that power level curves in Fig. 8 suggest also a major extension for locking range in our work.

106

Table 1. Comparison of frequency doublers operating at similar output frequency.

Reference	2010 [6]	2018 [5]	2014 [9]	2013 [2]	2010 [10]	**This work**
Process	130 nm CMOS	65 nm CMOS	180 nm CMOS	180 nm CMOS	180 nm CMOS	45 nm CMOS SOI
Topology	ILFD (varactor)	push-push balanced	casc. stacked current reuse	single balanced	casc. stacked current reuse	ILFD (mimcap)
-3 dB tuning range (GHz)	12 – 14.2 (16.8%)	22.5 – 24.8 (9.7%)	20 – 24 (18.2%)	15 – 36 (82.4%)	18 – 26 (36.4%)	14.6 – 19.2 (27.2%)
Input power (dBm)	5	0	0	5	0	0
Output power (dBm)	-5	5	-5	-5	4	-5
Fund. suppression (dBc)	45 – 53	44	32 – 53.8	33	30 – 50	36 – 55
3rd harm. suppr. (dBc)	57	–	–	–	15 – 25	40 – 54
Supply voltage	1.3	1.2	2.6	1	2.6	1
Power dissipation (mW)	5.2	31.2	13.9	4 – 11	18.2 – 20.8	5 – 11
Area (mm²)	0.083	0.35	0.57	0.32	0.29	0.12

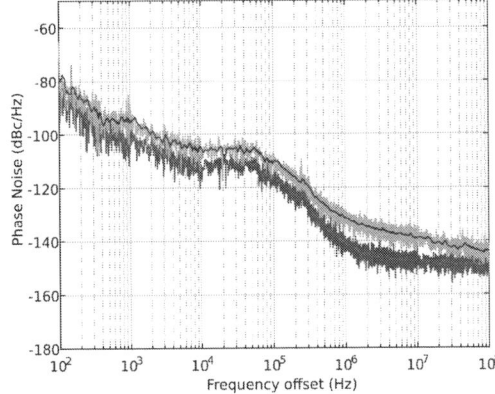

Fig. 9. Doubler phase noise measured at 18 GHz (yellow curve) compared to signal generator phase noise measured at 9 GHz (blue curve).

IV. CONCLUSION

The designed digitally controlled injection locked frequency doubler has a wide locking range while benefiting from the excellent phase noise performance inherent to injection locked multipliers. It can also operate with a relatively low input power level over a wide frequency range. The phase noise performance of the doubler has been demonstrated to be close to ideal with state-of-the-art fundamental and 3rd harmonic suppression. By using switchable MIM capacitors, locking range of the push-push injection locked doubler can be extended by 50%. Digital control with high-Q inductors and MIM capacitors enables to optimize the performance vs. tuning range with the on-resistance of the MOS switches even at frequencies above 10 GHz. The implemented doubler is capable of supporting wide bandwidth requirements of 5G mmW band solutions and can be further optimized for even wider tuning ranges by enhancing the digitally controlled capacitance matrix.

ACKNOWLEDGMENT

Nokia is acknowledged for financing the project, GlobalFoundries for chip processing and Keysight for providing measurement equipment. This work was supported in part by the Academy of Finland 6Genesis Flagship (grant no. 318927).

REFERENCES

[1] S. Chakraborty, L. E. Milner, X. Zhu, L. T. Hall, O. Sevimli, and M. C. Heimlich, "A k-band frequency doubler with 35-dB fundamental rejection based on novel transformer balun in 0.13- μm SiGe technology," *IEEE Electron Device Letters*, vol. 37, no. 11, pp. 1375–1378, Nov 2016.

[2] P. Tsai, Y. Lin, J. Kuo, Z. Tsai, and H. Wang, "Broadband balanced frequency doublers with fundamental rejection enhancement using a novel compensated marchand balun," *IEEE Transactions on Microwave Theory and Techniques*, vol. 61, no. 5, pp. 1913–1923, May 2013.

[3] N. Mazor, O. Katz, B. Sheinman, R. Carmon, R. Ben-Yishay, R. Levinger, and D. Elad, "A SiGe ku-band frequency doubler with 50% bandwidth and high harmonic suppression," in *2014 IEEE MTT-S International Microwave Symposium (IMS2014)*, June 2014, pp. 1–4.

[4] S. Saberi Ghouchani and J. Paramesh, "A wideband millimeter-wave frequency doubler-tripler in 0.13-μm CMOS," in *2010 IEEE Radio Frequency Integrated Circuits Symposium*, May 2010, pp. 65–68.

[5] S. Vehring and G. Boeck, "Truly balanced K-band push-push frequency doubler," in *2018 IEEE Radio Frequency Integrated Circuits Symposium (RFIC)*, June 2018, pp. 348–351.

[6] E. Monaco, M. Pozzoni, F. Svelto, and A. Mazzanti, "Injection-locked CMOS frequency doublers for μ-wave and mm-wave applications," *IEEE Journal of Solid-State Circuits*, vol. 45, no. 8, pp. 1565–1574, Aug 2010.

[7] G. Mangraviti, K. Khalaf, B. Parvais, K. Vaesen, V. Szortyka, G. Vandersteen, and P. Wambacq, "Design and tuning of coupled-LC mm-wave subharmonically injection-locked oscillators," *IEEE Transactions on Microwave Theory and Techniques*, vol. 63, no. 7, pp. 2301–2312, July 2015.

[8] B. Razavi, "A study of injection locking and pulling in oscillators," *IEEE Journal of Solid-State Circuits*, vol. 39, no. 9, pp. 1415–1424, Sep. 2004.

[9] S. Wang and C. Chang, "K-band CMOS frequency doubler with high fundamental rejection," *Electronics Letters*, vol. 50, no. 17, pp. 1211–1212, Aug 2014.

[10] J. Chen and H. Wang, "A high gain, high power K-band frequency doubler in 0.18 μm CMOS process," *IEEE Microwave and Wireless Components Letters*, vol. 20, no. 9, pp. 522–524, Sep. 2010.

Proceedings of the 16th European Microwave Integrated Circuits Conference

A W-Band Up-Conversion Mixer with Integrated LO Frequency Doublers in a 60 nm GaN Technology

Mingquan Bao[#], Robert Malmqvist[*], Rolf Jonsson[*], Jonas Hansryd[#], Kristoffer Andersson[#]

[#]Ericsson Research, Ericsson AB, Sweden

[*]Swedish Defence Research Agency (FOI), Sweden

{mingquan.bao, jonas.hansryd, kristoffer.andersson}@ericsson.com, {rma, roljon}@foi.se

Abstract — **A W-band up-conversion mixer with integrated frequency doublers is designed and manufactured in a 60 nm GaN-on-Si technology. The mixer is a single-balanced transconductance mixer. Via a 90^0 hybrid, the single-ended LO signal is split into I/Q signals fed separately to two frequency doublers, generating differential 2×LO signals. The circuit is characterized by on-chip probing. The measured lowest conversion loss is 6.4 dB. The 3-dB RF bandwidth is 8 GHz from 87.5 to 95.5 GHz and the 3-dB IF bandwidth is 4.0 GHz from 0.1 to 4.1 GHz. The output referred P_{1dB} is -8.5 dBm at 93 GHz RF frequency. The input power (LO) of the doubler is 17 dBm. The total DC power consumption of the mixer and the doublers is 620 mW. To the best of authors' knowledge, the mixer achieves the lowest conversion loss among the E/W-band GaN mixers published in the literature.**

Keywords — **frequency doubler, GaN, mixer, W-band.**

I. INTRODUCTION

Gallium nitride (GaN) devices have been applied increasingly in power amplifiers due to their superior high current/voltage capacity. An E-band and a W-band single-chip receivers in GaN technology have also been developed [1], [2]. In a transceiver, a mixer plays an important role, transferring a low frequency IF signal into a high frequency RF signal, or vice versa.

Only three E/W-band GaN mixers were published in literature [1], [8], [9]. Two of them were single-ended down-conversion mixers [1], [8]. The rest was a double-balanced down/up conversion mixer, but the performance of the mixer acting as an up-conversion mixer was not presented in the paper [9].

Resistive mixers are the most popular topology [1], [3]-[7], [9] which demonstrate a high linearity and a wide bandwidth. But the resistive mixer suffers a large conversion loss. At E/W-band, the conversion losses of the resistive mixers are 11-15 dB [9]; while the loss is 8 dB for a transconductance mixer [8].

Besides the problem of high conversion loss, at W-band, it is not easy to get a LO source with a power larger than 10 dBm. Therefore, a sub-harmonic mixer becomes attractive, which is driven by a LO with a half or a third of the frequency of a fundamental mixer. Unfortunately, the sub-harmonic mixer has an even higher conversion loss than a fundamental mixer. An alternative solution is utilizing frequency multipliers to convert a low frequency LO signal into a high frequency one [1], [10]-[12].

Most frequency multipliers based on GaN HEMTs operate at frequency lower than 4 GHz [10]-[12]. A 77 GHz frequency

doubler is reported in [1]. This E-band frequency doubler has an output power of 10 dBm with a conversion loss of 5 dB.

In this paper, a W-band frequency doubler utilizing cascode transistors is designed. The doubler is integrated with an up-conversion transconductance mixer in a single-chip.

II. CIRCUIT SCHEMATIC AND DESIGN

A. Circuit Schematic

The circuit schematic of the transconductance mixer and the frequency doublers are shown in Fig. 1. Transistor Q1 acts as a transconductance mixer. Both IF signal and 2×LO obtained from the frequency doubler are applied at the gate of Q1, where LO refers to the doubler's input signal. The output (RF) signal is extracted from the drain of Q1.

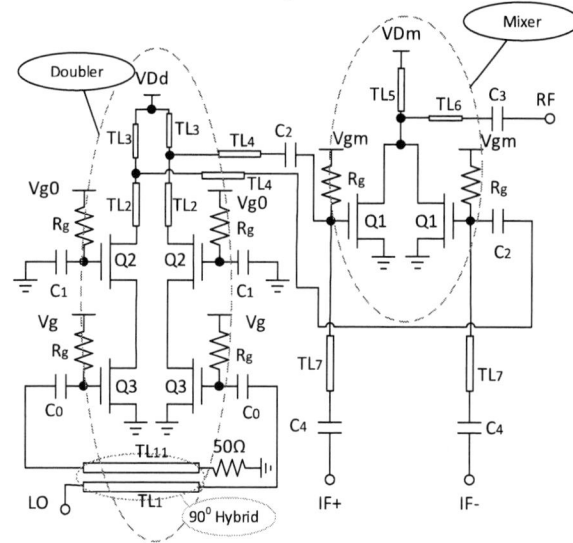

Fig. 1. Schematic of the transconductance mixer and the frequency doublers.

The frequency difference between RF and 2×LO is so small that it is difficult to remove the leakage of 2×LO signal at the RF port by utilising a filter. Therefore, a single-balanced mixer is designed, which consists of two identical mixers (Q1) driven by differential 2×LO. Meanwhile the outputs of two mixers are combined. Consequently, the 2×LO leakage from two mixers are differential signals and will be cancelled at the output (RF)

3−4 April 2022, London, UK

port. An impedance matching network consisting of transmission lines (TL_5/TL_6) and capacitor C_3 connects the drains and the RF port. This network is also a high pass filter, suppressing undesired LO leakage.

To add constructively the mixing products (RF signals) from two mixers, the input IF signals for two mixers are also the differential signals, which are applied at the mixers separately via two identical series LC circuits (TL_7 and C_4). This LC circuit has dual functions: 1) impedance matching for the IF signal; 2) blocking 2×LO signal.

To enhance the power of 2×LO, two identical frequency doublers are utilized, instead of a single frequency doubler plus a balun. The doublers are driven by quadrature I/Q signals separately. In this design, a Lange coupler (90^0 hybrid) is used, which consists of edge-coupled transmission lines (TL_1/TL_{11}). The lengths of TLs are a quarter-wavelength (732 μm). Capacitor C_0 is used for dc blocking and impedance matching.

Finally, the schematic of the frequency doubler will be discussed briefly. The common source configured Q3 is biased in class-B/C, to generate harmonics. The cascoded Q2 amplifies the 2nd harmonic. The gate of Q2 is AC grounded via small capacitor, C_1 (73 fF). A T-type network consists of TL_2, TL_3, as well as a series LC circuit (TL_4, and C_2), which acts as a high-pass filter for suppression of the 1st harmonic, as well as an impedance matching network.

B. Circuit design

The design process consists of two steps: (1) designing a stand-along mixer and a stand-along frequency doubler. All inputs and outputs are 50 Ω terminated. (2) connecting doubler's output port to the mixer's LO port and then designing an impedance matching network.

The devices' gate widths, bias voltages, as well as dimensions of the capacitors and of TLs in the matching networks are selected by harmonic balance simulations in Keysight ADS. EM simulations using Keysight Momentum are performed for all transmission lines and capacitors. In this design, the gate width of Q1 (in the mixer) is 2×26 μm, Q2 and Q3 (in the doubler) are 2×22 μm and 2×45 μm, respectively.

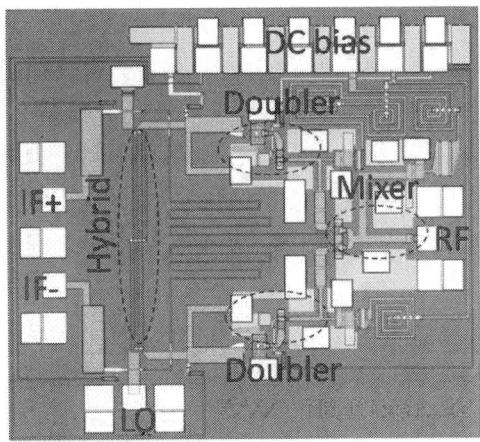

Fig. 2. Chip photo of the up-conversion mixer with the frequency doublers, white squares in figure are pads and ground vias. (size: 1.5×1.3 mm²).

C. Implementation

The designed mixer with integrated frequency doublers is fabricated in a 60 nm GaN-on-Si technology provided by OMMIC. The current gain cut-off frequency (f_T) is 130 GHz and the maximum drain current is 1.6 A/mm. The breakdown voltage VB_{CEO} is 21 V [13]. The substrate thickness is 100 μm. To suppress undesired higher-order transmission modes, grounded co-planar waveguides (GCPWs) are used in the mixer and the doublers, except the impedance matching networks for the low frequency IF and LO signals. The measured loss of GCPW (50 Ω) is less than 0.6 dB/mm at 2- 110 GHz [14].

The chip photos are shown in Fig. 2. Chip size is about 2.0 mm². White squares in figure are pads and ground vias.

III. MEASUREMENT RESULTS

A. Measurement Results

The mixer with integrated doublers is characterized by on-chip probing. The output powers are measured by a Keysight N9140B UXA signal analyser.

A stand-alone doubler, cutout from the integrated circuit, is characterized. The measured output powers of the fundamental and the second harmonic versus frequencies are shown in Fig. 3. The LO frequency is swept from 41.5 to 48.5 GHz, correspondingly, the 2×LO frequency is swept from 83 to 97 GHz. The input power is 17 dBm. The output power is almost saturated. In the measured frequency range, the maximum output power of 2×LO is 8 dBm and the maximum conversion gain is -9 dB.

The output power at the fundamental frequency (LO) is measured also. The fundamental rejection ratio varies between 2 to 12 dB. This relative low rejection ratio is attributed to the single-ended design. Even though a balanced doubler could have a better rejection ratio of undesired harmonics than a single-ended one, an extra balun in the balanced doubler could occupy a large chip area.

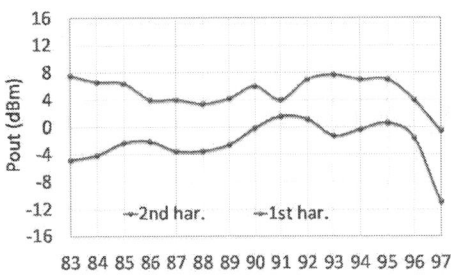

Fig. 3. The measured power of the fundamental and the 2nd harmonics versus the frequency of the second harmonic. Input power is 17 dBm.

DC supply voltage for the doubler is 10 V. The gate biases for the common source and cascode transistors are -2.3 V and 1.9 V, respectively. DC power consumption is 220 mW.

Now, the measurement results of the mixer with integrated doublers circuit will be presented. Fig. 4 plots measured conversion gains versus RF frequency, where 2×LO frequency is swept from 83 to 96 GHz, correspondingly, the frequency of

109

the doubler's input (LO) is swept from 41.5 to 48 GHz. The input power of the doubler is 17 dBm. The input IF signal has a power of -1 dBm at 1 GHz. For lower sideband (LSB), RF signal has a frequency of $f_{RF}= 2\times f_{LO}-f_{IF}$; while, for upper sideband (USB), $f_{RF}=2\times f_{LO}+f_{IF}$. As illustrated in Fig. 4, the maximum conversion gain is -6.4 dB (LSB). The 3-dB RF bandwidths are 8 GHz (from 87.5 to 95.5 GHz) for LSB, and 10.5 GHz (from 86.5 to 97 GHz) for USB.

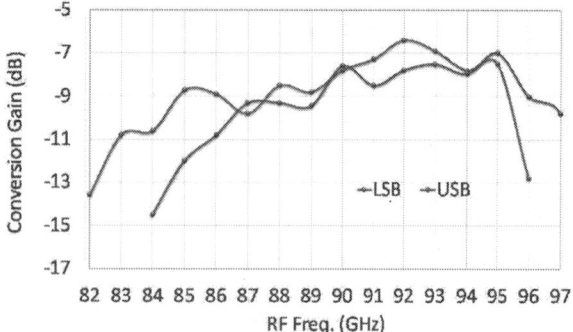

Fig. 4. The conversion gain versus RF frequency. P_{LO}=17 dBm, P_{IF}=-1 dBm, f_{IF}=1 GHz. $f_{RF}=2\times f_{LO}-f_{IF}$ (LSB), $f_{RF}=2\times f_{LO}+f_{IF}$ (USB).

In those measurements, DC supply voltage is 5.8 V for the mixer. The gate bias is -1.6 V. DC power consumption is 180 mW for the mixer, and 620 mW for the integrated circuit. The DC biases for the doublers are the same as those for the stand-alone doubler.

The mixer's conversion gain depends on IF frequency also, as shown in Fig. 5. The 3-dB IF bandwidths are 4.0 GHz (from 0.1 to 4.1 GHz) for LSB, and 3.5 GHz (from 0.1 to 3.6 GHz) for USB. In those measurements, $2\times$LO frequency is fixed at 92 GHz. The powers of LO and IF are 17 dBm and -1 dBm, respectively. IF frequency is swept from 0.1 GHz to 4.1 GHz. Most likely, the poor impedance matching at the IF port results in the ripple in the conversion gain.

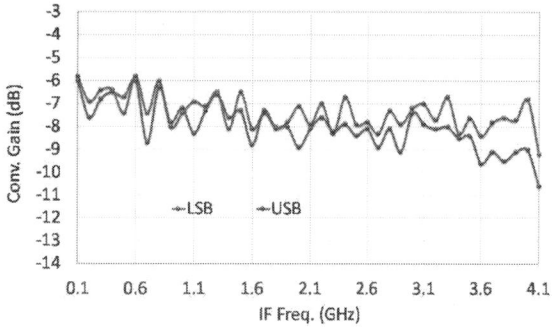

Fig. 5. Conversion gain versus IF frequency. $2\times f_{LO}$=92 GHz, P_{LO}=17 dBm, P_{IF}=-1 dBm. P_{IF}=-1dBm.

The mixer's conversion gain and output (RF) power are functions of the input (IF) power, as shown in Fig. 6, where P_{IF} is swept from -9 dBm to 2 dBm. The $2\times$LO and IF frequencies are 94 GHz and 1 GHz, respectively. The RF frequency is 93 GHz for LSB, and 95 GHz for USB. The input power of the doubler is 17 dBm. The measured output referred one dB

compression point, P_{1dB}, is -8.5 dBm. Probably, the $2\times$LO power of 8 dBm limits the mixer's linearity.

Furthermore, the mixer's conversion gain depends on the LO power, i.e., input power of the doubler, P_{LO}. As shown in Fig. 7, the conversion gain increases with increasing P_{LO}. When P_{LO} is larger than 15 dBm, the conversion gain becomes saturated. In these measurements, $2\times f_{LO}$ is equal to 94 GHz, IF frequency is fixed at 1 GHz, and IF power is -1 dBm.

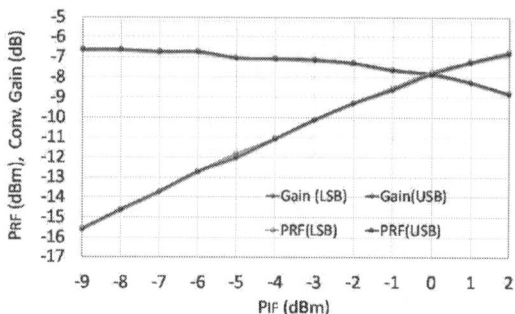

Fig. 6. Conversion gain and RF power versus the power of IF input. $2\times f_{LO}$=94 GHz, P_{LO}=17 dBm, f_{IF}=1 GHz. f_{RF}=93 GHz for LSB, and f_{RF}=95 GHz for USB.

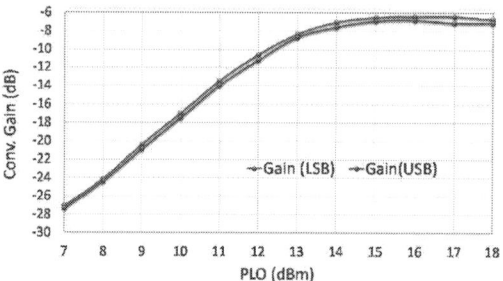

Fig. 7. Conversion gain versus the power of the LO, $2\times f_{LO}$= 94 GHz, f_{IF}=1GHz, P_{IF}=-1GHz.

The LO to RF port isolation is measured in the frequency range ($2\times f_{LO}$) from 83 to 96 GHz. As show in Fig. 8, the 2LO-to-RF isolation is larger than 23.5 dB which is a decent value for a mixer. The LO-to-RF isolation is larger than 13.3 dB which is inadequate. Fortunately, the LO-to-RF leakage signal would be blocked by the input impedance matching network at the cascaded power amplifier, because the frequency of the LO is far lower than RF frequency.

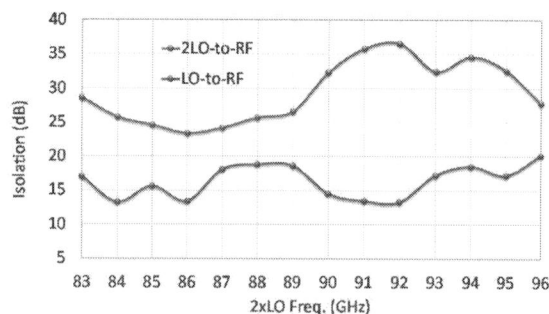

Fig. 8. Measured LO/RF port isolation.

B. Comparison of state-of-the-arts

The performance of the E/W-band GaN mixers published in the literature is listed in Table 1. The up-conversion mixer presented here achieves the lowest conversion loss. Required LO power is also lower, comparing with the mixer operating at 94 GHz [9]. The chip area is the smallest among the GaN mixers, considering two frequency doublers are integrated in the mixer chip.

IV. CONCLUSION

The performance of a W-band up-conversion mixer in GaN technology is demonstrated for the first time. The transconduce mixer topology applied here has benefits for reducing conversion loss, and the single-balance topology results in a decent 2LO-to-RF isolation. Furthermore, two frequency doublers are integrated in the mixer chip, to reduce the frequency of the input LO signal. Such a compact mixer with integrated doubles MMIC could be a building block for a full-integrated transmitter in GaN technology.

ACKNOWLEDGEMENTS

The European Union is acknowledged for the funding and support of the H2020 ICT project SERENA (Grant Agreement no.779305).

REFERENCES

[1] I. Kallfass, R. Quay, H. Massler, S. Wagner, D. Schwantuschke, C. Haupt, R. Kiefer, O. Ambacher, "A single-chip 77 GHz heterodyne receiver MMIC in 100 nm AlGaN/GaN HEMT technology", in *IEEE MTT-S International Microwave Symp.*, 2011.

[2] Robert Malmqvist, Rolf Jonsson, Mingquan Bao, Rémy LeBlanc, Koen Buisman, Christian Fager, Kristoffer Andersson, "A W-band single-chip receiver in a 60 nm GaN-on-silicon foundry process", submitted to *European Microwave Integrated Circuit Conference (EuMIC)*, 2021.

[3] J. Kang, A. Kurdoghlian, A. Margomenos, H. P. Moyer, D. Brown, and C. McGuire, "Ultra-wideband, high-dynamic range, low loss GaN HEMT mixer," *Electron. Lett.*, vol. 50, no. 4, pp. 295–297, Feb. 2014.

[4] K. Yuk, G. R. Branner and C. Cui, "Future directions for GaN in 5G and satellite communications", in *IEEE 60th International Midwest Symposium on Circuits and Systems (MWSCAS)*, 2017.

[5] M. S. Clements, A. V. Pham, J. S. Sacks; B. C. Henderson; S. E. Avery, "Comparison of highly linear resistive mixers in depletion and enhancement mode GaAs and GaN pHEMTs at Ka band", in *IEEE MTT-S International Microwave Symp.*, 2018, pp. 784-787.

[6] A. D. Padova, P. E. Longhi, S. Colangeli, W. Ciccognani, E. Limiti, "Design of a GaN-on-Si single-balanced resistive mixer for Ka-band satcom", *IEEE Microw. Compon. Lett.*, vol. 29, no. 1, pp. 56–58, Jan. 2019.

[7] N. T. Weerathunge; S. Chakraborty; S. J. Mahon; B. Wu; M. Heimlich, "Design of 30-56.5 GHz resistive single-ended mixer in 0.15 μm GaN/SiC technology", in *4th Australian Microwave Symposium (AMS)*, 2020

[8] I. Kallfass, G. Eren, R. Weber, S. Wagner, D. Schwantuschke, R. Quay and O. Ambacher, "High linearity active GaN-HEMT down-converter MMIC for E-band radar applications", in *Proc. of the 9th European Microw. Integrated Circuit Conf.*, 2014, pp. 128-130.

[9] A. Kurdoghlian, H. Moyer, H. Sharifi, D. F. Brown, R. Nagele, J. Tai, R. Bowen, M. Wetzel, R. Grabar, D. Santos, M. Micovic, "First demonstration of broadband W-band and D-band GaN MMICs for next generation communication systems", in *IEEE MTT-S International Microwave Symp.*, 2017, pp. 1126-1128.

[10] C. Wong, K. Yuk, G. R. Branner, S. R. Bahadur, "High power, wideband frequency doubler design using AlGaN/GaN HEMTs and filtering", in *41st European Microwave Conference*, 2011.

[11] V. Zomorrodian, R. A. York, "A MMIC frequency doubler using AlGaN/GaN HEMT technology", in *IEEE International Conference on Communication Problem-Solving (ICCP)*, 2011.

[12] M.-L. Chou, Yi-Qi Chiang, Hsien-Chin Chiu and Fan-Hsiu Huang, "A broadband and high power frequency tripler using 0.5 μm GaN-on-SiC HEMT technology", in *IEEE International Conference on Communication Problem-Solving (ICCP)*, 2015.

[13] W. Ciccognani, S. Colangeli, A. Serino, L. Pace, S. Fenu, P. E. Longhi, E. Limiti, J. Poulain, R. Leblanc, "Comparative noise investigation of high-performance GaAs and GaN millimeter-wave monolithic technologies", in *14th European Microwave Integrated Circuits Conference (EuMIC)*, 2019.

[14] R. Malmqvist, Rolf Jonsson, Anders Bernland, Mingquan Bao, Rémy LeBlanc, Koen Buisman, Christian Fager, Kristoffer Andersson, "E/W-band CPW-based amplifier MMICs fabricated in a 60 nm GaN-on-Silicon foundry process," in *15th European Microwave Integrated Circuits Conference (EuMIC)*, 2020.

Table 1. Performance of E/W-band GaN mixers

Ref.	GaN Technology	Circuit topology	Conversion	RF Freq (GHz)	Conv. Loss (dB)	IF freq. (GHz)	P_{1dB} (dBm)	LO power (dBm)	LO to RF isolation (dB)	Chip size (mm^2)
[1]	100 nm	Single-ended resistive mixer+doubler	down	77	--	0.4	--	15	---	2.5
[8]	100nm	Single-ended Trans-conductance mixer	down	75-81	8	0.4	13 (input referred)	13	13.4	2
[9]	40 nm	Double balanced resistive mixer	down/up	74-94	11-15	0.5 – 5.0	15 (input referred)	24	45	10.8
This work	**60 nm**	**Single-balanced Trans-conductance mixer + doubler**	**up**	**87.5 - 95.5**	**6.5-9.5**	**0.1-4.1**	**-8.5 (output referred)**	**17**	**23.5***	**2**

*2×f$_{LO}$ leakage

Gap in pagination due to unavailable paper.

Page 112

Proceedings of the 16th European Microwave Integrated Circuits Conference

SiGe BiCMOS Building Blocks for E- and D-Band Backhauling Front-Ends

G. Amendola[1], L. Boccia[1], F. Centurelli[2], P. Chevalier[3], A. Fonte[4], S. Karman[5], S. Levantino[5], A. Mazzanti[6], C. Mustacchio[1], A. Pallotta[7], I. Petricli[6], C. Samori[5], F. Tesolin[5], P. Tommasino[2], A. Traversa[4], A. Trifiletti[2]

[1]Dipartimento di Ingegneria Informatica, Modellistica, Elettronica e Sistemistica, University of Calabria, Rende, Italy

[2]Information Engineering, Electronics and Telecommunications Department, Sapienza University of Rome, Italy

[3]STMicroelectronics, 850 rue Jean Monnet, Crolles, France

[4]R&D Department, SIAE MICROELETTRONICA, Cologno Monzese (MI), Italy

[5]Dipartimento di Ingegneria Elettronica, Informazione e Bioingegneria, Politecnico di Milano, Milan, Italy

[6]Department of Electrical, Computer and Biomedical Engineering, University of Pavia, Italy

[7]STMicroelectronics, via Tolomeo 1, Cornaredo (MI), Italy

pasquale.tommasino@uniroma1.it

Abstract — **Millimeter-wave backhauling links are of primary importance in the architecture of 5G telecommunication systems. Their implementation demands highly integrated front-end chips capable of fully exploiting the most advanced semiconductor technologies' capabilities. This work reports the design and experimental validation of several fundamental building blocks of D- and E-band backhauling front-ends implemented using 55-nm SiGe BiCMOS semiconductor technology. Results include an E-band I-Q receiver, a frequency synthesizer, two D-band amplifiers and, a D-band power amplifier.**

Keywords — **Backhaul, BiCMOS integrated circuits, Point to point communication, SiGe, Wideband receiver front-end.**

I. INTRODUCTION

Backhaul point-to-point (P2P) radio links are undergoing rapid and significant growth, and new topologies and architectures of backhaul platforms have been developed. In particular, backhaul wireless networks had a huge increment in capacity and enhanced coverage. In this scenario, the introduction of 5G communication systems has offered the possibility to reach a data rate greater than 10 Gbps. This increasing demand in terms of data rate has driven the available spectrum's exploitation toward mmWaves. Two of them, the 71–76 and 81–86 GHz frequency bands, named E-band, and the 130-134, 141-148.5, 151.5-164, and 167-174.8 GHz frequency bands, named D-band, have been considered [1,3]. E-band has been allocated for gigabit-per-second P2P wireless communications. Its radio-based equipment offers a suitable solution for several reasons: i) the high capacity that can be reached -with 10 GHz of available spectrum-, ii) the possibility to have very large channel bandwidths -up to 2 GHz- and, iii) the links are often licensed under a "light license" process so that the licenses can be obtained quickly and cheaply by providing, at a fraction of the cost, the full benefits of traditional link licenses. D-band is also deemed a promising because, on the one hand, lower frequency bands

will get close to deployment saturation and, on the other hand, backhaul capacity requirements for mobile networks will increase in the next few years. To better exploit these new frequency bands and guarantee a massive deployment, it is necessary to reduce the costs and increase the performance. Silicon-based technologies and, in particular, SiGe BiCMOS technology have enabled a mixed-signal design that includes mm-wave, baseband, and digital functionalities based on the same process and embedded into a single chip.

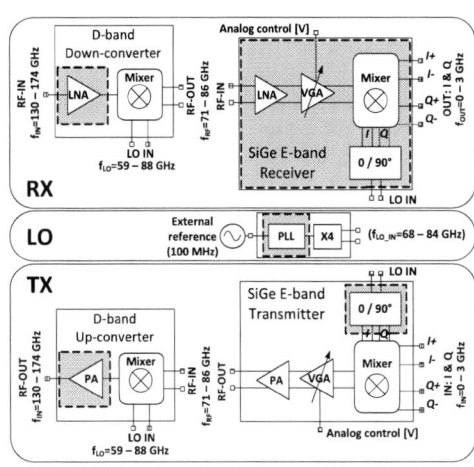

Fig. 1. E/D frontend block diagram. Highlighted blocks are the key components described in this paper.

In this paper, some key building blocks of a highly integrated SiGe BiCMOS D- and E-band I/Q transceiver are presented. The block diagram of the proposed transceiver is shown in Fig. 1. The proposed designs are based on a 55-nm SiGe BiCMOS semiconductor technology [4]. The proposed designs fully exploit the mixed-signal capabilities of the selected technology. In Section II, a brief overview of the

3–4 April 2022, London, UK

technology is shown. In Section III, the IC building blocks are presented, and their measured results are reported. Finally, in Section IV, the conclusions are drawn.

II. THE UNDERLYING TECHNOLOGY

SiGe BiCMOS technology has been widely used in high-speed optical modules and established today in the automotive radars market and it appears as the most appropriate choice for the infrastructure market, as the SiGe HBT addresses properly both optical link and mmWave IC requirements.

The performance of SiGe HBT is no longer the limiting factor for a mmWave transceiver front-end integration in applications with reduced output power, as in micro BTS or small cells, but rather the quality factor of the on-chip passive devices, such as inductors, capacitors, and transmission lines for matching and tuning, together with their accurate characterization in the mmWave frequency domain. The quality of on-chip passive devices is of primary importance for matching, tuning, filtering and frequency synthesis. Offering a capacitor combining high quality (i.e., high linearity, high Q factor, low leakage, large breakdown voltage) and high capacitance value is challenging. A solution allowed by the underneath CMOS node platform in ST's BiCMOS055 is to use a Metal-Oxide-Metal (MOM) structure (reaching ~2.5 $fF/\mu m^2$) for high-quality capacitors together with $5fF/\mu m^2$ Metal-Insulator-Metal (MIM). Back-end metal stack of the original digital 55nm CMOS technology has been modified to increase the Q-factor of inductors and reduce attenuation. The combination of: *i)* the high-frequency performance of the HBT, providing the high speed and gain that are critical for the high-frequency analog sections, *ii)* the CMOS technology excellent for building moderately complex low-power logic functions, to assist the analog front end in compensation, calibration, controlling and monitoring functionalities, and *iii)* the up to 9 metals Back End Of Line (BEOL) including the upper layers with thicker copper for improved quality factor at mmWave of the passive devices (inductors, capacitors and transmission lines), makes the ST's SiGe:BiCMOS055 perfectly fit for mixed-signal high-frequency IC chipsets, for applications in cellular network mmWave backhauling, front hauling, satellite communication and radar. There is a strong push to continuously improve SiGe HBT performances to provide the high speed and gain that are critical for the high-frequency analog sections. A first significant step has been done evolving from BiCMOS9MW (220/280 GHz f_T/f_{MAX}) to BiCMOS055 (320/370 GHz f_T/f_{MAX}). Further improving f_{MAX} requires moving to a new generation of architecture featuring an epitaxial intrinsic-to-extrinsic base link. Such an architecture is being developed in the new envisaged process, named BiCMOS055X, to target 400 GHz f_T and 600GHz f_{MAX} [5], which is mandatory for 6G. SiGe HBTs with lower f_T are available, when larger breakdown voltages (BVCEO) are required. RF switch device are also addressed in newly developed ST's BiCMOS process perimeters.

The building blocks for the D- and E-band transceiver presented in this paper have been designed on BiCMOS055 process and they will be redesigned in BiCMOS055X one.

III. BUILDING BLOCKS

A. E-band receiver

The E-band receiver chip is composed of a Variable Gain Low Noise Amplifier (LNA-VGA), a I/Q down-conversion mixer, and an E-band local oscillator 0-90° phase shifter.

I/Q E-band receiver characterization has been carried out at laboratories of Tor Vergata University, Rome, by using the test board designed by SIAE. Test-board includes custom E-band baluns and a commercial x6 frequency multiplier (see Fig. 2) to perform measurements in single-ended chip-on-board set-up (the I/Q E-band receiver has a fully differential input).

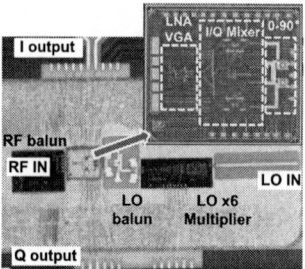

Fig. 2. Microphotograph of the chain for chip-on-board measurements of the E-band receiver. RX die size is (1.5 x 1.29) mm².

Furthermore, commercial 0 − 3 GHz baluns have been mounted at both I and Q base-band outputs, not shown in Fig. 2. Conversion gain measurements have been carried out by exploiting a Network Analyser with 2 single-ended RF sources, operating in mixer mode: one source drives the balun at the receiver input by GSG probe, the other one is connected by SMA coaxial connector at the I or Q output. A X-band generator has been used to provide the local oscillator signal at the x6 multiplier input. Two boards have been mounted and measured, each hosting a different E-band balun for the multiplier's interconnection to the on-chip local oscillator input: the two baluns show a 71 − 76 GHz and an 81 − 86 GHz operating band, respectively. S11 and conversion gain at minimum attenuation and with 0.5 GHz base-band frequency are shown in Fig. 3. It is worth noting that the peak receiver gain is shifted upward with respect to simulation. Finally, the conversion gain at 70.8 and 84.6 GHz LO frequency is shown in Fig. 4 for different VGA attenuations. More details on circuit design, and on performance with modulated signals can be found in [6].

B. D-band amplifiers

Two D-band amplifiers and one PA has been developed and and tested. Low noise amplifiers for the D-band extension module are implemented with cascode gain stages made of identical HBTs of 0.2x5.1 µm² emitter area, biased at 6.5 mA/µm² from a supply voltage V_{CC} = 2 V. At 150 GHz, a simpler common-emitter transistor features a Maximum Available Gain (MAG) of 5.5 dB only, while the cascode rises MAG to 12 dB. Amplifiers are designed with multiple stages, conjugately matched with input, output and interstage passive networks realized with shielded microstrip transmission lines.

114

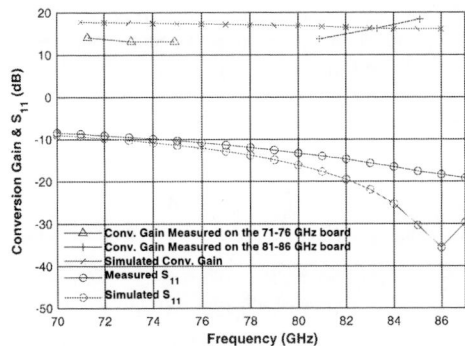

Fig. 3. E/D frontend block diagram. Highlighted blocks are the key components described in this paper.

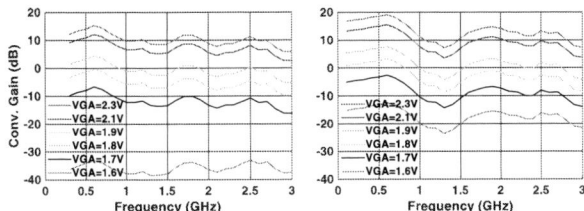

Fig. 4. Conversion gain @ several attenuation control voltages vs output frequencies at 70.8 and 84.6 GHz f_{LO}.

The networks are stagger tuned for bandwidth extension. The die photos of a 2-stage and 3-stage amplifier are shown in Fig. 5 (top). S-parameter measurements (reported in the plots of Fig. 5) are performed on-waver after probe tip calibration with custom TRL structures. The 2-stage amplifier features 20 dB gain at 144 GHz with -3 dB bandwidth on S_{21} of 66 GHz. The gain of the 3-stage design reaches 28.6 dB with 65 GHz, demonstrating 1.73 THz of gain-bandwidth product. The D-band power amplifier is designed exploiting HBTs in common-base, giving superior gain, output power, linearity and power efficiency than in common-emitter [7]. The common-base in the output stage and pre-driver are also exploited to adapt the supply current to the signal swing with current clamping, improving the power-added efficiency in back-off. The D-band power amplifier (with chip photo in Fig. 6) comprises 3 common-base stages and an input cascode stage, operating in class A, to rise the overall gain above 20 dB [8]. Large signal, continuous-wave measurements are reported in the plots in Fig. 6. The amplifier demonstrates the best performance at 135 GHz, with 24 dB power gain, a saturated output power, P_{sat}, of 17.6 dBm and output power at 1 dB gain compression point, P_{1dB}, of 16.8 dBm, less than 1 dB below P_{sat}. The PAE at the P_{1dB} is 17.1% and it is still 8.5% at 6 dB power back-off. The amplifier maintains P_{1dB} greater than 15 dBm with PAE of at least 15% from 125 to 145 GHz. In the same frequency range, the PAE at 6 dB back-off is above 6%.

C. Frequency Synthesizer

To mitigate the stringent noise requirements, the outputs of two PLLs driven by the same reference source are summed together, as shown in Fig. 7 [9].

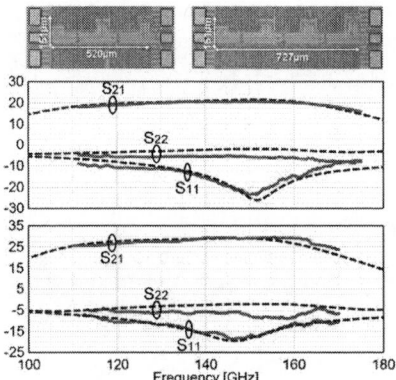

Fig. 5. 2-stage and 3-stage D-band signal amplifiers.

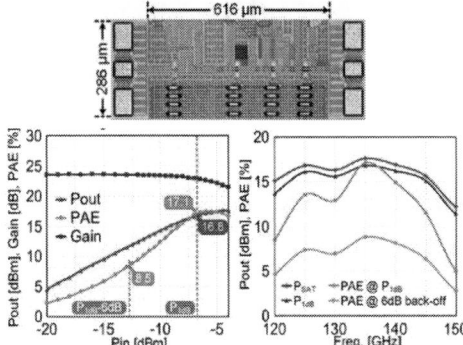

Fig. 6. D-band multi-stage power amplifier.

Assuming the two carriers are in-phase and have the same amplitude, while their noise is uncorrelated, the combined output phase is 3 dB lower than that of the individual carriers [10], [11]. Each core is based on a fractional-N digital PLL with bang-bang time-to-digital converter (BB-TDC) and a digital-to-time converter (DTC) on the reference path to cancel the $\Delta\Sigma$ quantization noise and to correct for any residual phase offset between the two PLLs. In practice, any mismatch in the DTCs of the two cores produces a phase offset and a reduced amplitude at the combiner output. Unlike [12], the offset detection circuit (ODC) senses the offset at the output of the PLLs and feeds it back into the two DTCs driven differentially. As the quality factor of LC resonators degrades at high frequency, each DCO runs at 10 GHz, and includes an intrinsic frequency multiplication by two. A subsequent frequency multiplier by four brings the output frequency to around 80 GHz. The DCO circuit shown in Fig. 7 is an evolution of the topology introduced in [13]. It uses a double nMOS/pMOS trans-conductor biased at $2V_{dd} = 2.0$ V, i.e. twice the core supply voltage, with two coupled tail resonators to boost the second-harmonic amplitude and a third coil coupled to them to extract the second-harmonic signal. The 20 GHz outputs of the two DCOs are then power-combined on-chip by means of another pair of coupled inductors.

The implemented frequency synthesizer whose die microphotograph is shown in Fig. 8 occupies an active area of 0.65 mm².

115

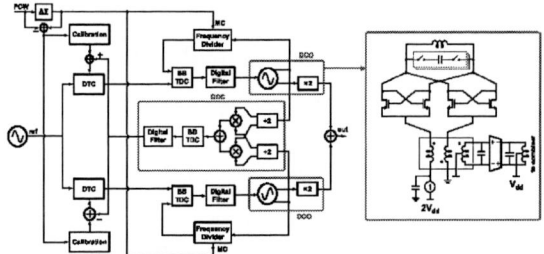

Fig. 7. Block diagram of the frequency synthesizer and DCO schematic.

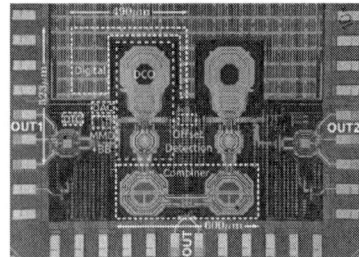

Fig. 8. Frequency synthesizer die micro-photograph.

The single DCO reaches a phase-noise-vs-power figure of merit (FoM) of -192.5 dB, thanks to the adopted topology. Enabling one or two cores, the PLL power consumption scales from 18.5 mW to 37.1 mW, and the phase noise at 1 MHz offset (measured at 4.855 GHz, after frequency division, and shown in Fig. 9) scales by 3 dB, from -125 dBc/Hz to -128 dBc/Hz. The latter is equivalent to -104 dBc/Hz at 1 MHz offset, for the 77.68 GHz carrier. The in-band noise, that is identical in both modes, is instead dominated by a correlated reference noise component. The rms integrated jitter from 1 kHz to 10 MHz is 206fs/174fs, with one/two cores activated. The reference spur level is about -64 dBc, while the in-band fractional spurs are below -45 dBc.

Fig. 9. Measured PLL phase noise in fractional-N mode (at 4.855GHz, after frequency division) with one-core and dual-core enabled.

IV. CONCLUSION

This work aims at demonstrating the impact which can be achieved in E- and D-band backhauling wireless systems when highly integrated mixed-signal front ends are implemented. Namely, the proposed study reports different key building blocks implemented using a 55-nm SiGe BiCMOS technology.

Simulated and experimental results are presented for and E-band I-Q receiver which show a conversion gain higher than 12 dB in the 71-76 GHz and 81-86 GHz bands. A frequency synthesizer was also reported showing that a single DCO reaches a phase-noise-vs-power figure of merit (FoM) of -192.5 dB. Two D-band amplifiers and one PA were also manufactured and tested. The first two feature 20 dB and 28 dB power gain around 140 GHz, and the PA presents 24 dB power gain and, saturated output power, P_{sat}, of 17.6 dBm at 135 GHz.

ACKNOWLEDGMENT

This work is supported by European ECSEL-JU/EU-H2020 under grant TARANTO "TowARds Advanced bicmos NanoTechnology platforms for rf and thz applicatiOns" no. 737454. The authors also wish to thank the research team of Prof. E. Limiti at Roma Tor Vergata University for conversion gain measurements carried out in their labs.

REFERENCES

[1] M. Giovanni and L. Frecassetti, "E-Band -Survey on Status of Worldwide Regulation", 1st edition: September 2020, ISBN No. 979-10-92620-32-2, ETSI, Sophia Antipolis, France, www.etsi.org.

[2] "ECC Recommendation ECC/REC/(18)01 on Radio frequency channel/block arrangements for Fixed Service in the band 130 174.8 GHz (D-band)", Electronic Communications Committee, 27 April 2018, www.cept.org.

[3] P. Nava, "Evolution of Fixed Services for wireless backhaul of IMT 2020 / 5G", ITU-R WP 5C, ITU-R Wireless Backhaul Workshop, 2020, www.itu.int.

[4] P. Chevalier et al., 'A 55 nm triple gate oxide 9 metal layers SiGe BiCMOS technology featuring 320 GHz fT / 370 GHz fMAX HBT and high-Q millimeter-wave passives', in 2014 IEEE International Electron Devices Meeting, Dec. 2014, doi: 10.1109/IEDM.2014.7046978.

[5] P. Chevalier et al., 'Nanoscale SiGe BiCMOS technologies: From 55 nm reality to 14 nm opportunities and challenges', in 2015 IEEE Bipolar/BiCMOS Circuits and Technology Meeting - BCTM, Oct. 2015, pp. 80–87, doi: 10.1109/BCTM.2015.7340556.

[6] G. Amendola et al., 'Compact E-Band I/Q Receiver in SiGe BiCMOS for 5G Backhauling Applications,' IEEE Transactions on Circuits and Systems II: Express Briefs, vol. 68, no. 9, Sept. 2021.

[7] E. Rahimi, J. Zhao, F. Svelto and A. Mazzanti, "High-Efficiency SiGe-BiCMOS E-Band Power Amplifiers Exploiting Current Clamping in the Common-Base Stage," in IEEE Journal of Solid-State Circuits, vol. 54, no. 8, pp. 2175-2185, Aug. 2019.

[8] I. Petricli, D. Riccardi and A. Mazzanti, "D-Band SiGe BiCMOS Power Amplifier With 16.8dBm P₁dB and 17.1% PAE Enhanced by Current-Clamping in Multiple Common-Base Stages," in IEEE Microwave and Wireless Components Letters, vol. 31, no. 3, pp. 288-291, March 2021.

[9] S. Karman et al., "A 18.9-22.3GHz Dual-Core Digital PLL with On-Chip Power Combination for Phase Noise and Power Scalability," to be published in Proc. of 2021 IEEE Radio-Frequency Circuits (RFIC) Symposium, 2021.

[10] S. Ek et al., "A 28-nm FD-SOI 115-fs Jitter PLL-Based LO System for 24–30-GHz Sliding-IF 5G Transceivers," IEEE J. of Solid-State Circuits, vol. 53, no. 7, pp. 1988-2000, Jul. 2018.

[11] S. Karman et al., "A Novel Topology of Coupled Phase-Locked Loops," IEEE Trans. on Circuits and Systems-I, vol. 68, pp. 989-997, Mar. 2021.

[12] A. Santiccioli et al., "A 98.4fs-Jitter 12.9-to-15.1GHz PLL-Based LO Phase-Shifting System with Digital Background Phase-Offset Correction for Integrated Phased Arrays," in IEEE Digest of IEEE Solid-State Circuits Conf. (ISSCC), paper 32.8, 2021.

[13] M. Garampazzi et al. "A 195.6dBc/Hz peak FoM P-N class-B oscillator with transformer-based tail filtering," IEEE J. of Solid-State Circuits, vol. 50, no. 7, pp. 1657-1668, Jul. 2015.

Proceedings of the 16th European Microwave Integrated Circuits Conference

A Superheterodyne 300 GHz Transmit Receive Chipset for Beyond 5G Network Integration

Iulia Dan[#], Christopher Grötsch[#], Laurenz John[*], Sandrine Wagner[*], Axel Tessmann[*], Ingmar Kallfass[#]

[#]University of Stuttgart ILH, Stuttgart, Germany
[*] Fraunhofer IAF, Fraunhofer Institute of Applied Solid State Physics, Freiburg, Germany

Abstract— This paper presents a compact solid-state, fully integrated transmitter and receiver chipset operating at 300 GHz fabricated using a 35 nm gate-length InGaAs metamorphic high-electron-mobility transistor technology. Both circuits integrate a fundamental frequency converter. The local oscillator path consists of a multiplier by three and a buffer amplifier. The transmitter uses a power amplifier as an output stage for increased output power. The first stage in the receiver is a low-noise amplifier. The transmitter achieves a high linearity: The output-referred 1- dB compression point lies at -3 dBm. The receiver has a conversion gain of around 10 dB without any IF post amplification and an estimated noise figure of 7.3 dB. The absolute 3 dB RF bandwidth of the system is 42 GHz, ranging from 288 GHz to 320 GHz. The local oscillator input frequency can vary between 72 and 75.5 GHz. The particularity of this chipset is the very high and wideband IF frequency range, from 75 to above 91 GHz. This superheterodyne architecture and the compatibility to IF systems composed of wireless links developed in the frame of 5G enables the integration into a real network, bringing 300 GHz communication networks a step closer to being implemented in applications like front- and back-hauling.

Keywords— 300 GHz wireless communication, broadband receiver, millimeter communication, submillimeter-wave monolithic integrated circuits (S-MMIC), metamorphic high electron mobility transistor technology (mHEMT)

I. INTRODUCTION

While 5G technologies and products become a growing and important part of wireless networks worldwide, the importance of terahertz (THz) communication in future and beyond 5G networks cannot be denied. The need for higher bandwidths is a direct consequence of the constantly growing demand for higher data rates. THz communication poses many challenges, not only in terms of device availability. Packaging at these high frequencies is a big concern as well as the processing of analog information delivered by the integrated devices. This is the reason why most successful wireless transmissions above 250 GHz [1]–[4] are using solid-state circuits and were carried out in a laboratory environment, taking advantage of powerful signal processing tools.

[3] presents the first, and up to now only, application-ready prototype implementing 300 GHz devices for back- and front-hauling, involving the combination of THz transmission and standard single-mode fibers. We have proposed another implementation of THz communication in a real-world application in [5]. For this solution of a future front- and back-haul network a superheterodyne architecture is necessary. The combination of commercially available 5G communication systems operating in V- and W-Band and a 300 GHz frontend,

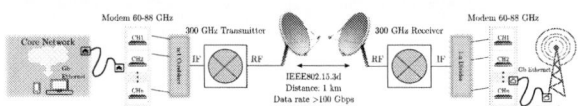

Fig. 1. Proposed solution for live integration of a 300 GHz frontend in a front- and back-haul network.

results in a wireless system compatible with the new IEEE 802.15.3d frequency standard [6], capable of being easily integrated in a live network. The work presented in [5] concentrates on the proof of concept for the proposed architecture and uses an existing radio frequency (RF) frontend designed for zero-intermediate frequency (IF) transmission. In this paper we are concentrating on the superheterodyne 300 GHz transmit receive chipset, with an IF operation range between 75 to 95 GHz. This chipset is the missing puzzle piece which completes the transmission experiment presented only as a proof-of-concept in [5].

II. PROPOSED SOLUTION FOR FUTURE FRONT- AND BACK-HAUL NETWORKS

The last portion of a communication network, the link between an edge cell and the core network, is realized at the moment by either wired or wireless links. The wired solutions are fast but very expensive and often impossible to deploy in remote areas. The future exponential growth of individual data rates means that the access cells have to be smaller and appear more frequently. The increasing number of cells will cause additional costs of optical fiber, which is already extremely high in an urban environment. Employing the solution pictured in Fig. 1 would provide the required capacity for this application. This system is based on a superheterodyne architecture and consists of two important components: the modem which provides the IF and the RF frontend. The modem communicates with the core network via Ethernet cable and realizes the first up-conversion to the frequency range between 60 and 88 GHz. According to the IEEE 802.15.3e-2017 standard [6] each modem channel can occupy up to 2.16 GHz of bandwidth. The IF can aggregate signals coming from one or more channels. The second up-conversion is realized by the 300 GHz transmitter. The RF signal is transmitted with data rates exceeding 100 Gbps using an antenna system over a variable distance up to 1 km to the base station where the 300 GHz RX and the modem down-convert the baseband signal.

3–4 April 2022, London, UK

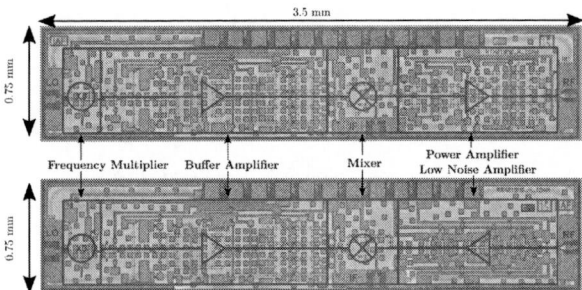

Fig. 2. Chip photographs of the fabricated transmitter and receiver MMIC.

Although 5G access networks extended the frequency spectrum above 6 GHz, where wireless transport links are currently operated, these products alone cannot support the beyond 5G requirements in large scale network deployment. In combination with a 300 GHz frontend like presented in Fig. 2 the ultra-high data rate demand can be satisfied. The 300 GHz transmitter and receiver presented in this work are based on millimeter-wave monolithic integrated circuits (MMICs) realized in a 35 nm gate-length InAlAs/InGaAs metamorphic high electron mobility transistor (mHEMT) technology. This mHEMT technology, described in detail in [7], is developed and optimized for high cut-off frequencies and ultra-low noise figure, featuring a transit frequency (f_T) of 515 GHz and a maximum frequency of oscillation (f_{max}) exceeding 1 THz.

The MMICs will be packaged into compact waveguide modules, similar to the ones presented in [4]. To enhance the maximum transmission distance different power amplifier modules can be connected after the 300 GHz TX like the one presented in [8].

The IF frequency is determined by the targeted application and ranges between 60 and around 90 GHz. Due to a frequency multiplier with a factor of three and a fundamental converter integrated on chip a local oscillator (LO) frequency in the range between 70 and 78 GHz is needed. The resulting RF range is between 270 and 320 GHz.

III. TX AND RX MMICs

Fig. 2 shows the chip photographs of the integrated transmitter and receiver, both with a chip size of 0.75 x 3.5 mm. Both chips integrate a multiplier by three, a buffer amplifier and a mixer. Furthermore, the transmitter has a power amplifier as a last stage. A low noise amplifier is the first stage in the receiver.

The input frequencies in V-Band (40 to 75 GHz) are very convenient because of a good availability of off-the-shelf components. Additionally it is easier to generate a high input power for the multiplier in V-band rather than W-band (75 to 110 GHz) and higher. The frequency multiplier-by-.three operates in compressed class-A mode. The buffer amplifier is needed to provide enough LO input power for the up- and down-converter. It consists of five gain stages with four parallel transistors in the last two stages. The fundamental frequency converter uses one mixer cell based on one common

source transistor. To achieve higher output powers a four-stage amplifier similar with the one used for the buffer stage is employed. To reduce the noise and make the down-converter suitable for THz communication a low noise amplifier (LNA) was designed and integrated in the receiver.

The DC power consumption of the transmitter is 357 mW, while the receiver consumes 281 mW. The difference is due to the lower power consumption of the LNA in comparison to the power amplifier. The overall functionality of the MMICs is verified by on-wafer measurements. The 300 GHz RF input signal of the receiver is synthesized using IAF in-house built frequency multiplier and amplifier modules. The LO signal around 70 GHz is generated using a commercially available frequency synthesizer with extension module and an IAF in-house built amplifier module. The same LO source is used for both transmitter and receiver. For the receiver the IF output signal at W-band frequencies is measured using a commercially available spectrum analyzer. Due to the lower cut-off frequency of the W-band waveguide, this measurement setup is limited to IF frequencies above 65 GHz. For the on-wafer measurement of the 300 GHz transmitter MMIC, the IF signal is generated using a commercially available frequency synthesizer with extension module. In this case, the IF frequency band is limited to frequencies above 70 GHz. The 300 GHz output power of the TX MMIC is measured using a commercially available power meter. All measurements are calibrated on waveguide level and the known probe losses are deembedded.

Fig. 3 shows the measured output power of the transmitter, P_{out}, versus IF input power, P_{IF}, and the comparison to the simulation. This measurement is conducted for an LO frequency of 73.5 GHz, an LO input power of −2 dBm and an IF frequency of 75 GHz. The $P_{1dB,out}$ is measured to be at -3 dBm, while the simulated output-related compression point lies at 2 dBm. Hence, a difference of 5 dB can be observed between measurement and simulation. If a 10 dB back-off from the 1-dB compression point is considered, which would allow the transmission of complex modulation formats like 64QAM, a distance of around 80 meters could be covered. For the link budget calculation an antenna system consisting of a transmitter and a receiver antenna with a total gain of around 90 dBi. To increase the distance even further the usage of additional power amplifier modules like the one presented in [8] is necessary.

Fig. 4 shows the measured output power of the transmitter, P_{out}, versus IF frequency, f_{IF}, and the comparison with the simulation. This measurement is conducted with an LO frequency of 73.5 GHz, an LO input power of −2 dBm and four different IF input powers: -15, -10, -5 and 0 dBm. The IF frequency is limited by the setup to 75 GHz. Very good accordance between measurement and simulation is observed for low input power, below −10 dBm. For increasing input powers the difference between measurement and simulation increases up to around 4 dB for an IF input of 0 dBm. Overall, it can be concluded that the transmitter is very broadband with over 25 GHz of IF bandwidth. The exact bandwidth cannot be

Table 1. Comparison of state-of-the-art transmitter and receiver chipsets with an operation frequency above 250 GHz.

Ref	Technology	RF Frequency Range (3 dB limits) in GHz	BB / IF Frequency Range in GHz	TX OP$_{1dB}$ in dBm	Receiver Noise Figure in dB	Power Consumption in mW
[9]	40 nm CMOS	260 - 270	0 - 20[1]	-5[2]	22.9	1790
[10]	80 nm InP-HEMT	272 - 302	2 - 32	-16	n.a.	n.a.
[11]	35 nm InGaAs mHEMT	270 - 314	0 - 22	-1	8.6	n.a
This work	35 nm InGaAs mHEMT	288 - 320	75 - 90	-3	7.3[1]	637

[1] according to simulation [2] estimation

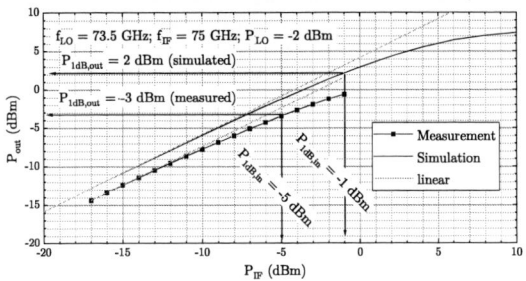

Fig. 3. Measurement results of the transmitted output power and comparison to simulation under the variation of the IF input power.

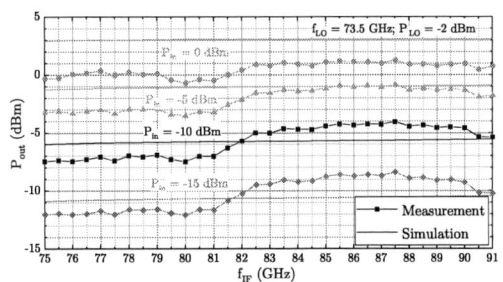

Fig. 4. Measurement results of the transmitted output power and comparison to simulation under the variation of the IF frequency.

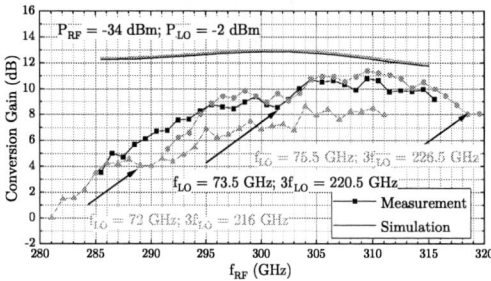

Fig. 5. Measurement results of the conversion gain of the designed receiver and comparison to simulation under the variation of the RF frequency.

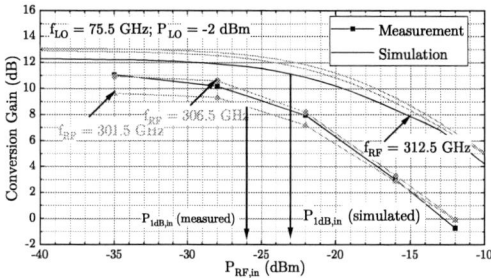

Fig. 6. Measurement results of the conversion gain of the designed receiver and comparison to simulation under the variation of the RF input power.

determined since it exceeds the measurement range.

For the integrated receiver similar measurements were conducted. The bandwidth of the receiver is dependent on the used LO frequency, as depicted in Fig. 5. For an LO of 72 GHz the 3-dB RF bandwidth calculates to 15 GHz, from 288 to 303 GHz. For an LO of around 73.5 GHz the maximum conversion gain is around 11 dB and the 3-dB RF bandwidth of 22 GHz between 293 and 315 GHz. The last measured LO frequency lies at 75.5 GHz. The possible measurement range for this LO frequency starts at only 291.5 GHz, limited by the measurement setup of the IF port. The 3-dB RF bandwidth covers the range from 295 GHz to 320 GHz. Compared to simulations, there is a good agreement, with deviations below 2 dB, for the LO frequencies of 73.5 GHz and 75.5 GHz in the RF frequency range between 303 and 313 GHz. Below 303 GHz and for the LO frequency of 72 GHz the deviation is higher, peaking at 8 dB for 285 GHz.

The linearity of the RX is plotted in Fig. 6. The measurement is conducted with a constant LO frequency of 75.5 GHz, at an input power of −2 dBm and three different RF frequencies: 301.5, 306.5 and 312.5 GHz. A relatively constant deviation of 2 to 4 dB is observed between measurement and simulation. A maximum conversion gain of around 13 dB was simulated, while the measured conversion gain only reaches a maximum of around 9 dB for an RF of 301.5 GHz and 11 dB for the other two RF frequencies. The measured input-referred 1-dB compression point is 3 dB lower than the simulated one, at −26 dBm RF input power.

An important measurement which cannot be conducted on-wafer due to the lack of available sources is the measurement of the noise figure of the receiver. Considering the simulated noise figure of the LNA, the overall noise figure of the receiver can be calculated using Friis' formula and lies at 7.2 dB which corresponds to measurements of similar circuits in that technology [12]. After the integration into a waveguide module, the noise figure can further degrade.

119

IV. COMPARISON TO STATE-OF-THE-ART

Table 1 shows the comparison to the state-of-the-art transmitter and receiver chipsets operating above 250 GHz. For this comparison all solid-state technologies that allow operation at these frequencies and the best performing chipset of the technology have been considered.

The MMIC chipset presented in this work is the only one operating with an IF input frequency above 75 GHz, supporting a superheterodyne transmission. This works exceeds the state of the art not only by architecture choice, but also by the lowest power consumption, a topic that gains more and more importance in the discussion about future networks. Although low-noise measurements cannot be conducted due to the lack of available sources, the receiver is expected to have a very low noise figure, similar to other 300 GHz receivers fabricated with the 35 nm InGaAs mHEMT technology [12]. In terms of linearity, especially important for the transmitter, measured by the output-referred 1-dB compression point, the presented TX MMIC is close to the state-of-the art. Only the transmitter fabricated in the same technology [11] shows a slightly better performance. Furthermore, the presented MMIC chipset operates at the highest reported RF range.

V. CONCLUSION

This paper presents a transmit and receive chipset fabricated in a 35 nm InGaAs mHEMT technology. The MMICs integrate a frequency multiplier by three, a buffer amplifier, a frequency converter and an RF power amplifier or a first RF low-noise amplifier. The LO input frequency lies between 72 and 75.5 GHz, the 3-dB bandwidth of the IF frequency is measured to exceed the range between 75 to 95 GHz and the RF frequency centers around 300 GHz, depending on the combination of LO and IF frequency. The superheterodyne transmission scheme allows the combination of wireless links developed in the frame of 5G, operating at frequencies above 60 GHz, with the RF system composed of the presented TX and RX. The proposed solution enables the integration of THz communication into a live application. The TX-RX chipset shows state-of-the-art high frequency performance in terms of bandwidth and linearity, and is therefore a promising candidate for future wireless high data rate communication systems in the lower THz band around 300 GHz.

ACKNOWLEDGMENT

The authors acknowledge and thank their colleagues from the Fraunhofer IAF epitaxy and technology department for their contributions during wafer growth and MMIC processing. This work has received funding from Horizon 2020, the European Union's Framework Programme for Research and Innovation, under grant agreement No. 814523. ThoR has also received funding from the National Institute of Information and Communications Technology in Japan (NICT).

REFERENCES

[1] H. Hamada, T. Tsutsumi, H. Matsuzaki, T. Fujimura, I. Abdo, A. Shirane, K. Okada, G. Itami, H. J. Song, H. Sugiyama, and H. Nosaka, "300-ghz-band 120-gb/s wireless front-end based on inp-hemt pas and mixers," *IEEE Journal of Solid-State Circuits*, vol. 55, no. 9, pp. 2316–2335, sep 2020.

[2] S. Lee, S. Amakawa, T. Yoshida, Y. Morishita, Y. Kashino, S. Hara, and M. Fujishima, "300-ghz cmos-based wireless link using 40-dbi cassegrain antenna for ieee standard 802.15.3d," in *2020 IEEE International Symposium on Radio-Frequency Integration Technology (RFIT)*. IEEE, sep 2020, pp. 136–138.

[3] C. Castro, R. Elschner, T. Merkle, and C. Schubert, "100 gbit/s terahertz-wireless real-time transmission using a broadband digital-coherent modem," in *2019 IEEE 2nd 5G World Forum (5GWF)*, Sep. 2019, pp. 399–402.

[4] I. Kallfass, I. Dan, S. Rey, P. Harati, J. Antes, A. Tessmann, S. Wagner, M. Kuri, R. Weber, H. Massler, A. Leuther, T. Merkle, and T. Kürner, "Towards MMIC-based 300 GHz Indoor Wireless Communication Systems," *IEICE Transactions on Electronics*, vol. E98-C, no. 12, pp. 1081–1090, Dec 2015.

[5] I. Dan, G. Ducournau, S. Hisatake, P. Szriftgiser, R. Braun, and I. Kallfass, "A terahertz wireless communication link using a superheterodyne approach," *IEEE Transactions on Terahertz Science and Technology*, vol. 10, no. 1, pp. 32–43, Jan 2020.

[6] *IEEE Standard for High Data Rate Wireless Multi-Media Networks Amendment 2: 100 Gbps Wireless Switched Point-to-Point Physical Layer*, IEEE-SA Standards Board Std.

[7] A. Leuther, A. Tessmann, H. Massler, R. Losch, M. Schlechtweg, M. Mikulla, and O. Ambacher, "35 nm metamorphic HEMT MMIC technology," in *Indium Phosphide and Related Materials, 2008. IPRM 2008. 20th International Conference on*, May 2008, pp. 1–4.

[8] L. John, A. Tessmann, A. Leuther, P. Neininger, T. Merkle, and T. Zwick, "Broadband 300-GHz Power Amplifier MMICs in InGaAs mHEMT technology," *IEEE Transactions on Terahertz Science and Technology*, pp. 1–1, 2020.

[9] S. Lee, R. Dong, T. Yoshida, S. Amakawa, S. Hara, A. Kasamatsu, J. Sato, and M. Fujishima, "An 80 Gb/s 300 GHz-Band Single-Chip CMOS Transceiver," in *2019 IEEE International Solid- State Circuits Conference - (ISSCC)*, Feb 2019, pp. 170–172.

[10] H. Hamada, T. Fujimura, I. Abdo, K. Okada, H. Song, H. Sugiyama, H. Matsuzaki, and H. Nosaka, "100-Gb/s InP-HEMT wireless transceiver using a 300-GHz fundamental mixer," in *2018 IEEE/MTT-S International Microwave Symposium - IMS*, June 2018, pp. 1480–1483.

[11] I. Kallfass, P. Harati, I. Dan, J. Antes, F. Boes, S. Rey, T. Merkle, S. Wagner, H. Massler, A. Tessmann, and A. Leuther, "MMIC chipset for 300 GHz indoor wireless communication," in *2015 IEEE International Conference on Microwaves, Communications, Antennas and Electronic Systems (COMCAS)*, Nov 2015, pp. 1–4.

[12] A. Tessmann, A. Leuther, S. Wagner, H. Massler, M. Kuri, H. . Stulz, M. Zink, M. Riessle, and T. Merkle, "A 300 GHz low-noise amplifier S-MMIC for use in next-generation imaging and communication applications," in *2017 IEEE MTT-S International Microwave Symposium (IMS)*, June 2017, pp. 760–763.

Implementation of Slow-Wave Thin-Film Microstrip Transmission Lines in a 35nm InGaAs Technology

Athanasios Gatzastras[#1], Hermann Massler[*], Arnulf Leuther[*], Sébastien Chartier[*], Ingmar Kallfass[#2]

[#]Institute of Robust Power Semiconductor Systems, University of Stuttgart, Germany
[*]Fraunhofer Institute for Applied Solid State Physics IAF, Freiburg, Germany
{[1]Athanasios.Gatzastras, [2]Ingmar.Kallfass}@ilh.uni-stuttgart.de

Abstract — This paper presents the implementation of slow-wave thin-film microstrip lines (TFMSL) and a comparison to conventional TFMSLs in the frequency ranges of 0-150 GHz and in H-band from 220-320 GHz fabricated in a 35nm InGaAs metamorphic high electron mobility (mHEMT) technology, whereby the implementation at lower THz frequencies represents a novelty. The use of slow-wave transmission lines has two positive effects. First, the insertion losses are reduced by the deplacement of the magnetic field under the signal line and second the phase velocity of the signal is reduced by implementing a slotted ground plane below the signal line. Both effects combined lead to a loss reduction per quarter wavelength up to 54 % at 110 GHz and up to 47 % in H-band in comparison to conventional TFMSLs.

Keywords — TFMSL, slow-wave line, H-band, 35nm InGaAs

I. INTRODUCTION

Since frequencies in the lower THz range from 251 GHz to 325 GHz are allocated by the new IEEE standard 802.15.3d-2017 [1], the demand of MMICs in that frequency range grows steadily. Furthermore, the decision by the World Radiocommunication Conference 2019 emphasizes the pursuits of developing applications like imaging and high-data rate communication systems in this frequency range [2]. With a rising number of applications and functionalities of analog circuits, the area need for those circuits is growing, too. A large proportion of the core area of those MMICs is often determined by passive circuits with a certain phase relation between their ports, like different kinds of couplers or power splitters. Since the electrical and therefore also the physical length of conventional transmission lines are determined by the electrical properties of the fabrication process, the approach of slow-wave transmission line types are known since 1994 [3]. This work picks up the topic of slow-wave transmission lines by introducing defects in the ground plane and applies it to TFMSL lines in a 35nm InGaAs technology in a frequency range up to 320 GHz.

The technology used in this work is a $In_{0.52}Al_{0.48}As/In_{0.8}Ga_{0.2}As$ mHEMT technology with a gatelength of 35 nm provided by the Fraunhofer IAF. These heterostructures are grown on 100 mm semi-insulating GaAs wafers by molecular beam epitaxy using a metamorphic buffer to adapt the lattice constant of GaAs wafers. The buffer is composed by a linear $In_xAl_{0.48}Ga_{0.52-x}As$ ($x = 0 \rightarrow 0.52$) [4]. The cutoff frequencies f_t and f_{max} are 517 GHz and

beyond 1 THz, respectively [5]. The metal stack consists of three metal layers (MET1-MET3) shown in Fig.1 for designing thin-film microstrip lines (TFMSL). The signal line is built by stacking MET2 and MET3 while MET1 is used as ground plane. The dielectric substrate is formed by a benzocyclobutene (BCB) layer with a relative permittivity of $\epsilon_r = 2.7$.

II. THE THIN-FILM TRANSMISSION LINE

Both types of the implemented TFMSLs are depicted in Fig. 1. Here, the dimensions of both signal lines are deliberately chosen to be identical for better comparison. The width of the lines is chosen to 3.4 µm targeting the 50 Ω characteristic impedance. Due to a fabrication grid of 250 µm and in order to prevent a capacitive coupling between the on-wafer probe-tips, the length of each line is chosen to 300 µm.

While Fig. 1a is showing a conventional TFMSL, Fig. 1b shows a slow-wave transmission line indicated by the slotted ground plane. In order to reach the highest possible slow-wave effect the slot width (w) in the ground plane and the slot spacing (b) of MET 1 depicted in Fig. 1b are set to 1 µm, respectively, as a result of 3D fullwave simulations.

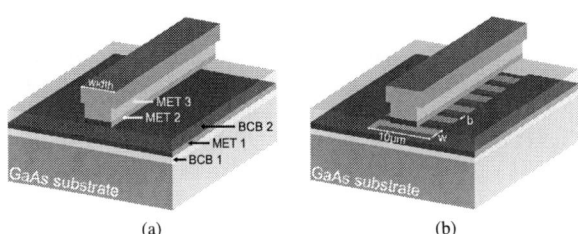

(a)   (b)

Fig. 1. Physical Layout of the implemented TFMSLs with (a) a conventional TFMSL with a continuously and (b) with a slotted ground plane.

Fig. 2 shows a photograph of the fabricated MMIC. In order to prevent a fabrication mismatch across the wafer and to reach representative measurement result, both lines are placed close to each other in one MMIC. Here, the slow-wave transmission line is the upper and the conventional TFMSL the lower one. The marked reference plane in the photograph indicates the boundary of the deembedded probe pad and the transmission line structure respectively.

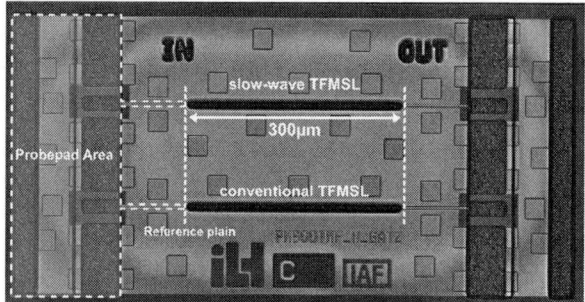

Fig. 2. Chip photograph of the implemented transmission lines.

III. THE WAVE PROPAGATION IN THE TFMSL

This section explains the field distribution of the slotted TFMSL and compares it to a conventional TFMSL with a continuous ground plane. To visualize the effect of the slotted ground plane this section shows the results of an EM simulation using Keysight's ADS Momentum. Fig. 3 collectively shows the distribution of the magnetic field in magnitude at a certain segment of each transmission line. The shown dimensions of the transmission lines and those of the slits are identical to the fabricated MMIC. While Fig. 3a and 3b show the top-view, Fig. 3c and 3d show a 3D side-view of both transmission line types.

From the respective views of the conventional microstrip transmission line it can be seen that the magnitude of the magnetic field has its maximum along the signal line and below in the ground plane as it is expected from theory. In contrast to that is the distribution of the magnetic field in a slow-wave TFMSL. While the magnitude in the signal line stays constant it can be observed that the magnetic field underneath the signal line is displaced orthogonally to the direction of the wave propagation in the signal line. This consequently means that by introducing slits the return currents are not able to flow directly below the signal line [6]. Instead, they are flowing sideways towards the continuous ground plane which leads to a lower power density right below the signal line and finally to lower insertion losses compared to conventional microstrip lines.

A. The Slow-Wave Effect

An additional effect of using transmission lines with a slotted ground plane is the slow-wave effect. Considering an equivalent circuit of a transmission line depicted in Fig. 4 the characteristic impedance Z_c and the phase velocity v_p can be derived with

$$Z_c = \sqrt{\frac{R' + L'}{G' + C'}} \quad \text{and} \quad v_p = \sqrt{\frac{1}{L'C'}} = \frac{c}{\sqrt{\mu_r \epsilon_r}} \quad (1)$$

respectively. It can be seen that by adding capacitance and inductance per unit length the phase velocity is reduced. Both

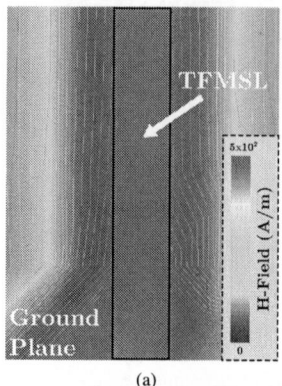 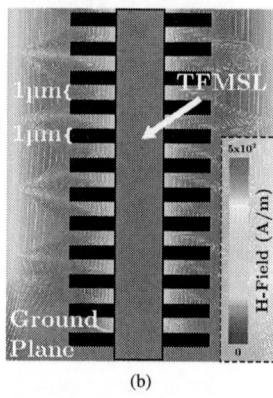

(a)   (b)

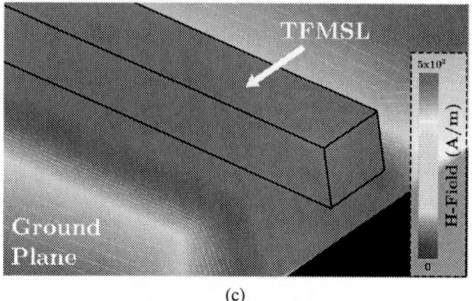

(c)

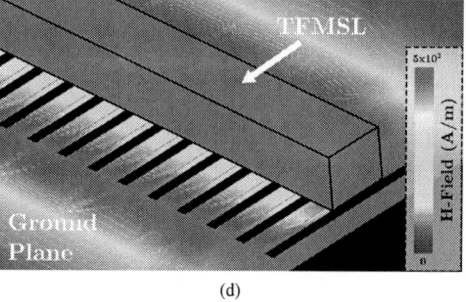

(d)

Fig. 3. Distrbution of the magnetic field in a conventional TFMSL and in a slow wave TFMSL based on a 3D EM simulation.

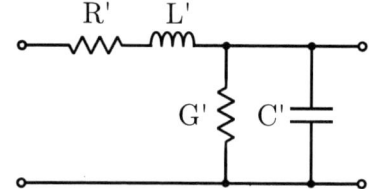

Fig. 4. Equivalent circuit of a transmission line element.

effects can be reached by introducing defects in the ground plane. Due to the fact that the return current is not able to flow directly below the signal line, additional inductance per unit length is introduced by the slits to the transmission line [6]. Whereas the capacitance per unit length of the transmission line is increased by the capacitive coupling between each slot. The goal of slow wave transmission lines is a reduction of the phase velocity of a signal leading into a reduction of the geometrical length of a transmission line for a certain wavelength. This means that the area need of passive on-chip structures depending on phase relations like couplers can be minimized.

IV. MEASUREMENT RESULTS

The following measurements are divided into two frequency ranges. The first range goes from 0 to 150 GHz and the second from 200 to 320 GHz. The on wafer S-parameter measurement setup was calibrated to the probe tip of the RF-probes by performing a thru-reflect-line calibration by using an impedance standard substrate (ISS). The probepads are deembedded by executing a full 3D EM simulation with CST Microwave Studio in the frequency range from 0 335 GHz. Fig. 5 shows the measured insertion loss of both lines with a linelength of 300 µm. The results in this frequency range show that the slotted transmission line has continuously lower insertion losses which confirms the theory described in chapter III and former investigations on slitted transmission lines [6],[7]. The property of the slitted transmission line

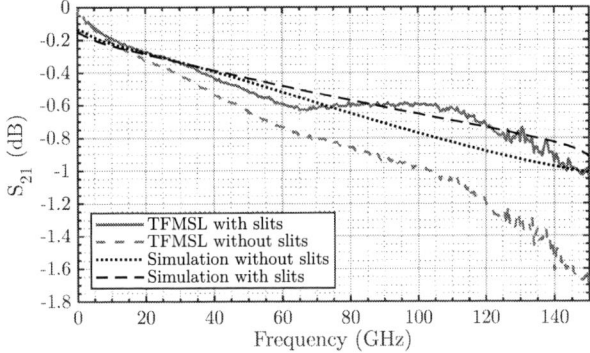

Fig. 5. Measured S21 parameter of the implemented TFMSLs over frequency

of reducing the phase velocity of an electromagnetic wave can be derived from Fig. 6. It can be observed that both phase responses are linear over frequency and have a different slope respectively. Thereby, the transmission line with a slotted ground plane has a higher slope than the conventional TFMSL.

The individual slopes can be determined graphically as follows:

$$\frac{\Delta Y_1}{\Delta X_1} = -0.77°/\text{GHz} \qquad \frac{\Delta Y_2}{\Delta X_2} = -0.64°/\text{GHz} \qquad (2)$$

The slope of the phase of the slotted TFMSL is due to the slow-wave effect about 17% higher in comparison to the one with a continuously ground plane which leads into a increased group delay by the same factor. Conversely, this means that the reduction the physical length for a certain electrical length by using slotted TFMSLs scales proportional with frequency.

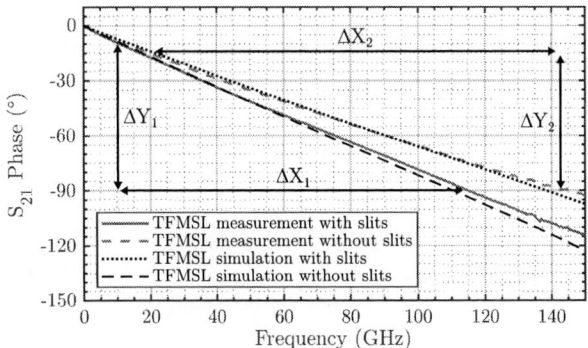

Fig. 6. Measured Phase of the S21 parameter over frequency.

It is derived that the length of e.g. a $\lambda/4$ transmission line at 145 GHz can be reduced over 20% by using slow-wave TFMSLs.

Fig. 7 shows the attenuation per quarter wavelength of the conventional TFMSL in comparison to those with a slotted ground plane for different slit spacings. This graph shows the impact of both effects described in the previous chapter III. Up to 54% less attenuation at e.g. 110 GHz can be achieved by introducing 1 µm slits in the ground plane of TFMSLs. The spacing between the slits has a moderate influence on the attenuation per quarter wavelength. This means that according

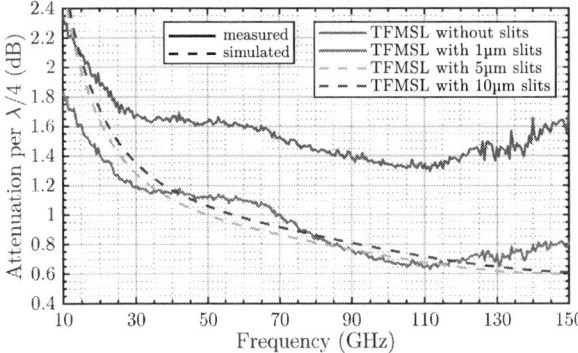

Fig. 7. Calculated attenuation per quarter wavelength of different TFMSLs.

to the simulation and the theory both effects are compensating each other. On the one hand the insertion losses become lower due to the displacement of the magnetic field, while on the other hand the slow-wave effect gets reduced by increasing the slit spacing. In total both effects lead to a similar attenuation per quarter wavelength over frequency.

A. Measurement Results in H-Band

In this paragraph the measurement results of both implemented TFMSL in H-band are introduced. From Fig.8 can be observed that in contrast to the previous results up to 150 GHz the insertion loss of the slow-wave TFMSL is no longer lower in comparison to the conventional TFMSL over this frequency range. Above a frequency of about 300 GHz

123

the insertion loss of the slow-wave TFMSL has a higher drop over frequency in comparison to the conventional TFMSL.

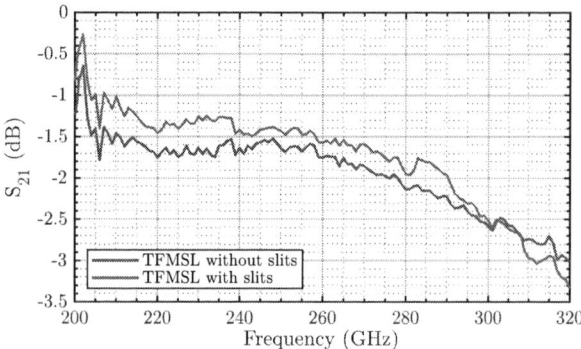

Fig. 8. Measured S21 parameter of the implemented TFMSLs in H-Band.

However, the courses of the respective phase depicted in Fig. 9 is as expected linear versus frequency and with

$$\frac{\Delta Y_1}{\Delta X_1} = -0.76°/\text{GHz} \qquad \frac{\Delta Y_2}{\Delta X_2} = -0.61°/\text{GHz} \quad (3)$$

almost congruent to those at lower frequencies. This means consequently that the reduction factor of the physical length of TFMSLs by using a slow-wave approach is proportional with frequency while a certain group delay and therefore the phase velocity are constant over frequency.

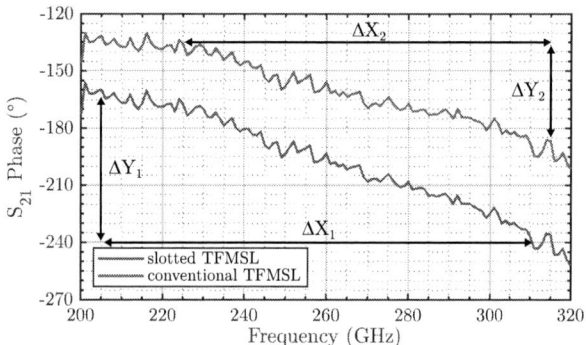

Fig. 9. Measured phase of the S21 parameter of the implemented TFMSLs in H-Band

Following Fig. 10 shows the calculated attenuation of a quarter wavelength line over frequency for both implementations. It can be observed that the attenuation referred to a quater-wavelength line is continuous lower at slow-wave lines over the total H-band with a maximum of 47 % at 220 GHz and minimal difference of 12 % at 320 GHz.

V. CONCLUSION

Since saving chip area is always a goal in integrated circuit design, this paper shows that the implementation of slow-wave transmission lines is very suitable for implementing signal lines and phase-dependent structures like power splitters or couplers in III-V technologies and over the entire

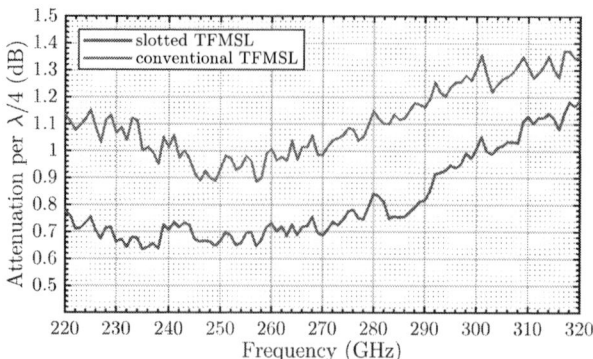

Fig. 10. Calculated attenuation per quarter wavelength of slotted TFMSL in comparison to a conventional TFSML.

millimeter-wave frequency range up to 300 GHz. Moreover, this work shows that slow-wave transmission lines will not only reduce the area of phase-related structures, but are also suitable for reducing the losses of those passive structures. The effect of the displacement of the magnetic field below the signal line and the slow-wave effect by introducing slits in the ground plane lead to a reduction of the attenuation up to 54 % at 110 GHz and 47 % at 220 GHz.

ACKNOWLEDGMENT

The authors would like to thank their colleagues from the Fraunhofer IAF epitaxy, technology and high-frequency department for excellent processing and characterization.

REFERENCES

[1] *IEEE Standard for High Data Rate Wireless Multi-Media Networks–Amendment 2: 100 Gb/s Wireless Switched Point-to-Point Physical Layer*, Std.

[2] *World Radiocommunication Conference 2019 - Final Acts*, 2019.

[3] R. Spickermann and N. Dagli, "Experimental analysis of millimeter wave coplanar waveguide slow wave structures on gaas," *IEEE Transactions on Microwave Theory and Techniques*, vol. 42, no. 10, pp. 1918–1924, 1994.

[4] A. Leuther, A. Tessmann, H. Massler, R. Losch, M. Schlechtweg, M. Mikulla, and O. Ambacher, "35 nm metamorphic hemt mmic technology," in *2008 20th International Conference on Indium Phosphide and Related Materials*, 2008, pp. 1–4.

[5] A. Leuther, A. Tessmann, M. Dammann, H. Massler, M. Schlechtweg, and O. Ambacher, "35 nm mhemt technology for thz and ultra low noise applications," in *2013 International Conference on Indium Phosphide and Related Materials (IPRM)*, 2013, pp. 1–2.

[6] J. J. Lee and C. S. Park, "A slow-wave microstrip line with a high-q and a high dielectric constant for millimeter-wave cmos application," *IEEE Microwave and Wireless Components Letters*, vol. 20, no. 7, pp. 381–383, 2010.

[7] H. Cho, T. Yeh, S. Liu, and C. Wu, "High-performance slow-wave transmission lines with optimized slot-type floating shields," *IEEE Transactions on Electron Devices*, vol. 56, no. 8, pp. 1705–1711, 2009.

Proceedings of the 16th European Microwave Integrated Circuits Conference

A 140 GHz to 170 GHz Active Tunable Noise Source Development in SiGe BiCMOS 55 nm Technology

Victor Fiorese[1,2], Joao Carlos Azevedo Goncalves[1], Simon Bouvot[1], Emmanuel Dubois[2],
Christophe Gaquiere[2], Guillaume Ducournau[2], François Danneville[2], Sylvie Lépilliet[2] & Daniel Gloria[1]

[1] STMicroelectronics, France

[2] IEMN – Institut d'Electronique de Microélectronique et de Nanotechnologie, France

victor.fiorese1@st.com

Abstract — **A new approach of millimeter wave (mmW) integrated active noise source (NS) is introduced for noise characterization up to 170 GHz. This NS is based on a diode biased in avalanche regime in BiCMOS 55 nm (B55) technology from STMicroelectronics. In order to increase the noise sensitivity of setup using this NS, a two-stage cascode Low Noise Amplifier (LNA) composed of 4 high speed NPN transistors is cascaded at its output, targeting high available Excess Noise Ratios (ENR_{av}). ENR_{av} levels have been extracted, showing tunable values ranged between 0 dB to 37 dB in the 140-170 GHz frequency range.**

I. INTRODUCTION

In the 5G development context, the increased need for accurate measurements at high frequencies (HF) has brought new concerns regarding testing setups. Indeed, the operating frequencies of recently developed transistors have exceeded 370 GHz for f_{max} [1], introducing the development of new mmW circuits. The design of such systems relies on accurate and reliable models of MOS or bipolar transistors that are obtained by small signal, large signal, and HF noise characterization. The usual frequency range of such noise models is limited to 110 GHz. Beyond this frequency limit, measurements are challenged by important interconnect losses and the difficulty to respect the minimum detectable signal (MDS) condition which is related to the noise receiver noise figure (NF_{RCV}). Moreover, by operating above 110 GHz comes the difficulty to generate a sufficient Excess Noise Ratio (ENR) that is directly used during the Y-factor measurement method. This method is commonly used for noise measurement and consists in determining the ratio of measured noise power when the NS is in ON state (P_{HOT}) and when it is in OFF state (P_{COLD}). The NS is based on a B55 noise diode [2] biased in avalanche regime for the ON state. Such noise sources can generate around 17 dB of ENR at 130 GHz [3]. Output noise power can prove to be insufficient to comply with noise receiver's sensitivity [4]. Indeed, the usual NF value for off-wafer noise receiver in these frequency ranges is 10 dB [5]. As suggested in [3], using a biased state instead of an unbiased state for noise source low state can help to validate the MDS condition. The drawback of such an approach is a decrease in the ENR maximum reachable levels, given that the delta between the two equivalent noise temperatures T_{hot} and T_{cold} corresponding respectively to the ON and OFF NS states is

smaller. Another solution can be the approach consisting in integrating a low noise amplifier (LNA) to have a constant output NS matching regardless of bias condition and higher ENR levels. Several works leveraging photodiodes [6], [7], NMOS transistors [8] or p-i-n diodes [9] showed conclusive results up to 40 GHz. In this work, the addressed frequency range is above 140 GHz. Limitations of HF noise measurement are detailed in section II while section III describes the solution proposed to ensure the MDS validation and a sufficient ENR level at the same time. Section IV describes the methodology used for ENR_{av} extractions. Finally, section V provides some results and prospects.

II. HIGH FREQUENCY NOISE MEASUREMENTS CHALLENGES OVERVIEW

A. DUT Noise factor measurement with Y-factor method

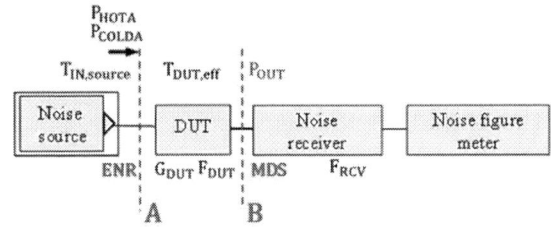

Fig. 1. Classical setup for NF measurement leveraging Y-factor method

The HF noise calibration is performed by connecting the noise source to the input of the noise receiver and making Y-factor measurement [4]. During the calibration, the noise figure (F_{RCV}) and gain (G_{RCV}) of the receiver are extracted allowing the calibration in plane B. Once the noise receiver is calibrated, the DUT can be characterized when inserted between noise source and noise receiver as shown in Figure 1. The noise figure of the DUT (F_{DUT}) can be determined using the Y-factor method [10]. To perform noise characterization above 110 GHz with sufficient accuracy, the MDS condition must be validated. This step follows the noise receiver calibration and is necessary to perform a noise characterization with sufficient responsivity and accuracy. This condition described in (1) is based on the equivalent noise temperature T_{DUT} and the available power gain G_{DUT} of the DUT.

125

3–4 April 2022, London, UK

$$P_{OUT} = G_{DUT}k_B(T_{DUT} + T_{IN})\Delta_f \geq 2F_{RCV}k_BT_a\Delta_f. \quad (1)$$

With k_B being the Boltzmann constant, $T_0 = 290K$, T_{IN} the equivalent temperature of the noise source, Δ_f the frequency bandwidth, T_a the ambient temperature, close to T_0. The MDS condition ensures that the output noise power signal of the noisy system (P_{OUT}) must exceed a certain value depending on the noise receiver NF in order to be detectable with a margin of 3 dB. Usually, at frequencies higher than 110 GHz, 10 dB is a relevant value for NF_{RCV}. In the case of a B55 high speed HBT noise characterization, the DUT contribution to G_{DUT} is not significant since this kind of component has roughly 1 dB of gain at 200 GHz. In the case of the Y-factor method, the two NS operating modes (ON and OFF) must validate the MDS criteria, the OFF state being the most challenging one.

B. The ENR level requirement

A noise source is always associated to an ENR level, defined as the noise power that a noise source can generate compared to the noise floor power of -174 dBm/Hz. This notion is introduced by the commonly used method of noise factor measurement involving the two operating states ON and OFF of a noise source, assigned to two noise temperatures T_{HOT} and T_{COLD}. From these two temperatures, a definition of the ENR can be introduced by (2), with T0=290 K.

$$ENR(dB) = 10log\left(\frac{T_{HOT} - T_{COLD}}{T_0}\right). \quad (2)$$

Commercial packaged noise sources have shown ENR levels of 12 dB up to 220 GHz [11] with a 2 dB flatness. Other works have developed on-wafer noise sources based on SiGe BiCMOS 55 nm noise diodes, showing monitorable ENR levels up to 260 GHz [12]. Such noise sources generate noise when biased in avalanche regime. By changing the biasing current value, several ENR levels can be reached, from 6 to 20 dB in the 130-170 GHz frequency range. Following the trend of these works, a new idea of an active noise source composed of a diode and a LNA is proposed. Such a solution can show a variable ENR level while respecting the MDS condition.

III. TEST STRUCTURES AND CHARACTERIZATION SETUP DESCRIPTION

A. Noise source and LNA description

The used NS is a SiGe BiCMOS 55 nm diode showing monitorable ENR levels up to 17 dB at 260 GHz when biased in avalanche regime [2]. The used LNA design and characterization well described in [4], has a 15 dB flat gain between 130 GHz and 160 GHz. It is structured as a classic cascode structure on two stages. Each stage is composed of two HBT NPN with an emitter width and length of respectively 0.2 µm and 5.56 µm. Since ENR measurements were performed between 130 GHz and 170 GHz, this LNA appears to be a relevant candidate for this study and is used as a buffer stage to enhance ENR levels. A layout view of this LNA is proposed in Figure 2.

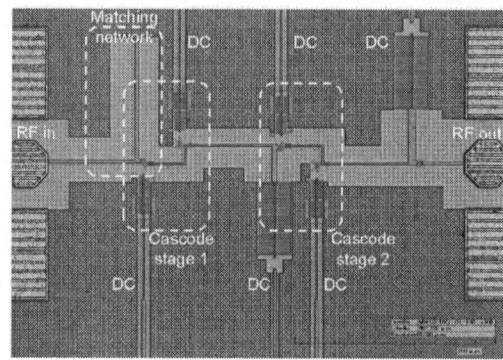

Fig. 2. Layout view of the LNA used in the present active noise source [4].

This LNA has been characterized in Sparams and noise in D band (140-170 GHz), showing a NF lower than 8 dB between 135 GHz and 170 GHz as described in Figure 3. The lower NF level corresponds to the NF of the LNA de-embedded of the RF pads.

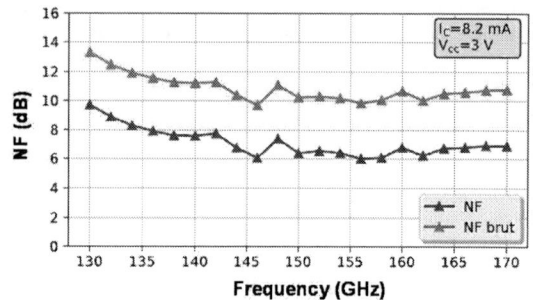

Fig. 3. NF measured in D-band of the 2 stage LNA [4].

B. ENR extraction when using a noise source and LNA setup

In order to assess the interest of adding a LNA to increase the ENR value, complementary extractions were performed in the setup illustrated on Figure 4:

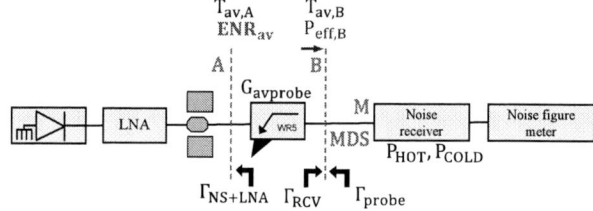

Fig. 4. Block diagram of the noise source and LNA ENR extraction setup

This setup has been used to extract ENR levels in several planes A and B to analyse the ENR level contribution of each element. To be able to extract ENR levels in plane A, measurements of the NS and LNA were performed in ON and OFF states. Besides this noise power level aspect, another advantage of using the chosen LNA is to greatly improve the output matching, regardless of NS bias current, due to its isolation of 35 dB in the 140-170 frequency range, as shown in Figure 5.

$$ENR(dB)=10log(\frac{T_{av,A,HOT}-T_{av,A,COLD}}{T_0}) . \qquad (3)$$

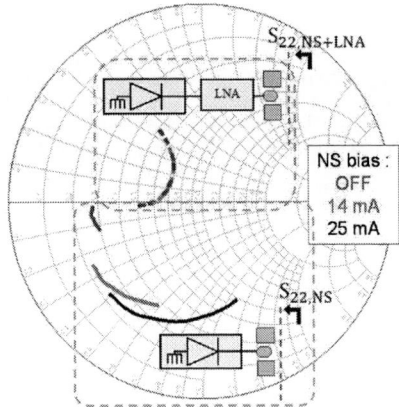

Fig. 5. Output matching comparison of NS (solid lines) and NS+LNA (dot lines) between 140 GHz and 170 GHz for 3 NS biasing.

IV. METHODOLOGY

In order to calculate the ENR, calibrated noise powers were measured in ON and OFF states at the input noise receiver plane [C]. To take into account the mismatch between the components, reflection coefficients and available gains calculation of the several blocks was made. Figure 2 shows the several parameters calculated to extract the ENR_{av} in planes A and B. At first, calculation of each reflection coefficient and available gain is done. This is possible due to the knowledge of the noise receiver and noise source reflection coefficients Γ_{RCV} and Γ_{NS+LNA} for each bias condition. Moreover, probe S-parameters are given by probe supplier and output pad is not de-embedded in this study. The first step consists in calculating Γ_{probe} and $G_{avprobe}$ with the knowledge of Γ_{NS+LNA} and Γ_{NS}, determined during S-parameters measurements. During this development, the noise receiver is assimilated to be a 50 Ω load termination. This permits to calculate the mismatch factor (M) [13], [14] used to transform effective noise powers into available noise powers, to be able to extract ENR_{av} levels. With this knowledge, the transform of relative measured effective noise powers (P_{eff}) compared to -174 dBm is done to obtain equivalent available noise temperatures at noise receiver input plane ($T_{av,B}$).

$$T_{av,B} = T_{eff,B}M = \frac{P_{eff,B}}{k_B}M. \qquad (4)$$

Then, the equivalent available noise temperatures at probe input plane ($T_{av,B}$) is calculated thanks to the knowledge of $G_{avprobe}$ as shown in equation (5):

$$T_{av,A} = \frac{T_{av,B}-T_a(1-G_{avprobe})}{G_{avprobe}} \quad \text{with } T_a= 296 \text{ K.} \qquad (5)$$

In this study, several bias states were used for NS and LNA. Extractions of ENR_{av} levels at plane A (see Figures 1 and 4) expressed in equation (2) have been done. $T_{av,A,COLD}$ is a bias state where NS is OFF and LNA is ON, $T_{av,A,HOT}$ is a bias state where NS is biased with currents varying from 0.25 mA to 25 mA and LNA is ON. It is to be highlighted that unlike conventional ENR definition, for which T_{COLD} is usually equal to the room temperature, used T_{COLD} values in our study are varying between 39000 K and 73000 K (see Figure 6).

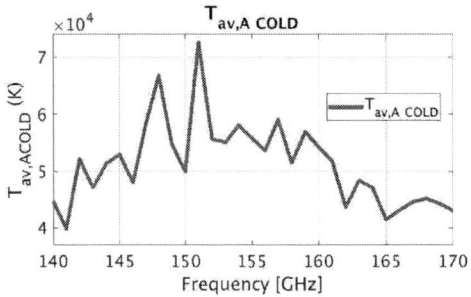

Fig. 6. Used values for $T_{av,A,COLD}$ (K) in probe input plane (A plane).

V. RESULTS

Extracted ENR_{av} levels are represented in Figure 7, showing highly monitorable powers up to 37 dB.

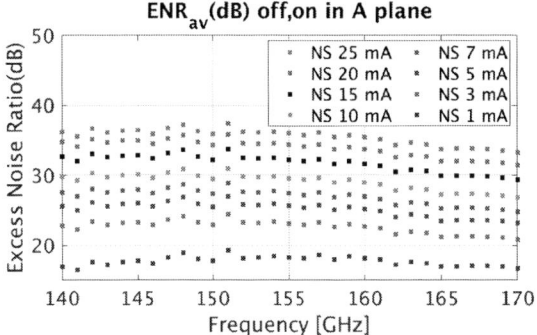

Fig. 7. Extracted ENR_{av} (dB) in probe input plane (A plane).

To our knowledge, these are the first results of such monitorable ENR_{av} levels at these frequencies using SiGe technology. Considering the MDS aspect, the use of NS in low state and LNA in high state shows a validation of the MDS condition with a 10 dB margin in the 140-170 GHz frequency band. When NS and LNA are set in low state, MDS condition can obviously not be satisfied as depicted in Figure 8.

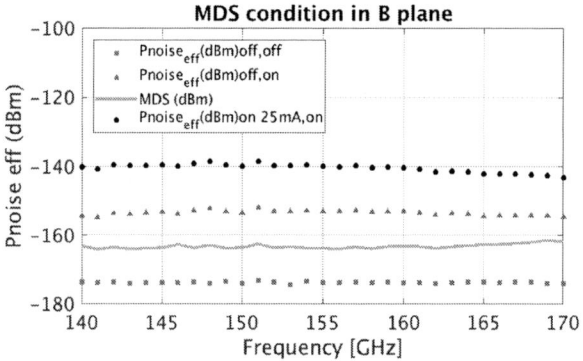

Fig. 8. MDS condition compared to Pnoise eff (dBm) in B plane

A mmW and sub-mmW noise sources state-of-the-art is presented Table 1.

Table 1. mmW and sub-mmW noise sources state-of-the-art

Technology	Frequency range (GHz)	ENR (dB)	Ref.
SiPho PD Ge	75-110	0-40	[7]
UTC-PD	0-20	0-62	[6]
BiCMOS 130 nm	0-40	15-30	[9]
Commercial	110-170	12	[11]
BiCMOS 55 nm	130-260	0-20	[2]
BiCMOS 90 nm	0-30	0-25	[8]
BiCMOS 55 nm	140-170	0-37	This work

VI. CONCLUSION

A new approach of in-situ active noise source integrated on silicon using a D-band LNA has been described. This way of operating induces higher monitorable ENR levels while respecting the MDS condition. A noise source low biasing LNA high biasing proves to be sufficient to validate this condition with a 10 dB margin. This active noise source shows a constant output matching in both high and low states of biasing. In the case of multi-impedance noise characterization of a DUT, an impedance tuner may be used in source-pull configuration. As a perspective, further studies will bring the noise measurement setup approach of this active noise source followed by an impedance tuner and a DUT in order to evaluate the multi-impedance method.

REFERENCES

[1] P. Chevalier et al., "SiGe BiCMOS Current Status and Future Trends in Europe," 2018 IEEE BiCMOS Compd. Semicond. Integr. Circuits Technol. Symp. BCICTS 2018, pp. 64–71, 2018.

[2] J. C. Azevedo Goncalves et al., "Millimeter-Wave Noise Source Development on SiGe BiCMOS 55-nm Technology for Applications up to 260 GHz," IEEE Trans. Microw. Theory Tech., vol. 67, no. 9, pp. 3732–3742, 2019.

[3] J. C. A. Goncalves, "Développement de bancs de caractérisation pour la mesure de bruit et la détection de puissance entre 130 GHz et 260 GHz," Université de Lille, 2019.

[4] S. Bouvot, "Contribution au BIST in-situ : Intégration sur silicium d'un banc de caractérisation en bruit en bande D," Université de Lille, 2018.

[5] M. Deng, "Contribution à la caractérisation et la modélisation jusqu'à 325 GHz de transistors HBT des technologies BiCMOS," Université de Lille, 2014.

[6] B. Vidal, "Broadband Photonic Microwave Noise Sources," IEEE Photonics Technol. Lett., vol. 32, no. 10, pp. 2020–2022, 2020.

[7] S. Oeuvrard et al., "On Wafer Silicon Integrated Noise Source Characterization up to 110 GHz based on Germanium-on-Silicon Photodiode," in 27th IEEE International Conference on Microelectronic Test Structures, ICMTS 2014, 2014, pp. 150–154.

[8] F. Alimenti, G. Tasselli, C. Botteron, P. A. Farine, and C. Enz, "Avalanche Microwave Noise Sources in Commercial 90-nm CMOS Technology," IEEE Trans. Microw. Theory Tech., vol. 64, no. 5, pp. 1409–1414, 2016.

[9] F. Alimenti, G. Simoncini, G. Brozzetti, D. D. Maistro, and M. Tiebout, "Millimeter-Wave Avalanche Noise Sources Based on p-i-n Diodes in 130 nm SiGe BiCMOS Technology: Device Characterization and CAD Modeling," IEEE Access, vol. 8, pp. 178976–178990, 2020.

[10] Keysight Technologies Appl. Note, "Fundamentals of RF and Microwave Noise Figure Measurements," pp. 1–44.

[11] "The SAGE Millimeter, Inc. website." [Online]. Available: https://www.eravant.com/110-to-170-ghz-12-db-enr-wr-06-waveguide-d-band-noise-source-with-faraday-isolator.

[12] H. Ghanem et al., "Modeling and Analysis of a Broadband Schottky Diode Noise Source Up To 325 GHz Based on 55-nm SiGe BiCMOS Technology," IEEE Trans. Microw. Theory Tech., pp. 1–10, 2020.

[13] C. K. S. Miller, W. C. Daywitt, and M. G. Arthur, "Noise Standards, Measurements, and Receiver Noise Definitions," Proc. IEEE, vol. 55, no. 6, pp. 865–877, 1967.

[14] L. P. Dunleavy, J. Randa, D. K. Walker, R. Billinger, and J. Rice, "Characterization and Applications of On-Wafer Diode Noise Sources," in IEEE Transactions on Microwave Theory and Techniques, 1998, vol. 46, no. 12, pp. 2620–2628.

Proceedings of the 16th European Microwave Integrated Circuits Conference

200 GHz Low Noise Amplifiers in 250 nm InP HBT Technology

Utku Soylu[*1], Ahmed S. H. Ahmed[*], Munkyo Seo[#], Ali Farid[*], Mark Rodwell[*]

[*]Department of Electrical and Computer Engineering, University of California, Santa Barbara, USA
[#]Department of Electrical and Computer Engineering, Sungkyunkwan University, South Korea
[1]utkusoylu@ucsb.edu

Abstract — We report 200 GHz InP DHBT low noise amplifiers in common base (CB) and common emitter (CE) topologies, together with a design procedure based on minimum noise measure. The CB design shows 7.4±0.7 dB noise figure over 196-216 GHz, 14.5 dB gain and -21.1 dBm Pin1dB at 200 GHz, and dissipates 9.2 mW, while the CE design shows 7.2±0.4 dB noise figure over 196-216 GHz, 13 dB gain and -18.2 dBm Pin1dB at 200 GHz, and dissipates 19.22 mW. To the authors' knowledge, these results demonstrate record noise figure for bipolar transistor amplifiers operating near 200 GHz.

Keywords —G-band, millimeter wave, noise figure (NF), noise measure (M), low noise amplifier (LNA), indium phosphide (InP), double heterojunction bipolar transistor (DHBT).

I. INTRODUCTION

There is increasing interest in the 200-300 GHz frequency band as the wide spectrum can support very high data rate wireless communications [1]. At these frequencies, greater integration scales have been demonstrated for receivers in SiGe and InP HBT than in III-V HEMT technologies, but HEMTs exhibit lower noise [2]-[8]. Reducing HBT LNA noise figure will make all-HBT 200-300 GHz receivers more competitive and, in hybrid receivers using a HEMT LNA and an HBT IC for the post-LNA and remaining receiver mm-wave stages, will minimize the required gain, hence number of stages, in the HEMT LNA. Here we present 200 GHz LNAs with low DC power consumption and record low noise for HBTs. We report common-base (CB) (Fig. 1) and common-emitter (CE) designs (Fig. 2). The CB design shows 7.4±0.7 dB noise figure over 196-216 GHz, 14.5 dB gain and -21.1 dBm P_{in1dB} at 200 GHz, and dissipates 9.2 mW. The CE design shows 7.2±0.4 dB noise figure over 196-216 GHz, 13 dB gain and -18.2 dBm P_{in1dB} at 200 GHz, and dissipates 19.2 mW.

In designing a multi-stage LNA for low total (cascaded) noise figure, the individual stages should be designed for lowest noise measure (M), not lowest noise figure (F), as this minimizes the total noise contribution of input and subsequent LNA gain stages. Passive element losses at 200GHz can significantly contribute to the LNA's total noise contribution, hence the LNA should be designed for minimum loss in the input matching network. The design procedure is therefore critical.

II. LOW NOISE AMPLIFIER DESIGN

LNAs were fabricated in the Teledyne 250 nm InP technology [1], which has four Au interconnect layers, 50 Ω/square thin film resistors and 0.3 fF/μm² MIM capacitors.

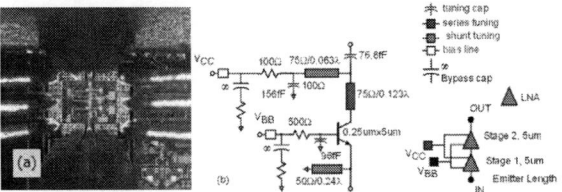

Fig. 1. Common-base amplifier: die photo (a) and amplifier circuit diagram (b). The die area, including DC routing and pads is 0.49 mm x 0.425 mm.

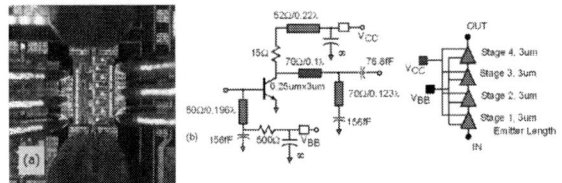

Fig. 2. Common-emitter amplifier: die photo (a) and amplifier circuit diagram (b). The die area, including DC routing and pads is 0.450 mm x 0.630 mm.

The HBT has a maximum 650 GHz power gain cut-off frequency (f_{max}), a maximum 3 mA/μm current density and the 4.5 V BV$_{\text{CEO}}$.

A. Noise Measure Technique

Transistor noise measure $M = (F-1)(1-G^{-1})^{-1}$ [10], where G is the gain, sets a lower bound for the cascaded noise figure F_{cascade} of a multi-stage LNA.

$$F_{\text{cascade}} = M + 1 = F + \frac{F-1}{G} + \frac{F-1}{G^2} + \cdots = \frac{F - G^{-1}}{1 - G^{-1}}$$

Further, if the embedding circuit is passive, lossless, and reciprocal, the minimum noise measure is independent of the surrounding circuit [10]. In particular, if passive element losses are negligible, the minimum F_{cascade} and M are identical in common-emitter and common-base configurations, and do not change if the stage is unilateralized or capacitively neutralized, or if gain is maximized using Singhakowinta's technique [11].

Given that F_{cascade} is independent of the stage configuration, we instead select the stage configuration based on either high feasible bandwidth or high gain per stage. The CB stage provides greater gain/stage, hence noise contributions associated with loss in the output matching network are reduced. Further, with greater gain per stage, fewer stages are required, reducing DC power. Because the CE stage has lower output impedance, its output matching network is more readily designed for wide bandwidth.

3–4 April 2022, London, UK

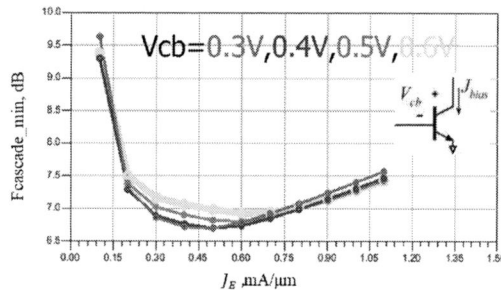

Fig. 3. Minimum $F_{cascade}$ as a function of emitter current density (J_E) and collector-base voltage (V_{CB}) for a 0.25 μm x 5 μm HBT.

B. Determining Bias Condition

Most widely-used RF computer-aided design programs compute F_{min} but not M_{min}, and compute F but not M as a function of source impedance. As will be subsequently described, Python scripts were written to compute these quantities from the output of the CAD simulation software. Given this, the first step in the design is to determine, from CAD simulation, the emitter current density and collector-base voltage giving the lowest M_{min}, hence the lowest $F_{cascade}$, at 200 GHz (Fig. 3). For the IC technology used, at 210 GHz, the simulated minimum $F_{cascade}$ is 6.7 dB with $J_E = 0.5$ mA/μm and $V_{CB} = 0.4V$.

C. Area Scaling, Base Capacitive Degeneration, and Input Matching Network

In common-emitter LNAs, an appropriate nonzero emitter inductive reactance $j\omega L_E$ allows input tuning for zero input reflection coefficient simultaneously with tuning for minimum $F_{cascade}$, doing so without increasing the minimum $F_{cascade}$. In common-base, a nonzero base capacitive impedance $1/j\omega C_{base}$ plays exactly the same role.

Subsequently, the HBT junction area is scaled, together with the DC current and the base capacitor, so that the source conductance for minimum noise measure is 20 mS (Fig. 4); this permits the input stage to be noise-matched to 50 Ω with a single inductive shunt element (Fig. 5), avoiding the added attenuation, hence the added noise, of a series matching element.

D. Displaying source impedance for minimum M

As with F and G_A (available gain), M can also be represented as a function of source reflection coefficient, displaying circles as contours of constant M [12]. Because widely-used RF computer-aided design programs do not provide this capability, a Python script was written to

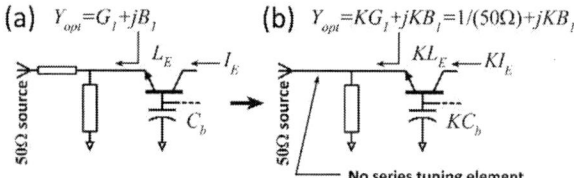

Fig. 4. (a) Input matching network without proper emitter junction area scaling (b) Input matching network with proper emitter junction area scaling.

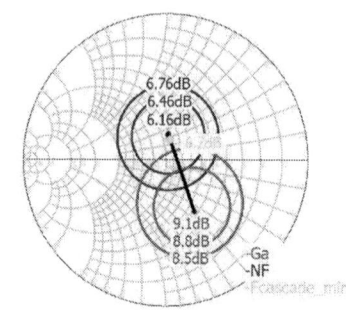

Fig. 5. CAD display, in Keysight ADS, of the contours of constant noise figure and available gain (G_A) in the plane of the source reflection coefficient. A Python script draws a line (black) between the centers of the F and G_A circles, computing the minimum noise measure along this line. Data is for an (0.25 x 5 μm^2) HBT in CB configuration with 200 fF base capacitance biased at $V_{CB}=0.4$ V and $J_E=0.5$ mA/μm.

approximately determine the minimum noise measure, and the associated source impedance, from the contours of constant F and G_A (Fig. 5). The script draws a line between the centers of the F and G_A circles, calculates M for each point on this line, and determines the point on the line having the smallest M.

If the F and G_A circles were exactly concentric, the impedance for minimum M would lie along the line so constructed, and this simple graphical algorithm would exactly determine M_{min} and the associated impedance. Graphically, we observe (Fig. 5) that the circles do not strongly deviate from this assumption. A more exact procedure would use the relationships of [12].

E. Output Matching Network

In a multistage LNA design, for lowest noise, each stage output is matched to the minimum noise measure impedance of the cascaded stage. This can be accomplished in a stage-by-stage design procedure in which each stage is designed to have minimum M for a 50 Ω external source impedance and maximum associated gain for a 50 Ω external load impedance.

LNA design must however balance noise against bandwidth and dynamic range. Consequently, the output tuning of the CB design was adjusted to increase the stage bandwidth and the 1dB gain compression point.

III. MEASUREMENT RESULTS

A. S-Parameter Measurements

Fig.1-2 show the chip micrographs. Measurements are performed on the 3-mil thinned die. S-parameters were measured using a Keysight network analyser with 220-325 GHz Oleson WR-03 frequency extender modules and 325 GHz GGB wafer probes. A short-open-load-thru (SOLT) calibration standard on an external substrate moves the reference plane to the probe tips.

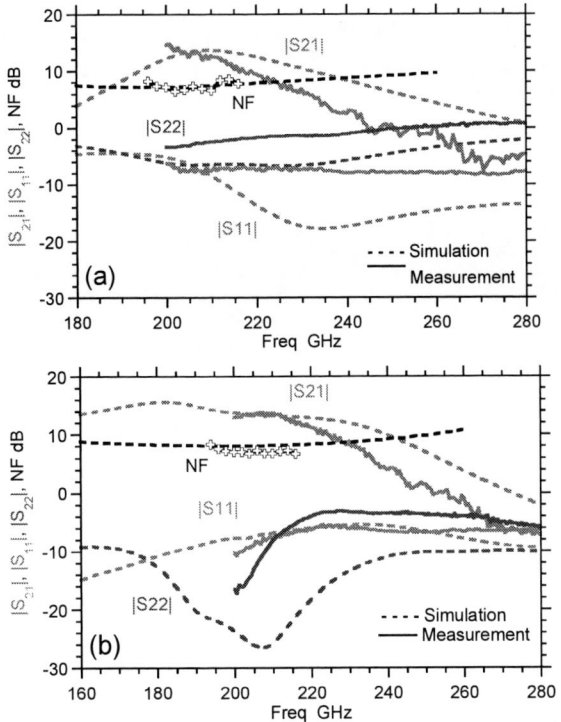

Fig. 6. Measured (solid) and simulated (dashed) S-parameters and NF: (a) common-base and (b) common-emitter amplifier.

Approximately 5% frequency downshift is observed between the amplifier's measured and simulated S-parameters (Fig. 6). The device model does not include base inductance, which is the main reason of the frequency shift. By adding extra series inductance to the base, the device model is adjusted.

The CB amplifier (Fig. 1) was biased at (V_{CCLNA}=1.5 V, I_{CCLNA}=6.03 mA, V_{BBLNA}=0.85 V, I_{BBLNA}=0.264 mA). The peak measured small-signal gain ($|S_{21}|$) is 14.5 dB at 200 GHz, in good agreement with simulation. The CE amplifier (Fig. 2) was biased at (V_{CCLNA}=1.6 V, I_{CCLNA}=11.68 mA, V_{BBLNA}=0.87 V, I_{BBLNA}=0.436 mA). The peak measured small-signal gain ($|S_{21}|$) is 13.69 dB at 206 GHz. The CB amplifier has a narrower bandwidth due to the higher CB output impedance.

B. Power Measurements

Fig. 7 shows the procedure for gain compression measurements [13], and the setup in the calibration and measuring phases are shown in Fig. 7. The ~200 GHz DUT excitation signal is generated by a synthesized microwave signal generator (N5183B) and an 8:1 VDI frequency multiplier. The signal is passed through a directional coupler to monitor the input power, and is passed through a G-band fixed attenuator to obtain power levels within the desired range to drive the LNA. By placing an Erickson power meter (PM4) at the attenuator output, and comparing its power measurement to that of the spectrum analyzer, the measurement of input power is thereby calibrated (Fig. 7a). The signal source is then

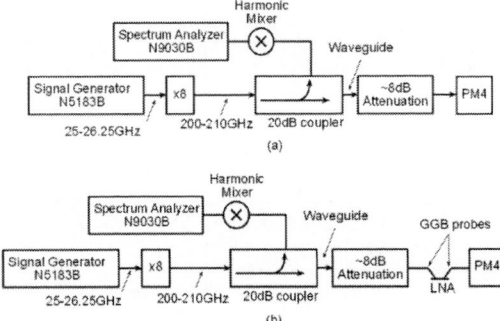

Fig. 7. Power measurement setup: a) calibration phase b) measurement phase.

connected, via GGB probes, to the amplifier input, and the amplifier output monitored by the PM4 power meter.

The CB and CE amplifiers are biased at (V_{CCLNA}=1.5 V, I_{CCLNA}=5.98 mA, V_{BBLNA}=0.864 V, I_{BBLNA}=0.268 mA) and (V_{CCLNA}=1.6 V, I_{CCLNA}=11.73 mA, V_{BBLNA}=0.87 V, I_{BBLNA}=0.522 mA) respectively. The CB and CE amplifiers have -21.1 dBm, and -18.2 dBm input referred P_{1dB} with 12.69 dB and 13.3 dB associated gain, respectively, at 200 GHz (Fig. 10a). At 210 GHz, the input referred P_{1dB} is -17.47 dBm (CB) and -19.08 dBm (CE) with 10.31 dB (CB) and 13.07 dB (CE) associated gain. The CB amplifier consumes 9.2 mW while the CE amplifier consumes 19.22 mW.

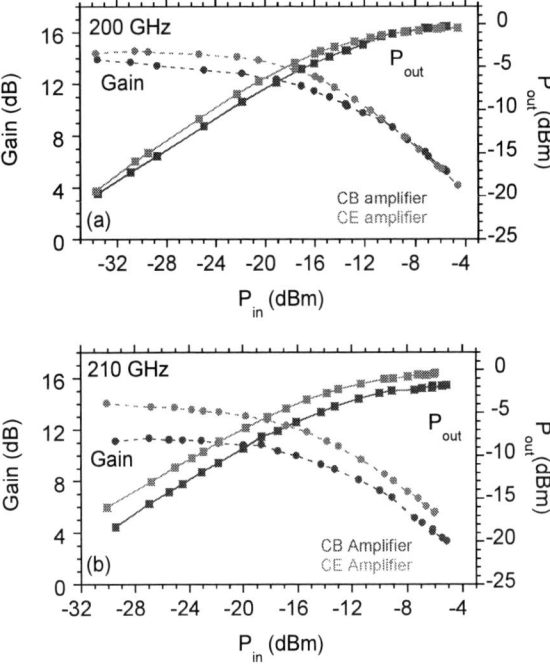

Fig. 8: Gain compression characteristics at (a) 200 GHz and (b) 210 GHz.

Table 1. Comparison of recently published >150 GHz low noise amplifiers

Ref.	Technology	Topology	Freq (GHz)	Gain (dB)	Gain/stage (dB)	NF (dB)	P_{DC} (mW)
[2]	250nm SiGe HBT	Cascode, diff. mode	156	26	8.7	8.5	-
[3]	50 nm mHEMT	CS	178-185	24.5	4.9	3.5	24
[4]	50 nm mHEMT	CS	206	16	4.0	4.8	-
[5]	32nm CMOS	CS	200-220	10-18	1.4-2.6	11	44.5
[6]	130nm Sige HBT	Cascode, diff. mode	220	18	6	16	151.2
[8]	250nm InP HBT	CE	265	24	4.8	10	81.7
[9]	250nm InP HBT	Cascode	288	8.4	8.4	11.2 at 300 GHz	-
This work	**250nm InP HBT**	CE	200	13	3.25	7.2	19.22
		CB	200	14.5	7.25	7.4	9.2

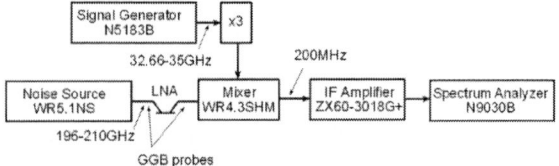

Fig. 9. Noise measurement setup.

C. Noise Measurements

The LNA noise figure is measured using the hot/cold Y-parameter method (Fig. 11). A VDI-WR5.1NS hot/cold noise source connected to the LNA input using a GGB 140-220 GHz wafer probe. The G-band probe loss is measured by landing probes on the through structure on the impedance standard substrate. The probe loss is found to be 2.0 dB loss at 200 GHz, and is deembedded from the measured noise figure. A ~20 dB low-noise post-amplifier (Mini Circuits, ZX60-3018G+) used to reduce the noise contribution from the spectrum analyzer. The subharmonic mixer's (VDI-WR4.3SHM) LO-signal was supplied by a QuinStar x3 WR-8.0 multiplier chain (QMM-933510030) and a signal generator (N5183B). The output noise power spectral density was measured at 200 MHz using a spectrum analyser (N9030B). The CB amplifier (Fig. 6a) shows 7.4±0.7 dB noise figure over 196-216 GHz, while the CE amplifier (Fig. 6b) shows 7.2±0.4 dB noise figure over 196-216 GHz.

Table 1 compares the performance of the low noise amplifier designed above 150 GHz. The amplifiers in Table 1 are the best reported in their respective transistor technology to the authors' knowledge.

IV. CONCLUSION

200 GHz low noise amplifier ICs in common-base and common emitter topologies were presented with record noise figure 7.4±0.7 dB over 196-216 GHz (CB) and 7.2±0.4 dB over 196-216 GHz (CE) in HBT technology.

ACKNOWLEDGMENT

This work was supported by ComSenTer, a JUMP program sponsored by the Semiconductor Research Corporation. The authors thank Teledyne Scientific & Imaging for the IC fabrication.

REFERENCES

[1] M. J. W. Rodwell et al., "100-340GHz Systems: Transistors and Applications," in Proc. IEEE IEDM, 2018, pp. 14.3.1-14.3.4

[2] Y. Zhao, E. Ojefors, K. Aufinger, T. F. Meister and U. R. Pfeiffer, "A 160-GHz Subharmonic Transmitter and Receiver Chipset in an SiGe HBT Technology," in IEEE Transactions on Microwave Theory and Techniques, vol. 60, no. 10, pp. 3286-3299, Oct. 2012.

[3] G. Moschetti et al., "A 183 GHz Metamorphic HEMT Low-Noise Amplifier With 3.5 dB Noise Figure," in IEEE Microwave and Wireless Components Letters, vol. 25, no. 9, pp. 618-620, Sept. 2015.

[4] A. Tessmann, A. Leuther, H. Massler, M. Kuri and R. Loesch, "A Metamorphic 220-320 GHz HEMT Amplifier MMIC," 2008 IEEE Compound Semiconductor Integrated Circuits Symposium, Monterey, CA, USA, 2008, pp. 1-4.

[5] Z. Wang, P. Chiang, P. Nazari, C. Wang, Z. Chen and P. Heydari, "A CMOS 210-GHz Fundamental Transceiver With OOK Modulation," in IEEE Journal of Solid-State Circuits, vol. 49, no. 3, pp. 564-580, March 2014.

[6] E. Ojefors, B. Heinemann and U. R. Pfeiffer, "Subharmonic 220- and 320-GHz SiGe HBT Receiver Front-Ends," in IEEE Transactions on Microwave Theory and Techniques, vol. 60, no. 5, pp. 1397-1404, May 2012.

[7] K. Eriksson, S. E. Gunnarsson, V. Vassilev and H. Zirath, "Design and Characterization of HH-Band (220–325 ∼GHz) Amplifiers in a 250-nm InP DHBT Technology," in IEEE Transactions on Terahertz Science and Technology, vol. 4, no. 1, pp. 56-64, Jan. 2014.

[8] J. Hacker et al., "THz MMICs based on InP HBT Technology," 2010 IEEE MTT-S International Microwave Symposium, Anaheim, CA, USA, 2010, pp. 1126-1129.

[9] M. Urteaga, Z. Griffith, M. Seo, J. Hacker, M. Rodwell, "InP HBT Technologies for THz Integrated Circuits", Proceedings of the IEEE, Vol. 105, No. 6, pp 1051-1067 June 2017.

[10] H. A. Haus and R. B. Adler, "Optimum Noise Performance of Linear Amplifiers," in Proceedings of the IRE, vol. 46, no. 8, pp. 1517-1533, Aug. 1958.

[11] A. Singhakowinta & A. R. Boothroyd, "Gain Capability of Two-port Amplifiers", International Journal of Electronics, Volume 21, Issue 6, 1966, pages 549-560.

[12] H. Fukui, "Available Power Gain, Noise Figure, and Noise Measure of Two-Ports and Their Graphical Representations," in IEEE Transactions on Circuit Theory, vol. 13, no. 2, pp. 137-142, June 1966.

[13] A. S. H. Ahmed, U. Soylu, M. Seo, M. Urteaga, J. F. Buckwalter and M. J. W. Rodwell., " A 190-210GHz Power Amplifier with 17.7-18.5dBm Output Power and 6.9-8.5% PAE.," in press, Proc. IMS2021.

Proceedings of the 16th European Microwave Integrated Circuits Conference

Output Power Limited Rugged GaN LNA MMIC

Evelyne Kaule[#1], Cristina Andrei[#], Matthias Rudolph[#*]

[#]Brandenburg University of Technology Cottbus-Senftenberg (BTU), Cottbus, Germany
Ulrich-L.-Rohde Chair for RF and Microwave Techniques
[*]Ferdinand-Braun-Institut gGmbH, Leibniz-Institut für Höchstfrequenztechnik (FBH), Berlin, Germany
[1]Evelyne.Kaule@b-tu.de

Abstract — **Rugged GaN HEMT low-noise amplifiers are in the receive path of T/R front ends are interesting for many applications. But the common concept of achieving ruggedness by reducing the gate DC current through a high resistance in the gate bias path might not be sufficient to control the output power under overdrive conditions. This paper focuses on the protection of the subsequent stages and suggests an AlGaN/GaN LNA MMIC circuit concept based on an output attenuator with the compact dimension of $4.35 \times 2.15\,\mathrm{mm}^2$. This provides an attenuation of up to 18 dB under input overdrive condition while gain, noise figure and output power in linear operation unaffected.**

Keywords — **Limit, GaN, HEMT, ruggedness, low-noise amplifier (LNA), MMIC, receiver.**

I. INTRODUCTION

Nowadays, GaN technology is well established in T/R-module front end applications, since it provides high output powers in the transmit path combined with low noise, high linearity and ruggedness in the receive path.

Rugged GaN LNAs commonly withstand pulsed input overdrive powers up to 43 dBm [1], [2]. Therefore, an input protection limiter is no longer required, allowing high-performance transceiver implementations on a single chip. For virtually all GaN LNAs, ruggedness is achieved by applying the gate bias voltage through a high ohmic resistor. In case of overdrive, the resistance will lower the gate voltage so that the HEMT achieves the class-C operation as described in [3]. Hence, most of the incident power is reflected and the output power is low. However, this concept alone may not be sufficient to protect the subsequent stages of the receiver path, e.g. when realized in CMOS. In order to operate reliably under overdrive conditions, a GaN LNA would be required to provide attenuation in excess of around 20 dB, which is, to the best of our knowledge, only addressed in the following works. In [4] and [5], a limiting circuit is presented, which consists of a HEMT connected between the LNA input and ground. It becomes conductive at high input voltage swings. The authors showed the implementation MMICs based on GaN. In [6], a circuit is integrated on the GaN MMIC that limits the drain supply current and thereby controls the maximum output power. Our previous work [7] proposes an adaptive drain DC supply, which switches the drain voltages to ground and an output attenuator becomes effective in case of overdrive condition. This output power limited LNA concept was realized in a hybrid implementation, which leads to a reduced bandwidth of the small-signal gain.

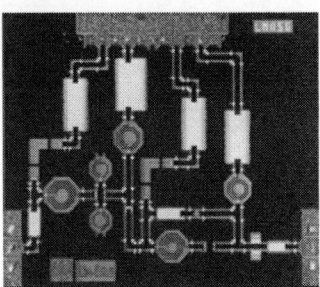

Fig. 1. Photography of the AlGaN/GaN LNA MMIC chip.

In the present work, the concept of an output attenuator, limiting the output power of a rugged GaN LNA in case of overdrive conditions, is taken up as a single chip implementation, in order to improve bandwidth and reduce size. The core of the output power limiting LNA MMIC is a rugged LNA MMIC based on AlGaN/GaN and an output attenuator realized by a GaN HEMT in common-gate configuration. The results obtained for this output power limited circuit are compared to an LNA MMIC featuring the LNA alone, to prove that the output protection is achieved without degradation of small-signal gain, noise figure, or linearity.

II. LNA MMIC

The core of the output power limiting LNA MMIC is the rugged LNA MMIC as shown in Fig. 1. The design is based on 0.15 μm-gate AlGaN/GaN HEMTs process line of the Ferdinand-Braun-Institut gGmbH (FBH) [8]. The LNA MMIC chip is $2.55 \times 2.15\,\mathrm{mm}^2$ in dimension and is developed as a two-stage amplifier in CPW technology, using $4 \times 50\,\mu\mathrm{m}$ transistors in order to obtain a minimum noise figure. Therefore, the first stage of the LNA MMIC is optimized for noise matching, while the second stage is adapted for high power. In order to improve noise performance and input return loss, source degradation is inserted in the first stage of the LNA MMIC, and the RC feedback in the second stage realizes an increased gain and a wideband operation. The bias network is designed to ensure stability and achieve gain over the C-band frequency range. In order to ensure the ruggedness, the gate bias network is fed through a high ohmic resistor [3].

3–4 April 2022, London, UK

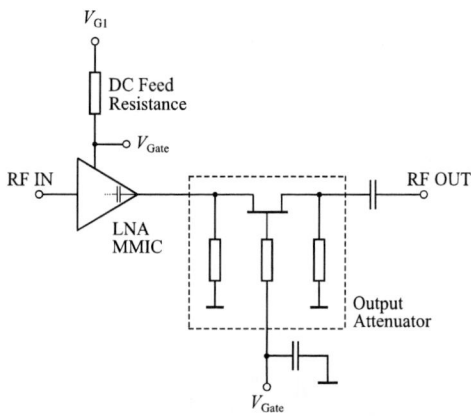

Fig. 2. Schematic of the output power limiting LNA MMIC.

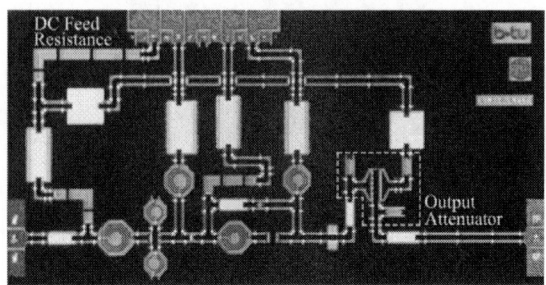

Fig. 3. Photography of the AlGaN/GaN output power limiting LNA MMIC chip.

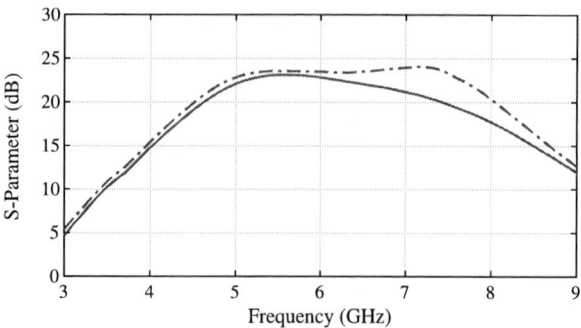

Fig. 4. Measured small-signal gain (S_{21}) of the output power limiting LNA (solid line) and of the LNA MMIC (dash-dotted line) at $V_{ds} = 16\,V$ and $I_{ds} = 50\,mA$.

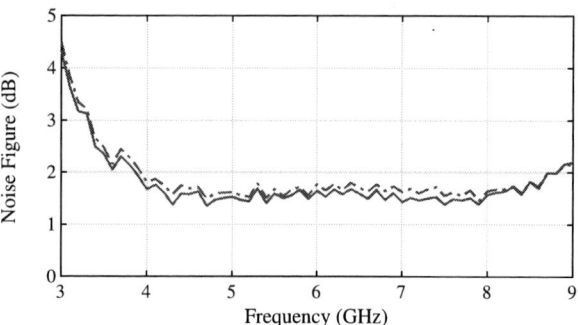

Fig. 5. Measured noise figure of the output power limiting LNA (solid line) and of the LNA MMIC (dash-dotted line) at $V_{ds} = 16\,V$ and $I_{ds} = 50\,mA$.

III. Output power limiting LNA MMIC

Due to the rugged LNA MMIC, an input attenuator is not necessary to protect the LNA in the overdrive condition. The LNA MMIC is rugged enough, however, it could not be sufficient to protect the subsequent stages. Furthermore, it is known from Friis' formula that the noise figure would be degraded by the attenuation inherent to a limiting circuit. Therefore, to limit the output power of an LNA in overdrive condition without degrading the small-signal gain, noise figure and the output power in linear range, the output power limiting LNA MMIC was developed.

Fig. 2 shows the circuit concept. The output attenuator connected to the rugged LNA MMIC output consists of a $12 \times 50\,\mu m$ HEMT in common-gate configuration, with resistors connected to drain and source providing 0 V DC voltage. If the LNA input is subject to high overdrive power, the first amplifier's stage gate supply voltage will drop since gate currents flow through the high ohmic DC feed resistance. This voltage change is used to switch the gate of the output attenuator in output power overdrive condition. The HEMT is conductive in small-signal operation, providing negligible attenuation. Once the output power increases and a DC gate current starts to flow at the first LNA stage, the output attenuator is switched off and limits the output signal. The output power limiting LNA MMIC requires no external control voltages and is completely independent, due to the fact that only the existing gate supply voltage of the first amplifier stage V_{G1} switches the output attenuator. Since for limiting merely a HEMT is used as a switch, the output power limiting LNA concept can be integrated as an MMIC with the dimension of $4.35 \times 2.15\,mm^2$, see Fig. 3.

IV. Measurement Results

In order to verify the concept, an output power limiting LNA MMIC was fabricated along with the same rugged LNA MMIC in conventional design without the limiter circuit. To characterize the small-signal parameters, a bias point of $V_{ds} = 16\,V$ and $I_{ds} = 50\,mA$ was selected for both stages, whereby the gain (S_{21}) and noise figure were measured, shown in Fig. 4 and Fig. 5, respectively. The solid lines refer to the output power limiting LNA MMIC, while the dash-dotted lines are the measurement results for the conventional rugged LNA MMIC. By integration of the output attenuator, the output power limiting LNA MMIC achieves the same bandwidth performance for the small-signal gain up to 6 GHz, providing a maximum gain of 23 dB at 5.5 GHz. The slight reduction in bandwidth and gain above 6 GHz is caused by the marginal change of the chip design, which was necessary to integrate the output attenuator and could be removed through an optimized redesign. Due to the output-side limiting concept, the noise figure remains unchanged, as shown in Fig. 5. In summary,

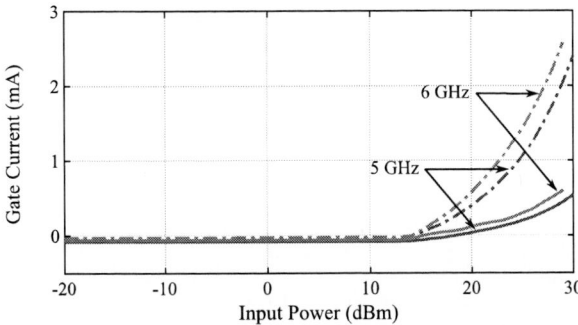

Fig. 6. Measured DC bias gate current of the first stage as a function of input power of the output power limiting LNA (solid line) and of the LNA MMIC (dash-dotted line) at 5 GHz and 6 GHz.

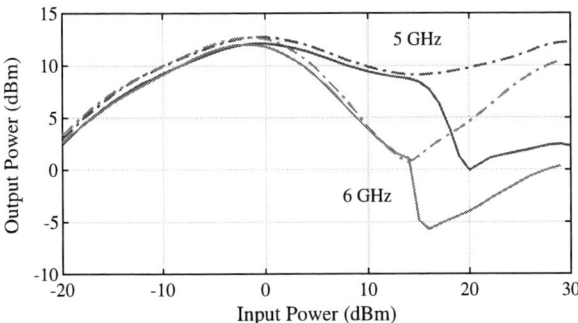

Fig. 7. Measured output power as a function of input power of the output power limiting LNA (solid line) and of the LNA MMIC (dash-dotted line) at 5 GHz and 6 GHz.

high small-signal gain and good noise properties in the C-band frequency range are achieved.

LNA ruggedness [3] and output power limiting is controlled by the flow of forward gate current under high input overdrive condition. As the input power increases, the gate current of the first stage starts to flow as soon as the input power exceeds about 14 dBm as shown in Fig. 6. The respective measured output powers as a function of input power are shown in Fig. 7, comparing the output power limited LNA MMIC (solid line) performance to the conventional rugged LNA MMIC (dash-dotted line) at the frequencies 5 GHz and 6 GHz. The limiting circuit reduces the output power in linear operation by only about 0.5 dB, as already observed in the small-signal gain measurement, shown in Fig. 4. Once the gate current of the first LNA stage starts flowing and the respective gate DC voltage drops, the output attenuator's HEMT is switched off. Thus, an additional attenuation of about 9.8 dB and 6.7 dB at the frequencies 5 GHz and 6 GHz, respectively, is achieved by the output attenuator alone, see Fig. 8. Thus, the output power limiting LNA MMIC provides an additional attenuation of about 12 dB and 18 dB as compared to the rugged LNA MMIC alone. Hence, the effectiveness and potential of the proposed output power limiting LNA MMIC concept is proven.

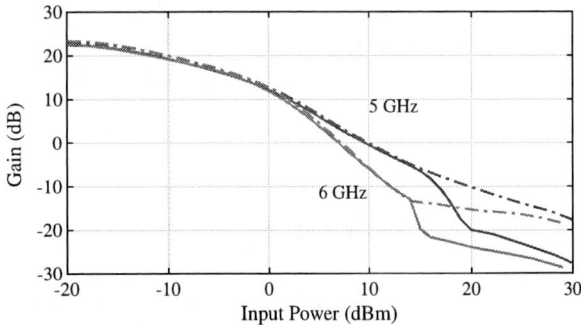

Fig. 8. Measured gain as a function of input power of the output power limiting LNA (solid line) and of the LNA MMIC (dash-dotted line) at 5 GHz and 6 GHz.

V. Conclusion

An on-chip circuit concept is presented to limit the output power of a rugged GaN LNA under input overdrive by using an output attenuator. The implementation reveals that the concept is capable of achieving an additional attenuation up to 18 dB under overdrive, leaving noise figure and small-signal gain virtually unchanged.

Acknowledgment

The authors thank our colleagues from BTU and FBH that assisted the research. Further thanks go to Ralf Doerner, Steffen Schulz and Jens Schmidt from FBH for performing the measurements.

References

[1] S. Colangeli, A. Bentini, W. Ciccognani, E. Limiti, and A. Nanni, "GaN-Based Robust Low-Noise Amplifiers," *IEEE Transactions on Electron Devices*, vol. 60, no. 10, pp. 3238–3248, Oct. 2013.

[2] C. Andrei, O. Bengtsson, R. Doerner, S. A. Chevtchenko, and M. Rudolph, "Robust Stacked GaN-based Low-Noise Amplifier MMIC for Receiver Applications," in *2015 IEEE MTT-S International Microwave Symposium*, May 2015, pp. 1–4.

[3] M. Rudolph, R. Behtash, R. Doerner, K. Hirche, J. Würfl, W. Heinrich, and G. Tränkle, "Analysis of the Survivability of GaN Low-Noise Amplifiers," *IEEE Transactions on Microwave Theory and Techniques*, vol. 55, no. 1, pp. 37–43, Jan. 2007.

[4] P. Schuh and R. Reber, "Robust X-Band Low Noise Limiting Amplifiers," in *2013 IEEE MTT-S International Microwave Symposium Digest (MTT)*, June 2013, pp. 1–4.

[5] C. Yağbasan and A. Aktuğ, "Robust X-Band GaN LNA with Integrated Active Limiter," in *2018 48th European Microwave Conference (EuMC)*, Sept. 2018, pp. 1205–1208.

[6] R. Rieger, A. Klaassen, P. Schuh, and M. Oppermann, "GaN Based Wideband T/R Module for Multi-Function Applications," in *2015 European Microwave Conference (EuMC)*, Sept. 2015, pp. 514–517.

[7] E. Kaule, C. Andrei, S. Gerlich, R. Doerner, and M. Rudolph, "Limiting the Output Power of Rugged GaN LNAs," in *2019 49th European Microwave Conference (EuMC)*, Oct. 2019, pp. 794–796.

[8] S. A. Chevtchenko, P. Kurpas, R. Lossy, and J. Würfl, "Investigation and Reduction of Leakage Current Associated with Gate Encapsulation by SiN$_x$ in AlGaN/GaN HFETs," in *2011 International Conference on Compound Semiconductor Manufacturing Technology, CS MANTECH 2011*, May 2011.

Proceedings of the 16th European Microwave Integrated Circuits Conference

A highly linear 79 GHz Low-Noise Amplifier for Civil-Automotive Radars in 22 nm FD-SOI CMOS with -6 dBm iP$_{1dB}$ and 5 dB NF

Songhui Li[#1], David Fritsche[#], Laszlo Szilagyi[#], Xin Xu[#], Quang Huy Le[*], Defu Wang[*], Thomas Kämpfe[*], Corrado Carta[#], Frank Ellinger[#,$]

[#]Chair for Circuit Design and Network Theory, TU Dresden, Dresden, Germany
[*]Fraunhofer IPMS, Dresden, Germany
[$]CeTi, Center for Tactile Internet, Dresden, Germany
[1]songhui.li@tu-dresden.de

Abstract—This paper presents a highly linear 79 GHz differential low-noise amplifier (LNA) for civil-automotive radars operating at the predefined frequency range from 77 GHz to 81 GHz. The circuit is optimized for frequency-modulated continuous-wave (FMCW) radar application, which typically require a very high input-referred 1 dB-compression point (iP$_{1dB}$). A reconfigurable differential common-source stage with capacitive neutralization is employed together with a common-gate stage in cascode configuration as the core of the LNA. The performance of the circuit can be easily adjusted within the gain-NF-P$_{1dB}$ trade-off boundaries by changing the voltage at the back-gate terminal of the common-source stage, thus tailored to the application specific requirements. Passive baluns are placed at input and output to characterize the differential circuit with the available single-ended laboratory instrumentation. The LNA is implemented in a 22 nm FD-SOI CMOS technology. Its core is very compact with an area of 0.04 mm^2. The fabricated chip is experimentally characterized in the lab, and it shows a peak gain of 8.7 dB at 80 GHz. From 75 GHz to 85 GHz, the measured input referred P$_{1dB}$ (iP$_{1dB}$) is about -6 dBm, and the minimum noise figure (NF) is 5 dB. Compared with the state-of-the-art for LNAs operating in a similar frequency range, the presented circuit shows the highest iP$_{1dB}$ and has the most compact circuit core, together with an excellent NF and a moderate gain, resulting in the best figure-of-merit.

Keywords—CMOS integrated circuits, Silicon-on-insulator, broadband amplifiers, differential amplifiers, Low-noise amplifiers, millimeter wave integrated circuits.

I. INTRODUCTION

The increasing interest in the research and development of autonomous vehicles motivates significantly the study of sensing and radar systems, which are indispensable in self-driving cars. Compared with other sensing techniques like camera or lidar, the radar is the most reliable one with no dependency on visibility and high adaptability to bad weather. Continuous-wave (CW) radar is preferred since it supports constant envelope modulations, maximizing the transmitter efficiency. In CW radars, the transmitter spillover to receiver is inevitable, since the transmitter and receiver are working concurrently and in the same frequency band. The magnitude of the leakage from transmitter to receiver can be several orders higher than the target reflections. This constrains severely the linearity of the receiver and the

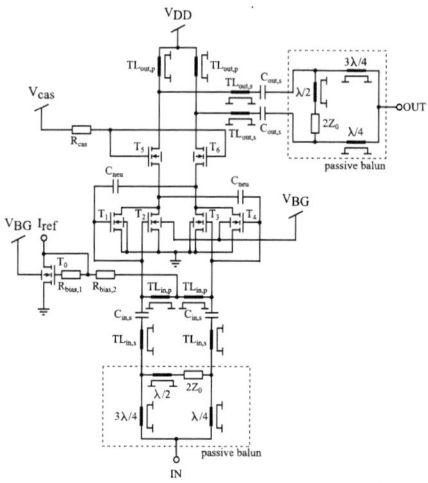

Fig. 1. Schematic of the differential amplifier with on-chip passive baluns at input and output.

dynamic range of the analog-to-digital converter [1]. To reduce the effect of the transmitter spillover, system-level cancellation techniques have been introduced such as in [1]. At the circuit level, a highly linear receiver, in particular an LNA, with high P$_{1dB}$ is desired, because it has a high tolerance on the transmitter spillover. Currently, novel LNA designs in advanced CMOS technologies operating around 79 GHz have been published. The 28 nm CMOS design in [2] has shown a gain of 29.6 dB. NF of 4 dB with DC power consumption of 10.8 mW have been achieved in 22 nm FinFET in [3]. The state-of-the-art of figure-of-merit (FoM) has been reported in [4] with a 22 nm Fully-Depleted Silicon-On-Insulator (FD-SOI) technology. However, the iP$_{1dB}$s in these designs are less than -20 dBm, making them not well suited for CW radar applications.

In this paper, we present the design of a differential LNA in a 22 nm FD-SOI technology (22FDX$^®$) with reported f_t / f_{max} of 347 GHz / 371 GHz [5]. The circuit is designed for a FMCW civil-automotive radar application with regulated

3–4 April 2022, London, UK

frequency range from 77 GHz to 81 GHz. To characterize the differential circuit, passive baluns are placed at input and output. With the effect of the balun, the characterized circuit shows the state-of-the-art of P_{1dB}, comparable gain and NF. By introducing the back-gate control, the trade-off between gain, NF and P_{1dB} is adjustable after fabrication, offering the flexibility to tailor the LNA performance to the application requirements.

II. CIRCUIT IMPLEMENTATION

The schematic of the LNA is shown in Fig. 1. A reconfigurable differential common-source stage ($T_{1,2}$ and $T_{3,4}$) as in [6] is implemented. As one of the key features of the 22FDX® technology, the threshold voltage of a transistor can be tuned with the voltage of the back-gate (BG) terminal. Exploiting this feature, the class of operation of the transistors $T_{2,4}$ can be adjusted while maintaining the same DC gate-source voltage as $T_{1,3}$. Thus the gate of all transistors $T_{1...4}$ can be joined together and no additional large DC-blocking capacitor is required, which makes the circuit significantly more compact. It is noteworthy that the drain current of the transistors are less sensitive to the voltage change of the BG than that of the gate. Compared to the normal gate bias, the BG biasing makes the circuit more tolerant to bias voltage variations. Enjoying this feature, $T_{2,4}$ can be easily biased either in the same region as $T_{1,3}$ or more towards their cut-off region. Analogous to a combination of amplifiers operating in different classes such as a Doherty amplifier, high linearity can be expected from this structure. Additionally, the trade-off between gain, NF, and P_{1dB} can be adjusted even after fabrication.

A neutralization capacitor C_{neu} is cross-connected to the differential common-source stage to cancel the parasitic gate-to-drain capacitance and improves the reverse isolation. Transistors $T_{5/6}$ are added as common-gate stage cascodes in order to further strengthen the reverse isolation, easing the design of the input and output matching networks. Resistor R_{cas} is used to improve the common-mode rejection. The transistor size is chosen such, that the differential source impedance of $100\,\Omega$ is included in the noise-circle (NC) with minimum noise, which is at the same time the input matching, target for optimizing the gain. As a part of the input matching network, the parallel stubs $TL_{in,p}$ are used as well for DC biasing. Aiming at a high P_{1dB}, the output matching network is optimized for 1-dB-compression-point load-pull. Parallel stubs $TL_{out,p}$ are introduced for DC supply. The virtual ground of the differential structure and its symmetric parasitics ease the routing of the supply power.

In order to characterize the differential circuit with the available ground-signal-ground (GSG) RF probes up to 110 GHz, an on-chip passive balun based on the Wilkinson divider [7] is placed at input and output. The balun consists of a $100\,\Omega$-resistor, $\frac{\lambda}{4}$, $\frac{\lambda}{2}$ and $\frac{3\lambda}{4}$ transmission lines as depicted in Fig. 1. The 180° phase offset is achieved by the different transmission line length of $\frac{\lambda}{4}$ and $\frac{3\lambda}{4}$ in the two branches.

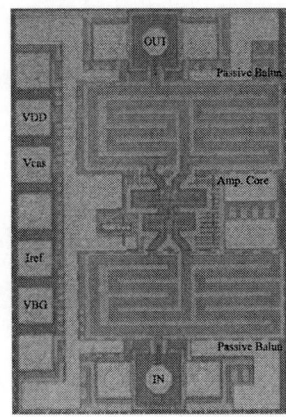

Fig. 2. Photograph of the LNA chip (total area: $600\,\mu m \times 850\,\mu m$, core area: $220\,\mu m \times 180\,\mu m$).

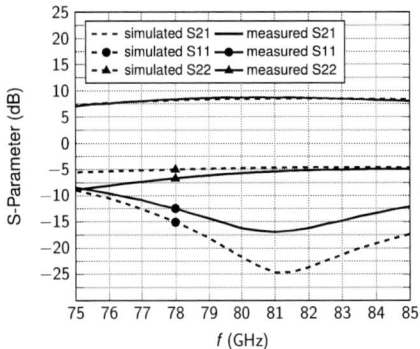

Fig. 3. S-parameter of the LNA with baluns while $V_{BG} = 0\,V$.

The amplitude is balanced by the $\frac{\lambda}{2}$ transmission line and the isolation resistor of $100\,\Omega$.

The chip photo is shown in Fig. 2. The whole chip size is $600\,\mu m \times 850\,\mu m = 0.51\,mm^2$, where the most area is taken by pads and the passive baluns. The core of the differential amplifier including matching networks is only $220\,\mu m \times 180\,\mu m = 0.04\,mm^2$. The RF input and output are at the bottom and top edge of the chip. Four supply and DC biasing pads (V_{DD}, V_{cas} I_{ref} and V_{BG}) are placed at the right edge and the corresponding DC voltage/current are fed to the circuit through zero-Ohm transmission lines.

III. MEASUREMENT RESULTS

The circuit is experimentally characterized in the lab. It is biased with $V_{DD} = 1.6\,V$, $V_{cas} = 1.2\,V$ and $I_{ref} = 0.4\,mA$. By setting different V_{BG}, from 0 V to 2 V the circuit performance can be tuned continuously. At these two boundaries, the average DC currents are 22 mA and 12 mA with V_{BG} of 0 V and 2 V, respectively.

The S-parameters are characterized with a network analyzer and corresponding frequency up-converter. The S-parameter measurements of the differential LNA with passive baluns, while $V_{BG} = 0\,V$ and $V_{BG} = 2\,V$, are shown

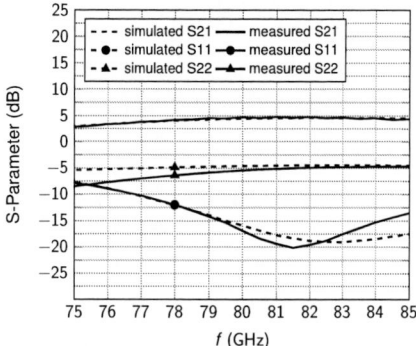

Fig. 4. S-parameter of the LNA with baluns while $V_{BG} = 2$ V.

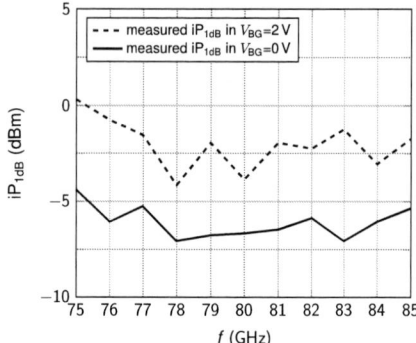

Fig. 5. Measured iP$_{1dB}$ of the LNA with baluns while $V_{BG} = 0$ V and $V_{BG} = 2$ V.

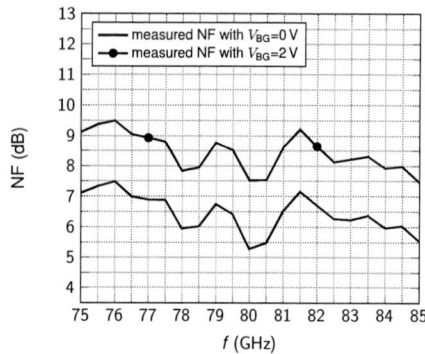

Fig. 6. Measured NF of the differential LNA with balun while $V_{BG} = 0$ V and $V_{BG} = 2$ V.

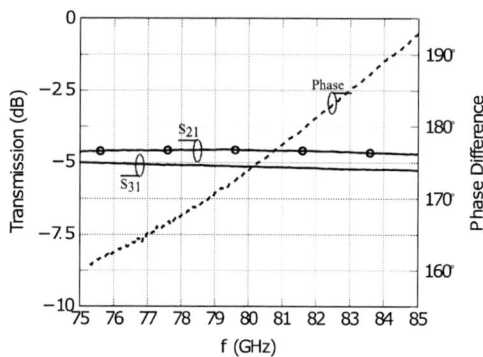

Fig. 7. Measured parameters of the baluns: transmission and phase difference from 75 GHz to 85 GHz.

in Fig. 3 and Fig. 4 in the frequency range from 75 GHz to 85 GHz. The simulation results of the LNA with baluns are compared correspondingly as well. The peak gain is 8.7 dB at 80 GHz and 4.8 dB at 81 GHz with $V_{BG} = 0$ V and $V_{BG} = 2$ V, respectively. In this frequency range, the gain drop is less than 2.5 dB. The input matching is better than -10 dB and the output matching is better than -5 dB. The simulation and measurement are consistent. Remarkably, the input and output matchings are not significantly affected by changes in V_{BG}.

To characterize the P$_{1dB}$, the input signal is generated with a frequency multiplier succeeded by a tunable attenuator, while the output signal is connected to a power meter. The extracted iP$_{1dB}$ of the characterized chip from 75 GHz to 85 GHz at the boundary values of V_{BG} are shown in Fig. 5. The measured iP$_{1dB}$ at $V_{BG} = 0$ V and $V_{BG} = 2$ V is frequency dependent, around -6 dBm and -2.5 dBm, respectively.

The noise characterization is performed with the W-band measurement setup described in [8]. A calibrated noise source is used initially for the noise receiver characterization, and the cold-source technique is applied for the measurement. The measured NF from 75 GHz to 85 GHz is shown in Fig. 6. With V_{BG} of 0 V, the minimum NF is 5.5 dB at 85 GHz, while at $V_{BG} = 2$ V the minimum NF is 7.5 dB at 85 GHz.

The used balun was also fabricated as test structure and

characterized independently. The results are shown in Fig. 7. From 75 GHz to 85 GHz, the transmission of the balun is between -4.6 dB and -5.1 dB, while the phase difference is from 175° to 185°. Aside from the 3 dB coming from power division, the insertion loss of the balun is around -1.6 dB and -2.1 dB in the two branches.

To estimate the performance of the differential circuit, the effect of the passive baluns was de-embedded as in [9]. The gain and NF after de-embedding with $V_{BG} = 0$ V are shown in Fig. 8. The simulation results of the differential LNA core without baluns are depicted here for comparison. The de-embedded peak gain is 12.4 dB at 80.5 GHz. The minimum NF is around 4 dB. An iP$_{1dB}$ around -8 dBm is expected.

IV. CONCLUSION

A highly linear 79 GHz differential LNA was designed and implemented in GLOBALFOUNDRIES 22FDX® FD-SOI technology. Its high P$_{1dB}$ is desired by civil-automotive CW radar applications because of the high tolerance on the transmitter-to-receiver leakage. To achieve the high P$_{1dB}$, a reconfigurable highly linear LNA core was designed. Common-source stages operating in different classes are combined for high linearity, while common-gate stages in

Table 1. Comparison with state-of-the-art LNAs around 80 GHz

Ref.	[1]	[2]	[3]	[4]	This work direct measurement	This work with balun de-embedded
G_{peak} (dB)	16.5	29.6	20	20	8.7	12.5
f_{centre} (GHz)	80	82.3	73.5	77	80	80.5
NF_{min} (dB)	5.2	6.4	4	4.6	5.5	4
iP_{1dB} (dBm)	−15.5	−28.1	−22.8	−27.4	−6	−8
P_{DC} (mW)	26.1	31.3	10.8	9	35.2	35.2
Core area (mm^2)	0.14	0.25	0.15	0.35	0.04	0.04
FoM	0.021	0.013	0.032	0.011	0.021	0.053
Tech.	28 nm CMOS	28 nm CMOS	22 nm FinFET	22 nm FD-SOI	22 nm FD-SOI	22 nm FD-SOI

$$FoM = \frac{Gain[lin.] \times iP_{1dB}[mW]}{P_{DC}[mW] \times (NF[lin.]-1)}$$

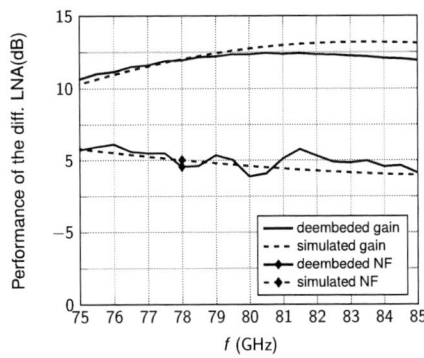

Fig. 8. Performance of the differential LNA with balun effect de-embedded with different V_{BG} values.

cascode configuration are implemented to increase the reverse isolation. A load-pull output matching strategy was taken place for a high P_{1dB}. With the proper size of the transistors and the input matching strategy, the circuit was designed also with a very good NF and moderate gain. Additionally, the trade-off of the gain, NF, and iP_{1dB} of the presented circuit can be adjusted easily through BG voltage after fabrication to fit application specifications. Passive baluns were placed for characterization at the input and output of the differential circuit. The direct measurement results, as well as the results after de-embedding the balun effects are compared with state-of-the-art for LNAs around 80 GHz in Table 1. The presented circuit shows the highest iP_{1dB} both with or without the passive balun. The core area of the differential amplifier is the smallest reported. The directly measured NF is moderate towards comparable with the state-of-the-art, after de-embedding. Compared to the FoMs presented in [3] and [4] we have added the iP_{1dB} and removed the bandwidth since the frequency range of civil-automotive radars is predefined. The FoM is shown in linear scale. It can be seen that the FoM of the presented circuit is the highest reported.

ACKNOWLEDGMENT

This work was funded via subcontract from GLOBALFOUNDRIES Dresden Module One within the framework Important Project of Common European Interest (IPCEI), by the Federal Ministry for Economics and Energy (BMWi) and by the State of Saxony.

REFERENCES

[1] A. Medra, D. Guermandi, K. Vaesen, S. Brebels, A. Bourdoux, W. Van Thillo, P. Wambacq, and V. Giannini, "An 80 GHz Low-Noise Amplifier Resilient to the TX Spillover in Phase-Modulated Continuous-Wave Radars," *IEEE J. Solid-State Circuits*, vol. 51, no. 5, pp. 1141–1153, May 2016.

[2] M. Vigilante and P. Reynaert, "On the Design of Wideband Transformer-Based Fourth Order Matching Networks for *E*-Band Receivers in 28-nm CMOS," *IEEE Journal of Solid-State Circuits*, vol. 52, no. 8, pp. 2071–2082, 2017.

[3] W. Shin, S. Callender, S. Pellerano, and C. Hull, "A Compact 75 GHz LNA with 20 dB Gain and 4 dB Noise Figure in 22nm FinFET CMOS Technology," in *2018 IEEE Radio Frequency Integrated Circuits Symposium (RFIC)*, 2018, pp. 284–287.

[4] L. Gao, E. Wagner, and G. M. Rebeiz, "Design of E- and W-Band Low-Noise Amplifiers in 22-nm CMOS FD-SOI," *IEEE Transactions on Microwave Theory and Techniques*, vol. 68, no. 1, pp. 132–143, 2020.

[5] S. N. Ong, S. Lehmann, W. H. Chow, C. Zhang, C. Schippel, L. H. K. Chan, Y. Andee, M. Hauschildt, K. K. S. Tan, J. Watts, C. K. Lim, A. Divay, J. S. Wong, Z. Zhao, M. Govindarajan, C. Schwan, A. Huschka, A. Bellaouar, W. LOo, J. Mazurier, C. Grass, R. Taylor, K. W. J. Chew, S. Embabi, G. Workman, A. Pakfar, S. Morvan, K. Sundaram, M. T. Lau, B. Rice, and D. Harame, "A 22nm FDSOI Technology Optimized for RF/mmWave Applications," in *2018 IEEE RF Integr. Circuits Symp. (RFIC)*, June 2018, pp. 72–75.

[6] S. Li, M. Cui, X. Xu, L. Szilagyi, C. Carta, W. Finger, and F. Ellinger, "An 80 GHz Power Amplifier with 17.4 dBm Output Power and 18% PAE in 22 nm FD-SOI CMOS for Binary-Phase Modulated Radars," in *2020 Asia-Pacific Microwave Conference (APMC)*, December 2020.

[7] Jong-Sik Lim, Hoe-Sung Yang, Young-Taek Lee, Sungwon Kim, Kwang-Seok Seo, and Sangwook Nam, "A 77 GHz CPW Balun using Wilkinson Structure," in *34th Eur. Microw. Conf. (EuMC), 2004.*, vol. 1, Oct 2004, pp. 377–380.

[8] Q. H. Le, D. K. Huynh, D. Wang, Z. Zhao, S. Lehmann, T. Kämpfe, and M. Rudolph, "W-Band Noise Characterization with Back-Gate Effects for Advanced 22nm FDSOI mm-Wave MOSFETs," in *2020 IEEE Radio Frequency Integrated Circuits Symposium (RFIC)*, 2020, pp. 131–134.

[9] O. Garcia-Perez, V. Gonzalez-Posadas, L. E. Garcia-Munoz, and D. Segovia-Vargas, "A De-embedding Method to Characterize Differential Amplifiers Using Passive Baluns," in *2011 41st European Microwave Conference*, 2011, pp. 377–380.

Proceedings of the 16th European Microwave Integrated Circuits Conference

Highly Linear D-Band Low-Noise Amplifier with 8.5dB Noise Figure in InP-DHBT Technology

M. Hossain[1], R. Doerner[1], H. Yacoub[1], T.K. Johansen[2], W. Heinrich[1], V. Krozer[1]

[1]Ferdinand-Braun-Institut, Leibniz-Institut für Höchstfrequenztechnik (FBH), Germany

[2]Technical University of Denmark (DTU), Denmark

maruf.hossain@fbh-berlin.de

Abstract — This paper presents a D-band low noise amplifier (LNA) using an 0.5 µm InP-DHBT technology. The LNA circuit design is based on a 2-way combined cascode unit power cell. The measured LNA exhibits 10 to 16 dB small signal gain and 11 to 8.5 dB noise figure (NF) in the frequency range from 140 to 170 GHz. The dc power consumption is only 103 mW and results in a power-added efficiency (PAE) of 11 % at 146 GHz. The output 1-dB compression point (OP_1dB) reaches 10 dBm at 146 GHz. The chip area is only 1.6 × 1.1 mm². To the best knowledge of the authors, the performance combination in terms of low NF, high linearity as well as PAE beyond 140 GHz is unique so far.

Keywords — InP double heterojunction bipolar transistor (DHBT), monolithic microwave integrated circuit (MMIC), low noise amplifier (LNA), noise figure (NF), transferred-substrate process (TS).

I. INTRODUCTION

Nowadays, high-speed communication and imaging systems require high-performance circuit functionalities at 100 GHz and beyond. Performance targets include in particular a combination of low noise figure, high linearity (OP_1dB), and high efficiency. The low noise amplifier (LNA) is a key building block for such applications in the receiver frontend. These characteristics are required in order to deal with strong interferer signals as well as to achieve high sensitivity of the receiver. Among the frequencies beyond 100 GHz, presently particularly the D-band (ranging from 110-170 GHz) is of interest, e.g., for wireless communications [1, 2].

In the past years, significant progress has been made in pushing low noise amplifiers to the mm-wave frequency range, using various semiconductor technologies. Several low noise amplifiers for D-band have been published [3, 4, 5], which show relatively good noise figure, but they suffer from low values for linearity as well as efficiency. Furthermore, so far InP based MMICs provide a different kind of transceivers building blocks at D-band such as power amplifier (PA), Up/down-converters, switches, sources. Having all of these system components in one technology allows compact integration. The main motivation behind this work is to demonstrate this, showing a highly linear and high-performance LNA.

The work presented here serves the communication system being developed in the frame of the EU project 'ULTRAWAVE'. The project aims to develop a dense ultra-capacity backhaul that enables 5G cell densification by exploiting the frequency range from 141 GHz to 150 GHz in D-band [6].

The paper is organized as follows: Sec. II briefly describes the technology and Sec. III discusses the circuit design. Section IV presents the realized circuit and the measurement results, followed by a discussion, while Sec. V contains the conclusions.

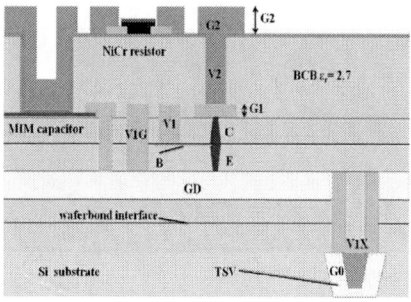

Fig. 1. Vertical layer stack of InP-DHBT technology.

II. INP-DHBT TECHNOLOGY

The low noise amplifier circuit is fabricated using the FBH transferred-substrate (TS) InP-DHBT MMIC process. Fig. 1 shows the vertical layer stack of this technology. The transistors have 0.5 µm wide emitters and achieve ft/fmax values of ~360/490 GHz with BVCEO = 4 V. The BCB stack above the Si substrate includes three gold metal layers (G1, GD, G2) with 2 µm, 2.5 µm, and 4.5 µm thickness, respectively, MIM capacitors with a sheet capacitance of 0.3 fF/µm² and NiCr thin film resistors (sheet resistance of 25 Ω/sq.). In order to suppress substrate modes, a high resistivity Si substrate with through-silicon vias (TSVs) is used as the host wafer for the wafer bonding process. For further details of the MMIC process see [7].

III. CIRCUIT DESIGN

The schematic diagram of the low noise amplifier including its bias network is shown in Fig. 2. The circuit is designed based on cascode topology. The amplifier consists of two identical cascode power unit cells combined on-chip, in order to achieve sufficient gain in the D-band. Each cascode unit cell comprises a double-finger HBT with an emitter size

140

3–4 April 2022, London, UK

of 0.5 × 6 µm² and a ft/fmax of 360/490 GHz. (see Fig. 2). The input and output matching networks are designed to achieve highest small- and large-signal gain as well as low noise figure.

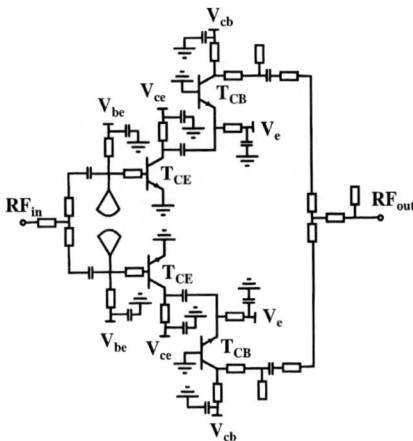

Fig. 2. Schematic diagram of the D-band LNA.

All the passive components, such as MIM capacitors and the NiCr on-chip resistors, are used for DC-coupling purposes as well as to enhance electrical stability. The splitter and the combiner at the input and the output of the LNA are designed to be identical. The circuit is realized in a thin-film microstrip line environment.

IV. MEASURED RESULTS

The chip photograph of the realized D-band LNA amplifier is shown in Fig. 3. The chip area is 1.6 × 1.1 mm². The circuit was characterized on-wafer under both small and large-signal excitation in two different setups. First, the small-signal performance was measured. For this purpose, a ground-signal-ground (GSG) 50 µm pitch WR-5 waveguide probe with a vector network analyzer were used. The system was calibrated applying the multiline Thru-Reflect-Line (mTRL) approach to place the reference planes at the probe pad.

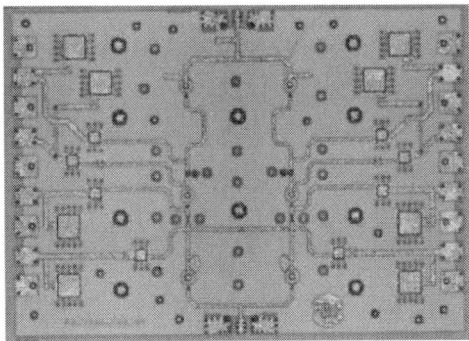

Fig. 3. Chip photograph of the D-band LNA.

Fig. 4 shows the measured (solid lines) and the simulated (dash lines) S-parameter results of the circuit. The forward small-signal gain S21 has an average value of more than 10 dB within 140-170 GHz. The measured input reflection coefficient S11 remains below -10 dB, while the output reflection coefficient S22 is -3 dB. The poor output matching performance is expected because of the inherent high output impedance of the cascode amplifier. But one can improve this by adding an output buffer stage which will not affect the noise figure, rather increase the gain of the LNA. However, one can clearly see that the measured and the simulated small-signal results agree well each other.

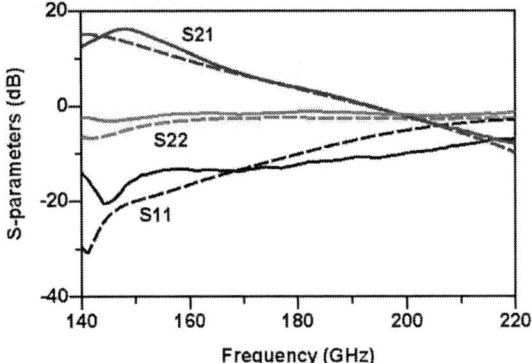

Fig. 4. Measured (solid) and simulated (dash) s-parameters of the D-band LNA.

The second step was to measure large-signal behavior of the circuit. The input power of the circuit is fed by a WR-5 module which includes a WR-10 ×6 multiplier, an electronic attenuator, an amplifier and a WR-5 doubler, followed by a ground-signal-ground (GSG) probe. To detect the output power of the circuit, a WR-5 GSG waveguide probe was connected to a WR5-to-WR10 taper, which connects to the input of a power sensor and an Erickson PM4 power meter. After that, the input and the output power at the probe tips is estimated by de-embedding the probe losses provided by the vendor.

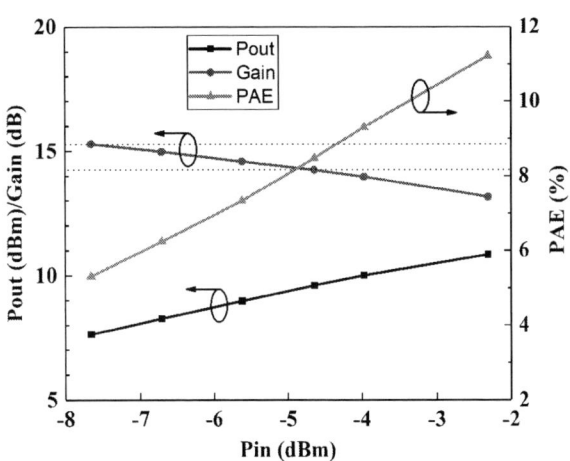

Fig. 5. Measured large-signal performances of the D-band LNA.

141

Table 1. State-of-the-art of D-band (110-170 GHz) low noise amplifier in different technologies.

Reference	Technology	Frequency [GHz]	Gain [dB]	NF [dB]	P_{DC} [mW]	P_{1dB} [dBm]	P_{sat} [dBm]	PAE [%]
[9]	130 nm SiGe HBT	108-150	12-27	5.5-7[a]	12	-	-	-
[10]	50 nm GaAs mHEMT	110-150	22[a]	3.4[d]	57.6	-	-	-
[13]	130 nm SiGe HBT	130	17.5	7.7	31.5	-	-	-
[8]	65nm CMOS	126	14.4	12	22.6	-	-	-
[11]	130 nm SiGe HBT	132-160	21	<9.5[b]	14.5	-	-	-
[12]	130 nm SiGe HBT	70-140	17-25	7-9[a]	54	-	-	-
[5]	GaN	107-114	28-31	7.5-9.5	3600	10.9[c]	18.5[c]	1.7[c]
This work	**500 nm InP DHBT**	**140-170**	**10-16**	**11-8.5**	**103**	**10**	**>11**	**11**

[a]estimated from graph, [b]simulation, [c]120 GHz, [d]mean noise figure

Fig. 5 provides the measured large-signal performance of the LNA at 146 GHz. For an input power of -2 dBm, an output power of 11 dBm is achieved with 103 mW DC power consumption, which corresponds to 11 % Power Added Efficiency (PAE). The output 1-dB compression point (OP1dB) is around 10 dBm. One can see from the curves that the saturated output power is not reached, which is due to limited available input power from the test setup.

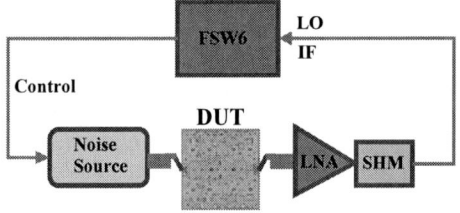

Fig.6. Noise figure (NF) measurement setup.

The on-wafer 50-Ohm noise figure measurement setup for the D-band in Fig. 6 comprises a Rohde & Schwarz FSW67 spectrum analyzer complemented by an external Radiometer Physics (RPG) G band mixer SMA-220, an RPG G band low-noise amplifier and an ELVA ISSN-06 D-band noise source. So, the frequency range for this setup effectively covers 140-170GHz.

The system calibration is performed as a two-step procedure employing waveguide and on-wafer through connections. The relatively large losses introduced by the waveguide to on-wafer transition between the noise source and the DUT results in an increased confidence interval for the results and are estimated to be ± 0.5 dB.

The noise figure (NF) measurements for the frequency range from 140 GHz to 170 GHz are plotted in Fig. 7. The measured NF shows nearly constant behaviour with frequency, with an average value of 9.5 dB and a variation of ±1 dB within this frequency range.

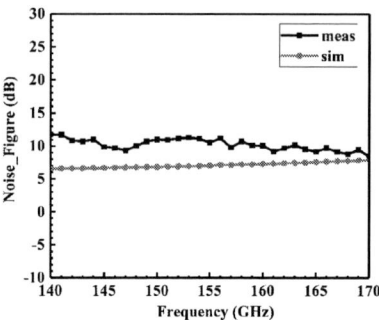

Fig. 7. Measured and simulated noise figure as a function of frequency.

As can be seen, measured and simulated NF values exhibit a difference of about 5 dB at the lower frequency end, while these deviations decrease continuously with growing frequency and above 160 GHz both values agree quite well. This can be explained by the noise modelling of the HBTs.

Table 1 benchmarks the performance of our circuit against other published D-band (110-170 GHz) LNAs in different technologies. The noise figure of our LNA is comparable with other published LNAs in different technologies. At the same time, our LNA shows highest linearity (OP$_{1dB}$) among the few data available and the same is true for PAE. Only LNA from [5] exhibits a slightly higher OP_1dB, but at 120 GHz frequency. Overall, although dedicated III-V HEMT technology achieves better noise figure performance in this frequency range, our LNA demonstrates unique performance when combining noise figure, linearity and power added efficiency.

V. CONCLUSION

This paper presents a D-band low noise amplifier, based on 500-nm transferred-substrate (TS) InP-DHBT technology. The LNA shows 11 to 8.5 dB noise figure from 140 to 170 GHz and the output 1-dB compression point (OP_1dB) is around 10 dBm. It delivers +11 dBm output power with 11 %

power added efficiency at 146 GHz. These results demonstrate the high performance a single-chip transceiver module realized using the InP-DHBT process at D-band frequencies can achieve.

ACKNOWLEDGMENT

The authors gratefully acknowledge financial support by the DLR project MIMIRAWE and EU H2020 project ULTRAWAVE (contract no. 762119). Also, this work benefitted from funding by the German Federal Ministry of Education and Research (BMBF) under the project reference 16FMD02 (Forschungsfabrik Mikroelektronik Deutschland).

REFERENCES

[1] A. Karakuzulu, M. H. Eissa, D. Kissinger and A. Malignaggi, "Full D-Band Transmit–Receive Module for Phased Array Systems in 130-nm SiGe BiCMOS," in IEEE Solid-State Circuits Letters, vol. 4, pp. 40-43, 2021.

[2] S. Koch, M. Guthoerl, I. Kallfass, A. Leuther and S. Saito, "A 120–145 GHz Heterodyne Receiver Chipset Utilizing the 140 GHz Atmospheric Window for Passive Millimeter-Wave Imaging Applications," in IEEE Journal of Solid-State Circuits, vol. 45, no. 10, pp. 1961-1967, Oct. 2010.

[3] E. Aguilar, A. Hagelauer, D. Kissinger and R. Weigel, "A low-power wideband D-band LNA in a 130 nm BiCMOS technology for imaging applications," 2018 IEEE 18th Topical Meeting on Silicon Monolithic Integrated Circuits in RF Systems (SiRF), Anaheim, CA, USA, 2018, pp. 27-29.

[4] R. Cleriti et al., "D-band LNA using a 40-nm GaAs mHEMT technology," 2017 12th European Microwave Integrated Circuits Conference (EuMIC), Nuremberg, Germany, 2017, pp. 105-108.

[5] R. Weber et al., "A Beyond 110 GHz GaN Cascode Low-Noise Amplifier with 20.3 dBm Output Power," 2018 IEEE/MTT-S International Microwave Symposium - IMS, Philadelphia, PA, USA, 2018, pp. 1499-1502.

[6] http://ultrawave2020.eu/

[7] N. G. Weimann, et al., "Transferred-Substrate InP/GaAsSb Heterojunction Bipolar Transistor Technology With fmax ~ 0.53 THz," in IEEE Transactions on Electron Devices, vol. 65, no. 9, pp. 3704-3710, Sept. 2018.

[8] K. Katayama, K. Takano, S. Amakawa, T. Yoshida and M. Fujishima, „14.4-dB CMOS D-band low-noise amplifier with 22.6-mW power consumption utilizing bias-optimization technique," *2016 IEEE International Symposium on Radio-Frequency Integration Technology (RFIT)*, Taipei, Taiwan, 2016, pp. 1-3.

[9] A. Ç. Ulusoy et al., "A SiGe D-Band Low-Noise Amplifier Utilizing Gain-Boosting Technique," in IEEE Microwave and Wireless Components Letters, vol. 25, no. 1, pp. 61-63, Jan. 2015.

[10] R. Weber, H. Massler and A. Leuther, "D-band low-noise amplifier MMIC with 50 % bandwidth and 3.0 dB noise figure in 100 nm and 50 nm mHEMT technology," 2017 IEEE MTT-S International Microwave Symposium (IMS), Honolulu, HI, USA, 2017, pp. 756-759.

[11] B. Zhang, Y.-Z. Xiong, L. Wang, S. Hu, and L.-W. Li, "Gain-enhanced 132–160 GHz low-noise amplifier using 0.13 m SiGe BiCMOS," Electron. Lett., vol. 48, no. 5, pp. 257–259, Mar. 2012.

[12] G. Liu and H. Schumacher, "Broadband Millimeter-Wave LNAs (47–77 GHz and 70–140 GHz) Using a T-Type Matching Topology," in IEEE Journal of Solid-State Circuits, vol. 48, no. 9, pp. 2022-2029, Sept. 2013.

[13] B. Zhang et al., "130-GHz gain-enhanced SiGe low noise amplifier," 2010 IEEE Asian Solid-State Circuits Conference, Beijing, China, 2010, pp. 1-4.

C-Band Low-Noise Amplifier MMIC with an Average Noise Temperature of 44.5 K and 24.8 mW Power Consumption

Felix Heinz[1], Fabian Thome, Arnulf Leuther, Oliver Ambacher

Fraunhofer IAF, Fraunhofer Institute for Applied Solid State Physics, Germany
[1] felix.heinz@iaf.fraunhofer.de

Abstract — This paper reports on a 4–8 GHz (C-band) low-noise amplifier monolithic microwave integrated circuit in 50 nm metamorphic high electron mobility transistor technology aimed for the use in large scale systems. The two-stage design exhibits a small-signal gain of 31 dB and a room temperature average noise temperature of 44.5 K between 4 and 8 GHz with a minimum of 38.4 K. The power consumption of the amplifier at optimal noise bias is only 24.8 mW. To the best of the authors' knowledge, the amplifier has the lowest noise temperature reported among monolithic microwave integrated circuits that do not utilize an off-chip input matching network in C-band.

Keywords — High-electron-mobility transistors (HEMTs), metamorphic HEMTs (mHEMTs), low-noise amplifiers (LNAs), monolithic microwave integrated circuits (MMICs), C-band.

Fig. 1. Chip micrograph of the C-band MMIC in 50 nm metamorphic HEMT technology. The chip occupies an area of $(2.5 \times 1)\,\text{mm}^2$.

I. INTRODUCTION

The C-band (4–8 GHz) is widely used in manifold applications such as radar, radio-astronomy, quantum computing and in parts for mobile communication. These systems utilize broadband low-noise amplifiers (LNAs) directly attached to the receiver input to keep the system noise temperature at the level of the LNA.

Nowadays, the lowest noise temperatures both at room temperature and under cryogenic conditions are achieved by high electron mobility transistors (HEMTs) utilizing an InGaAs-channel. Such HEMTs can be realized either directly lattice matched on InP substrate or on GaAs substrate where a metamorphic buffer is used to adapt the different lattice constants (metamorphic HEMTs). Besides the technology, circuit design mainly determines the noise performance of an amplifier.

State-of-the-art noise performance at C-band is achieved by hybrid amplifiers [1], [2] or by hybrid-integrated circuits which utilize an off-chip input matching network [3], [4]. The lower noise temperature of hybrids compared to fully integrated solutions at these frequencies originates from the low-loss off-chip substrates that can be used for matching, especially for the input matching of the first amplifier stage. The matching networks of monolithic designs usually have higher losses at these frequencies due to the thickness of the substrate which consequently limits the width of transmission lines.

However, MMIC processes offer the possibility to easily shrink the size of the amplifiers compared to

hybrid approaches. Furthermore, monolithic co-integration of different circuits on a single multi-functional chip is only possible for monolithic designs. This is of special importance for large systems that might need to scale towards a high number of LNAs, as for example multiple-input-multiple-output (MIMO) receiver arrays or read-out circuits for future quantum computing systems with a high number of quantum-bits (Qubits). When a fully error corrected quantum-computer shall be realized the number of Qubits needs to be scaled to approximately one million. The read-out of such a large system will need novel cryogenic multiplexing techniques and still the number of (cryogenic) read-out LNAs will need to scale according to the number of Qubits. Therefore, further scaling of a quantum computer towards such high number of Qubits seems hardly feasible with today's hybrid solutions. Thus, a MMIC solution can help to ease the scaling of future quantum computing systems, since they are smaller in size and can be integrated into larger multi-functional chips.

This work focuses on the design of an LNA MMIC in 50 nm metamorphic HEMT (mHEMT) transistor technology [5] targeting the C-band. Besides the requirement for lowest noise temperature, the power consumption of the MMIC shall be minimized to enable easier scaling of a large read-out array. Circuit design techniques that allow for low noise temperature and low power consumption are discussed. Finally, a detailed investigation of the power-consumption of the MMIC taking into account how it trades with the noise performance is presented.

II. 50-NM METAMORPHIC HEMT TECHNOLOGY

This section provides a brief overview on the 50 nm metamorphic HEMT technology [5] of the presented LNA. A linearly graded InAlGaAs buffer layers adapts the lattice constant of the 100 mm semi-insulating GaAs wafers to the value of InP. On top, the $In_{0.8}Ga_{0.2}As/In_{0.53}Ga_{0.47}As$ composite-channel is formed between $In_{0.52}Al_{0.48}As$ barrier-layers. A saturation doped $In_{0.53}Ga_{0.47}As$ cap-layer allows for low resistive ohmic contacts, which are important for low noise operation.

The 50 nm gate is structured by electron beam lithography with a four layer resist. After wet-chemical recess etching, the PtTiPtAu gate is formed by electron beam evaporation. A 450 nm wide metal strip of the first metalization layer is placed on the head of the T-gate to further reduce the gate line resistance, which is mandatory for low-noise applications. The gate is encapsulated in low-k BCB for reduced parasitic capacitances compared to a SiN layer.

The process supports full MMIC design with SiN-MIM capacitors, NiCr thin film resistances and a plated Au-layer in air-bridge technology. After front-side processing the wafers are thinned down to 50 μm thickness to suppress parasitic substrate modes. Through substrate vias are dry etched and a 2.7 μm plated Au-layer forms the backside metalization.

III. LNA MMIC DESIGN

The LNA design targets lowest noise performance over the whole the C-band (4 – 8 GHz) without the use of matching networks on external substrates. Furthermore, the power consumption of the MMIC shall be kept as low as possible. Following design goals that need to be traded against each other conclude from this: Broadband amplification, low noise, low power consumption and unconditional stability.

The LNA comprises two stages in common-source configuration with inductive source feedback applied to both stages. A low-loss grounded coplanar waveguide environment is used for interconnections and matching. Fig. 1 shows a chip- micro-graph of the processed LNA MMIC. The circuit occupies an area of of $(2.5 \times 1)\,mm^2$.

Matching of the first stage is most critical for low noise performance of every LNA since losses before the first amplification stage directly contribute to the amplifier noise temperature. Furthermore, the input of the first stage HEMT needs to be simultaneously matched for lowest noise and power. A common technique for simultaneous noise and power matching is inductive source degeneration. Symmetrical source-lines are connected to the HEMT to realize the desired source inductances. This reduces the quality factor of the input matching network and allows for broadband matching. However, the length of the source line trades with the amplifier gain and has to be chosen carefully. For wide-band performance inductive source degeneration is applied to both amplifier stages.

On-chip spiral inductors realized in airbridge technology are used for matching. The use of the airbridge-layer to realize inductors is motivated by reduced capacitive coupling between

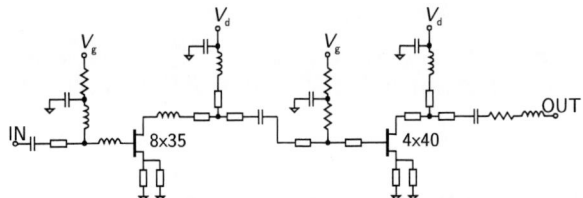

Fig. 2. Simplified schematic of the C-band MMIC.

the turns of the inductor compared to a inductor directly placed on the substrate. This results in a higher resonance frequency of the inductors making it easier to keep the target band resonance free. Furthermore, the increased cross-section of the galvanic metal compared to the first metal layer yields low ohmic losses. This allows for lowest possible loss of the on-chip matching networks. Losses before the first amplifier stage are crucial especially for fully monolithic solutions. The on-chip matching networks induce higher losses than equivalent structures realized on much thicker substrates as used for hybrid amplifiers. Therefore, as few components as possible have to be used for input matching. Good matching with a low number of components can be achieved when the gate width of the first stage HEMT is chosen properly. At C-band frequencies only large absolute gate width allows for matching with very small and therefore low-loss input matching networks. However, the minimum noise temperature of a transistor increases with the width of the transistor fingers due to linear scaling of the gate line resistance with the transistor-finger width. This problem can to some extend be overcome by parallelization of multiple short fingers to obtain the desired absolute gate width. However, massive parallelization leads to very wide devices which are hard to integrate into the chosen waveguide environment and tend to behave non-ideal. It has been found that for the desired frequency regime a parallelization of eight fingers is uncritical while providing only one sixty-fourth of effective gate line resistance compared to a single finger device of the same absolute gate width.

Although increasing the absolute gate width of the first stage eases low-loss matching, it violates the requirement for low power consumption. The absolute current needed to obtain the same gain scales linearly with the absolute gate width making smaller devices favorable. Therefore, the power consumption which is determined by the gate width needs to be traded with a low-loss matching network. The first stage HEMT has been chosen to be of an absolute gate width of $W_g = 8 \times 35\,μm$ to achieve low noise performance in combination with low power consumption. Since the first stage provides sufficient gain, the matching of the interstage and output is less critical in terms of noise performance [6]. Therefore, the size of the second stage HEMT is chosen much smaller ($W_g = 4 \times 40\,μm$) to further reduce the power consumption of the MMIC. The design avoids resistors in the drain-to-source path to prevent additional power dissipation. A simplified schematic of the LNA is shown in Fig. 2.

145

IV. MEASUREMENT RESULTS

On-wafer S-parameter and noise temperature measurements have been performed in one contact using a Keysight PNA-X vector network analyzer with integrated noise receiver. The noise temperature measurement is done using the vector corrected cold-source method [7]. Fig. 3 shows the measured S-parameters and noise temperature of the LNA at optimal noise biasing conditions ($V_{d1} = 0.6$ V, $I_{d1} = 100$ mA/mm). Furthermore, the design-simulation is shown in Fig. 3 and an excellent fit between the model and the measurement is found.

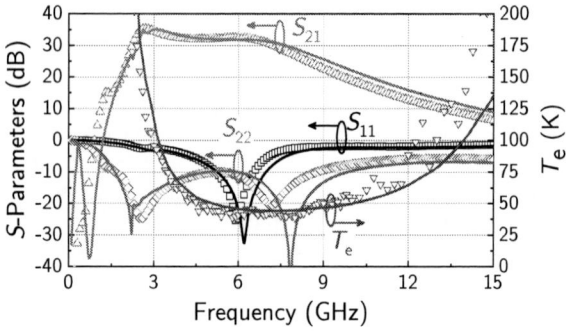

Fig. 3. On-wafer S-parameter and noise temperature measurements (symbols) of the LNA MMIC and the corresponding simulation model (lines) when the LNA is operated at optimum noise bias ($V_{d1} = 0.6$ V, $I_{d1} = 100$ mA/mm) at room temperature.

The LNA exhibits an average small-signal gain of more than 31 dB in C-band ranging from 28 to 32 dB. The input matching $|S_{11}|$ is subject to be traded for noise performance especially under consideration of lossy on-chip matching networks but is still better than -10 dB between 5 and 7 GHz. The output match $|S_{22}|$ is better than -10 dB between 2 and 9 GHz. The average noise temperature of the amplifier between 4 and 8 GHz is 44.5 K with a minimum of 38.4 K. All results are given at optimal noise bias which is reached at a first stage bias voltage of $V_{d1} = 0.6$ V and a drain current density of $I_{d1} = 100$ mA/mm. The overall chip dissipates a power of only 24.8 mW at this optimal noise bias point.

Besides the optimal noise performance it is of special interest how the amplifier performance changes when the bias is varied. The dependence of the small-signal gain and the dependence of the noise temperature on the first stage bias (which is most critical for noise performance) is investigated. The S-parameters and the noise temperature are measured at different first stage bias points when the second stage bias is fixed at $V_{d2} = 0.5$ V and $I_{d2} = 100$ mA/mm. Fig. 4 shows the dependence of the small-signal gain on the first stage bias and Fig. 5 shows the corresponding dependency of the noise temperature on the bias point. The circles denote the bias points where the actual measurements took place and the contour is obtained by interpolation between these points.

It has been found that the small-signal gain is relatively robust against changes in the first stage bias. Only for very low

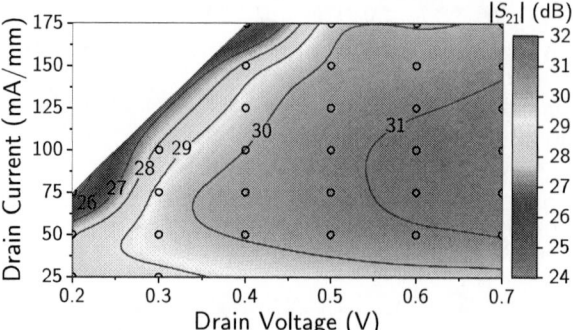

Fig. 4. Measured small-signal gain averaged between 4–8 GHz of the LNA MMIC dependent on the first stage bias point. The points denote the bias-points where the actual measurements have been executed and the contour plot is obtained by interpolation between the measured bias points.

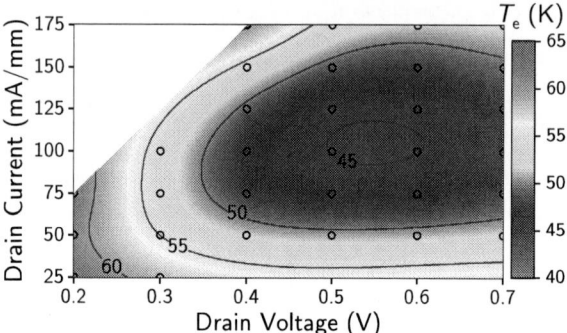

Fig. 5. Measured noise temperature averaged between 4–8 GHz of the LNA MMIC dependent on the first stage bias point. The points denote the bias-points where the actual measurements have been executed and the contour plot is obtained by interpolation between the measured bias points.

voltages and currents the gain performance degrades notably. However, the noise temperature shows a clear optimum around a drain current density of $I_{d1} = 100$ mA/mm and a drain voltage ranging between approximately $V_{d1} = 0.5$ V to $V_{d1} = 0.6$ V. The increase in noise temperature when varying the bias point around this minimum is relatively small, only at bias points far from the optimum the noise performance starts to degrade significantly. This minimum originates from the trade-off between higher current and voltage bias leading to higher device gain, but also increasing the channel noise power. Higher gain suppresses output related noise as e.g. the channel noise [8] but at some point the increase in channel noise power cannot be compensated anymore. This leads to the observed optimal noise bias point. The observed bias dependency allows to identify further useful operating points. At a first stage bias point of $V_{d1} = 0.4$ V and $I_{d1} = 50$ mA/mm the power consumption reduces to 13.6 mW (55 % of the optimal noise bias point) while the average noise temperature over the full band only increases to 50.4 K and a small-signal gain of 30.1 dB is maintained. Even at an extreme low power bias (9.4 mW) of $V_{d1} = 0.2$ V and $I_{d1} = 25$ mA/mm an average noise temperature of 62.3 K with a small-signal gain of 28 dB can be reached, which is still competitive. However, one would rather find a more balanced first to second stage bias ratio if such low powers shall be considered.

V. COMPARISON TO STATE-OF-THE-ART

Table 1 summarizes the results and compares them to state-of-the-art room temperature C-band LNAs. In the upper part, hybrid LNAs and hybrid-ICs (IC with external input matching network) are listed. MMIC LNAs are given below.

Although, hybrid LNAs based on InP HEMTs demonstrate the lowest noise temperatures today [2] it can be seen that the full MMIC solution presented in this paper is competitive to some hybrids in terms of noise [1]. Furthermore, it is able to even demonstrate a lower noise temperature compared to some hybrid-IC solutions [4]. The dc-power consumption per amplifier stage at optimal noise bias is the lowest among sub 100 K noise temperature LNAs and can be further reduced to 13.6 mW by adaption of the first stage bias with just a slight loss in noise performance. To the best of the authors' knowledge, the presented LNA has the lowest noise performance among MMIC LNAs at room temperature.

Table 1. State-of-the-Art Room Temperature C-band LNAs.

Ref.	Techn.	Freq. (GHz)	P_{dc} (mW)	Gain (dB)	T_e (K)	Topology
[1]	130 nm InP HEMT	$4-8$	56.25	45	45	3-stage hybrid
[2]	100 nm InP HEMT	$4-8$	60.75	43	32	3-stage hybrid
[3]	100 nm InP HEMT	$0.3-14$	100	40.7	60.7	3-stage hybr.-IC
[4]	100 nm mHEMT	$4-12$	90.2	38	47.8	3-stage hybr.-IC
[9]	110 nm InAs/InSb HEMT	$4-8$	17.6	$25-29$	150	3-stage hybrid
[10]	150 nm mHEMT	$4-8$	240	28	65	2-stage MMIC
[11]	150 nm mHEMT	$4-12$	30	27	190	2-stage MMIC
This Work	50 nm mHEMT	$4-8$	24.8	31	44.5	2-stage MMIC

VI. CONCLUSION

In this work, we present a LNA MMIC with an average noise temperature of 44.5 K over the whole C-band. The 2-stage amplifier exhibits on average 31 dB of gain in the $4-8$ GHz range. The design focuses on a monolithic solution without the use of external matching networks to allow for scalable integration into larger readout or receiver systems that can hardly be achieved by hybrid designs. Low power dissipation is a key requirement besides noise performance. A study on circuit design techniques to reduce the power consumption of MMIC LNAs is given. A trade-off between small-size input matching networks that demand for larger gate width with as short as possible gate width transistors allows

for low noise and low power consumption in a monolithic design. The presented MMIC dissipates only 24.8 mW of dc power at optimal noise bias. Furthermore, an extensive study on the bias dependency of the gain and noise temperature of the LNA shows that the power consumption of the MMIC can further be reduced to 13.6 mW (55% of power consumption at optimal noise bias) with just a slight degradation of the noise performance to 50.4 K.

Although, hybrid LNAs demonstrate lower noise temperatures, the present MMIC is getting close to hybrid performance and already competes with some integrated amplifiers that utilize an external input matching network. To the best of the authors' knowledge, the LNA demonstrates the lowest effective noise temperature among monolithic LNAs at room temperature.

ACKNOWLEDGMENT

The project has received funding from the European Union's Horizon 2020 research and innovation programme under Grant agreement number 871764.

We express our gratitude to our colleagues in the Fraunhofer IAF technology department for their excellent contributions during epitaxial growth and wafer processing.

REFERENCES

[1] J. Schleeh, H. Rodilla, N. Wadefalk, P. Nilsson, and J. Grahn, "Cryogenic noise performance of InGaAs/InAlAs HEMTs grown on InP and GaAs substrate," *IEEE J. Solid-State Circuits*, vol. 91, pp. 74 – 77, 2014.

[2] E. Cha, N. Wadefalk, G. Moschetti, A. Pourkabirian, J. Stenarson, and J. Grahn, "A 300-μW Cryogenic HEMT LNA for Quantum Computing," in *IEEE MTT-S Int. Microw. Symp. Dig.*, 2020, pp. 1299–1302.

[3] E. Cha *et al.*, "0.3–14 and 16–28 GHz Wide-Bandwidth Cryogenic MMIC Low-Noise Amplifiers," *IEEE Trans. Microw. Theory Tech.*, vol. 66, no. 11, pp. 4860–4869, 2018.

[4] B. Aja Abelan *et al.*, "4–12- and 25–34-GHz Cryogenic mHEMT MMIC Low-Noise Amplifiers," *IEEE Trans. Microw. Theory Tech.*, vol. 60, no. 12, pp. 4080–4088, 2012.

[5] A. Leuther, A. Tessmann, H. Massler, R. Aidam, M. Schlechtweg, and O. Ambacher, "450 GHz amplifier MMIC in 50 nm metamorphic HEMT technology," in *Proc. Int. Conf. Indium Phosphide Rel. Mater.*, 2012, pp. 229–232.

[6] H. T. Friis, "Noise Figures of Radio Receivers," *Proceedings of the IRE*, vol. 32, no. 7, pp. 419–422, 1944.

[7] "High-Accuracy Noise Figure Measurements Using the PNA-X Series Network Analyzer, Application Note 1408-20," Agilent Technologies, 2013.

[8] M. W. Pospieszalski, "Modeling of noise parameters of MESFETs and MODFETs and their frequency and temperature dependence," *IEEE Trans. Microw. Theory Tech.*, vol. 37, no. 9, pp. 1340–1350, 1989.

[9] G. Moschetti *et al.*, "Cryogenic InAs/AlSb HEMT Wideband Low-Noise IF Amplifier for Ultra-Low-Power Applications," *IEEE Microw. Compon. Lett.*, vol. 22, no. 3, pp. 144–146, 2012.

[10] M. Kelly, I. Angelov, J. P. Starski, N. Wadefalk, and H. Zirath, "4-8 GHz Low Noise Amplifiers using metamorphic HEMT Technology," in *Proc. Eur. Microw. Integr. Circuits Conf.*, 2006, pp. 118–121.

[11] C. Chiong, W. Tzeng, Y. Hwang, W. Wong, H. Wang, and M. Chen, "Design and measurements of cryogenic MHEMT IF low noise amplifier for radio astronomical receivers," in *Proc. Eur. Microw. Integr. Circuits Conf.*, 2009, pp. 1–4.

Proceedings of the 16th European Microwave Integrated Circuits Conference

Transient Field-Plate Thermometry Demonstrated on a 20-W X-Band GaN Power Amplifier

Simon J. Mahon, Melissa C. Gorman, Michael C. Heimlich

Macquarie University, NSW 2109, Australia

simon.mahon@mq.edu.au

Abstract — **Transient field-plate thermometry is demonstrated experimentally on a 20-W X-band GaN power. The technique is similar to gate-resistance thermometry but more convenient to implement in a power amplifier layout close to the heat source, to maintain accuracy, and without increasing MMIC dimensions. The new technique is used to extract the thermal time constants for heating and cooling under pulsed operation.**

Keywords — **Gallium nitride, MMICs, power amplifier, temperature measurement.**

I. INTRODUCTION

Thermal characterisation of microwave and millimetre-wave high power amplifiers is critical for radar and other pulsed applications. Monitoring transistor temperature in real-time facilitates the dynamic adjustment of pulse widths, biases and input signal levels to improve performance and reliability.

Gate-resistance thermometry (GRT) has been demonstrated as a useful technique for the measurement of temperatures in single, stand-alone devices for both gallium arsenide (GaAs) and gallium nitride (GaN) technology [1-2].

In [3], the use of GRT was demonstrated in a 10-W, X-band power amplifier; however, the GRT function could not be implemented on an active transistor. Two disadvantages follow: (i) the GRT is moved some distance from the heat generating transistors, reducing measurement accuracy, and (ii) the amplifier needs to be made larger to allocate space for the GRT.

In this paper, we present the first application of an alternate approach, field-plate thermometry (FPT), to measure the transient temperature of 20-W, X-band GaN power amplifier. A single FPT structure is embedded in the centre transistor of both the input and output stages to sense temperature as the amplifier bias and pulsed drive level are varied.

In Section II, the design of the field-plate thermometer and 20-W amplifier are described. Calibration of the FPT, and comparison to GRT, are discussed in Section III. The measured RF and thermal performance are presented in Section IV and the results are discussed in Section V.

II. DESIGN

A. Field-Plate Thermometer

FPT is similar in concept to GRT. Both use a two- or four-wire technique to measure the change in resistance of a thin strip of metal close to the location of heat generation in the transistor, which is slightly below the surface of the semiconductor under the gate, offset towards the drain. This resistance change is compared to a prior calibration to determine the transistor's rise in temperature.

GRT offers the advantage of measurement directly on the semiconductor surface, however, it is difficult to measure changes in the resistance of an operational gate without impacting RF performance; hence, it is often positioned, as a separate device, up to 100 μm away from the active transistor (e.g. 85 μm away in [3]), reducing measurement accuracy.

FPT solves this problem by measuring the resistance change in the active transistor's field-plate which is not intentionally carrying RF signal, with the small penalty of sensing the temperature an additional micrometre, or less, further above the heat source. FPT is incorporated into the active transistor.

B. X-band Power Amplifier

The two stage X-band power amplifier is shown in Fig. 1. The first stage consists of two 8×125 μm transistors and the second eight 8×125 μm transistors. The fabrication process is WIN's 0.15 μm GaN-on-SiC, NP15.

There are two field-plate thermometers, one in the centre of each stage. Both use the two-wire technique with the MMIC ground providing the return path. A close up of the FPT in stage two is shown in Fig. 2 where it is part of an RF-active transistor.

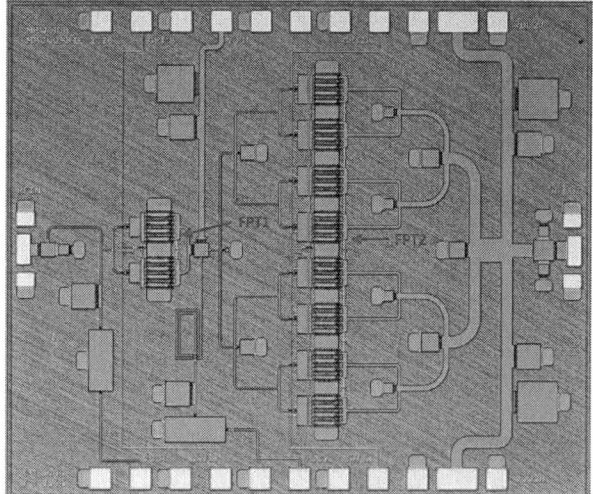

Fig. 1. Photograph of the two-stage X-band power amplifier with the locations of the two FPT devices indicated. MMIC dimensions are 3.0 mm x 2.5 mm.

3–4 April 2022, London, UK

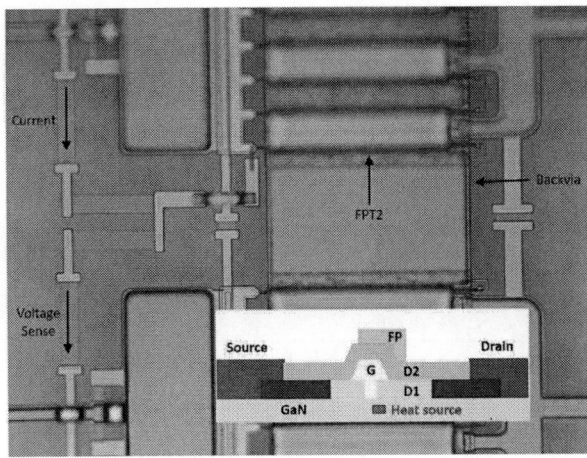

Fig. 2. Close up of the field-plate thermometer connexion in stage two. A sketch of the transistor cross-section is inset and valid for both ordinary transistors and FPT2. The two dielectric layers, D1 and D2, are less than 1 μm thick so both the gate, G, and the field-plate, FP, are close to the heat source. The source and drain ohmic is dark blue and connecting metal dark green.

The field-plate in Fig. 2 is shorted to the source, and hence ground, on the right. On the left, the large value resistors ensure that it is an approximate open at X-band, allowing it to perform as a normal field-plate at RF, but it allows the dc to pass for the 2-wire resistance and, hence, temperature measurement.

III. FPT CALIBRATION

The combined GRT and FPT structure in Fig. 3 is used to calibrate the change in resistance with the change in temperature as the measurement system is heated from room temperature to 200°C.

Three of these combined structures were measured and the results are presented in Fig. 4, along with the GRT calibration data from [3]. All nine measurements are in close agreement. The fit to FPT1, shown in the figure, is used in Section IV.

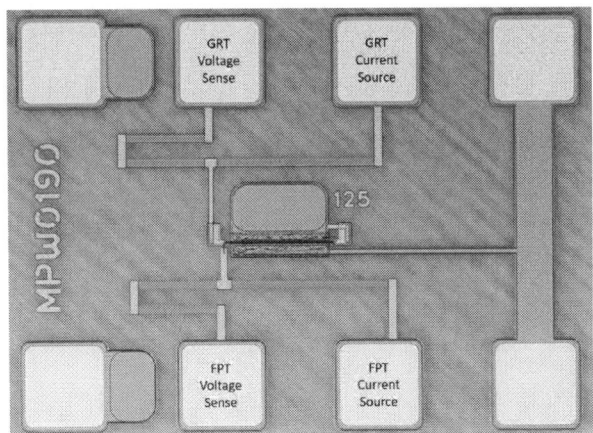

Fig. 3. Combined gate-resistance and field-plate thermometry test structure.

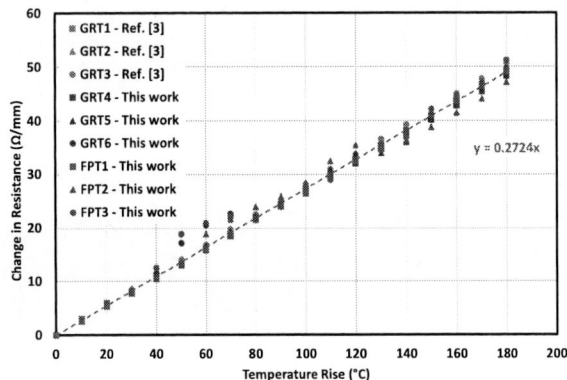

Fig. 4. Change in resistance as a function of temperature rise. Three GRT samples from [3] (red), three GRT from this work (blue) and three FPT samples from this work (green) are shown. The fit to FPT1 is also shown.

IV. AMPLIFIER RF AND THERMAL MEASUREMENTS

A. RF Measurements

The amplifier is measured at room temperature with a 1 kHz dc and RF pulse. The dc pulse starts at -100 μs and the RF pulse at -90 μs. The dc and RF turn off at 0 μs. The drain bias is 30 V. Three gate biases are used: -1.5, -2.0 and -2.5 V.

The output power of the amplifier is shown in Fig. 5. The maximum output of 43.3 dBm (21.1 W) is obtained at 10.5 GHz with the -1.5 V bias. At this same frequency, the maximum output power is 43.1 dBm (20.3 W) and 42.7 dBm (18.5 W) at biases of -2.0 and -2.5 V, respectively.

In Fig. 6, the output power, gain and power-added efficiency (PAE) are plotted as a function of input power at these three gate biases. The maximum PAE of 20.9% is obtained at 27 dBm input power and a gate bias of -1.5 V. The maximum PAE at -2.0 and -2.5 V, is 20.5% and 20.4%, respectively, both obtained at 30 dBm input power.

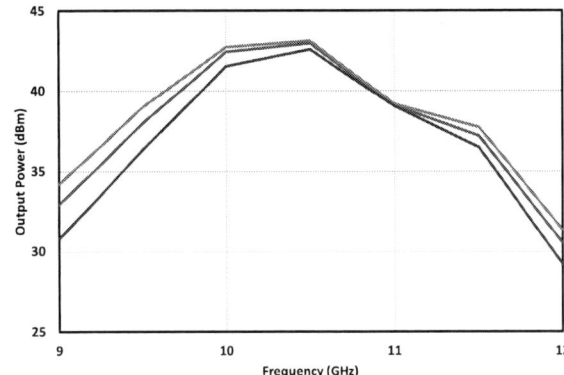

Fig. 5. Amplifier output power as a function of frequency with 30 dBm input. $V_d = 30$ V, $V_g = -2.5$ V (blue), -2.0 V (green) and -1.5 V (red).

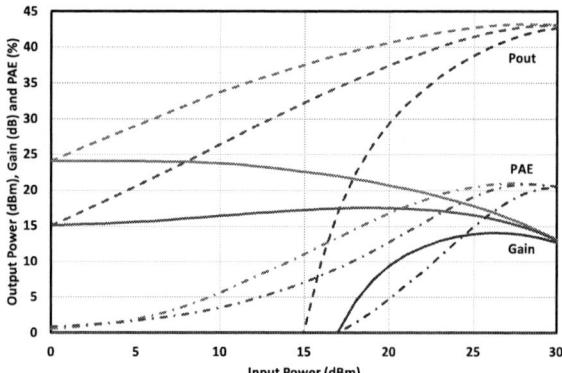

Fig. 6. Output power (dashed lines), gain (solid) and power-added efficiency (dot-dashed) as a function of input power at 10.5 GHz. V_d = 30 V, and V_g = -2.5 V (blue), -2.0 V (green) and -1.5 V (red).

B. Transient Thermal Measurements

Figs 7 and 8 illustrate the temperature profile in the two stages under the three different gate bias conditions, V_g = -2.5, -2.0 and -1.5 V, and 30 dBm input power.

With the higher gate bias of -1.5 V, the transistors draw current and temperature increases as soon as the dc is applied. For the more negative biases, the temperature does not start to rise until the RF is applied 10 μs later as the dc alone causes insufficient current flow to noticeably heat the transistor.

The temperature rise, as a function of input power, is shown in Fig. 9 for stage 1 of the amplifier and Fig. 10 for stage 2. The bias is Vd = 30 V and Vg = -2.0 V.

The maximum temperature rise occurs at the end of the pulse. It is 109°C and 135°C for stages 1 and 2, respectively, with an input power of 30 dBm (output of 43 dBm).

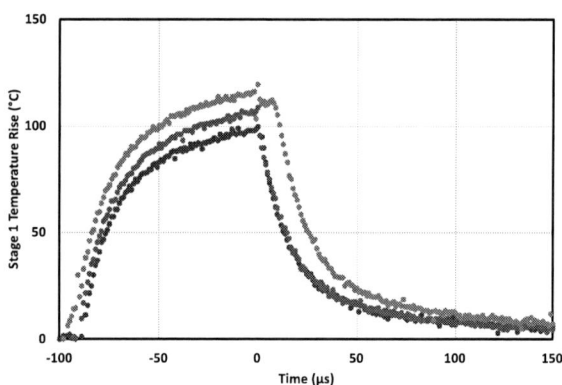

Fig. 7. Stage 1 temperature rise versus time for 10.5 GHz and 30 dBm input power with V_d = 30 V and V_g = -2.5 V (blue), -2.0 V (green) and -1.5 V (red).

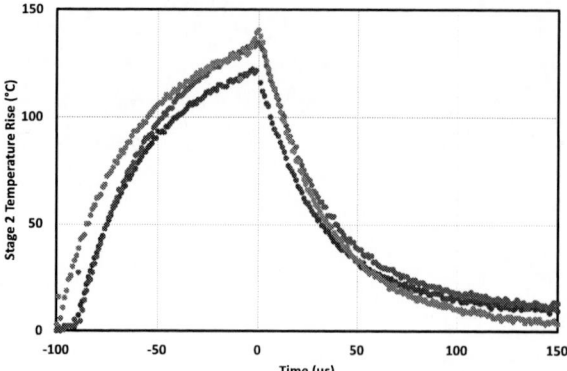

Fig. 8. Stage 2 temperature rise versus time for 10.5 GHz and 30 dBm input power with V_d = 30 V and V_g = -2.5 V (blue), -2.0 V (green) and -1.5 V (red).

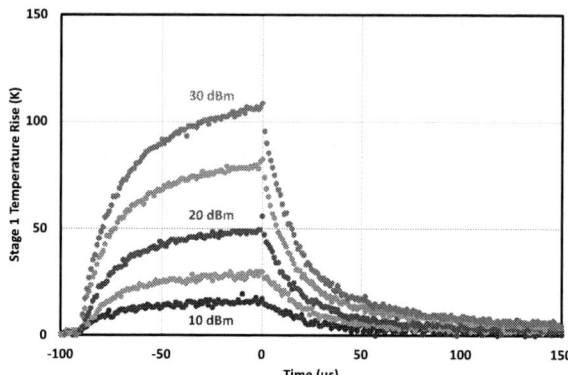

Fig. 9. Stage 1 temperature rise versus time for 10.5 GHz input powers from 10 to 30 dBm (5 dB steps). DC pulse starts at -100 μs, RF pulse starts at -90 μs. DC and RF turn off at 0 μs. V_d = 30 V, V_g = -2.0 V.

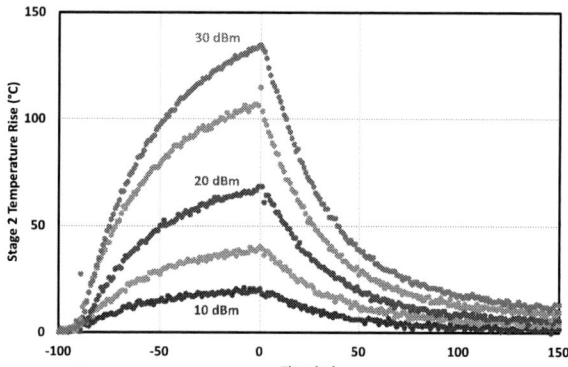

Fig. 10. Stage 2 temperature rise versus time for 10.5 GHz input powers from 10 to 30 dBm (5 dB steps). DC pulse starts at -100 μs, RF pulse starts at -90 μs. DC and RF turn off at 0 μs. V_d = 30 V, V_g = -2.0 V.

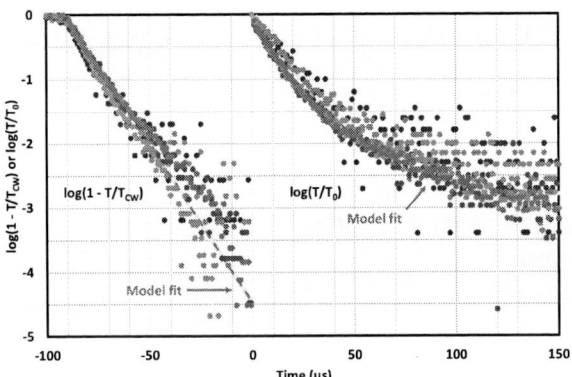

Fig. 11. Normalised stage 1 temperature plotted on a logarithm scale for input powers: 10 dBm (blue), 15 dBm (cyan), 20 dBm (green), 25 dBm (orange) and 30 dBm (red). $V_d = 30$ V, $V_g = -2.0$ V. Model fit for 30 dBm is also shown.

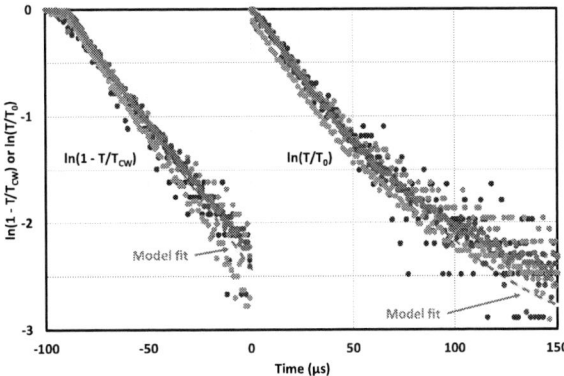

Fig. 12. Normalised stage 2 temperature plotted on a logarithmic scale for input powers: 10 dBm (blue), 15 dBm (cyan), 20 dBm (green), 25 dBm (orange) and 30 dBm (red). $V_d = 30$ V, $V_g = -2.0$ V. Model fit for 30 dBm is also shown.

A model is fitted to the data from Figs 9 and 10. A single time constant is sufficient to model transistor heating during the pulse, -90 µs < t < 0, but two time constants are required to accurately model the cooling period, t > 0.

$$\Delta T = \Delta T_{CW} \left(1 - e^{\frac{t+90}{\tau_1}} \right) \qquad (1)$$

$$\Delta T = \Delta T_{t=0} \left(e^{\frac{-t}{\tau_1}} + \alpha . e^{\frac{-t}{\tau_2}} \right) \qquad (2)$$

where the values for τ_1 the temperature rise to CW operation, ΔT_{CW}, and α are listed in Table 1. The time constant for the slow cooling, τ_2, is 1000 µs. Figs 11 and 12 show the data from Figs 9 and 10 and the model at 30 dBm.

V. DISCUSSION

In [3], a 10-W, X-band amplifier was used to demonstrate gate-resistance thermometry. In the same NP15 process, and under CW drive at a bias of $V_d = 30$ V, $V_g = -2.0$ V, a maximum temperature of 152°C was measured with a GRT device positioned in the centre of the second stage but 85 µm away from the nearest active transistor gate. Here, for a amplifier with similar output power per unit periphery (2.6 W/mm), under

Table 1. Thermal time constant, τ_1, CW temp. rise and α from Figs 9 and 10.

Stage	Input (dBm)	τ_1 (µs)	ΔT_{CW} (°C)	α
1	10	17	16	0.12
	15	19	29	0.10
	20	20	50	0.085
	25	17	78	0.095
	30	20	108	0.071
2	10	34	22	0.044
	15	38	44	0.063
	20	39	75	0.059
	25	33	114	0.21
	10	37	148	0.19

the same bias, a CW temperature rise of 148°C has been extracted at an active transistor gate (Table 1), implying a transistor temperature of about 173°C in a room temperature laboratory. The 21°C difference is consistent with the effect of positioning the GRT far from the heat source in that amplifier but it is complicated by the more compact layout in this work.

Transient measurements of GaAs devices [2] suggest faster rise times than observed here; however, this may be due to the poorer thermal conduction of GaAs compared to GaN-on-SiC. In [2], a two-step temperature rise for the 8-finger transistor was also reported. This was not observed here which may be due to the placement of the FPT in a transistor adjacent to a heat spreading backvia (Fig. 2).

The transistors here heat and initially cool at the same rate; however, they then experience a slower cooling rate some 50 µs after the pulse ends. This slower cooling time constant, τ_2, equals the pulse repetition period. Overall, the second stage of the amplifier heats and cools at about half the rate of the first.

VI. CONCLUSION

Transient thermometry of a 20-W X-band GaN power amplifier has been demonstrated experimentally using the new technique of field-plate thermometry. The method is similar to gate-resistance thermometry but more convenient to implement in a power amplifier without placing it away from the heat source, reducing accuracy, and increasing MMIC dimensions. This new approach has been used to extract the thermal time constants for heating and cooling under pulsed operation.

ACKNOWLEDGEMENT

The authors thank Andrew Jones and Gerry McCulloch for measurements and WIN Semiconductor Corp., Taiwan, for providing access to their NP15 process.

REFERENCES

[1] B. K. Schwitter, A. E. Parker, S. J. Mahon and M. C. Heimlich, "Transient gate resistance thermometry demonstrated on GaAs and GaN FET," 2016 IEEE MTT-S International Microwave Symposium (IMS), 2016, pp. 1-4, doi: 10.1109/MWSYM.2016.7540035.

[2] B. K. Schwitter, A. E. Parker, S. J. Mahon and M. C. Heimlich, "Characterisation of GaAs pHEMT Transient Thermal Response," 2018 13th European Microwave Integrated Circuits Conference (EuMIC), 2018, pp. 218-221, doi: 10.23919/EuMIC.2018.8539961.

[3] S. J. Mahon et al., "Real-time. In-circuit Temperature Sensing of an X-Band GaN Power Amplifier," 2020 15th European Microwave Integrated Circuits Conference (EuMIC), 2021, pp. 37-40, doi: 10.1109/EuMIC48047.2021.00021.

Proceedings of the 16th European Microwave Integrated Circuits Conference

A 27dBm Ku-Band SiGe Power Amplifier Working up to 90°C with High Robustness to the 2:1 SWR

B. Coquillas[1,2], E. Kerhervé[1], S. Redois[3], AC. Amiaud[2], L. Roussel[2], B. Louis[3], E. Itcia[3], T. Merlet[2]

[1]IMS Laboratory, University of Bordeaux, CNRS UMR 5218, France

[2]THALES LAS, France

[3]THALES DMS, France

benjamin.coquillas@ims-bordeaux.fr

Abstract — **This paper presents a compact Ku-Band SiGe cascode power amplifier (PA) with a saturated output power (Psat) higher than 27dBm up to 90°C and a highly robust power combination to the Standing Wave Ratio (SWR). Improved transistor and twisted coupler layout techniques are combined in a balanced architecture. As a proof of concept, the PA was implemented with a 0.13-μm SiGe BiCMOS technology. The measured performances at 18GHz show a Psat of 27.4dBm and 27.1dBm and a peak-PAE of 25.8% and 24.5% at 30°C and 90°C, respectively. With a 2:1 SWR the PAE and Psat drop less than 7.2% and 1.5dB, respectively.**

Keywords — **defense industry, design methodology, directional coupler, power amplifier, temperature measurement**

I. INTRODUCTION

The application of active scanning antenna module is already mastered and used in several defense equipments. This application raises two major technological challenges for small-scale on-board radiofrequency sensors: the integration of all antenna circuits and the control of the on-board electronics cost. The PA is an elementary block in the communication chain of these antenna circuits. It is subject to these strong technological and economic constraints.

The PA survey from Georgia Tech presents the limits of the state-of-the-art performances for the low cost SiGe technology with a maximum output power level of 1W in the 6-20 GHz frequency range [1]. SiGe is an alternative technology to the GaAs. In the Ku-band, the SiGe 130nm state of the art offers very few compromise solutions between watt-level output power and more than 30% of power added efficiency (*PAE*) [2]. In addition, the temperature constraint for on-board front-end module (FEM) and the FEM robustness to the antenna active SWR are poorly informed for Ku-band transmitters. The use of a cascode topology with a low base impedance termination and a minimum voltage-current waveform overlap extends the V_{CE} swing on the upper SiGe HBT in the cascode to operate beyond BV_{CB0}, boosting the output power and the *PAE* [3]-[4].

In this paper two power cells (PCs) are combined in a balanced PA architecture. The block diagram is shown in Fig. 1. Two twisted 90° hybrid directional couplers perform the power combination and improve the PA robustness to the SWR antenna variations. The circuit is implemented from [5] in a 130nm BiCMOS SiGe technology from STMicroelectronics, with many relevant measurements. The

PA achieves a *peak-PAE* of 25.8% and a *Psat* of 27.4dBm at 18GHz and 30°C. With a 2:1 SWR variation, the *peak-PAE* and *Psat* drops are less than 7.2% and 1.5dBm, respectively.

To the authors' knowledge, this is the first Ku-band SiGe power amplifier with an efficient 2:1 SWR robustness.

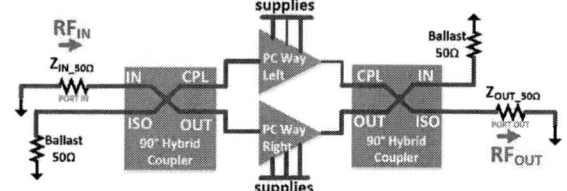

Fig. 1. Block diagram of the designed balanced PA

II. POWER AMPLIFIER THEORY AND DESIGN

A. Description of the cascode power cell

The sizing and the biasing of the PC are fully explained in [5]. Each PC is made of two nx-CBEBC transistors (T1, T2) with 5 emitter fingers of 13μm each, for a total emitter area of 35.1μm² per PC.

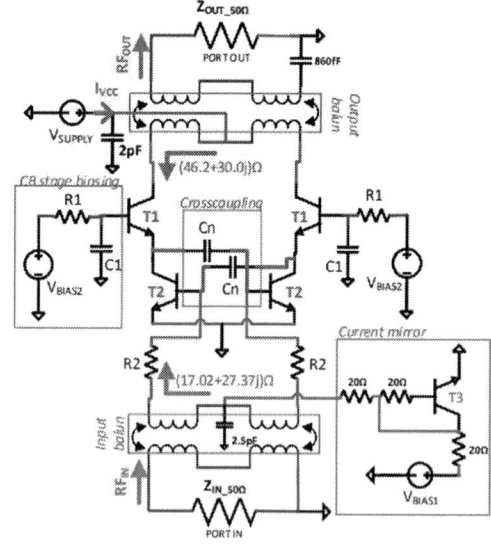

Fig. 2. Schematic of the elementary power cell

152

3–4 April 2022, London, UK

B. Figure of merit for self-contained architectures

The phenomenon of the SWR is illustrated in Fig. 3.

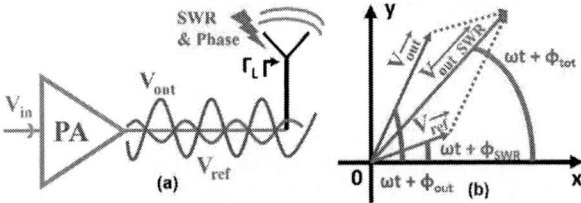

Fig. 3. (a) SWR phenomenon (b) Fresnel diagram of the Vout$_{SWR}$

The linear gain $Gain_{SWR}$ of the output signal $V_{out_{SWR}}$ influenced by a given SWR is calculated from the Fresnel diagram, using the Pythagoras theorem (1). The theoretical drop $\Delta Gain_{SWR}$ between the minimum and the maximum values of $Gain_{SWR}$ only depends on the SWR (2). Γ_L is the reflection coefficient. ϕ_{out} is the phase of the output signal. ϕ_{SWR} is the phase of V_{ref}, the signal reflected because of the SWR. ϕ_{tot} is the difference between ϕ_{SWR} and ϕ_{out} (Fig. 3b).

$$Gain_{SWR} = 20\, log \sqrt{[\,(V_{out}/V_{in})^2 + (V_{ref}/V_{in})^2 + 2 \times}$$
$$(V_{out} \times V_{ref}/V_{in}^2) \times cos(\phi_{tot})} \tag{1}$$

$$\Delta Gain_{SWR} = Gain_{SWR(\phi tot=180°)} - Gain_{SWR(\phi tot=0°)}$$
$$= 20\, log[\sqrt{(1 + \Gamma_L^2 + 2\Gamma_L)/(1 + \Gamma_L^2 - 2\Gamma_L)}]$$
$$= 20\, log(SWR). \tag{2}$$

A balanced architecture can be considered as a shield which reduces the SWR seen by the PC (Fig. 4). To quantify this shield effect, a new figure of merit (FOM) is introduced. SSC_{SWRa} means Shield ratio for Self-Contained architectures, defined for an applied SWR. The SWR_a is the SWR applied to the PA. The SWR_s is the SWR sensed by the PC (3)-(4).

$$SWR_S = 10^{\frac{\Delta Gain_{measured}}{20}} \tag{3}$$

$$SSC_{SWR_a} = SWR_a / SWR_s. \tag{4}$$

Table 1 presents the theoretical drop of the linear gain $\Delta Gain_{SWR}$ and the $SSC_{2:1}$ with different values of sensed SWR.

Table 1. Linear gain drop, SSC2:1 with different values of SWR

FOM \ SWR$_S$	1:1	1.2:1	1.3:1	2:1	4:1
$\Delta Gain_{SWR_S}$	0dB	1.6dB	2.3dB	6dB	12dB
$SSC_{2:1}$	2	1.7	1.5	1	0.5

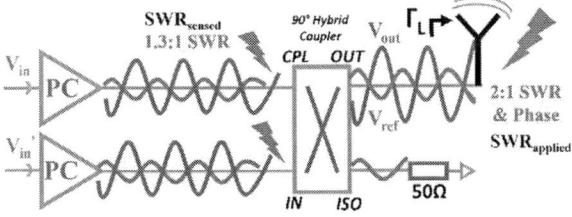

Fig. 4. Shield effect of the hybrid coupler with two power cells (PC)

C. Implementation of the twisted coupler

A twisted coupler topology is used for the power combination. To validate the proof of concept of a SWR robustness at 18GHz, a twisted coupler with 3 branches and 7 twists is designed. The design methodology of the twisted coupler is illustrated in Fig. 5.

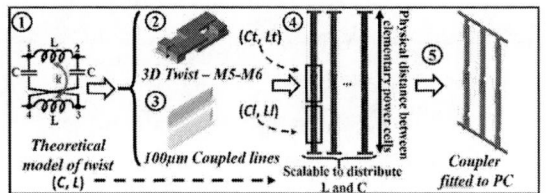

Fig. 5. Design process of the twisted coupler

The calculation ① of the initial inductance and capacitance (L=630pH, C=126fF) is based on [6], with a coupling factor k of 0.6. Electromagnetic (EM) simulations give the EM characterization of a single twist (Lt=25pH, Ct=260fF) ② and a 100µm-coupled line (Ll=58pH, Cl=70fF) ③. Then the number of 7 twists and the length of 520µm of coupled line are determined to fit the initial values (L, C) and to obtain a coupler centered at 18GHz ④. The chosen form factor fits the space available between the 2 PCs ⑤. The manufactured coupler is depicted in Fig. 6.

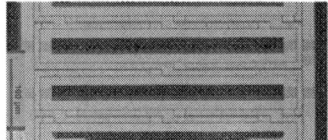

Fig. 6. Microphotography of the SiGe twisted coupler

III. IMPROVED TRANSISTOR LAYOUT TECHNIQUES

Two specific drawing techniques are used for the tracks around the transistors (Fig. 7). First, the "stair" topology minimizes the parasitic coupling capacitances induced by the metal levels of the collector and the emitter facing each other. Second, the double access at the base of the transistor reduces the parasitic effects and facilitates the track layout.

The choice of a small size of transistors for T1 and T2 and the isolation of their bases directly impact the layout of the circuit (Fig. 8). The positive charges resulting from the impact ionization [3]-[4] are evacuated in the very close C1 capacitors.

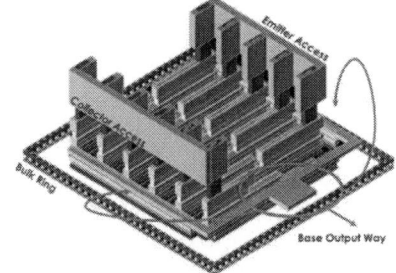

Fig. 7. Intrinsic transistor view (layout) with a "stair" topology

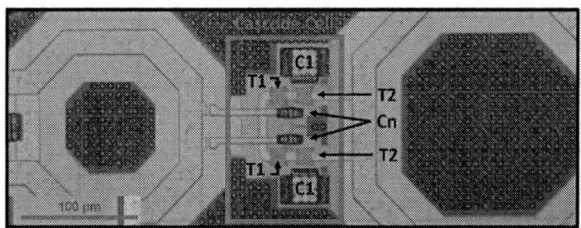

Fig. 8. Microphotography of the cascode power cell

IV. MEASUREMENT RESULTS

The compact twisted balanced PA was manufactured with the SiGe 130nm BiCMOS9MW from STMicroelectronics. The microphotography of the PA is shown in Fig. 13. The chip die size is 1.2mm² including the pads. The active core size is 0.4mm². Measurements are performed on chip with a 3.3 to 4.2V supply voltage (V_{SUPPLY}) and a 1.1V biasing voltage (V_{BIAS1}). The total quiescent DC current is 65.7mA. The measured S-parameters are shown in Fig. 9. The small-signal gain exhibits a relative bandwidth at 30°C of 27.8% and 47.2% at -1dB (BW_{1dB}) and -3dB (BW_{3dB}), respectively.

The RF performances of the PA are depicted in Fig. 10. With a V_{SUPPLY} of 3.3V and at 3dB of gain compression the PA achieves 25.2dBm of P_{OUT} and 27.2% of PAE. It also achieves 26.1dBm of $Psat$. With a V_{SUPPLY} up to 4.2V, it achieves a $Psat$ of 27.4dBm and a PAE at 3dB of gain compression of 25%.

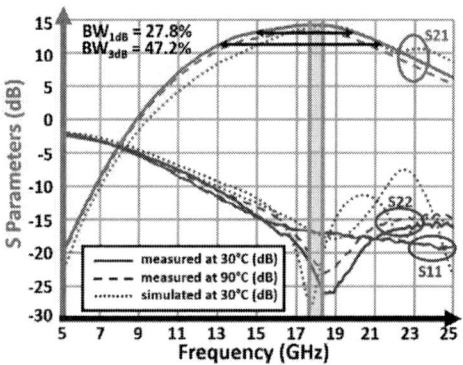

Fig. 9. S-parameters of the balanced power amplifier (V_{SUPPLY} = 3.3V)

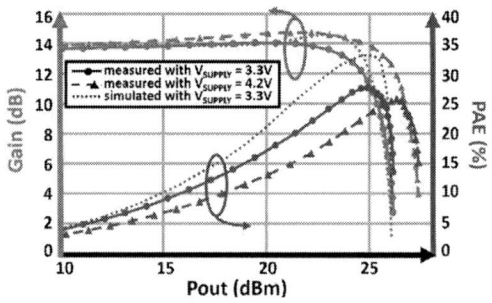

Fig. 10. Linear gain and PAE vs P_{OUT} of the balanced power amplifier at 18GHz and 30°C

The measured $Psat$ at 30°C (Fig. 11) is stable over the 16-19GHz frequency range and is very close to the simulated $Psat$. The $peak$-PAE is centered at 18GHz which shows a good output matching considering the RF performances up to 90°C.

The measured $Psat$ and $peak$-PAE of the PC and the PA with a 2:1 SWR at 18GHz and 30°C are compared in Fig. 12 and summarized in Table 2. The maximum $peak$-PAE, linear gain and $Psat$ drops of the PA are respectively limited to 7.2%, 1.9dB and 1.5dB. Whereas the elementary PC performs a drop of 15.7% $peak$ PAE, 3.4dB linear gain and 3dB $Psat$.

According to the Tables 1 and 2, the value of the $SSC_{2:1}$ is between 1.5 and 1.7, which means that this balanced architecture demonstrates a drop of the impact of the SWR from 2:1 to less than 1.3:1. These results justify the use of the balanced architecture with the twisted hybrid couplers in the case of a 2:1 SWR.

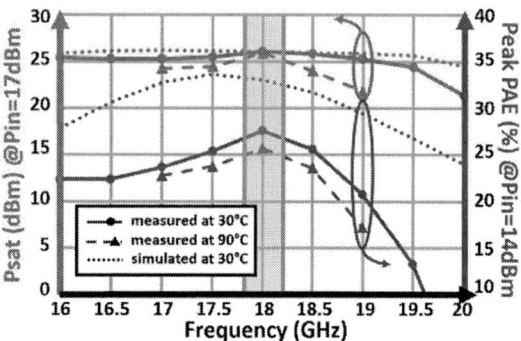

Fig. 11. Psat and peak-PAE vs frequency of the balanced power amplifier at V_{SUPPLY} = 3.3V

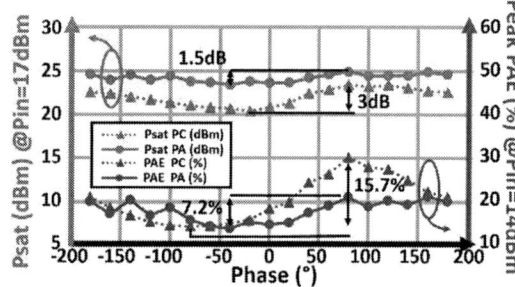

Fig. 12. Power cell (PC) and power amplifier (PA) 2:1 SWR sensitivity at 18GHz, 30°C, V_{SUPPLY} = 3.3V

Table 2. Summary of 2:1 SWR drops for elementary power cell and power amplifier @30°C, 18GHz, V_{SUPPLY} = 3.3V

Performances	Delta gain	Delta peak-PAE	Delta Psat
Elementary PC	3.4dB	15.7%	3dB
Balanced PA	1.9dB	7.2%	1.5dB

These measured performances allow to validate the architecture chosen to achieve a power amplifier with good levels of PAE and output power at 18GHz up to 90°C.

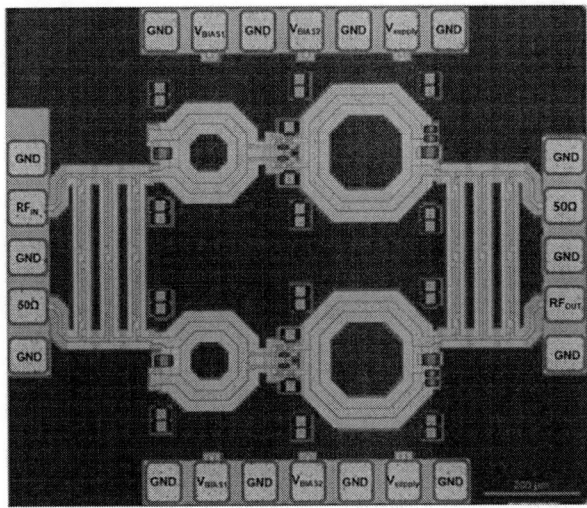

Fig. 13. Microphotography of the SiGe Ku-band PA

Compared with the reported Ku-band silicon-based PAs in Table 3, the proposed PA demonstrates the first balanced architecture reported with both a saturated power higher than 27dBm and a high robustness to the 2:1 SWR. It also demonstrates for an elementary power cell the highest *PAE* at 3dB compression point (28.7%), the highest *Psat* (23.6dBm) and the highest linear gain (14.9dB) at 30°C for a chip area of 0.4mm². It finally demonstrates the first PA implemented to achieve similar performances at 30°C and 90°C.

Compared to [7] presenting a high linear gain at 12.1GHz and a higher *PAE_{3dB}* with a 0.35μm SiGe technology, our proposed elementary PC achieves at the same temperature a slightly higher *Psat* for a higher frequency and a significant reduced size. Compared to [8] working at 14GHz with a 0.25μm SiGe technology, our elementary PC achieves with the same chip area, at a higher temperature and at a higher frequency a similar linear gain and a similar saturated power with a higher *PAE_{3dB}*. Compared to [2] presenting a very high *Psat* and a higher *PAE_{3dB}* at 14GHz with a 0.25μm SiGe technology, our proposed PA achieves at a higher frequency a better linear gain for a significant reduced size.

The biasing of the PC and the balanced PA are identical. Despite the losses of the power combiner, a similar *PAE* is obtained because the couplers of the balanced PA lead to a slightly different output matching which improves the *PAE*.

No similar PAs are reported for any frequencies upper than 14GHz in the Ku-band, neither for a 0.13 μm SiGe technology, nor on measurement temperatures above 30°C, nor with a 2:1 SWR among all reported Ku-band SiGe PAs.

V. CONCLUSION

A 0.13μm BiCMOS SiGe compact power amplifier operating at 18GHz and up to 90°C with a strong robustness to the 2:1 SWR is implemented. Two power cells are placed in a balanced architecture using an innovative combination of transistor and coupler layout techniques.

This power amplifier offers a twisted power combination for saturated power levels greater than 27dBm up to 90°C in the Ku-band, with a high robustness to the 2:1 SWR and with efficiency levels higher than 25%.

ACKNOWLEDGMENT

This work was supported by the ANRT association. The authors thank B. Plano (INP Bordeaux), M. De Matos and M. Potereau (IMS laboratory) for their technical supports.

REFERENCES

[1] H. Wang *et al.,* "Power Amplifiers Performance Survey 2000-Present," [Online]. Available: https://gems.ece.gatech.edu/PA_survey.html

[2] Y. Chen, *et al.,* "A 1-Watt Ku-band power amplifier in SiGe with 37.5% PAE," in *2016 IEEE Radio Frequency Integrated Circuits Symposium (RFIC)*, 2016, pp. 324-325.

[3] I. Ju and J. D. Cressler, "An X-band inverse class-F SiGe HBT cascode power amplifier With harmonic-tuned output transformer" in *2017 IEEE Radio Frequency Integrated Circuits Symposium (RFIC)*, 2017, pp. 390-393.

[4] S. Redois *et al.,* "Quasi Inverse Class-F X-Band Highly Efficient Power Amplifier with 51.8% Peak PAE in SiGe," in *2019 49th European Microwave Conference (EuMC)*, 2019, pp. 832-835.

[5] B. Coquillas *et al.,* "Ku-Band Cascode Balanced SiGe Power Amplifier with High Robustness to Active SWR" in *2021 19th IEEE International New Circuits And Systems Conference (NEWCAS)*, 2021, pp. 1-4.

[6] V. Knopik *et al.,* "Integrated scalable and tunable RF CMOS SOI quadrature hybrid coupler" in *2017 12th European Microwave Integrated Circuits Conference (EuMIC)*, 2017, pp. 159-162.

[7] S. Gerlich and P. Weger, "Highly Efficient Packaged 11–13 GHz Power Amplifier in SiGe-Technology With 37.3% of PAE," in *IEEE Microwave and Wireless Components Letters*, vol. 23, no. 10, pp. 539-541, Oct. 2013.

[8] Y. S. Noh *et al.,* "A Compact Ku-Band SiGe Power Amplifier MMIC With On-Chip Active Biasing," in *IEEE Microwave and Wireless Components Letters*, vol. 20, no. 6, pp. 349-351, June 2010.

Table 3. Comparison with the state-of-the-art of the Ku-band SiGe-based power amplifiers (‡graphically estimated)

References	SiGe (μm)	Freq. (GHz)	V_SUPPLY (V)	Pout 3dB / Psat (dBm)	PAE 3dB (%)	Gain (dB)	Chip Area (mm²)	PA Architecture
[7] (30°C)	0.35	12.1	1.8	22.5‡ / 23.4	35‡	20.0‡	1.71	Differential 2 stages
[8] (30°C)	0.25	14	4	23.0‡ / 24.45	27.5‡	15.8‡	0.38	One-stage cascode
[2] (30°C)	0.25	14	5	30 / 29	36‡	12.8	9.45	16-way output comb.
This work (30°C)	0.13	18	3.3	23.4 / 23.6	28.7	14.9	0.38	1 PC alone
		18	3.3	25.2 / 26.1	27.2	13.5	1.2	1 balanced PA
		18	4.2	26.6 / 27.4	25.0	13.7		
This work (90°C)	0.13	18	3.3	23.0 / 24.0	27.1	14.5	0.38	1 PC alone
		18	3.3	24.8 / 25.9	25.4	13.3	1.2	1 balanced PA
		18	4.2	26.3 / 27.1	24.1	12.8		

Proceedings of the 16th European Microwave Integrated Circuits Conference

A 4 GBaud 5 Vpp Pre-Driver for GaN based Digital PAs in 22 nm FDSOI using LDMOS

Frowin Buballa[#], Sebastian Linnhoff[#], Thomas Hoffmann[*], Andreas Wentzel[*], Wolfgang Heinrich[*], Friedel Gerfers[#]

[#]Mixed Signal Circuit Design, Technische Universität Berlin, Germany

[*]Ferdinand-Braun-Institut gGmbH (FBH), Leibniz-Institut für Höchstfrequenztechnik, Germany

{f.buballa, sebastian.linnhoff, friedel.gerfers}@tu-berlin.de, {thomas.hoffmann, wentzel, heinrich}@fbh-berlin.de

Abstract — This paper presents a 4 GBaud pre-driver chip for GaN-based RF digital transmitter chains providing voltage swings up to 5 V_{pp} at typical capacitive load impedances (0.25 pF). The compact monolithic microwave integrated circuit (MMIC) driver has been fabricated using a 22 nm FDSOI technology featuring 6.5 V LDMOS to achieve high-voltage swings and high switching speeds at the same time. The chip exhibits an area of 710x760 μm^2 and consumes about 488 mW of DC power. Standalone driver characterization and tests in combination with a real GaN-MMIC have been conducted. Using a high ohmic (1 kΩ), ultra-compact self-made probe the driving signal of the proposed pre-driver has been tested at the interconnection between MOS and GaN devices. For a 0.35 V_{pp} 1 GHz PWM 50 % duty-cycle differential input signal the driver IC achieves 4.1 V_{pp} of output voltage amplitude with rise/fall times of 200 ps and 150 ps, respectively, at the gate of the 4x125 μm first-stage GaN-HEMT.

Keywords — CMOS, LDMOS, 22nm FDSOI, power amplifiers, digital, GaN, driver amplifier, digital Tx.

I. INTRODUCTION

Over the recent years there has been an increased demand for high-power, high-speed switching transistors for the next generation of wireless base stations, as well as power electronics and microwave power amplification applications. Gallium Nitride (GaN) based high electron-mobility transistors (HEMTs) have proven to be a suitable technology to fulfill the challenging requirement of high switching speed combined with a high breakdown voltage [1], [2]. However, depending on the application, interface issues with typically employed state of the art devices arise. While GaN based HEMTs require a switching voltage of several Volt (4 to 5 V), state of the art CMOS-technologies operate with supply core voltages of 1 V and below. Therefore a suitable pre-driver is necessary, which is the focus of this work. Recent publications engaging with this topic feature BiCMOS technologies or make use of a monolithic integrated GaN gate driver [1]-[3] . Employing BiCMOS technologies shows promising results, however these designs only achieve switching speeds up to several MHz [4].With a monolithic integration of GaN-HEMT and driver, data rates up to the GHz-range are achievable [3]. However, GaN designs consume a large area and are therefore expensive, compared to state of the art CMOS technology implementations.This work presents the first design of a ultra-wideband pre-driver operating up to 4 GBaud driving a GaN-HEMT, completely designed in a 22 nm CMOS technology, by using laterally diffused

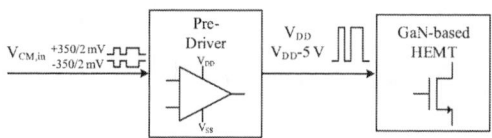

Fig. 1. Signal chain of the amplifier design. For the demonstration, the digital input-signal will be generated by laboratory equipment

MOSFET (LDMOS) devices, available in this technology, with a breakdown voltage up to 6.5 V. The presented pre-driver is designed to deliver the required swing of 5 V_{pp} while providing data rates up to 4 GBaud, thereby providing high data rates while avoiding the need for expensive GaN integration. In this paper the LDMOS driver design is discussed in Section II, including a brief introduction of the high-voltage LDMOS device available in 22nm FDSOI. In Section III measurements results, which include the GaN-HEMT load, are discussed.Section IV concludes the paper.

II. LDMOS PRE-DRIVER DESIGN

A data rate of 4 GBaud with rise- and fall- times below 100 ps is targeted. Therefore a bandwidth beyond 3 GHz is desirable. For the design the capacitive load of 250 fF primarily from the GaN-HEMT and the additional impedance of the bondwire connecting the amplifier to the GaN device has to be taken into account. The pre-driver is designed to be driven by an differential pulse-width modulated (PWM) input signal with a 350 mV_{pp} amplitude.

A. 22 nm FDSOI technology

The 22 nm FDSOI technology features a special high-voltage LDMOS transistor which is able to reliably operate at supply voltages of up to 6.5 V. As this voltage range exceeds the requirement for a 5 V output swing, while still enabling highest switching speeds in the GS/s range, the pre-amplifier circuit can be substantially simplified. Figure 2 shows the cross-section of the nLDMOS. The drain of the transistor is enveloped by two additional wells, compared to a common MOS transistor. The n-type channel is created in a lightly-doped n-type region resulting in desirable features such as low on-resistance and a high voltage isolation. With an f_T of 40 GHz it is a excellent candidate for a high-speed high-voltage

3–4 April 2022, London, UK

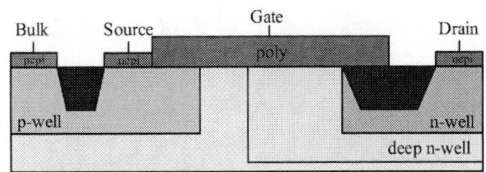

Fig. 2. Cross-section of the nLDMOS [5]

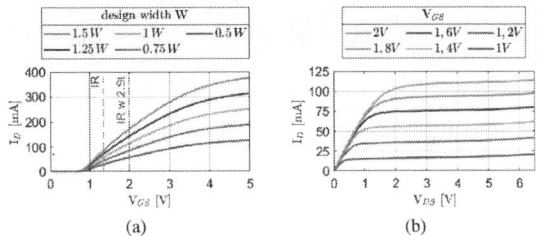

Fig. 3. (a) Simulated input characterization for the different nLDMOS widths, comparable to the utilized width W. Input Range (IR) is marked; (b) Simulated output characterization for the nLDMOS for different input voltages

switching applications [5]. Figure 3a shows the input- and 3b the output characterization n-type LDMOS for different widths input voltages respectively, to give a brief overview of the LDMOS performance.

B. Driver Design

Enabled by the LDMOS, a class-A architecture featuring a fully differential nLDMOS pair with a resistive load is chosen. This architecture enables both, a sufficient slew rate to achieve the desired rise- and fall-times and low output impedance to achieve a high bandwidth given a 250 fF load. However, for this current-mode logic design as shown in Figure 4, the input devices (M6/M7) switch between low- and high-current mode. To enable the maximum output swing, while only a $350\,mV_{pp}$ input swing is provided, a very large input device is required, which significantly lowers the bandwidth due to the introduced gate capacitance. To circumvent this limitation the design features an additional amplifier stage which provides sufficient amplification to generate an input-signal with a $>1\,V_{pp}$ signal swing for the 2nd stage. First and second stage use different supply voltages of 2 V for the first and 6.5 V for the second stage. The first stage is implemented using thick oxide 1.8 V I/O devices, while the second stage features the previously discussed LDMOS devices to enable the desired output swing. To further expand the bandwidth, as well as improve the rise- and fall-times, inductive peaking is introduced via the inductors $L_{1.st}$ and $L_{2.st}$. Figure 5a shows the simulated bandwidth of the driver for SS- and FF-corner as well as TT-corner with and without inductive peaking. Figure 5b shows the eye-diagram for a 4 GBaud PRBS input signal across corner based on the extracted layout, including a bond wire model. In TT-corner, the driver achieves rise- and fall-times (10 %-90 %) of t_{rise}=90 ps and t_{fall}= 95 ps. To address the variation of the resistance across process-corner and temperature an on-chip tracking loop

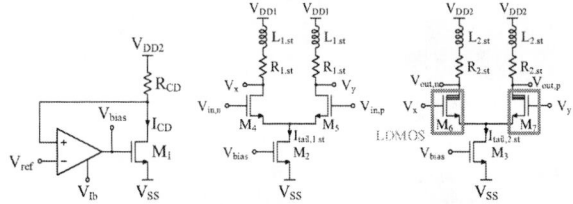

Fig. 4. Overall schematic of the proposed design, including the 2 amplifier stages and the voltage regulator

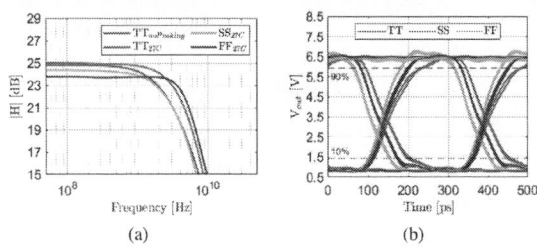

Fig. 5. (a) Simulated bandwidth of the amplifier across corner; (b) Simulated eye diagram of single ended output for 250 ps bit-period PRBS-input signal

is implemented to adaptively adjust the bias voltages of the current sources.

III. MEASUREMENT

A. CMOS driver chip

A demonstrator was fabricated in 22 nm FDSOI technology measuring 710 µm x 790 µm. The MMIC PA takes an area of $81\,\mu m^2$ including $66\,\mu m^2$ for the inductors. The functionality of the chip was tested using wafer-level probing, therefore omitting the additional inductive load of the bondwire and the capacitive load of the GaN-HEMT. Figure 6b shows the block diagram of measurement setup, including bias tees to adjust the input common mode and DC-blocks at the output, due to the voltage level limitation of the used real-time oscilloscope. Since the driver is designed for a capacitive load, the output resistance of the driver forms a a resistive divider with the $50\,\Omega$ input impedance of the oscilloscope, decreasing the output amplitude. Figure 7 shows the measurement results for a 6 GBaud PRBS input signal. Comparing the difference between the one-and zero-voltage to the simulated results, it can be calculated, that the eye height is decreased by a factor $k = 2.7$. Therefore it can be assumed, that the achieved eye-height of 1.62 V corresponds to an eye-height of 4.3 V for a pure capacitive load.

B. Digital Tx chain: CMOS and GaN MMIC

To check the performance of the fabricated CMOS pre-driver on the real capacitive input load of around 0.25 pF of a typical first-stage GaN-HEMT the driver MMIC is connected to a GaN chip via bond wires. For reasons of simplicity an available GaN-MMIC from FBH is selected. It has been published in [1]. The GaN-MMIC is a 2-stage circuit which was fabricated using the FBH 0.25 µm GaN-HEMT process (GaN-on-SiC). Figure 8a shows the schematic and 8b the photograph of the used GaN chip.

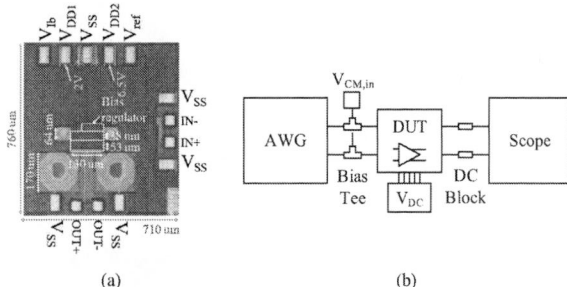

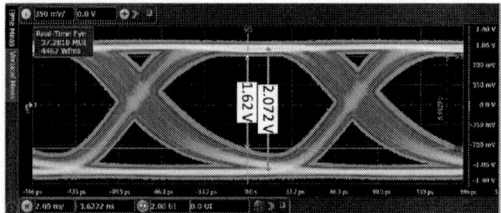

(a)   (b)

Fig. 6. (a) photograph of the fabricated driver chip; (b) block diagram of the setup for the $50\,\Omega$ measurement to confirm the functionality

Fig. 7. Measured eye diagram with a 50Ohm input impedance sampling oscilloscope for a 6 GBaud data rate

The GaN PA chip uses a push-pull configuration of two $8\times250\,\mu m$ GaN-HEMTs ($T_{1/2}$ in Fig. 8a) in the final-stage switching the output voltage at the node RF_{out} on and off between V_{dcin} (up to 40 V) and 0 V. The driver is an active pull-up circuit simply connecting the high-side driver (drain of T_{D3}) to the switching node RF_{out}. More details on the module and chip can be found in [1].

To be able to better classify and interpret the measurement results, it is important to look at the actual (capacitive) input impedance of the GaN transistors in the first stage of the GaN chip ($T_{D1/2}$ in Fig. 8a). T_{D1} and T_{D2} have a gate-width of $2\times125\,\mu m$ and $4\times125\,\mu m$, respectively. According to parameters of FBH GaN-HEMT model for the appropriate wafer run and transistor sizes T_{D2} is the best choice to check in terms of driving input signal. This is due to the fact that the input capacitive load ($C_{gs} + C_{gd}$) is around 300 fF for T_{D2} and thus very close to the specified 250 fF (cp. Sec. II). T_{D1} shows approximately 140 fF of input capacitance.

CMOS driver as well as GaN converter chip are combined together on a copper heatsink and connected via gold bond wires. The CMOS chip is glued while the GaN MMIC is soldered on the heat sink. A PCB around the chips provides DC biasing and RF input and output connection. The compact final demonstrator is shown in Fig. 9a.

To gain a clear picture about the driving capabilities of the proposed CMOS pre-driver circuit (Sec. II) it is important to probe directly at the input of the GaN chip, i.e., in this case on the input pad of T_{D2} (see above and Fig. 8a). As the pad is small ($140 \times 140\,\mu m^2$) and additionally a connecting bond wire is placed on it, a very precise and small probe is required. For this purpose, a self-made high ohmic ($\approx 1\,k\Omega$) probe was built. It uses two very thin former DC needles at the end of

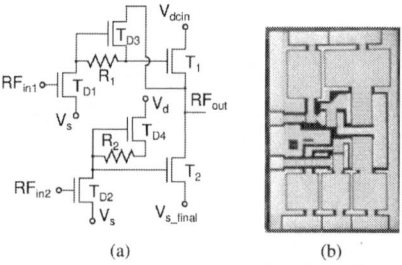

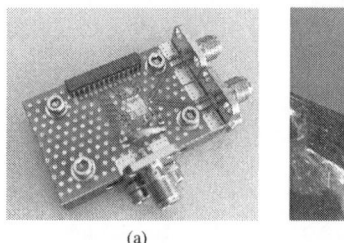 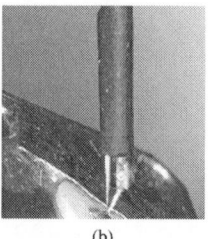

(a)   (b)

Fig. 8. Schematic (a) and chip photograph (b) of used 2-stage GaN MMIC to check driving capability of proposed CMOS chip; chip area: $1.6 \times 2.4\,mm^2$.

(a)   (b)

Fig. 9. (a) Fabricated hybrid module including CMOS pre-driver and GaN converter chip as well as DC bias circuitry on a PCB; size: $40 \times 25 \times 14\,mm^3$; (b) self-made highly precise high ohmic ($1\,k\Omega$) probe.

a coaxial cable to form GND and signal port. In terms of the signal line first two $500\,\Omega$ 02016 resistors from Vishay's CH series are connected in series via soldering and the output pin of the second resistor is connected to the signal line needle for on-chip/pad probing. Consequently, with the scope input impedance of $50\,\Omega$ this results in a voltage division ratio of 1:21. The probe has been characterized with a network analyzer from 10 MHz to 8 GHz and in this range the attenuation varies by only 0.3 dB. Thus, 3-dB bandwidth is higher than 8 GHz. Additionally, it is important to note that the probe impedance varies around the targeted $1\,k\Omega$ value from minimum $720\,\Omega$ (at 5.5 GHz) to maximum $1230\,\Omega$ (at 7.1 GHz). Fig. 9b shows the built-up probe.

All measurements have been conducted with PWM input signals generated by a Keysight M8195A AWG. The voltage amplitude of the two input PWM signals of the driver MMIC is $350\,mV_{pp}$ each. Bias-Tees in each input branch set the required DC potential for the two input gates of the driver (see Fig. 4), similar to the measurement setup in Figure 6b. A 70 GHz bandwidth real-time scope from Tektronix is used to capture the voltage signal from the high-ohmic probe in time-domain. It is important to note that, unfortunately, the output signals of the GaN chip could not be measured as it has been damaged after saving measured data from the probe connection.

First, a 50 % duty-cycle PWM signal with a bit rate of 2 GBit/s (1 GHz signal frequency) has been checked at the input of T_{D2} (cp. Fig. 8a). Fig. 10 shows the measured output voltage in time-domain.

Considering the characteristics of the high ohmic probe (varying impedance vs. frequency) and influences of the bond wire nearby one can derive from Fig. 10 a clean input signal for T_{D2}.

158

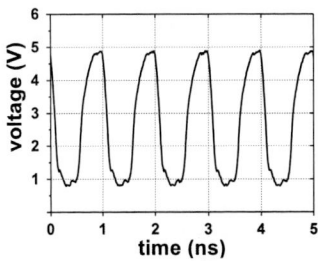

Fig. 10. Measured output voltage waveform vs. time for 1 GHz 50 % duty-cycle PWM input signal; probed with 1 kΩ self-made probe at input of T_{D2} (RF$_{in2}$ in Fig. 8a).

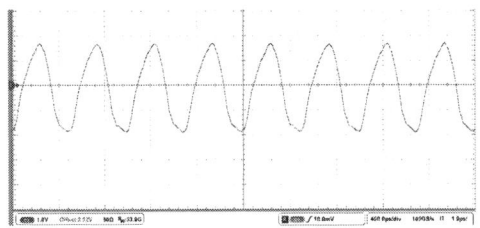

Fig. 11. Screenshot of measured output voltage waveform (1.1 V – 4.7 V) vs. time (0 – 4 ns) of proposed CMOS driver MMIC at the gate of T_{D2} (see Fig. 8a); input signal: 50 % duty-cycle PWM; signal frequency: 2 GHz.

With the bias conditions from Sec. II. the pre-driver delivers a voltage amplitude of around 4.1 V$_{pp}$ between 0.8 – 4.9 V at the gate of T_{D2}. This amplitude is enough to safely switch the transistor on and off. One has to mention that the effective V$_{gs}$-swing of the GaN-HEMT T_{D2} has been set between -4 V and 0 V. This has been realized by shifting up all potentials of the GaN chip (see Fig. 8a) according to the required input bias points, i.e., in this case by +11 V. Also, there is still room for higher output amplitude of the CMOS driver chip, but for this first test it was always operated in a safe region. In addition, rise- and fall-times have been determined from the signal in Fig. 10. While rise times (10 %-90 %) are in the range of 200 ps fall times (10 %-90 %) of around 150 ps have been measured. These values are almost double the simulated ones and 130 % increased compared to 50 Ω characterization of the CMOS driver chip alone. These values clearly need improvement.

The direct influence of these higher rise- and fall-times on the waveform of the driving signal of the GaN-HEMT is visualized with a voltage waveform screenshot for a 50 % duty-cycle PWM input signal with a bit rate of 4 Gbit/s. The resulting signal frequency is 2 GHz. Fig. 11 shows the waveform. Please note that only the shape of the output waveform is important here.

One can clearly see that the voltage waveform for such a high bit rate suffers from the observed rise- and fall-times. Since one period for this signal is 500 ps in time and pulses have a width of 250 ps in the 50 % duty-cycle case they are distorted to a triangular shape. Moreover, a lower output voltage amplitude is achieved (3.6 V$_{pp}$). This can be critical in terms of fully switching on/off the GaN-HEMTs which then decreases efficiency of the whole GaN circuit. The reasons for these deviations are currently being intensively investigated and will

be considered in the next CMOS design process. But one also has to state that for the first CMOS-based GaN pre-driver circuit with new nLDMOS option the results are good and very encouraging. Future work will focus on making the driver even faster to realize clean edges up to signal frequencies of 6 GHz for 5G applications. But not only for this, also for other sub 6 GHz applications like e.g. in VHF DC/DC converters, where line-up efficiency and costs are very important, the proposed driver approach represents a suitable candidate.

IV. CONCLUSION

A new CMOS-based driver circuit for GaN-HEMTs operating in switching mode has been proposed in this paper. It consist of two amplifier stages making use of LDMOS devices and inductive peaking. The driver has a DC power consumption of 488 mW. The driving capabilities have been checked with an existing GaN-MMIC from FBH which was fabricated on the in-house 0.25 µm GaN-on-SiC process run. By using a self-made high ohmic (1 kΩ) probe from FBH the input signals at the real capacitive input load (around 0.3 pF) of the GaN-HEMT used (4x125 µm gate-width) have been tested directly at the gate of the GaN-transistor. The driver chip delivers for a 1 GHz 50 % duty-cycle PWM input signal a voltage amplitude of 4.1 V$_{pp}$ at the gate of the GaN transistor. Rise- and fall-times of 200 ps and 150 ps have been measured, respectively. Thus, this first version of a CMOS driver chip for GaN-based transmitter chains is capable to amplify GHz-range signals on a capacitive load up to 0.3 pF. Future work will focus on improvement of rise- and fall-times to drive GaN circuits up to 6 GHz for e.g. 5G communication. Also, the reduction of the power consumption is a center topic for the next generation of the driver.

ACKNOWLEDGMENT

The authors would like to thank Ravi Subramanian and Mentor Siemens for Analog FastSPICE (AFS) support. They would also like to thank GLOBALFOUNDRIES for the University Multi Project Wafer Program.

REFERENCES

[1] A. Wentzel *et al.*, "A highly efficient ghz switching gan-based synchronous buck converter module," in *2019 European Microwave Conference in Central Europe (EuMCE)*, 2019, pp. 18–21.

[2] Y. Zhang *et al.*, "Very high frequency pwm buck converters using monolithic gan half-bridge power stages with integrated gate drivers," *IEEE Transactions on Power Electronics*, vol. 31, no. 11, pp. 7926–7942, 2016.

[3] T. Hoffmann *et al.*, "Broadband driver amplifier with voltage offset for gan-based switching pas," in *2020 IEEE/MTT-S International Microwave Symposium (IMS)*, 2020, pp. 273–276.

[4] A. Seidel and B. Wicht, "A fully integrated three-level 11.6nc gate driver supporting gan gate injection transistors," in *2018 IEEE International Solid - State Circuits Conference - (ISSCC)*, 2018, pp. 384–386.

[5] C. Schippel *et al.*, "A 22nm fdsoi technology with integrated 3.3v/5v/6.5v rfldmos devices for iot soc applications," in *2018 IEEE SOI-3D-Subthreshold Microelectronics Technology Unified Conference (S3S)*, 2018, pp. 1–3.

Proceedings of the 16th European Microwave Integrated Circuits Conference

400-Watt S-band Power Amplifier MMIC

A.P. de Hek[1], G. van der Bent[2], F.E. van Vliet[3]

TNO, The Hague, The Netherlands

{[1]peter.dehek, [2]gijs.vanderbent, [3]frank.vanvliet}@tno.nl

Abstract — **A state-of-the-art, second harmonically tuned, 400 W S-band MMIC power amplifier has been developed. This amplifier delivers in the 2.8 – 3.3 GHz band an output power of 400 W and a PAE between 50-55%. The obtained amplifier performance is the result of the structured design approach discussed in this paper.**

Keywords — **High Power Amplifier, Microwave Monolithic Integrated Circuit (MMIC), S-band, PAE.**

I. INTRODUCTION

There is an ever−increasing need for higher output power levels for S-band radars to be able to increase the probability of detection. At the same time these increased power levels will come with increased DC current levels. Since phased array radar systems can contain a couple of thousand elements, it becomes necessary to reduce the supply current by using a higher drain voltage. For this design the NP45-11 technology as described in section II is used, which allows for operation up to 55 V. An MMIC implementation was selected to be able to increase the gain of the amplifier to more than 20 dB. In doing so, the drive level will remain compatible with standard components like for instance a phase driver; a recent example can be found in [1]. Such components are used to control the transmit beam and at the same time generate sufficient power to drive the power amplifier into compression. The design of the power amplifier is discussed in section III. A summary of the obtained measurement results is given in section IV. Finally, in section V a comparison with the state-of-the-art is made and several conclusions have been drawn.

II. TECHNOLOGY.

For the design the NP45-11 technology of WIN Semiconductors has been used.

Fig. 1. Photograph of 10x500 um transistor.

This is a 0.45 μm Power GaN/SiC HEMT technology with an Enhanced Moisture Ruggedness (EMR) layer, which enables application of the amplifier in a plastic package. The technology is suitable for operation up to C-band.

The transistor layout utilizes no airbridges to make the EMR layer feasible. Therefore, the transistor layout has a via-hole between every pair of two gate fingers, see Fig. 1.

The process includes TaN resistors, inductors, MIM capacitors, via-holes and two interconnection metals.

III. AMPLIFIER DESIGN

A. Design procedure

The amplifier design is performed in the following steps:

- Determine the optimum load impedance, output power density and PAE of a transistor
- Establish amplifier architecture
- Stabilise selected transistors
- Design matching networks
- Stabilise amplifier
- Perform thermal evaluation and check lifetime

These steps will be discussed into more detail in the following sections.

B. Transistor characterization

Fundamental and second harmonic load-pull measurements have been performed on a transistor with 10 gate fingers which each have a 500 μm gate width. The results, at the load impedance used in the amplifier design, are depicted in Fig. 2.

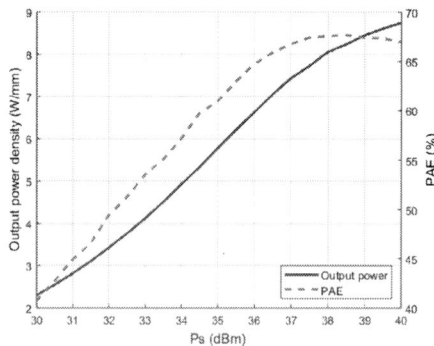

Fig. 2. Measured output power and PAE of a 10x500 um NP45-11 transistor measured at f=3.2 GHz, Vd=48 V, Vg=-2.8 V, PW=10 us, DC=10%, T=25 °C, load reflection coefficients: Γ_{f0}=-0.54+0.36i and Γ_{2f0}=1.

3−4 April 2022, London, UK

At 48 V an output power density of ~9 W/mm and a PAE of 67% has been measured. Application at 55 V would result in ~11 W/mm and a similar PAE. The large-signal power gain is 17 dB, which is sufficiently high to be able to apply a transistor with these dimensions in a power amplifier.

The sensitivity to second harmonic load variations has been measured, see Fig. 3. In the amplifier matching network design care has been taken that the second harmonic phase of the load reflection coefficient is between 0-100°.

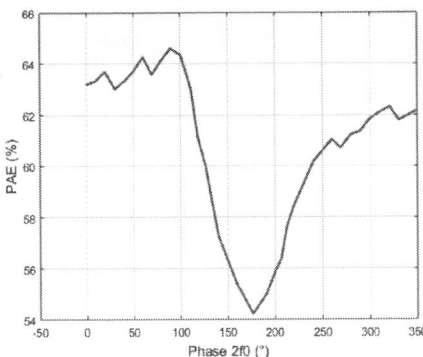

Fig. 3. Second harmonic load dependency of PAE with $|\Gamma_{2f0}|=1$ measured at f=3.2 GHz, Vd=48 V and Γ_{f0}=-0.59+j0.30.

C. Amplifier architecture

Based on the measured output power density it was decided to put eight 10x500 μm transistors in parallel. Considering an output matching network (OMN) loss of 0.7 dB, an output power of 375 W is expected. To be able to guarantee that sufficient margin is available to drive the output stage into compression two 10x250 um transistors have been used in the input stage, see Fig. 4.

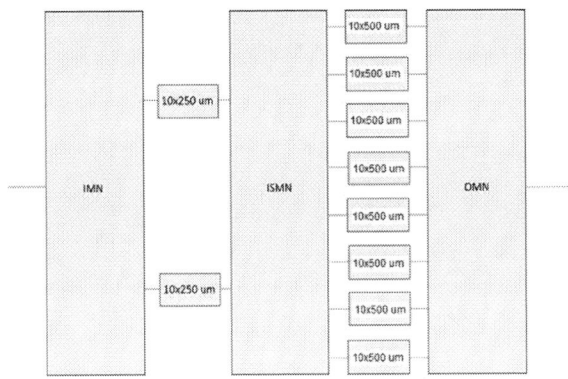

Fig. 4. Amplifier architecture.

D. Stability transistor

Before the design of the matching networks is performed, first the transistors are made unconditionally stable by applying a parallel RC network [2] at the gate of the transistors. The values of this RC network are established by first determining the minimum required series resistance, over bias, frequency and temperature that is needed to guarantee

unconditional stability. This resistance value will vary over frequency and an RC network is used to create the necessary frequency dependent resistance.

E. Matching networks design

The optimum load impedance and the transistor input impedance, determined from the before discussed load-pull measurement results, are used for the design of the matching networks. The used impedances are listed in Table 1.

Table 1. Model of input and output impedance transistor.

R$_{IN}$ (Ω.mm)	C$_{IN}$ (pF/mm)	R$_{LOAD}$ (Ω.mm)	C$_{LOAD}$ (pF/mm)
1.58	12.00	185.00	0.39

For the output matching design, the following strategy has been used. First two transistors have been combined. It has been investigated whether combining four transistors would be a better strategy, as it could have resulted in a more compact layout. However, the lower input impedance of the network after combining would have led to more losses.

Secondly the imaginary part of the resulting load impedance of the transistors has been compensated by adding a parallel inductor after the combination point. This inductor is also used to bias the transistors [3].

Thirdly all transistors are combined with low-pass networks. The resulting network has a loss of around 0.7 dB. The metal widths have been dimensioned such that that they can support duty cycles up to 30%.

For the design of the interstage matching and input matching networks the influence of the stability network at the gate must be taken into consideration. The capacitor present in this network will reduce the obtainable bandwidth. For the design of these networks the procedure outlined in [3] has been used.

The layout of the interstage matching network has been kept as symmetrical as possible, to be able to reduce the amplitude unbalance at the input of the second stage transistors to a minimum. This is necessary to prevent reduction of the obtainable frequency band.

F. Stability amplifier

The overall stability of the amplifier has been checked under both small-signal and large-signal conditions with the help of the method described in [4]. No instabilities were found. Resistors have been placed in between the transistors to be able to further guarantee odd mode stability.

G. Thermal evaluation

The temperature of the amplifier has been simulated for the case in which the chip is soldered with AuSn in a QFN7x7 package. The simulated performance for a 100 μs pulse width is depicted in Fig. 5. A junction temperature of 225 °C is taken as maximum for reliable operation. From the depicted results it can be concluded that the maximum duty cycle that can be used depends on the backside temperature of the QFN package.

For a temperature of 100 °C the maximum duty cycle is 30%, which corresponds to the choices in metal width.

For monitoring the backside temperature of the amplifier, a temperature sensor has been integrated with the help of TaN resistors at both sides of the output stage.

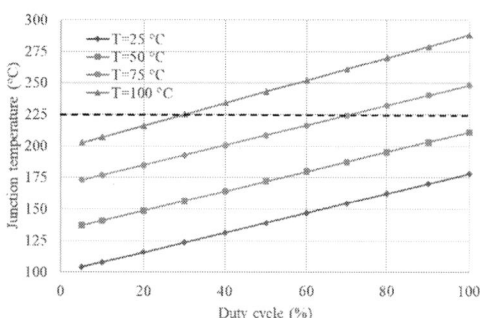

Fig. 5. Simulated temperature output stage for a pulse width of 100 us as function of duty cycle and backside temperature QFN package.

IV. MEASUREMENTS

The realised amplifier is depicted in Fig. 6.

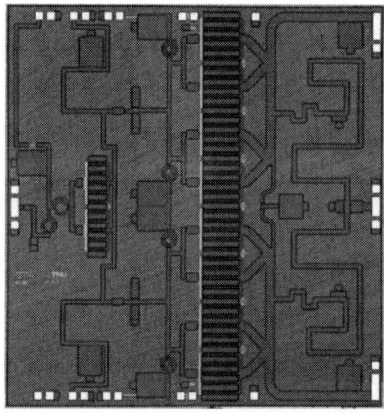

Fig. 6. Photograph of 400 W amplifier, chip size: 5.25 x 5.4 mm^2.

Two (add wafer diameter) wafers have been processed. On-wafer measurements have been performed on one halve wafer. In total 28 amplifiers were fabricated on this halve wafer. Twenty-six amplifiers were fully functional. The measured output power and PAE have been depicted in Fig. 7. The results show over the 2.9 - 3.3 GHz band an output power between 390 – 410 W with a variation between chips of ±0.1 dB. A PAE between 49-55% with a variation between chips of ± 1% point has been measured.

Samples from the second wafer have been soldered on a Cu/Mo carrier. The measured output power and PAE are comparable to the on-wafer results depicted in Fig. 7. Both the output power and PAE have been measured as function of the applied drain voltage. The results depicted in Fig. 8 show that

the PAE varies only with ± 1.5% point over a large part of the measured voltage range. The output power increases, as expected, almost according to a straight line with increasing supply voltage.

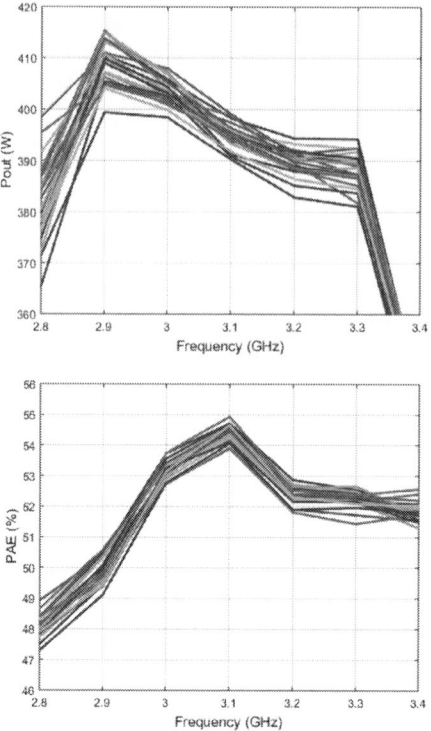

Fig. 7. Measured output power and PAE at Vd=55 V, Idq0=1 A, Ps=30 dBm, PW=2 us, DC=1%, and T=25 °C.

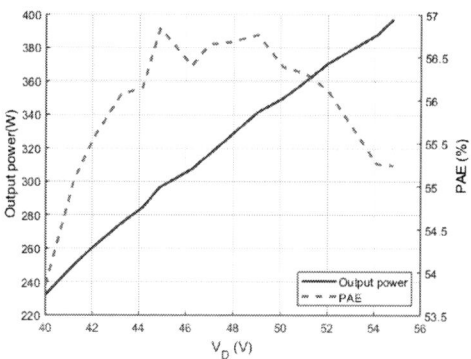

Fig. 8. Measured output power and PAE as function of drain voltage measured at f=3.1 GHz, Vg=-3 V, Ps=30 dBm, PW=2 us, DC=1%, and T=25 °C.

The measured temperature dependency is shown in Fig. 9. For the output power a reduction of 0.32 W/°C and for the PAE a value of -0.045 %/°C over temperature are found.

The dependency of pulse width is depicted in Fig. 10. As expected, both the output power and PAE are reduced with increasing pulse width. A thermal simulation of the amplifier shows a temperature increase, between a pulse width of 5 and 100 µs, of ~50 °C. This would, from a thermal point of view, correspond with a reduction of respectively 16 W in output power and 2.25% in PAE. The measured results show a reasonably good correspondence for the PAE. However, for the output power a reduction of 25 W is measured.

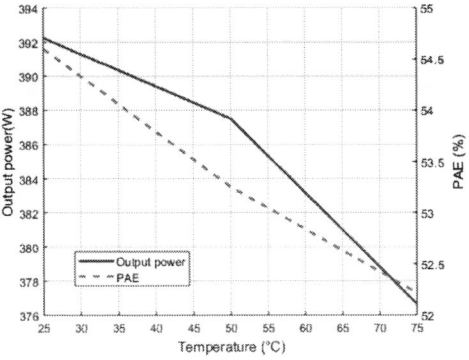

Fig. 9. Measured output power and PAE as function of temperature measured at f=3.1 GHz, Vd=55 V, Vg=-3 V, Ps=30 dBm, PW=2 us, DC=1%, and T=25 °C.

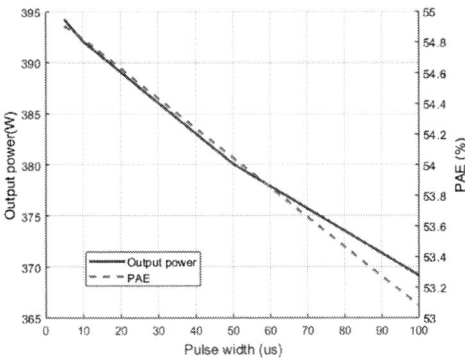

Fig. 10. Measured output power and PAE as function of pulse width measured at f=3.1 GHz, Vd=55 V, Vg=-3 V, Ps=30 dBm, DC=1%, and T=25 °C.

V. CONCLUSIONS

In this paper a structured design approach of an S-band power amplifier MMIC has been discussed. The realised amplifier has a measured average output power of 400 Watt

and a PAE of 53% in the 2.9-3.3 GHz frequency band, to the knowledge of the authors, the highest output power realized on-chip in this frequency band has been obtained, as can been seen from the results listed in Table 2, where a comparison with the state-of-the art has been made. The results show that 5 dB more output power has been realized on a MMIC with 26 dB of gain. This high gain makes it possible to drive the amplifier directly with already commercially available products.

ACKNOWLEDGMENT

The authors like to thank WIN Semiconductors for making the transistor samples available that are used in the load-pull measurements on which this amplifier design is based.

REFERENCES

[1] M. Van Heijningen, J. Essing and F. E. Van Vliet, "X-Band GaAs Phase Driver MMIC Optimized for GaN-Based Phased-Array Radar Transmit Chain," 2018 13th European Microwave Integrated Circuits Conference (EuMIC), Madrid, Spain, 2018, pp. 118-121, doi: 10.23919/EuMIC.2018.8539936.

[2] D. Teeter, A. Platzker and R. Bourque, "A compact network for eliminating parametric oscillations in high power MMIC amplifiers," 1999 IEEE MTT-S International Microwave Symposium Digest (Cat. No.99CH36282), Anaheim, CA, USA, 1999, pp. 967-970 vol.3, doi: 10.1109/MWSYM.1999.779547.

[3] Hek, de, A. P. (2002). Design, realisation and test of GaAs-based monolithic integrated X-band high-power amplifiers. Technische Universiteit Eindhoven. https://doi.org/10.6100/IR55615.

[4] M. Ohtomo, "Stability analysis and numerical simulation of multidevice amplifiers," in IEEE Transactions on Microwave Theory and Techniques, vol. 41, no. 6, pp. 983-991, June-July 1993, doi: 10.1109/22.238513.

[5] Qorvo," TGA2813, 3.1-3.6 GHz 100 W GaN Power Amplifier", data sheet 2013.

[6] A. Alexander and J. Leckey, "A 120 watt GaN power amplifier MMIC utilizing harmonic tuning circuits for S-band applications," 2015 IEEE MTT-S International Microwave Symposium, Phoenix, AZ, USA, 2015, pp. 1-3, doi: 10.1109/MWSYM.2015.7167045.

[7] G. van der Bent, A. P. de Hek, F. E. van Vliet and Z. Ouarch, "Single-Chip 100-Watt S-band Power Amplifier in 0.25 µm GaN HEMT MMIC Technology," 2020 15th European Microwave Integrated Circuits Conference (EuMIC), Utrecht, Netherlands, 2021, pp. 21-24, doi: 10.1109/EuMIC48047.2021.00017.

[8] P. K. Singh, K. Suman, S. K. Gedela, K. Bantupalli, K. Y. Varma and R. S. N. Gongo, "100 W High Power Amplifier MMIC in 0.45 µm GaN Technology," 2019 14th European Microwave Integrated Circuits Conference (EuMIC), Paris, France, 2019, pp. 1-4, doi: 10.23919/EuMIC.2019.8909397.

Table 2. Overview of state-of-the-art of S-band MMIC power amplifiers.

Ref.	Technology	Frequency (GHz)	Pout (W)	PAE (%)	Gain (dB)	Vd (V)	Size (mm²)
[5]	Qorvo TQGaN25	3.1-3.4	115	55	22	30	5.4x6.7
[6]	GSC	2.8-3.5	125	47	18	40	6.0x5.1
[7]	UMS GH25-10	2.9-3.3	100	57	24	30	6.0x5.0
[8]	WIN NP45-11	2.8-3.4	110	40	21	50	5.8x3.3
This work	WIN NP45-11	2.9-3.3	400	53	26	55	5.3x5.4

A 41.5 dBm Broadband AlGaN/GaN HEMT Balanced Power Amplifier at K-Band

S. Samis[*], C. Friesicke[+], T. Maier[+], R. Quay[+], A.F. Jacob[*]

[*]Institute of High-Frequency Technology, Hamburg University of Technology, Germany
[+]Fraunhofer Institute for Applied Solid State Physics (IAF), Germany
stanislav.samis@tuhh.de

Abstract—This paper describes the design and characterization of a broadband balanced power amplifier (BPA) MMIC at K-band. The utilized technology is the $0.25\,\mu$m AlGaN/GaN HEMT process provided by Fraunhofer IAF. The BPA is designed as a two-stage power amplifier (PA) with a total gate width (TGW) of 4.32 mm. More than 41.5 dBm of output power with a power-added efficiency (PAE) greater than $21\,\%$ are demonstrated between 21 and 23 GHz. Furthermore, the BPA exhibits a maximum saturated output power of 17 W associated with $22\,\%$ of PAE at 22.5 GHz. A peak PAE of $30\,\%$ is realized at a drain supply voltage of 28 V.

Keywords—Balanced power amplifier, K-band, MMIC, gallium nitride, satellite communication.

I. INTRODUCTION

Due to the ever increasing demand of higher data rates, today's satellite communication systems use extensively the K-band for data downlink. The performance of the transmitter RF front-end is commonly dominated by the high power amplifier (HPA). Although traveling-wave tube amplifiers (TWTAs) are still commonly used because of their outstanding power and efficiency characteristics, the choice of the HPA technology depends strongly on the particular application. In distributed antenna structures, such as active phased arrays or multiple-feed-per-beam (MFB) systems, where the needed output power per antenna element is in the range of several Watts, solid-state power amplifiers (SSPAs) provide many application-specific advantages. Compared to TWTAs, SSPAs can be directly integrated into the antenna structure and placed close to the individual radiating elements. Consequently, the power distribution takes place in front of the amplifier modules, where the low power levels allow for more loss in the beam-forming network.

The balanced configuration approach is a widely used power combining technique first introduced by Eisele [1]. Fig.1 is a schematic representation of the BPA. The structure consists of two identical amplification modules (PA) embedded between two quadrature hybrid couplers at the in- and output. The signal P_{in} entering at the RF input port is first divided by the input coupler into two equal components with a $90°$ phase offset. After amplification both signals are recombined by the second coupler at the RF output. Due to the phase properties of the coupler, the BPA can significantly improve the in- and output reflection and is thus very useful especially in broadband applications. In this work,

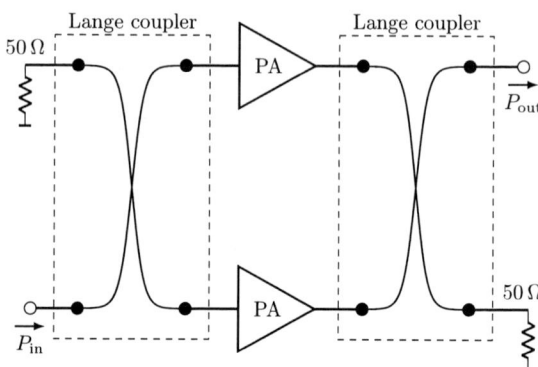

Fig. 1. Schematic representation of the BPA.

the balanced configuration is employed in order to reach more than 41.5 dBm of output power at K-band.

The work is structured as follows. Following a classification of the achieved results and their comparison to the published benchmarks in Section II, Section III introduces the employed AlGaN/GaN technology. Section IV and V present the realization of the Lange coupler and BPA, respectively. Afterwards, a discussion of the measured results and a conclusion are given in Sections VI and VII, respectively.

II. STATE OF THE ART

The realization of power amplifier MMICs exceeding 10 W of output power at K-band is nowadays still a great challenge. A comparison of published state-of-the-art K-band

Table 1. Comparison of published state-of-the-art GaN power amplifiers at K-band.

Freq. (GHz)	P (W)	Gain (dB)	PAE (%)	Gate length (μm)	Stages -	Ref. -
17.0-20.2	10	22	30	0.1	3	[3]
17.2-20.2	10	18	35	0.2	2	[4]
17.3-20.2	9	24	28	0.1	3	[5]
17-21	10	18.5	32	0.15	2	[6]
18.75	15.8	19	36	0.2	2	[7]
21-23	**14**	**20**	**21**	**0.25**	**2**	**BPA**

MMICs based on the GaN technology is illustrated in Tab.1. In [3]-[6] good efficiency is demonstrated between 17 and 21 GHz. However, the achieved output power over the reported bandwidth is in the range of 10 W. Moreover, 15.8 W of output power associated with 36 % of PAE was obtained at a single frequency of 18.75 GHz in [7]. The BPA presented here exhibits more than 41.5 dBm (14 W) of output power between 21 and 23 GHz. Additionally, the maximum saturated output power of 17 W with 22 % of PAE was measured at 22.5 GHz. To the best of the authors' knowledge this is the highest output power obtained from a single K-band GaN on SiC MMIC. The results presented here therefore demonstrate state-of-the-art performance with regard to the combination of output power, bandwidth, and efficiency at K-band.

III. 250 nm ALGAN/GAN HEMT TECHNOLOGY

The amplifier MMICs realized in this work are fabricated on a 4-inch semi-insulating SiC substrate wafer with a thickness of 100 μm. The European AlGaN/GaN HEMT process (GaN25) with a gate length of 0.25 μm was provided by the Fraunhofer Insitute of Applied Solid-State Physics (IAF) [8]. The epitaxial structures are grown by metal organic chemical vapor deposition (MOCVD). The gates are defined by electron-beam lithography. The technology features MIM-capacitors, nickel chromium (NiCr) thin-film resistors and square via-holes. The GaN25 process was extensively analyzed at X-band. However, At K-band the technology is already close to its frequency limits. The cut-of frequencies f_T and f_{max} are 28 GHz and 40 GHz, respectively. A fully implemented Keysight ADS design kit including a passive microstrip line library and active HEMT models is available for circuit simulation.

IV. LANGE COUPLER DESIGN

The in- and output coupler used in this work is an interdigitated 3-dB coupler described by Lange [9]. It consists of four parallel lines (fingers) with alternate fingers tied together. Fig. 2 shows the schematic of the employed Lange coupler. The geometry is determined by three parameters w,

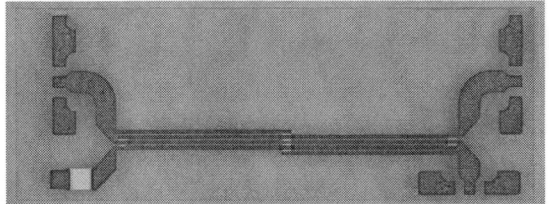

Fig. 3. Chip photograph of the realized Lange coupler.

s, and l. The length of the coupling region is controlled by l and corresponds approximately to a quarter wavelength at the center frequency of the operating band. All fingers have the same width w and are spaced by s. The connections between the interdigitated fingers are realized by cross-over elements consisting of air bridges and underpasses. The optimal parameters of the Lange coupler were extracted by the 3D EM simulator Ansys-HFSS. Fig. 3 shows the chip photograph of the manufactured Lange coupler as a three port. The termination of the isolated port was realized by a grounded 50 Ω NiCr thin-film resistor.

The S-parameters of the Lange coupler were extracted on-wafer in a 50 Ω environment. The measured transmissions of the direct (S_{31}) and the coupled paths (S_{21}) are depicted in Fig. 4. Both curves overlap between 17 and 21 GHz. The coupler demonstrates a high degree of symmetry around 20 GHz. At this frequency the amplitude imbalance $\varepsilon = ||S_{31}| - |S_{21}||$ corresponds to 0.2 dB. Furthermore, ε is less than 0.5 dB in the frequency range from 14.5 to 24.5 GHz. The results achieved for the phase offset between the two output ports are very convincing, too. Fig. 5 illustrates the phase difference $\Delta\phi = \phi_{31} - \phi_{21}$ between the direct (ϕ_{31}) and the coupled (ϕ_{21}) connection. The Lange coupler exhibits an almost perfect quadrature response up to 25 GHz. The absolute deviation from 90° is in the range of ±1°. However, $\Delta\phi$ increases when moving towards higher frequencies and exceeds 95° beyond 28.6 GHz. A further important figure of merit of the coupler is the net loss, since it directly impacts the BPA performance, especially at the output. As can be calculated from the extracted S-parameters, the net loss is less

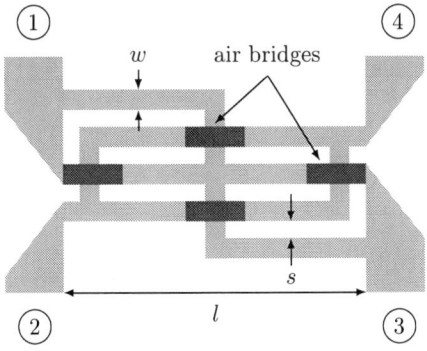

Fig. 2. Schematic of the microstrip Lange coupler.

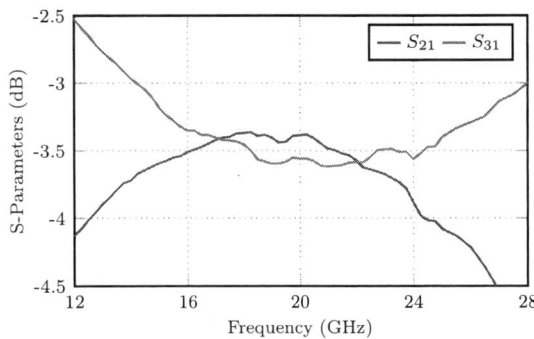

Fig. 4. Measured transmission response of the direct (S_{31}) and the coupled (S_{31}) paths.

165

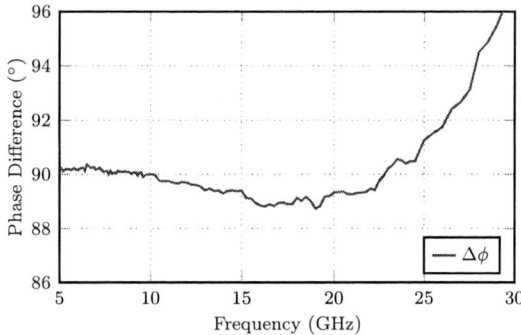

Fig. 5. Measured phase difference between the direct (ϕ_{31}) and the coupled (ϕ_{31}) paths.

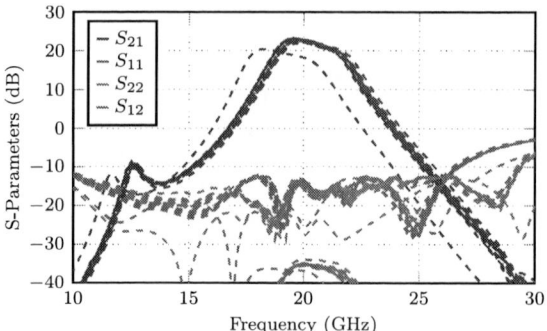

Fig. 7. Wafer mapping of the measured (solid) and simulated (dashed) S-parameters of the balanced power amplifier at $V_{DS} = 28\,V$ drain voltage and $I_{DS} = 100\,mA/mm$ of quiescent drain current.

than 0.5 dB within the operating band of the realized coupler. Furthermore, the measured isolation of the Lange coupler is better than 17 dB up to 24 GHz.

V. BPA REALIZATION

Each PA module combines four $6 \times 90\,\mu m$ HEMTs at the output. This yields a TGW of 2.16 mm in the final-stage. In order to reach 20 dB of small-signal gain, a driver-stage based on two similar transistors ($6 \times 90\,\mu m$) was added at the input to form a staging ratio of $1 : 2$. Additionally, the amplifier employs the concept of via sharing, where the adjacent HEMTs are connected by a common via. Consequently, in order to improve the odd-mode stability, a NiCr thin-film resistor is mounted between the connected HEMTs at both the input and the output. The main target of the PA was to obtain a good combination between the power and efficiency over a bandwidth of at least 2 GHz. The output matching network was designed as a reactive $4 : 1$ combining structure, which transforms the $50\,\Omega$ impedance to the optimum load of the individual HEMTs. Nonlinear load-pull simulation was performed in order to identify the optimum load, which maximizes the power and efficiency at the output of the utilized transistor. Interstage and input matching networks were realized as band-pass filters, which provide a flat gain characteristic over the targeted bandwidth. Finally, an RC

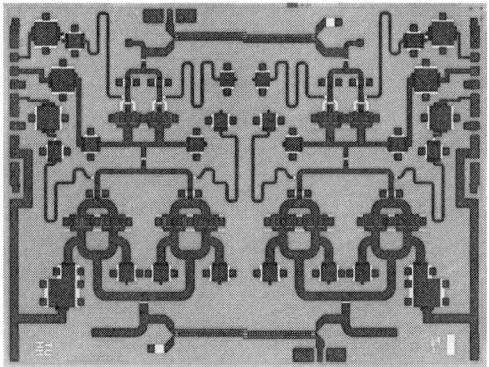

Fig. 6. Chip photograph of the BPA MMIC.

stabilization network is placed directly at the input of the driver to ensure unconditional stability.

The overall BPA was realized, combining two of the PA modules using the designed Lange coupler at the in- and output. The structures were integrated on a single MMIC occupying a chip area of $4.0 \times 5.5\,mm^2$. Consequently, the TGW of the complete BPA in the final-stage is doubled and corresponds to 4.32 mm. The photograph of the fabricated MMIC is depicted in Fig. 6. Similar to the individual coupler test structure, the isolation port was terminated using a $50\,\Omega$ NiCr resistor following by a grounded via. Since there is no direct DC routing between both amplifier modules, the supply voltage has to be applied from both sides of the chip.

VI. MEASURED RESULTS OF THE BPA

For the characterization of the BPA, small- and large-signal on-wafer measurements were performed. The linear S-parameters were extracted using an automated wafer probing system and a CW stimulus. The measurements were carried out with a drain bias of 28 V and a quiescent current of 100 mA/mm. Fig. 7 shows the comparison between the simulated (dashed) and measured (solid) performance using a strict selection of 11 functional cells on a wafer. Due to manufacturing tolerances and simulation inaccuracies, the measured response of the small-signal gain ($|S_{21}|$) is shifted by approximately 2 GHz towards higher frequencies compared to simulation. Additionally, the measured $|S_{21}|$ is higher than predicted by the model. It exceeds 20 dB between 19 and 21 GHz and reaches its maximum of about 22 dB at 19.5 GHz. As expected, the BPA demonstrates broadband in- and output matching, due to the balanced configuration. The associated in- and output reflection coefficients ($|S_{11}|, |S_{22}|$) are less than $-10\,dB$ up to 26 GHz. Moreover, in the operating frequency band the magnitude of both coefficients varies between -13 and $-21\,dB$.

The power performance of the BPA was measured in the pulsed-mode with a $100\,\mu s$ pulse width and a duty cycle of 10 %. Fig. 8 shows the obtained results for swept input power at $I_{DS} = 50\,mA/mm$ and two different drain supply voltages. For both operating points the best performance is

166

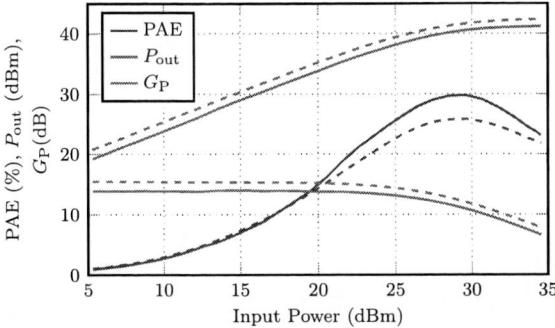

Fig. 8. Power sweep performed on the BPA for $I_{DS} = 50\,\text{mA/mm}$ of quiescent drain current and $V_{DS} = 28\,\text{V}$ (solid), and $V_{DS} = 35\,\text{V}$ (dashed) drain voltage at 22.5 GHz.

achieved at 22.5 GHz. A peak PAE of 30 % is measured at an input power of 29 dBm (3 dB compression) and 28 V drain bias. Due to the relatively low quiescent currents, the peak PAE is already achieved at about 3 dB compression, which additionally indicates that the BPA operates in the range of good linearity. The associated output power and large-signal gain are 11 W (40.5 dBm) and 11 dB, respectively. The maximum saturated output power of 13 W with 25 % of PAE was observed at about 7 dB compression. As depicted in Fig. 8 (dashed), the same power sweep was performed with a 20 % larger drain current. At $V_{DS} = 35\,\text{V}$ and a compression of about 3 dB, the peak PAE is reduced by 4 % and drops to 26 %. However, the associated output power and gain increase and reach 15 W and 12 dB, respectively. Furthermore, the BPA exhibits 17 W of output power and 22 % of PAE, when it is driven by 34 dBm of input power (8 dB compression). At this point, the power gain is still 8 dB.

The saturated output power, the PAE, and the large-signal gain over frequency at an input power of $P_{in} = 29\,\text{dBm}$ and $V_{DS} = 35\,\text{V}$ are reported in Fig. 9. Over the 21 − 23 GHz frequency band, the PAE exceeds 21 % and reaches its maximum of 26 % at 22.5 GHz. The large-signal gain shows a slight roll-off when moving towards higher frequencies. From

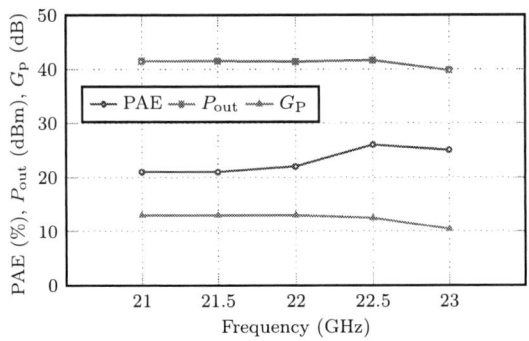

Fig. 9. Frequency sweep performed on the BPA for $I_{DS} = 50\,\text{mA/mm}$ of quiescent drain current and $V_{DS} = 35\,\text{V}$ drain voltage at an input power of $P_{in} = 29\,\text{dBm}$.

approximately 13 dB up to 22.5 GHz it drops to 10.4 dB at 23 GHz. This behavior is directly reflected by the associated output power, which exceeds 14 W between 21 and 22.5 GHz. However, at 23 GHz the 14 W output power level is reached, when the BPA is driven by $P_{in} = 33\,\text{dBm}$. At this input power, the amplifier exhibits 26 % of PAE and 8 dB of large-signal gain.

VII. CONCLUSION

A state-of-the-art balanced power amplifier MMIC utilizing a 0.25 μm AlGaN/GaN HEMT process technology has been presented. The MMIC operates at K-band and demonstrates a linear gain in the range of 20 dB. The in- and output reflection remains below −10 dB up to 26 GHz. The large-signal measurements show a saturated output power of 17 W and a peak PAE of 30 % at 22.5 GHz. For a drain supply voltage of $V_{DS} = 35\,\text{V}$, the output power of the BPA exceeds 14 W (41.5 dBm) between 21 and 23 GHz. In this frequency range the associated PAE varies between 21 and 26 %.

ACKNOWLEDGMENT

This work was supported and funded by the German Space Management (DLR) on behalf of the German Ministry for Economic Affairs and Energy (BMWi) under research contact 50YB1809.

REFERENCES

[1] K. Eisele, R. Engelbrecht, and K. Kurokawa, "Balanced Transistor Amplifiers for Precise Wideband Microwave Applications," *in IEEE International Solid-State Circuit Conference*, pp. 18-19, Feb. 1965

[2] Sechi, F., Bujatti, M., *Solid-State Microwave High-Power Amplifiers*, Artech House, Boston London, 2009.

[3] P. Colantonio and R. Giofré, "A GaN-on-Si MMIC Power Amplifier with 10 W Output Power and 35 % Efficiency for Ka-Band Satellite Downlink." in *2020 15th European Microwave Integrated Circuits Conference (EuMIC)*, pp. 29-32, Utrecht, Netherlands, 2021.

[4] S. Din, A. M. Morishita, N. Yamamoto, C. Brown, M. Wojtowicz and M. Siddiqui, "High-Power K-Band GaN PA MMICs and module for NPR and PAE," *2017 IEEE MTT-S International Microwave Symposium (IMS)*, pp. 1838-1841, Honolulu, 2017.

[5] R. Giofré, P. Colantonio, F. Costanzo, F. Vitobello, M. Lopez and L. Cabria, "A 17.3-20.2-GHz GaN-Si MMIC Balanced HPA for Very High Throughput Satellites," in *in IEEE Microwave and Wireless Components Letters*, pp. 1-4, 2021.

[6] L. Marechal et al., "10 W K Band GaN MMIC Amplifier Embedded in Waveguide-Based Metal Ceramic Package," in *2019 14th European Microwave Integrated Circuits Conference (EuMIC)*, pp. 184-187, Paris, 2019.

[7] J. J. Sowers and S. Tabatabaei, "K-Band Doherty and Class AB Gallium Nitride MMIC Power Amplifiers for Space Applications," *2019 IEEE Asia-Pacific Microwave Conference (APMC)*, pp. 1047-1049, Singapore, 2019.

[8] P. Waltereit *et al.*, "GaN HEMT and MMIC Development at Fraunhofer IAF: Performance and Reliability," *Physica Status Solidi (a)*, vol. 206, no. 6, pp. 1215-1220, 2009.

[9] J. Lange, "Interdigitated Stripline Quadrature Hybrid," *IEEE Trans. Microw. Theory Tech.*, vol. MTT-57, pp. 1150-1151, Dec. 1969.

Gap in pagination due to unavailable paper.

Page 168

Proceedings of the 16th European Microwave Integrated Circuits Conference

120 GBd SiGe-Based 2:1 Analog Multiplexer Module for Ultra-Broadband Transmission Systems

C. Schmidt[#1], T. Tannert[*], J.H. Choi[#], C. Caspar[#], D. Pech[#], S. Wünsch[#], G. Ropers[#], J. Schostak[#],
V. Jungnickel[#], R. Freund[#], M. Grözing[*], M. Berroth[*]

[#] Fraunhofer Heinrich-Hertz-Institute, Berlin, Germany
[*] University of Stuttgart, Stuttgart, Germany
[1]christian.schmidt@hhi.fraunhofer.de

Abstract — A 2:1 analog multiplexer (AMUX) module based on a SiGe-HBT IC is presented. The AMUX module offers a 6-dB analog signal path bandwidth of 61 GHz. We show a 120 GBd 4-level pulse amplitude modulation (PAM) signal with the AMUX module fed by two CMOS-digital-to-analog converters (DACs), each with 19 GHz analog 3-dB bandwidth only, thus demonstrating high-baud-rate signal generation with bandwidth-limited CMOS DACs enabled by advanced digital signal processing.

Keywords — analog integrated circuits, multiplexing, digital-analog conversion, integrated circuit packaging.

I. Introduction

The demand for ever increasing capacity in optical communication networks requires the ability to generate electrical signals with ever increasing symbol rate. The analog bandwidth of deep-submicron CMOS-based digital-to-analog converters (DACs) is not expected to increase substantially with new CMOS technology nodes [1]. By interleaving CMOS-DACs with a broadband circuit in SiGe or InP technology, both the analog bandwidth and the sampling rate can be enhanced.

During the last years, multiple research activities originated in the field of DAC interleaving with analog multiplexers (AMUX). Integrated circuits (ICs) using InP bipolar transistors have shown up to 160 GBd PAM-4 eye diagrams with 91 GHz bandwidth in a packaged module and even 168 GBd for an optical 16-QAM transmission [2]-[5]. However, well-established silicon processes using SiGe bipolar devices enable more complex circuits with better yield and lower costs. Therefore, they are much more promising for large scale commercialization. SiGe-based AMUX ICs [6]-[8] have shown remarkable results, including the demonstration of more than 67 GHz analog signal path bandwidth [6], 100 GBd PAM-4 signal generation together with a 92 GS/s arbitrary waveform generator (AWG) [7] and 120 GBd PAM-4 signal generation together with ultra-broadband SiGe DACs [8].

We demonstrate a 120 GBd PAM-4 signal generated with an AMUX module based on a SiGe-HBT IC supported by advanced digital signal processing (DSP) and using commercially established 19 GHz analog bandwidth CMOS DACs, which are better suited for commercial application than SiGe DACs due to their lower power consumption and possible monolithic co-integration with a digital signal processor. The 2:1 AMUX IC design and the AMUX module design are both presented and discussed. S-parameter measurement results indicate their performance capabilities. Finally, the time-domain performance is evaluated and a 120 GBd signal with an AMUX-DAC, enabled by advanced DSP, is demonstrated.

II. 2:1 AMUX IC

The AMUX IC is fabricated in a 130 nm SiGe-BiCMOS technology with f_T = 300 GHz and f_{max} = 500 GHz [9]. Details of the circuit are presented in [6]. Fig. 1 shows a block diagram (a) and a chip photograph (b) of the AMUX-IC with a size of 1.6 x 1 mm^2. The inner pads are considered for flip-chip mounting to dissipate heat and provide stability. The AMUX core, that performs the interleaving of the two inputs, is based on a Gilbert-cell topology shown in Fig. 1c. The two differential input voltages feed two linearized transconductance stages.

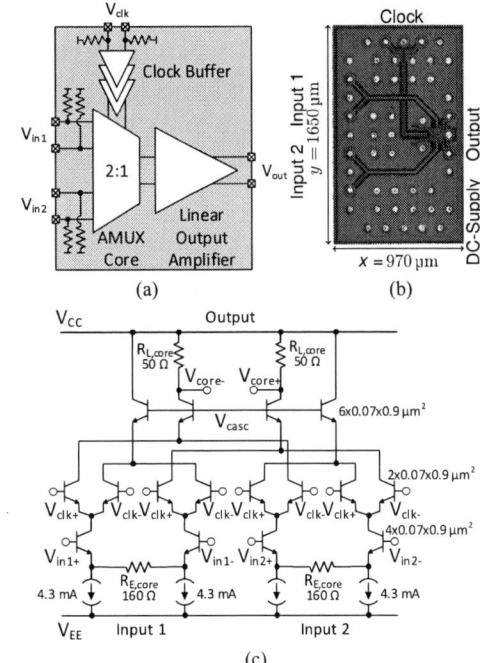

Fig. 1. AMUX IC block diagram (a), IC photo (b), AMUX core schematic (c). (a): ©2017 IEEE [6].

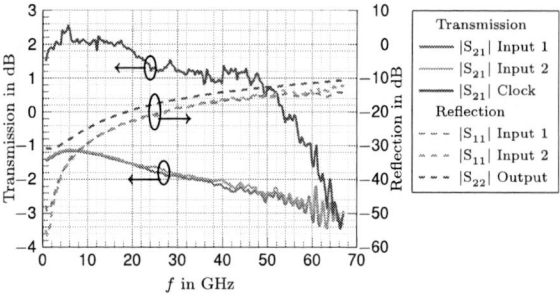

Fig. 2. On-chip measured differential S-parameters.

The transconductor currents to pass to the output are selected by a sampling clock, which controls a four-path differential current switch. The selected transconductor currents are switched to the output, whereas the non-selected currents are switched to a dummy load. At the output of the AMUX there is a linear amplifier, which enables a differential output voltage swing of up to 1 V with linear gain.

Differential S-parameter measurements are performed on-wafer. The signal path's transmission and reflection parameters for the differential input power at +1 dBm are given in Fig. 2.

The signal path's transmission amplitude drops by 2.2 dB within the measurement range of 67 GHz, i.e. the signal path's 3-dB bandwidth exceeds 67 GHz. The inactive signal path has an attenuation of more than 29 dB. The reflection parameters at the 100 Ohm differential signal inputs and output stay below -10 dB. The S-parameters are independent of signal power, which indicates a good linearity. For the clock path, a 3-dB bandwidth of approximately 60 GHz is measured at 0 dBm differential input power. This indicates proper switching operation of the AMUX IC at sampling rates of up to 120 GS/s. The power consumption of the AMUX IC is 1.06 W, which is negligibly dependent of the clock rate.

III. 2:1 AMUX MODULE

The AMUX IC is bonded onto an RF interposer using flip-chip technology as shown in Fig. 3. The AMUX module includes six G3PO connectors for two differential data inputs and a differential clock input, two 1 mm connectors for the differential output signal, and a DC connector.

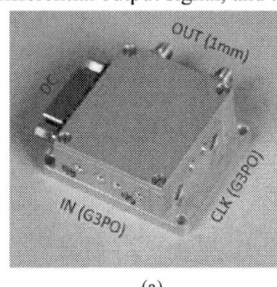
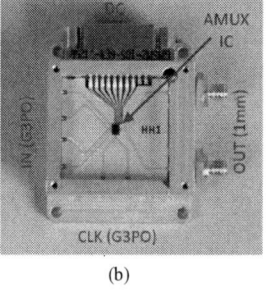

(a)   (b)

Fig. 3. Photographs of the AMUX module (size: 30x26x13 mm³): (a) 2x differential inputs (G3PO), 1x differential output (1 mm), 1x differential clock (G3PO), (b) open housing with flip-chipped AMUX IC on interposer.

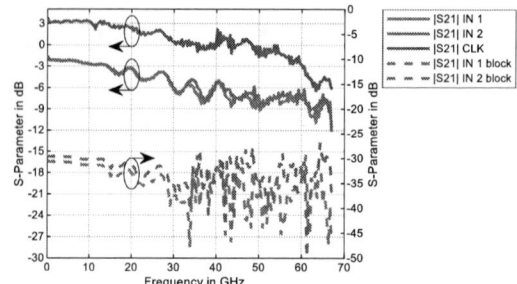

Fig. 4. AMUX module S-parameter measurement results.

The RF interposer with conductor-backed coplanar waveguides and via-holes is made of quartz material with ε=3.78 and 254 μm thickness. The longest transmission line at the output is 16.6 mm and shows 2.9 dB insertion loss at 110 GHz, < -14 dB reflections, and +/- 3 ps group delay distortion as measured on-wafer.

The equal-delay RF input transmission lines maintain a phase difference of < 5° up to 70 GHz in 3D electromagnetic simulation.

The module is characterized in the frequency-domain using the Rohde&Schwarz network analyzer ZVA60. The network analyzer is configured to operate in the true-differential mode.

Fig. 4 shows the measured differential S-parameters of the AMUX module. The 3-dB and 6-dB analog bandwidth of the transparent data paths are 29 and 61 GHz, respectively. The attenuation for the blocked data paths is > 30 dB. The S-parameters are independent of signal power, which indicates a good linearity. The S-parameters of the differential clock path are similar to the curve of the data paths, whereby the 3-dB and 6-dB analog bandwidth are 34 and 60 GHz, respectively.

IV. 2:1 AMUX-DAC

To assess the performance of the 2:1 AMUX module in the time domain, it is combined with two DACs. Then, the so-called AMUX-DAC can generate signals, which exceed both the sampling rate and the bandwidth of the individual DACs.

A. Measurement Setup

The AMUX module performance is evaluated with the measurement setup shown in Figs. 5-6. The two differential AMUX input signals are generated using two 8-bit, 90 GS/s, 19 GHz 3-dB analog output bandwidth Socionext CMOS-DACs. Analog phase shifters ensure 90° offset with respect to the symbol rate. One of the AMUX's differential output ports is connected via a bias-T to the digital storage oscilloscope Tektronix DPO77002SX with 70 GHz analog bandwidth and 200 GS/s. The other output port is terminated.

The clock network is fed from the frequency generator Agilent E8257D, whose output signal is split and divided to generate both the DAC clock and the differential AMUX clock.

The AMUX performance is evaluated at two different clock frequencies: 30 and 60 GHz, corresponding to AMUX input baud rates of 30 GBd and 60 GBd and AMUX output baud rates (f_C) of 60 GBd and 120 GBd, respectively.

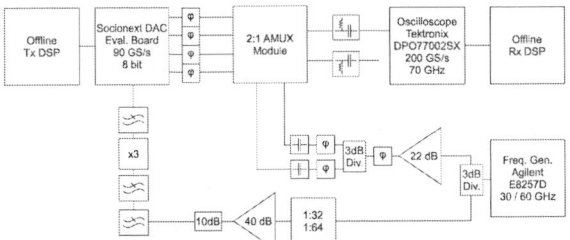

Fig. 5. AMUX-DAC measurement setup.

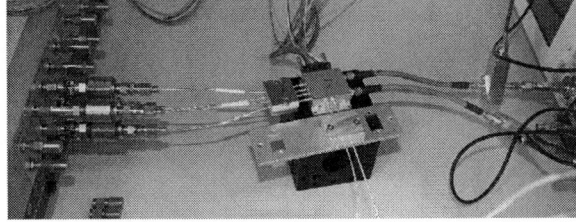

Fig. 6. Photograph of the AMUX-DAC measurement setup: the DACs (left), the AMUX module (center), and the scope (right).

The CMOS DAC conversion rate is fixed at $f_S = 90$ GS/s, resulting in a combined AMUX-DAC sampling rate of 180 GS/s and providing 3-fold and 1.5-fold oversampling at the 30 GBd and 60 GBd AMUX inputs, respectively.

In Fig. 6, a photograph of the measurement setup is provided. The AMUX's data input ports, clock input port, data output port and DC supply are connected from the left, the bottom, the right, and the top, respectively. The CMOS-DAC 19" rack mount and the oscilloscope input are visible on the left- and right-hand side, respectively.

B. Digital Signal Processing

In order to assess the AMUX-DAC performance, a pulse amplitude modulation (PAM)-signal is generated digitally. The signal is passed through the DSP pipeline visualized in Fig. 7. The DSP generates the individual signals for the DACs.

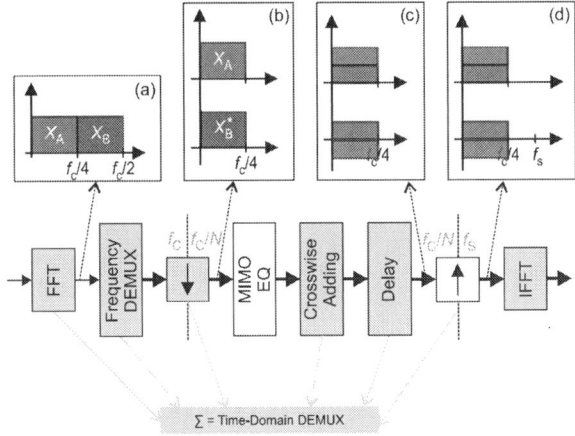

Fig. 7. AMUX-DAC digital signal processing block diagram (bold arrows denote multiple parallel signals). At distinct points in the DSP pipeline, the spectral representation is visualized for the exemplary case of two DACs.

It pre-compensates the CMOS DACs', AMUX's and RF components' linear impairments. The operations at distinct points in the DSP pipeline are shown in the frequency domain for the exemplary case of two DACs in Fig. 7a-d.

To perform the pre-compensation for signal paths having (slightly) different frequency responses, the equalization cannot be performed before digital time-domain demultiplexing of the combined signal into sub-signals nor after the demultiplexing [10]. Rather, the equalization needs to be performed during the time-domain demultiplexing, which is only possible by performing this operation in the frequency domain. In the frequency domain, multiple processing steps are required to represent the time-domain demultiplexing operation. Hence, the multiple-input multiple-output (MIMO) equalization can be performed in between these processing steps as described mathematically in detail in [10] and [4].

The processing is performed as follows: the combined digital signal with the sampling rate f_C is Fourier-transformed first (Fig 7a). Then, it is split with a frequency demultiplexer (e.g. a diplexer) into two signals with sampling rate f_C. These signals are then downsampled to f_C/N, whereby $N=2$ in this case. This way, the second signal is implicitly downconverted to baseband due to bandpass sampling (Fig. 7b). A MIMO equalizer pre-compensates the linear impairments of the DACs, the AMUX and the RF wires as well as the mismatches between the individual signal paths. The signals are then added and delayed, completing the corresponding time-domain demultiplexing operation (Fig. 7c). Both sub-signals are upsampled to the DAC sampling rate f_S (Fig. 7d) and converted back to the time-domain.

To calibrate the AMUX-DAC pre-compensating MIMO equalizer, its coefficients are estimated. First, a least-squares (LS) MIMO channel estimation is performed. Second, the MIMO filter coefficients are obtained by calculating the LS pseudo-inverse of the estimated MIMO channel, i.e. the equalizer coefficients are derived as described in [10].

At the receiver, the offline DSP consists of a low-pass filter, clock phase recovery, channel estimation, and bit error counting.

C. Results

In Figs. 8a-b, the measurement results are depicted for an AMUX output symbol rate of 2x 30 GBd = 60 GBd. Open eyes and distinct PAM levels can be observed at PAM4 and PAM8, respectively. For both PAM-4 and PAM-8 an imbalance between even and odd eyes is observed, which may be attributed to the difference of the two analog channels. A BER of 6.7e-6 and 1.1e-2 was achieved for PAM-4 and PAM-8, respectively. The eye diagrams are generated from the recorded oscilloscope samples after low-pass filtering and clock phase recovery.

In Figs. 8c-f, the measurement results are depicted for 2x 60 GBd = 120 GBd. Open eyes and distinct PAM levels can be observed at PAM2 and PAM4 in Figs. 8e-f, respectively. A BER of 7.6e-6 and 2.9e-3 was achieved for PAM-2 and PAM-4, respectively.

The frequency response of the uncompensated AMUX-DAC, which has been used for calibration, is visualized in Fig. 8c.

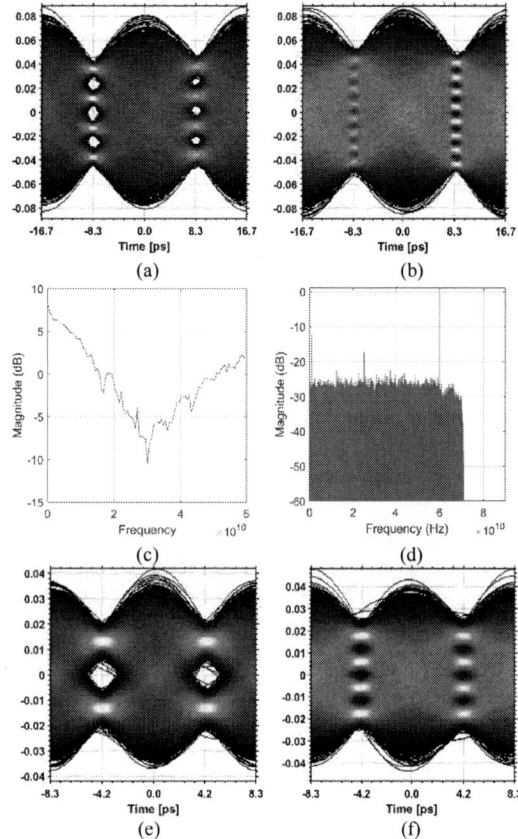

Fig. 8. AMUX-DAC measurement results: (a,b) 60 GBd eye diagrams for PAM-4 and PAM-8, (c) AMUX-DAC frequency response, (d) measured spectrum, (e,f) 120 GBd eye diagrams for PAM-2 and PAM-4.

The frequency response magnitude drops by 17 dB from DC to half of the signal bandwidth; then, it increases again by 12 dB. This behaviour becomes comprehensible, if we think of the AMUX as a combination of buffer and mixer [4], [10].

It both passes the input signal in the baseband and simultaneously upconverts the input signal to the AMUX's clock frequency, combining both parts at its output. The resulting AMUX-DAC frequency response is a combination of the DAC frequency response, the AMUX frequency response, and the response of interconnecting cables.

After applying the Tx DSP pre-processing, the resulting received spectrum is visualized in Fig. 8d. Clock spurs appear at 22.5 and 60 GHz, which can be attributed to the DAC clock and the AMUX clock, respectively. Beyond 70 GHz, the oscilloscope's low-pass cut-off is present.

V. CONCLUSION

We presented a 120 GBd 2:1 AMUX module based on SiGe-HBT technology supported by advanced DSP. The module consists of a metal package with G3PO and 1 mm connectors. Its data path has a 6-dB analog bandwidth of 61 GHz. We demonstrate 120 GBd 4-level pulse amplitude modulation AMUX-DAC signal generation based on an

AMUX module, two commercially available CMOS-DACs each with only 19 GHz analog 3-dB bandwidth and advanced DSP. This paper shows how the bandwidth limitations of CMOS D/A converters can be overcome by using a broadband SiGe-AMUX module in combination with advanced digital signal preprocessing.

ACKNOWLEDGMENT

This work was supported in part by Deutsche Forschungsgemeinschaft (DFG, grant no. 423436357).

REFERENCES

[1] H. J. Lee, S. Rami, S. Ravikumar, V. Neeli, K. Phoa, B. Sell, Y. Zhang, "Intel 22nm FinFET (22FFL) Process Technology for RF and mm Wave Applications and Circuit Design Optimization for FinFET Technology", *IEEE International Electron Devices Meeting (IEDM)*, 2018, doi: 10.1109/IEDM.2018.8614490.

[2] D. Ferenci, M. Grözing, M. Berroth. "A 25 GHz Analog Multiplexer for a 50 GS/s D/A-Conversion System in InP DHBT Technology" in *IEEE Compound Semiconductor Integrated Circuit Symposium (CSICS)*, Waikoloa, HI, 2011, doi 10.1109/CSICS.2011.6062440.

[3] M. Nagatani, H. Wakita, H. Yamazaki, H. Nosaka, K. Kurishima, M. Ida, Y. Miyamoto, "A 128-GS/s 63-GHz-bandwidth InP-HBT-based analog-MUX module for ultra-broadband D/A conversion subsystem", *IEEE MTT-S International Microwave Symposium (IMS)*, 2017, doi: 10.1109/MWSYM.2017.8058858.

[4] M. Nagatani, H. Wakita, H. Yamazaki, Y. Ogiso, M. Mutoh, M. Ida, ... H. Nosaka, "A Beyond-1-Tb/s Coherent Optical Transmitter Front-End Based on 110-GHz-Bandwidth 2:1 Analog Multiplexer in 250-nm InP DHBT", *IEEE Journal of Solid-State Circuits*, 2020, 55, (10), pp. 2301 – 2315, doi 10.1109/JSSC.2020.2989579.

[5] R. Hersent, A. Konczykowska, F. Jorge, M. Riet, C. Mismer, V. Nodjiadjim, B. Duval, J-Y. Dupuy, "Analog-Multiplexer (AMUX) circuit realized in InP DHBT technology for high order electrical modulation formats (PAM-4, PAM-8)," in *International Microwave and Radar Conference (MIKON)*, Warsaw, Poland, 2020, pp. 222–224, doi 10.23919/MIKON48703.2020.9253772.

[6] T. Tannert, X. Q. Du, D. Widmann, M. Grözing, M. Berroth, C. Schmidt, C. Caspar, J. H. Choi, V. Jungnickel, R. Freund, "A SiGe-HBT 2:1 Analog Multiplexer with more than 67 GHz Bandwidth" in *IEEE Bipolar/BiCMOS Circuits and Technology Meeting (BCTM)*, 2017, pp. 146-149, doi 10.1109/BCTM.2017.8112931.

[7] H. Ramon *et al.*, "A 100-GS/s Four-to-One Analog Time Interleaver in 55-nm SiGe BiCMOS," *IEEE J. Solid-State Circuits*, vol. 56, no. 8, pp. 2539–2549, 2021, doi: 10.1109/JSSC.2021.3057575.

[8] M. Collisi, M. Möller, "A 120 GS/s 2:1 Analog Multiplexer with High Linearity in SiGe-BiCMOS Technology" in *IEEE BiCMOS and Compound semiconductor Integrated Circuits and Technology Symposium (BCICTS)*, 2020.

[9] H. Rücker, B. Heinemann, and A. Fox, "Half-Terahertz SiGe BiCMOS technology," in *IEEE 12th Topical Meeting on Silicon Monolithic Integrated Circuits in RF Systems*, 2012, pp. 133–136.

[10] C. Schmidt, *Interleaving Concepts for Digital-to-Analog Converters* 1st ed., Wiesbaden, Germany: Springer Vieweg, 2020, doi: 10.1007/978-3-658-27264-7.

Proceedings of the 16th European Microwave Integrated Circuits Conference

A 7-30 GHz, 80-dBΩ Noise-Optimized, Bandpass-Like TIA in 130 nm SiGe BiCMOS Technology for Quasi-Coherent Optical Receivers

Tom K. Johansen[#], Guillermo Silva Valdecasa[#*], Monika Kupska[*], Jose A. Altabas[*], Omar Gallardo[*], Michele Squartecchia[*], Jesper B. Jensen[*]

[#]Department of Electrical Engineering, Technical University of Denmark, Denmark
[*]Bifrost Communications Aps, Denmark

tkj@elektro.dtu.dk

Abstract — In this paper, a high-gain, noise-optimized bandpass-like transimpedance amplifier (TIA) chip is presented. The TIA chip is implemented in a 130 nm SiGe BiCMOS technology and consists of a 2nd order high-pass filter (HPF), shunt-feedback TIA stage, Gilbert cell based variable gain amplifier (VGA) having an ac-coupled Cherry-Hooper load circuitry allowing for low supply voltage and finally a cascode based output buffer. The TIA chip is optimized for operation with a commercial photodetector (PD) in a quasi-coherent optical receiver. The experimental results of the fabricated TIA chip shows a 7-30 GHz bandwidth and differential transimpedance of 80-dBΩ. An evaluation board containing the TIA chip is employed in a 10 Gbps transmission experiment and a state-of-the-art sensitivity for quasi-coherent receivers is achieved.

Keywords — Access network, Optical receivers, SiGe BiCMOS, Transimpedance amplifier (TIA).

I. INTRODUCTION

With the emergence of ever more bandwidth-intensive applications, such as streaming, Internets of Things (IoT), cloud computing, 5G mobile and augmented reality, the exponential growth in data traffic experienced during the recent years is expected to continue well into the future. To keep up with the data traffic demand for end-users, passive optical networks (PONs) are gaining popularity as a low-cost solution. The next-generation of PONs (NG-PON2) offer new, innovative ways to be fast and flexible [1]. The NG-PON2 standard, however, puts high demanding requirements on the optical network units (ONUs) and optical line terminals (OLTs), in particular on the receiver side.

The quasi-coherent receiver has been proposed as a high sensitivity, low complexity and low cost solution allowing high split ratio in NG-PON2 and similar networks [2, 3]. These receivers have better sensitivity than direct detection receivers and do not employ power hungry DSP algorithms as in coherent receivers [4]. In the quasi-coherent receiver, an intensity modulated signal from the transmitter is being mixed with the signal from a local oscillator (LO) laser. The mixing takes place in a photodetector (PD), downconverting the data to an intermediate frequency (IF). The IF is equal to the difference between the signal wavelength (λ_S) and the LO wavelength ($\lambda_{LO} = \lambda_S + IF$) and will typically fall within the microwave frequency range. Following amplification in the

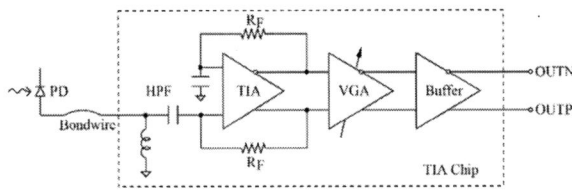

Fig. 1. Architecture of TIA chip.

electrical domain, a high-speed envelope detector restores the original data in real time.

TIAs for quasi-coherent receivers separate themselves from other TIAs developed for direct detection and coherent detection schemes, see e.g. [5, 6], in the sense that they must be of the bandpass type, provide sufficient transimpedance gain to effectively drive the high-speed envelope detectors and provide lowest possible input referred noise to comply with the reach and split ratio requirements of access networks [7].

In this paper, a high-gain, ultra-low noise bandpass-like SiGe BiCMOS TIA for quasi-coherent receivers is reported. The organization of the paper is as follow. In section II the circuit design of the bandpass-like TIA is described. Section III discusses the noise optimization procedure followed for the TIA + PD assembly. The experimental verification of the TIA chip using both small-signal measurements and employment in a transmission experiment is presented in section IV. Finally, section V concludes the paper.

II. CIRCUIT DESIGN

The architecture of the TIA chip is shown in Fig.1. It consists of a 2nd order high-pass filter (HPF), TIA stage, VGA and output buffer. It is co-designed for use with a commercial photo-detector (PD) with 37 GHz electrical bandwidth. The specifications demands a passband range from approximately 10 GHz to 30 GHz in order to accommodate various transmitter frequencies and to have flexibility in the selection of the LO laser wavelength. The differential transimpedance gain should be larger than 5KΩ. The TIA chip is implemented in the 130 nm SiGe:C BiCMOS SG13S technology from IHP. The SG13S technology offers HBTs with f_T/f_{max} of 240/330 GHz and static current gain β of 900 [9]. The simplified schematic of the TIA chip is shown in Fig. 2. In the following,

3-4 April 2022, London, UK

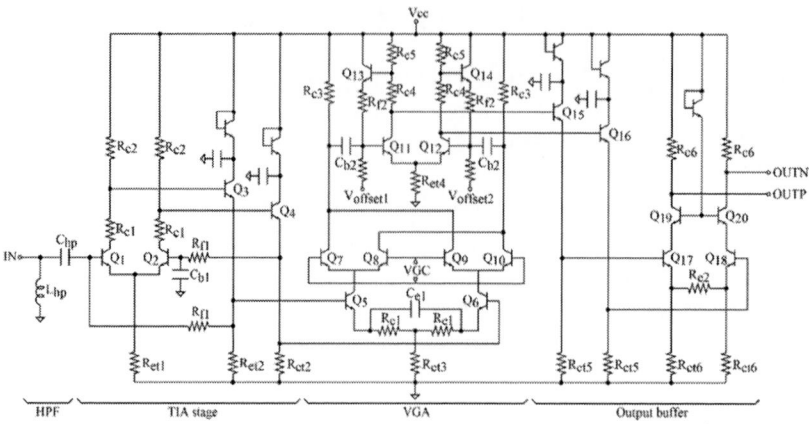

Fig 2. Simplified schematic of TIA chip.

the design of the various blocks of the TIA chip will be described.

A. Input Filter

The purpose of the input filter is two-fold. At first, it should eliminate the direct detection term coming from the mixing process in the PD. Secondly, it should be able to sink the DC current from the PD. The values, L_{HP} and C_{HP}, of the 2nd order high-pass filter are selected so that the lower cut-off frequency is ~7 GHz and optimized using EM simulation to avoid deteriorating effects in the passband of the TIA.

B. TIA Stage

The TIA stage uses a shunt feedback topology with emitter followers due to its favourable bandwidth, gain and noise behaviour. The transimpedance limit dictates that the product of the bandwidth squared and the transimpedance resistor R_{f1} should be constant for a given technology [5, 11]. As the transimpedance resistor R_{f1} should be as large as possible to lower the noise contribution from the TIA stage the bandwidth of this stage is reduced. This reduction in bandwidth can be recovered in the following amplifying stages by introducing controlled amount of peaking as proposed in [5]. To further lower the noise contribution from the TIA stage, multifinger devices with $8 \times 120 \times 840$ nm^2 emitter area are used for Q_1-Q_4 and current sources are implemented using resistors, R_{et1} and R_{et2}. The collector load resistances are each split into two resistances, R_{c1} and R_{c2}. This is necessary to stay well below the breakdown voltage of the transistors Q_1 and Q_2 but may introduce a small noise penalty. The TIA stage also provides single-ended to differential conversion by ac grounding the base of transistor Q_2 through capacitor C_{b1}.

C. VGA Stage

The VGA stage is formed around a Gilbert cell where the biasing of the transistor pair Q_8-Q_9 relative to Q_7-Q_{10} provides the gain control. The reduced bandwidth of the TIA stage is recovered by emitter degeneration of the transistor pair Q_5 and Q_6 by R_{e1} and C_{e1}, The bandwidth can be further extended by employing a Cherry-Hooper load circuit [9, 10]. Due to the limited voltage headroom imposed by the 3.3 V power supply it is proposed to ac couple the output of the Gilbert cell to the input of the load circuitry consisting here of transistors Q_{11}-Q_{14}. Provided that the Gilbert cell load resistance, R_{c3}, is larger than the input impedance looking into the Cherry-Hooper shunt feedback circuit the full benefit can be obtained even at a reduced voltage headroom.

D. Output Buffer

The output buffer consists of a pair of emitter followers, Q_{15} and Q_{16}, followed by a Cascode gain cell formed by transistors Q_{17}-Q_{20}. The collector resistor, R_{c6}, is selected as 50 Ω for improved output matching. The emitter degeneration resistor R_{e2} is selected as a trade-off between small-signal transimpedance gain and large-signal voltage drive capability.

E. Layout Consideration

A chip microphotograph of the TIA chip is shown in Fig. 3. The chip size is 1200×1156 um^2 including pads. A high degree of symmetry is observable in the layout. In particular, a dummy high-pass filter is included at the input for added symmetry. The core of the amplifier has been compacted as much as possible to keep bandwidth high. The RC parasitics associated with the core layout have been extracted using the extractor in Cadence and used in the final optimization.

Fig. 3. Microphotographs of TIA chip. The size is 1200×1156 μm^2.

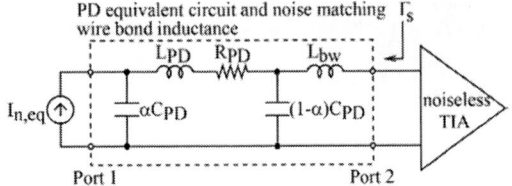

Fig. 4. Conceptual model used for optimization of equivalent input noise current of TIA + PD assembly.

III. NOISE OPTIMIZATION OF TIA + PD ASSEMBLY

The equivalent input noise current of a TIA is often reported with the input open-circuited. While this may be used to benchmark the noise performance of the TIA in itself, it does not well represent the noise performance of the TIA in an optical receiver front-end with a given PD. To optimize the noise performance of the optical receiver front-end, the interaction between the TIA input, the PD and any packaging parasitics or noise matching network between them must be included into the analysis. Fig. 4 shows conceptually the model view we propose in order to optimize the optical receiver front-end. The PD equivalent circuit and noise matching wire bond inductance are combined into a general two-port network. The total noise current spectral density in units $\frac{A}{\sqrt{Hz}}$. referred to the intrinsic photodetector can be determined as

$$I_{n,eq} = \frac{\sqrt{4kT_0G_dF}}{|H_{pd}|} \quad (1)$$

where K is Boltzmann's constant, T_0 the reference temperature of 290K, G_d is the conductance looking back into the two-port from the TIA and is most easy determined from the ABCD parameters as $G_d = \text{Re}\{C/D\}$, F is the noise figure which can be determined once the source reflection coefficient, Γ_s, is known and $H_{pd}=1/D$ the current referral function [11]. This equivalent input noise current referred to the intrinsic photodetector is what should be minimized for an optimum optical receiver noise performance.

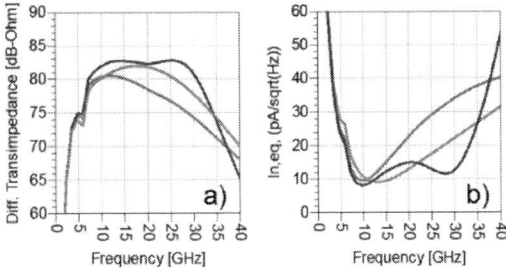

Fig. 5. Simulated differential transimpedance a) and equivalent input noise current spectral density b). The red curve is for a wire bond inductance of L_{bw}=0.0 nH; the blue curve is for an optimum wire bond inductance of L_{bw}=0.52 nH while the magenta curve represents the TIA result alone. The PD equivalent circuit assumes C_{PD}=90 fF, α=2/3, L_{PD}=60 pH and R_{PD}=15 Ω.

In order to illustrate how the above analysis can be used in the noise optimization of the TIA + PD assembly three situations are compared for a representative PD equivalent circuit as shown in Fig. 5. In the first case, the noise matching

wire bond inductance is set to zero. This is shown to lead to a strong increase in the equivalent input noise current and drop in the transimpedance gain at high frequency. Selecting the wire bond inductance to have a rather large value of L_{bw}=0.52 nH minimize the equivalent input noise current. It is observed that this noise optimum wire bond inductance does not lead to excessive peaking in the transimpedance gain. Finally, the results for the TIA alone are shown as a reference. It is seen that the equivalent input noise current will fall between the two extremes values of the wire bond inductance.

IV. EXPERIMENTAL VERIFICATION

A. Small-Signal Characterization

The TIA chip was at first characterized in the electrical domain via on-wafer measurements of the small-signal S-parameters. Due to the very high gain of the TIA an input power below -40 dBm was used during calibration and measurements. The supply voltage was set at 3.3 V and the current draw was 78 mA. Fig. 6 shows the S-parameters measured between the input and one of the two outputs with the unused output terminated to 50 Ω through an external bias-T. In general, a very good agreement between measurements and simulations is observed. Interchanging the two output arms leads to identical results. The gain is also shown for a few selected gain control voltages.

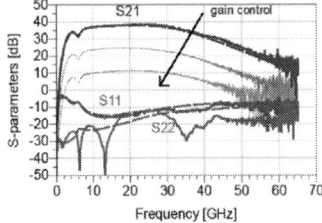

Fig. 6. Measured (solid lines) versus simulated (dashed lines) single-ended S-parameters. The gain is also shown for different gain control settings.

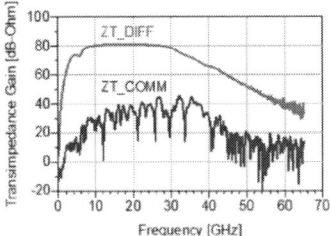

Fig. 7. Measured differential-mode (ZT_DIFF) and common-mode (ZT_COMM) transimpedance gain.

The differential and common-mode transimpedance can be calculated from the measured S-parameters. In the calculation the effect of the PD and noise matching bondwire is included. The differential transimpedance has a maximum of 80.8 dBΩ and a -3 dB bandwidth from 6.9 GHz to 30.5 GHz. The common-mode transimpedance remains at least 35 dB below the differential transimpedance which indicates an excellent balance between the two outputs of the TIA.

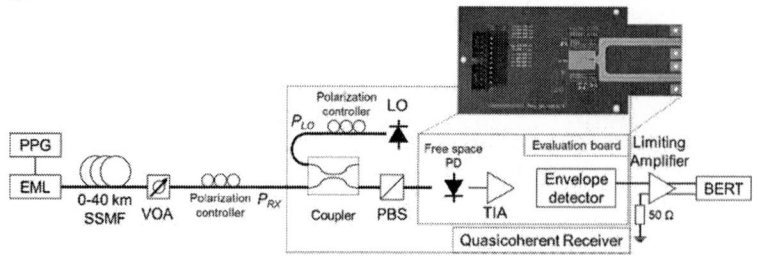

Fig. 8.Experimental setup for testing the SiGe BiCMOS TIA chip in a quasi-cohernt receiver.

B. Transmission Experiment

The developed TIA chip is employed in a quasi-coherent receiver similar to that of [2]. The experimental setup is shown in Fig. 8. The quasi-coherent receiver consists of an optical coupler, where the signal and the LO are joined, and a polarization beam splitter (PBS), where the two polarizations of the signal and the LO are divided. The signal and the LO are mixed in the PD, downconverting the data to an IF centred around 17 GHz. The LO optical power (P_{LO}) is 0 dBm. The PD has a responsivity of 0.8 A/W and an electrical bandwidth of 37 GHz. After the PD, the signal is amplified using the developed TIA and downconverted to baseband using a proprietary high-speed envelope detector. The data signal is generated with a commercially available externally modulated laser (EML), which is modulated with a 10 Gbps non-return to zero signal generated with a pulse pattern generator (PPG). The signal is propagated through different spans of standard single mode fiber (SSMF), which varies between 0 km (optical back-to-back -BTB) and 40 km. Errors in the transmission are recorded using the bit-error-rate-tester (BERT). The resulting BER vs. received power (P_{RX}) are shown in Fig. 9. The sensitivity of the quasi-coherent receiver is defined as the P_{RX} where the BER is equal to 10^{-3}, which is the FEC limit of the NG-PON2 standard. The quasi-coherent receiver sensitivity with the proposed TIA for back-to-back (BTB) transmission is -27.5 dBm, as can be seen in Fig. 9. The power penalty due to dispersion is below 0.25 dB for 10 km SSMF, 0.5 dB for 20 km SSMF and 0.75 dB for 40 km SSMF.

These sensitivities should be compared to those reported using a commercial TIA in reference [2, Fig. 4]. We find that the sensitivity of the quasi-coherent receiver with the developed SiGe BiCMOS TIA chip using an LO power of only 0 dBm is actually 1.5 dB better than the one that uses a LO optical power of 6.5 dBm in [2]. This clearly verifies the low-noise performance of the developed TIA chip.

V. CONCLUSION

In this paper, a high-gain, ultra-low noise bandpass-like SiGe BiCMOS TIA with 80 dBΩ differential transimpedance gain developed for quasi-coherent receivers has been reported.

ACKNOWLEDGMENT

This work was supported by the INCOM project (ID: 8057-00059B), partly funded by Innovation Fund Denmark.

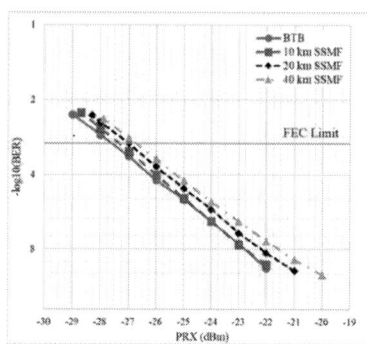

Fig. 9. BER vs. received power for back-to-back, 10, 20, and 40 km SSMF.

REFERENCES

[1] Broadband Forum Market Update. (2019) The Future of Passive Optical Networking is Here (MU-437). [Online]. Available: http://www.broadband-forum.org/marketing/download/MU-437.pdf

[2] J. A. Altabas *et al.*, "Real-Time 10 Gbps Polarization Independent Quasicoherent Receiver for NG-PON2 Access Networks," in *Journal of Lightwave Technology*, vol. 37, no. 2, pp. 651-656, 15 Jan.15, 2019.

[3] J. A. Altabas, O. Gallardo, G. S. Valdecasa, M. Squartecchia, T. K. Johansen and J. B. Jensen, "DSP-Free Real-Time 25 GBPS Quasicoherent Receiver With Electrical SSB Filtering for C-Band Links up to 40 km SSMF," in *Journal of Lightwave Technology*, vol. 38, no. 7, pp. 1785-1788, 1 April, 2020

[4] V. A. Thomas, S. Varughese, and S. E. Ralph, "Quasicoherent Receivers for Access Neworks Using Fullwave Rectification Based Envelope Detection," CLEO 2019, pages 1–2.

[5] I. G. Lopez et al., "100 Gb/s Differential Linear TIAs With Less Than 10 pA/√Hz in 130-nm SiGe:C BiCMOS," in IEEE Journal of Solid-State Circuits, vol. 53, no.2, pp. 458-469, Feb. 2018.

[6] J.-Y. Dupuy, F. Jorge, M. Riet, V. Nodjiadjim, H. Aubry, and A. Konczykowska, "59-dBΩ 68-GHz Variable Gain-Bandwidth Differential Linear TIA in 0.7-μm InP DHBT for 400-Gb/s Optical Communication Systems," 2015.IEEE Compound Semiconductor Integrated Circuit Symposium (CSICS), pp. 1–4.

[7] G. S. Valdecasa, J. B. Jensen, M. Didriksen and T. K. Johansen., "A 2-38 GHz Linear GaAs pHEMT TIA for Quasi-Coherent Optical Receiver," Proc. 14th European Microwave Integrated Circuits Conference (EuMIC), 2019, pp. 160-3.

[8] H. Rücker et al., "A 0.13 μm SiGe BiCMOS technology featuring f_T/f_{max} of 240/330 GHz and gate delays below 3 ps," in IEEE Journal of Solid-Stage Circuits, Vol. 45, No. 9, pp. 1678– 1686, Sept. 2010.

[9] C. D. Holdenried, J. W. Haslett, and M. W. Lynch, "Analysis and Design of HBT Cherry-Hooper Amplifiers With Emitter-Follower Feedback for Optical Communications," in IEEE Journal of Solid-Stage Circuits, Vol. 39, No. 11, pp. 1959– 1967, Nov. 2004.

[10] H. Rein, and M. Möller, "Design Considerations for Very-High-Speed Si-Bipolar IC's Operating up to 50 Gb/s," in IEEE Journal of Solid-Stage Circuits, Vol. 31, No. 8, pp. 1076– 1090, Aug. 1996.

[11] E.. Säckinger, Analysis and Design of Transimpedance Amplifiers for Optical Receivers, 1st edition. John Wiley & Sons, 2018.

Proceedings of the 16th European Microwave Integrated Circuits Conference

Multi-Phase Clock Path Circuit up to 57 GHz Including 5 bit Programmable Phase Interpolators for Time-Interleaved Broadband Data Converters in a 28 nm FD-SOI CMOS Technology

Daniel Widmann, Tobias Tannert, Xuan-Quang Du, Markus Grözing, Manfred Berroth

University of Stuttgart, Institute of Electrical and Optical Communications Engineering, Germany
{daniel.widmann, tobias.tannert, markus.groezing, manfred.berroth}@int.uni-stuttgart.de

Abstract — Clock paths in mixed-signal integrated circuits are critical building blocks possibly determining the entire circuit performance. A precisely controllable clock phase is highly desirable e.g. for monolithic, ultra high-speed data converters with time-interleaving, i.e. digital-to-analog (DAC) and analog-to-digital (ADC) converters, to adjust the time-interleaved converter channels' timing. More precisely, these converters use the means of analog multiplexing at the DAC outputs or analog demultiplexing at the ADC inputs, respectively. A broadband and low jitter clock path for frequencies up to 57 GHz is presented including 5 bit programmable phase interpolators at half input frequency with a phase delay resolution of about 1.25 ps realized in a 28 nm FD-SOI CMOS technology. A combination of current mode logic and CMOS logic is used in the proposed circuit.

Keywords — CMOS integrated circuits, phase shifters, mixed analog digital integrated circuits, analog-digital conversion, digital-analog conversion.

I. INTRODUCTION

Mixed-signal integrated data converter circuits, such as time-interleaved digital-to-analog converters (DACs) and analog-to-digital converters (ADCs) with ultra high conversion rates, demand challenging clock path circuits possibly determining the entire circuit performance. First, frequency dividers for different circuit parts are required. Secondly, several clock signal properties like common mode level, duty cycle and magnitude have to be optimized. Finally, for proper timing in clocked, time-interleaved front-end systems, the clock phase relations have to be precisely adjustable to omit intersymbol interference and ensure ideal sampling instants. Especially DACs that use the means of analog multiplexing at the DAC outputs [1] or a set of ADCs that use demultiplexing at their inputs [2], respectively, require precise relative clock timing (skew) of the single converters' clocks. Additionally, jitter is a critical aspect determining effective number of bits (ENOB) with increasing signal frequency. Here, a low jitter clock path up to 57 GHz including a 5 bit programmable phase interpolator realized in a 28 nm fully-depleted silicon-on-insulator (FD-SOI) CMOS technology [3] allowing forward body-biasing is presented. It is a key element for paving the way to higher sampling rates of monolithic, time-interleaved data converters in CMOS. The input circuit parts at highest frequencies (f_{clk} and $f_{clk}/2$) are

realized in (inductively peaked) current mode logic (CML). After level conversion, the design passes into common CMOS logic.

II. CLOCK PATH CONCEPT

Fig. 1 shows a block diagram of the entire clock system. Although the real circuit provides two equivalent branches of $f_{clk}/2$ channels, only one channel is discussed here. As an example, in a time-interleaved data converter, f_{clk} is used for the high-speed output multiplexer (MUX) whereas the other clocks on lower frequency domains provide clock signals for the digital MUXs at its input. Next to differential amplifiers (DAs), source followers (SFs) and different frequency dividers in different architectures, an offset control circuit, a start circuit, a programmable phase interpolator as well as a common mode control circuit are key components of the system. After the first clock division and phase rotation, a level conversion takes place providing the signals for the much more compact CMOS logic part with less power consumption.

III. CML PART

The CML part uses inductive peaking techniques in most components. The clock path starts with a passive offset control circuit depicted in Fig. 2. This circuit allows to control the common mode voltages of the two input clock signals and thus the duty cycle of the differential clock signal. On the right hand side ($R_3...R_5$, C_3 and L), a DC bias network sets the common mode voltage V_{cm} of both clock signals. On the left hand side, the external offset control voltage is shown which allows detuning the two common mode voltages for precise alignment. The resistor values are chosen to achieve $50\,\Omega$ termination single-ended (SE) ($100\,\Omega$ differential) in total.

Next, a cascade of DAs with inductive series as well as shunt peaking (see e.g. [4]) amplifies the signal before reaching a start circuit (see Fig. 3). This chain of amplifiers ensures limitation even for small input clock power levels. The start process is of particular importance for the first frequency divider in the CMOS logic part in the clock domain $f_{clk}/2$. Since it is run at its upper frequency limit, the clock signal has to be initialized and started at a dedicated phase. Moreover, transient settling effects at the start can cause short pulses

3–4 April 2022, London, UK

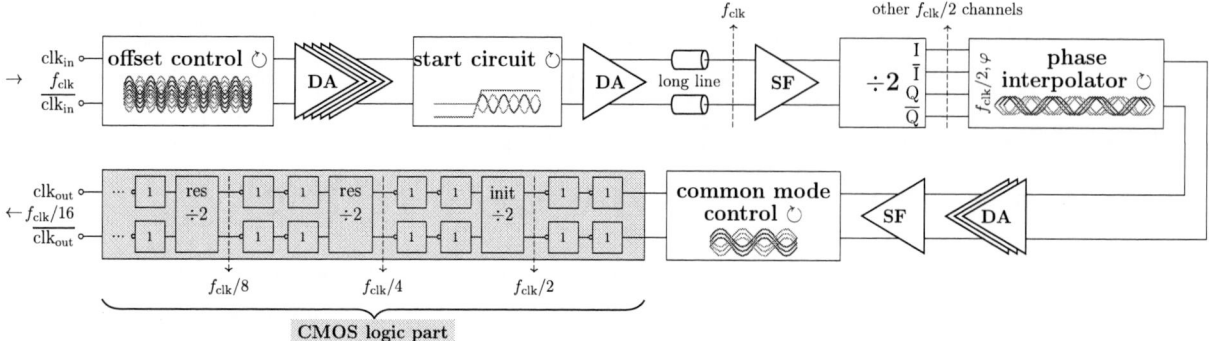

Fig. 1. Block diagram of the clock path (DA: differential amplifier, SF: source follower). The gray part is realized in common CMOS logic whereas the other parts are realized in CML, mostly with inductive peaking. Dashed arrows indicate the junctions in different clock domains where the clock signal can be retrieved for the data converter. Externally adjustable circuit parts are marked with ↻. In the CMOS logic part, different frequency dividers with initialization (*init*) or synchronous start (*res*) are used.

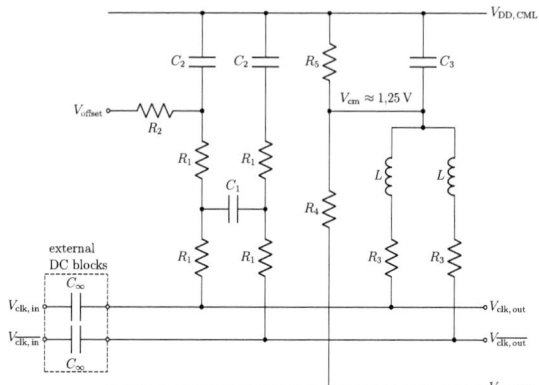

Fig. 2. Passive offset and duty cycle control circuit.

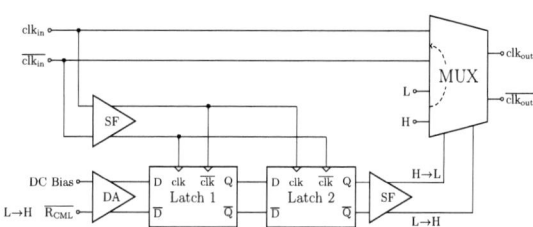

Fig. 3. CML start circuit. Latches are defined in such a way that they are opaque for clk = H. On start, the inductively peaked MUX switches from static levels to the clock input signal (dashed arrow).

leading to a nonfunctional static state of the divider at $f_{\text{clk}}/2$ and thus to a breakdown of divider activity. The circuit in Fig. 3 is responsible for both, a proper phase for initialization as well as a sampling of the reset signal $\overline{R_{\text{CML}}}$ to control the length of the first clock half-pulses.

The frequency divider consists of two CML latches with initialization transistors in negative feedback configuration and modified by shunt peaking.

Another key component is the 5 bit programmable phase interpolator, similar to e.g. [5]. Its core is a weighted adder which is shown in Fig. 4. In the following, the phase

interpolator's output delay is not considered in the output phase. Assuming ideal, single-tone signals and a perfectly linear system, the output signal of the weighted adder depending on the input amplitudes $\hat{V}_{\text{clk1}}$ and $\hat{V}_{\overline{\text{clk2}}}$ can be described as the following phasor:

$$\underline{V}_{\text{clk, out}} \sim \alpha \cdot \hat{V}_{\text{clk1}} - j \cdot (1 - \alpha) \cdot \hat{V}_{\overline{\text{clk2}}} \quad . \tag{1}$$

Without loss of generality, the phase difference between $\underline{V}_{\text{clk1}}$ and $\underline{V}_{\overline{\text{clk2}}}$ is assumed to be $+\pi/2$. The weighting factor α can be tuned by three programmable current sources with binary weighting. A programmable register allows setting the logical inputs sel_0, sel_1 and sel_2. For binary weighting, the current sources in cascode configuration are adapted in their transistor widths (w, $2w$, $4w$) and for symmetry reasons, the resistor values (R_0, $2R_0$, $4R_0$) are scaled accordingly. Ideally, the output signal's phase referred to $\underline{V}_{\text{clk1}}$ assuming the same amplitudes for the input signals is

$$arg\left\{\underline{V}_{\text{clk, out}}\right\} = -arctan\left(\frac{1 - \alpha}{\alpha}\right) \tag{2}$$

with

$$\alpha = \left(\text{sel}_2 \cdot 2^2 + \text{sel}_1 \cdot 2^1 + \text{sel}_0 \cdot 2^0\right) / \left(2^3 - 1\right) \quad . \tag{3}$$

$\alpha \in [0, 1]$ and for the logical selection inputs, it holds $\text{sel}_i \in \{0, 1\}$. Generally, the output phase $\phi_{\text{clk, out}}$ of $\underline{V}_{\text{clk, out}}$ can be expressed as

$$\phi_{\text{clk, out}} = \frac{\sum_{i=0}^{N-1} \left(\text{sel}_i \cdot 2^i \cdot \phi_{\text{clk1}} + (1 - \text{sel}_i) \cdot 2^i \cdot \phi_{\overline{\text{clk2}}}\right)}{2^N - 1} . \tag{4}$$

Here, a $N = 3$ (bit) weighted adder is discussed. The weighted adder can only interpolate phases between those of the input signals (e.g. $[-\pi/2, 0]$), their phases included. I.e., only phases of one quadrant can be reached. For a full phase control, two MUXs with another two control inputs sel_3 and sel_4 are required in front of the weighted adder enabling the choice of the desired quadrant by adaption of the input signals (see Fig. 5). Therefore, all four phases (I, $\overline{\text{I}}$, Q, $\overline{\text{Q}}$) provided by the frequency divider are required. It has to be mentioned that the phase interpolator works best for sinusoidal signals which is

178

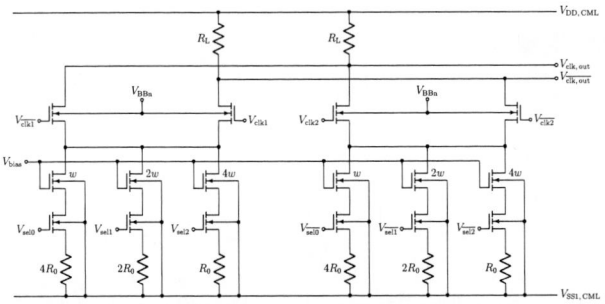

Fig. 4. Schematic of the weighted adder. Programmable current sources with binary weighting of currents realize the weighting function.

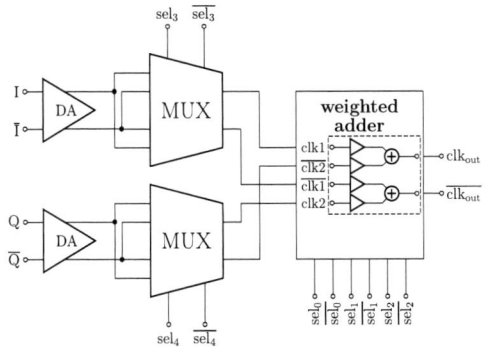

Fig. 5. Block diagram of the 5 bit programmable phase interpolation circuit. The choice of connections is defined by layout reasons.

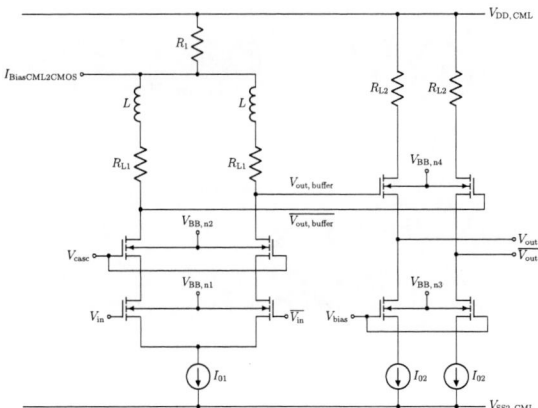

Fig. 6. Common mode control for CML to CMOS interface.

especially given for high frequencies where this circuit part is not in limitation anymore.

Finally, the interface between the CML and the CMOS logic part is of special importance. The common mode level and thus the duty cycle referred to the switching point of the CMOS logic part has to be adapted. Fig. 6 shows the common mode shift in two steps. First, an amplifier stage with an additional load resistor $R_1 = 10\,\Omega$ shifts the common mode output level by $R_1 \cdot I_{01}$. An extra control input $I_{\text{BiasCML2CMOS}}$ allows detuning the shift for precise level adaption at this critical interface. Secondly, a source follower for each clock signal causes another fixed level shift of a gate-source voltage.

After common mode level adaption, architecture is changed to CMOS logic for power consumption and chip area reasons.

IV. CMOS Logic Part

The CMOS logic part starts with two inverters for voltage level regeneration providing a limited signal to the first frequency divider. There are two types of frequency dividers in this part as depicted in [6]. Their basic latch structures are the same. However, the first frequency divider at $f_{\text{clk}}/2$ is driven at its frequency limit which is why initialization and a correct start clock phase are required. Minimum initialization transistors in combination with the proper initial phase delivered by the CML part are essential at the start. As mentioned before, the first clock pulses are critical and

there is only little tolerance for short pulses caused by transient settling effects. Going beyond leads to a common mode shift away from $\sim V_{\text{DD}}/2$ and the divider stops action ending in a nonfunctional static state. To omit a critical voltage drop at start due to the sudden current demand through supply lines with inherent inductances and bonding wires, the frequency dividers have their own supply voltage with large block capacitors not being disturbed by other circuit parts and are switched on sequentially.

At the lower frequency domains, resettable frequency dividers are used [6] with AND/NAND gates implemented as differential cascode voltage switch logic (DCVSL) gates. Their input reset signal is sampled and operation starts synchronously to the input clock signal. In case of several $f_{\text{clk}}/2$ channels are required (see "other $f_{\text{clk}}/2$ channels" in Fig. 1), a synchronous start by sampling of the reset signal is important for synchronization of the different channels. Therefore, the initialization and reset concept enables a multi-clock, multi-phase system that can start a clocked CMOS converter with a large power consumption jump on startup.

V. Measurement Results

The successive start sequence is given in Table 1 and run automatically on-chip by analog delay elements for the global reset signal. External supply voltages vary from $0.9\,\text{V}$ to $1.15\,\text{V}$ (overdrive) for different CMOS logic parts. For CML parts, the externally applied positive voltage is $V_{\text{DD, CML}} = 1.7\,\text{V}$, the bottom ones are $V_{\text{SS1, CML}} = 0.0\,\text{V}$ and $V_{\text{SS2, CML}} = -0.7\,\text{V}$, respectively. On the one hand, a slight overdrive to the transistors is applied. On the other hand, the voltage drops caused by series resistances on the supply lines are compensated by the given voltage values. The estimated total power consumption including two equivalent branches of $f_{\text{clk}}/2$ channels for the $f_{\text{clk}}/16$ signal not considering the CMOS inverters and the $50\,\Omega$ output drivers for the $f_{\text{clk}}/16$ signal is about $1.1\,\text{W}$. The CMOS frequency dividers' contribution is less than $6\,\%$.

Fig. 7 shows the $f_{\text{clk}} = 57\,\text{GHz}$ SE measurement results of the output clock signal clk_{out} at $f_{\text{clk}}/16 = 3.5625\,\text{GHz}$

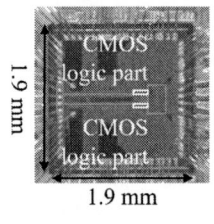 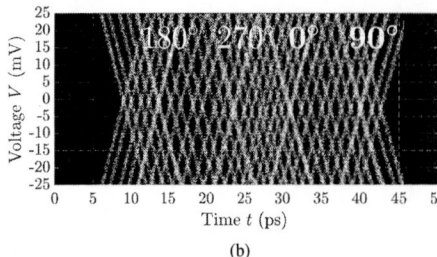 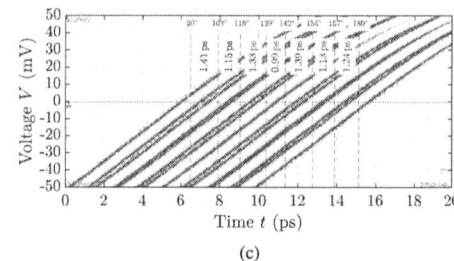

(a)	(b)	(c)

Fig. 7. (a) Photograph of the bonded die in the RF board cavity. In blue color, the CML part including GSSG input is shown. The small CMOS logic parts are shown in white color. Next to the clock path, the die also contains other circuits not presented here. (b) and (c) show the measurement results of the divided SE output clock signal clk_{out} at $f_{clk}/16 = 3.5625\,\text{GHz}$ for (b) all phase interpolator positions within $360°$ and (c) the phase positions within one quadrant in a closer view. For measurement reasons, rising and falling edges appear together in (b).

Table 1. Stepwise start sequence.

Step	Part	Signal
1	divider $f_{clk} \to f_{clk}/2$	init.: on $\to$ off
	divider $f_{clk}/2 \to f_{clk}/4$	init.: on $\to$ off
2	start circuit f_{clk}	MUX: static $\to$ clock
3	divider $f_{clk}/4 \to f_{clk}/8$	reset: on $\to$ off (sampled)
4	divider $f_{clk}/8 \to f_{clk}/16$	reset: on $\to$ off (sampled)

for different phase interpolator settings. Measurements were done using a subsampling oscilloscope with a sampling module bandwidth of $70\,\text{GHz}$ and a phase reference module driven at f_{clk} for low jitter measurements. Fig. 7b shows all phase interpolator positions inside a phase window of $360°$ referred to the $f_{clk}/2$ domain. In principle, a 5 bit programmable phase interpolator can generate 32 different phases. As the four phase positions at the corners of the four quadrants fall together, 28 different positions can be observed in one cycle resulting in a phase delay resolution of $\Delta t = \frac{2}{28 \cdot f_{clk}} \approx 1.25\,\text{ps}$ or an angle resolution referred to $f_{clk}/2$ of $\Delta\phi = 360°/28 \approx 12.9°$, respectively. Due to asymmetries, the two corresponding values at each quadrant corner slightly differ which leads to broader curves at these positions. A closer view of all positions within one quadrant is shown in Fig. 7c revealing phase delay resolutions of about $1.0\,\text{ps} - 1.4\,\text{ps}$. Furthermore, a low RMS jitter of $\sim180\,\text{fs}$ can be determined. It is quite comparable to the value of $(<)150\,\text{fs}$ in [7] discussing a state-of-the-art FinFET DAC. However, it is determined differently. It has to be mentioned that this value represents the one of the whole system including CMOS logic parts (also parts for different chip functionality not discussed here) that dynamically load the supply voltage and additionally the one of the measurement setup. Consequently, the on-chip jitter at the CML part might be even less. The lowest operation frequency verified by measurements is $2\,\text{GHz}$. However, this value is limited by the measurement setup.

VI. CONCLUSION

A low jitter clock path up to $57\,\text{GHz}$ including a 5 bit programmable phase interpolator realized in $28\,\text{nm}$ FD-SOI CMOS technology is presented. The key aspects are the separation of the circuit into a partially inductively peaked

CML and a common CMOS logic part as well as a clock startup concept considering a successive start sequence with initialization and reset parts. Furthermore, separated supply voltage domains with large block capacitors omitting critical voltage drops have been implemented to ensure proper startup operation of all sensitive circuit parts. The presented clock path enables time-interleaving of ultra high-speed CMOS integrated sampling systems, i.e. broadband data converters.

ACKNOWLEDGMENT

This project is funded by the *Deutsche Forschungsgemeinschaft* (DFG, German Research Foundation) – 276016065.

We thank our partners from Nokia Bell Labs Stuttgart for their extensive help in packaging of this integrated circuit. Special thanks to Fred Buchali and Peter Klose for their support.

REFERENCES

[1] K. Schuh, Q. Hu, M. Collisi *et al.*, "100 GSa/s BiCMOS Analog Multiplexer Based 100 GBd PAM Transmission over 20 km Single-Mode Fiber in the C-Band," in *European Conference on Optical Communications (ECOC)*, 2020, pp. 1–4.

[2] X.-Q. Du, M. Grözing, A. Uhl *et al.*, "A 112-GS/s 1-to-4 ADC front-end with more than 35-dBc SFDR and 28-dB SNDR up to 43-GHz in 130-nm SiGe BiCMOS," in *IEEE Radio Frequency Integrated Circuits Symposium (RFIC)*, 2019, pp. 215–218.

[3] N. Planes, O. Weber, V. Barral *et al.*, "28nm FDSOI technology platform for high-speed low-voltage digital applications," in *Symposium on VLSI Technology (VLSIT)*, 2012, pp. 133–134.

[4] J. S. Walling, S. Shekhar, and D. J. Allstot, "Wideband CMOS Amplifier Design: Time-Domain Considerations," *IEEE Transactions on Circuits and Systems I: Regular Papers*, vol. 55, no. 7, pp. 1781–1793, 2008.

[5] S. Sidiropoulos and M. Horowitz, "A semidigital dual delay-locked loop," *IEEE Journal of Solid-State Circuits*, vol. 32, no. 11, pp. 1683–1692, 1997.

[6] D. Widmann, M. Grözing, and M. Berroth, "High-Speed Serializer for a 64 GS s^{-1} Digital-to-Analog Converter in a 28 nm Fully-Depleted Silicon-on-Insulator CMOS Technology," *Advances in Radio Science*, vol. 16, pp. 99–108, 2018.

[7] R. L. Nguyen, A. M. Castrillon, A. Fan *et al.*, "8.6 A Highly Reconfigurable 40-97GS/s DAC and ADC with 40GHz AFE Bandwidth and Sub-35fJ/conv-step for 400Gb/s Coherent Optical Applications in 7nm FinFET," in *IEEE International Solid- State Circuits Conference (ISSCC)*, vol. 64, 2021, pp. 136–138.

Proceedings of the 16th European Microwave Integrated Circuits Conference

A DC to 20 GHz Variable Gain Amplifier with Tunable Input Matching in 22 nm FDSOI Technology

Seyyedmohsen Seyyedrezaei[1], Manu Viswambharan Thayyil, Corrado Carta, Frank Ellinger

Chair for Circuit Design and Network Theory, Technische Universität Dresden, Germany

[1]seyyedmohsen.seyyedrezaei@tu-dresden.de

Abstract — The design and characterization of a broadband radio-frequency variable-gain baseband amplifier, enhanced with tunable input-impedance matching and implemented in a 22 nm fully-depleted silicon-on-insulator technology is presented. The proposed amplifier utilizes the back-gate control of the transistors offered in the technology for gain tuning. Using a tunable nFET-based resistor, the amplifier features an adaptive input impedance and can be matched to various circuits, such as a mixer with low to moderate output impedance to serve as a final stage to drive a 50 Ω load. Measurement results show an achieved gain variability of about 9.2 dB with a maximum gain of 13.7 dB. The designed amplifier obtained a 3 dB bandwidth of nearly 20 GHz covering multiple industrial, scientific and medical (ISM) frequency bands. The circuit consumes a power of 5.3 mW from a 1 V power supply. The chip area including pads amounts to 0.277 mm².

Keywords — Baseband, Broadband amplifiers, CMOS integrated circuits, Feedback amplifiers, Impedance matching, Inductive peaking, Noise figure, Tunable circuits and devices.

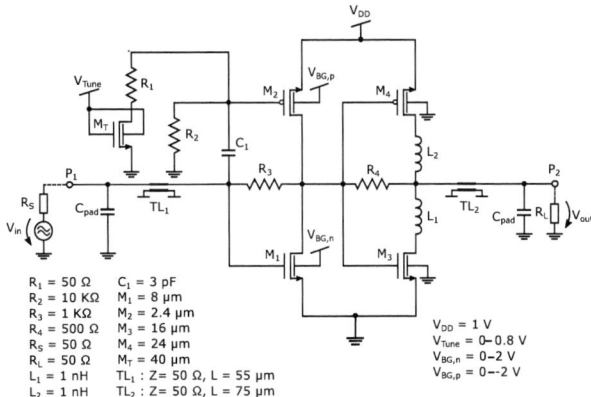

Fig. 1. Schematic of the designed BVGA.

I. INTRODUCTION

Many of recent applications such as positioning and localization radars, communication and identification systems require the design and implementation of silicon-based monolithic microwave integrated circuits (MMICs). On one hand, the precision of such systems is highly dependent on the bandwidth (BW), while on the other hand, preserving a required gain of the system over a high bandwidth leads to higher power consumption and degradation of the power efficiency of the system. An amplifier providing a variable gain over a large bandwidth, consuming low DC power with tunable input-impedance and low noise figure (NF) is desirable to drive a 50 Ω load for various system architectures such as zero intermediate-frequency (IF) and super-heterodyne receivers.

Although variable gain amplifiers (VGAs) such as [1], [2], [3], and [4] have achieved a moderate to large range of gain tunability, the highest bandwidth among them belonging to [3] is 4 GHz. The power consumption of [1] is relatively high compared to others due to the need for input and output buffers and a four-stage VGA core. Even though [5] has obtained a high gain-tunability over a bandwidth of nearly 15 GHz, the maximum achieved gain is -7.5 dB. Distributed amplifiers such as [6] and [7] deliver a high gain over a wide bandwidth, but the consumed power and area are substantially higher compared to [1]–[5]. Furthermore, the gain variability is not pursued in [6] and [7].

This work addresses these issues and presents a broadband variable gain amplifier (BVGA) design which uses the back-gate biasing in order to achieve a moderate range of gain tunability. Utilizing a tunable nFET-based resistor at the input, the amplifier is enhanced with an adaptive input impedance and can be used as a final stage for different circuits with different output impedances. Moreover, an inductive peaking is used to broaden the bandwidth of the circuit enabling the amplifier to drive a 50 Ω load for systems with operating frequency from 0 to around 20 GHz. The implementation is performed in a commercially available 22 nm fully-depleted silicon-on-insulator technology from GLOBALFOUNDRIES. The employed metal stack has 10 copper layers and one aluminum layer on the top. The design uses both pFETs and nFETs with f_t/f_{max} of 275 GHz / 299 GHz and 347 GHz / 371 GHz, respectively [8]. The fully-depleted transistors are isolated from the low-resistivity substrate by a thin buried oxide layer, decreasing the capacitive parasitics. The technology also provides a back-gate node to be able to tune the threshold voltages of the transistors.

The circuit design of the BVGA is discussed in Section II. Characterization setup and the measured results are presented in Section III. Section IV summarizes the work, highlights the key features of the design, and compares the results with the state-of-the-art (SoA).

3–4 April 2022, London, UK

II. Circuit Design

The schematic of the designed BVGA is shown in Fig. 1. The design is based on the cascade of two feedback amplifier stages, with input port P_1 and output port P_2. The transconductance g_m required for amplification is provided by minimum length transistors M_{1-4}. nFET M_1 in combination with pFET M_2 forms the first stage, and nFET M_3 together with pFET M_4 forms the second stage. Using the g_m of the pFETs besides nFETs, the topology allows current reuse and hence better efficiency.

In order to match the input impedance of the amplifier to different impedances, the diode-connected transistor M_T serves as a tunable nFET-based resistor. When M_T is turned on with $V_{Tune} \approx 500\,\text{mV}$, the input source resistance $R_s = 50\,\Omega$ forms a voltage divider with the equivalent parallel combination of $R_{in,eq} = (R_1 + R_{on,M_T}) || R_2 || R_{in,eq1} \approx 50\,\Omega$. Where $R_{in,eq1} \approx 300\,\Omega$ is the real part of the input impedance of the first amplifier stage calculated according to [9] as, $Z_{in,1} = R_3/(1 - a_{v,1})$. Here $R_3 = 1\,\text{k}\Omega$ is the first stage feedback resistance and $|a_{v,1}|$ is the magnitude of voltage gain, given by $|a_{v,1}| = ((g_{m,M1} + g_{m,M2})R_{o,eq1}) \approx 2.2$. The sizes of M_1 and M_2 are chosen to obtain $g_{m,M1} = 15\,\text{mS}$ and $g_{m,M2} = 3\,\text{mS}$, and the corresponding widths are $W_{M1} = 8\,\mu\text{m}$ and $W_{M2} = 2.4\,\mu\text{m}$. $R_{o,eq1}$ is the first stage equivalent parallel output resistance given by $R_{o,eq1} \approx R_3 || R_{in,eq2} \approx 125\,\Omega$. Similar to the first stage design, the equivalent input resistance of the second stage $R_{in,eq2}$ is calculated from the real part of second stage input impedance, $Z_{in,2} = R_4/(1 - a_{v,2}) \approx 140\,\Omega$. Similarly, the corresponding second stage magnitude of voltage gain $|a_{v,2}| = (g_{m,M3} + g_{m,M4})R_{o,eq2} \approx 2.4$, with $R_{o,eq2} \approx R_4 || R_L \approx 45\,\Omega$ is calculated. M_3 and M_4 are chosen to have widths $W_{M3} = 24\,\mu\text{m}$ and $W_{M4} = 16\,\mu\text{m}$ to obtain $g_{m,M3} = 28\,\text{mS}$ and $g_{m,M4} = 24\,\text{mS}$, respectively. The metal-oxide-metal capacitance $C_1 \approx 3\,\text{pF}$ enables the independent biasing of $M_{1,2}$, and facilitates the control of input impedance using V_{Tune}.

Due to the relatively large size of M_3 and M_4 required to drive a $50\,\Omega$ load resistance and the parasitic capacitance of the feedback resistors, the output capacitance of $C_{out,2} \approx 200\,\text{fF}$, limits the bandwidth of the circuit. Peaking inductors $L_{1,2} \approx 1\,\text{nH}$ help to improve the bandwidth by forming a synthetic transmission line [10] at the output, together with the pad capacitance $C_{pad} \approx 30\,\text{fF}$ and $C_{out,2}$. The circuit is interfaced to the pads using grounded coplanar waveguide transmission lines $TL_{1,2}$ having width $W = 5.5\,\mu\text{m}$, spacing to ground wall $S = 7.25\,\mu\text{m}$ and a characteristic impedance of $50\,\Omega$. The corresponding lengths are $55\,\mu\text{m}$ and $75\,\mu\text{m}$, respectively.

The magnitude of the effective gain of the circuit a_v is the product of the individual stage gains and the losses due to the input voltage divider: it can be written as, $|a_v| = \frac{R_{in,eq}}{R_s + R_{in,eq}}|a_{v,1}||a_{v,2}| \approx 4.5$. Varying V_{Tune} from 0 V to 800 mV varies the input impedance $R_{in,eq}$ from approximately $270\,\Omega$ to $42\,\Omega$, enabling the circuit to have a tunable input impedance. Additionally, varying the back-gate bias $V_{BG,n}$ of M_1 from 0 V to 2 V, the first stage gain is tunable to a lesser extent,

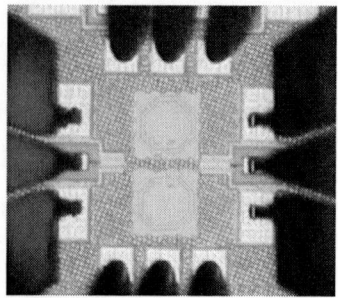

Fig. 2. Chip micrograph of the BVGA.

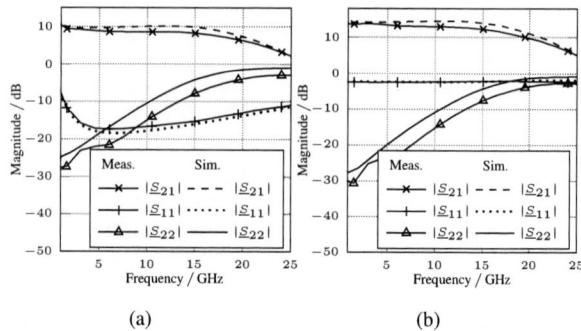

Fig. 3. Measured and simulated small signal parameters for (a) $50\,\Omega$ mode; (b) high-impedance mode.

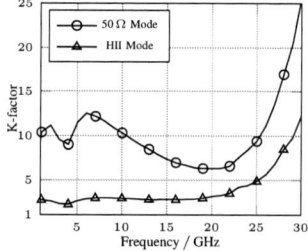

Fig. 4. Measured K-factor for different input impedance modes.

without impacting the input impedance. This enables the use of the circuit in applications like IF amplifiers following down-conversion mixers where a high ohmic input impedance is needed, and the amplifier needs to drive a $50\,\Omega$ output load.

III. Characterization

The chip micrograph of the fabricated BVGA is shown in Fig. 2. The occupied area of the chip is $530\,\mu\text{m} \times 523\,\mu\text{m} = 0.277\,\text{mm}^2$ including the pads. The BVGA draws a current of 5.3 mA from a supply of 1 V corresponding to total power consumption of 5.3 mW. The characterization is done for two modes, the $50\,\Omega$ input-impedance mode with $V_{Tune} = 0.8\,\text{V}$ and the high input-impedance (HII) mode with $V_{Tune} = 0\,\text{V}$. Small and large signal measurements of the chip are performed on a wafer prober using a R&S® ZVA67 vector network analyzer and 40 GHz GSG probes. The S-parameters are measured from 20 MHz to 25 GHz. Fig. 3(a) and 3(b) show the magnitude of

182

Table 1. Comparison with State-Of-The-Art CMOS Variable Gain Amplifiers.

Ref.	Technology (CMOS)	Gain (dB)	BW$_{-3dB}$ (GHz)	iP_{1dB} (dBm)	NF (dB)	P_{DC} (mW)	Area (mm^2)	FoM[a]	Tunable Input Impedance
[1]	65 nm	-39.4 to +20.2	4	-17 to -30	10 to 27	26	0.316[b]	6.9	No
[2]	65 nm	6 to 44	1	-13 to -43[c]	18 to 26	6[e]	0.027[d]	10.7[e]	No
[5]	28 nm	-7.5 to -40.5	15	-8 to -5	N/A	9.5	0.026[d]	N/A	Yes
[3]	65 nm	-19 to 21	4	-8 to -13	17 to 47	3.5[e]	0.012[d]	20.4[e]	No
[4]	65 nm	2 to 24	2.2	-2 to -22	24 to 29	9.96 / 3.48[e]	0.01[d]	4	No
This Work	22 nm FD-SOI	4.5 to 13.7	20	-8.6 to -16.7	8.3 to 13.1	5.3	0.277[b] / 0.044[d]	36.3	Yes

[a] $FoM = \frac{Gain_{max} \times BW \times Gain\ Range}{P_{DC} \times NF_{max}}$ [b] Including pads [c] Estimated data from the reported paper [d] Only core [e] Excluding output buffer

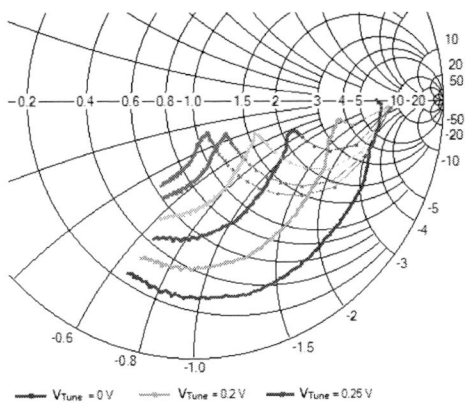

Fig. 5. Measured input impedance against different V_{Tune}.

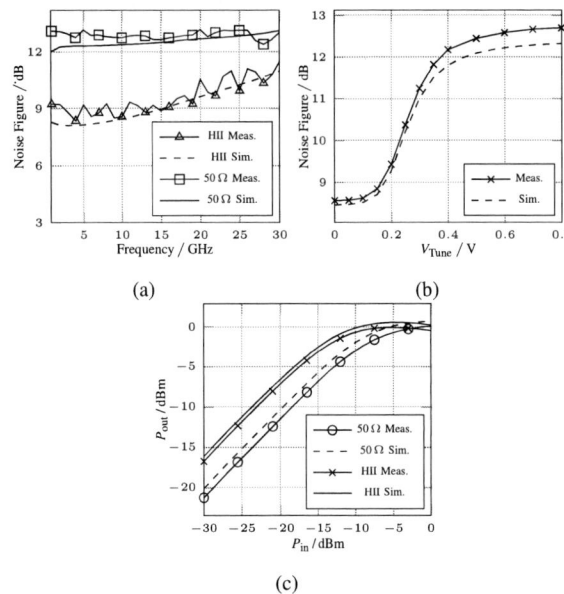

Fig. 6. Measured and simulated (a) noise figure for different modes; (b) noise figure against tuning voltage (V_{Tune}) at 10 GHz; (c) output power as a function of input power levels from -30 to 0 dBm for two impedance modes.

measured and simulated reflection coefficients $|\underline{S}_{11}|$, $|\underline{S}_{22}|$ and small signal gain $|\underline{S}_{21}|$ of the fabricated BVGA for 50 Ω and HII modes, respectively. For 50 Ω mode, the 3-dB bandwidth is between 1 GHz and 20 GHz with a peak gain of 9.3 dB. The $|\underline{S}_{11}|$ is below -10 dB from 1 GHz to 28.2 GHz, and the $|\underline{S}_{22}|$ is below -10 dB up to 13.3 GHz. For the HII mode, the 3-dB bandwidth is 18.5 GHz with a peak gain of 13.5 dB. The $|\underline{S}_{22}|$ is below -10 dB up to 13.4 GHz. The measured reverse isolation is better than 33 dB, and as shown in Fig. 4, the circuit is unconditionally stable for both modes. The trends of the measured S-parameters are similar to the simulated ones with a slight deviation which can be due to the temperature, process variations and additional losses.

Fig. 5 depicts the measured input impedance of the BVGA against tuning voltage. As can be seen, the real part of the input impedance can be changed up to 86% from approximately 50 Ω for $V_{Tune} = 1$ V to around 345 Ω for $V_{Tune} = 0$ V. By varying the input impedance, the NF of the BVGA can be tuned as required by the application. The noise figure is measured using a R&S® FSW67 spectrum analyzer in combination with a NC346V noise source. The results of NF measurement against simulation are shown in Fig. 6(a) for both input impedance modes. The minimum NF of 8.3 dB and 12.3 dB are measured for HII and 50 Ω modes, respectively. Fig. 6(b) shows a 3.9 dB noise figure tunability can be obtained for a 0.4 V change in V_{Tune} and at the expense of around 4 dB power gain.

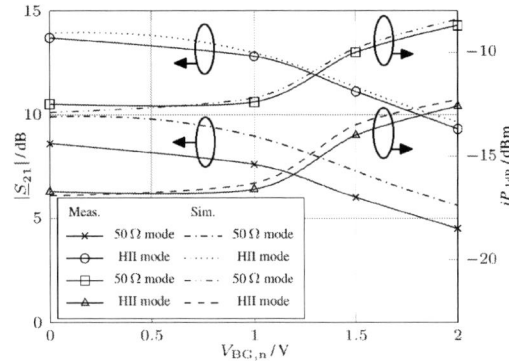

Fig. 7. Measured and simulated gain and input 1 dB compression point against back-gate bias of nFET transistor ($V_{BG,n}$) for different modes.

The large signal measurement of the fabricated BVGA in the middle of the band, 10 GHz, is shown against simulation results in Fig. 6(c). The measured values align well to the simulations. The measured input-referred 1 dB compression point (iP_{1dB}) for the 50 Ω input impedance mode is -12.3 dBm with a saturated output power of 0.5 dBm. For the high input impedance mode, the input-referred 1 dB compression point is -16.7 dBm with a saturated output power of about 0 dBm. Fig. 7 shows the results for gain and linearity in terms of $|\underline{S}_{21}|$ and input-referred 1 dB compression point against $V_{BG,n}$. The measured results show a total gain tunability of 9.2 dB and 8 dB change in input-referred 1 dB compression point.

IV. CONCLUSION

This research work presented a single-ended broadband variable gain amplifier implemented in 22 nm FDSOI technology with a large bandwidth from DC to 20 GHz. The design features a gain tuning using back-gate control of the transistors offered in the employed technology. Utilizing an impedance tunability at the input, the input impedance of the presented BVGA can be adapted and matched to the different impedances and the amplifier can drive a 50 Ω load for various systems over a frequency range of 0 to 20 GHz. The real part of the input impedance of the designed BVGA can be changed about 85% from 50 Ω to 345 Ω using a tunable nFET-based resistor. Thanks to a current reuse topology using both pFETs and nFETs, the circuit consumes low DC power. An overview of the state-of-the-art for baseband VGAs is given in Table 1. As shown, this design has the combination of the highest reported bandwidth with the lowest NF whereas having one of the lowest reported consumed power. The design also has the best reported figure of merit (FoM), to the knowledge of the authors.

ACKNOWLEDGMENT

The authors would like to thank GLOBALFOUNDRIES for fabricating the circuit. This work is supported by Bundesministerium für Bildung und Forschung (BMBF) within the framework of project REGGAE.

REFERENCES

[1] T. B. Kumar, K. Ma, and K. S. Yeo, "A 4 GHz 60 dB Variable Gain Amplifier With Tunable DC Offset Cancellation in 65 nm CMOS," *IEEE Microwave and Wireless Components Letters*, vol. 25, no. 1, pp. 37–39, 2015.

[2] L. He, L. Li, X. Wu, and Z. Wang, "A Low-Power Wideband dB-Linear Variable Gain Amplifier With DC-Offset Cancellation for 60-GHz Receiver," *IEEE Access*, vol. 6, pp. 61 826–61 832, 2018.

[3] L. Kong, H. Liu, X. Zhu, C. C. Boon, C. Li, Z. Liu, and K. S. Yeo, "Design of a Wideband Variable-Gain Amplifier With Self-Compensated Transistor for Accurate dB-Linear Characteristic in 65 nm CMOS Technology," *IEEE Transactions on Circuits and Systems I: Regular Papers*, vol. 67, no. 12, pp. 4187–4198, 2020.

[4] H. Liu, C. C. Boon, X. He, X. Zhu, X. Yi, L. Kong, and M. C. Heimlich, "A Wideband Analog-Controlled Variable-Gain Amplifier With dB-Linear Characteristic for High-Frequency Applications," *IEEE Transactions on Microwave Theory and Techniques*, vol. 64, no. 2, pp. 533–540, 2016.

[5] E. Sobotta, R. Wolf, N. Joram, and F. Ellinger, "RF variable gain amplifier with linear control and automatic matching in 28 nm CMOS," in *2015 IEEE International Conference on Microwaves, Communications, Antennas and Electronic Systems (COMCAS)*, 2015, pp. 1–5.

[6] L. Gao, Q. Ma, and G. M. Rebeiz, "A 1–17 GHz Stacked Distributed Power Amplifier with 19–21 dBm Saturated Output Power in 45nm CMOS SOI Technology," in *2018 IEEE/MTT-S International Microwave Symposium - IMS*, 2018, pp. 454–456.

[7] M. M. Tarar, T. Beucher, S. Qayyum, and R. Negra, "Efficient 2–16 GHz flat-gain stacked distributed power amplifier in 0.13m CMOS using uniform distributed topology," in *2017 IEEE MTT-S International Microwave Symposium (IMS)*, 2017, pp. 27–30.

[8] S. N. Ong, S. Lehmann, W. H. Chow, C. Zhang, C. Schippel, L. H. K. Chan, Y. Andee, M. Hauschildt, K. K. S. Tan, J. Watts, C. K. Lim, A. Divay, J. S. Wong, Z. Zhao, M. Govindarajan, C. Schwan, A. Huschka, A. Bellaouar, W. LOo, J. Mazurier, C. Grass, R. Taylor, K. W. J. Chew, S. Embabi, G. Workman, A. Pakfar, S. Morvan, K. Sundaram, M. T. Lau, B. Rice, and D. Harame, "A 22nm FDSOI Technology Optimized for RF/mmWave Applications," in *2018 IEEE Radio Frequency Integrated Circuits Symposium (RFIC)*, 2018, pp. 72–75.

[9] M. V. Thayyil, S. Li, N. Joram, and F. Ellinger, "A 4–32 GHz SiGe Multi-Octave Power Amplifier With 20 dBm Peak Power, 18.6 dB Peak Gain and 156% Power Fractional Bandwidth," *IEEE Microwave and Wireless Components Letters*, vol. 29, no. 11, pp. 745–748, 2019.

[10] S. Voinigescu, *High-Frequency Integrated Circuits*, 1st ed. Cambridge: Cambridge University Press, 2013.

Proceedings of the 16th European Microwave Integrated Circuits Conference

Noise Modeling of GaN/AlN HEMT

Sanaul Haque[#1], Frank Schnieder[*], Oliver Hilt[*], Ralf Doerner[*], Frank Brunner[*], Matthias Rudolph[*#]

[#]Ulrich L. Rohde Chair of RF and Microwave Techniques, Brandenburg University of Technology, Cottbus, Germany
[*]Ferdinand-Braun-Institut gGmbH, Leibniz-Institut für Höchstfrequenztechnik, Berlin, Germany
[1]sanaul.haque@b-tu.de

Abstract — AlN outperforming GaN as a buffer layer material makes GaN/AlN HEMTs promising for future high power and low noise applications. However, information on the noise performance and the noise modeling of GaN/AlN HEMTs is hardly available in the literature. This paper demonstrates that GaN/AlN HEMT technology at its early development stage presents similar noise parameters and competitive noise performance compared to an existing conventional GaN HEMT. This work also shows that the common Pucel and Pospieszalski noise models are valid and well applicable to the GaN/AlN HEMT devices without any modification.

Keywords — HEMTs, noise, semiconductor device modeling, semiconductor device noise.

I. INTRODUCTION

Gallium Nitride (GaN) High Electron Mobility Transistors (HEMTs) have emerged as the semiconductor device technology of choice for high power, high frequency and high temperature applications due to GaN's wide bandgap, high breakdown field, excellent transport properties and reasonable thermal conductivity [1]. However, GaN HEMT has its shortcomings, namely gate lag and drain lag effects causing current collapse [2] and dispersion [3]. The gate lag effect, caused by electron trapping in the surface states of the barrier layer, can be suppressed by surface passivation. On the contrary, the drain lag effect is a consequence of electron trapping in GaN buffer layer trap states introduced by the inevitable doping with either Fe [4] or C [5] for better electron confinement in the GaN channel and less leakage current in the GaN buffer. To counter this issue, Aluminium Nitride (AlN), which being a wider bandgap and higher resistivity material does not require the trap-prone doping, has been effectively utilized as the buffer layer in [6], [7] exhibiting significant suppression of current collapse and leakage current as well as enhancement of RF performance and breakdown voltage. Recently, a similar approach of replacing GaN with AlN in the buffer layer has been made to fabricate GaN/AlN HEMTs at Ferdinand-Braun-Institut (FBH).

The power handling capability of GaN HEMTs provides the opportunity for rugged and compact receiver and/or transceiver design without any protection circuitry [8], [9]. To assist low noise design in GaN/AlN HEMT technology, it is imperative to have the knowledge of noise performance and a noise model of these devices, which to the best of the authors' knowledge have not been published before. This paper reports on the noise performance and the noise modeling of GaN/AlN HEMT devices for the first time. As the technology is still under development, the devices at the current state do not

yet provide their full potential. However, the devices already show competitive performance compared to an established state-of-the-art GaN process. Like our previous study on GaN HEMT [10], this work is able to present the validity of the common Pucel [11] and Pospieszalski [12] noise model for the prediction of noise parameters of a typical GaN/AlN HEMT device.

II. OVERVIEW OF NOISE MODELS

Pucel noise model describes the intrinsic FET noise by two noise sources, one at the gate-source terminal and the other at the drain-source terminal of the intrinsic device. The drain-source noise generator represents the noise in the channel. This channel noise, which can be seen as a fluctuation of charge, is then coupled to the gate terminal mostly by a gate-source capacitance (C_{gs}) and is represented by the gate-source noise generator. Due to the coupling, these two noise sources are correlated. The short circuit gate and drain noise currents from the Pucel model are found to be [11]

$$\langle i_g^2 \rangle = 4kT_oBR\frac{(\omega C_{gs})^2}{g_m} \tag{1}$$

$$\langle i_d^2 \rangle = 4kT_oBPg_m \tag{2}$$

$$\langle i_g^* i_d \rangle = C\sqrt{\langle i_g^2 \rangle \langle i_d^2 \rangle} \tag{3}$$

where $\langle i_x^2 \rangle$, k, T_o, B and g_m are mean-square current, Boltzmann constant, reference temperature (290 K), bandwidth and transconductance, respectively. P, R and C are noise model parameters where C, the correlation coefficient of both noise sources, is expected to be a negative imaginary number.

The Pospieszalski noise model, on the other hand, explains the intrinsic FET noise by assigning an equivalent temperature T_g and T_d to the intrinsic gate resistance (R_{gs}) and drain resistance (R_{ds}), respectively. Unlike the Pucel noise model, the Pospieszalski model assumes no correlation between the two noise sources represented by the noise model parameters T_g and T_d. The short circuit noise currents can be derived from the Pospieszalski model as [12]

$$\langle i_g^2 \rangle = 4kT_gBR_{gs}\left|\frac{j\omega C_{gs}}{1+j\omega C_{gs}R_{gs}}\right|^2 \tag{4}$$

$$\langle i_d^2 \rangle = 4kB\left(\frac{T_d}{R_{ds}}+T_gR_{gs}\left|\frac{g_m}{1+j\omega C_{gs}R_{gs}}\right|^2\right). \tag{5}$$

3–4 April 2022, London, UK

III. Transistor Technology

The epitaxial layers of the $Al_{0.31}Ga_{0.69}N/GaN/AlN$ HEMT (thickness of 10 nm/148 nm/1350 nm) were grown by MOCVD on a 4" semi-insulating SiC substrate. No compensation doping was used in the AlN buffer layer to prevent related dispersion effects. A 2 nm GaN cap was added on top of the structure in order to prevent Al oxidation. It is important to notice that the GaN channel layer is relaxed and not compressively strained according to the AlN buffer layer lattice constant. The AlGaN barrier layer is tensile strained with respect to the GaN channel as typical for GaN HEMTs. The semiconductor stack is passivated by 2 nm in-situ SiN and 150 nm PECVD-deposited SiN. A 150 nm long gate trench is opened in the SiN. Using FBH Ir sputter-gate technology, the gate trench is covered with a 380 nm thick and 400 nm long Ir/Ti/Au metal stack to form the Schottky-type gate metal. Device isolation was achieved by N^+ ion implantation. A 2DEG sheet resistance of 440 Ω/sq was determined using the van der Pauw method. Ohmic contact definition relies on a Ti/Al-based metallization. After rapid thermal annealing, an ohmic contact resistance of 0.36 $\Omega \cdot$mm was obtained. Device on-state resistance is 2.9 $\Omega \cdot$mm and maximum drain current is 0.9 A/mm. The technology provides transistors with maximum intrinsic transconductance, f_t and f_{max} around 600 mS/mm, 60 GHz and 150 GHz, respectively. The device used in this work has 4x50 μm gate width, 0.5 μm gate-source distance and 2 μm gate-drain distance.

IV. Noise Model Parameter Extraction Procedure

The parameter extraction procedure closely follows the extraction routine of [13]. At first, the extrinsic and intrinsic small signal equivalent circuit parameters are determined from the cold-FET and bias dependent S-parameter measurement, as in [14]. Using the relationship in [15], a noise correlation matrix in chain representation for the entire FET device is then extracted from the measured noise parameters (i.e. minimum noise figure (NF_{min}), noise resistance (R_n) and optimum source reflection coefficient (Γ_{opt})). The extrinsic parameters are deembedded (as in [15]) from this correlation matrix to get the noise contribution from the intrinsic device. The updated matrix elements in Y representation are proportional to $\langle i_g^2 \rangle$, $\langle i_g i_d^* \rangle$, $\langle i_g^* i_d \rangle$ and $\langle i_d^2 \rangle$ [15]. Using (1) - (5), the noise model parameters are extracted at each measured frequency. A single value of each noise model parameter is then determined by averaging over a suitable frequency range that yields good agreement between measured and modeled noise parameters.

V. Measurement and Modeling of Transistor

Three transistors from different wafer positions were measured for cold-FET, S-parameter and noise parameter data and then modeled accordingly. Due to similar characteristics of all the transistors, noise modeling of only one transistor is presented here. The noise measurement was performed using a source-pull system at two bias points i.e. I_{ds} of 27 mA and 39 mA (roughly 15% and 22% of $I_{ds,max}$, respectively) at drain voltage of 15 V in the frequency range of 4 - 26.5 GHz.

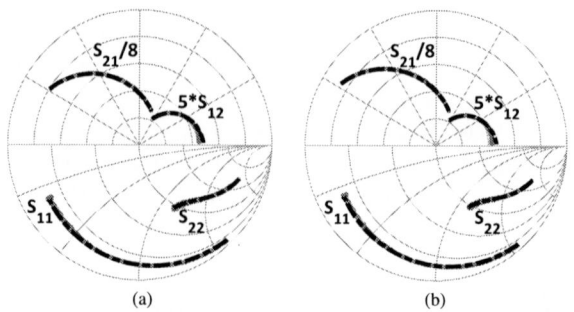

Fig. 1. Measured (red solid line) and simulated (black dash-dot line) S-parameters of GaN/AlN HEMT at (a) I_d = 27 mA and (b) I_d = 39 mA

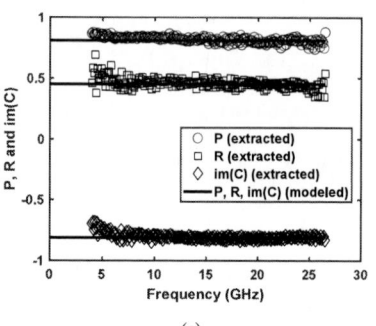

(a)

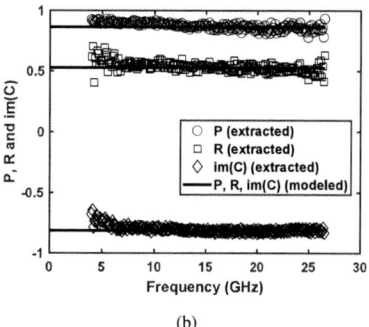

(b)

Fig. 2. Pucel noise model parameters of GaN/AlN HEMT at (a) I_d = 27 mA and (b) I_d = 39 mA

A. Noise Model Parameter Extraction

Parameter extraction begins with the extraction of small signal parameters. Fig. 1 shows a good agreement between measured and modeled S-parameters, except minor mismatch in high frequency S_{12} at both bias conditions.

The extracted and modeled Pucel noise model parameters (P, R and C) at both bias currents are shown in Fig. 2. The moderate spread of extracted parameters at each frequency helps in reliable modeling of the parameters by averaging over frequency. The values of parameters P and R are real, while parameter C is negative imaginary, as expected.

Fig. 3 displays the extracted and modeled Pospieszalski noise model parameters (T_g and T_d) at both bias conditions. The extracted T_g values vary between 350 K and 700 K.

186

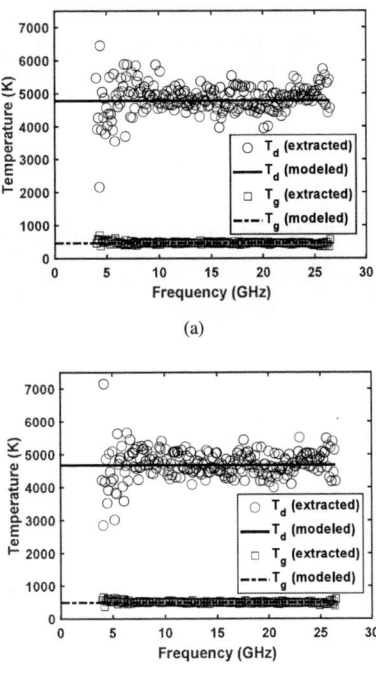

(a)

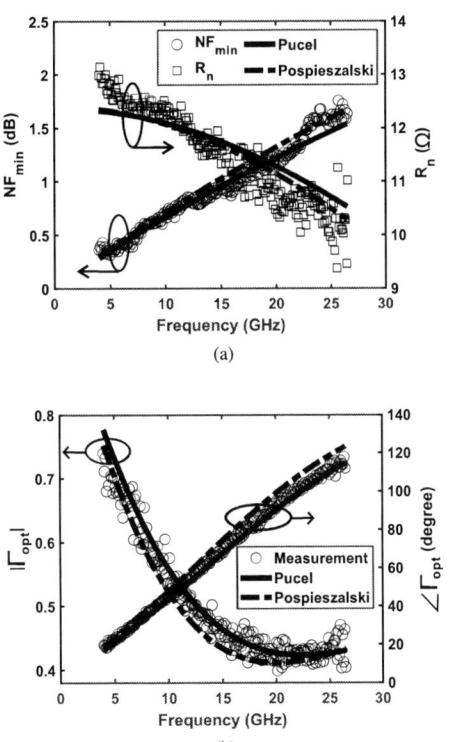

(b)

Fig. 3. Pospieszalski noise model parameters of GaN/AlN HEMT at (a) I_d = 27 mA and (b) I_d = 39 mA

(a)

(b)

Fig. 4. Measured (symbols) and simulated (lines) noise parameters of GaN/AlN HEMT at I_d = 27 mA

It is worth mentioning that the extraction of T_g directly depends on R_{gs} which has high uncertainty in the extraction procedure [10]. The majority of T_d values for both bias points are found to be in the range of 4000 K to 5500 K.

B. Noise Model Validation

The ultimate test of a noise model is to accurately predict the noise parameters at different frequencies. Fig. 4 and Fig. 5 illustrate the simulated noise parameters from both noise models in comparison with the measurements at bias current of 27 mA and 39 mA, respectively. In general, the simulated noise parameters from the models closely match the corresponding ones from the measurement at both bias conditions. Especially at high bias condition, the predictions of NF_{min} and Γ_{opt} from both the models (Fig. 5a and Fig. 5b) are similar and align with the measurement. At low bias, the Pucel model expects slightly less NF_{min} at higher frequencies compared to the Pospieszalski model, but still predicts the measurement well (Fig. 4a). $|\Gamma_{opt}|$ and $\angle\Gamma_{opt}$ of the Pucel model fit better than those of the Pospieszalski model to the measurement at low bias current (in Fig. 4b). The prediction trend of R_n (in Fig. 4a and Fig. 5a) at both bias points is identical where both the models have slightly gradual slope compared to the measurement.

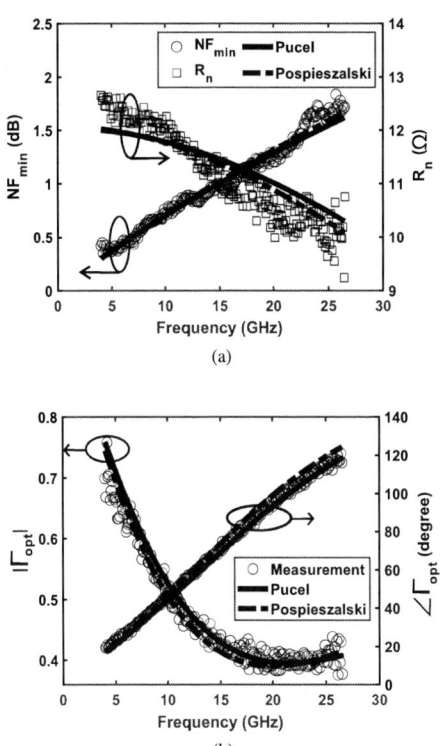

(a)

(b)

Fig. 5. Measured (symbols) and simulated (lines) noise parameters of GaN/AlN HEMT at I_d = 39 mA

C. Comparison between GaN and GaN/AlN HEMT

To benchmark GaN/AlN HEMT's noise parameters, it is compared with a contemporary FBH GaN HEMT of identical physical dimensions, as presented in Fig. 6. All the parameters are well comparable. It should be noted that although having a similar contact and sheet resistance, the GaN/AlN HEMT has higher NF_{min} than the GaN HEMT and the difference increases with frequency. Considering the early stage of development and the room for improvement and optimization in process technology and layer epitaxy, GaN/AlN HEMTs already show promising results for future low noise and high power design.

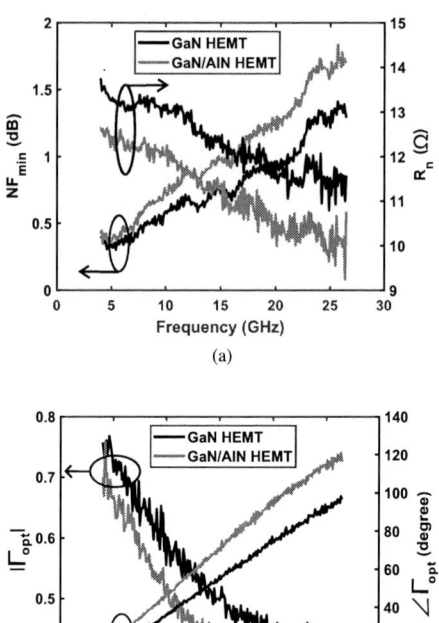

(a)

(b)

Fig. 6. Comparison of measured noise parameters of GaN HEMT ($V_d = 20$ V, $I_d = 40$ mA) and GaN/AlN HEMT ($V_d = 15$ V, $I_d = 39$ mA)

VI. CONCLUSION

GaN/AlN HEMTs promise to be the next big step of GaN HEMT technology. This paper, to be the best of the authors' knowledge, presents for the first time the measured noise parameters and noise modeling of GaN/AlN HEMT devices. The detailed parameter extraction and noise parameters validation with measurement prove the applicability of both Pucel and Pospieszalski noise model to these devices without considering gate leakage current, generation-recombination and trap assisted noise. Comparison with GaN HEMTs confirms higher NF_{min}, but still competitive noise performance of GaN/AlN HEMTs

which can be expected to improve even further as the technology gets more mature and the devices offer their full potential.

ACKNOWLEDGMENT

Financial support by the German Ministry of Education and Research (BMBF) within the program ForMikro-LeitBAN, grant no. 16ES1112, and under the project reference 16FMD02 (Forschungsfabrik Mikroelektronik Deutschland), is gratefully acknowledged.

REFERENCES

[1] L. Eastman and U. Mishra, "The toughest transistor yet [GaN transistors]," *IEEE Spectrum*, vol. 39, no. 5, pp. 28–33, May 2002.

[2] C. Nguyen, N. X. Nguyen, and D. E. Grider, "Drain current compression in GaN MODFETs under large-signal modulation at microwave frequencies," *Electronics Letters*, vol. 35, no. 16, pp. 1380–1382, 1999.

[3] E. Kohn, I. Daumiller, P. Schmid, N. Nguyen, and C. Nguyen, "Large signal frequency dispersion of AlGaN/GaN heterostructure field effect transistors," *Electronics Letters*, vol. 35, no. 12, pp. 1022–1024, 1999.

[4] S. Heikman, S. Keller, S. P. Denbaars, and U. K. Mishra, "Growth of Fe doped semi-insulating GaN by metalorganic chemical vapor deposition," *Applied Physics Letters*, vol. 81, no. 3, pp. 439–441, 2002.

[5] C. Poblenz, P. Waltereit, S. Rajan, S. Heikman, U. K. Mishra, and J. S. Speck, "Effect of carbon doping on buffer leakage in AlGaN/GaN high electron mobility transistors," *Journal of Vacuum Science & Technology B: Microelectronics and Nanometer Structures Processing, Measurement, and Phenomena*, vol. 22, no. 3, pp. 1145–1149, 2004.

[6] Z. Y. Fan, J. Li, M. L. Nakarmi, J. Y. Lin, and H. X. Jiang, "AlGaN/GaN/AlN quantum-well field-effect transistors with highly resistive AlN epilayers," *Applied Physics Letters*, vol. 88, no. 7, 2006.

[7] Y. L. Fang, Y. M. Guo, J. Y. Yin, B. Wang, Z. R. Zhang, J. Li, W. L. Lu, N. Gao, and Z. H. Feng, "High Breakdown Field AlGaN/GaN HEMT with AlN Super Back Barrier," in *2018 1st Workshop on Wide Bandgap Power Devices and Applications in Asia (WiPDA Asia)*, 2018, pp. 217–220.

[8] M. Rudolph, R. Behtash, R. Doerner, K. Hirche, J. Würfl, W. Heinrich, and G. Tränkle, "Analysis of the survivability of GaN low-noise amplifiers," *IEEE Transactions on Microwave Theory and Techniques*, vol. 55, no. 1, pp. 37–43, Jan. 2007.

[9] J. P. Janssen, M. Van Heijningen, G. Provenzano, G. C. Visser, E. Morvan, and F. E. Van Vliet, "X-band robust AlGaN/GaN receiver MMICs with over 41 dBm power handling," in *2008 IEEE Compound Semiconductor Integrated Circuits Symposium*, 2008, pp. 1–4.

[10] M. Rudolph, R. Doerner, E. Ngnintendem, and W. Heinrich, "Noise modeling of GaN HEMT devices," in *2012 7th European Microwave Integrated Circuit Conference*, 2012, pp. 159–162.

[11] R. A. Pucel, H. A. Haus, and H. Statz, "Signal and noise properties of gallium arsenide microwave field-effect transistors," in *Advances in Electronics and Electron Physics*, 1975, vol. 38, pp. 195–265.

[12] M. Pospieszalski, "Modeling of noise parameters of MESFETs and MODFETs and their frequency and temperature dependence," *IEEE Transactions on Microwave Theory and Techniques*, vol. 37, no. 9, pp. 1340 – 1350, Sep. 1989.

[13] P. Heymann, M. Rudolph, H. Prinzler, R. Doerner, L. Klapproth, and G. Bock, "Experimental evaluation of microwave field-effect-transistor noise models," *IEEE Transactions on Microwave Theory and Techniques*, vol. 47, no. 2, pp. 156–163, Feb. 1999.

[14] G. Dambrine, A. Cappy, F. Heliodore, and E. Playez, "A new method for determining the FET small-signal equivalent circuit," *IEEE Transactions on Microwave Theory and Techniques*, vol. 36, no. 7, pp. 1151–1159, Jul. 1988.

[15] H. Hillbrand and P. Russer, "An efficient method for computer aided noise analysis of linear amplifier networks," *IEEE Transactions on Circuits and Systems*, vol. 23, no. 4, pp. 235–238, Apr. 1976.

Efficient TCAD Temperature-dependent Large-Signal Simulation of a FinFET Power Amplifier

E. Catoggio, S. Donati Guerrieri, F. Bonani, G. Ghione

Dipartimento di Elettronica e Telecomunicazioni, Poltecnico di Torino, Italy

{eva.catoggio, simona.donati, fabrizio.bonani, giovanni.ghione}@polito.it

Abstract— We show a complete temperature-dependent analysis of a low power FinFET-based class A amplifier for small-cell applications based on an efficient approach to the temperature-dependent physics-based analysis of electron devices in Large Signal (LS) nonlinear conditions. The method extends the Green's Function (GF) approach, already developed for the device LS noise and technological sensitivity, to calculate the LS device response to the temperature variation from a nominal, "cold" condition with a negligible numerical overhead with respect to the other GF-based analyses. T-dependent TCAD simulations are applied to assess the robustness of the FinFET-based power amplifier against device heating and load variations. Temperature variations dominate over load sensitivity, showing more than 1 dB output power loss and a PAE reduction from 28% to 23%. The proposed approach represents a first step towards the development of physically sound, temperature dependent, LS circuit models of nonlinear stages.

Keywords— Semiconductor devices, Nonlinear device models, TCAD simulations, Harmonic Balance

I. INTRODUCTION

Physics-based device simulations represent an ideal environment to accurately model the behavior of the active device in RF/microwave circuits, as they keep trace of the underlying technological and physical parameters. With the ever increasing capability of computation machines, the frequency domain analysis of electron devices operated in highly nonlinear conditions has proved to be a fairly manageable task even within TCAD simulators, especially using the Harmonic Balance formalism for mixed-mode simulations, where the device physical equations need to be solved concurrently with the external harmonic tuning circuits [1]–[3]. The outcome of Large Signal (LS) TCAD simulations can be also successfully integrated into circuit-level simulators through X-parameters [4], and further coupled to accurate Electro Magnetic (EM) simulations of the passive structures [5], thus making the physically based analysis of a complete nonlinear stage, such as a power amplifier, an attractive circuit simulation scenario. Nonetheless, to be successfully used for circuit analysis, the physics-based models must be able to predict the *sensitivity* of the nonlinear stage performance towards the variations of: 1) multiple physical/technological parameters (e.g. doping, geometry or material parameter spread); 2) variations of the embedding circuit (effectively corresponding to a load-pull analysis around the nominal load conditions); 3) temperature variations. Previously, we developed efficient numerical algorithms, based on the efficient computation of the Green's Functions of the device linearized

physical model, to model the first two sources of variation [6]–[8]. In this work we address the problem of temperature variations, especially relevant in the scenario of power devices (e.g. GaAs or GaN based HEMTs) and nanoscale devices (e.g. FinFETs) [9], [10]. While self-consistent physics-based electro-thermal (ET) simulations, would be the ideal choice at the TCAD level [11], they meet fundamental limitations due to the numerical burden, especially in the nonlinear case. Full ET simulations can be simplified defining a device ("junction") temperature parametrizing the electrical model and self-heating is then accounted for through a self-consistent solution of the T-dependent electrical model coupled to an external thermal circuit. Hence, in this work we address the problem of T-dependent nonlinear simulations, where in the physical model a unique equivalent lattice temperature describes the overall device heating [12].

Our in-house code, allowing for the Harmonic Balance Large Signal analysis of electron devices, has been exended to account for the temperature dependency of the material properties, finally allowing for T-dependent LS simulations. The Conversion Matrix Green Function capability, originally developed for fast and numerically efficient LS noise, LS variability and synthetic load pull analyses, has been extendend accordingly [12]. The GF approach allows now for the fast and numerically efficient T-dependent LS analysis, starting from a single "cold" simulation. The variations with temperature can be easily coupled with other concurrent technological or load variations, finally allowing for an global assessment a nonlinear stage reliability.

In this paper, the novel code is used to simulate a FinFET-based power amplifier (PA) as a function of both temperature and load variations (akin to synthetic load-pull), showing that the GF approach is capable to accurately reproduce the LS stage performance up to 50 K temperature increase. We demonstrate that the stage is dominated by thermal degradation, showing more than 1 dB output power loss and PAE reduction from 28% to 23%.

II. EFFICIENT T-DEPENDENT TCAD LS ANALYSIS

Consider an active device with N-ports connected to an N-port external load, as in Fig. 1 (left). The device constitutive equations collectively represent a physics-based model, e.g. the drift-diffusion equations, discretized over the device volume, and coupled to the external circuit, here represented by the equivalent load and power sources [13]. The system is

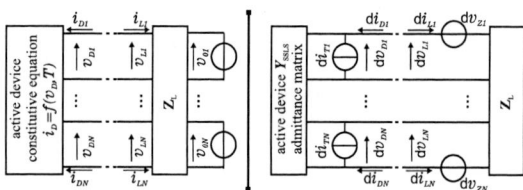

Fig. 1. Schematic representation of the linearized device including temperature induced and load-induced variations

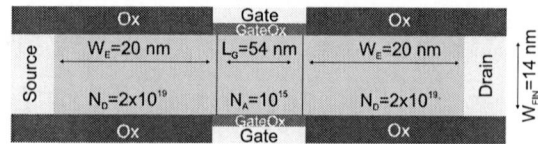

Fig. 2. Double Gate structure representing the cross-section of each elementary fin of the power cell.

linearized around the LS steady-state (nominal load $\mathbf{Z}_L$ and "cold" temperature T_0), to account for a temperature variation δT in the physical model and a load variation $\delta \mathbf{Z}_L$, yielding:

$$\delta \vec{i}_D = \frac{\partial f(\vec{v}_D, T)}{\partial \vec{v}_D}\bigg|_0 \delta \vec{v}_D + \frac{\partial f(\vec{v}_D, T)}{\partial T}\bigg|_0 \delta T$$

$$= \mathbf{Y}_{SSLS}\,\delta \vec{v}_D + \delta \vec{i}_T; \tag{1}$$

$$\delta \vec{v}_L = \mathbf{Z}_L\,\delta \vec{i}_L + \delta \mathbf{Z}_L\,\vec{i}_L\big|_0 = \mathbf{Z}_L\,\delta \vec{i}_L + \delta \vec{v}_Z \tag{2}$$

where subscript "0" refers to the LS working point, "D" to the device ports, and "L" to the load ports . In (1)-(2), $\mathbf{Y}_{SSLS}$ is the Small-Signal Large-Signal (SSLS) device admittance matrix, computed from SSLS TCAD analysis. The impressed generators $\delta \vec{i}_T$ and $\delta \vec{v}_Z$ collectively represent the equivalent terminal effect of δT and $\delta \mathbf{Z}_L$. The linearized model allows for the representation in Fig. 1 (right).

With $\delta \vec{v}_L = \delta \vec{v}_D$ and $\delta \vec{i}_L = -\delta \vec{i}_D$, (1)-(2) yield

$$\delta \vec{i}_D = (\mathbf{I} + \mathbf{Y}_{SSLS}\mathbf{Z}_L)^{-1}\left(\delta \vec{i}_T + \mathbf{Y}_{SSLS}\delta \vec{v}_Z\right) \tag{3}$$

Generators $\delta \vec{v}_Z$ are directly computed from the nominal LS solution using (2). Impressed generators $\delta \vec{i}_T$, instead, are computed using the in-house TCAD simulator by means of the Conversion Matrix Green's Functions (CGF), with negligible numerical overhead with respect to the computation of the nominal LS working point: the details can be found in [6], [12]. For this work, the in-house code was extended to account for all temperature dependencies of the physical model, including mobility, velocity saturation and carrier statistics.

III. FinFET Power Amplifier

FinFET technology, primarily developed for digital applications, is being actively investigated for its possible applications in analog stages, being the natural evolution of the RF CMOS technology [14], [15]. For the exploitation of these extremely miniaturized devices in analog circuits, the impact of parasitics and the difficult thermal management are the primary concerns, calling for accurate nonlinear, self-heating oriented models. In the RF 5G scenario, small and medium power amplifiers are among the most challenging stages: here we propose the thermal analysis of a small power class A power amplifier operating at the frequency of 70 GHz. A unit cell made of a multifinger device (10 fingers of 30 fins each) with a fin height of 25 nm, corresponding to a total gate periphery of 15 μm, is used for the development of the power stage. More cells can be then combined to increase the output power.

The cross-section of each elementary fin of the power cell is represented in Fig. 2. This elementary structure was analyzed with our in-house 2D TCAD simulator to assess the power cell degradation with increasing temperature.

A preliminary device analysis was performed to select the DC bias $V_G = 0.675$ V and $V_D = 0.6$ V and the optimum load $Z_{opt} = 53 + j6\Omega$ according to the load-line approach. Notice that Z_{opt} has been calculated at the nominal "cold" temperature $T_0 = 300$ K, while at higher temperatures the device will exhibit a de-tuning due to the change of the ouput characteristics with T. The device was then simulated in large-signal conditions, with $n_H = 10$ harmonics and increasing input power from back-off to 2 dB gain compression. Higher harmonics are supposed to be shorted at this high operating frequency. In this preliminary simulation campaign, the input port has been left unmatched and terminated with a 50 Ω/mm impedance.

A. T-dependent LS Analysis

We first address the stage temperature dependency with the nominal load condition, i.e. setting $\delta \mathbf{Z}_L = 0$ in (2). At each input power, the CGFs are calculated at $T_0 = 300$ K and the drain current variation with T is evaluated according to (3) for 5 temperatures, spanning the interval $[310-350]$ K. GF results are verified against the reference solution, corresponding to repeated LS analyses with varying temperature (incremental method, INC). With the 5 temperatures under test, the simulation time of the INC analysis is roughly 5 times the one required for the GF approach (in this example, around 25 hours INC vs. 5 hours GF). All results show an excellent accuracy of the GF approach in all operating conditions.

Figs. 3 and 4 show the output power and gain of the power cell with increasing T. The power performance exhibits a noticeable degradation, with up to 1 dB less output power and gain at $T = 350$ K. The thermal sensitivity is higher in back-off and limited in compression: in fact, the variation of the output power above the 1 dB compression point is due to the knee voltage walk-out with T, which is in any case quite limited, see also Fig. 5 (solid line). Fig. 6 shows the stage efficiency (left) and its variation with respect to the "cold" case (right) as a function of the input power. The thermal sensitivity depends again on the input power having a maximum value at 1 dB compression, but a significant efficiency reduction is found in a wide range of output power (roughly from 5 dB OBO to saturation). This needs to be taken carefully into account in the design of quasi-linear stages, often operated with modulated signals whose average value is well in back-off.

190

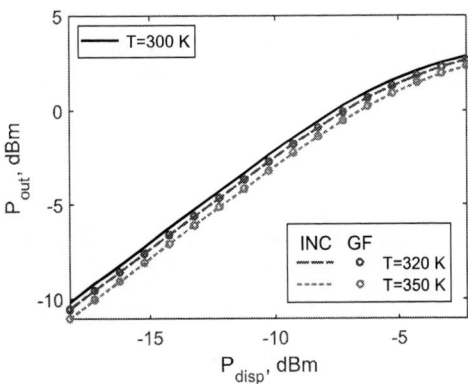

Fig. 3. $P_{in} - P_{out}$ plot for the class A PA as a function of temperature. Lines: incremental simulations at $T = 300$ K (solid), $T = 320$ K (dashed) and $T = 350$ K (dotted). Symbols: GF approach.

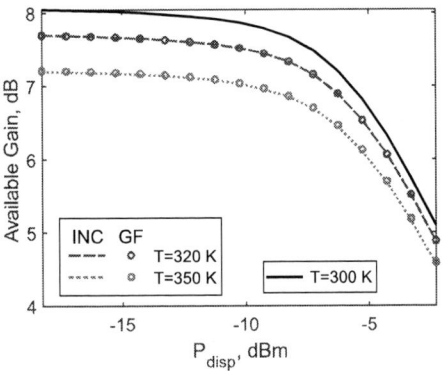

Fig. 4. Available Gain for the class A PA at varying temperature. Lines: incremental simulations at $T = 300$ K (solid), $T = 320$ K (dashed) and $T = 350$ K (dotted). Symbols: GF approach.

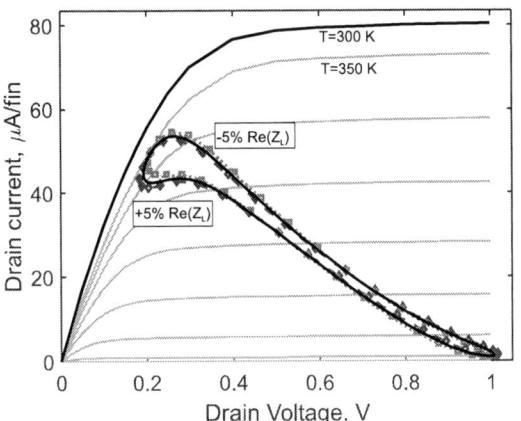

Fig. 5. Dynamic load lines at 2 dB gain compression, $T = 350$ K and varying the load condition. Black solid line: nominal load. Dotted line and squares: -5% variation of the nominal load real part. Dashed line and diamonds: $+5\%$ variation of the nominal load real part. Lines: incremental simulations. Symbols: GF approach. Output characteristics at $T = 350$ K are in grey, while the black output characteristic shows the "cold" case $T = 300$ K at the largest gate voltage.

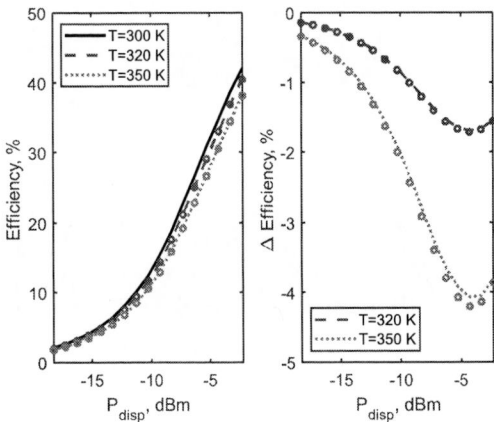

Fig. 6. Drain efficiency (left) and its variation with respect to the nominal temperature (right) as a function of the input power. Lines: incremental simulations at $T = 300$ K (solid), $T = 320$ K (dashed) and $T = 350$ K (dotted). Symbols: GF approach.

B. T-dependent, Load-dependent Analysis

We now extend the previous analysis to assess the overall power stage robustness against the concurrent variations of temperature and load termination (e.g. due to the variability of the matching network). Varying the real part of the optimum load by ± 5 % with respect to the nominal case (a range in line with EM simulations, see e.g. [5]), doubles the simulation time for the INC approach, while the GF analysis following (3) requires a negligible time overhead. Fig. 5 shows the dynamic load lines of the PA unit cell at 2 dB gain compression. The device is driven more harshly into compression by the increasing load (blue dashed lines), whereas the compression is lower in the opposite case, but the output swing is in any case reduced by the T-induced knee voltage increase. The accuracy of the time domain waveforms (DLL) shows that the harmonics are also well reproduced by the proposed GF T-dependent LS analysis. The $P_{in} - P_{out}$ and PAE reported in Fig. 7 also show examples of the concurrent load (here $+5\%$) and T variations.

Turning to the relative importance of the two variations, Fig. 8 reports the P_{out} and PAE variation with respect to their *nominal value* (i.e., $T = 300$ K and nominal load). In this figure, lines represent the variations due to temperature only, while the error bars represent the spread expected because of the $\pm 5\%$ load variations. Noticeably, the detailed variations depend on the input power: with larger load ($+5\%$), for example, P_{out} in back-off exhibits a lower reduction with respect to the nominal load (higher power and higher gain), while the opposite is true at higher input power since the compression is greater (see again Fig. 5). Furthermore the load sensitivity is lower in compression, again due to the reduced effect of the load termination at the knee. PAE has the highest load sensitivity close to the onset of gain compression (here around -7 dBm), but variations with temperature are in any case dominating, leading to a significant PAE reduction in a wide range of output power (as already noticed, roughly

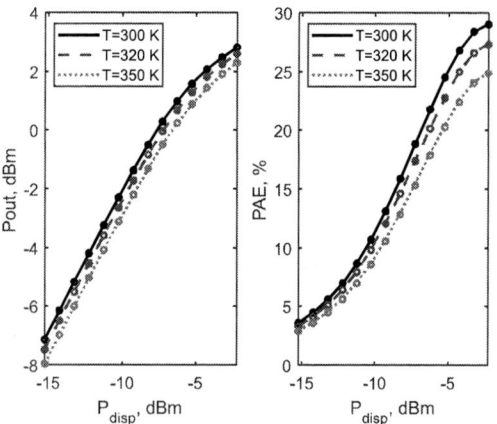

Fig. 7. Output power (left) and PAE (right) with varied load (+5% real part) and at different temperatures. Solid lines: incremental simulations. Symbols: GF approach with concurrent T and load variations according to (3).

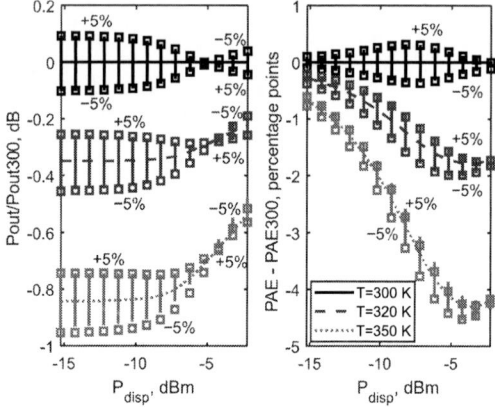

Fig. 8. Variation of the output power (left) and PAE (right) with temperature and load. Lines: incremental simulations with nominal load and $T = 300$ K (solid), $T = 320$ K (dashed) and $T = 350$ K (dotted). Error bars denote the expected spread due to ± 5 % variation of $\mathrm{Re}(Z_L)$ evaluated with the incremental analysis. Symbols: spread due to ± 5 % variation of $\mathrm{Re}(Z_L)$ following the GF approach with concurrent temperature and load variations.

from 5 dB OBO to saturation). All these intermixed effects are correctly reproduced by the GF analysis. Overall, the stage exhibits a spread of roughly 1 dB for P_{out} and 5 percentage points for PAE.

IV. CONCLUSION

A novel TCAD approach for the efficient temperature dependent analysis of nonlinear stages has been validated in a FinFET-based power amplifier along with the effect of concurrent load variations. The thermal degradation is shown to affect all the operating conditions, including the back-off case. The proposed approach demonstrates that LS TCAD analysis is mature for the assessment of nonlinear circuits and represents a first step towards the development of physically sound, temperature dependent, LS circuit models.

ACKNOWLEDGMENT

This work has been supported by the Italian MIUR PRIN 2017 Project "Empowering GaN-on-SiC and GaN-on-Si technologies for the next challenging millimeter-wave applications (GANAPP)"

REFERENCES

[1] B. Troyanovsky, Z. Yu, and R. W. Dutton, "Large signal frequency domain device analysis via the harmonic balance technique," in *Simulation of Semiconductor Devices and Processes*. Springer Vienna, 1995, pp. 114–117.

[2] F. M. Rotella, G. Ma, Z. Yu, and R. W. Dutton, "Design optimization of RF power MOSFET's using large signal analysis device simulation of matching networks," in *Simulation of Semiconductor Processes and Devices 1998*. Springer Vienna, 1998, pp. 26–29.

[3] J. Fang, J. Moreno, R. Quaglia, V. Camarchia, M. Pirola, S. Donati Guerrieri, C. Ramella, and G. Ghione, "3.5 GHz WiMAX GaN doherty power amplifier with second harmonic tuning," *Microwave and Optical Technology Letters*, vol. 54, no. 11, pp. 2601–2605, 2012.

[4] S. Donati Guerrieri, F. Bonani, and G. Ghione, "Linking X parameters to physical simulations for design-oriented large-signal device variability modeling," in *2019 IEEE MTT-S International Microwave Symposium (IMS)*. IEEE, jun 2019.

[5] S. Donati Guerrieri, C. Ramella, F. Bonani, and G. Ghione, "Efficient sensitivity and variability analysis of nonlinear microwave stages through concurrent TCAD and EM modeling," *IEEE Journal on Multiscale and Multiphysics Computational Techniques*, vol. 4, pp. 356–363, 2019.

[6] S. Donati Guerrieri, F. Bonani, F. Bertazzi, and G. Ghione, "A unified approach to the sensitivity and variability physics-based modeling of semiconductor devices operated in dynamic conditions—Part I: Large-signal sensitivity," *IEEE Transactions on Electron Devices*, vol. 63, no. 3, pp. 1195–1201, mar 2016.

[7] ——, "A unified approach to the sensitivity and variability physics-based modeling of semiconductor devices operated in dynamic conditions.—Part II: Small-signal and conversion matrix sensitivity," *IEEE Transactions on Electron Devices*, vol. 63, no. 3, pp. 1202–1208, mar 2016.

[8] S. Donati Guerrieri, M. Pirola, and F. Bonani, "Concurrent efficient evaluation of small-change parameters and green's functions for TCAD device noise and variability analysis," *IEEE Transactions on Electron Devices*, vol. 64, no. 3, pp. 1269–1275, mar 2017.

[9] C. Prasad, "A review of self-heating effects in advanced CMOS technologies," *IEEE Transactions on Electron Devices*, vol. 66, no. 11, pp. 4546–4555, nov 2019.

[10] F. Bonani, V. Camarchia, F. Cappelluti, S. Donati Guerrieri, G. Ghione, and M. Pirola, "When self-consistency makes a difference," *IEEE Microwave Magazine*, vol. 9, no. 5, pp. 81–89, oct 2008.

[11] A. Benvenuti, W. M. Coughran, and M. R. Pinto, "A thermal-fully hydrodynamic model for semiconductor devices and applications to III-v HBT simulation," *IEEE Transactions on Electron Devices*, vol. 44, no. 9, pp. 1349–1359, 1997.

[12] E. Catoggio, S. Donati Guerrieri, and F. Bonani, "Efficient TCAD thermal analysis of semiconductor devices," *IEEE Transactions on Electron Devices*, 2021.

[13] S. Donati Guerrieri, F. Bonani, and G. Ghione, "A comprehensive technique for the assessment of microwave circuit design variability through physical simulations," in *2017 IEEE MTT-S International Microwave Symposium (IMS)*. IEEE, jun 2017.

[14] J.-P. Raskin, "FinFET versus UTBB SOI — a RF perspective," in *2015 45th European Solid State Device Research Conference (ESSDERC)*. IEEE, sep 2015.

[15] A. M. Bughio, S. Donati Guerrieri, F. Bonani, and G. Ghione, "Multi-gate FinFET mixer variability assessment through physics-based simulation," *IEEE Electron Device Letters*, vol. 38, no. 8, pp. 1004–1007, aug 2017.

Gap in pagination due to unavailable paper.

Pages 193-196

Proceedings of the 16th European Microwave Integrated Circuits Conference

Trap Characterization in InAlN/GaN and AlN/GaN based HEMTs with Fe- and C-doped Buffers

Emmanuel Dupouy, P. Vigneshwara Raja, Florent Gaillard, Raphaël Sommet, Jean-Christophe Nallatamby

XLIM Laboratory, CNRS UMR 7252, University of Limoges, France

raphael.sommet@xlim.fr, jean-christophe.nallatamby@unilim.fr

Abstract — Electrically active traps in InAlN/GaN and AlN/GaN based high-electron mobility transistors (HEMTs) with different buffer doping impurities are characterized by drain current transient (DCT) spectroscopy. A single deep-level trap at $E_C - 0.63$ eV ($\sigma = \sim 6 \times 10^{-15}$ cm^2) is identified in the InAlN/GaN HEMT with Fe-doped buffer layer. The DCT experiments on InAlN/GaN HEMT (C-doped buffer) reveal an electron trap at $E_C - 0.14$ eV with a capture cross-section of 4×10^{-18} cm^2. In AlN/GaN HEMT with C-doped buffer, a trap signature at $E_C - 0.33$ eV is observed with lower capture cross-section value of 2×10^{-20} cm^2. The DCT spectra of the C-doped HEMT devices have shown Arrhenius-like behaviour in the emission time constants, suggesting that charging/discharging dynamics of these traps is not governed by the defect bands, instead the carrier exchange may occur between the particular trap level and the conduction band through SRH recombination.

Keywords — AlN/GaN HEMT, C-doping, electron trap, Fe-doping, InAlN/GaN HEMT, current transient.

I. INTRODUCTION

The InAlN/GaN HEMT has offered several advantages over the existing AlGaN/GaN HEMT technology, i.e. lattice matched heterostructure, high spontaneous polarization, absence of inverse piezoelectric polarization, and thin InAlN barrier layer [1]. The AlN/GaN HEMT is another promising alternative to the AlGaN/GaN system due to their superior properties such as high spontaneous polarization and wide bandgap (6.2 eV) of AlN, ultra-thin AlN barrier (< 5 nm), and high 2DEG density ($\sim 6 \times 10^{13}$ cm^{-2}) [2]. Hence, the InAlN/GaN and AlN/GaN based HEMT devices could be used in the future RF and microwave power applications. To improve the carrier confinement in the 2DEG and to eliminate the buffer leakage and punch-through currents, the GaN buffer region is intentionally doped with compensational impurities viz. iron (Fe) and carbon (C) to achieve semi-insulating buffer layer [3]. However, the compensational dopants introduce electrically active traps and promote charge trapping, current collapse, drain-lag and gate-lag events, thus deteriorating the dynamic performance of the HEMT [3], [4].

In comparison with the Fe-doping in buffer, the C-doping technology is not yet matured enough to the industry standards. In fact, the C-doping concentration can be controlled in a step-like fashion, thereby ensuring precise control of the doping profile in the buffer [4]. The C-doping in GaN has an amphoteric nature, i.e. creates acceptor-like or donor-like states according to the Fermi level. The C-doping density around 10^{18} cm^{-3} provide net n-type buffer layer [4]. While, high C density ($\sim 10^{19}$ cm^{-3}) is necessary to achieve high breakdown voltage,

such high concentrations may lead to an effective p-type buffer region and also augment the current collapse [4], [5]. Moreover, the C-doping produces highly localized defect states (known as dislocations) in the buffer, which significantly deteriorate the HEMT characteristics. So the identification of C-doping induced traps in the buffer requires a special emphasis. Note that, deep-level traps in the AlGaN/GaN HEMTs are well documented in the literature. On the other hand, limited reports are available for the traps present in InAlN/GaN and AlN/GaN based HEMTs. In this work, electrically active traps in the InAlN/GaN and AlN/GaN HEMTs with Fe- and C-doped buffers are characterized by drain current transient (DCT) spectroscopy. The identified trap signatures may be useful to the GaN material community for controlling C-doping incorporation in the buffer.

II. EXPERIMENT

For initial trap characterization, InAlN/GaN HEMT with Fe-doped buffer (HEMT1-InAlN:Fe) was considered; the layered structure contains GaN cap layer, InAlN barrier layer, and Fe-doped GaN buffer layer grown on silicon carbide (SiC) substrate. Second device, InAlN/GaN HEMT structure (HEMT2-InAlN:C) includes GaN cap, InAlN barrier, and C-doped GaN buffer on SiC substrate. The AlN/GaN HEMT device (HEMT3-AlN:C) consists of AlN barrier, GaN channel layer, C-doped GaN buffer layer on SiC substrate. The HEMTs have the gate length of 0.1 μm and feature nitride passivation. Further device details cannot be given because of the intellectual property rights.

The DCT measurement test-bench developed at XLIM laboratory has been used and it utilizes time domain analysis to investigate the time evolution of charge carriers associated with the trapping and de-trapping process. The DCT experiments were carried out by using 2 pulse generators (Agilent HP 81110A and HP8114A) and two scope DPOs (Tektronix DP07054) to monitor drain current. Both the gate and drain terminals were pulsed (double pulse condition) in the DCT characterization. During the initial trap-filling phase, HEMT devices were pulsed to the OFF-state bias point ($V_{GF} = -6$ V, $V_{DF} = 10$ V) for 1 ms time duration, to populate the traps in the HEMT. Afterwards, the devices were switched to an ON-state bias point (V_{GM}, V_{DM}) and the drain current transient spectra were measured over eight decades of time (1 μs to 100 s) to examine the transient recovery due to the charge detrapping from a particular trap level. The double pulse scheme employed in the DCT measurement is illustrated in Fig. 1. Low drain

3–4 April 2022, London, UK

voltage ($V_{DM} \leq 7$ V) was considered in the ON-state bias points to eliminate self-heating effects in the detrapping transients. To identify the trap parameters, the DCT measurements were conducted at different temperatures (30°C to 130 °C) by varying the sample chuck temperature.

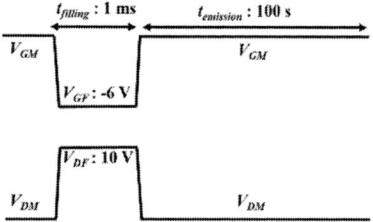

Fig. 1. Double pulsing scheme employed in the DCT characterization.

III. RESULTS AND DISCUSSION

A. HEMT1-InAlN:Fe

Figure 2(a) shows the drain current transient spectra obtained for the InAlN/GaN HEMT with Fe-doped buffer (HEMT1-InAlN:Fe) at the bias point ($V_{GS} = -1.6$ V, $V_{DM} = 7$ V) for different temperatures of 30 °C to 105 °C. A sharp rise in I_{DS} is noticed in the DCT spectra, due to the emission of trapped electrons from an active electron trap (labelled as F1) located below the conduction band edge [1], [6]. To estimate the emission time constant (τ_n), the DCT spectra were fitted by stretched multi-exponential function [1], [6] and the derivative spectra [$\partial I_{DS}/\partial \log_{10}(t)$] of the fitted DCT data for the HEMT1-InAlN-Fe1 are plotted in Fig. 2(b). The emission time constant of the detrapping transient is about 0.01 s at 30 °C. It is noticed from Fig. 2 that the DCT spectrum shifts towards lower time constants (τ_n) with the rise in temperature (T), indicating that the electron emission process from the trap is thermally activated [6], i.e. carrier emission rate follows the Arrhenius relationship. Accordingly, from the DCT characteristics at various temperatures, $ln(\tau_n T^2)$ versus ($1/kT$) plot is made for the trap F1 (see Fig. 3) and the trap signatures are computed by using the following Arrhenius expression [1], [6]

$$\ln(\tau_n T^2) = \frac{E_a}{kT} - \ln\left(\frac{\sigma_n N_C v_{th}}{g T^2}\right) \quad (2)$$

where E_a is the trap activation energy, σ_n is the capture cross-section, k is the Boltzmann constant, N_C is the effective density states in the conduction band, g is the degeneracy factor and v_{th} is the carrier thermal velocity. The activation energy (0.63 eV) and capture cross-section ($\sim6 \times 10^{-15}$ cm^2) of the trap F1 are determined from the slope and intercept values of the Arrhenius plot by using Eqn. 2.

Figure 3(a) depicts the DCT derivative spectra of the HEMT1-InAlN:Fe device acquired at two different bias points ($V_{GM} = -1.6$ V, $V_{DM} = 1$ V) and ($V_{GM} = -1.6$ V, $V_{DM} = 7$ V) for the various temperatures of 30 °C, 45 °C and 60 °C. In both cases of Fig. 3(a), the same gate voltage is maintained, but the drain voltage is changed from 1 V to 7 V. The emission time constant at 30 °C extracted from the first and second cases of

Fig. 3(a) (indicated by arrow) is ~0.05 s and ~0.01 s, respectively. This indicates that the carrier emission time constant decreases (emission rate increases) when the drain voltage is increased from 1 V to 7 V, due to the field-assisted carrier emission through the Poole-Frenkel effect [1]. Nevertheless, Arrhenius analyses for these two cases of DCT spectra yield the same trap parameters ($E_a = 0.63$ eV, $\sigma_n = \sim6 \times 10^{-16}$ cm^{-2}) for the trap F1, as observed from Fig. 3(b). It is worth recalling that positive peaks in the DCT derivative spectrum reveal the presence of electron trap in the device [1], [6]. So, a discrete deep-level trap at $E_C - 0.63$ eV (F1) is identified in the HEMT1-InAlN:Fe device. This outcome supports the reported works in the literature that the Fe-doping induces acceptor-type traps in the GaN buffer region with the activation energy of ~0.6 eV [3], [4]. Hence, the electron trap F1 at $E_C - 0.63$ eV is attributed to the Fe-doping related point defects in the GaN buffer layer.

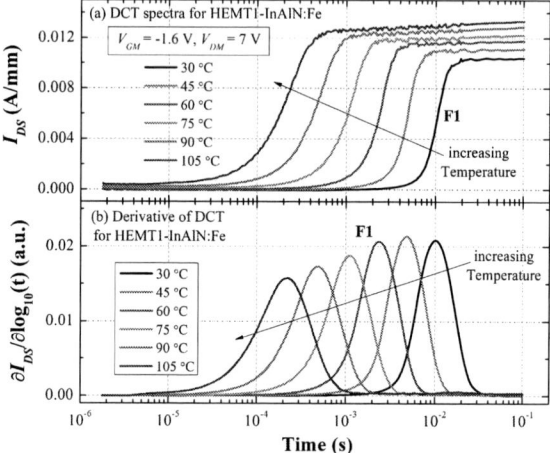

Fig. 2. (a) Drain current transient (DCT) spectra for InAlN/GaN HEMT with Fe-doped buffer at different temperatures (30 °C to 105 °C). (b) Derivative spectra of the fitted DCT data for the HEMT1-InAlN:Fe device.

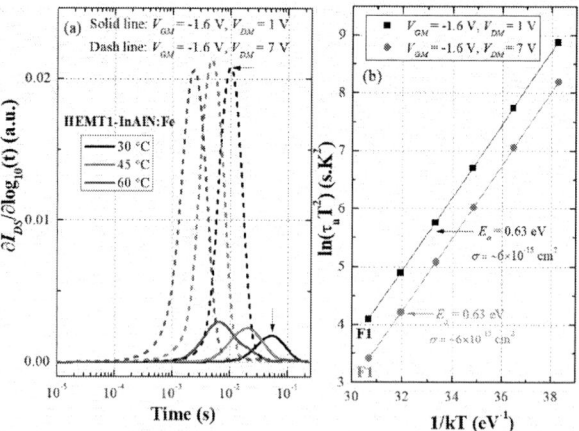

Fig. 3. (a) Derivative spectra for HEMT1-InAlN:Fe acquired at two different bias points ($V_{GM} = -1.6$ V, $V_{DM} = 1$ V) and ($V_{GM} = -1.6$ V, $V_{DM} = 7$ V) for various temperatures. (b) Arrhenius plots obtained from these two bias points.

B. HEMT2-InAlN:C

Figure 4 illustrates the DCT derivative spectra for InAlN/GaN HEMT with C-doped buffer (HEMT2-InAlN:C) attained at the bias point (V_{GM} = -2.2 V, V_{DM} = 1 V) for increasing temperatures (from 30 °C to 110 °C). The positive peaks C1 in the DCT derivative spectra reveal the I_{DS} transient recovery due to the electron emission from an active electron trap [1], [6]. It is noted from Fig. 4 that the emission time constant for C1 decreases with the increasing temperature. The time constant (τ_n) for the trap C1 is extracted at different temperatures and the $ln(\tau_n T^2)$ versus ($1/kT$) for C1 is plotted in the inset of Fig. 4. The activation energy (~0.14 eV) and capture cross-section (4×10^{-18} cm^2) for the trap C1 is determined from the Arrhenius calculation as per the Eqn. 2. From the DCT spectra, the emission time constant of the trap C1 at $E_C - 0.14$ eV is extracted as 11.6 s at 30 °C, which is higher than that for the Fe-induced trap F1 at $E_C - 0.63$ eV (τ_n = 0.01 s). In comparison with the deep-level trap F1 (refer Fig. 2), slow detrapping transient responses are perceived in the Fig. 4 for the shallow-level trap C1, due to the adopted bias point and their low capture cross-section of 4×10^{-18} cm^2. Nonetheless, it can be stated that the emission time constant depends not only on trap activation energy, but also on the capture cross-section [1].

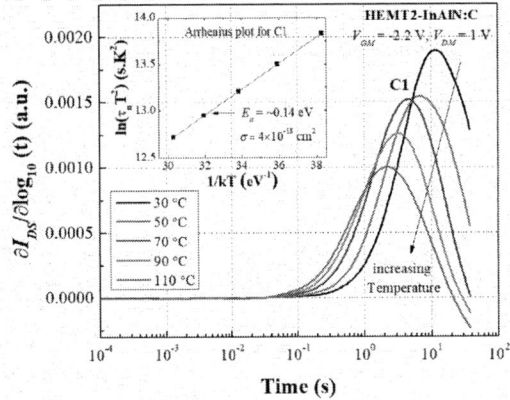

Fig. 4. (a) DCT derivative for InAlN/GaN HEMT with C-doped buffer acquired for different temperatures (30 °C to 110 °C). The inset shows Arrhenius plot for trap C1 identified in the HEMT2-InAlN:C device.

As expected, the variation of $ln(\tau_n T^2)$ with respect to ($1/kT$) is found to be linear (straight line) in the temperature range of 30 °C to 110 °C, as visualized from the inset of Fig. 4. Therefore, the emission rate of carrier from the trap C1 follows the Arrhenius law ($exp(-E_a/kT)$) and Shockley-Read-Hall (SRH) recombination statistics. Koller *et. al.* [5] observed a non-Arrhenius charging/discharging behavior with $exp(aT)$ temperature dependence in the capacitance transient spectroscopy for the highly C-doped GaN layers (carbon concentration $\geq 10^{18}$ cm^{-3}). They concluded that the charging and discharging process is governed by the charge transport between the defect band (DB) and the C-produced defect level [5]. The defect bands are associated with the C-induced

dislocations in the buffer. If the trapping/detrapping kinetics are governed by other non-thermal emission mechanisms such as hopping or trap assisted tunnelling, then the weak temperature dependent DCT properties are expected along with a computation of lower thermal activation energy [4]. In this work, typical Arrhenius-like behavior is observed in the emission time constant of the trap C1 at $E_C - 0.14$ eV, suggesting that the charging/discharging dynamics is carried out through the conventional carrier exchange (as per SRH recombination) between the particular trap energy at E_C - 0.14 eV and the conduction band. Although the energy position of the trap C1 at $E_C - 0.14$ is located close to the well-known donor-level at $E_C - 0.11$ eV considered in the autocompensation C_N-C_{Ga} defect model [7]. The trap parameters of C1 ($E_C - 0.14$ eV, 4×10^{-18} cm^2) is better matched with the donor-level state at $E_C - 0.15$ eV (2.3×10^{-18} cm^2) observed in the unintentionally carbon doped GaN films [8].

C. HEMT3-AlN:C

The DCT derivative spectra for the AlN/GaN HEMT with C-doped buffer (HEMT3-AlN:C) acquired at the bias point (V_{GM} = -0.32 V, V_{DM} = 1 V) for different temperatures (50 °C to 130 °C) are plotted in Fig. 5. The positive peaks C2 in the transient spectra indicate the electron trapping phenomena in the HEMT3-AlN:C device. As the temperature increases, the transient spectrum moves to lower τ_n. Arrhenius analysis for C2 yields the activation energy of 0.33 eV and the capture cross-section of 2×10^{-20} cm^2. The energy level of the trap C2 ($E_C - 0.33$ eV) is higher to the trap C1 ($E_C - 0.14$ eV) identified in the HEMT2-InAlN:C. The trap C2 can be associated to an extended defect with localized states [9]. Verzellesi *et al.* [7] also detected similar trap energy 0.37 eV (1.1×10^{-21} cm^2) in the AlGaN/GaN HEMT with C-doped buffer.

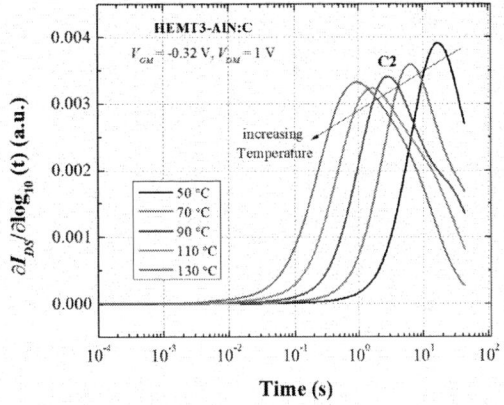

Fig. 5. DCT derivative spectra for the HEMT3-AlN:C device acquired at the bias point (V_{GM} = -0.32 V, V_{DM} = 1 V) for different temperatures (50 °C to 130 °C) reveal the electron trap C2 at E_C - 0.33 eV.

Figure 6 shows the DCT derivative spectra for the HEMT3-AlN:C device obtained at the bias point (V_{GM} = -0.52 V, V_{DM} = 6 V) for different temperatures (50 °C to 130 °C). The positive peaks C2* are noticed in the Fig. 6 and the corresponding $ln(\tau_n T^2)$ versus ($1/kT$) plot is displayed in Fig. 7.

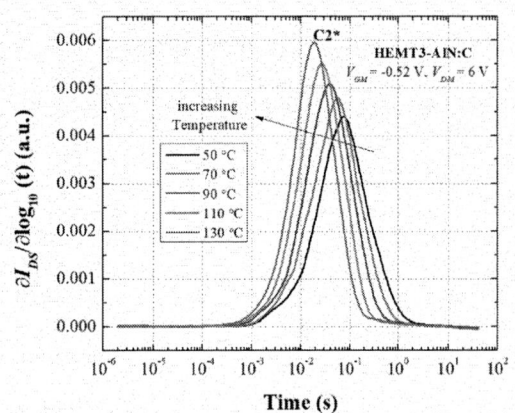

Fig. 6. DCT derivative spectra for AlN/GaN HEMT with C-doped buffer (HEMT3-AlN:C) acquired at the bias point (V_{GM} = -0.52 V, V_{DM} = 6 V) for different temperatures (50 °C to 130 °C) reveal the trap C2* (0.16 eV).

Another activation energy of 0.16 eV with a capture cross-section of 5×10^{-20} cm^2 is determined for the same previous electron trap C2 from the Arrhenius relation. We can conclude that electric field influences trap releasing through Poole-Frenkel effect, decreasing activation energy as long as drain voltage increases. This fact shows the importance to make the DCT measurements at low drain voltage / electric field. It has two advantages: first, it decreases effect of electric field in trap releasing, second, it decreases self-heating effect in transistor that could influence trap releasing dynamics too. Furthermore, Fig. 7 shows the typical Arrhenius-like properties in the emission time constant of the trap C2. As a result, the charging and discharging kinetics are essentially governed by the conventional carrier transport between the trap C2 and the conduction band through the SRH recombination statistics, and are not associated with the C-induced defect bands [5].

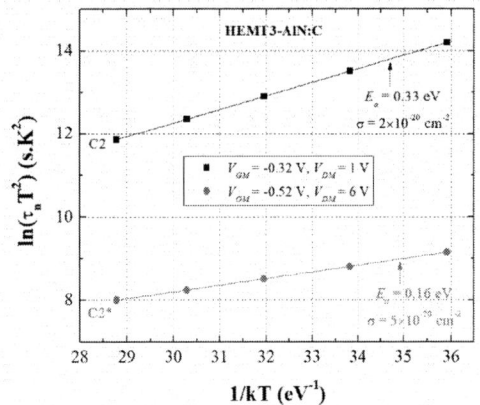

Fig. 7. Arrhenius plots for the electrically active traps C2 and C2* identified in the HEMT3-AlN:C device at different bias points.

IV. CONCLUSION

The electrically active traps in the InAlN/GaN and AlN/GaN based HEMTs with Fe- and C-doped buffers are identified by the DCT characterization. The deep-level trap F1 (E_C − 0.63 eV, 6×10^{-15} cm^2) is identified in the HEMT1-InAlN:Fe device. A single electron trap C1 (E_C − 0.14 eV, 4×10^{-18} cm^2) is observed in the HEMT2-InAlN:C device, while another trap C2 (E_C − 0.33 eV, 2×10^{-20} cm^2) is detected in the AlN/GaN HEMT with C-doped buffer. Typical Arrhenius-like behavior is observed in the emission time constant of the traps (C1 and C2) identified in the C-doped samples; this suggests that charging and discharging kinetics are essentially governed by the carrier exchange between the particular trap level and the conduction band, and are not associated with the defect bands.

ACKNOWLEDGMENT

We greatly appreciate and acknowledge the Agence Nationale de la Recherche (ANR) and Direction Générale de l'Armement (DGA) France for funding under contract ANR-17-ASTR- 0007-01 (COMPACT project).

REFERENCES

[1] P. V. Raja, M. Bouslama, S. Sarkar, K. R. Pandurang, J.-C. Nallatamby, N. DasGupta, and A. DasGupta, "Deep-level traps in AlGaN/GaN and AlInN/GaN based HEMTs with different buffer doping technologies," *IEEE Trans. Electron Devices*, vol. 67, no. 6, pp. 2304-2310, Jun. 2020,.

[2] N. K. Subramani, A. K. Sahoo, J.-C. Nallatamby, R. Sommet, N. Rolland, F. Medjdoub, and R. Quéré, "Characterization of parasitic resistances of AlN/GaN/AlGaN HEMTs through TCAD-based device simulations and on-wafer measurements," *IEEE Trans. Microw. Theory Techn.*, vol. 64, no. 5, pp. 1351-1358, May 2016.

[3] M. Meneghini, I. Rossetto, D. Bisi, A. Stocco, A. Chini, A. Pantellini, C. Lanzieri, A. Nanni, G. Meneghesso, and E. Zanoni, "Buffer traps in Fe-doped AlGaN/GaN HEMTs: Investigation of the physical properties based on pulsed and transient measurements," *IEEE Trans. Electron Devices*, vol. 61, no. 12, pp. 4070-4077, Dec. 2014.

[4] J. Bergsten, M. Thorsell, D. Adolph, J.-T. Chen, O. Kordina, E. Ö. Sveinbjörnsson, and N. Rorsman, "Electron trapping in extended defects in microwave AlGaN/GaN HEMTs with carbon-doped buffers," *IEEE Trans. Electron Devices*, vol. 65, no. 6, pp. 2446-2453, Jun. 2018.

[5] C. Koller, G. Pobegen, C. Ostermaier, and D. Pogany, "Effect of carbon doping on charging/discharging dynamics and leakage behavior of carbon-doped GaN," *IEEE Trans. Electron Devices, vol. 65, no. 12*, pp. 5314-5321, Dec. 2018.

[6] M. Meneghini, G. Meneghesso, and E. Zanoni, "Trapping and degradation mechanisms in GaN-based HEMTs," in *Gallium Nitride (GaN) Physics, Devices, and Technology*, Boca Raton, USA: CRC Press, 2016, pp. 327-361.

[7] G. Verzellesi, L. Morassi, G. Meneghesso, M. Meneghini, E. Zanoni, G. Pozzovivo, S. Lavanga, T. Detzel, O. Häberlen, and G. Curatola, "Influence of buffer carbon doping on pulse and AC behavior of insulated-gate field-plated power AlGaN/GaN HEMTs," *IEEE Electron Device Lett.*, vol. 35, no. 4, pp. 443-445, Apr. 2014.

[8] A. Armstrong, A. R. Arehart, D. Green, U. K. Mishra, J. S. Speck, and S. A. Ringel, "Impact of deep levels on the electrical conductivity and luminescence of gallium nitride codoped with carbon and silicon," *J. Appl. Phys.*, vol. 98, no. 5, Art. no. 053704, Sep. 2005.

[9] H. Wang, P.-C. Hsu, M. Zhao, E. Simoen, A. Sibaja-Hernandez, et J. Wang, "Investigation of Defect Characteristics and Carrier Transport Mechanisms in GaN Layers With Different Carbon Doping Concentration," *IEEE Trans. Electron Devices*, vol. 67, no. 11, p. 4827-4833, Nov. 2020.

Mechanisms of Buffer and Surface Traps in GaN HEMTs for Low Frequency Y21 and Y22 parameters

Tomohiro Otsuka[#], Yutaro Yamaguchi [#], Masaomi Tsuru [#], Toshiyuki Oishi[*]

[#] Mitsubishi Electric Corporation, Japan

[*]Saga University, Japan

Otsuka.Tomohiro@bp.MitsubishiElectric.co.jp

Abstract — **Characteristics in low frequency Y-parameters for GaN HEMTs is important to investigate device physics of traps and construct equivalent circuit model to improve linearity in power amplifier such as Doherty. We have clarify effects of buffer and surface traps to Im(Y21) and Im(Y22) by device simulation and small signal equivalent circuit model. Results of the device simulation including the buffer trap, surface trap and self-heating model can explain experimental data adequately. Peak at lowest frequency appeared in only Im(Y21) and has positive value. This peak originates from the surface trap. The peaks appeared in both Im(Y21) and Im(Y22) with different polarity originate from the buffer traps. Furthermore, the self-heating effects generate the peak in both Im(Y21) and Im(Y22) with the same polarity. The signals of the traps in Y-parameters induce from drain current changed by ionized traps. In the case of the buffer in Im(Y21), the ionized traps decreased at the drain side in GaN layer as gate voltage change to larger value. These negative charges decrease due to acceptor type trap and the drain current induced by the buffer trap increases. Therefore, the signal for the buffer trap appear as the peak with negative polarity at the frequency determined from trap properties.**

Keywords — **GaN, HEMT, buffer trap, surface trap, low frequency Y-parameters, TCAD device simulation, small signal equivalent circuit.**

I. INTRODUCTION

Due to the excellent material properties, such as high breakdown electric field, a high electron saturation velocity and high thermal conductivity, gallium nitride based high electron mobility transistors (GaN HEMTs) are used for high-power and high-frequency amplifiers in radio-frequency (RF) systems, like a radar and communication system. There are still some issues for GaN HEMTs to improve the amplifier performances. One of the issues is analyzation and reduction of traps existing various location in GaN HEMTs [1]. These traps cause current collapse and degrade the efficiency in microwave characteristics. Low frequency Y-parameter measurements have been investigated on the traps in GaN HEMTs [2, 3]. In previous works, the imaginary part of output admittance (Im(Y_{22})) were mainly studied to detect the buffer traps. In addition, self-heating effect was reported to generate a peak in Im(Y_{22}) around MHz [3]. Frequency dependent circuit parameters such as Y-parameters are useful to design circuits. The characteristics in low frequency region become important to improve advanced power amplifiers such as Doherty [4, 5]. Because Y21 is a parameter from input to output signal and includes variation of drain current (I_d) in the numerator, it is considered to be important among Y-parameters. However, it is not clear on the relationship between Y21 and trapping effect.

II. EXPERIMENTAL METHOD AND SIMULATION MODEL

Figure 1(a) shows a schematic cross-sectional structure of GaN HEMTs studied in this work. AlGaN barrier layer, GaN layer were fabricated on SiC substrates. GaN layer plays a role as channel and buffer. Iron (Fe) atoms were doped in GaN layer to suppress drain leakage current. SiN passivation layer was formed to cover surface between each electrode. Gate Schottky electrode was with T shaped. Gate width was 0.1 mm.

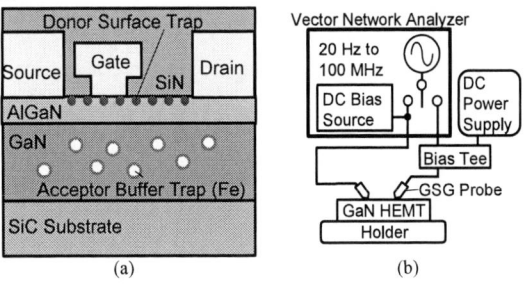

Fig. 1. (a) Schematic cross-sectional structure of GaN HEMTs, (b) Experimental setup for measurement of Y parameters.

Device simulation was carried out using a technology computer-aided design (TCAD) (Silvaco ATLAS). In this work, buffer trap, surface trap model and self-heating effect were considered. The buffer trap with acceptor type corresponds to Fe doping in GaN layer. The energy level below the conduction band, concentration and capture cross-section were 0.6 eV, 1.0E18 cm^{-3} and 8.0E−14 cm^2, respectively. On the other hand, the surface trap with donor type is set at the surface with thickness of 1 nm between the source and the drain electrodes. Their energy level from the conduction band, concentration and capture cross-section were 0.3 eV, 5.0E 12cm^{-2} and 5.0E−23 cm^2, respectively. In addition, a self-heating effect is important for low frequency measurement, because a thermal time constant for GaN HEMT is μs order (MHz order in frequency domain).

Experimental data of Y parameters were obtained by an on-wafer probing system and the vector network analyser (Keysight E5601B) as shown in Fig.2(b). S-parameters were measured for the frequency from 20 Hz to 100 MHz at R.T. and converted to Y parameters. GaN HEMTs were biased at the on-state condition though bias tees (gate bias (V_g) of 0 V, drain bias (V_d) of 20 V). We have selected the imaginary parts of Y-parameters for analyzation of the trapping effects, because the characteristics have Gaussian-like curve with a peak. In addition, there is no signal in Im(Y_{11}) and Im(Y_{12}) due to little gate current.

Figure 2 shows a small signal equivalent circuit model used in this work. Equivalent circuit model is essential to design a power amplifier. A buffer trap circuit with R-C components was implemented between the source and the drain, and corresponds to the GaN region between the source and the drain in Fig. 1(a). The buffer trap related current source was inserted between the drain and the source, and controlled by voltage (V_{dtrap}) applied to the buffer trap capacitance (C_{dtrap}). The surface traps were implemented in the similar way to the buffer trap circuits and represented by green squares in Fig. 2. R-C circuit for the surface trap was implemented between the gate and the source, and corresponds to the surface in Fig. 1. The suffixes of "d" and "g" in the trap parameters represent the buffer traps and the surface traps, respectively. A current source for the surface trap was expressed by $K_{gtrap}g_mV_{gtrap}$ (the degradation of transconductance by traps). In this work, this current source was controlled only by voltage applied to C_{gtrap} located between the gate and the source. By using the equivalent circuit model, frequency and magnitude for the peak of Im(Y_{21}) are as follows.

$$f_{peak@Im(Y_{21})} = \frac{1}{R_{gtrap}C_{gtrap}} \tag{1}$$

$$Im(Y_{21})@peak = \frac{K_{gtrap}g_m}{2} \tag{2}$$

In similar way, frequency and magnitude for the peak of Im(Y_{22}) were obtained by replacing "gtrap" to "dtrap" in Eq. (1) and (2).

$$f_{peak@Im(Y_{22})} = \frac{1}{R_{dtrap}C_{dtrap}} \tag{3}$$

$$Im(Y_{22})@peak = \frac{K_{dtrap}g_m}{2} \tag{4}$$

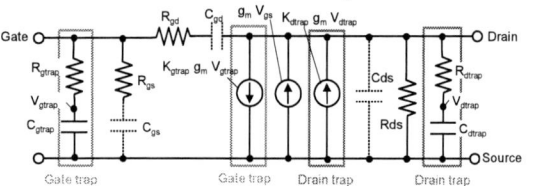

Fig. 2. Small signal equivalent circuit with trap for GaN HEMTs.

III. EXPERIMENTAL RESULT

Figure 3 shows experimental results of Im(Y_{21}) and Im(Y_{22}) for GaN HEMTs biased at the on-state condition ($V_g = 0$ V, $V_d = 20$ V) as function of frequency. Each signal has Gaussian distribution like shape and peak at a frequency. We categorized these peaks into groups with similar peak

frequency, because trap time constant was mainly determined by the trap energy level and the capture cross-section. The signal named as "f1" has the peak with positive polarity at the frequency around 100 Hz and was appeared only in Im(Y_{21}). The signal named as "f2" were appeared at frequency around 10 KHz order in both Im(Y_{21}) and Im(Y_{22}). Polarities for each signal are different, (the negative peak for Im(Y_{21}) and the positive peak for Im(Y_{22})). The signals named as "f3" have the peaks at 4 MHz and the same positive polarity in Im(Y_{21}) and Im(Y_{22}). We have investigated on the origin for these three type of signals by device simulation and equivalent circuit model.

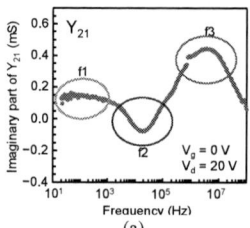 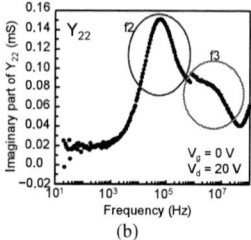

(a)   (b)

Fig.3. Experimental results of imaginary parts of (a) Y21 and (b) Y22

IV. INFLUENCE OF TRAPS IN EQUIVALENT CIRCUIT

Firstly, trapping effects were considered in terms of the equivalent circuit of Fig. 2. I_d is represented using three current sources as follow.

$$I_d = g_mV_g - K_{gtrap}g_mV_{gtrap} - K_{dtrap}g_mV_{dtrap} \equiv g_mV_g - I_{d\_trap} \tag{5}$$

where I_{d_trap} is drain current induced by the surface and buffer traps. The contribution of the buffer traps was introduced by the derivative of I_d with respect to V_g or V_d. In the same way to the buffer traps, the contribution of the surface trap was obtained. As results, the following equations were obtained.

$$\frac{\Delta I_{d\_trap}}{\Delta V_g} = -K_{gtrap}g_m \tag{6}$$

$$\frac{\Delta I_{d\_trap}}{\Delta V_d} = -K_{dtrap}g_m \tag{7}$$

By inserting Eq. (5) into Eq. (2), Im(Y_{21})@peak is as follow.

$$Im(Y_{21})@peak = -\frac{1}{2}\frac{\Delta I_{d\_trap}}{\Delta V_g} \tag{8}$$

Therefore, the amplitude indicates the drain current change due to the traps when V_g changes slightly. In the same way, Im(Y_{22})@peak is shown as follow:

$$Im(Y_{22})@peak = -\frac{1}{2}\frac{\Delta I_{d\_trap}}{\Delta V_d} \tag{9}$$

Because the traps change their charges as according to potential, the trap induced drain current changes by ionized trap density in GaN HEMTs. From Eqs. (8) and (9), the information of the magnitude to Im(Y_{21}) and Im(Y_{22}) can be known by variation of the ionized density for each traps as the slightly change of V_g or V_d.

V. DEVICE SIMULATION

As first step of device simulation study, we calculated about GaN HEMT with one type of traps (buffer or surface) without the self-heating effect in order to make clear the role of the individual trap. The biases are the same as the experimental ($V_g = 0$ V, $V_d = 20$ V). The trap parameters are changed intentionally from values used in Sec. II in order to make each trap effect clear. Figure 4 shows Im(Y_{21}) and Im(Y_{22}) depending on frequency for GaN HEMTs with only buffer trap. Gaussian-distribution like signals were appeared in both Im(Y_{21}) and Im(Y_{22}). Two curves have a peak at the same frequency and opposite polarity. The peak in Im(Y_{21}) has negative value (negative peak), while the peak in Im(Y_{22}) has positive value (positive peak). From Fig. 4, the effects of the buffer traps is observed in not only Im(Y_{22}) but also Im(Y_{21}) and the magnitude of Im(Y_{21}) was larger than that of Im(Y_{22}).

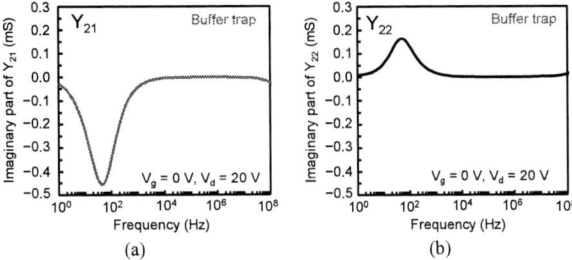

Fig. 4. Device simulation results for only buffer trap : (a) Im(Y_{21}) and (b) Im(Y_{22}) depending on frequency

Figure 5 shows Im(Y_{21}) and Im(Y_{22}) depending on frequency in the case of GaN HEMT with only surface trap. No peaks were appeared in Im(Y_{22}) unlike the case of only the buffer traps as shown in Fig. 4(b). On the other hand, a peak was appeared with positive value in Im(Y_{21}), which is opposite polarity as compared with the case of only buffer trap (Fig. 4(a)).

From Figs. 4 and 5 in the TCAD simulation results, there is the different characteristics between buffer trap and surface trap. Therefore, we can distinguish clearly two types of traps using Im(Y_{21}) and Im(Y_{22}) depending on frequency. If there are a negative peak in Im(Y_{21}) and positive peak in Im(Y_{22}) at the same frequency, this peak is originated from the buffer trap. On the other hand, if there is a positive peak in only Im(Y_{21}), this peak is originated from the surface trap.

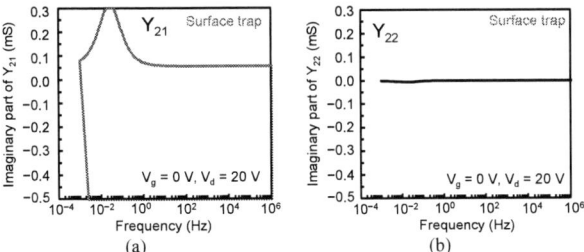

Fig. 5. Device simulation results for only surface trap : (a) Im(Y_{21}) and (b) Im(Y_{22}) depending on frequency.

Figure 6 shows two-dimensional (2D) plot of ionized buffer trap density in the case of GaN HEMTs with only the buffer trap. Ionized buffer traps with negative charge were distributed under the gate region with high density. The negative charges elevate the conduction band energy, which results to the conduction current decrease.

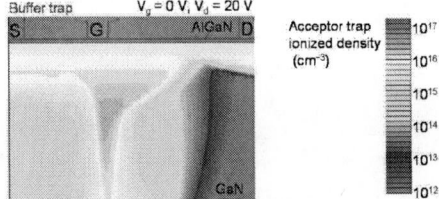

Fig. 6. Two-dimensional plot for ionized buffer trap density

In order to investigate an impact of the buffer trap to Im(Y_{21}) and Im(Y_{22}), we have calculated the ionized trap density profiles at different voltage (V_g or V_d) by 1 V from the bias ($V_g = 0$ V, $V_d = 20$ V). Figure 7 shows the distribution of the ionized buffer trap density ratio along the lateral direction in GaN layer at depth of 250 nm from AlGaN/GaN interface. Fig. 7(a) and (b) are the case of V_g change and V_d change, respectively. All buffer traps were ionized in GaN layer under gate electrode. The ratio of the ionized buffer trap decreases from center position toward the source or drain side. The ratio of the ionized buffer trap varied at drain side region according to the change of the voltage. In the case of V_g change (Fig. 7(a)), the region decreasing ionized buffer trap density moved to center position as V_g increases. This indicated that the number of the ionized buffer trap (negative charge) decreased and the conduction band energy decreases as V_g increase. Therefore, the drain current induced by the buffer trap increased as V_g increase. From Eq. (8), the magnitude of Im(Y_{21}) at the peak frequency have negative value. This consists with the result in Fig. 4(a) with the observation of the negative peak. On the other hand, in the case of V_d change, area existing the ionized buffer trap expanded toward the drain side as V_d increase as shown in Fig. 7(b). From Eq. (9) and Fig. 7(b), Im(Y_{22}) was found to have the positive peak. Actually, the positive peak was observed for the device simulation shown in Fig. 4(b). In addition, variation of the ionized buffer trap density for the case of V_g change was larger than that for the case of V_d change. This difference was observed as the magnitude of the peak for Im(Y_{21}) and Im(Y_{22}) in Fig. 4.

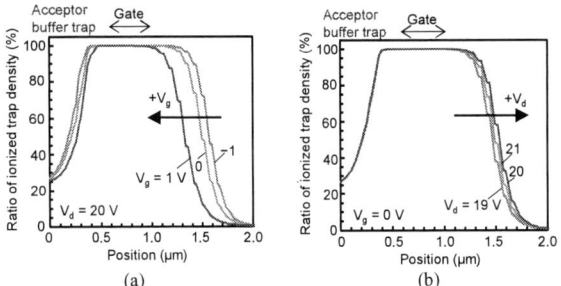

Fig. 7. Ionized buffer trap lateral profile in case of (a) Vg and (b) Vd changes with variation of 1 V.

203

Next, we calculated the surface donor trap case. Figure 8 shows 2D plot of ionized surface trap density at the surface in AlGaN layer. Many traps were ionized at the surface under gate. Because the ionized surface traps have positive charge due to the donor-type, the conduction band energy decreases and increases the drain current.

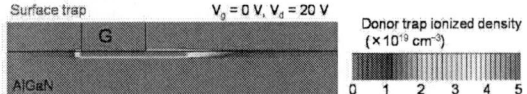

Fig. 8. Two-dimensional plot for ionized surface trap density.

Figure 9shows the distribution of the ionized surface trap density ratio along the lateral direction from the source to the drain at the surface. Figures 9(a) and (b) show the case for change of V_g and V_d, respectively. The variation of the voltage change is from −1 to 1 V. Width of the ionized surface trap under gate decrease as V_g increase as shown in Fig. 9(a). Because total number of the positive charges (ionized surface traps) decrease for V_g increase, the trap-induced drain current decreased. Therefore, the magnitude of Im(Y_{21}) for the surface trap has positive value from Eq. (8). On the other hand, V_d change didn't give impact to the profile of the ionized surface trap density as shown in Fig. 9(b) because the potential from the gate is dominated near the surface. Therefore, there is no surface trap signal in Im(Y_{22}).

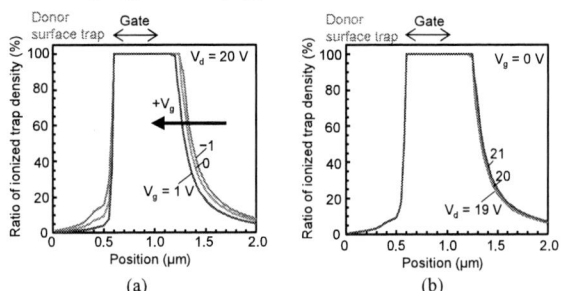

Fig. 9. Ionized surface trap lateral profile in case of (a) V_g and (b) V_d changes with variation of 1 V.

VI. TRAPPING MECHANISM

Figure 10 shows the Im(Y_{21}) and Im(Y_{22}) depending on frequency by including the buffer trap, the surface trap and the self-heating model. The simulation results agree with the experimental data in Fig. 3. The peaks "f1", "f2" and "f3" in Fig. 3. were found to correspond to the surface trap, the buffer trap and the self-heating, respectively. There are three peaks in Im(Y_{21}). The peak at lowest frequency is originated by the surface trap because of no observation in Im(Y_{22}). The peaks at middle frequency is originated by the buffer traps because of observation in both Im(Y_{21}) and Im(Y_{22}). There are two kinds of peaks for the buffer trap and the self-heating in Im(Y_{22}). From previous study, the peaks at highest frequency are originated from the self-heating [3]. The heat capacitance and thermal conductivity are different from each material such as SiC and GaN. This variation of the thermal parameters spreads the peak width for the self-heating effect.

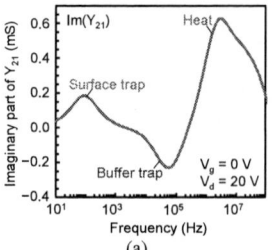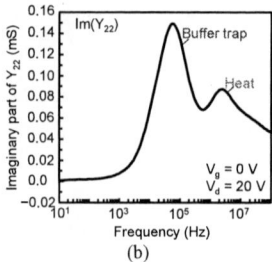

Fig. 10. Device simulation results with buffer trap, surface trap and self-heating model for (a) Im(Y_{21}) and (b) Im(Y_{22}) as function of frequency

VII. CONCLUSION

We have clarified the origin of the peaks in Im(Y_{21}) and Im(Y_{22}) depending on frequency for GaN HEMTs. There are three type of peaks in low frequency region between 20 Hz to 100 MHz for on-state condition. These peaks were categorized into the buffer trap, surface trap and self-heating using device simulation and small signal equivalent circuit model. The bias change in measurement of Y-parameters change the ionized traps density. The change of the drain current induced by ionized traps density appear as peak in Im(Y_{21}) and Im(Y_{22}). In the case of the buffer trap, V_g increase with variation of 1 V decrease the negative charged traps at the drain side in GaN layer between the source and the drain. This change increased the drain current and generated the negative peak in Im(Y_{21}). The buffer traps generated the positive peak in Im(Y_{22}) in similar way. Regarding the surface trap, the signal was appeared in Im(Y_{21}). There is no signal in Im(Y_{22}), because the ionized trap density profile didn't move by V_d change dur to the strong influence of the potential from the gate. We can distinguish the kind of origin for the peak by these feature and peak frequency determined by trap energy level and capture cross-section. The information obtained in this work are useful to analyze trap properties and location, and construct an equivalent circuit model.

REFERENCES

[1] D. Bisi, M. Meneghini et al., "Deep-Level Characterization in GaN HEMTs-Part I: Advantages and Limitations of Drain Current Transient Measurements," IEEE Trans. on Electron Devices, vol. 60, no. 10, pp. 3166-3175, Sep. 2013.

[2] N. K. Subramai et al., "Identification of GaN Buffer Traps in Microwave Power AlGaN/GaN HEMTs Through Low Frequency S-Parameters Measurements and TCAD-Based Physical Device Simulations" IEEE Journal of the Electron Devices Society, Volume: 5, Issue: 3, pp. 175-181, May 2017.

[3] T. Otsuka et al., "Study of Self-heating Effect of GaN HEMTs with Buffer Traps by Low Frequency S-parameters Measurements and TCAD Simulation," 2019 IEEE BiCMOS and Compound semiconductor Integrated Circuits and Technology Symposium (BCICTS), 2019, pp. 1-4.

[4] F. M. Barradas et al., "Compensation of Long-Term Memory Effects on GaN HEMT-Based Power Amplifiers," IEEE Trans. Microw. Theory Tech.,, vol. 65, no. 9, pp. 3379-3388, Mar. 2017.

[5] Y. Yamaguchi et al., "Modeling of frequency dispersion at low frequency for GaN HEMT," 2014 Asia-Pacific Microwave Conference, 2014, pp. 798-800.

A Low Phase Noise Phase-Locked Loop With Short Settling Times for Automotive Radar

Tobias T. Braun[#1], Marcel van Delden[#2], Christian Bredendiek[*3], Jan Schoepfel[#4], Nils Pohl[#*5]

[#]Ruhr-University Bochum, Germany
[*]Fraunhofer Institute for High Frequency Physics and Radar Techniques (FHR), Germany
{[1]Tobias.T.Braun,[4]Jan.Schoepfel,[5]Nils.Pohl}@rub.de, [2]Marcel.VanDelden@est.rub.de, [3]Christian.Bredendiek@fhr.fraunhofer.de

Abstract — **This paper presents a millimeter-wave (mm-wave) synthesizer for use in the automotive band. The main focus is achieving low phase noise in conjunction with short settling times. By using a reference frequency of 1 GHz, the synthesizer achieves a phase noise of lower than -105 dBc/Hz at an offset frequency of 100 kHz over the entire automotive band. The corresponding jitters, integrated from offset frequencies of 10 kHz to 10 MHz, are lower than 61 fs. Simultaneously, with a loop bandwidth of 5 MHz, a settling time of 880 ns is determined for a frequency step with an error band of 10 ppm. To improve linearity, the use of a static prescaler inside the loop was omitted. To the best of the author's knowledge, this has not yet been achieved regarding a synthesizer using a fundamental VCO inside the automotive band.**

Keywords — **Fractional N-synthesizer, frequency modulated continuous wave (FMCW), phase-frequency detector (PFD), programmable frequency divider, SiGe bipolar integrated circuit (IC).**

I. INTRODUCTION

Reliable sensors are mandatory to push towards a higher level of automation in ground transportation and increase traffic safety. Frequency-modulated continuous-wave (FMCW) radar sensors qualify as such because of their ability to measure the distance and velocity of obstacles and other vehicles reliably, throughout a broad range of weather conditions.

Fractional-N frequency synthesizers can generate the required FMCW signals with high linearity while maintaining a small division ratio. On the one hand, several key parameters of such a synthesizer, like the center frequency and output bandwidth, are defined by the regulations for automotive applications to 76.5/79 GHz and 1/4 GHz, respectively [1]. On the other hand, the design of the synthesizer itself dictates critical aspects such as phase noise, chirp linearity, and settling times, thus determining the performance of the radar system.

Especially short settling times, which allow for a higher modulation rate of the output signal, are essential in automotive applications. The signal processing of the fast chirp sequence determines the distance of the target via a fast Fourier transform (FFT). A second FFT of this intermediate frequency (IF) allows for the measurement of the relative velocity by extracting the phase difference between chirps. The phase difference must be smaller than π to be unambiguous. Therefore, the maximum velocity measurable unambiguously is limited to

$$v_{\max} = \frac{\lambda}{4T_C}, \tag{1}$$

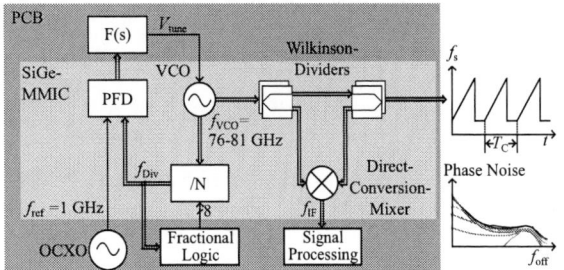

Fig. 1. Block diagram of the proposed fractional-N frequency synthesizer utilizing a reference frequency (f_{ref}) of 1 GHz while neglecting the use of a static prescaler to improve the state of the art regarding linearity, phase noise and settling time.

with the wavelength of the electromagnetic wave λ and the chirp repitition rate T_C [2]. E.g. concerning a velocity of 130 km/h of the vehicle a chirp repetition rate of $T_C \approx 13\,\mu s$ is necessary when taking two opposing vehicles into account. Thus, faster chirps, combined with minimum idle time inbetween them, are desirable for automotive applications to measure the occurring relative velocities.

II. PLL FUNDAMENTALS

The proposed transceiver shown in Fig. 1 is based on a phase-locked loop (PLL). It consists of a voltage-controlled oscillator (VCO), a fractional-N divider, a phase-frequency detector (PFD), and a loop filter (F(s)). Additionally, a static prescaler at the VCO's output is often used to obtain a manageable input frequency for the programmable frequency divider.

To generate faster chirps the PLL's settling time t_{settle} has to be reduced. It can be approximated as

$$t_{\mathrm{settle}} = \frac{-2\ln\left(\frac{f_{\mathrm{err}}\sqrt{1-\zeta^2}}{f_{\mathrm{step}}}\right)}{2\pi \cdot B_{\mathrm{loop}}} \tag{2}$$

with the allowed frequency error f_{err}, the size of the frequency step f_{step}, the PLL's damping factor ζ and the loop bandwidth B_{loop} in accordance with [3]. While ζ can be optimized to reduce t_{settle}, it depends entirely on the PLL's phase margin, thus being limited by stability constraints. Thusly, to decrease t_{settle}, the loop filter bandwidth has to be increased. However, since B_{loop} also affects phase noise it is restricted by the phase noise generated by the $\Sigma\Delta$-modulator at the very least. The $\Sigma\Delta$-modulator's noise is given by

$$\mathcal{L}_{\mathrm{SDM}}(f_{\mathrm{off}}) = (2\pi)^2 \cdot \left(2\sin\left(\frac{\pi \cdot f_{\mathrm{off}}}{f_{\mathrm{ref}}}\right)\right)^{2\cdot(n-1)} \cdot \left(\frac{1\,\mathrm{Hz}}{12 f_{\mathrm{ref}}}\right), \quad (3)$$

where f_{off} is the offset frequency and n is the order of the $\Sigma\Delta$-modulator [3]. To achieve adequate suppression of $\mathcal{L}_{\mathrm{SDM}}$, B_{loop} is limited to approximately $\frac{f_{\mathrm{ref}}}{20}$. While this serves as an upper limit of the loop bandwidth, to minimize integrated phase noise, the intersection point of the open-loop phase noise of the VCO and the closed-loop phase noise of the complete PLL is often regarded as a starting point [3]. Because the intersection point is in the VCO's $1/f^2$ region, its phase noise can be calculated as

$$\mathcal{L}_{\mathrm{VCO}}(f_{\mathrm{off}}) = 10\log\left(\frac{F k_{\mathrm{B}} T}{2 P_{\mathrm{s}}} \cdot \left(\frac{f_{\mathrm{s}}}{2 Q f_{\mathrm{off}}}\right)^2\right), \quad (4)$$

where P_{s} is the VCO's output power, F is its noise figure and Q is the quality factor. The PLL's in-loop flat noise is typically determined by the phase noise of the reference, the frequency divider, and the PFD, summarized by $\mathcal{L}_{\mathrm{flat}}$. The contribution to the PLL's phase noise improves quadratically for a higher f_{ref} due to the lower multiplier N. Contrarily the PFD's additive noise increases linearly for a higher f_{ref} in relation to the frequency f_{char} at which the PFD was characterized [4]. As a result, the PLL's noise floor can be calculated as

$$\mathcal{L}_{\mathrm{PLL}}(f_{\mathrm{off}}) = \mathcal{L}_{\mathrm{flat}} + 20\log\left(\frac{f_{\mathrm{s}}}{f_{\mathrm{ref}}}\right) + 10\log\left(\frac{f_{\mathrm{ref}}}{f_{\mathrm{char}}}\right). \quad (5)$$

By equating (4) and (5) the offset frequency corresponding to the intersection point of the VCO's open loop noise and the PLL's noise floor can be calculated as

$$f_{\mathrm{cross}} = \sqrt{\frac{F k_{\mathrm{B}} T f_{\mathrm{ref}} f_{\mathrm{char}}}{8 P_{\mathrm{s}} Q^2 \cdot 10^{\frac{\mathcal{L}_{\mathrm{flat}}}{10}}}} \Rightarrow f_{\mathrm{cross}} \propto \sqrt{f_{\mathrm{ref}}}. \quad (6)$$

This proportionality helps to combine low phase noise with short settling times. Increasing f_{ref} allows for a higher loop bandwidth not only based on the $\frac{f_{\mathrm{ref}}}{20}$ limit explained earlier. It also improves the PLL's noise floor and thusly reduces the negative effect of a high B_{loop} on the integrated phase noise.

Regarding the chirp linearity mentioned in Section I, a higher reference frequency is also beneficial, because more sampling points are used to synthesize the chirps [5]. Furthermore, the linearity of state-of-the-art PLLs is often hindered by a low input frequency of the programmable frequency divider. During fractional operation the $\Sigma\Delta$-modulation at the programmable frequency divider continuously generates a phase error at the input of the PFD. The magnitude of this phase error is proportional to the value of the static prescaler R, necessitated by the limits of the programmable divider's input frequency. Therefore, minimizing R is beneficial for the linearity of the chirps due to the decreased phase error at the PFD's input.

The transceiver presented in this work will take advantage of those conditions by using a reference frequency of

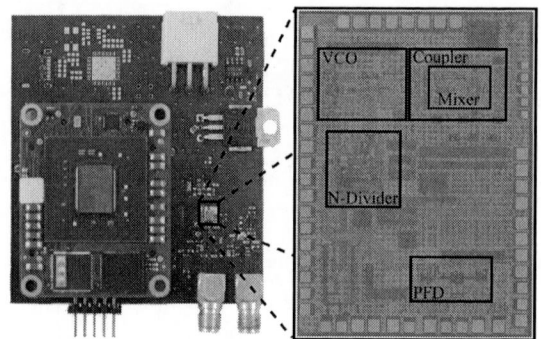

Fig. 2. Photograph of the realized MMIC containing the mixer, coupler, VCO, programmable frequency divider and PFD, mounted on the PCB.

$f_{\mathrm{ref}} = 1\,\mathrm{GHz}$ while using the fundamental VCO-frequency directly as the input frequency of the programmable frequency divider, without the use of a static prescaler. To the best of the author's knowledge, this has not been achieved before, regarding a fundamental VCO in the automotive band.

III. Synthesizer Design

The proposed synthesizer was realized on a monolithic microwave integrated circuit (MMIC), which has a size of $1964\mathrm{x}1448\,\mu\mathrm{m}^2$, and is shown in Fig. 2.

An essential part of the implemented PLL is the fully programmable frequency divider similar to [6]. By using merged emitter-coupled logic in conjunction with inductive peaking, a maximum input frequency of 94 GHz is achieved. Thus, a static frequency division inside the PLL can be omitted. Additionally, the programmability of every integer division factor $12 \leq N \leq 259$ allows for high flexibility of the reference frequency.

To operate at a reference frequency of 1 GHz, a highly linear PFD with a high input frequency is used, comparable to [5]. The PFD's gain linearity is required, because, for fractional operation, the $\Sigma\Delta$-modulation generates significant phase errors at the PFD's input, even when the PLL is locked. Using a highly linear PFD prevents the quantization noise of the $\Sigma\Delta$-modulation folding back to the carrier similar to [7]. To achieve the required operating frequencies, the PFD is also implemented in merged emitter-coupled logic, like the programmable divider. The output stage is realized with open-collector outputs designed to drive an external differential active low-pass filter without the use of a charge pump (CP). A CP is conventionally used to drive the loop filter. It allows for a passive loop filter and, thus, a fully integrated PLL. However, the use of an integrated loop filter significantly limits the achievable transfer functions and tuning voltages. Additionally, the mismatch of the up and down currents at these high input frequencies would considerably degrade the linearity of the PFD and lead to high dynamic mismatch noise in the PLL [5].

Moreover, the MMIC contains a transceiver. To realize a monostatic radar system, it consists of a mixer, a coupler,

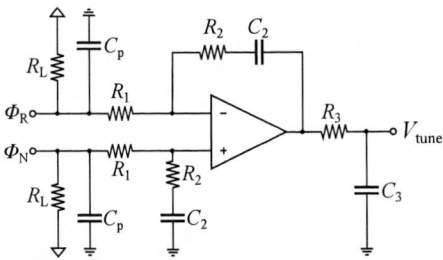

Fig. 3. Schematic of the loop filter, applying the differential output signal of the PFD directly, without the use of a CP.

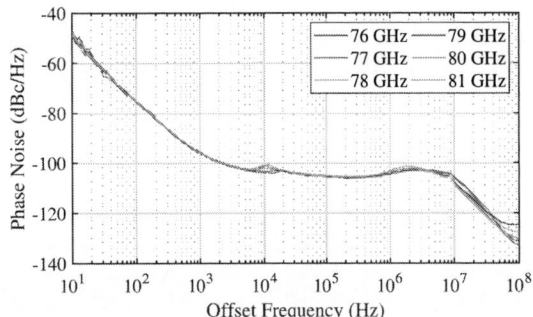

Fig. 5. Measured phase noise at the transceiver's output over the range of the automotive band.

and a VCO covering the E-band similar to [8]. The VCO offers a tuning range of 31 GHz. The applicable tuning voltage ranges from 1.1 V to 9.3 V. With an output power of 7 dBm provided by the VCO, the transceiver delivers a TX output power of approximately -1 dBm, while simultaneously driving the LO-port of the direct-down-conversion mixer. The VCO achieves a minimum phase noise of -99 dBc/Hz at an offset frequency of 1 MHz.

The entire MMIC draws a current of approximately 340 mA from a 3.3 V power supply.

For the reasons mentioned above, the transceiver renounces the use of a CP while using an external active loop filter. Fig. 3 presents the filter's architecture. Based on the fundamentals explained in II, a loop bandwidth of 5 MHz is used to decrease settling time without sacrificing phase noise. The phase margin is chosen to be $\varphi_m = 60\,°$ at a center frequency of $f_s = 78$ GHz. This ensures a sufficient phase margin across the required bandwidth to guarantee stability while compromising between phase noise and settling time. Since the loop gain variation over the automotive bandwidth is 1:1.7, no additional compensation is necessary to stabilize the PLL for its designated application.

IV. MEASUREMENT RESULTS

To validate the synthesizer's functionality, the bare die of the MMIC was mounted on a printed circuit board (PCB). The PCB provides the supply voltage, the divider value, the reference frequency and the loop filter.

The reference signal of 1 GHz is provided by an oven-controlled crystal oscillator (OCXO) (Axtal AXIOM1000). The phase noise is measured by using a phase noise and signal source analyzer (R&S FSWP) in conjunction with a W-band extender module (VDI WR10SAX). The output signal of the transceiver is directly probed and fed into the W-band extender. A field-programmable gate array (FPGA) module provides the necessary divider value. The result of the phase noise measurement at a carrier frequency of 78 GHz is depicted in Fig. 4. Moreover, Fig. 4 presents the simulated phase noise contributions of the PLL components at 78 GHz, which are based on corresponding phase noise measurements of the individual components. Overall, the simulation and measurement are in good agreement.

Close to the carrier, the phase noise is dominated by the OCXO. For higher offset frequencies inside the loop bandwidth, the frequency divider has the highest influence, after the PFD dominates the phase noise between 200 Hz and 1 kHz. Inside the loop bandwidth, the phase noise is very low compared to the state of the art with approximately -105 dBc/Hz. This results in a jitter of only 53 fs, integrated from offset frequencies of 10 kHz to 10 MHz. Fig. 5 presents the measured phase noise at different carrier frequencies covering the entire automotive band. Within this band, the

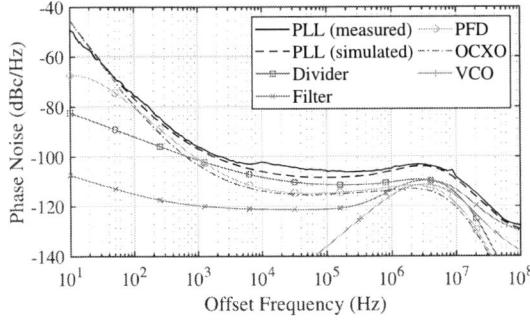

Fig. 4. Measured and simulated phase noise at f_s=78 GHz. The simulated contributions are calculated based on measurements of the individual components.

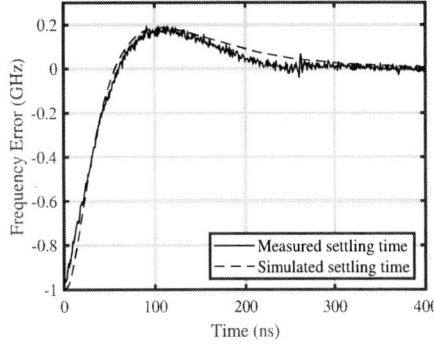

Fig. 6. Measured and simulated settling time of the transceiver for a frequency step from 77 GHz to 78 GHz.

Table 1. State-of-the-art FMCW-synthesizers with high relevance to this work.

Ref.	This work	[9]	[5]	[10]	[11]
Year	2022	2018	2016	2017	2018
Technology	130nm SiGe	130nm SiGe	130nm SiGe	22nm CMOS	65nm CMOS
f_s (GHz)	78	79	80	77	24
f_{ref} (MHz)	1000	125	167	160	52
PN @ 10 kHz (dBc/Hz)	-102	-63	-97	-65	-89
PN @ 1 MHz (dBc/Hz)	-104	-97	-94	-95	-99
RMS time jitter (fs)	53	687*	70	583*	212*
B_{loop} (kHz)	5000	140	1400	250	200
Calculated t_{settle} (µs)	0.88	26.2†	2.6†	-	-

*Graphically determined by integrating the phase noise measurements from 10 kHz to 10 MHz.
†Best case calculation for a 10 ppm step utilizing (2), based on the specified B_{loop} with optimum ζ.

variation of the in-loop phase noise over the carrier frequencies is less than 3 dB. The integrated jitter for each frequency of the automotive band is less than 61 fs.

To demonstrate the low settling times enabled by the high loop bandwidth, a frequency step from 77 GHz to 78 GHz is conducted. To observe the PLL's control characteristics, the tuning voltage is measured, and the corresponding frequency error is calculated via the measured tuning characteristic of the VCO. Fig. 6 shows the results of this measurement in comparison with the simulated frequency error.

The simulation and measurement are in good agreement. However, the small deviation comes down to the component tolerances of the resistors and capacitors used in the loop filter. Concerning simulations and a threshold frequency error of ± 10 kHz, corresponding to 10 ppm of the frequency step, the settling time is approximately 880 ns.

Compared to recently published FMCW-synthesizers shown in Table 1, the RMS time jitter of the realized transceiver is lower than the state of the art. This is a direct result of the fundamentals discussed in Section II. By increasing f_{ref}, the PLL's in-band phase noise is significantly reduced, resulting in a lower RMS time jitter, while using a higher loop bandwidth. Thusly enabling a short settling time, which ultimately allows for a high modulation rate. Further comparisons regarding the modulation rate, as well as chirp linearity will be investigated in future works, when the presented MMIC is measured in conjunction with the $\Sigma\Delta$-modulation.

V. CONCLUSION

We presented a mm-wave synthesizer based on a PLL for automotive radar combining low phase noise and short settling times. It omits a static prescaler and utilizes a high reference frequency of $f_{ref} = 1$ GHz. Thus, a loop bandwidth of 5 MHz was realized without sacrificing phase noise performance. In-loop phase noise of less than -105 dBc/Hz, and a jitter of less than 61 fs was achieved over the entire automotive band. Simultaneously, a short settling time of 880 ns was demonstrated for a frequency step with an error band of 10 ppm.

ACKNOWLEDGMENT

The authors would like to thank Infineon Technologies AG for fabricating the chips. This work is partially funded by the Deutsche Forschungsgemeinschaft (DFG, German Research Foundation) – Project-ID 287022738 – TRR 196" and partially funded by the German Federal Ministry of Education and Research (BMBF) within the project SeeYou.

REFERENCES

[1] "The european table of frequency allocations and applications in the range 8.3 kHz to 300 GHz (ECA table)," Mar. 2019, ECC.

[2] V. Winkler, "Range Doppler detection for automotive FMCW radars," in Proc. European Radar Conf, Oct. 2007, pp. 166–169.

[3] D. Banerjee, PLL Performance, Simulation, and Design, 5th Edition. Dog Ear Publishing, 2017.

[4] P. V. Brennan and I. Thompson, "Phase/frequency detector phase noise contribution in PLL frequency synthesiser," Electronics Letters, vol. 37, no. 15, pp. 939–940, 2001.

[5] G. Hasenaecker. M. van Delden, T. Jaeschke, N. Pohl, K. Aufinger, and T. Musch, "A SiGe fractional- N frequency synthesizer for mm-wave wideband FMCW radar transceivers," IEEE Transactions on Microwave Theory and Techniques, vol. 64, no. 3, pp. 847–858, Mar. 2016.

[6] M. van Delden, N. Pohl, K. Aufinger, and T. Musch, "A 94 GHz programmable frequency divider with inductive peaking for wideband and highly stable frequency synthesizers," in Proc. 12th European Microwave Integrated Circuits Conf. (EuMIC), Oct. 2017, pp. 9–12.

[7] D. Cherniak, C. Samori, R. Nonis, and S. Levantino, "PLL-based wideband frequency modulator: Two-point injection versus pre-emphasis technique," IEEE Transactions on Circuits and Systems I: Regular Papers, vol. 65, no. 3, pp. 914–924, Mar. 2018.

[8] C. Bredendiek, S. Hansen, G. Briese, and N. Pohl, "A full E-band single-channel SiGe transceiver MMIC for monostatic FMCW radar systems," in Proc. 15th European Microwave Integrated Circuits Conf. (EuMIC), Jan. 2021.

[9] J. Vovnoboy, R. Levinger, N. Mazor, and D. Elad, "A dual-loop synthesizer with fast frequency modulation ability for 77/79 GHz FMCW automotive radar applications," IEEE Journal of Solid-State Circuits, vol. 53, no. 5, pp. 1328–1337, May 2018.

[10] A. R. Fridi, C. Zhang, A. Bellaouar, and M. Tran, "A low power fully-integrated 76-81 GHz ADPLL for automotive radar applications with 150 MHz/us FMCW chirp rate and -95dBc/Hz phase noise at 1 MHz offset in FDSOI," in Proc. IEEE RFIC, 2019, pp. 327–330.

[11] D. Cherniak, L. Grimaldi, L. Bertulessi, R. Nonis, C. Samori, and S. Levantino, "A 23-GHz low-phase-noise digital bang–bang PLL for fast triangular and sawtooth chirp modulation," IEEE Journal of Solid-State Circuits, vol. 53, no. 12, pp. 3565–3575, 2018.

Proceedings of the 16th European Microwave Integrated Circuits Conference

A Passively-Coupled 39.5 GHz Colpitts Quadrature VCO in SiGe HBT Technology

Janis Wörmann[#1], Aleksey Dyskin[*], Sébastien Chartier[#†], and Ingmar Kallfass[#2]

[#]Institute of Robust Power Semiconductor Systems (ILH), University of Stuttgart, Stuttgart, Germany
[*]Department of Electrical Engineering, Technion-Israel Institute of Technology, Haifa, Israel
[†]is now with Fraunhofer Institute for Applied Solid State Physics (IAF), Freiburg, Germany
{[1]janis.woermann, [2]ingmar.kallfass}@ilh.uni-stuttgart.de

Abstract — **A quadrature voltage controlled oscillator (VCO) using Colpitts topology is presented. The quadrature voltage controlled oscillator (QVCO) facilitates low phase noise due to an innovative coupling mechanism between two differential cores. The passive capacitive coupling operates in-phase, utilizing injection locking technique and is unambiguous in its phase position. A prototype monolithic microwave integrated circuit (MMIC) using heterojunction bipolar transistors in 0.13 μm SiGe BiCMOS technology achieves a measured differential output power of 0 dBm at the output of each core and a phase noise of -104 dBc/Hz at 1 MHz offset, with 39.5 GHz center frequency of oscillation. The frequency tuning range is 6.4% and power efficiency is 1.25%.**

Keywords — **Injection-locked oscillators, Voltage-controlled oscillators, Silicon Germanium, Heterojunction bipolar transistors, MMICs, K_a-band.**

I. INTRODUCTION

Quadrature signal sources are indispensable components of modern communications and radar systems, where they allow, among other things, the use of direct down-conversion topologies and complex modulation formats in which the phase of a signal is used to encode information (coherent modulation). The low-noise, unambiguous and reliable generation of frequency-variable quadrature signals sets the lower bound for system performance as it is usually a core of the signal chain.

A variety of Colpitts voltage controlled oscillator (VCO) topologies exist. A design meant to improve tuning range (TR) and output voltage swing renounces on collector resistors and uses a transformer to decouple the output signal from transistor bases [1]. The work in [2] improves phase noise (PN) by adding emitter-inductance for a higher varactor quality-factor (Q-factor) and decoupling of noise from the current source.

Several methods for generating a quadrature signal are well known. The following conventional methods differ from the proposed method by a higher component count and lower radio frequency (RF) performance leading to higher power consumption and higher costs. Passive methods such as RC-CR polyphase filters suffer from limited bandwidth, inherent loss and worsen the PN performance [3], [4]. Another method is frequency division by two of a signal with twice the desired frequency, which requires an oscillator at twice the frequency and thus is not preferred in high frequency range [3]. A different method is to lock two identical oscillator cores with

each other to achieve 90° phase shift between them. Here, one approach is to extract the second harmonic in each core and then invert and insert this signal to the other core[5], [6]. Phase inversion at the second harmonic corresponds to 90° phase relation between the fundamental signals. This mechanism is simple but faces ambiguity of the phase relation (±90°) and amplifies the second harmonic artificially. For the latter method yet another approach is to extract a 90° phase-shifted current or voltage signal e.g. by using a cross-coupled buffer or additional cascade stages [3]. This signal is then coupled directly into the other core and inversely. The coupling mechanism is simple as well, resulting in fewer unwanted harmonic components, but comes with an increased number of transistors and possible need for capacitive gate decoupling, which results in increased noise [7], [8], [9]. Direct coupling of two cores is suggested in [10], using intrinsic *bulk-to-x* capacitances of metal oxide semiconductor field effect transistor (MOSFET) and varactor diodes, which are part of the oscillator core. The coupling is achieved by cross-connecting the intrinsic bulk terminals. This technique avoids additional coupling elements, but shows impact on the coupling characteristic when tuning the varactors and faces increased (flicker)-noise, since it is based on MOSFET technology by design. The coupling proposed in this work improves on the latter method.

The generation of quadrature signals should generally feature low complexity, high stability over time, bandwidth, spectral purity, output power at a given power supply and especially low noise. The PN of oscillators is proportional to the number of active devices, among other factors. Therefore, a quadrature coupling technique is presented, to passively couple two VCO cores via injection locking, keeping the number of active devices at a minimum, and thus to achieve minimum phase noise possible. The proposed coupling aims to be robust against environmental variations and is, most importantly, unambiguous in phase relation by design. To the authors' knowledge, the proposed passive, capacitive coupling method presented here is novel and the quadrature voltage controlled oscillator (QVCO) is the first in K_a-band, providing high, unbuffered output power in the order of 0 dBm, while achieving the stated low PN.

3–4 April 2022, London, UK

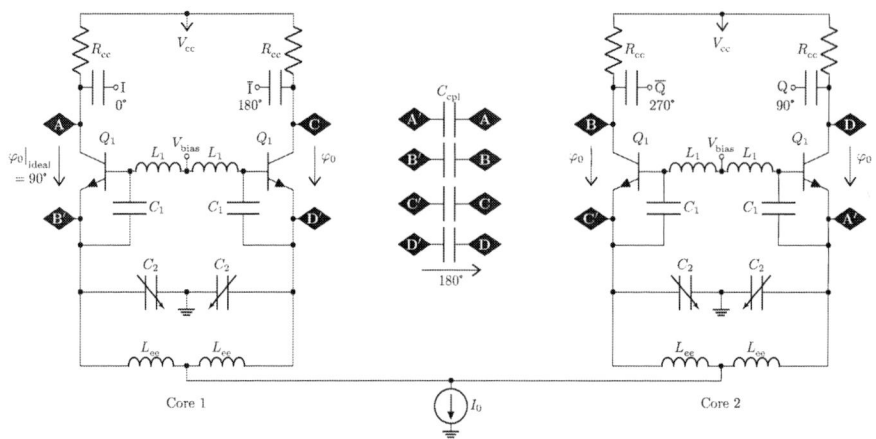

Fig. 1. Two identical differential Colpitts cores, coupled to form the QVCO. The phase of each output in degrees relative to output I is indicated, as well as a relative phase shift over particular components given.

II. CIRCUIT DESIGN

The QVCO is fabricated in $0.13\,\mu m$ SiGe BiCMOS technology (*SG13S*) from IHP, which offers npn-heterojunction bipolar transistors (HBTs) with f_T of $250\,GHz$ and f_{max} up to $340\,GHz$, $1.7\,V$ collector-emitter breakdown voltage and 7 metal layers. Inductors can be realized on the upper two thick metal layers.

As shown in Fig. 1, the proposed QVCO consists of two identical differential oscillator cores using Colpitts topology, whereas each core is formed by two single ended VCO branches in back-to-back configuration. Colpitts topology is chosen for the oscillator as it promises lowest PN and power consumption at a given bias current [1]. The theory of a Colpitts VCO is well known [3], [11] and therefore not addressed in this paper.

A single-ended VCO-branch consists of the transistor Q_1, inductor L_1 implemented using a transmission line (TL) and capacitors C_1 and C_2. Each branch is degenerated by means of an inductor L_{ee} and using a current source I_0 to suppress common mode oscillations. The virtual ground nodes on the symmetry axis of both differential cores are used for frequency tuning via varactor C_2 and biasing of the transistors via V_{bias}. In contrast to cross-coupled oscillator topologies using HBT no capacitive decoupling of the transistor bases (from the opposite collector) for base biasing is necessary. The output signals are decoupled at the collector nodes without additional output buffer stages.

Both cores are coupled in a way to achieve 90° phase shift between the two differential in-phase and quadrature (IQ)-outputs, using injection locking. For this, an innovative coupling technique is introduced.

A phase shift φ_0 across transistors Q_1 will emerge when the collector is loaded mainly with resistive load, while the emitter is loaded mainly with capacitive load, as indicated in Fig. 1. Ideally, the voltage signal phase at the emitter node would lag 90° to the one at the collector i.e. $\varphi_0|_{ideal} = 90°$. By using coupling capacitors C_{cpl}, each emitter node is coupled to a collector node of the corresponding branch of the other core.

An injection-current over each C_{cpl} is fed into the other core, forming a closed loop (360°) over all four coupling-paths. The capacitive quadrature coupling is mathematically confirmed.

An injected current signal into an oscillator in proximity of its resonance frequency will lock this oscillator to the injected frequency. In case of matching injected and resonance frequency, this locking will settle with zero phase shift between both frequency signals. It is known that the quality factor of a resonator tank reaches its high value if an injected current is in phase with the tank voltage [12]. As the injected current in each branch of the QVCO already shows a 90° shift, the presented direct quadrature coupling acts in-phase, which means that it has no negative effect on the noise behavior by means of quality factor degeneration.

Both cores lock unambiguously and in quadrature even in case of a non-ideal, inherent phase shift $\varphi_0|_{nonideal} \neq 90°$ over the transistor, if one assumes the branches to be free-running. The use of distinct coupling capacitors promises high robustness. Symmetric design is strictly applied to achieve low phase error.

III. RESULTS

Frequency domain measurements of the device under test (DUT) were performed on-chip. Fig. 2 shows a photograph of the oscillator MMIC. For frequency domain measurements only one differential probe is attached and the other differential output left open. This asymmetric loading is simulated and used for correcting the power measurement in post-processing. Simulation results at the same operating points are displayed as dashed lines, accounting for extracted parasitics from the layout. Bondpad parasitics are not included.

Variation of the varactor tuning voltage V_{tune} changes the effective resonator tank capacitance and thus the resonance frequency. Measurement results are given in Fig. 3. The frequency differs approximately $1\,GHz$ between measurement and simulation. The use of TL models is probably one reason for this, since inductance deviations of several percent can accumulate with ease without verification by EM simulation.

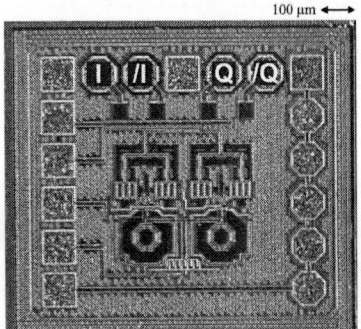

100 μm ◄──►

Fig. 2. Chip-photo of the 39.5 GHz quadrature oscillator. Full chip dimensions are 800 μm x 700 μm and oscillator dimensions are ≈ 350 μm x 400 μm. 3-port spiral inductors L_{ee} and 3-port inductors L_1 implemented as TL of each core can be clearly seen. Outputs are labeled on the north side of the chip.

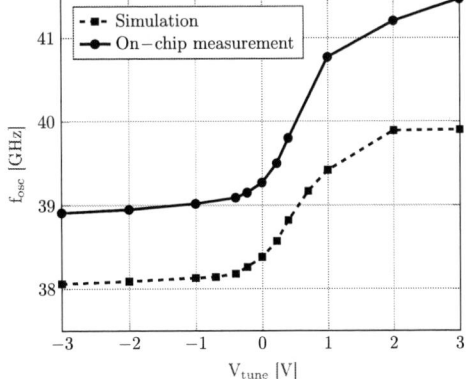

Fig. 3. Frequency measurement and simulation results over V_{tune}, at nominal OP, $V_{cc} = 1.8$ V.

Still, the covered tuning range as well as the gradients of frequency over tuning voltage are in a good agreement. In measurement a maximum absolute tuning range of 2.55 GHz is achieved. With a nominal frequency of 39.5 GHz this yields 6.4% relative tuning range.

Varying V_{tune} is not expected to have significant effects on the output power, however a PN degrading can occur. The differential varactor consists of two diodes with a common-collector. Applying a voltage V_{tune} larger than ≈ −0.7 V, the diodes begin conducting. Since this leads to lower effective Q-factor of the resonator tank due to current leakage, a decrease of PN and output power is expected.

As depicted in Fig. 4 the measurements show small dependency of output power over tuning range. Above 1 V tuning voltage the measured power suffers a noticeable loss. This may be a result of increased loss of the measurement setup, as the frequency is already well above 40 GHz, which is exceeding the specifications of probe and connectors.

Accurate de-embedding of the DUT could not be implemented due to unavailability of the required calibration substrate. Therefore, a rough power correction based on the typical loss of the on-chip probe and cables is applied only. Further, the simulated output power is extracted at

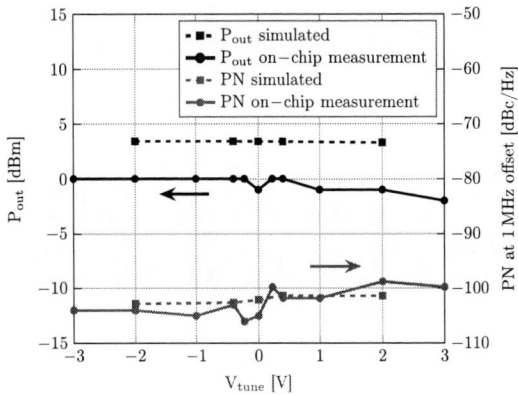

Fig. 4. Differential output power and PN measurement and simulation results over V_{tune}, at nominal OP, $V_{cc} = 1.8$ V.

the oscillator core outputs with ideal matching, as it does not account bondpad parasitics, which is presumably one reason for the shown deviation between simulated and on-chip measured power.

Only a small PN dependency on V_{tune} can be confirmed. Below −0.7 V a current at the tuning voltage supply of the varactor diodes in the mA-range is measured, which implies a leakage of current from the resonator tank, as expected. Decreasing frequency counteracts the loss, since the Q-factor is inversely proportional to the frequency. In contrast to the power measurements, inaccurate correction of the losses of the measurement setup does not affect the measured PN, since PN is specified relative to the carrier power.

Time-domain measurements measured at all four outputs simultaneously prove the robustness of the quadrature coupling even at limit values of the operating points (OPs). The 90°-locking of the two cores is always obtained in these OPs.

For comparison the following figure of merit (FOM) is applied

$$\text{FOM} = L(\Delta f) - 20 \log_{10}\left(\frac{f_{osc}}{\Delta f}\right) + 10 \log_{10}\left(\frac{P_{DC}}{P_{out}}\right) \quad (1)$$

with f_{osc} [Hz] the center frequency of oscillation, Δf [Hz] the frequency offset for phase noise $L(\Delta f)$ [dBc / Hz], output power P_{out} [W] and DC supply power P_{DC} [W].

Table 1 summarizes state-of-the-art QVCOs reported so far. It can be seen, that there are oscillators which sport better performance but come with some important limitations. The differential HBT based VCOs reported in [1] and [2] yield high output power and comparable PN, but large supply voltage is needed and most importantly only a single differential output is provided. The complementary metal oxide semiconductor (CMOS) pendant [13] yields significantly lower output power as well as moderate PN, since is given at 4 MHz offset.

Only one quadrature VCO [9] is a suitable candidate for direct comparison, as the others are not intended to provide an output signal in the similar power range. As no output power is stated in [6] and [8], 1 mW P_{RF} is assumed to perform

211

Table 1. Summary and comparison with state-of-the-art oscillators. *(Italic for poorly comparable works.)*

Ref.	Type	Coupling	f_{osc} [GHz]	TR [%]	V_{DC} [V]	P_{DC} [mW]	P_{out} [dBm]	η [%]	PN [dBc/Hz]	FOM [dBc/Hz]
[1] '13	VCO SiGe HBT	n.a.	74	4.8	2.5	65	2	2.43	-99	-180
[2] '03	VCO SiGe HBT	n.a.	43	26	5.5	280	6.5	1.6	-110	-184
[13] '04	VCO CMOS SOI	n.a.	40	15	1.5	11.3	-8	1.4	-97 ****	-158
[6] '19	QVCO CMOS	superharm., tail impedance	40	18.4	0.75	8.4	n.a.	n.a.	-94.3	*-177* **
[8] '07	QVCO CMOS	transistor, cross-coupled	41	24.4	1.2	7.8	n.a.	n.a.	-104	*-187* **
[9] '05	QVCO CMOS SOI	transistor, cross-coupled	39.5	12.5	1.5	81	-6.5 *	0.55	-87 ***	-147
[10] '09	*QVCO CMOS*	*bulk cap., passive*	*5.35*	*18*	*1.8*	*8*	*n.a.*	*n.a.*	*-128.7*	*n.a.*
This work	QVCO SiGe HBT	direct cap., passive	39.5	6.4	1.8	160	0 *	1.25	-104	-177

* per channel   ** with assumed $P_{out} = 1\,mW$ *** at 3 MHz offset   **** at 4 MHz offset

FOM calculation. While [9] yields doubled TR and draws only half DC-power, only low output power with low efficiency is generated, along with moderate PN, given at 3 MHz offset. The QVCO of [10] is cited as using comparable passive capacitive coupling via isolated bulk in CMOS. However, the achieved PN cannot be used directly for comparison since the operating frequency is much smaller.

The DC-power draw of $160\,mW$ of the presented Colpitts QVCO is reasonable, since the yielded efficiency of $\approx 1.25\%$ (with $2\mathrm{x}\ 0\,\mathrm{dBm}$ output power) is in the same order as of other designs. As the FOM suggests, when considering the last mentioned restrictions, the oscillator can compete with comparable topologies. To the authors' knowledge, the proposed oscillator is the first K_a-band quadrature VCO that provides high unbuffered output power at the given order while achieving the specified low phase noise performance.

IV. Conclusion

A Colpitts quadrature VCO in SiGe HBT $0.13\,\mu m$ technology is presented. The oscillator's measured center frequency is $\approx 39.5\,GHz$ sporting a TR of 6.4% and low PN of $-104\,dBc/Hz$ at 1 MHz offset. The output power of $\approx 0\,dBm$ at $100\,\Omega$ differential load per IQ channel is obtained without an output buffer. This, and a proposed capacitive injection lock coupling mechanism minimizes the generated PN, as no additional active devices are introduced. Further on, the passive coupling operates in-phase, not degrading the resonator Q-factor. By design, the coupling is circumventing issues and constraints of other common coupling techniques. Compared to its counterparts, the reported QVCO shows superior performance in terms of phase noise and demonstrates robustness of quadrature signal generation to environmental conditions.

Acknowledgment

The authors would like to thank the professors and colleagues from the Institutes of Semiconductor Engineering (IHT), Smart Sensors (IIS), Electrical and Optical Communications Engineering (INT), and Radio Frequency Technology (IHF) (all from the University of Stuttgart) and the Institute of Radio Frequency Engineering and Electronics, KIT Karlsruhe (IHE) for generously helping with assembly and measurement equipment.

References

[1] G. Sapone, E. Ragonese, A. Italia, and G. Palmisano, "A 0.13um SiGe BiCMOS Colpitts-Based VCO for W-Band Radar Transmitters," *IEEE Transactions on Microwave Theory and Techniques*, vol. 61, no. 1, pp. 185–194, 2013.

[2] H. Li and H.-M. Rein, "Millimeter-wave VCOs with wide tuning range and low phase noise, fully integrated in a SiGe bipolar production technology," *IEEE Journal of Solid-State Circuits*, vol. 38, no. 2, pp. 184–191, 2003.

[3] L. B. Oliveira, J. R. Fernandes, I. M. Filanovsky, M. Ismail, M. M. Silva, and C. J. Verhoeven, *Analysis and Design of Quadrature Oscillators*, ser. Analog Circuits and Signal Processing Series. Berlin and Heidelberg: Springer, 2008. [Online]. Available: http://nbn-resolving.de/urn:nbn:de:1111-20080919672

[4] T. H. Lee, *The design of CMOS radio-frequency integrated circuits*, 2nd ed. Cambridge: Cambridge Univ. Press, 2004. [Online]. Available: http://www.loc.gov/catdir/description/cam051/2003061322.html

[5] I. Nasr, M. Dudek, R. Weigel, and D. Kissinger, "A 33% tuning range high output power V-band superharmonic coupled quadrature VCO in SiGe technology," in *2012 IEEE Radio Frequency Integrated Circuits Symposium*. IEEE, 17.06.2012 - 19.06.2012, pp. 301–304.

[6] L. Zhang, N.-C. Kuo, and A. M. Niknejad, "A 37.5–45 GHz Superharmonic-Coupled QVCO With Tunable Phase Accuracy in 28 nm CMOS," *IEEE Journal of Solid-State Circuits*, vol. 54, no. 10, pp. 2754–2764, 2019.

[7] M. J. Hemmati and A. Hakimi, "A CMOS quadrature VCO with optimized Colpitts topology for low-voltage applications," *Microelectronics Journal*, vol. 72, pp. 32–39, 2018.

[8] M. Usama and T. A. Kwasniewski, "A 40 GHz Quadrature LC VCO and Frequency Divider in 90-nm CMOS Technology," in *2007 IEEE International Symposium on Circuits and Systems*. IEEE, 27.05.2007 - 30.05.2007, pp. 3047–3050.

[9] F. Ellinger and H. Jackel, "38-43 GHz quadrature VCO on 90 nm VLSI CMOS with feedback frequency tuning," in *IEEE MTT-S International Microwave Symposium Digest, 2005*. IEEE, 12-17 June 2005, pp. 1701–1703.

[10] E. Ebrahimi and S. Naseh, "A new low-phase noise direct-coupled CMOS LC-QVCO," *IEICE Electronics Express*, vol. 6, no. 18, pp. 1337–1344, 2009.

[11] F. Ellinger, *Radio Frequency Integrated Circuits and Technologies*, second edition ed. Berlin, Heidelberg: Springer-Verlag Berlin Heidelberg, 2008.

[12] J. van der Tang, P. van de Ven, D. Kasperkovitz, and A. van Roermund, "Analysis and design of an optimally coupled 5-GHz quadrature LC oscillator," *IEEE Journal of Solid-State Circuits*, vol. 37, no. 5, pp. 657–661, 2002.

[13] N. Fong, J. Kim, J.-O. Plouchart, N. Zamdmer, D. Liu, L. Wagner, C. Plett, and G. Tarr, "A low-voltage 40-GHz complementary VCO with 15% frequency tuning range in SOI CMOS technology," *IEEE Journal of Solid-State Circuits*, vol. 39, no. 5, pp. 841–846, 2004.

Proceedings of the 16th European Microwave Integrated Circuits Conference

30-46 GHz 1.5dB IL Negative Gate Control SPDT with 24.5dBm IP1 in 130nm CMOS

Sumeet Londhe[1], Noam Bar-Helmer[2], Samuel Jameson[2], and Eran Socher[1]

[1]School of Electrical Engineering, Tel Aviv University, Israel

[2]Rafael, Israel

Abstract—In this paper a novel way of improving IP1 of SPDT switches is presented. IP1 of a SPDT generally degrades with power due to the leakage to the isolated port. Adding a negative control to the off state of the switch improves IP1 performance considerably while lowering the insertion loss. Compared to traditional series/shunt SPDT designs the negative gate control shunt SPDT provides wider bandwidth, lower insertion loss, higher isolation and higher IP1. The negative gate control SPDT outperforms traditional SPDTs and state of the art achieving an insertion loss of 1.5dB with wide band isolation (20–50GHz) better than 23dB and an IP1 of 24.5dBm in low-cost 130nm CMOS.

Keywords—5G, CMOS, array, efficiency, SPDT, IP1.

I. INTRODUCTION

MMW systems are becoming more important for industrial, wireless communication and military applications. Compact antenna dimensions and higher resolution is obtained by working at higher frequencies, especially frequencies like 24 GHz, 35 GHz and 94GHz are important for their low atmospheric loss. As demand for 5G rises, efficient ways of using silicon die frontends is necessary. SPDTs allow the reuse of the on-package antenna for both TX-RX saving space and packaging costs. On the other hand, SPDTs limit the performance of the TX by limiting the power injected into the antenna and also that of the RX by introducing insertion loss [1,2]. Increasing the power handling capacity of the SPDT and lowering its insertion loss are key to system performance.

This paper presents design and measured results of Ka-band series, shunt stacked and the novel negative gate control SPDT in low-cost 130nm CMOS. The pros and cons of each SPDT are described in the paper. A new way of improving performance compare to traditional tradeoffs is proposed and demonstrated in the paper, by implying negative voltage to shunt devices.

II. CIRCUIT DESIGN

Three SPDT configurations were designed and analyzed in order to optimize for a particular parameter in this work: series, shunt stack and the new negative gate shunt SPDT, which provides better performance tradeoff. Series SPDT circuits have a high insertion loss (IL) as the signal travels through the switch and a bandlimited high isolation [1]. Shunt switches have broadband insertion and isolation response due to lack of resonating elements [1,4]. For transistor based SPDTs IP1 depends upon the gate bias. Higher ON/OFF voltage increases the IP1. This upper limit is determined by process node and it is not reliable to increase this beyond the recommended value.

Floating gate circuits were reported to improve IP1. It was shown that the channel modulation of the OFF state limits the

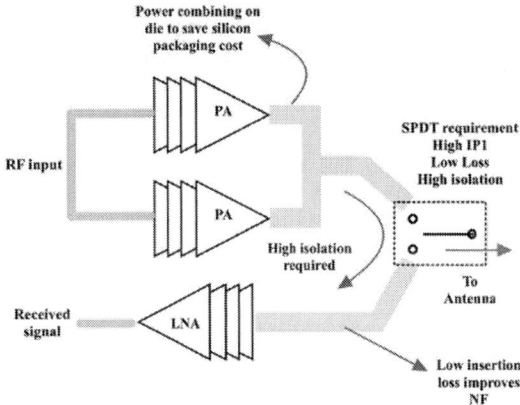

Fig. 1 SPDT requirements and use in a 5G frontend block diagram

IP1 performance of the SPDT [1]. In our design a novel way of improving IP1 without compromising the integrity of the process is demonstrated.

Each SPDT is designed with a maximum Gmax approach (minimization of IL). Once the optimum size for minimum isolation is selected, the OFF capacitance of the switch is resonated out to improve isolation. The input and output matching networks are designed to have minimum loss. All circuits were designed using 1.2V DNW transistors and passives were simulated using 3D Momentum.

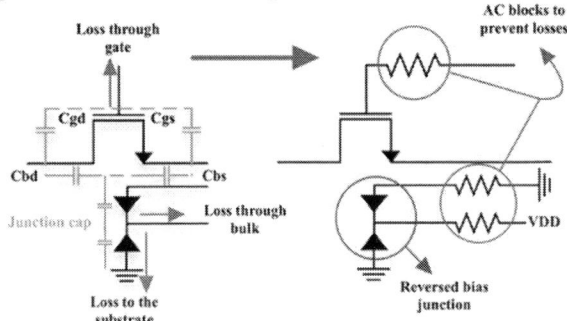

Fig. 2. Transistor model as a switch

A. Transistor Switch Model

Fig. 2 presents a simplified model for parasitic losses in a transistor as a switch. In ON state losses primarily arise due to channel resistance which improves with larger size. As the size of the transistor increases the parasitic capacitance start to leak signal to the metal gate and to the substrate. Even though loss

3–4 April 2022, London, UK

through the channel resistance drops, the parasitic loss dominates as size increases [2].

Separating the bulk of the transistor from the substrate using deep N-well transistors and using high resistor to block the leakage improves the performance of the switch. High resistance at the gates avoids breakdown of the transistor gates and avoids leakage through the gate. The deep N-well separates the bulk and the substrate minimising loss to the substrate. Optimising the N-well area to lower capacitance while providing isolation is essential for performance.

B. Series SPDT switch

In the series topology, transmit or receive signals are delivered or blocked by the channel low or high impedances, controlled by the gate voltage. Large size transistors (W> 90um) are required to satisfy IL<2dB in 130nm CMOS. These large transistors provide high parasitic capacitance in OFF state. This parasitic capacitance leads to lower isolation as it provides a low impedance path for the signal from the input to the OFF channel even though the channel resistance might be high. An inductor is added between the input and output nodes of the channel to resonate with the OFF capacitance of the channel which improves the isolation considerably for this configuration limited by Q of the inductor. Fig.3 shows the schematic of the switch. The series SPDT achieves an insertion loss of 2dB with isolation >30dB for narrow bandwidth and IP1 of 18dBm.

C. Shunt stacked SPDT

Shunt SPDT are widely used for their wideband and low insertion loss characteristics. Each channel is connected to the ground via a shunt transistor chains. IL can be better since the signal path is normally open, and shunt transistors are used to block signal delivery by absorbing its power. On the antenna port, a lumped λ/4 transmission line is used to transform the ground connection into an open. Small size transistors are required in shunt configuration to lower parasitic capacitors. However, a small single shunt transistor has lower power handling capacity [1]. A larger stacked transistor improves power handling capacity as it lowers the gate voltage modulation of the primary shunt transistor. Fig.4 shows the schematic of this shunt stacked switch. This SPDT configuration achieves an insertion loss of 1.8dB with a broadband isolation >15dB and an IP1 of 17dBm.

D. Negative gate control shunt SPDT

Shunt SPDT circuits have previously demonstrated low insertion loss capabilities with adequate isolation characteristics. Lack of sharp resonating inductors/capacitors provide a relatively wide band response. The problem with shunt SPDT lies with its power handling capacity. Shunt SPDTs produce poor results with respect to IP1 characteristics.

As the input power increases V_{GS} of the shunt transistors in both OFF and ON state starts to increase. When biased at 0/1.2V control voltage at high input power channel modulation of the OFF stage causes leakage to ground. The larger signal opens up the shunt transistors connecting the signal line to ground and increases insertion loss drastically [1]. Changing the

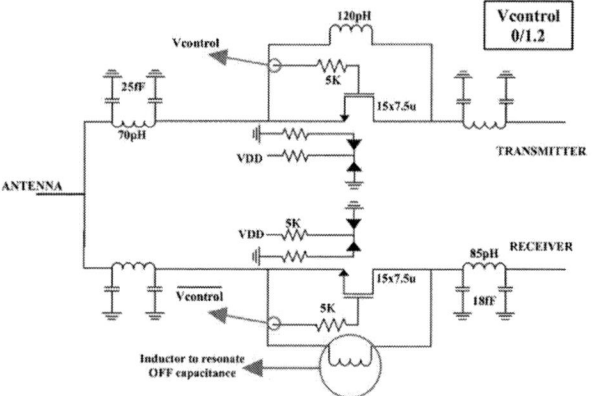

Fig. 3. Traditional series SPDT schematic

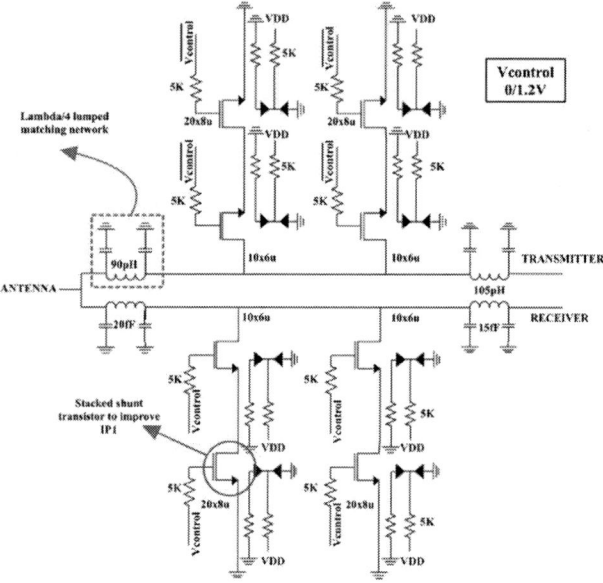

Fig. 4. Shunt stacked SPDT

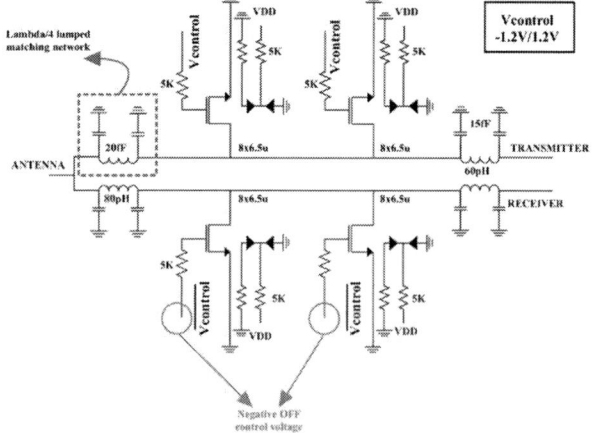

Fig. 5. Negative gate control SPDT

the control voltage to 1.2/-1.2 V solves this problem. Fig.6 shows the V_{GS} swing at the gate of the ON and OFF transistors at 20 dBm input power. Even though V_{GS} signal of the OFF state is high the bias point is low enough so that the transistor does not generate channel inversion.

Fig.5 shows schematic of the negative gate controlled SPDT. Negative bias at the gates introduces capacitance at the gate-bulk junction. Adding a high resistance at the bulk node lowers the effect of this capacitance. Lower size shunt transistors can be used in this case to improve IL further. Fig.6 shows the voltage swing at the gates of the ON and OFF transistors at 20dBm input power. The OFF state transistor is far from entering linear mode. This design achieves an insertion loss of 1.5dB with isolation of >25dB and IP1 of 24dB.

III. MEASUREMENT

The IC was manufactured in Tower-Jazz 130nm 7 Metal CMOS process. The S-parameters for the chips were measured using Agilent E8361C vector network analyzer. The large signal power sweep measurement was done using Agilent E8257D signal generator along with a wide band PA to provide input power large enough to compress the SPDT, and Agilent E4448A spectrum analyzer. Chips from multiple sections of the wafer were tested to compare performance and all measured using on-chip probing.

A. S-parameter measurement analysis

The core area of the series SPDT is 317 x 264 um. The S-parameters for the series SPDT are shown in Fig 7. The series SPDT achieve minimum IL of 2.2dB with a 1dB bandwidth of 16 GHz (28 GHz-44 GHz) measured at 1.2V. The maximum isolation achieved is 23dB. The isolation is narrow band as a result of the resonance of off capacitance. Isolation is better than 20dB between 36GHz-43GHz. The input and output return loss are >15dB across the bandwidth.

The core area of the shunt SPDT is 310 x 400um. The S-parameters for the shunt stacked SPDT are shown in Fig 8, achieving minimum IL of 2dB with a 1dB bandwidth of >25GHz (23 GHz-50 GHz) measured at 1.2V. The maximum isolation achieved is 18.5dB. The isolation is wide band as it does not depend on resonance. Isolation is better than 16dB between 20GHz-50GHz. The input and output return loss are >15dB in the bandwidth.

The core area of the negative shunt SPDT is 190 x 310um. The S-parameters for the negative shunt SPDT are shown in Fig 9, achieving record minimum IL of 1.5dB with a 1dB bandwidth of >16GHz (30GHz-46GHz) measured at 1.2V. The maximum isolation achieved is 28dB. The isolation is wide band again since no resonance is used. Isolation is better than 24dB between 20GHz-50GHz.The input and output return loss are >15dB in the bandwidth.

B. Power handling measurement analysis

Fig. 10 shows insertion loss and isolation measured for the SPDTs up to 28dBm input power. Power measurements were done at 37 GHz for all SPDTs. The limiting factor for IP1 in SPDT is generally the leakage from the isolated transistor

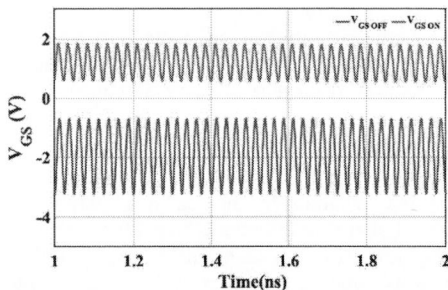

Fig. 6. V_{GS} ON and OFF state at 20dBm input power for the negative gate control SPDT

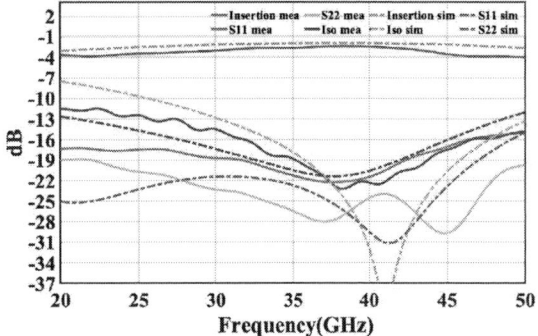

Fig. 7. Measured vs simulated s-parameters of series SPDT

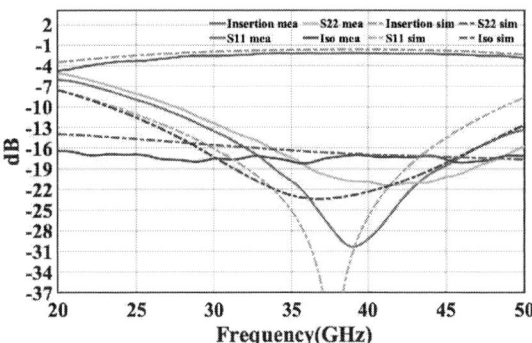

Fig. 8. Measured vs simulated S-parameters shunt stacked SPDT

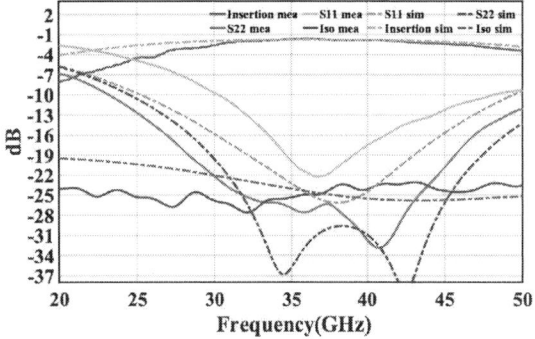

Fig. 9. Measured vs simulated S-parameters Negative gate controlled SPDT

branch that starts conducting at higher input power. As the input power increases the swing at the gate voltage increases which modulated the channel resistance. As shown from Fig. 10 as input power increases the isolation starts degrading and leaking power to the isolated side increasing the insertion loss. The series and shunt stacked SPDTs suffer from this phenomenon at lower input power. On the other hand, the negative gate control SPDT holds the isolation for higher input powers which improves the IP1 significantly.

The measured IP1 for the shunt stacked and series SPDT is 17dBm and for the negative gate control SPDT is 24.5dBm. The series and shunt stacked SPDT has isolation better than 15dB up to 15dBm of input power. The negative gate control SPDT has isolation better than 20 dB until 27dBm input power. All SPDTs were measured at 1.2V control voltage.

IV. CONCLUSION

In this paper multiple configurations of SPDTs were designed, fabricated and measured. Traditional series and shunt SPDTs have bandwidth and IP1 limitation respectively. Adding a stacked transistor to a traditional shunt SPDT improves its IP1 considerably with an increase in insertion loss. The series SPDT provides a narrow band high isolation better than 20dB between 36-43 GHz while the shunt SPDT provides wide band isolation better than 15dB. The series SPDT has an insertion loss of 2.2dB which is higher than the insertion loss of the shunt stacked SPDT of 2dB as a result of the signal passing through the conducting channel during ON state. These design show performance on par with state of the art as shown in Table 1. The limitations of the above traditional SPDTs are alleviated by adding a negative gate control voltage. The negative gate control voltage shunt SPDT uses smaller transistors improving isolation and insertion while increasing the IP1 of the SPDT. The negative gate control SPDT as an insertion loss of 1.5dB and isolation better than 24dB with an IP1 of 24.5dB. Compared to the state of the art this SPDT has the best performance in all fields across all CMOS technology nodes while being designed in a low cost 130nm technology node.

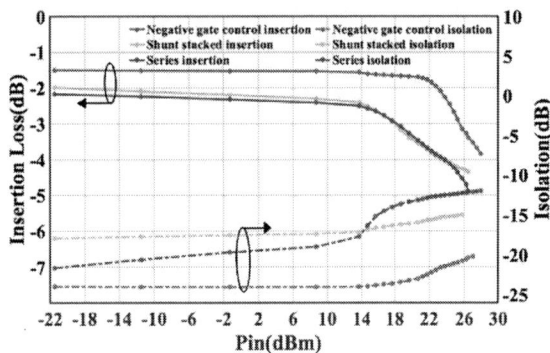

Fig. 10. Measured power handing performance of the demonstrated SPDT's

Table 1. Comparison to State of Art

Ref	Units	This Work	TMTT '08 [1]	SiRF '20[3]	TMTT '17[4]	TMTT '20[5]
Process		130nm CMOS	130nm CMOS	65nm SOI	130nm SOI**	45nm SOI
BW (1dB)	GHz	30-46	35-40	24-31	0-50	0-50
Insertion	dB	1.5	2.2	3.5	1.7-2.1	1#
Isolation	dB	25	>32	21	27	25
IP1	dBm	24.5	22	19.5	14	21
Size	mm²	0.05	0.06	-	0.04	-

** high resistivity trap rich # De-embedded pads

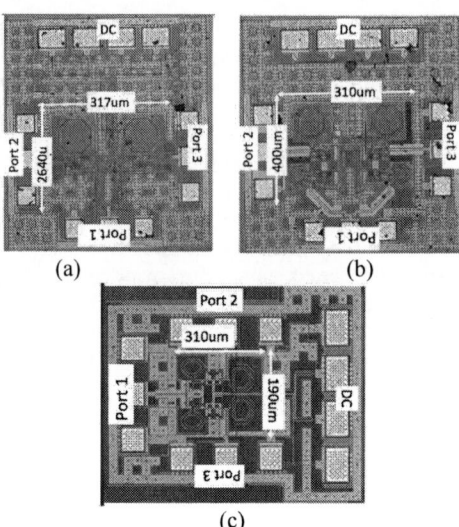

Fig. 11. (a) Series SPDT (b) Shunt SPDT (c) Negative gate control SPDT

REFERENCES

[1] B. Min and G. M. Rebeiz, "Ka-Band Low-Loss and High-Isolation Switch Design in 0.13-μm CMOS," in *IEEE Transactions on Microwave Theory and Techniques*, vol. 56, no. 6, pp. 1364–1371, June 2008, doi: 10.1109/TMTT.2008.921749.

[2] S. Londhe, N. Shmilovitz, S. Avner, N. Bar-Helmer, S. Jameson and E. Socher, "34–42GHz CMOS Transceiver Frontend for Versatile Arrays," *2020 15th European Microwave Integrated Circuits Conference (EuMIC)*, Utrecht, Netherlands, 2021, pp. 73–76, doi: 10.1109/EuMIC 48047.2021.00030.

[3] T. Despoisse, N. Deltimple, A. Ghiotto, M. D. Matos and P. Busson, "Low-Loss Ka-band SPDT Switch Design Methodology for 5G Applications in 65 nm CMOS SOI Technology," *2020 IEEE 20th Topical Meeting on Silicon Monolithic Integrated Circuits in RF Systems (SiRF)*, San Antonio, TX, USA, 2020, pp. 5–8, doi: 10.1109/ SIRF46766.2020.9040177.

[4] B. Yu et al., "Ultra-Wideband Low-Loss Switch Design in High-Resistivity Trap-Rich SOI With Enhanced Channel Mobility," in *IEEE Transactions on Microwave Theory and Techniques*, vol. 65, no. 10, pp. 3937–3949, Oct. 2017, doi: 10.1109/TMTT.2017.2696944.

[5] M. Lokhandwala, L. Gao and G. M. Rebeiz, "A High-Power 24–40-GHz Transmit-Receive Front End for Phased Arrays in 45-nm CMOS SOI," in *IEEE Transactions on Microwave Theory and Techniques*, vol. 68, no.11, pp. 4775–4786, Nov. 2020, doi: 10.1109/TMTT.2020. 2998011.

A Highly Linear SiGe BiCMOS Gilbert-Cell based Downconversion Mixer for 5G Applications

Mir Hassan Mahmud[#], Abdurrahman Burak[#], Can Çalışkan[#], Tahsin Alper Ozkan[#], Ali Bahadir Ozdol[#], Melik Yazici[#], Yasar Gurbuz[#]

[#]Faculty of Engineering and Natural Sciences, Sabanci University, Orhanli, Tuzla, 34956, Istanbul, Turkey
yasar@sabanciuniv.edu

Abstract — In this brief a high linearity, low loss, downconversion mixer requiring low LO drive power is presented for use in 5G beamforming applications. The emitter degenerated Gilbert-cell mixer with a four-way power combiner, baluns, and output filter is implemented using a 130-nm SiGe BiCMOS technology. The measured input 1-dB compression point of the fully integrated mixer is 6.4 dBm with a conversion gain of -12.9 dB. The mixer consumes 29.7 mW during small-signal operation and has a core area of 0.99 mm². The relatively low loss and high linearity performance of the core mixer makes it uniquely suitable for use in 5G receivers.

Keywords — mixer, 5G mobile communication, MIMO, BiCMOS integrated circuits, beam steering, baluns, linearity.

I. INTRODUCTION

Next generation (5G) communication systems aim to address the ever-increasing demand for tens of Gbps data-rates through the deployment of Enhanced Mobile Broadband (eMBB) services [1]. This will be facilitated by utilizing the wide bandwidth available at mmWave frequencies such as the band from 24.25-to-27.50 GHz. Recently, various beamforming transceivers have been presented to overcome the path loss of at mmWave frequencies [2], [3]. In these works, Si-based technologies are favored over III-V counterparts, because of their integration capability. Among Si-based technologies, SiGe heterojunction bipolar transistors (HBTs) also offer improved gain and noise performances, compared to CMOS [4].

The linearity of a downconversion mixer has a higher priority than conversion gain (CG), and noise figure (NF), because the mixer should handle the high RF power at the receiver output. Due to the inherent trade-off between CG and linearity, typically mixers with low CG are preferred. The NF of a mixer has lower importance in beamforming applications since the RF front-end typically has a large enough gain to greatly suppress the noise added by the mixer.

Various mixer topologies have been reported for 5G applications, in order to improve the linearity [5], [6], [7]. Active mixers can provide a CG and require lower local oscillator (LO) power than passive mixers. On the other hand, passive mixers typically exhibit superior linearity performance. Active or passive double-balanced mixers offer higher port-to-port isolation than their single-balanced and unbalanced counterparts. Hence, double-balance mixers are preferred for reduced feedthrough [8]. High LO drive power, which increases overall DC power consumption, and high

conversion loss (CL) make passive mixers (such as [5], [6]) less attractive for beamforming applications [2]. On the other hand active mixers require additional techniques to be applied in order to meet the high linearity requirements, which can increase their complexity and power consumption [7]. So, for 5G applications there still exists a need for a low complexity, highly linear, low loss mixer with low to moderate LO and dc power requirements.

In this work a Gilbert-cell based downconversion mixer exhibiting high linearity and low loss while utilizing low LO power is presented for 5G applications. The mixer utilizes an emitter degeneration linearization technique. It is implemented by using the IHP 130-nm SiGe BiCMOS technology featuring HBTs with f_T/f_{MAX} of 250/340 GHz. The paper is organized as follows. Section II describes the design considerations and methodology, while section III presents the measurements results. Finally, a conclusion is provided in section IV.

II. CIRCUIT DESIGN

The presented mixer is aimed at European 5G applications with an RF band from 24-to-28 GHz. To relax the image rejection requirements, an intermediate frequency (IF) of 3 GHz is selected. With the RF and IF set, the required LO is defined as 21-to-25 GHz.

The active, double-balanced Gilbert-cell mixer in Fig. 1 is selected to attain low loss and high isolation. It is then optimized for high linearity using inductive degeneration. The mixer includes a four-way Wilkinson Power Combiner (WPC) to interface a multi-channel beamforming front-end. As a fully integrated solution, three on-chip baluns are integrated to convert the single-ended RF and LO inputs to differential inputs and the differential output to a single-ended output. An on-chip low pass filter (LPF) at the IF output is used to attenuate mixing spurs and feedthrough.

A. Gilbert Cell Core

The linearity of a Gilbert-cell mixer is limited by the transconductance stage (g_m) transistor pair Q_1 and Q_2 in Fig. 1. Emitter degeneration improves the input third order intermodulation intercept point (IIP$_3$) of a transconductance stage at the cost of degraded CG and NF performance[9]. The intermodulation products (IM$_3$) depend on the degeneration impedance Z_E as shown in (1) [9].

$$|\text{IM}_3| \propto |1 + s\text{C}_{je}(\text{Z}_B(s) + \text{Z}_E(s))| \tag{1}$$

3–4 April 2022, London, UK

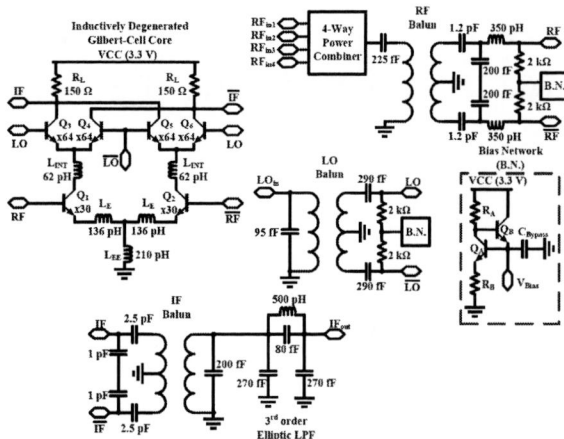

Fig. 1. Schematic of the emitter degenerated Gilbert-cell with 4-way combiner and baluns.

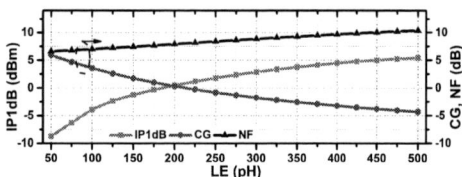

Fig. 2. Effect of inductor L_E on IP1dB, CG, and NF

In (1) C_{je} is the base-emitter junction capacitance of the transistor and Z_B includes all the impedances at the base of the transistor. From (1) it can be inferred that $sC_{je}Z_E$ is a negative real number for an inductor with impedance sL_E. So, the inductor L_E can be used in the emitter to decrease the magnitude of the IM_3 products, and hence improve the linearity of the mixer. Additionally, inductive degeneration introduces significantly lower noise when compared to resistive degeneration [8]. While degeneration improves the linearity, it negatively affects the CG of the mixer, as shown by the voltage CG of the degenerated Gilbert Cell in (2) [10].

$$A_v = -\frac{2}{\pi}\left(\frac{g_m}{1 + j\omega_{RF}L_E g_m}\right)R_L \qquad (2)$$

The effect of L_E on the input compression point (IP$_{1dB}$), CG, and NF is simulated as in Fig. 2 using ideal lumped components. To compensate for losses in the passive structures an L_E of 136 pH is selected to provide a positive CG for the mixer core while still achieving good IP$_{1dB}$ performance. The DC operating point of the g_m stage is set to have 1.6 V across it (maximum allowable voltage). The voltage drop over the switching quad (Q_3 to Q_6) is set equal to that across the load resistance, R_L to obtain the required output signal swing. R_L is selected to be 150 Ω in order to balance the trade-off between voltage drop and CG, as well as to enable IF matching with on-chip components. The base bias of Q_1/Q_2 is determined as 850 mV considering the power consumption, linearity, and CG trade-off. Large devices are selected for the devices Q_3 to Q_6 in order to improve NF [8] and provide

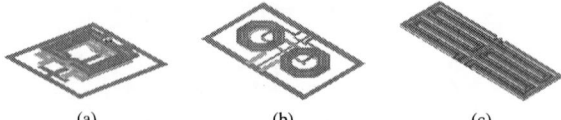

Fig. 3. 3D view of (a) RF Balun; (b) IF Balun; (c) WPC

an impedance suitable for output matching. L_E modifies the differential input impedance of the RF port, and aids in the input matching to 50 Ω [8]. Matching networks are used in the RF and LO ports to obtain the appropriate impedance for the baluns (50 Ω differential). The IF port is matched by using series and shunt capacitors and the IF balun. The shunt capacitor at the output also filters out high frequency mixing spurs. The integrated bias networks in Fig. 1 allow the mixer the operate from a single supply of 3.3 V. The tail current source is replaced by an inductor, L_{EE} to improve the headroom. Moreover, L_{EE} in conjunction with L_{INT} improves the second harmonic suppression [10]. L_{INT} also enhances CG flatness over the RF bandwidth [11].

A third order elliptic π-type low-pass filter is used at the IF output to obtain high attenuation in the stop band with reduced number of required filter sections.

B. Balun and Four-Way Wilkinson Power Combiner Design

A balun, shown in Fig. 3a, is used at the RF input along with a similar one at the LO input. A passive balun was selected over an active balun in order to improve the overall linearity of the mixer and reduce the dc power consumption. The broadside coupled transformer balun is selected since it is more compact than the Marchand balun. The two top metal layers in the technology were used for the primary and secondary coils of the transformer to minimize the insertion loss. The balun utilizes almost identical coils of approximately 410 pH inductance with a quality factor of around 18 each. The coils are offset about their centers to provide low amplitude and phase imbalance over the required bandwidth [12]. The balun occupies an area of about 0.014 mm^2. A similar balun is used for converting the single-ended LO signal into a differential signal.

The balun at the IF output, shown in Fig. 3b, requires larger coils of 4.8 nH at the differential port and 1.0 nH at the single-ended port due to the lower operating frequency. Hence, the coils were unwrapped in order to optimize the dimensions of the balun with respect to the overall floorplan. The IF balun occupies around 0.12 mm^2.

The outputs of the four-channel receivers are combined by using a distributed four-way WPC. For this purpose, two-way WPCs are cascaded to improve the bandwidth compared to lumped counterparts. Grounded coplanar transmission lines (T-lines) with multiple bends are used for the quarter-wave sections of the WPC as in Fig. 3c. Two-way WPCs are then cascaded using grounded coplanar T-lines to form the four-way WPC.

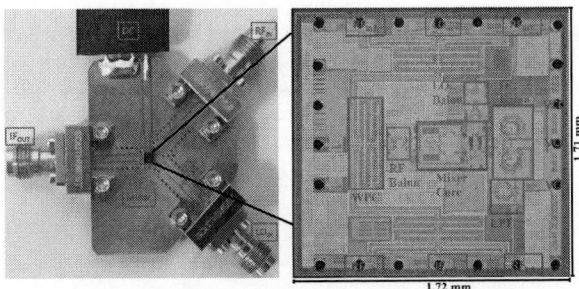

Fig. 4. Chip micrograph with the bonded PCB view

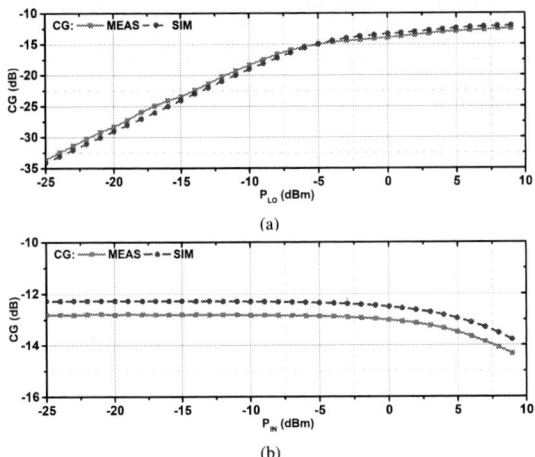

Fig. 5. Simulated and measured (a) CG vs. LO power (b) CG vs. RF input power at 26 GHz

III. MEASUREMENT RESULTS

The chip is flip-chip bonded as in Fig. 4 by utilizing in-house Au-stud bump capabilities to a Rogers RO5870 PCB. The measurement results include PCB T-line losses, and so for proper comparison with the simulation results the effects of these lines must be included. For a fair comparison, the effects of the PCB T-lines are added to the mixer's simulations. The measured mixer consumes 29.7 mW from a 3.3 V supply.

The CG of the mixer saturates with 6 dBm LO power as in Fig. 5a. It should be noted that the LO balun together with the interconnects and PCB traces have an insertion loss of around 4 dB. Hence the LO power at each of the switching quad is around -1 dBm. Fig. 5b demonstrates that the measured and simulated CG and IP_{1dB} of the mixer at 26 GHz differs by about 0.7 dB. The average CG over the 24 - 28 GHz bandwidth is -12.9 ± 1 dB from Fig. 6a. The NF of the mixer is measured as 23.3 dB (average) over the desired band. It is dominated by the loss of the passive components at the ports of the mixer.

The reflection coefficients at the RF and IF ports are better than -10 dB as in Fig. 7a and Fig. 7b respectively. The LO reflection coefficient shows a shift in the peak from 23 GHz to 26 GHz. Port-to-port isolations are all better than about 30 dB as in Fig. 8.

The measured IP_{1dB} is within 1 dB of the simulated values and its minimum value within the band is 5.4 dBm at 24 GHz

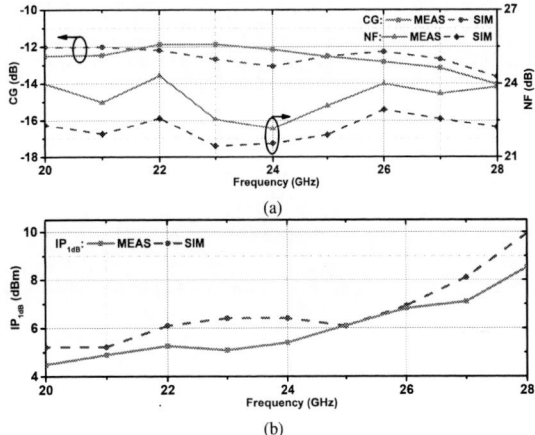

Fig. 6. Simulated and measured (a) CG, NF vs. frequency (b) IP_{1dB} vs. frequency

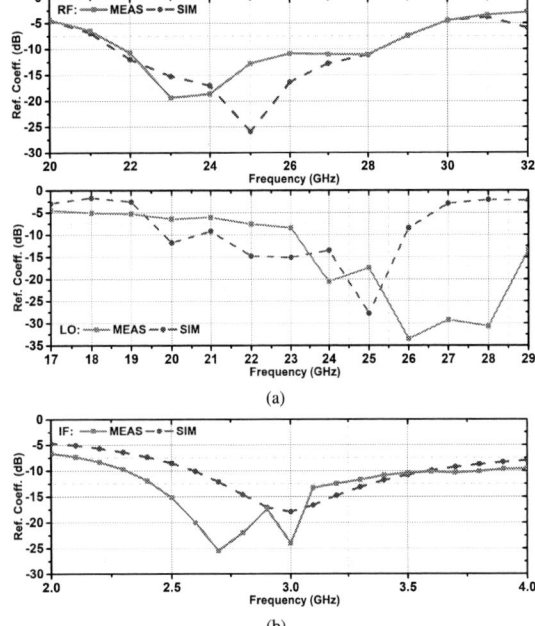

Fig. 7. Simulated and measured (a) RF, LO reflection coefficient (b) IF reflection coefficient

from Fig. 6b . The average IP_{1dB} is 6.4 dBm over the desired band. Measurements verify that the applied inductive emitter degeneration technique leads to improved linearity.

The measurements of a standalone two-way WPC in Fig. 9a and Fig. 9b are used to de-embed the performance of the core mixer. In Table 1 the simulation results of the core mixer without the four-way WPC shows that the core mixer exhibits a low CL of 2.05 dB and a high IP_{1dB} of -2.95 dBm. The NF of the core mixer is 11.2 dB, which shows that the overall NF of the mixer is dominated by the WPC.

The measurements are coherent with the simulations. The core mixer in the presented work achieves a low loss

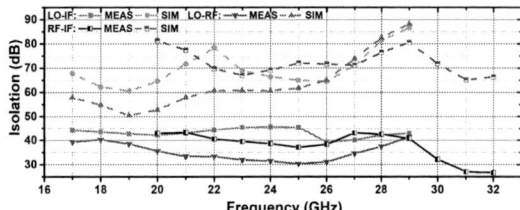

Fig. 8. Simulated and measured port-to-port isolations

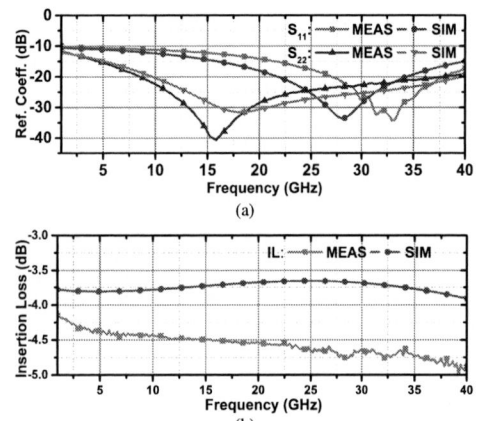

(a)

(b)

Fig. 9. Simulated and measured (a) Two-way WPC reflection coefficient and (b) Two-way WPC insertion loss

compared to the passive implementations of [5] and [6] while requiring lower or the same LO power. Compared to the active single-balanced mixer of [13], this work attains better linearity and isolation but consumes more DC power. It must be noted that this is a fully integrated implementation and hence the presented mixer can be used to meet the criteria set for linearity, CG, NF, and LO drive requirements for 5G beamforming applications as a compact solution.

IV. CONCLUSION

An emitter degenerated Gilbert cell is implemented to obtain high linearity by utilizing 130-nm SiGe BiCMOS technology. It includes integrated four-way Wilkinson power combiner, baluns, and LPF. Hence the presented work can be readily integrated with 5G RF beamforming front-ends. High linearity is achieved through inductive emitter degeneration of a Gilbert-Cell mixer. The fully integrated mixer has an IP_{1dB} of 6.4 dBm, a CL of 12.9 dB, and an NF of 23.3 dB. The measurements are for the flip-chip bonded mixer and thus the results represent the overall performance of the mixer ready for integration with a front-end. The presented mixer is suitable for 5G beamforming applications due to the high linearity and low loss performance it achieves while requiring a low LO drive power.

ACKNOWLEDGMENT

The authors would like to thank IHP Microelectronics for IC fabrication facilities and Ali Kasal for assistance with PCB fabrication and flip-chip bonding.

Table 1. Comparison of 5G Downconversion Mixers

	This Work	This Work (sim. w/o WPC)	[5]	[6]	[13]	[14]
Tech.	130-nm SiGe	130-nm SiGe	65-nm CMOS	180-nm CMOS	180-nm SiGe	180-nm SiGe
RF (GHz)	24-28	24-28	26-30	27-40	20-32	22-39
CG (dB)	-12.9	-2.05	-8	-15.2	3	-8
IP_{1dB} (dBm)	6.4	-2.95	9.4[1]	-5	-8.05[1]	-1.5[1]
P_{DC} (mW)	29.7	29.7	0	0	18	0
P_{LO} (dBm)	6	6	12[2]	6	5	3
SSB NF (dB)	23.3	11.2	–	–	10.5	14
Min. Iso. (dB)	30	51	35	60	15	20

[1]Estimated $IP_{1dB} = IIP_3 - 9.6$ [2]Assumed 10% of power consumption of LO driver.

REFERENCES

[1] "Technical Specification Group Services and System Aspects: Release 15 Description; Summary of Rel-15 Work Items (Release 15)," 3rd Generation Partnership Project, Tech. Rep., Oct. 2018.

[2] K. Kibaroglu, M. Sayginer, and G. M. Rebeiz, "A 28 GHz transceiver chip for 5G beamforming data links in SiGe BiCMOS," in *2017 IEEE Bipolar/BiCMOS Circuits and Technology Meeting (BCTM)*, Oct 2017, pp. 74–77.

[3] B. Sadhu, Y. Tousi, J. Hallin, S. Sahl, S. K. Reynolds, Ö. Renström, K. Sjögren, O. Haapalahti, N. Mazor, B. Bokinge, G. Weibull, H. Bengtsson, A. Carlinger, E. Westesson, J. Thillberg, L. Rexberg, M. Yeck, X. Gu, M. Ferriss, D. Liu, D. Friedman, and A. Valdes-Garcia, "A 28-GHz 32-Element TRX Phased-Array IC With Concurrent Dual-Polarized Operation and Orthogonal Phase and Gain Control for 5G Communications," *IEEE Journal of Solid-State Circuits*, vol. 52, no. 12, pp. 3373–3391, Dec 2017.

[4] J. D. Cressler, *Fabrication of SiGe HBT BiCMOS Technology*. Boca Raton, FL: CRC Press, 2008.

[5] H. A. Ameen, K. Abdelmonem, M. A. Elgamal, M. A. Mousa, O. Hamada, Y. Zakaria, and M. A. Y. Abdalla, "A 28 GHz four-channel phased-array transceiver in 65-nm CMOS technology for 5G applications," in *2017 29th International Conference on Microelectronics (ICM)*, Dec 2017, pp. 1–4.

[6] T. Tsai, I. Huang, J. Tsai, A. Alshehri, A. Almalki, A. Saved, and T. Huang, "A Ka-Band Sub-Harmonically Pumped Mixer Using Diode-Connected MOSFET for 5G MM-Wave Transceivers," in *2018 Asia-Pacific Microwave Conference (APMC)*, Nov 2018, pp. 488–490.

[7] C. Choi, J. H. Son, O. Lee, and I. Nam, "A +12-dBm OIP3 60-GHz RF Downconversion Mixer With an Output-Matching, Noise- and Distortion-Canceling Active Balun for 5G Applications," *IEEE Microwave and Wireless Components Letters*, vol. 27, no. 3, pp. 284–286, March 2017.

[8] Keng Leong Fong and R. G. Meyer, "Monolithic RF active mixer design," *IEEE Transactions on Circuits and Systems II: Analog and Digital Signal Processing*, vol. 46, no. 3, pp. 231–239, March 1999.

[9] K. L. Fong and R. G. Meyer, "High-frequency nonlinearity analysis of common-emitter and differential-pair transconductance stages," *IEEE Journal of Solid-State Circuits*, vol. 33, no. 4, pp. 548–555, April 1998.

[10] S. Voinigescu, *High-Frequency Integrated Circuits*. New York, NY: Cambridge University Press, 2013.

[11] D. Alldred, B. Cousins, and S. P. Voinigescu, "A 1.2V, 60-GHz radio receiver with on-chip transformers and inductors in 90-nm CMOS," in *2006 IEEE Compound Semiconductor Integrated Circuit Symposium*, Nov 2006, pp. 51–54.

[12] J. R. Long, "Monolithic transformers for silicon RF IC design," *IEEE Journal of Solid-State Circuits*, vol. 35, no. 9, pp. 1368–1382, Sep. 2000.

[13] M. El-Nozahi, E. Sánchez-Sinencio, and K. Entesari, "A 20–32-GHz Wideband Mixer With 12-GHz IF bandwidth in 0.18-μm SiGe Process," *IEEE Transactions on Microwave Theory and Techniques*, vol. 58, no. 11, pp. 2731–2740, Nov 2010.

[14] V. Issakov, H. Knapp, M. Wojnowski, A. Thiede, and W. Simbürger, "A 22–39 ghz passive mixer in sige:c bipolar technology," in *2010 IEEE MTT-S International Microwave Symposium*, 2010, pp. 1012–1015.

Proceedings of the 16th European Microwave Integrated Circuits Conference

37.2-to-42.0 GHz VCO with -93.4 dBc/Hz Phase Noise for FMCW Radar in 22 nm FDSOI

Laszlo Szilagyi[*#1], Songhui Li[#], Xin Xu[#], Paolo Valerio Testa[*#], Andres Seidel[#], Corrado Carta[#], Frank Ellinger[#$]

[*]GLOBALFOUNDRIES, Dresden, Germany
[#]Technische Universität Dresden, Germany
[$]CeTi, Centrum for Tactile Internet, Dresden, Germany
[1]laszlo.szilagyi@globalfoundries.com

Abstract — This paper presents a voltage-controlled oscillator (VCO) realized in a 22 nm FDSOI CMOS technology for frequency-modulated continuous-wave (FMCW) radars. Several different circuit techniques applied to reduce phase noise, increase the tuning range and output power are described and validated with experimental results. Furthermore, carefully designed, reversed varactors are implemented to achieve a constant tuning slope, thus minimum FMCW-chirp error. The tuning range is 12.1 %, from 37.2 to 42 GHz, whith a tuning slope error of < 17 % and < 1 % for a 0.6 GHz automotive radar chirp. Meanwhile a phase noise of -93.4 dBc/Hz at 1 MHz offset is measured at 42 GHz. With 45.3 mW DC power consumption, 2.5 dBm output power are measured on a differential load, yielding a DC-to-RF efficiency of 3.6 %. With 0.015 mm^2 the VCO is very compact. It performs remarkably well in every parameter when compared to the state of the art, especially the area, phase noise, dc-to-RF efficiency and provides a very linear control slope.

Keywords — Cross-coupled VCO, FDSOI, FD-SOI, FMCW radar, low phase-noise, oscillator, VCO.

I. INTRODUCTION

Milimeter-wave radar systems have established themselves among the surroundings sensor for autonomous and assisted driving: even the lowest level of assistance require radar system. Short-range systems can be used for proximity detection, partly to replace ultrasonic sensors, while medium-range radars measure velocity and distance to farther targets. The 77 GHz frequency modulated continuous-wave (FMCW) radar has already become a common choice for the automotive industry. At present, specialised high-frequency processes are used for the front-end of these systems [1]. However, the digital signal processor (DSP) represents a significant part of radar systems and, for a low power consumption, higher integration and lower area, it needs to be integrated in high-performance processes such as the 22 nm fully depleted silicon on insulator (FDSOI). By avoiding complex packaging, the integration of analog mm-wave frontends in the same chip as the DSP circuitry contributes significantly to cost reduction and – to a certain extent – power efficiency. For CMOS designs to compete with those in Si-Ge or III-V, new circuit techniques and a careful design optimization are necessary.

An important component of this civil automotive radar system is the chirp-generation block including a voltage controlled oscillator (VCO). In 77 GHz radars a chirp with 0.6 to 1.2 GHz bandwidth needs to be generated between 76 and 81 GHz. The generation of low phase-noise signal at mm-wave frequency is challenged by the losses in the resonant tank of the oscillator. An established solution consists in realizing a VCO at a suitable sub-harmonic of the target frequency and in combination with a multiplier [2]: the design presented in this paper relies on a frequency doubler. A VCO for such a sub-system is the key subject of this work, covering in detail techniques to reduce its phase noise, linearize VCO control slope (k_{VCO}) increase the tuning range and output power.

II. DESIGN OF THE OSCILLATOR

Standard automotive radar will have a frequency sweep in the 76-81 GHz frequency band [1], meaning 38-40.5 GHz in the present system. This results in a 6.0 % tuning range with sufficient margin to be added for process and temperature variation (PVT), as well as to avoid using the edges of the control characteristic, which has excessive deviation from k_{VCO}. Figure 1 shows a conventional varactor configuration, geometry optimized for the target frequency and a high quality factor (Q), that results in a 216 % capacitance tuning range. V_{tank} is assumed to be biased at 0.8 V. Simulations show that the previously mentioned tuning range cannot be met with this

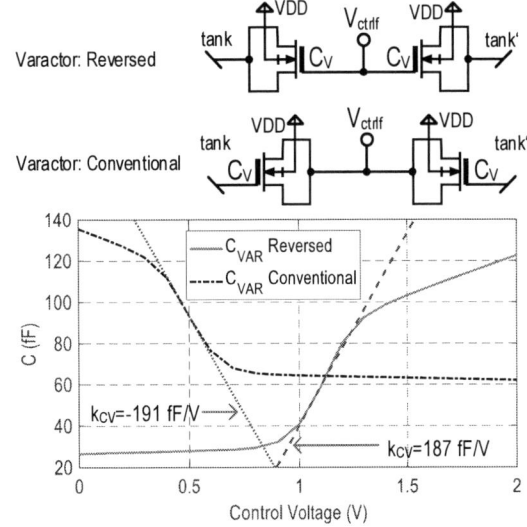

Fig. 1. Varactor: reversed and conventional. Simulated capacitance vs. voltage.

3–4 April 2022, London, UK

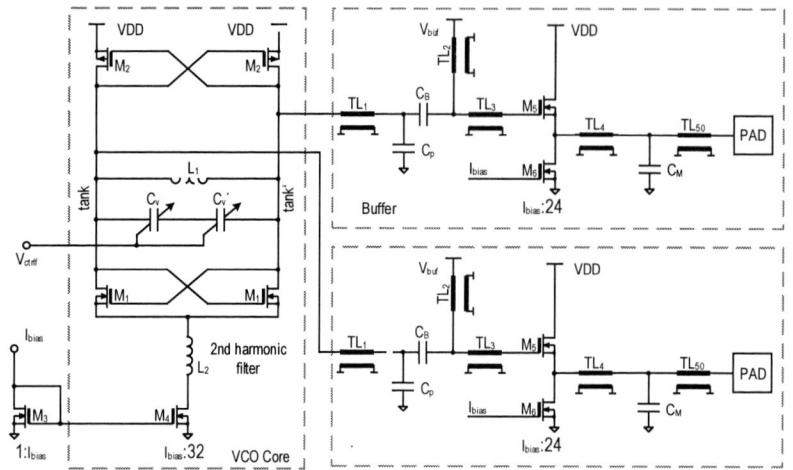

M₁	12 μm/18 nm
M₂	12 μm/18 nm
M₃	22 μm/250 nm
M₄	704 μm/250 nm
M₅	24 μm/18 nm
M₆	264 μm/250 nm
TL₁	20 μm
TL₂	307 μm
TL₃	15 μm
TL₄	290 μm
C_B	468 fF
C_p	135 fF
C_M	117 fF

C_v:
l/w: 100 nm/500 nm
gates×diffusions: 8×7

L₂:
L=166 pH, Q=17 @80 GHz

Fig. 2. Schematic of the VCO and buffers.

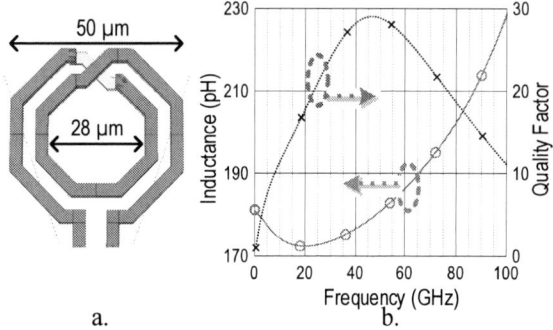

Fig. 3. (a) Inductor L_1 and (b) its simulated inductance and quality factor.

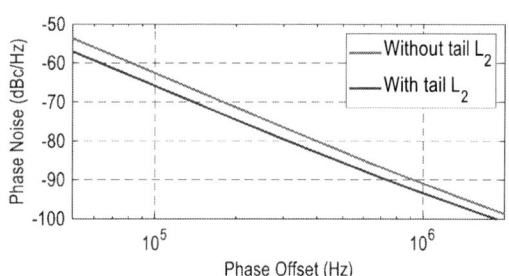

Fig. 4. Simulated phase noise without and with the tail inductor L_2.

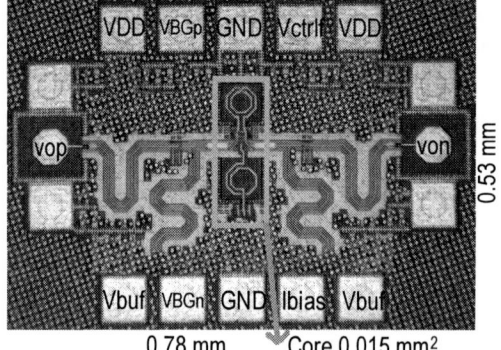

Fig. 5. Chip micrograph of the VCO.

varactor when also considering the existing fixed capacitances. A possible solution is to increase its size, however this has several disadvantages among which Q deterioration and excessive increase of k_{VCO}. This can cause significant phase noise as result of amplitude modulation to phase modulation (AM-PM) conversion [2]. Optimal k_{VCO} should be chosen $<0.1 \cdot f_0$ [3]. Given the structure of the FDSOI process, the varactor can be connected reversely, as depicted by Fig. 1, resulting a similar k_{VCO}, but a higher capacitive tuning of 480 %.

The schematic of the VCO is presented in Fig. 2. It consists

of the VCO core and its output buffers. Since the 22FDX technology offers pMOS of performance comparable to nMOS [4], both pMOS and nMOS cross-coupled pairs were used to compensate the losses of the tank. This technique improves the tuning range of the VCO, since the varactor voltage can be set between a positive and a negative potential, referred to the fixed bias point of *VDD*/2. Moreover, the output swing of the core is is twice what achievable with the nMOS-only VCO at the same current [3]. Figure 3(a) shows the geometry of the inductor in the resonant tank, L_1. It is implemented in the topmost copper of the metal stack; electromagnetic simulations - in Fig. 3(b) - return a peak Q of 28 at about 40 GHz. The current mirror, formed by M_3 and M_4, allows a precise control over biasing the core. Furthermore, given the high impedance at the output of the current source, the sensitivity of the core to supply noise is reduced. A further measure to improve phase noise is the use of inductor L_2 in order to filter out the second harmonic as described in [5]. This has an inductance of 166 pH and a Q of 17 at 80 GHz. Although in [5] the improvement in phase noise is 7 dB, in this design only 3.7 dB could be achieved as the simulated plots of Fig. 4 show. The reason

222

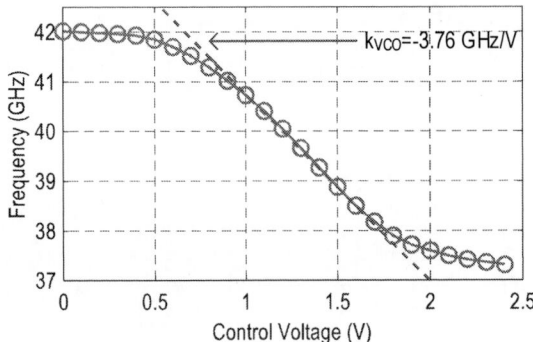

Fig. 6. Measured control slope, tuning range.

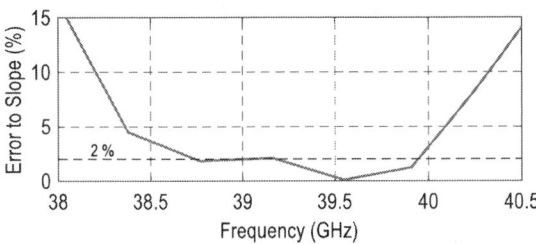

Fig. 7. Measured error to the linear control slope.

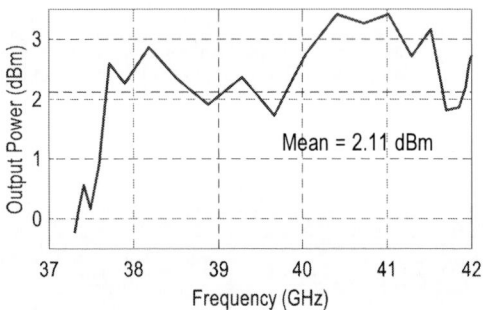

Fig. 8. Measured differential output power.

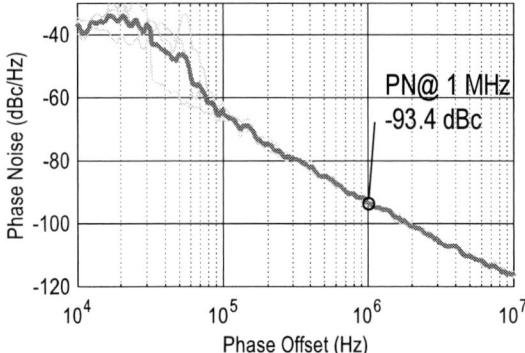

Fig. 9. Measured phase noise at 42 GHz.

is that the benefit of this technique depends on the Q of the filter inductor and at 80 GHz this is severely limited by increasing-with-frequency skin effect and substrate losses.

Common-drain stages are used as output buffers. These isolate the tank from the load and provide sufficient output power to drive the frequency doubler and divider [6] circuits. Ac coupling between the core and buffers was chosen. By doing so, the output power can be adjusted independently from the core bias. In this way, the buffer can be supplied even from a different VDD than the core. Transmission lines (TL) TL_1 to TL_3 together with C_B form the bias network to M_5, ac de-coupling and power matching between the core and buffers. These TL have a characteristic impedance (Z_0) of 62 Ω. The output of the VCO is matched to 50 Ω with TL_4 and C_M.

III. MEASUREMENT RESULTS

The VCO was implemented in the 22 nm FDSOI of GLOBALFOUNDRIES (22FDX®). As shown in Fig. 5, the pad-limited die area is 0.78×0.53 mm², where the core area is 0.015 mm². With 1.6 V VDD the complete circuit consumes 45.3 mW of which 17.4 mW the core. The tuning curve is presented in Fig. 6. The lower and upper limits are 37.2 and 42 GHz respectively, resulting a tuning range of 12.1 % around the center frequency of 39.6 GHz. Multiplied by 2, the output frequency of the VCO covers the requirement for 77 GHz radar. The control curve has a slope of -3.76 GHz/V. As it can be seen in Fig. 7 the error from this slope is less than 17 % over the frequency band of interest, therefore better than recommended value of <20%, critical for designing a stable phase locked loop (PLL) [3]. Furthermore, the error is less

than 1 % for 0.6 GHz sections of the slope that ensure linear chirps of the FMCW radar.

The output power level was measured with a thermal power sensor. Losses in the probes, cables and dc-blocks were de-embedded from the measurements. As it can be seen in Fig. 8, the differential output power does not vary significantly over frequency, averaging 2.11 dBm over the complete tuning range. An average dc-to-RF efficiency (η) of 3.6 %, including the buffers, is reached.

Phase-noise (PN) measurements were performed on the free-running VCO using a spectrum analyzer. Locking on the free-running oscillator was relaxed by interposing a /8 divder from [6] and de-embedding the phase noise improvement of the dividers $\Delta PN = 20 log_{10}(8)$ [3]. The red line in Fig. 9 shows the average of the measurements with $V_{ctrlf} = 0$ V, thus $f_{osc} = 42$ GHz. The phase noise value at 1 MHz offset of -93.4 dBc/Hz, falls into the $1/f^2$ region. PN measurements at a smaller offset than 100 kHz could not be measured reliably for the free-running VCO, due to the sensitivity of the analyzer.

Table 1 compares the proposed design with other recently published VCOs. This design has a well balanced performance in comparison to the state of the art. This design should be compared first and foremost to [15], [16] and [17], in the same technology. Within this subset, this VCO is among the lowest PN, lowest core consumption and nearly the highest figure of merit (FOM) implementations. The core area is the second smallest reported, after [17]; however, that design operates at a much higher frequency, which allows for more compact

Table 1. Comparison with State of the Art of CMOS VCO circuits

Ref.	Tech. (nm)	f_0 (GHz)	TR (%)	Output Power (dBm)	PN@1MHz (dBc/Hz)	Core Area (mm^2)	Core P. (mW)	Supply (V)	η (%)	FOM (dBc/Hz)	FOM$_T$ (dBc/Hz)	FOM$_A$ (dBc/Hz)
[7]	180	71	10.8	-15	-74.83	n/a	24.3	1.8	0.1	-158.1	-158.7	n/a
[8]	65	40	7.1	4.6	-96	n/a	26	0.9	4.1	-173.9	-170.9	n/a
[9]	65	77.4	10.2	n/a	-100	0.1	61.5	1.3	n/a	-179.7	-178.0	-189.7
[2]	65	77	14.5	6.2	-88	n/a	190	1	2.2	-162.9	-166.2	n/a
[10]	65	65	7.9	n/a	-105.9	0.15	61.2	n/a	n/a	-183.6	-181.6	-191.8
[11]	45	83.4	9.7	n/a	-91.2	0.06	5.7	1	n/a	-182.1	-181.8	-194.3
[12]	45	60.5	19	-20.3	-101.7	0.1	40	1	0.03	-181.3	-187.0	-191.3
[13]	28	64	11.1	n/a	-92.5	0.162	3.15	0.9	n/a	-183.6	-184.5	-191.5
[14]	28	60	16	-23	-87	0.02	11	0.9	0.02	-172.1	-176.2	-189.1
[15]	22	60	34	n/a	-84.3	0.015	7.5	0.7	n/a	-172.3	-181.7	n/a
[16]	22	60.6	17	15.1	-92.1	0.094	31.7	0.8/1.65	20.1	-170.4	-175.0	-180.7
[17]	22	64.3	25	-0.4	-90.0	0.0063	20	1	4.6	-173.3	-181.3	-195.3
This work	22	39.6	12.1	2.11	-93.4	0.015	17.4	1.6	3.6	-172.9	-174.6	-191.2

$$FOM = PN(\Delta f) - 20log\left(\frac{f_0}{\Delta f}\right) + 10log\left(\frac{P_{dc}}{1mW}\right),\ FOM_T = FOM - 20log\left(\frac{TR(\%)}{10}\right),\ FOM_A = FOM + 10log\left(A(mm^2)\right).$$

inductors.

IV. CONCLUSION

A VCO at half frequency for a 77 GHz FMCW civil automotive radar is implemented in 22FDX®. Cross-coupled nMOS and pMOS pairs are used for the core, resulting in a measured tuning range of 12.1 %, from 37.2 to 42 GHz. The linearity of the tuning curve is better than $< 17\%$ for the frequency band of interest and $< 1\%$ for a 0.6 GHz automotive radar chirp Two techniques were employed to reduce phase noise: the addition of a tail current and second-harmonic filtering. Those result in a measured phase noise of -93.4 dBc/Hz at 1 MHz offset and 42 GHz oscillation. This second method reduces phase noise by nearly 4 dB. Ac-coupled output buffers provide in average 2.11 dBm output power on a differential load, resulting in a dc-to-RF efficiency of 3.6 %. This design meets the requirements for automotive radar chirp generation; with one of the smallest area among all designs and lowest phase noise among 22 nm implementations reported, it offers a well balanced performance ranking high in all metrics.

ACKNOWLEDGMENT

This work was funded via subcontract from Globalfoundries Dresden Module One within the framework Important Project of Common European Interest (IPCEI), by the Federal Ministry for Economics and Energy (BMWi) and by the State of Saxony. We would like to acknowledge the CeTi project for fruitful discussions regarding millimetre-wave IC design and tactile systems.

REFERENCES

[1] T. Fujibayashi et al., "A 76- to 81-GHz multi-channel radar transceiver," IEEE Jour. of Solid-State Circ., vol. 52, no. 9, pp. 2226–2241, Sep. 2017.

[2] V. P. Trivedi, K. To, and W. M. Huang, "A 77GHz CMOS VCO with 11.3GHz tuning range, 6dBm output power, and competitive phase noise in 65nm bulk CMOS," in Proc. IEEE RFIC Symp, Jun. 2011, pp. 1–4.

[3] B. Razavi, Design of CMOS Phase-Locked Loops: From Circuit Level to Architecture Level. Cambridge University Press, 2020.

[4] P. V. Testa, L. Szilagyi, C. Carta, and F. Ellinger, "A complementary ring mixer driven by a single-ended LO in 22-nm FD-SOI CMOS for K and Ka-bands," IEEE Open Journal of Circuits and Systems, vol. 2, pp. 293–303, 2021.

[5] E. Hegazi, H. Sjoland, and A. A. Abidi, "A filtering technique to lower LC oscillator phase noise," IEEE Journal of Solid State Circuits, vol. 36, no. 12, pp. 1921–1930, 2001.

[6] L. Szilagyi et al., "A divide-by-4 and 8 Circuit for 77 GHz radar in 22 nm FD-SOI CMOS," in Proc. 4th Australian Microw. Symp. (AMS), 2020, pp. 1–2.

[7] Y. Chang, Y. Chiang, and C. Yang, "A V-band push-push VCO with wide tuning range using 0.18 μm CMOS process," IEEE Microwave and Wireless Comp. Lett., vol. 25, no. 2, pp. 115–117, Feb. 2015.

[8] V. P. Trivedi and K. To, "A novel mmwave CMOS VCO with an AC-coupled LC tank," in Proc. IEEE RFIC Symp, Jun. 2012, pp. 515–518.

[9] X. Yi, C. C. Boon, G. Feng, and Z. Liang, "An eight-phase in-phase injection-coupled VCO in 65-nm CMOS technology," IEEE Microwave and Wireless Comp. Lett., vol. 27, no. 3, pp. 299–301, Mar. 2017.

[10] J. Zhang et al., "An ultralow phase noise eight-core fundamental 62-to-67-GHz VCO in 65-nm CMOS," IEEE MWCL, vol. 29, no. 2, pp. 125–127, Feb. 2019.

[11] J. Xu, K. He, Y. Wang, R. Zhang, and H. Zhang, "A 79.187.2 GHz 5.7-mW VCO with complementary distributed resonant tank in 45-nm SOI CMOS," IEEE MWCL, vol. 29, no. 7, pp. 477–479, Jul. 2019.

[12] J. Rimmelspacher, R. Weigel, A. Hagelauer, and V. Issakov, "A quad-core 60 GHz push-push 45 nm SOI CMOS VCO with 101.7 dBc/Hz phase noise at 1 MHz offset, 19 % continuous FTR and 187 dBc/Hz FoMT," in Proc. ESSCIRC 2018, Sep. 2018, pp. 138–141.

[13] T. Forsberg, J. Wernehag, A. Nejdel, H. Sjland, and M. Trmnen, "Two mm-wave VCOs in 28-nm UTBB FD-SOI CMOS," IEEE Microw. and Wireless Comp. Lett., vol. 27, no. 5, pp. 509–511, May 2017.

[14] V. Issakov, F. Padovan, J. Rimmelspacher, R. Weige, and A. Geiselbrechtinger, "A 52-to-61 GHz push-push VCO in 28 nm CMOS," in Proc. 48th Eur. Microwave Conf. (EuMC), Sep. 2018, pp. 1009–1012.

[15] C. Zhang and M. Otto, "A wide range 60 GHz VCO using back-gate controlled varactor in 22 nm FDSOI technology," in Proc. IEEE SOI-3D-Subth. Microele. Tech. Unif. Conf. (S3S), Oct. 2017, pp. 1–3.

[16] M. Cui, C. Carta, and F. Ellinger, "Two 60-GHz 15-dBm output power VCOs in 22-nm FDSOI," IEEE Microwave and Wireless Components Letters, vol. 31, no. 2, pp. 184–187, 2021.

[17] Z. Tibenszky, D. Fritsche, C. Carta, and F. Ellinger, "An efficient wide tuning range 0.4 dBm 65 GHz NMOS VCO on 22 nm FD-SOI CMOS," in Proc. IEEE Int. Symp. Radio-Frequency Integration Technology (RFIT), 2020, pp. 37–39.

Proceedings of the 16th European Microwave Integrated Circuits Conference

A 100 GHz Class-F-Like InP-DHBT PA with 25.4% PAE

Amit Shrestha[1], Ralf Doerner[1], Hady Yacoub[1], Tom K. Johansen[2], Wolfgang Heinrich[1], Viktor Krozer[1],
Matthias Rudolph[1,3], Andreas Wentzel[1]

[1]Ferdinand-Braun-Institut gGmbH, Berlin, Germany

[2]Technical University of Denmark (DTU), Kgs. Lyngby, Denmark

[3]Ulrich L. Rohde Chair of RF and Microwave Techniques, Brandenburg University of Technology (BTU), Cottbus, Germany

Amit.Shrestha@fbh-berlin.de

Abstract — **This paper presents for the first time a class-F-like W-band power amplifier in InP DHBT technology. It reaches a power added efficiency (PAE) of 25.4% at 8.8 dBm output power (Pout) at 101 GHz. At the same frequency, the best PAE/Pout trade-off is achieved with 24%/10.9 dBm. The switch-mode PA applies a two-finger 0.5 μm InP-DHBT and delivers a saturated output power of 12.5 dBm and a maximum large-signal gain of 5 dB. To the authors' best knowledge this is the first class-F-like W-band PA to date.**

Keywords — **power amplifier, class-F PA, W-band PA, InP DHBT, switch-mode, power-added efficiency.**

I. INTRODUCTION

Recently, the W-band spectrum is becoming more and more important in commercial, military and 5G/6G applications for high capacity and low-cost wireless data transmission [1]. Higher output power, gain, bandwidth as well as proof of MIMO system operation and functionality with multi-element arrays at highest frequencies have the highest priority in such systems [2]. However, the efficiency (PAE) of the PAs is still a bottleneck. The PA is a major power dissipation contributor in the front-end, thus significantly affecting overall efficiency. To overcome this problem is the focus of this work. There are several approaches reported for efficiency enhancement in PA at mm-waves. Up to now, the switched-mode concept such as class-F and class-E [3]-[4] seem to be the most promising solution with regard to efficiency at W-band and beyond. However, it is a major challenge to fine-tune class-E/F networks for W-band, due to significant parasitics and degrading accuracy of the models at the harmonics involved. An important factor is also the proper modelling of the high current regime in the saturation region of the transistor.

In W-band, highest power-added efficiency (PAE) of more than 40% has been published [3] using a 130 nm SiGe technology. But when high power is required at the same time, e.g., for the final-stage in future high-speed communication front-ends, InP-technology is a more promising candidate. Consequently, the motivation of this paper is to evaluate the potential of switch-mode W-band PA designs in terms of efficiency using FBH 0.5 μm InP-DHBT process line for the application in next-generation (6G) communication systems.

The paper is organized as follows: Section II describes FBH InP technology used while Section III explains the basics of class-F PA design and EM-circuit co-simulated results. In section IV, the measurements of the realized MMIC are discussed. Section V concludes the paper.

II. FBH TRANSFERRED-SUBSTRATE InP-DHBT TECHNOLOGY

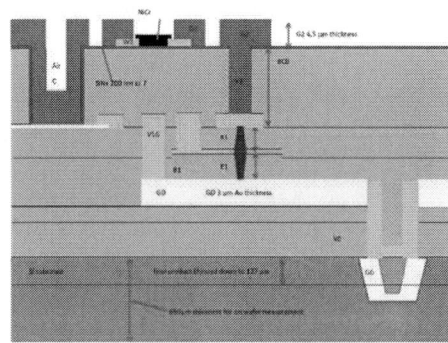

Fig. 1. Cross-section of FBH InP-DHBT transferred substrate technology [5].

The proposed class-F PA was fabricated using FBH InP-DHBT technology [5]. The MMIC process uses InP-DHBT structures transferred to a Si host wafer. This reduces the extrinsic base-collector capacitance C_{bcx} and hence improves the cut-off frequencies of operation. Technology layers (Fig. 1) include thin-film SiNx capacitors and NiCr resistors along with gold-filled via interconnects between 3 layers of routable interconnect metals. BCB is used as a dielectric interlayer. For example, the 2-finger InP-DHBTs with an emitter area of $2 \times 0.5 \times 6$ μm^2 achieve cutoff frequencies f_T/f_{max} of 425/475 GHz at 40 mA collector current and 2 V collector voltage. Breakdown voltage $V_{BD, CEO}$ exceeds 4 V. Heat dissipation across the Si substrate is facilitated by the addition of deep gold-filled vias (VX via), which provide the thermal connection to the top surface of the Si substrate.

III. CLASS F PA DESIGN

Class-F theory demands for short-circuit load termination at even-order harmonics and open load termination at odd harmonics which is to be provided by the output matching network. This leads to waveform engineering of collector voltage (V_C) and collector current (I_C) in such a way that when voltage is high the current is low or minimized and when the

3–4 April 2022, London, UK

voltage is low the current is high or maximized thus creating a non-overlapping V_C and I_C waveforms [6]. This lowers power dissipation due to the active transistor. Note that from the conventional bias point definition of PA, the class-F bias point may fall into deep class-AB or class-B. However, the special feature of class-F is that the transistor is operated as a switch. The introduction of odd harmonics clips the voltage waveform approaching a square wave. This enhances the amplitude of the fundamental signal up to a fixed voltage swing just before when output voltage starts clipping. Therefore, the efficiency of the PA is increased compared to the non-switched case.

A. Harmonically tuned output matching network (OMN)

In this 100 GHz class-F PA, the 2nd and 3rd harmonics fall at 200 GHz and 300 GHz respectively, controlling 4th or 5th harmonics is even more challenging. However, it is worth noting that the introduction of only 3rd harmonics to the fundamental signal increases the amplitude of the fundamental signal (V_{PK}) already by 12.5% compared to the usual fundamental output voltage (V_O) (i.e. $V_{PK} / V_O = 1.125$). On the other hand, adding all remaining odd harmonics would increase the fundamental output voltage by only 26 % (i.e. $V_{PK} / V_O = 1.26$). The problem is that tuning more harmonics needs extra matching components. This might be counterproductive because it increases losses, not only for the higher harmonics but also the fundamental. Thus, a trade-off is required. In our case, the OMN is tuned up to the 3rd harmonic, which is accomplished applying large-signal load-pull simulations.

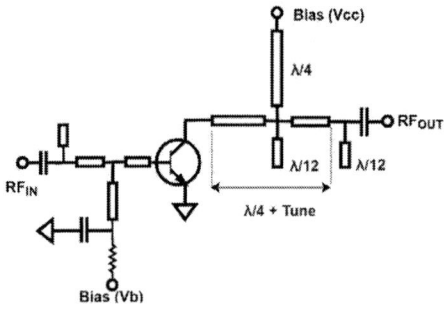

Fig. 2. Schematic of proposed class-F-like PA including IMN and OMN.

At first, a deep class-AB bias point is chosen and load termination is introduced. For the output matching network (OMN), distributed transmission lines are employed. This is because transmission lines lead to simple topologies, can be controlled better and have lower losses compared to lumped components. PAE is tuned carefully, varying the phase of the 2nd and 3rd harmonic reflection coefficient along the unity circle, keeping the fundamental signal close to 50 Ω load. The quarter-wave collector bias line is also incorporated in the OMN. EM simulation shows that no extra component for stability is required. The OMN includes shunt stubs for better harmonic control and tuning lines for compensating the device's parasitic passive components effects on tuning [7]. An important aspect is that series tuning and shunt tuning behave different and generally series tuning is preferred, when the impedance falls,

while shunt tuning is preferred when operating on the voltage. Shunt tuning requires shunt lines, which is not a huge impact, whereas series tuning is very difficult at these frequencies if the line impedance ratio is limited.

The ideal class-F does not always show the best results. Our observation on simulated results also found that the maximally flat square voltage waveform performed better compared to sharper square voltage waveform. Therefore, tuned harmonic terminations are needed rather than the OMN according to ideal class-F. The input matching network corresponding to the optimized load impedance is also designed. The resistor at the base avoids low-frequency instability of the PA. The input matching network (IMN) plays an important role, especially it maintains a certain gain of the PA. The complete class-F PA schematic is presented in Fig. 2.

B. Simulation results

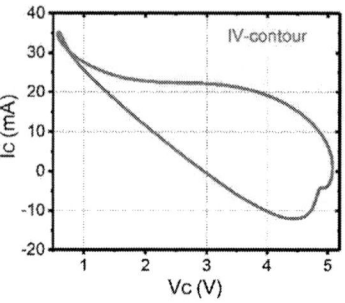

Fig. 3. Simulated current-voltage contour of proposed InP-DHBT class-F-like PA at the intrinsic collector port.

The PA schematic from Fig. 2 was transferred into a layout and EM-circuit co-simulation supports the final design. Fig. 3 shows simulated current and voltage at the intrinsic ports of the transistor which is accessible via the used FBH HBT model. Usual commercial (HBT) models do not provide this access to the intrinsic ports which is an important tool for waveform engineering. The corresponding I-V contour reaches up to 5.2 V, beyond the breakdown voltage ($V_{BD, CEO}$) of the transistor (4.5 V). In this PA, the base of the transistor in operation is not open, allowing a higher breakdown voltage limit than $V_{BD,CEO}$. The PA is operated in deep class-AB bias giving rise to high current dynamics. The collector current (I_C) is significantly low at high collector voltage (V_C). The lower V_C is limited by the knee voltage (turn on resistance) of the transistor. Ideally, zero turn-on resistance would allow the large-signal to be operated with a wider contour leading to the higher delivered output power and efficiency. Also, such a condition is ideal for pure switch-mode PA since collector current would be high with zero collector voltage. The technology used shows a knee voltage of ~ 0.5 V (see Fig. 4) limiting the amount of possible energy loss reduction with the switch-mode approach. Several on-going technology tests focus on reducing the knee voltage (e.g., using antimonide) as it is a major issue to be solved for highly-efficient switch-mode PA design. Also, from Fig. 4 we can derive that even at the intrinsic node in simulation it is not possible to avoid a strong overlap between I_C and V_C.

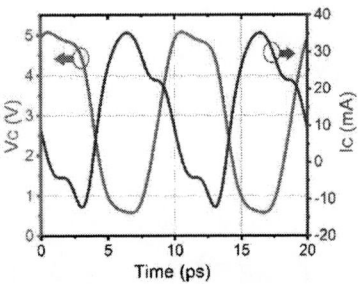

Fig. 4. Simulated current-voltage waveform at the intrinsic-collector plane.

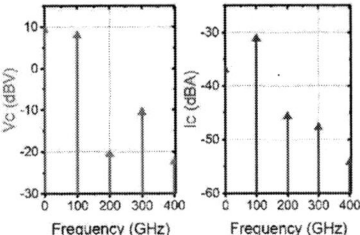

Fig. 5. Simulated spectrum of current-voltage at the intrinsic-collector port.

At the collector output, the 2nd harmonic of the voltage is suppressed while the 3rd harmonic is increased (see Fig. 5, left), creating a square wave-like shape as required for class-F. In terms of current the 2nd harmonic is suppressed too much while 3rd harmonic is not suppressed enough resulting in non-ideal class-F current spectrum characteristics (see Fig. 5, right).

After optimizing the OMN the PAE of 32%, P_{sat} of 14 dBm and gain of 5 dB is obtained at 101 GHz frequency (see Fig. 6). As could be expected, both voltage and current waveforms differ from ideal class-F curves. However, it is the best compromise that could be achieved with the matching elements.

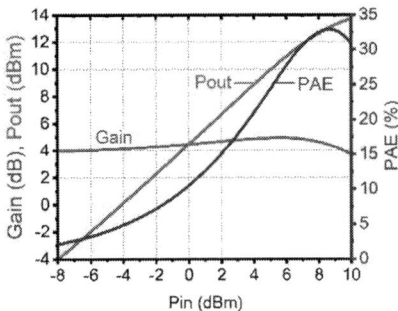

Fig. 6. Simulated PAE, output power (Pout), and gain versus input power (Pin) at 101 GHz (with $V_{CC} = 2.9$ V and $V_b = 0.87$V).

The layout is kept as simple as physically possible (see Fig. 7). Meander lines are not used both in RF and DC lines. Squeezing transmission lines into a small area showed an adverse effect on the PAE of the PA during EM simulations. Base and collector biasing are fed from opposite sides of the chip to restrict any coupling. The two shunt bypass capacitors are added to each bias line. Finally, Fig. 7 shows a photograph of the fabricated class-F PA chip exhibiting an area of 1.3×1.6 mm².

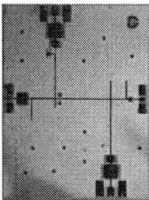

Fig. 7. Chip photo of fabricated InP class-F-like PA; area: 1.3 x 1.6 mm².

IV. MEASUREMENTS

A. S-parameter

First, the small-signal on-wafer measurements were carried out and plotted in Fig. 8. Here, to check for maximum gain, the PA is biased at class-A like high-gain setting with $V_b = 1.1$ V and $V_{CC} = 2.2$ V. The measurement set up has the capability to measure the scattering parameter up to 220 GHz in a single sweep. The measured reflection coefficients S11 and S22 are less than -10 dB and -7.5 dB, respectively, around 100 GHz. A small-signal gain of 6 dB is obtained at 101 GHz (S21, Fig. 8).

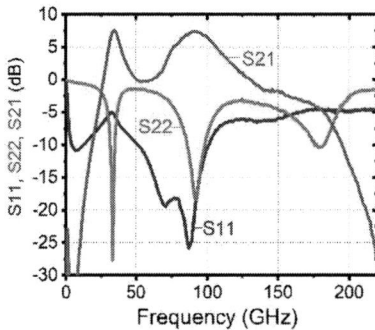

Fig. 8. Measured input and output reflection coefficient (S11, S22) and small-signal gain (S21) vs. frequency; bias: Vcc = 2.2 V, $V_b = 1.1$ V.

B. Power measurements

The large-signal measurement characteristics of the class-F-like PA are plotted in Fig. 9 for different bias points. According to best PAE results, which was the major focus of this work, 101 GHz has been chosen as frequency of operation.

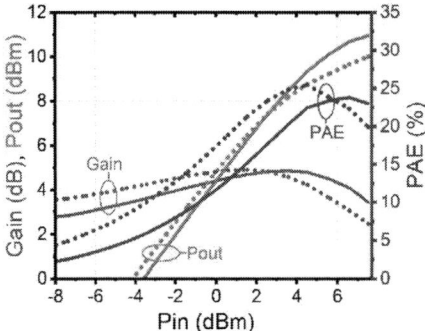

Fig. 9. Measured Pout, large-signal gain and PAE vs. Pin; bias1: Vcc = 2.9 V, Vb = 0.9 V (solid lines), bias2: Vcc = 2.85 V, Vb = 0.87 V (dotted lines).

Table 1. State-of-the-art solid state single-chip W-band PA comparisons

Reference	Technology	PA topology	Frequency (GHz)	Peak PAE (%)	Pout (dBm)	Psat (dBm)	Gain (dB)
[8]	250 nm InP HBT	2-stage, single-ended, CE	110	22.5	20	-	10
This work	**500 nm InP HBT**	**One stage, class-F-like tuned**	**101**	**25.4 / 24**	**8.8 / 10.9**	**11 / 12.5**	**5**
[4]	800 nm InP HBT	One stage, digital, overdriven sinus	95	31	12.1	14.4	5.6
[3]	130 nm SiGe HBT	Cascode, one-stage, class-E tuned	93	40.4	16.7	19.3	10.3
[10]	45 nm SOI nFET	Stacked (multidrive), single stage	91	14	19	19.2	6
[8]	250 nm InP HBT	2-stage, single-ended, CE	81	39.2	20	-	10.5

The realized class-F-like PA reaches a maximum PAE of 25.4% at 8.8 dBm output power for biasing condition $V_{CC} = 2.85$ V and $V_b = 0.87$ V (see Fig. 9, dotted curves). However, the best trade-off between Pout (10.9 dBm) and PAE (24%) has been reached for a slightly different bias point ($V_{CC} = 2.9$ V and $V_b = 0.9$ V). It is shown in the solid line curves (Fig. 9). The measured saturated output power is 12.5 dBm which is 1.5 dBm less compared to the simulated one. The proposed PA was also measured in a wider range from 92 to 103 GHz with peak PAE between 23 and 25%, Pout between 9 and 12 dBm and 4-5 dB gain. Note that several biasing conditions are tested to achieve the optimum performance of the PA. However, the optimum biasing conditions for measured and simulated results are similar. The drop in PAE compared to simulation stems from very sensitive non-overlapping condition of V_C and I_C waveforms. Also, intrinsic transistor gain is slightly lower than expected from simulations, which influences the performance as well. This is even more true if we compare with measured collector efficiencies of more than 40% indicating that the lack in gain is a major drawback. This will be subject of further investigations in future designs.

Finally, Table 1 presents performance comparison to state-of-the-art W-band single-chip PAs realized in different technologies. We can state that the proposed PA achieved state-of-the-art results for InP-based switch-mode configurations around 100 GHz. Above 100 GHz, it reaches even the highest PAE up to now. With regard to maximum PAE in W-band the results need to be improved which is to be expected when increasing the gain. In summary we can state that the performance obtained is promising and serves as a good starting point for further highly- efficient switch-mode PA designs which significantly improve energy efficiency in future high-speed communication systems.

V. CONCLUSION

In this paper, we present for the first time a class-F-like power amplifier in W-band at 101 GHz realized in FBH 0.5 μm InP-DHBT technology. The PA achieves the maximum peak PAE of 25.4%. Saturated maximum output power is 12.5 dBm and gain is 5 dB. This work demonstrates the potential of energy-efficient switch-mode PAs like class-F at W-band frequencies, despite the design challenges associated with the harmonic trapping matching networks. Ongoing work will focus on increased gain, minimized knee voltage and also on energy-efficiency enhancements at power back-off which is a major challenge in high-speed communication systems.

ACKNOWLEDGMENT

The authors would like to thank German Research Foundation (DFG) for funding under funding number WE 6288/2-1. Moreover, this project was supported by the German BMBF within the "Forschungsfabrik Mikroelektronik Deutschland (FMD)" framework under ref. 16FMD02.

REFERENCES

[1] J. Schellenberg, A. Tran, Lani Bui, A. Cuevas and E. Watkins, "37 W, 75–100 GHz GaN power amplifier," 2016 IEEE MTT-S International Microwave Symposium (IMS), San Francisco, CA, USA, 2016, pp. 1-4, doi: 10.1109/MWSYM.2016.7540195.

[2] H. Hamada et al., "300-GHz-Band 120-Gb/s Wireless Front-End Based on InP-HEMT PAs and Mixers," in IEEE Journal of Solid-State Circuits, vol 55, no. 9, pp. 2316-2335, Sept. 2020, doi: 10.1109/JSSC.2020.3005818.

[3] P. Song et al., "A Class-E Tuned W-Band SiGe Power Amplifier With 40.4% Power-Added Efficiency at 93 GHz," in IEEE Microwave and Wireless Components Letters, vol. 25, no. 10, pp. 663-665, Oct. 2015, doi: 10.1109/LMWC.2015.2463231.

[4] A. Wentzel et al., "An Efficient W-Band InP DHBT Digital Power Amplifier," International Journal of Microwave and Wireless Technologies, EuMW 2016 Special Issue, Vol. 9, Issue 6, pp. 1241 - 1249, July 2017.

[5] N. G. Weimann et al., "Transferred-Substrate InP/GaAsSb Heterojunction Bipolar Transistor Technology With $f_{max} \sim 0.53$ THz," in IEEE Transactions on Electron Devices, vol. 65, no. 9, pp. 3704-3710, Sept. 2018, doi: 10.1109/TED.2018.2854546.

[6] T. Sharma, R. Darraji, F. Ghannouchi and N. Dawar, "Generalized Continuous Class-F Harmonic Tuned Power Amplifiers," in IEEE Microwave and Wireless Components Letters, vol. 26, no. 3, pp. 213-215, March 2016, doi: 10.1109/LMWC.2016.2524989.

[7] Young Yun Woo, Youngoo Yang and Bumman Kim, "Analysis and experiments for high-efficiency class-F and inverse class-F power amplifiers," in IEEE Transactions on Microwave Theory and Techniques, vol. 54, no. 5, pp. 1969-1974, May 2006, doi: 10.1109/TMTT.2006.872805

[8] Z. Griffith, M. Urteaga, P. Rowell and R. Pierson, "71–95 GHz (23–40% PAE) and 96–120 GHz (19–22% PAE) high efficiency 100–130 mW power amplifiers in InP HBT," 2016 IEEE MTT-S International Microwave Symposium (IMS), San Francisco, CA, USA, 2016, pp. 1-4, doi: 10.1109/MWSYM.2016.7540041.

[9] P. Song, A. Ç. Ulusoy, R. L. Schmid, S. N. Zeinolabedinzadeh and J. D. Cressler, "W-band SiGe power amplifiers," 2014 IEEE Bipolar/BiCMOS Circuits and Technology Meeting (BCTM), Coronado, CA, USA, 2014, pp. 151-154, doi: 10.1109/BCTM.2014.6981303.

[10] A. Agah, J. A. Jayamon, P. M. Asbeck, L. E. Larson and J. F. Buckwalter, "Multi-Drive Stacked-FET Power Amplifiers at 90 GHz in 45 nm SOI CMOS," in IEEE Journal of Solid-State Circuits, vol. 49, no. 5, pp. 1148-1157, May 2014, doi: 10.1109/JSSC.2014.230829.

Proceedings of the 16th European Microwave Integrated Circuits Conference

A 117.5-130 GHz 22.1 dBm 11.5% PAE DAT Based Power Amplifier in InP 130 nm HBT Technology

Linsheng Zhang[*], Vinay Iyer, Jay Sheth, Linli Xie, Robert M. Weikle, Steven M. Bowers

Department of Electrical and Computer Engineering, University of Virginia, USA
[*]lz4wa@virginia.edu

Abstract — **An 8-way combined power amplifier is demonstrated in InP 130nm HBT technology. Each power unit adopts a stacked topology to overcome breakdown voltage limitations. A distributive active transformer is implemented as a low-loss power combiner, with directly matched impedance for each power unit by utilizing capacitive tuning at the output. The proposed amplifier achieves a maximum output power of 22.1 dBm with 11.5% PAE at 120 GHz, and obtains a 3 dB small signal bandwidth from 100 GHz to 140 GHz.**

Keywords — **power amplifiers, Millimeter wave integrated circuits, Heterojunction bipolar transistors.**

I. INTRODUCTION

High power (>20dBm) and high frequency (>100GHz) amplifiers are critical in applications ranging from radar exciters, drivers for diode mixers, to transmitters, and imaging devices. Higher power can extend detection range for imaging and radar applications, and a higher power-added efficiency (PAE) improves the overall efficiency of the entire system. It is also extremely attractive to be able to cover the entire frequency band desired with a single amplifier. Silicon-based processes typically have limited breakdown voltage when the dimension is scaled to achieve higher f_t and f_{max}. InP process have high power heterojunction bipolar transistors (HBTs) with higher f_t/f_{max} compared to silicon-based processes due to the wide bandgap of the material, and hence are an attractive solution for high-efficiency amplifiers beyond 100 GHz. As a trend of scaling down the size of the device for higher f_t and f_{max} needed to obtain THz amplifier design, the breakdown voltage becomes lower. Thus, efficient power combing techniques become more attractive for advanced technology nodes.

Several high-power amplifiers have been implemented beyond 100 GHz using the 250 nm InP process [1]-[5]. [1] showed greater than 30% PAE with a single-stage design, but with limited gain. To further increase gain and output power, multistage power combined topologies have been implemented to achieve as high as 24 dBm output power. A low loss Wilkinson power combiner was used in [2], [3]. Customized transmission line based combiners [5] were implemented to further reduce loss. However, the Psat was limited due to the 4:1 combining. 8:1 transmission line combiner was proposed to further increase the output power [4], however the 3 dB bandwidth became narrower with the addition of four more power amplifier (PA) cells.

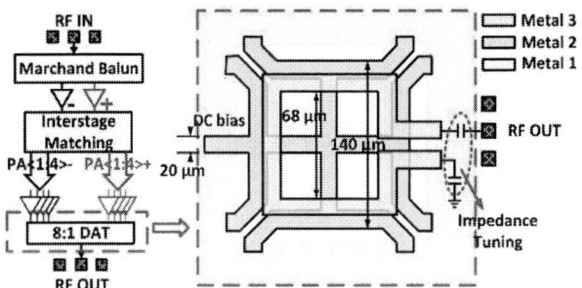

Fig. 1. Architecture of the two stage PA with 8-way DAT combiner.

To achieve a good balance between power and bandwidth, this paper proposes a high-power PA with an 8-way The Distributive Active Transformer (DAT) power combining technique in InP 130 nm HBT process. The DAT power combiner is directly matched to the load-pull impedance of the PA to reduce loss from additional matching networks.

II. CIRCUIT ARCHITECTURE

A. DAT design and implementation

Fig. 1 shows the block diagram of the power amplifier. The input is converted to differential through a balun and fed to the driver stage. Each driver drives four power cells.

To increase the output power beyond the individual device limit, an efficient power combiner is required. The DAT architecture has been implemented up to 71 GHz [6] and shows the benefit of low loss, compact core area, wide bandwidth, and ability to achieve impedance transformation [6], [7]. The main challenge of scaling the DAT to above 100 GHz are the imbalance between the different inputs because of the distributed effect of the transformer and conversion from differential output to single-ended output.

The inter-winding capacitance between primary and secondary contributes to the imbalance between different inputs [8]. To reduce the inter-winding capacitance, the 1:1 ratio stacked transformer structure was proposed using Metal 2 and 3. Ideally, such a transformer presents 1/8 of the load impedance to each of the 8 unit cells. Considering the inductance of the primary, the electromagnetic (EM) simulated impedance shows additional imaginary part with small real part as the red lines shown in Fig. 2 (a). However, the stacked PA

3–4 April 2022, London, UK

designed in the InP process has an optimal load-pull impedance close to 25+25j. To reach this target impedance, two series capacitors can be added at the output of the DAT. They act as parasitic capacitance and push the self-resonance frequency lower. As shown in Fig. 2 (a), the input impedance thus gets closer to the target impedance after adding two series capacitors at the outputs of the DAT.

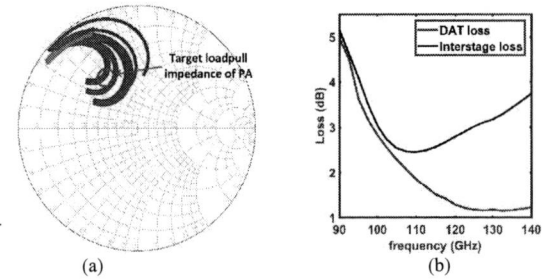

(a)   (b)

Fig. 2. (a) Simulated input impedance of DAT from 90 GHz to 140 GHz. Red lines are without series capacitance and blue lines are with two series capacitors at the DAT output. (b) Simulated DAT loss and interstage loss including the metal loss and impedance mismatch.

Three metal layers are available for interconnection in the InP 130 nm technology. Considering that the width of the microstrip line on Metal 3 is wider than that on Metal 2 for same impedance, Metal 2 was used for primary and Metal 3 was used for secondary. The width of primary and secondary are both 20 μm for low metal loss. They are offset by 16 μm horizontally to reduce capacitive coupling. The inner radius of the secondary is 34 μm to keep the DAT compact. The Metal 1 ground underneath the DAT is cut out to reduce the eddy current loss and parasitic capacitance to the ground.

The differential pairs create virtual ground at the center of each of the four sections of the primary at the fundamental frequency. It enables the VCC of each power unit to be supplied from these nodes directly without additional quarter-wave bias lines. The base bias of the Common Emitter (CE) stage was provided through a quarter-wave transmission line on Metal 2. The base bias of the Common Base (CB) stage was through the input matching with quarter-wave transmission line on Metal 3. The whole DAT was electromagnetically simulated using HFSS FEM simulator, and achieved a simulated loss of 1.16 dB at 135 GHz as shown in Fig. 2 (b). In addition, the simulated loss including metal loss and impedance mismatch is smaller than 1.3 dB from 120 GHz to 140 GHz.

B. Unit cell design

The driver stage and power stage have similar core designs to facilitate reuse. Fig. 2 shows the schematic of the unit cells. The size of the HBT is 4×6μm. Each unit cell is implemented using stacked topology with capacitance at the base of the CB stage to add swings at both collectors in phase, and maximize the voltage swing on the load. A series transmission line is used as the inter-stage matching between the CE and CB stage. The InP 130nm HBT technology has breakdown voltage BV_CEO ~ 3.5V, and a typical beta ~18. f_t/f_{max} is 521GHz/1.15THz [9].

Compared to the power stage unit, the driver unit has an additional input matching network to match to 50 Ω balun. An additional 5 Ω resistor is added at the emitter of the driver stage to suppress the common mode oscillation. It increases the dc power consumption but PAE drops less than 1% in simulation.

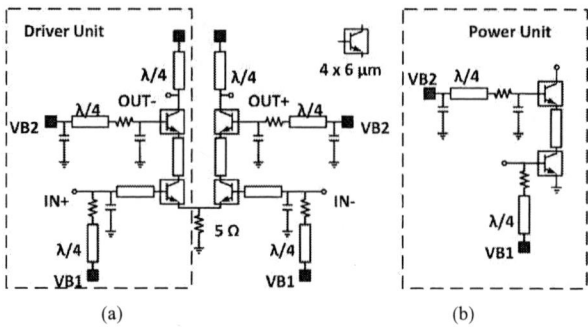

(a)   (b)

Fig. 3. Architecture of the unit cell in (a) driver stage and (b) power stage.

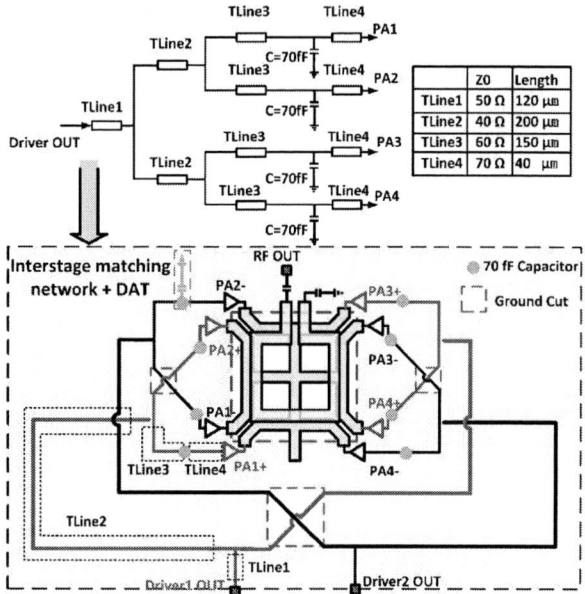

Fig. 4. Schematic and diagram of the inter-stage matching between the driver stage and power stage.

C. Inter-stage matching and power splitter design

The input power splitter is a transmission line based design. Fig. 4 shows the schematic details. The two differential signals are each power split into four and feed to the power stage. It directly matches the output of the driver to the input of the power stage, and also maintains the balance between the differential signals due to the symmetrical design. The impedance of the microstrip lines on Metal 2 and Metal 3 are difficult to match because of the different heights to the ground. So, at three crossovers of the microstrip lines, Metal 1 ground is cut out to improve the signal balance. As shown in Fig.2(b), the simulated loss is ~2.5dB to 4 dB including the metal loss and impedance mismatch.

D. Stability

Each unit PA was designed to be unconditionally stable at the nominal bias. Common mode oscillation was identified at the driver stage in simulation and suppressed by adding a 5 Ω emitter resistance. The whole design was EM simulated to ensure unconditional stability at nominal bias using the K factor and Δ, as shown in Fig. 5 (a).

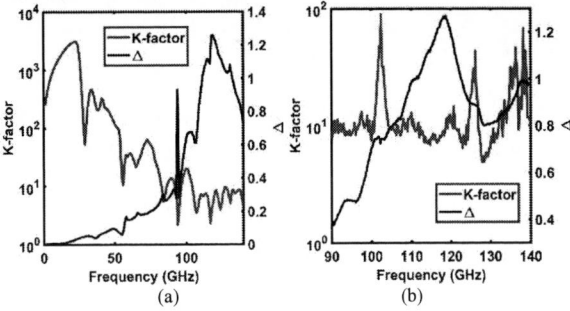

Fig. 5. (a) Simulated K-factor and Δ from 1 GHz to 140GHz and (b) Measured K-factor and Δ from 90 to 140 GHz (right).

Thermal runaway could be a potential issue for the HBTs. Ballast resistors were added on the base dc bias lines to mitigate thermal runaway and also help with stability.

III. MEASUREMENT RESULTS

A. Small signal measurement

The die micrograph is shown in Fig. 6. The whole chip is 1.86 mm x 1.56 mm including pads. To feed the DAT, a power splitter is needed which occupies ~5x more area than the DAT core itself, and thus dominates the chip size. The input balun is designed using Metal 3 as ground, and there are vias at the transition of the Metal 1 and Metal 3 to connect the ground. Driver unit and power unit each have two base biases and one VCC. Due to the 5 Ω emitter resistance, the base bias for the driver is 300 mV higher than that for the power stage, and VCC becomes 600mV higher for the same reason. The VCC of the power stage is 3.55 V with a 215 mA DC current. The driver stage is biased at $VCC_{Dr} = 4.1V$ with 63 mA DC current.

S parameters of the PA was measured by using the Keysight PNA-X Vector Network Analyzer along with the WR8 VDI extender. An additional waveguide attenuator was inserted at the input to maintain small signal power levels level. Formfactor WR-8 Infinity GSG probes were used for RF input and output. MPI Multi-Contact Probes were used to provide DC bias. An on-die TRL cal standard set was used for calibration. Due to equipment limitations, the S parameters have been measured from 90 GHz to 140 GHz. The simulated and measured S parameters are shown in Fig. 7. The peak S21 is

17.77 dB at 133.75 GHz. The 3 dB bandwidth is across 100 GHz to 140 GHz. The measured K-factor is plotted in the Fig. 5 (b) to confirm the in band stability.

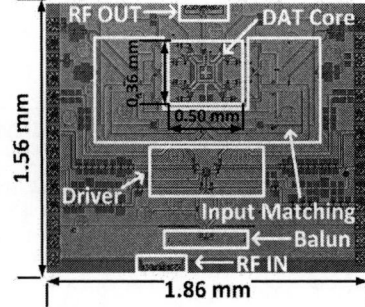

Fig. 6. Die micrograph (1.86 mm × 1.56 mm) of the proposed 8-way DAT combined PA.

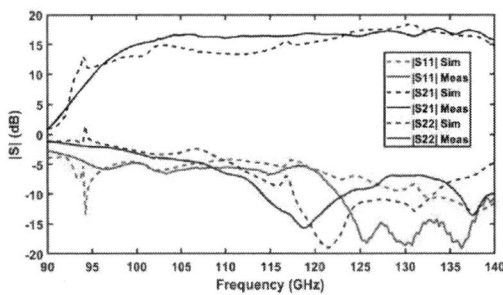

Fig. 7. Simulated and measured S parameter from 90 GHz to 140 GHz.

B. Large signal measurement

Large-signal measurements are performed with an input signal generated from the Keysight signal generator followed by a VDI WR8 tripler and a VDI WR 9 power amplifier to drive the PA. A VDI PM5 was used to measure the power at the output.

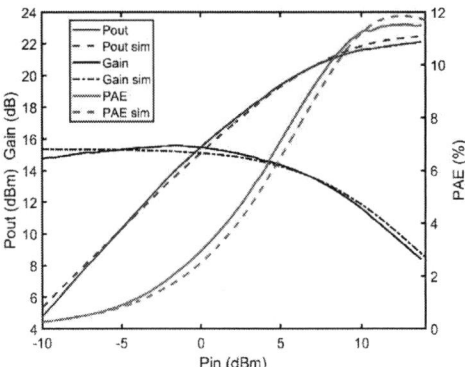

Fig. 8. Measured and Simulated output power, gain and PAE versus input power at 120 GHz.

231

Table 1. Table of comparison

Ref	Psat 1 dB bandwidth (GHz)	Technology	Topology	Peak Psat (dBm)	Peak PAE (%)	Peak S21 (dB)	Small signal bandwidth (GHz)	Size (mm²)
[1]	80-105	250nm InP	1 stage CE stack	20.1	34	6.5	45#	0.52
[2]	88-104	250nm InP	3 stage 2 way combined	20.8	24.7	29.5	16.5	1.15
[3]	55-135	250nm InP	4 stage 2 way combined	21.3	8.9	27.3	76.3	1.19
[3]	110-150	250nm InP	5 stage 4 way combined	24.0	7.0	29.4	35.6	1.89
[4]	125-151#	250nm InP	3 stage 8 way combined	23	17.8	22.8	20	1.34
[5]	125-150	250nm InP	3 stage 4 way combined	20.5	20.8	20.3	43	0.69
This work	117.5-130*	130nm InP	2 stage 8 way combined	22.1	11.5	17.8	40*	2.90

Graphically estimated. * The upper limit of the bandwidth is limited by equipment.

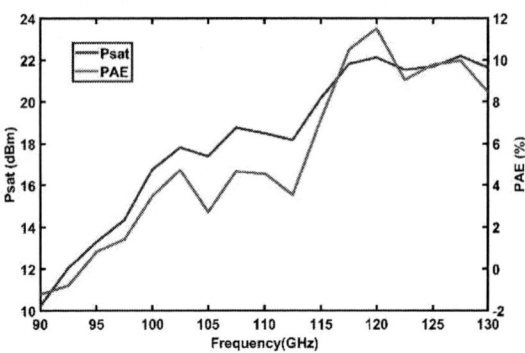

Fig. 9. Measured saturated power and PAE versus frequency.

Fig. 8 shows the large signal measurement vs the available input power at 120 GHz. The saturated output power is 22.1 dBm with 11.5% PAE and 8.6 dB large signal gain. The output 1 dB compression is 20.3 dBm with 8.7% PAE and 13.8 dB gain.

Large signal performance was measured from 90 GHz to 130 GHz, limited by the available equipment. As shown in Fig. 9, the PA has larger than 20 dBm saturated output power from 115 GHz to 130 GHz with PAE higher than 7%.

Table 1 shows state of the art in InP PA that achieve greater than 20 dBm output power around 120 GHz. Compared to [1],[2],[5], the proposed PA achieved higher output power. Compared to [3], the proposed PA has a higher PAE. The proposed PA showed double the small-signal bandwidth of [4] with comparable output power.

IV. CONCLUSION

A two-stage efficient high-power PA has been demonstrated at 120 GHz in a InP 130 nm HBT process. It combines the power from 8 power units with DAT topology. A driver stage provides additional gain. The DAT based PA adopts a stacked transistor structure with compact geometry to minimize the imbalance caused by interwinding capacitance. In addition, capacitive tuning has been implemented at the output of the DAT to achieve direct matching to the unit PA. It achieves 22.1 dBm peak output power with 11.5 % PAE. The 3 dB bandwidth is across 100 GHz to 140 GHz. It utilizes a power combing solution to achieve high output power, efficiency, and bandwidth.

ACKNOWLEDGMENT

The authors would like to thank Virginia Diodes, Inc (VDI) for assistance with measurements, the National Ground Intelligence Center (NGIC) for funding, and Dr. Herbert Zirath, Professor, Chalmers University for technical discussions. The authors would like to thank Teledyne Scientific & Imaging for the IC fabrication.

REFERENCES

[1] Z. Liu, T. Sharma and K. Sengupta, "80–110-GHz Broadband Linear PA With 33% Peak PAE and Comparison of Stacked Common Base and Common Emitter PA in InP," in *IEEE Microwave and Wireless Components Letters*, vol. 31, no. 6, pp. 756-759, June 2021.

[2] Z. Griffith et al. "A W-Band SSPA With 100–140-mW Pout , >20% PAE, and 26–30-dB S21 Gain Across 88–104 GHz," *IEEE Microwave and Wireless Components Letters*, vol. 30, no. 2, pp. 189-192, Feb. 2020.C

[3] Z. Griffith, M. Urteaga, and P. Rowell, "A 140-GHz 0.25-W PA and a 55-135 mW PA: High-gain, broadband power amplifier MMICs in 250-nm InP HBT," in *IEEE MTT-S Int. Microw. Symp. Dig.*, Boston, MA, USA, Jun. 2019.

[4] A. S. H. Ahmed, M. Seo, A. A. Farid, M. Urteaga, J. F. Buckwalter and M. J. W. Rodwell, "A 200mW D-band Power Amplifier with 17.8% PAE in 250-nm InP HBT Technology," *2020 15th EuMIC*, 2021, pp. 1-4.

[5] A. S. H. Ahmed, M. Seo, A. A. Farid, M. Urteaga, J. F. Buckwalter and M. J. W. Rodwell, "A 140GHz power amplifier with 20.5dBm output power and 20.8% PAE in 250-nm InP HBT technology," *2020 IMS*, 2020, pp. 492-495.

[6] Y. Jen, J. Tsai, T. Huang and H. Wang, "Design and Analysis of a 55–71-GHz Compact and Broadband Distributed Active Transformer Power Amplifier in 90-nm CMOS Process," in *IEEE Transactions on Microwave Theory and Techniques*, vol. 57, no. 7, pp. 1637-1646, July 2009.

[7] I. Aoki, S. D. Kee, D. B. Rutledge and A. Hajimiri, "Distributed active transformer-a new power-combining and impedance-transformation technique," in *IEEE Transactions on Microwave Theory and Techniques*, vol. 50, no. 1, pp. 316-331, Jan. 2002.

[8] H. T. Nguyen, D. Jung and H. Wang, "4.9 A 60GHz CMOS Power Amplifier with Cascaded Asymmetric Distributed-Active-Transformer Achieving Watt-Level Peak Output Power with 20.8% PAE and Supporting 2Gsym/s 64-QAM Modulation," *2019 IEEE ISSCC*, 2019, pp. 90-92.

[9] M. Urteaga, Z. Griffith, M. Seo, J. Hacker and M. J. W. Rodwell, "InP HBT Technologies for THz Integrated Circuits," in Proceedings of the IEEE, vol. 105, no. 6, pp. 1051-1067, June 2017.

Proceedings of the 16th European Microwave Integrated Circuits Conference

A 47-50 GHz 3 W MMIC Power Amplifier Using 100 nm Gallium Nitride Technology

Seifeddine Fakhfakh[#], Guillaume Callet[#], Estelle Byk [#], Laurent Favede[#],
Aleksandra Malko[*], Sandra Riedmueller[*], Pierre Denis [*], Hervé Blanck[*], Marc Camiade[#]

[#]United Monolithic Semiconductors SAS, France
[*]United Monolithic Semiconductors GmbH, Germany

Abstract — **This paper presents the design and test results of a 47-50 GHz power amplifier MMIC based on a 100 nm gate length in-house GaN technology under development. The Q-band HPA device was designed using a preliminary design kit inherited from GH15 equation set, validated with Load-Pull at 45 GHz and extrapolated to higher frequencies. The MMIC exhibits an output power of 4 W at 50 GHz and greater than 3W over the 47-50 GHz band with Power Added Efficiency (PAE) higher than 25% and a linear gain higher than 21 dB at V_{DD}=15 V. According to the author's knowledge, the presented GaN HPA results are the best among the ones published for frequencies up to 50 GHz. This paper contributes therefore to the state of the art.**

Keywords — **HEMTs, MMIC, power amplifiers, Gallium Nitride, Q-band.**

I. INTRODUCTION

In recent years, with increasing traffics on commercial wireless communication systems, many studies have started to evaluate the potentialities of using new set of frequencies such as Q/V-band [1]-[2] with the aim of to free up spectrum for user traffic. The most critical components for such systems is the high power MMIC amplifiers based on Gallium Arsenide (GaAs) and Gallium Nitride (GaN) which are the most popular technologies usually adopted for microwave systems. However, comparing them in terms of power capability and ruggedness, the GaN technology is nowadays the preferred. It allows to implement solid state power amplifiers (SSPAs) which can replace the traveling wave tube amplifier (TWTA) in several applications [3]. The GaN technology is also becoming the component of choice to provide the needed RF power at millimeter-wave frequencies.

Thus we investigated a GaN HEMT technology dedicated to microwave and millimeter wave circuits, using a 100 nm gate length and a 70 GHz cutoff frequency. In order to validate the non-linear model extracted and to assess the preliminary performances of this technology in development, a first investigation step was done on a broadband Ka-band HPA which demonstrated state-of-the-art performance. This MMIC exhibits an output power greater than 8 W over 26-35 GHz with a minimum of PAE of 26.5% and a linear gain higher than 23 dB at V_{DD}=15 V. Then a second investigation step was done in 47-50 GHz band as detailed in this paper.

This contribution focuses on a recent benchmark for the state of the art performance achieved by a single MMIC chip for a Q-band MMIC PA in this bandwidth.

This paper is organized as follows: in section II, the capability of the technology used is pictured by the performances of a single device of 4 fingers of 50 μm width. Section III presents the results of the HPA designed in order to demonstrate the capability of the technology up to Q-band. Finally, the conclusion and comparison with the state of the art are given in Section IV.

II. SINGLE DEVICE PERFORMANCES

In order to address applications in mm-wave frequency bands, the downsizing of the gate associated with the specific epitaxy definition are the main focus.

The HPA design presented hereafter relies on a 100nm gate length GaN based High Electron Mobility Transistor. The silicon carbide substrate (SiC) is thinned down to 70 μm.

In order to reach the targeted performances, the technology requires specific features such as the removal of the source field plate which limits the performances at higher frequencies [4]. But efforts are also made on passives with two different capacitor densities 175 and 350 pF/mm^2 compatible with capacitors over via-holes in order to improve compactness and reduce parasitic effects.

The [S]-parameter characterization of a 4x50 μm device up to 50 GHz is presented in Fig. 1 at quiescent biasing $V_{ds} = 15V$, $I_{ds} = 150mA/mm$.

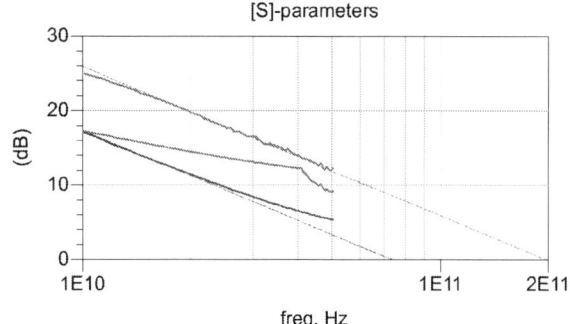

Fig. 1. [S]-parameters measurements of 4x50 μm device at $V_{ds} = 15V$, $I_{ds} = 150mA/mm$. In red Mason Gain, in green the maximum gain and in blue the current gain $|H21|^2$.

The device measured achieves a maximum gain of 10.3 dB at 45 GHz, a F_t=70 GHz and Fmax around 200 GHz.

3–4 April 2022, London, UK

This small signal campaign has been completed by a Load-pull characterization in order to extract the preliminary design kit and also to evaluate the performances of the single device used for the design of the demonstrator presented in Fig. 2.

The 4x50µm has been evaluated using 45 GHz loadpull bench. On the optimum load for PAE, the device delivers a peak PAE of 41% with an associated output power density of 2.3 W/mm and 7.2 dB power gain. The optimum output power load allows to reach a 3.1 W/mm power density.

These technology features in development allow to compete with industrial performance records [5]-[6] at nominal biasing Vds=15 V, output power higher than 3 W/mm and a peak PAE (above 40%) up to 50 GHz.

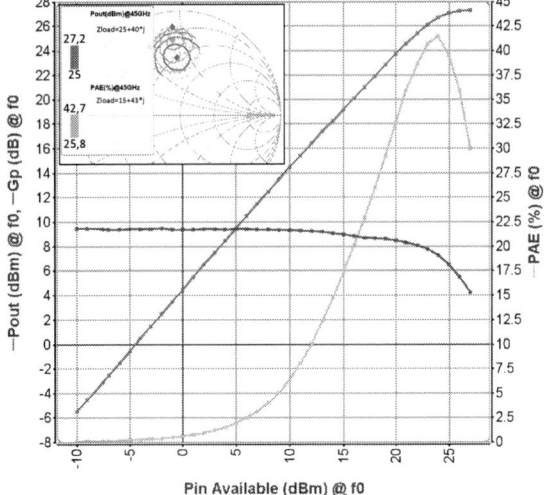

Fig. 2. Load-pull characterization performed at 45 GHz on 4x50 µm device at Vds=15 V, Ids=150 mA/mm. Top left figure gives the contours measured for optimum impedances; Power performances: Pout in red, PAE in green and power gain in blue obtained at Zopt-Pout.

III. Q-BAND POWER AMPLIFIER

A Q-band power amplifier has been designed to achieve following performances: 47-50 GHz bandwidth, greater than 20 dB small signal gain, greater than 3 W saturated output power and higher than 25% PAE. The design aims to evaluate the realizable short gate length transistor performance in an MMIC in the frequency range up to 50 GHz.

Firstly, the design starts with the choice of circuit topology. Taking into account the losses of the various matching networks, a 3 stage topology has been chosen to reach the 20 dB small signal gain. Considering the output combiner loss (about 0.9 dB), 8 transistors with a geometry of 4x50 µm are necessary to reach the requested output power.

In order to drive enough the second and final stages and to minimize low AM/PM distortion, a 1:2 staging ratio was adopted leading to a 2:4:8 unit cell layout. The MAG/MSG transition of the 4x50 µm transistor was evaluated in order to

check that it is not located in our band of interest as this can generate design difficulties and potential instabilities.

Before starting with the MMIC design, load-source pull simulations were carried out on single device at the fundamental and second harmonic in order to determine appropriate loads of the non-linear model using ADS simulation. Consequently, the output-load for the selected transistor has been chosen to maximize the trade-off between PAE and output power criteria in deep AB class of operation. All passive networks (output combiner and interstage splitters) are designed to provide this optimum load to the unit transistor cell.

The matching circuits were designed in the following order: output, interstage, and input with high pass and low pass elements to provide a broad band of operation. They are tuned for the best overall circuit performance. This design is fully matched to 50Ω at the input and output which is the most important matching network in determining the output power and efficiency since it includes the power combining structure.

The tuning capacitors were chosen in order to avoid their resonance inside the frequency range of interest. This is helpful to make the circuit less sensitive.

The Dielectric High Density (DHD) capacitors (350 pF/mm^2) available in the new process with three levels of metallization have been used in this presented chip version in order to validate its functionality. It should be noted that there was no difference on chip size in comparison with the version based on the use of classic capacitors (due to the tile constraints). Therefore, the chip area of $4.52mm \times 3.3mm$ ($14.9mm^2$) including the dicing street can be easily reduced. The final layout and related topology are shown on Fig. 3. The selected bias point for the design is $V_{DD} = 15V$ for the drain voltage and $420mA$ of the quiescent current for the PA corresponding to $150mA/mm$ on the transistor.

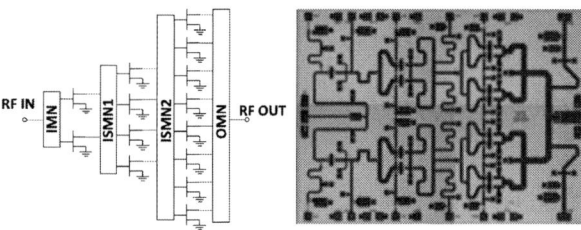

Fig. 3. Schematic (on the left) and photograph (on the right) of the 3W Q-band Power Amplifier MMIC ($4.52mm \times 3.3mm$).

As we can see on the right of Fig. 3, the layout of the chip is completely symmetrical with ground pad placed next to each DC pad to avoid low-frequency oscillations. Except V_{D1} and V_{D2} the biases supplies can be provided from the north, south or both sides for flexibility of insertion into different power modules.

Keysight's ADS and Momentum software were used for the electric design and Electro-Magnetic design respectively. The EM blocks for passive networks are simulated from 0 to 120 GHz (second harmonic included) which is a compromise between number of harmonics and models validity, knowing that active devices have furthermore a low gain beyond 120 GHz. For reasons of accuracy, the via hole and transistors access in microwave mode conditions were integrated with the appropriate mesh and excitation ports. A special care is taken regarding the low impedance of gate access.

At the end of the design, stability analysis was checked and ensured for the full HPA as well as for each transistor cell (internal loops) for multiple different process variants to make sure there was sufficient margin. To avoid odd-mode and parametric oscillations, specific linear and non-linear analysis were done for potentially adding new resistors in the network. Thereafter, the circuit sensitivity (active and passive elements) was checked using Monte-Carlo simulations as well as multi-biasing simulations.

IV. MEASURED PERFORMANCES

This section details the results of the characterization obtained by means of On-Wafer (OW) measurements of the Q-band HPA. It consists in low and large-signal characterizations under pulsed condition of $25\mu s$ duration and 10% duty cycle to limit the thermal dissipation.

A. Small-signal response

In order to show the sensitivity of the designed HPA, 35 chips have been measured on [S]-parameter from 0.5 to 67 GHz with a step of 0.25 GHz. The nominal bias point is fixed at (15V; 150mA/mm) consistent with the one used for design and a carrier plate backside at room temperature. These measurements are shown on Fig. 4 with no issue observed inside or outside the useful Q-band. The isolation (S12 parameter) is higher than 40dB in the frequency band of interest as we can see in the top left of Fig. 4.

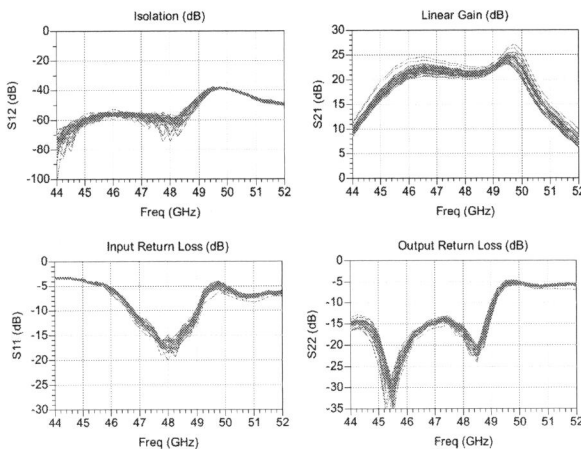

Fig. 4. Measured small-signal performance of the designed Q-band HPA MMIC at (15V/420mA) at room temperature. The measurement pulse condition is $25\mu s - 10\%$.

This characterization result shows that the spread between measured chips is quite low with an average measured linear gain (at the top right of Fig. 4) exceeding 21 dB over the entire broadband 47-50 GHz. Input and output return losses are displayed on the bottom left and the bottom right of the Fig. 4 respectively. A worst case of almost -5dB is observed on input and output return loss.

Next results deal with the OW power measurements. The nominal biasing point is similar to simulation conditions and S-parameters test previously performed.

B. Large-signal results

Following wide band [S]-parameters measurements, power test has been carried out on-wafer from 46 to 51 GHz, the highest frequency covered by the designed HPA.

Fig. 5 depicts the comparison between measured (red) and simulated (blue) large signal performances (PAE, Gain and output power) of the designed Q-band MMIC versus input power. This measure was carried out at (15 V; 150 mA/mm) and a single drive frequency of 50 GHz which represents the higher frequency covered by the amplifier.

It has to be noted that the measurement has been realized in Pulsed condition whereas the model is extracted in continuous waveforms (CW) conditions. This difference explains the shift observed between measurement and simulation.

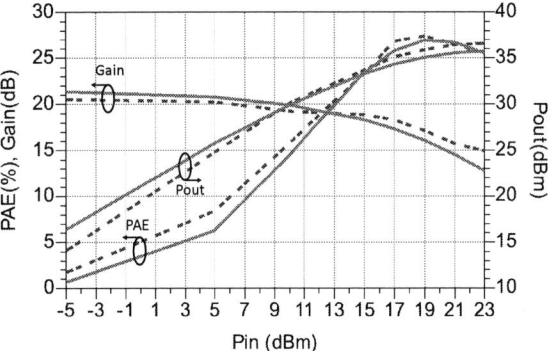

Fig. 5. Comparison between measured (red) and simulated (blue) circuit performance vs. input power at 50 GHz for (15 V/42 0mA). The measurement pulse condition is $25\mu s - 10\%$.

From Fig. 5 we can conclude that the input power of 19 dBm appears as the best trade-off for input power and PAE performances at (15 V; 150 mA/mm) biasing point for which the measured PAE peaks at 27% with an output power of 35 dBm at a gain compression of 5 dB.

The comparison between five measured samples (red) and simulated (blue) large signal performances is presented on Fig. 6 as a function of the frequency at 19 dBm of input power.

The PA exhibits between 3 and 4 W of output power at the nominal biasing point (15 V; 150 mA/mm) for the frequency bandwidth 47-50 GHz with a PAE greater than 25% (28% at 47 GHz and 27% at 50 GHz).

The performance of the designed HPA was compared with other published, available products and demonstraors HPAs in Table 1. This benchmarks presents two or more stages HPAs reported under both pulsed and CW operating conditions. Until now, the only published HPA operating in the same frequency

235

range is the dual band of G.Lv [7] which works at 29.5 and 47 GHz. Thus, this work is a compelling design for Q-band applications requiring 47-50 GHz frequency bandwidth, with state-of-the-art gain, power added efficiency and output power.

Even if other HPAs are realized to operate in Q-band, this MMIC is precursor comparing to them since it operates at higher frequencies with state-of-the-art PAE.

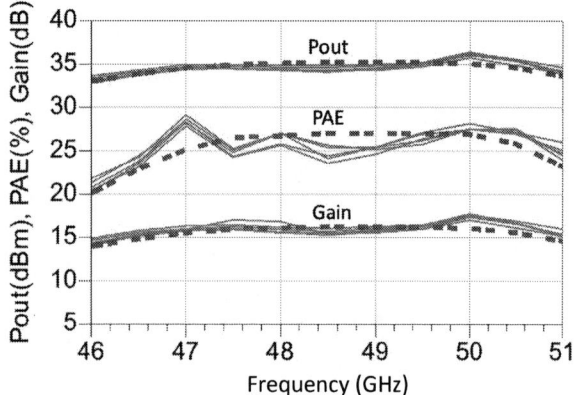

Fig. 6. Comparison between measured (red) and simulated (blue) circuit performance vs. frequency at (15 V/420 mA) and room temperature. The measurement pulse condition is $25\mu s - 10\%$.

Table 1. Q-BAND MMIC PA BENCHMARKS

Ref (#)	Freq (GHz)	Gain (dB)	Pout (dBm)	PAE (%)	VDS (V)	Size (mm²)
[7]	47	15.5	22.7	40	4	0.98
[8]	41-46	17	33	NA	6	15.1
[9]	47.2-50.2	≥ 16	37	≥ 20	12	10.36
[10]	37.5-42.5	22	≥ 36.5	30	8	12.25
[11]	37.5-42.5	NA	33	≥ 14	15	NA
[12]	47.5-51.4	≥ 17	35.5	11	28	3.92
This work	47-50	≥ 21	≥ 34.5	≥ 25	15	14.85

V. CONCLUSION

The design of a Q-band power amplifier MMIC using a 100 nm GaN technology non-linear model is presented. This model is scalable in respect of number of gate fingers and gate width, shows good agreement with the experimental data up to 50 GHz. Performances of the HPA are also detailed and compared to other published HPA MMICs. Under pulsed operation, on-wafer measurements demonstrate a small signal gain exceeding 21 dB over a 47-50 GHz frequency band. In the same frequency range, the HPA exhibits an output power in excess of 3 W at 5 dB gain compression, with a peak of 4 W at 50 GHz and a PAE between 25 and 28%. To the author's best knowledge, this is the first demonstration of 3 W Q-band power amplifier MMICs up to 50 GHz, they contribute therefore to the state of the art.

REFERENCES

[1] M. Aloisio, P. Angeletti, F. Coromina, and R. De Gaudenzi, "Technological challenges of future broadband telecommunication

satellites in q/v-band," in *2012 IEEE International Conference on Wireless Information Technology and Systems (ICWITS)*, 2012, pp. 1–4.

[2] R. Emrick, P. Cruz, N. B. Carvalho, S. Gao, R. Quay, and P. Waltereit, "The sky's the limit: Key technology and market trends in satellite communications," *IEEE Microwave Magazine*, vol. 15, no. 2, pp. 65–78, 2014.

[3] R. Giofrè, P. Colantonio, L. Gonzalez, L. Cabria, and F. De Arriba, "A 300w complete gan solid state power amplifier for positioning system satellite payloads," in *2016 IEEE MTT-S International Microwave Symposium (IMS)*, 2016, pp. 1–3.

[4] J. Dufraisse, G. Callet, O. Jardel, E. Chartier, N. Sarazin, S. Piotrowicz, M. Di Forte Poisson, P. Bouysse, R. Quéré, and S. L. Delage, "Characterizations of inaln/aln/gan transistors for s-band applications," in *2011 6th European Microwave Integrated Circuit Conference*, 2011, pp. 140–143.

[5] Y. Cao, V. Kumar, S. Chen, Y. Cui, S. Yoon, E. Beam, A. Xie, J. Jimenez, A. Ketterson, C. Lee, D. Linkhart, and A. Geiler, "Qorvo's emerging gan technologies for mmwave applications," in *2020 IEEE/MTT-S International Microwave Symposium (IMS)*, 2020, pp. 570–572.

[6] J. Moron, R. Leblanc, F. Lecourt, and P. Frijlink, "12w, 30ghz power amplifier mmic using a commercially available gan/si process," in *2018 IEEE/MTT-S International Microwave Symposium - IMS*, 2018, pp. 1457–1460.

[7] G. Lv, W. Chen, X. Chen, and Z. Feng, "An energy-efficient ka / q dual-band power amplifier mmic in 0.1- μ m gaas process," *IEEE Microwave and Wireless Components Letters*, vol. 28, no. 6, pp. 530–532, 2018.

[8] *Qorvo*, https://www.qorvo.com/products/p/TGA4046.

[9] *Ommic*, https://www.microwavejournal.com/ext/resources/pdfdownloads/DSSF2019/OMMIC-2019-DSSF.pdf.

[10] F. Costanzo, R. Giofrè, A. Salvucci, G. Polli, and E. Limiti, "A 4w 37.5-42.5 ghz power amplifier mmic in gan on si technology," in *2018 14th Conference on Ph.D. Research in Microelectronics and Electronics (PRIME)*, 2018, pp. 137–140.

[11] F. Costanzo, R. Giofrè, G. Polli, A. Salvucci, and E. Limiti, "A q-band mmic power amplifier in gan on si technology for space applications," in *2018 International Workshop on Integrated Nonlinear Microwave and Millimetre-wave Circuits (INMMIC)*, 2018, pp. 1–3.

[12] *Northrop*, https://www.northropgrumman.com/wpcontent/uploads/MicroelectronicsAPN319.

Proceedings of the 16th European Microwave Integrated Circuits Conference

A D-Band Power Amplifier with 12dBm P1dB, 10% Power Added Efficiency in InP-DHBT Technology

M. Hossain[1], T. Shivan[1], R. Doerner[1], S. Seifert[1], H. Yacoub[1], T.K. Johansen[2], W. Heinrich[1], V. Krozer[1]

[1]Ferdinand-Braun-Institut, Leibniz-Institut für Höchstfrequenztechnik (FBH), Germany
[2]Technical University of Denmark (DTU), Denmark
maruf.hossain@fbh-berlin.de

Abstract — **This paper presents a D-band power amplifier (PA) using 800 nm InP-DHBT technology. It consists of a driver stage with a 2-way combined cascade unit power cell. Measurements show 12 dB small signal gain. The output 1-dB compression point (OP1dB) and the saturated output power (Psat) occur around 12 dBm and 14 dBm, respectively. The maximum dc power consumption is 212 mW and results in a power-added efficiency (PAE) of up to 10% at 145 GHz. The chip area is only 1.5x1.2 mm². This amplifier demonstrates very similar Psat, OP1dB, and PAE values as compared to 250 nm InP-DHBT technologies.**

Keywords — **InP double heterojunction bipolar transistor (DHBT), monolithic microwave integrated circuit (MMIC), power amplifier (PA), transferred-substrate process (TS).**

I. INTRODUCTION

There is an increasing interest in millimeter-wave (mm-wave) frequencies for high-speed communications as well as improved performance of sensors and imaging systems. Operating frequencies beyond 100 GHz are attractive particularly for D-band (ranging from 110-170 GHz) wireless communications [1, 2]. The power amplifier (PA) in the transmit chain is a key building block for such applications. Performance targets include a combination of high output power, linearity (OP1dB) and high efficiency.

In the past years, significant progress has been made in pushing power amplifiers to the mm-wave frequency range, using various semiconductor technologies. Several power amplifiers have been published [4, 5, 6] at D-band which deliver high output power, but they are relatively power hungry. Therefore, the main motivation behind this work is to reach the combined performance data mentioned above.

The work presented here serves the communication system being developed in the frame of an EU project 'ULTRAWAVE'. The project aims to develop a dense ultra-capacity backhaul that enables 5G cell densification by exploiting the frequency range from 141 GHz to 150 GHz in D-band [3].

The paper is organized as follows: Sec. II briefly describes the technology and Sec. III discusses the circuit design. Section IV presents the realized circuit and the measurement results, followed by a discussion, while Sec. V contains the conclusion.

II. THE InP-DHBT TECHNOLOGY

The power amplifier circuit is fabricated using a monolithic transferred-substrate (TS) InP-DHBT process and transistors with 0.8x6 µm² emitter width. Fig. 1 shows the vertical layer stack of this technology. The InP transistors are transferred and bonded to a Si substrate in a wafer-level bonding process using benzocyclobutene (BCB). The InP DHBTs have 0.8 µm wide emitters and achieve ft and fmax values of about 350 GHz with BVCEO = 4 V. The BCB stack above the Si substrate includes three gold metal layers (G1, GD, G2) with 2 µm, 2.5 µm, and 4.5 µm thickness, respectively, MIM capacitors with a sheet capacitance of 0.3 fF/µm² and NiCr thin film resistors (sheet resistance of 25 Ω/sq.). In order to suppress substrate modes, a high resistivity Si substrate with through-silicon vias (TSVs) is used as the host wafer for the wafer bonding process. For further details of the MMIC process see [7].

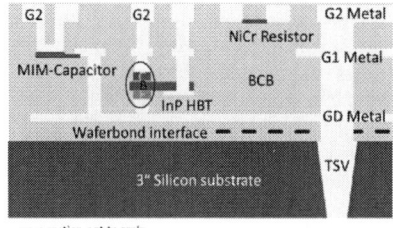

Fig. 1. Vertical layer stack of InP-DHBT technology.

III. CIRCUIT DESIGN

The schematic diagram of the power amplifier including its bias network is shown in Fig. 2. The circuit is designed based on a common emitter (CE) single-finger HBT with an emitter size of 0.8 × 6 µm² and a ft/fmax of 350/350 GHz. The amplifier consists of a driver stage with two identical power unit cells combined on-chip, in order to achieve sufficiently high output power in the D-band. The power unit cell is designed combining a single-finger driver stage with a 2-finger power stage.

In order to increase maximum available gain at desired frequencies, two single-finger HBTs have been used to form a double finger HBT by using a small piece of transmission line (see Fig. 2).

3–4 April 2022, London, UK

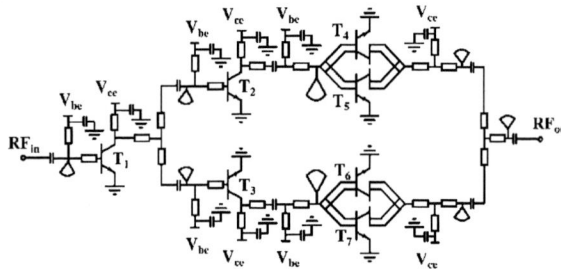

Fig. 2. Schematic diagram of the D-band power amplifier.

All the passive components, such as MIM capacitors and the NiCr on-chip resistors, are used for DC-coupling purposes as well as enhanced electrical stability. The conventional 50-ohm input/output matching technique is used. The splitter and the combiner at the input and the output of the PA are designed to be identical. In order to make the layout as compact as possible DC biasing of the HBTs is realized through the $\lambda/4$ short-circuited matching stubs. This also improves the overall stability factor. The circuit is realized in a thin-film microstrip line environment.

IV. MEASURED RESULTS

The chip photograph of the realized D-band power amplifier is shown in Fig. 3. To improve the heat sinking of the HBTs, large thermal vias are introduced close to HBT's as well as free part of the layout area (see Fig. 3). The chip area is 1.5x1.2 mm². The circuit was characterized on-wafer under both small and large-signal excitation in two different setups. First, the small-signal performance was measured. For this purpose, a ground-signal-ground (GSG) 50 μm pitch WR-5 waveguide probe with a vector network analyzer was used. The system was calibrated with multiline Thru Reflect Line (mTRL) to place the reference plane at the probe tips.

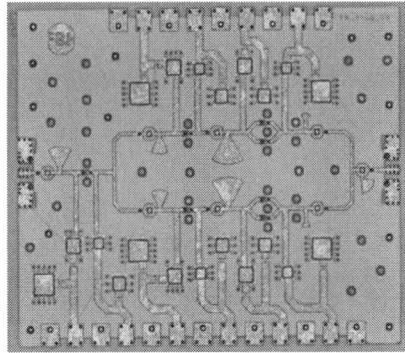

Fig. 3. Chip photograph of the D-band power amplifier.

In Fig. 4, the measured (solid lines) and simulated (dotted lines) S-parameters of the circuit are plotted. The forward small-signal gain, (S21) has an average value of more than 10 dB within 140-150 GHz. The output and the input reflection coefficients (S22, S11) remain below -15 dB within the desired frequency band. Measured and simulated small signal

results agree reasonably well with each other. It is important to note that one could design cascaded PA to improve on the gain and linear output power by tuning the first PA to a different bias condition.

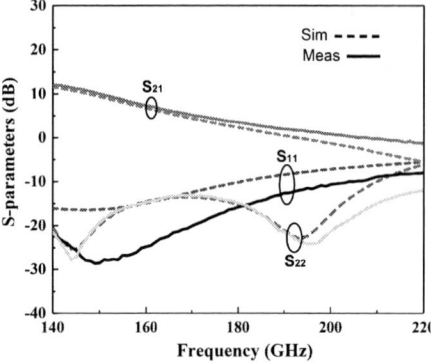

Fig. 4. Measured (solid) and simulated (dashed) S-parameters of the D-Band power amplifier.

The second step was to measure the large signal behavior of the circuit. The input power of the circuit is fed by a WR-5 module which includes WR-10 X6 multiplier, attenuator, amplifier and WR-5 doubler, followed by a ground-signal-ground probe. To extract the output power of the circuit, a WR-5 GSG waveguide probe was connected to a WR5-to-WR10 taper, which connects to the input of a power sensor and an Erickson PM4 power meter. After that the input and the output power at the probe tips is estimated by de-embedding the probe losses provided by the vendor.

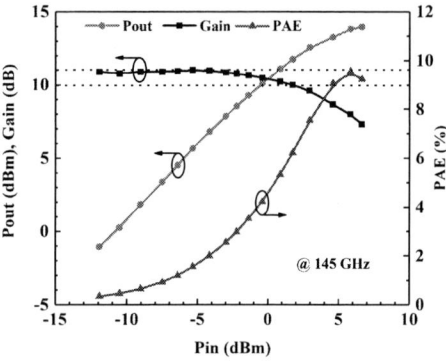

Fig. 5. Measured large-signal performance of the D-Band power amplifier.

The measured large-signal performance of the amplifier at 145 GHz is provided in Fig. 5. A saturated output power of 14 dBm is achieved with 212 mW DC power consumption, which corresponds to 9.5 % Power Added Efficiency (PAE). In Fig.5, one can also see that the measured output 1-dB compression point (OP1dB) is 12 dBm. The results shown in Fig. 5 are achieved with a safe biasing point. One can obtain a saturated output power exceeding 15 dBm by increasing the base currents (Ib), which will decrease the total PAE.

Table 1. State-of-the-art of power amplifier at 130 GHz and beyond on different technologies.

Reference	Technology	Frequency [GHz]	Gain [dB]	PDC [mW]	P1dB [dBm]	Psat [dBm]	PAE [%]
[6] IMS'19	250 nm InP DHBT	150	[a]26	3460	[a]21	[a]23	[a]6.1
[8] TMTT'05	250 nm InP DHBT	150 165	10 6	188	[a]10	11 9	4 5
[9] MWCL '17	130 nm SiGe HBT	160	27	-	[a]11	[b]14	5.7
[10] BCTM '16	130 nm SiGe HBT	162	35.4	504	10.3	14	4.8
[11] IMS'12	65 nm CMOS	150	16	115.2	7.6	12.2	12.1
[12] ISSCC'09	65 nm CMOS	150	8.2	25.5	1.5	6.3	9.5
[13] ESSCIRC'19	130 nm SiGe HBT	170	30		15.6	18	4
[14] RFIC'20	250 nm InP DHBT	130	7	-	14.4	15.3	32
[15] TMTT'19	100 nm GaN HEMT	145	25	-	-	17	1
[16] EuMIC'13	50 nm InP mHEMT	140	-	-	-	5	-
[17] IMS'09	100 nm InP mHEMT	144	10	162	11.4	-	10
This work	**800 nm InP DHBT**	**145**	**11**	**212**	**12**	**14**	**10**

[a]estimated from the graph, [b]differential mode

In order to further investigate the circuit in terms of bandwidth, it was measured by varying the input frequency with the maximum input power available from the test setup. Considering the waveguide system and probe loss, one obtains an actual output power of +14 dBm at the circuit output, at the desired frequency mentioned above. In Fig. 6 the measured output power of the circuit is plotted versus frequency. As can be seen, more than +10 dBm output power is achieved between 141 GHz and 170 GHz.

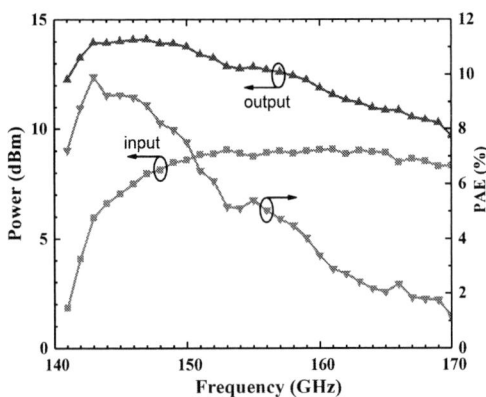

Fig. 6. Measured input and output power and power added efficiency (PAE) as a function of frequency of the D-Band power amplifier.

Table 1 benchmarks the performance of our power amplifier against other published D-band power amplifiers on different technologies for frequencies in the 140 GHz band and beyond. As can be seen, when looking for the combined performance in terms of linear output power and efficiency, only 3 of the published amplifiers reach OP1dB values above 10 dBm and a PAE of 10% and more, i.e., [14], [17] and our work. Particularly, the PA according to [14] delivers higher output power and PAE at 130 GHz, but with a 250 nm

technology node, compared to an 800 nm technology node in our case.

V. CONCLUSION

This paper presents a D-band power amplifier, based on an 800 nm transferred-substrate (TS) InP-DHBT technology. It delivers +14 dBm saturated output power (Psat) with 10 % power added efficiency at 145 GHz. The output 1-dB compression point (OP1dB) shows +12 dBm. This demonstrates that also an InP DHBT process with a relatively relaxed geometry node can almost reach the performance level of a 250 nm node state-of-the-art process.

ACKNOWLEDGMENT

The authors gratefully acknowledge partial financial support by the DLR project MIMIRAWE and EU H2020 project ULTRAWAVE (contract no. 762119). This work was also partly funded by the German Federal Ministry of Education and Research (BMBF) under the project reference 16FMD02 (Forschungsfabrik Mikroelektronik Deutschland).

REFERENCES

[1] Y. Zhao, et al., "A 160GHz Subharmonic Transmitter and Receiver Chipset in an SiGe HBT Technology," Microwave Theory and Techniques, IEEE Transactions on, vol. 60, no. 10, pp. 3286–3299, 2012.

[2] S. Carpenter, Z. S. He and H. Zirath, "A direct carrier I/Q modulator for high-speed communication at D-band using 130nm SiGe BiCMOS technology," 2017 12th European Microwave Integrated Circuits Conference (EuMIC), Nuremberg, 2017, pp. 265-268.

[3] http://ultrawave2020.eu/

[4] S. Daneshgar and J. F. Buckwalter, "Compact Series Power Combining Using Subquarter-Wavelength Baluns in Silicon Germanium at 120 GHz," in IEEE Transactions on Microwave Theory and Techniques, vol. 66, no. 11, pp. 4844-4859, Nov. 2018, doi: 10.1109/TMTT.2018.2867467.

[5] H. Lin and G. M. Rebeiz, "A 112–134 GHz SiGe amplifier with peak output power of 120 mW," 2014 IEEE Radio Frequency Integrated Circuits Symposium, Tampa, FL, 2014, pp. 163-166, doi: 10.1109/RFIC.2014.6851686.

[6] Z. Griffith, M. Urteaga and P. Rowell, "A 140-GHz 0.25-W PA and a 55-135 GHz 115-135 mW PA, High-Gain, Broadband Power Amplifier MMICs in 250-nm InP HBT," 2019 IEEE MTT-S International Microwave Symposium (IMS), Boston, MA, USA, 2019, pp. 1245-1248, doi: 10.1109/MWSYM.2019.8701019.

[7] N. G. Weimann, et al., "Transferred-Substrate InP/GaAsSb Heterojunction Bipolar Transistor Technology With fmax ~ 0.53 THz," in IEEE Transactions on Electron Devices, vol. 65, no. 9, pp. 3704-3710, Sept. 2018, doi: 10.1109/TED.2018.2854546.

[8] K. Paidi et al., "G-band (140-220 GHz) and W-band (75-110 GHz) InP DHBT medium power amplifiers," in IEEE Transactions on Microwave Theory and Techniques, vol. 53, no. 2, pp. 598-605, Feb. 2005, doi: 10.1109/TMTT.2004.840662.

[9] M. Furqan, F. Ahmed, B. Heinemann and A. Stelzer, "A 15.5-dBm 160-GHz High-Gain Power Amplifier in SiGe BiCMOS Technology," in IEEE Microwave and Wireless Components Letters, vol. 27, no. 2, pp. 177-179, Feb. 2017, doi: 10.1109/LMWC.2016.2646910.

[10] J. Al-Eryani et al., "A 162 GHz power amplifier with 14 dBm output power," 2016 IEEE Bipolar/BiCMOS Circuits and Technology Meeting (BCTM), New Brunswick, NJ, 2016, pp. 174-177, doi: 10.1109/BCTM.2016.7738965.

[11] Zuo-Min Tsai et al., "A 1.2V broadband D-band power amplifier with 13.2-dBm output power in standard RF 65-nm CMOS," 2012 IEEE/MTT-S International Microwave Symposium Digest, Montreal, QC, 2012, pp. 1-3, doi: 10.1109/MWSYM.2012.6258394.

[12] M. Seo et al., "A 1.1V 150GHz amplifier with 8dB gain and +6dBm saturated output power in standard digital 65nm CMOS using dummy-prefilled microstrip lines," 2009 IEEE International Solid-State Circuits Conference - Digest of Technical Papers, San Francisco, CA, 2009, pp. 484-485, doi: 10.1109/ISSCC.2009.4977519.

[13] M. Kucharski, H. J. Ng and D. Kissinger, "An 18 dBm 155-180 GHz SiGe Power Amplifier Using a 4-Way T-Junction Combining Network," ESSCIRC 2019 - IEEE 45th European Solid-State Circuits Conference (ESSCIRC), Cracow, Poland, 2019, pp. 333-336, doi: 10.1109/ESSCIRC.2019.8902847.

[14] K. Ning, Y. Fang, M. Rodwell and J. Buckwalter, "A 130-GHz Power Amplifier in a 250-nm InP Process with 32% PAE," 2020 IEEE Radio Frequency Integrated Circuits Symposium (RFIC), Los Angeles, CA, USA, 2020, pp. 1-4, doi: 10.1109/RFIC49505.2020.9218351.

[15] M. Ćwikliński et al., "D-Band and G-Band High-Performance GaN Power Amplifier MMICs," in IEEE Transactions on Microwave Theory and Techniques, vol. 67, no. 12, pp. 5080-5089, Dec. 2019, doi: 10.1109/TMTT.2019.2936558.

[16] T. Merkle, S. Koch, A. Leuther, M. Seelmann-Eggebert, H. Massler and I. Kallfass, "Compact 110–170 GHz amplifier in 50 nm mHEMT technology with 25 dB gain," 2013 European Microwave Integrated Circuit Conference, Nuremberg, 2013, pp. 129-132.

[17] I. Kallfass et al., "A 144 GHz power amplifier MMIC with 11 dBm output power, 10 dB associated gain and 10 % power-added efficiency," 2009 IEEE MTT-S International Microwave Symposium Digest, 2009, pp. 429-432, doi: 10.1109/MWSYM.2009.5165725.

Proceedings of the 16th European Microwave Integrated Circuits Conference

A 28-GHz-Band GaN HEMT MMIC Doherty Power Amplifier Designed by Load Resistance Division Adjustment

Ryo Ishikawa, Takuya Seshimo, Yoichiro Takayama, Kazuhiko Honjo

The University of Electro-Communications, Japan

{r.ishikawa, y.takayama honjou}@uec.ac.jp

Abstract — A 28-GHz-band GaN HEMT MMIC Doherty power amplifier has been developed by using 0.15-μm GaN HEMT MMIC technology. The Doherty amplifier was designed by adaptively adjusting a load resistance division to a carrier amplifier (CA) and a peaking amplifier (PA) according to an output power balance between the CA and PA. The fabricated GaN HEMT MMIC Doherty amplifier exhibited a maximum drain efficiency of 54% and a maximum power-added efficiency of 44% at 29.1 GHz, with a saturation output power of 30 dBm. In addition, a drain efficiency of 33% was achieved at 29.1 GHz on a 9-dB output back-off condition.

Keywords — Wireless communication, Quasi-millimeter wave, Doherty power amplifier, GaN HEMT MMIC.

I. INTRODUCTION

To achieve a higher data rate and greater data capacity, operation frequencies are increased up to millimeter-wave regions for 5th-generation (5G) wireless communication systems. In addition, the peak-to-average power ratio on used digital-modulated signals also increases to about 10 dB [1]. For this issue, millimeter-wave power amplifiers become a key device to determine a system performance. Doherty power amplifiers (DPA) [2], [3] are commonly used for the wireless communication systems, owing to their wide dynamic range property. Some methods to enlarge the dynamic range were reported at a few gigahertz band [4], [5]. Since a transistor capability degrades and circuit loss increases at the millimeter-wave region, gain and efficiency decrease, and matching circuits have to be simplified. As a result, it is difficult for a DPA with a complex circuit to obtain a sufficient performance.

As one of method to decrease the number of circuit elements, we have proposed a DPA concept in which both matching and load modulation are simultaneously treated [6]. The proposed concept was applied to discrete DPA modules [7], [8] and a 4-GHz-band GaN HEMT MMIC DPA [9]. However, at the millimeter-wave region, it is difficult to realize a suitable circuit that fits to the theoretical condition. Therefore, a more practical design algorithm with a readjustment process is newly applied to a 28-GHz-band DPA design, by simplify and clarify a design target.

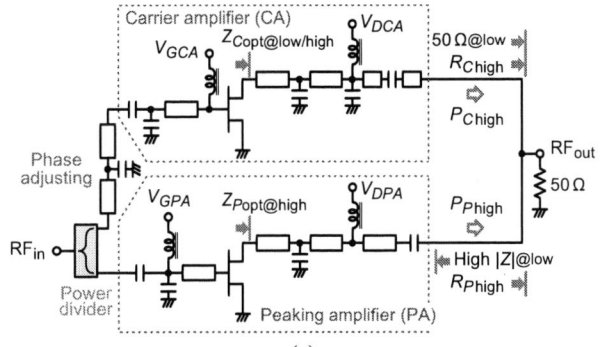

(a)

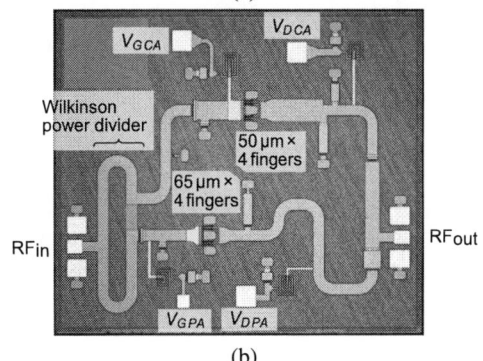

(b)

Fig. 1. (a) Constructed 28-GHz-band Doherty power amplifier (DPA) circuit, and (b) fabricated GaN HEMT MMIC DPA (2540 × 2140 μm).

II. DESIGN ALGORITHM FOR ADJUSTING LOAD RESISTANCE DIVISION TO CA AND PA

Fig. 1 shows a constructed DPA circuit and a fabricated GaN HEMT MMIC DPA. The MMIC was fabricated by using GaN MMIC technology (0.15-μm GaN/SiC HEMT process; WIN Semiconductors Corp.). The transistor sizes were determined, as shown in Fig. 1, to obtain a few hundred milliwatts at the back-off condition. A design kit provided by the foundry was used for the design in the circuit simulator (ADS; Keysight Technologies).

The DPA consists of a carrier amplifier (CA) and a peaking amplifier (PA). The CA and PA were designed based on the dual-power-level impedance optimization technique [9]. In this

241

3–4 April 2022, London, UK

technique, transistors with parasitic elements are assumed since a transistor model usually includes them. The CA and PA matching circuits were simply constructed with transmission lines and shunt capacitors, as shown in Fig. 1. At the CA output matching, a two-section low-pass-form circuit was used to adjust for the dual-power-level conditions.

The proposed design algorithm is shown in Fig. 2. First, optimum impedance values at "low (back-off)" and "high (saturation)" power levels for the GaN HEMTs are estimated by a load-/source-pull simulation. From this simulation, the saturation powers for the CA and PA were obtained ($P_{C\,high}$ and $P_{P\,high}$ shown in Fig. 1 (a), respectively). And load resistances for the CA and PA are derived as follows [9]:

$$R_{P\,high}=50\left(1+\frac{P_{C\,high}}{P_{P\,high}}\right),\ R_{C\,high}=50\left(1+\frac{P_{P\,high}}{P_{C\,high}}\right).\quad (1)$$

The load resistances are used to design the CA and PA output matching circuits [9], as initial values. Here, the matching circuit for the PA is designed with a compromise for the infinite output impedance at the PA off-state (@low). On the other hand, for the CA, there is no guarantee that the matching circuit always fulfills the optimum condition due to the circuit reciprocal theorem for a two-port circuit [9]. However, the optimum impedance values have a margin to maintain performances. So, if estimated saturation powers after the circuit designs are deviated from those on the load-/source-pull simulation results, replaced load resistances are used instead of the initial ones. If it is acceptable, the design goes to the next step; if not, the saturation power has to be changed by adjusting the drain voltage and/or gate width. In this paper, only V_{DCA} was adjusted so that the Z_{Copt} shift was as small as possible. On the other hand, in a theoretical Doherty

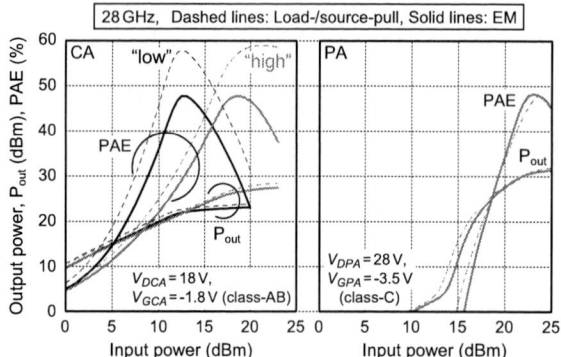

Fig. 3. Simulated CA and PA characteristics. For the characteristics by using electro-magnetic (EM) simulation results of the matching circuits, $R_{P\,high}=72\,\Omega$ and $R_{C\,high}=166\,\Omega$ were set.

operation, a constant power gain is maintained by increase the CA gain at "low", owing to the load modulation. However, in actual cases, it is difficult to keep the constant power gain, since the impedance variation of the load modulation cannot be strictly controlled caused by the influence of parasitic circuit elements. Here, $Z_{C\,opt@low}$ was also carefully selected for the CA gain.

Fig. 3 shows simulated CA and PA characteristics. This was the final result derived from the adjustment. From the characteristics by the load-/source-pull simulation, $R_{P\,high}=74\,\Omega$ and $R_{C\,high}=154\,\Omega$ were estimated from output power values of $P_{C\,high}=720\,mW$ and $P_{P\,high}=1500\,mW$, at an input power of 25 dBm. On the other hand, for the characteristics by using electro-magnetic (EM) simulation results of the matching circuits, the output power values were decreased to $P_{C\,high}=580\,mW$ and $P_{P\,high}=1350\,mW$, due to incomplete matching circuit and insertion loss, though the deviation was decreased by the feed-back adjustment. From these values, load resistances were changed to $R_{P\,high}=72\,\Omega$ and $R_{C\,high}=166\,\Omega$. As a result, similar characteristics were obtained, though CA efficiency was decreased due to the insertion loss. After the CA and PA design, a conventional Wilkinson power divider and phase adjusting circuit were designed as shown in Fig. 1.

Fig. 4 shows simulated DPA characteristics. In the simulation, an EM simulation result of the whole circuit except for GaN HEMTs was used. From these results, good efficiency performances was estimated, especially at an output power back-off (OBO) region.

III. Measured Results for Fabricated GaN HEMT MMIC DPA

To check the simulation validity, the simulated and measured S-parameters are shown in Fig. 5. About 1 GHz shift was observed in S_{21}, due to an inaccuracy in the simulation models, though the magnitude values became similar.

The measured gain and efficiency characteristics for the fabricated GaN HEMT MMIC DPA are shown in Fig. 6. The fabricated MMIC was measured by using an on-wafer probing

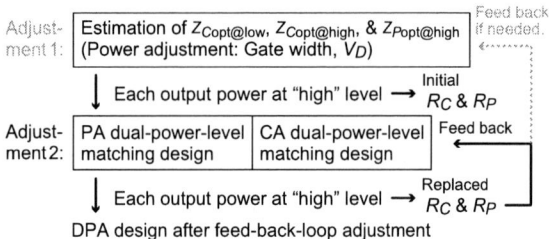

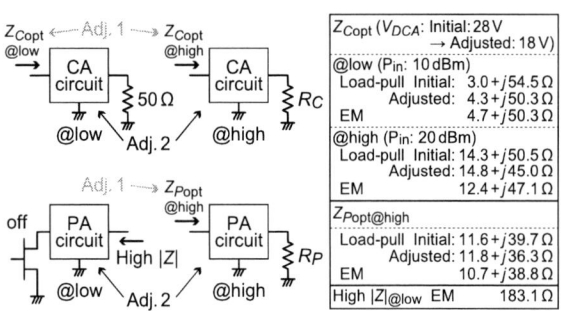

Fig. 2. Proposed carrier amplifier (CA) and peaking amplifier (PA) design algorithm, and estimated impedance values.

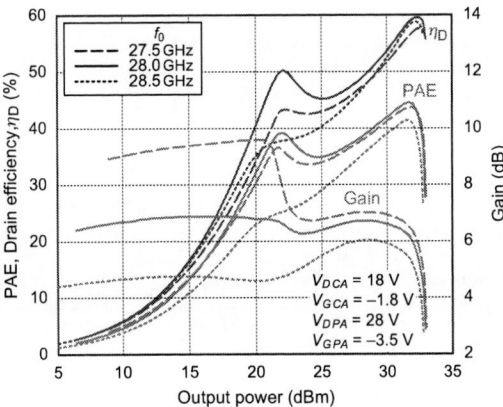

Fig. 4. Simulated DPA characteristics.

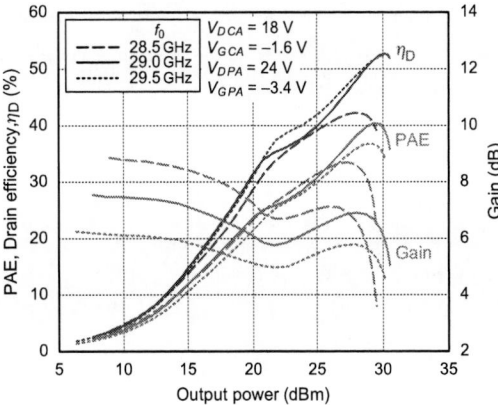

Fig. 6. Measured gain and efficiency characteristics for the fabricated GaN HEMT MMIC DPA.

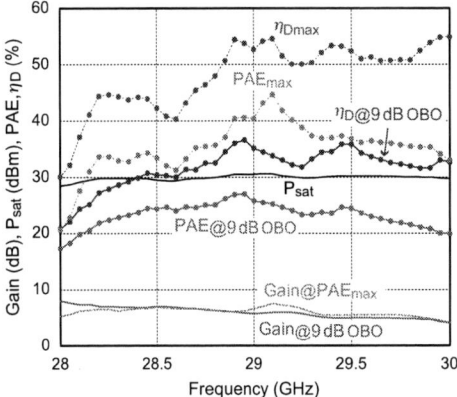

Fig. 7. Measured frequency characteristics for the fabricated DPA.

system. Any influence from probe and cable insertion losses was removed by calibration. In the measurement, DC bias voltages were adjusted to obtain better performance. From Fig. 6, similar characteristics to those in the simulated results were obtained, though the efficiency decreased due to loss, the saturation power decreased by the PA bias voltage decrease, and the operation frequency shifted as predicted from Fig. 5.

The measured frequency characteristics for the fabricated DPA are shown in Fig. 7. From Fig. 6, a maximum drain efficiency (η_D) of 54% and a maximum power-added efficiency (PAE) of 44% at 29.1 GHz were achieved with a saturation output power of 30 dBm. In addition, an η_D of 33% and a PAE of 25% were achieved at 29.1 GHz with a 9-dB OBO condition, where the OBO position was estimated from each saturation power. The efficiency reduction at the OBO compared with the simulated one is attributed to an increase of power leak to the PA output side due to the insufficient PA infinite output impedance. As a bandwidth, more than 23% PAE was maintained from 28.4 to 29.5 GHz with a 9-dB OBO condition. Finally, the gain characteristic must be further increased, and this is the future issue.

The measured input and output spectrum for a 256 QAM modulated signal is shown in Fig. 8. The modulated signal was created with random IQ baseband signals, just to check an adjacent channel leakage ratio (ACLR) performance. From

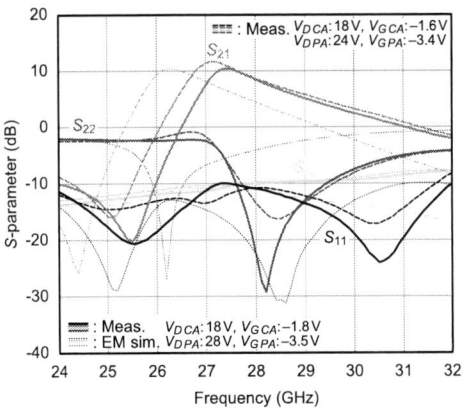

Fig. 5. Simulated and measured S-parameters.

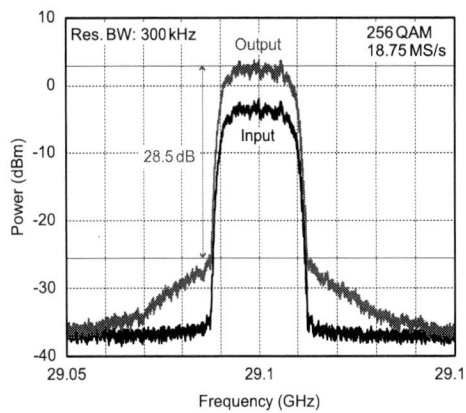

Fig. 8. Measured input and output spectrum for a 256 QAM modulated signal.

243

Table 1. Performance summary and comparison with previous works

	Process	Frequency	$\eta_{D\max}$	PAE$_{\max}$	PAE@OBO	P_{sat}
[10]	0.15 μm GaN HEMT	26 GHz	27.5%	21.7%	20% @6-dB	32 dBm
[11]	100 nm GaN HEMT	28 GHz	39%	36.2%	30% @6-dB	33 dBm
[12]	0.15 μm GaN HEMT	28.5 GHz	NA	25.5%	19.5% @8-dB	35.6 dBm
[13]	0.15 μm GaN HEMT	30 GHz	NA	39.8%	NA	37.6 dBm
[14]	0.15 μm GaN HEMT	28.5 GHz	NA	29%	19% @9-dB	39 dBm
[15]	0.15 μm GaN HEMT	27.5 GHz	NA	43%	23% @9-dB	36 dBm
This work	0.15 μm GaN HEMT	29.1 GHz	54%	44%	25% @9-dB	30 dBm

Fig. 8, at an average output power of about 21 dBm, an ACLR of 28.5 dB was maintained without a digital pre-distortion (DPD). It is expected that the ACLR is improved by the DPD. A performance summary and comparison with previous works are listed in Table I.

IV. CONCLUSION

A quasi-millimeter wave GaN HEMT MMIC Doherty power amplifier (DPA) was developed based on a proposed load resistance division adjustment. By adaptively adjusting the load resistance division to the carrier amplifier (CA) and peaking amplifier (PA) according to an output power balance between the CA and PA, a more practical design could be carried out. The fabricated GaN HEMT MMIC DPA exhibited a maximum drain efficiency of 54% and a maximum power-added efficiency (PAE) of 44% at 29.1 GHz, with a saturation output power of 30 dBm. In addition, a drain efficiency of 33% was achieved at 29.1 GHz with a 9-dB output back-off condition. As a bandwidth, more than 23% PAE was maintained from 28.4 to 29.5 GHz with a 9-dB OBO condition.

REFERENCES

[1] F. Balteanu, "Linear front end module for 4G/5G LTE advanced applications," in *Proc. 2018 European Microw. Conf.*, Sept. 2018, pp. 251–254.

[2] W. H. Doherty, "A new high efficiency power amplifier for modulated waves," *Proc. IRE*, vol. 24, no. 9, pp. 1163–1182, Sep. 1936.

[3] P. Colantonio, F. Giannini, R. Giofrè, and L. Piazzon, "The AB-C Doherty amplifier, Part I: Theory," *Int. J. RF and Microwave Computer-Aided Engineering*, vol. 19, no. 3, pp. 293–306, May 2009.

[4] X. H. Fang and K.-K. M. Cheng, "Extension of high-efficiency range of Doherty amplifier by using complex combining load," *IEEE Trans. Microw. Theory Tech.*, vol. 62, no. 9, pp. 2038–2047, Sep. 2014.

[5] M. Özen, K. Andersson, and C. Fager, "Symmetrical Doherty power amplifier with extended efficiency range," *IEEE Trans. Microw. Theory Tech.*, vol. 64, no. 4, pp. 1273–1284, Apr. 2016.

[6] Y. Takayama, and K. Honjo, "Proposals of compact broadband microwave Doherty amplifiers with generalized output-combining condition," *IEICE Tech. Rep.* (in Japanese), vol. 112, no. 312, MW2012-115, pp. 7–12, Nov. 2012.

[7] S. Watanabe, Y. Takayama, R. Ishikawa, and K. Honjo, "A miniature broadband Doherty power amplifier with a series-connected load," *IEEE Trans. Microw. Theory Tech.*, vol. 63, no. 2, pp. 572–579, Feb. 2015.

[8] T. Seshimo, Y. Takayama, R. Ishikawa, and K. Honjo, "Harmonic-tuned high-efficiency GaN HEMT Doherty power amplifier based on two-power-level impedance optimization," in *Proc. 2019 Asia Pacific Microw. Conf.*, Dec. 2019, pp. 375–377.

[9] R. Ishikawa, Y. Takayama, and K. Honjo, "Fully integrated asymmetric Doherty amplifier based on two-power-level impedance optimization," in *Proc. 2018 European Microw. Conf.*, Sept. 2018, pp. 1221–1224.

[10] R. Guo, H. Tao, and B. Zhang, "A 26 GHz Doherty power amplifier and a fully integrated 2 × 2 PA in 0.15 μ m GaN HEMT process for heterogeneous integration and 5G," in *Proc. 2018 IEEE MTT-S Int. Wireless Symp.*, May 2018, DOI: 10.1109/IEEE-IWS.2018.8401017.

[11] R. Giofrè, A. D. Gaudio, W. Ciccognani, S. Colangeli, and E. Limiti, "A GaN-on-Si MMIC Doherty power amplifier for 5G applications," in *Proc. 2018 Asia Pacific Microw. Conf.*, Nov. 2018, pp. 971–973.

[12] K. Nakatani, Y. Yamaguchi, Y. Komatsuzaki, and S. Shinjo, "Millimeter-wave GaN power amplifier MMICs for 5G application," in *Proc. 2019 IEEE Int. Symp. Circuits and Systems*, May 2019, DOI: 10.1109/ISCAS.2019.8702133.

[13] B. Schmukler, K. M. Bothe, S. Ganguly, T. Alcorn, J. Gao, C. Hardiman, E. Jones, D. Namishia, F. Radulescu, J. Barner, J. Fisher, D. A. Gajewski, S. T. Sheppard, and J. W. Milligan, "A high efficiency, Ka-band, GaN-on-SiC MMIC with low compression," in *Proc. 2019 IEEE BiCMOS Compound semiconductor Integr. Circuits Technol. Symp.*, Nov. 2019, DOI: 10.1109/BCICTS45179.2019.8972749.

[14] D. Wohlert, B. Peterson, T. R. M. Kywe, L. Ledezma, and J. Gengler, "8-watt linear three-stage GaN Doherty power amplifier for 28 GHz 5G applications," in *Proc. 2019 IEEE BiCMOS Compound semiconductor Integr. Circuits Technol. Symp.*, Nov. 2019, DOI: 10.1109/BCICTS45179.2019.8972750.

[15] M. Bao, D. Gustafsson, R. Hou, Z. Ouarch, C. Chang, and K, Andersson, "A 24–28-GHz Doherty power amplifier with 4-W output power and 32% PAE at 6-dB OPBO in 150-nm GaN technology," *IEEE Microw. Wireless Compon. Lett.*, vol. 31, no. 6, pp. 752–755, Jun. 2021.

Proceedings of the 16th European Microwave Integrated Circuits Conference

Field-Plate Mixer

Simon J. Mahon, Michael C. Heimlich

Macquarie University, NSW 2109, Australia

simon.mahon@mq.edu.au

Abstract — **A novel GaN FET field-plate mixer is presented. The local oscillator (LO) is connected to the FET's field-plate which is disconnected from the transistor source. The RF/IF input is connected to the gate (RF for down-, IF for up-conversion) and the IF/RF output to the drain. Over the 18-40 GHz range, the mixer cell has an average measured down-conversion loss of 8.8 dB, up-conversion loss of 8.2 dB, IP1 of 10.8 mW/mm and IIP3 of 69.7 mW/mm; values that compare well with the state-of-the-art for passive FET mixers. Good agreement with simulation is demonstrated.**

Keywords — **GaN, MMICs, mixers, field-plates.**

I. INTRODUCTION

Linearity is a key performance parameter for model mm-wave systems, including radar and communications. Resistive FET mixers, which apply LO to the FET gate and RF and IF to the drain through filters, offer excellent linearity and have been incorporated into single-chip GaAs MMIC receivers and up-converters [1]. Recently, the superior power performance of GaN amplifiers has led to interest in passive and active GaN FET mixers [2-4].

This paper presents a novel enhancement to the FET mixer concept for GaN FETs with source-connected, field-plate technology. The connectivity of the field-plate is changed. The field-plate is disconnected from the source and connected to the LO instead. The field-plate remains isolated from the FET's electrodes and semiconductor surface by dielectric layers. For down-conversion, the incoming RF is connected to the gate and the IF to the drain. For up-conversion, the incoming IF is connected to the gate and the RF to the drain. All three ports are unmatched to allow the core performance to be examined.

This manuscript is organised as follows: In Section II, the fabrication and simulation of the filed-plate mixer is discussed. Measurements and simulation of down- and up-conversion loss, and up-conversion single-tone IP1 (input-referred, P1dB compression point) and two-tone IIP3 (input-referred, third-order intercept point) linearity over the 18-40 GHz range are presented in Section III. Discussion of measurement vs simulation and comparison of the results with selected FET mixer publications is the topic of Section IV before some concluding remarks are made.

II. FIELD-PLATE MIXER

A. Fabrication

In a GaN FET, source field-plates are thin metal lines fabricated above the gate passivation over the gate electrode and gate-drain gap to increase gate-drain breakdown and output power. In the WIN Semiconductor's 0.15-µm NP15 GaN

process used here, one end is open-circuited and the other connected to the FET's source metal as illustrated in Fig. 1.

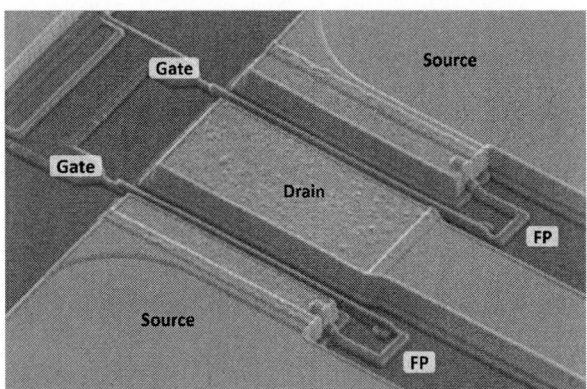

Fig. 1. Photograph of a standard 2x50 µm GaN FET in WIN Semiconductor's 0.15-µm NP15 process. Each field-plate ("FP") is fabricated above, and slightly to the drain side, of the gate. It is separated by dielectric layers from both the gate electrode and the GaN semiconductor below. The left end is open circuited and the right is connected to the source and, hence, ground. Photograph courtesy of WIN Semiconductor.

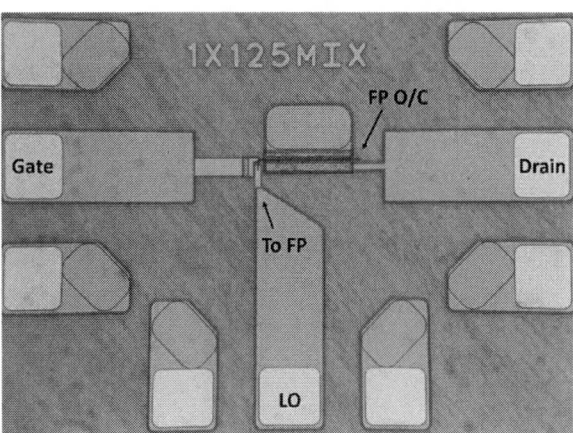

Fig. 2. Field-plate mixer photograph. RF/IF input at the gate (RF for down-conversion, IF for up-conversion), IF/RF output at the drain, LO connected to the field-plate on the left. The right-hand end of the field-plate is an open circuit.

In this work, the field-plate is disconnected from the source at one end (becoming an open circuit) and its other, previously open circuited, end is connected to a local oscillator (LO) thereby creating a mixer. A photograph of the mixer, fabricated in NP15 using a 1x125 µm FET, is shown in Fig. 2.

245

3–4 April 2022, London, UK

In down-conversion mode, the incoming RF is connected to the gate and the outgoing IF to the drain. In up-conversion these roles are reversed, the incoming IF is connected to the gate and the outgoing RF to the drain.

B. Simulation

The mixer is simulated using the single-finger [5] implementation of the MQFET model [6]. Fig. 3 shows the meshed structure in AXIEM EM software tool. The inset illustrates the passage of the gate (blue), to which the internal gate-source and gate-drain ports are attached for the single-finger MQFET model, underneath the field-plate (cyan).

The MQFET model had been extracted for power amplifier use with a traditional source-connected field-plate. For use here, the field-plate capacitance was deactivated inside the model to avoid double counting its effect with the LO field-plate modelled explicitly here in the EM simulation.

The LO signal is applied to the field-plate which is isolated from both the gate and the GaN semiconductor and is terminated in an open circuit. In the simulation, the proximity of the field-plate to the gate and drain couples the LO signal to both these ports of the intrinsic non-linear transistor model. Direct modulation of the semiconductor, through the dielectric layers, may occur but it is not considered in this simulation.

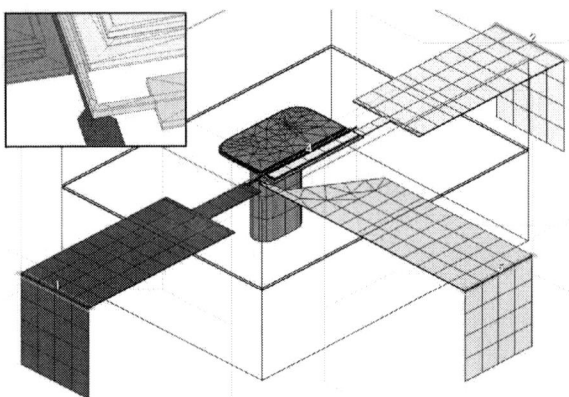

Fig. 3. Field-plate mixer simulation in Axiem EM software. RF/IF input at the gate (blue), IF/RF output at the drain (yellow), LO connected to the field-plate (cyan). The source is connected to ground through a backvia (magenta).

III. MEASUREMENTS

A. Conversion Loss

The measured down-conversion loss for lower and upper side band (LSB and USB) operation over an LO range of 18 to 40 GHz, with a 1 GHz IF, is shown in Fig. 4 along with the simulated value for the LSB case (USB simulation is similar).

Fig 5 shows the LSB and USB up-conversion measurements along with the LSB simulation values.

The LO signal induces a DC potential on the drain of the FET therefore the mixer operates in a hybrid between passive and low-bias active modes. The measured and simulated drain potential for the LSB up-conversion case is shown in Fig. 6.

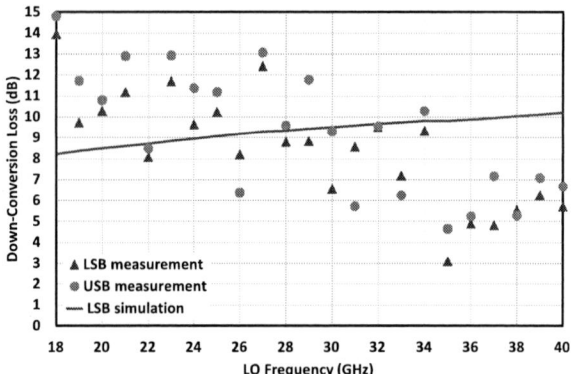

Fig. 4. Measured LSB and USB down-conversion loss vs LO frequency at 15 dBm. IF = 1 GHz and -20 dBm. (Av. meas. conv. loss = 8.8 dB.)

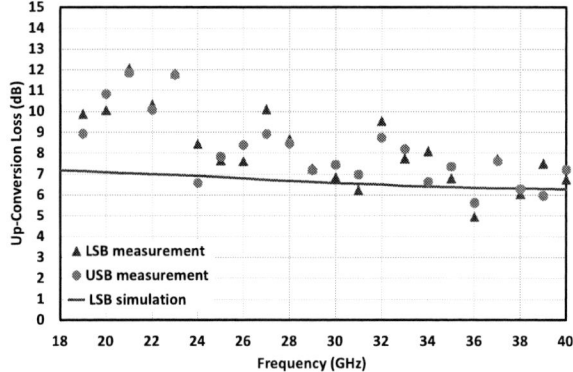

Fig. 5. LSB and USB up-conversion loss vs LO frequency. LO = 15 dBm. IF = 1 GHz and -20 dBm. Gate bias = -2 V. (Av. meas. conv. loss = 8.2 dB)

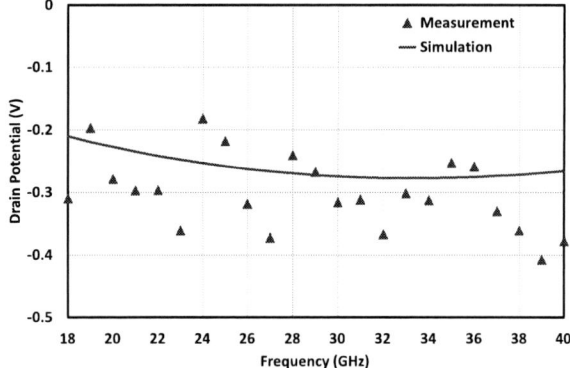

Fig. 6. Measured and simulated drain-source voltage for LSB up-conversion vs LO frequency. LO = 15 dBm. IF = 1 GHz and -20 dBm. Gate bias = -2 V.

Fig 7(a). shows the dependence of up-conversion loss on gate bias and LO power level. Optimal conversion loss is obtained with an LO power level around 15 dBm and a gate bias near -2.0 V which is approximately the turn-off voltage for NP15. The induced drain potential is also dependent on gate bias and LO level - see Fig. 7(b).

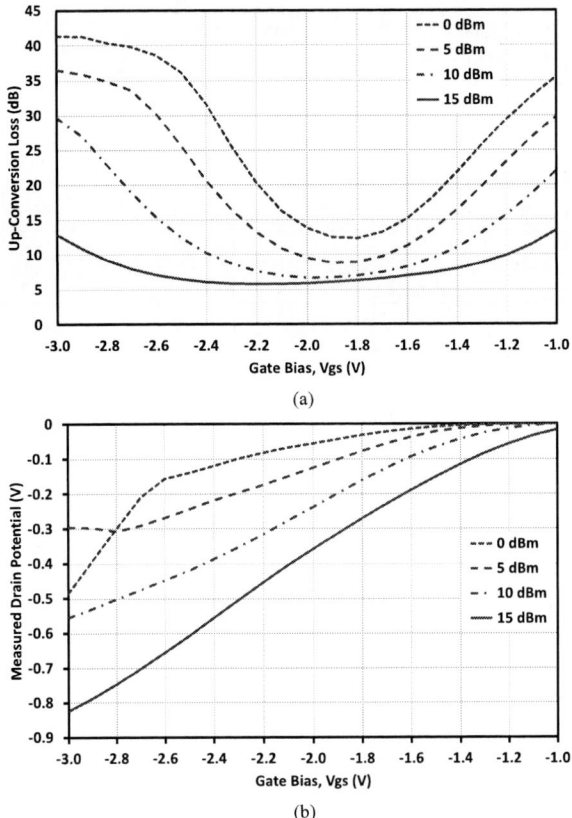

(a)

(b)

Fig. 7. Measured (a) LSB up-conversion loss and (b) drain potential vs gate bias and LO powers of 0, 5, 10 and 15 dBm at 38 GHz.

B. IP1 and IIP3

Measurement of P1 and IP3 in down-conversion mode was inhibited by the failure of a mm-wave instrumentation amplifier. Measurements in up-conversion mode, with a 1 GHz IF, were not affected.

In Fig. 8, the measured LSB and USB IP1 and the simulated IP1 for the LSB case (USB is similar) are shown. The average measured IP1 is 1.3 dBm (10.8 mW/mm).

The variation in LSB and USB IP1 with LO power is shown in Fig. 9 along with the LO leakage (this new topology, like other FET mixers, requires a balun to suppress LO leakage).

Fig. 10 presents the LSB and USB input-referred third-order intercept point (IIP3) measured with two -20 dBm signals at 0.9945 and 1.0055 GHz (i.e. 11 MHz tone spacing centred at 1.0 GHz). The simulated LSB IIP3 is also shown. The average measured IIP3 is 9.4 dBm (69.7 mW/mm).

Finally, Fig. 11 presents the useful metric of measured IIP3 - IP1, which has an average value of 8.1 dB across the band.

IV. DISCUSSION

A. Measurement vs Simulation

The simulated conversion loss is in good agreement with the measured up- and down-conversion loss shown in Figs. 4 and 5. The LO induces a measurable DC potential at the drain

terminal which may act to improve the conversion gain. This effect is captured by the model and is confirmed by measured data, as seen in Fig. 6; however, comparing Figs 7(a) and (b) suggest that it is a second-order effect and that the mixer's operation is fundamentally passive.

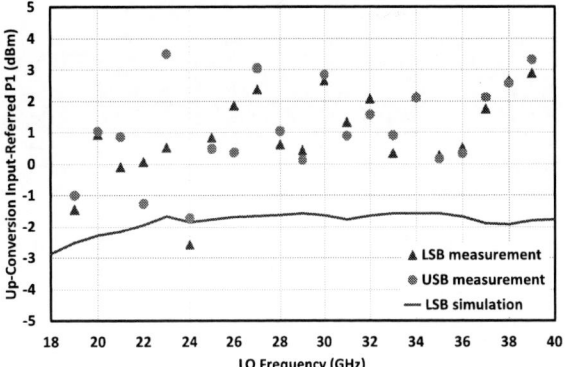

Fig. 8. Measured LSB and USB up-conversion input-referred P1 compression point, and simulated LSB, vs LO frequency. LO = 15 dBm. IF = 1 GHz. Gate bias = -2 V. (Average measured input IP1 = 1.3 dBm)

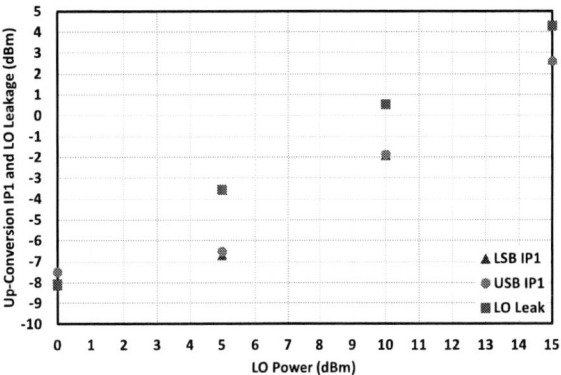

Fig. 9. LSB and USB up-conversion input-referred P1 compression point and LO leakage vs LO power at 38 GHz. IF = 1 GHz. Gate = -2 V.

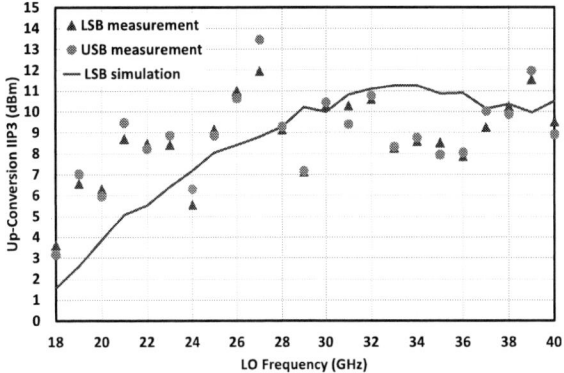

Fig. 10. LSB and USB up-conversion input-referred IP3 vs frequency. LO = 15 dBm. IF = 1 GHz. Gate bias = -2 V. (Av. meas. IIP3 = 9.4 dBm).

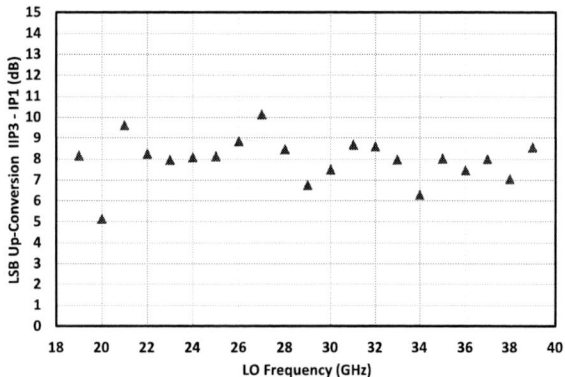

Fig. 11. LSB up-conversion input-referred IP3 – P1 vs frequency. LO = 15 dBm. IF = 1 GHz. Gate bias = -2 V. (Av. meas. = 8.1 dB)

The model underestimates the field-plate mixer's single-tone IP1 linearity (Fig. 8) but the simulation agrees quite well with the two-tone IIP3 measurements shown in Fig. 10. This remains a topic of research.

Overall, despite the IP1 discrepancy, good agreement has been demonstrated between measurement and simulation. This validates the model assumption that the LO signal is mainly coupled to the gate and drain, and hence into the FET, rather than by direct modulation of the semiconductor as would be the case for a dual-gate FET.

B. Comparison with Published Work

Table 1 compares the performance of the unmatched novel mixing device presented here with selected GaAs and GaN resistive-FET MMIC mixers operating in the lower mm-wave range. The single-tone (IP1) and two-tone (IP3) data are scaled by width. Size is not compared as the other mixers also contain matching circuitry, LO baluns and image-reject structures.

The linearity of the GaAs mixers in [1] can be extracted from the indicated measurement graphs. In up-conversion, the IIP3 is approximately 63 mW/mm, whereas, in down-conversion, a higher value of 1000 mW/mm is obtained.

In [2], a down-conversion mixer was designed using both 0.15 μm GaN, giving a IP1 of 15.8 mW/mm measured at 28 GHz, and with 0.25 μm GaN, giving an IIP3 of 25.1 mW/mm at 28 GHz.

The active E-band down-conversion mixer in [3] demonstrates excellent single-tone linearity with an IP1 of 109.6 mW/mm measured at 77 GHz. It uses a 15 volt drain supply which makes comparison with the passive mixers challenging.

The single-balanced, down-conversion mixer presented in [4] achieves IP1 of 15.8 mW/mm and IIP3 of 200 mW/mm at 29 GHz in the centre of the mixer's 26-31 GHz band.

The work presented here archives a IP1 of 10.5 mW/mm and IIP3 of 66.1 mW/mm averaged over the 18-40 GHz band. Both these results are in up-conversion mode which, as the GaAs results in [1] suggest, may be exceeded by the down-conversion results when they become available. Regardless, this novel, unmatched, field-plate mixer compares well with the state-of-the-art for GaAs and GaN mixers.

Table 1. Measured linearity for selected mm-wave passive (applied Vd = 0 V) and active (applied Vd > 0) GaAs and GaN resistive-FET mixers.

Ref.	Tech.	Freq (GHz)	Vd (V)	IP1 (mW /mm)	IIP3 (mW /mm)	Up / Down
[1] Figs 17-20	0.15 μm GaAs	32-45	0	-	~63	Up
[1] Figs 25-26	0.15 μm GaAs	32-45	0	-	~1000	Down
[2]	0.15 μm / 0.25 μm GaN	26-30	0	15.8	25.1	Down
[3]	0.10 μm GaN	75-81	15	109,6	-	Down
[4]	0.10 μm GaN	26-31	0	15.8	200	Down
This work	0.15 μm GaN	18-40	0	10.8	69.7	Up

V. CONCLUSION

A novel GaN FET field-plate mixer has been presented. The local oscillator (LO) is connected to the FET's field-plate which is disconnected from the transistor source. Over the 18-40 GHz range, the mixer cell has an average measured down-conversion loss of 8.8 dB, up-conversion loss of 8.2 dB, IP1 of 10.8 mW/mm and IIP3 of 69.7 mW/mm; values that compare well with the state-of-the-art for passive FET mixers. Good agreement with simulation has been demonstrated.

ACKNOWLEDGMENT

The authors thank Andrew Jones and Gerry McCulloch for measurements, Evgeny Kuxa for model extraction and WIN Semiconductor Corp., Taiwan, for providing access to their NP15 process.

REFERENCES

[1] S. J. Mahon et al., "Broadband integrated millimeter-wave up- and down-converter GaAs MMICs," in IEEE Transactions on Microwave Theory and Techniques, vol. 54, no. 5, pp. 2050-2060, May 2006, doi: 10.1109/TMTT.2006.872793.

[2] M. Do, M. Seelmann-Eggebert, R. Quay, D. Langrez and J. Cazaux, "AlGaN/GaN mixer MMICs, and RF front-end receivers for C-, Ku-, and Ka-band space applications," The 5th European Microwave Integrated Circuits Conference, Paris, 2010, pp. 57-60.

[3] I. Kallfass et al., "High linearity active GaN-HEMT down-converter MMIC for E-band radar applications," 2014 9th European Microwave Integrated Circuit Conference, Rome, 2014, pp. 128-131, doi: 10.1109/EuMIC.2014.6997808.

[4] A. De Padova, P. E. Longhi, S. Colangeli, W. Ciccognani and E. Limiti, "Design of a GaN-on-Si Single-Balanced Resistive Mixer for Ka-band Satcom," in IEEE Microwave and Wireless Components Letters, vol. 29, no. 1, pp. 56-58, Jan. 2019, doi: 10.1109/LMWC.2018.2880315.

[5] S. J. Mahon et al., "LNA Design Based on an Extracted Single Gate Finger Model," 2010 IEEE Compound Semiconductor Integrated Circuit Symposium (CSICS), Monterey, CA, 2010, pp. 1-4, doi: 10.1109/CSICS.2010.5619678.

[6] A.E. Parker, "Advances in nonlinear characterization of millimetre-wave devices for telecommunications," Proc. of SPIE, vol. 6798, no. 12, pp. 67 980L (1–11), 4-7 Dec. 2007.

Proceedings of the 16th European Microwave Integrated Circuits Conference

17.6 dB Variable-Gain and Variable-Bandwidth Upconverter in 65 nm CMOS for 60 GHz Bands

Oner Hanay, David Bierbuesse, Renato Negra

Chair of High Frequency Electronics, RWTH-Aaachen University, Germany

{oner.hanay, david.bierbuesse, rnegra}@hfe.rwth-aachen.de

Abstract — This paper presents a variable-gain and variable-bandwidth IF-to-millimetre-wave upconversion mixer in 65 nm standard CMOS technology for multistandard transmitter topologies. The fabricated mixer occupies only 0.28 mm^2 of chip area including LO- and RF-baluns, the LO-buffer and pads. The return loss of the differential IF input is better than -12 dB for up to the 3rd harmonic of the input signal centred at 5 GHz. The 3-dB IF-bandwidth and the gain are tunable by the quality-tuned inductive peaking stage. The maximum measured conversion gain is 17.6 dB at an LO power level of +1 dBm and an IF-bandwidth of 3.2 GHz. The highest IF-bandwidth is 8.1 GHz while providing a conversion gain of 5.2 dB. The 3-dB RF-bandwidth spans from 55 to 64.5 GHz, whereas the measured conversion gain between 52.5 to 67 GHz is above 10 dB. The output-referred 1-dB compression point is -2.9 dBm with an IIP3 of -12 dBm. The total power consumption is 98.46 mW, without the LO-buffer it is 47.92 mW. The LO-to-RF isolation is better than 26 dB in the entire bandwidth.

Keywords — Mixer, upconverter, variable gain, 60 GHz, transmitters.

I. INTRODUCTION

Currently, 5G standards utilise communication in the already congested sub-6-GHz range to ensure rural coverage at reduced bandwidths and the millimetre-wave (mmW) range around 28 GHz for urban high data-rate communication [1], [2]. However, the unlicensed ISM band at 60 GHz provides wider available bandwidth which can be used in order to overcome the capacity restrictions of the 28 GHz band. In combination with the significant channel losses due to the resonance of O_2 in the air, the 60 GHz frequency band is crucial for high-density femto-cells in overpopulated urban areas [3], [4]. Modern hand-held devices support several wireless communication standards utilising a wide carrier frequency range which leads to the introduction of carrier frequency relocation from several hundred megahertz up to the mmWs. In order to implement the carrier frequency relocation, multistandard transmitter architectures are required in order to achieve manageable production cost. In this regard, conventional, *i.e.* RF-DAC-based, direct transmitters operating in the sub-6 GHz region are extended with additional upconverters to implement the carrier frequency selection [5]. Therefore, the output of the RF-DAC can directly be fed to an upconversion mixer's IF input which needs to provide a suitable input impedance as well as good matching at least up to the 3rd harmonic, since the impedance directly affects the linearity of the preceding RF-DAC [6]. Furthermore, the upconversion circuits shall provide sufficient

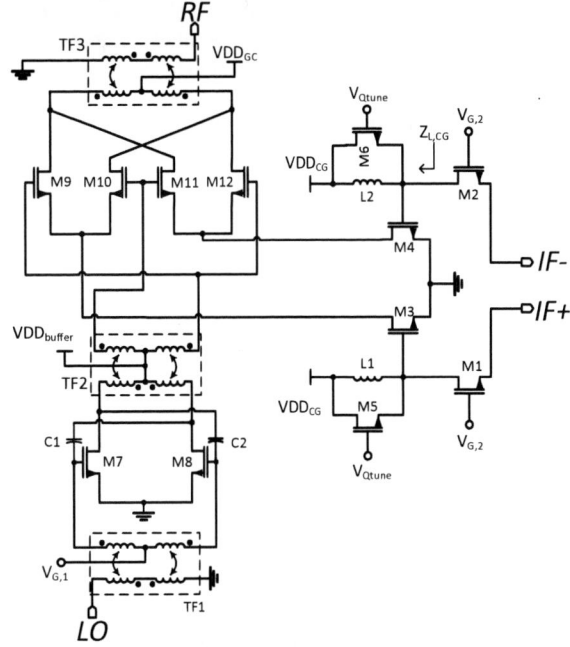

Fig. 1. Simplified schematic of the proposed standard CMOS IF-to-mmW upconverter with integrated LO-Buffer, RF, and LO baluns.

IF-bandwidth in the range of several GHz [4] and the RF output is required to cover the complete 60 GHz ISM/WiGig (IEEE802.11ad) bands from 57 to 66 GHz. The conversion gain is one of the most important properties, especially, when the output is in the mmW-range. A high conversion gain relaxes the constraints on the preceding RF-DAC and the following amplification stages. The degree of integration of the complete transmitter needs to be as high as possible in order to reduce the production cost. Thereby, the feasibility of high performance and chip-area-efficient standard CMOS upconverters is essential such that they can be integrated with the complete transmitter.

The Gilbert-cell is a popular mixer topology where both the IF and LO signals are fed to the gates of the common-source transistors or to the gates of the Gilbert-quad transistors. In terms of broadband matching, especially in this particular case where the IF input is centred at 5 GHz and shall be matched up to the 3rd harmonic, it is not beneficial to feed the signals

3–4 April 2022, London, UK

to the gates. In [7] the IF input is directly fed to the gates of the common-source stage and, hence, the input matching is provided for low fractional bandwidth. To overcome the bandwidth limitation, [8] introduces a resistive feedback input stage which in turn leads to a trade-off between conversion gain and the input impedance. In [9] a SiGe process is used to achieve outstanding conversion gain at high output power. Nevertheless, the adoption of a SiGe process increases the production cost significantly for high volume applications.

In this paper, the proposed mixer uses a common-gate IF input stage to enable broadband impedance matching from DC up to 18 GHz. The common-gate stage is connected via quality-tuned inductive peaking to the common-source transistors of the Gilbert-cell as shown in Fig. 1. The quality-tuning is used to control the gain versus the 3-dB IF-bandwidth. Moreover, the transformer based LO- and RF-baluns are integrated on-chip to improve the impedance matching. The measured performance of the mixer reveals outstanding conversion gain up to 17.6 dB for a standard CMOS upconversion circuit with reasonable LO pumping power in the mmW-range, together with broadband IF input matching while retaining the required operation bandwidth.

II. PROPOSED DESIGN CONSIDERATIONS

A. Gain-bandwidth tuning and input stage

Fig. 1 presents the schematic of the proposed mixer. The common-gate stage at the IF input provides a precisely controllable broadband matching due to its input impedance dominated by $1/g_m$. The pseudodifferential common-gate stage does not deliver any current gain. Thus, it is used as impedance transformer establishing voltage gain. The load impedance, $Z_{L,CG}$, seen by the common-gate stage is set by the LC shunt resonance formed by the gate capacitance of the transistors, $M2$ or $M3$, and the shunt inductances, $L1$ or $L2$. Considering $M5$ and $M6$ as controllable resistances, the quality factor of the inductances and, hence, the bandwidth of the resonators are tunable. The increased bandwidth goes along with a reduced magnitude of the resonator impedance. In combination with the impedance transforming common-gate stage, the bandwidth extension results in lower voltage gain. Fig. 2a shows the magnitude of $Z_{L,CG}$ at different tuning voltages based on EM-simulated passives and routing.

The transistors $M5$ and $M6$ are reverse biased for $V_{Qtune} = 0\,\mathrm{V}$ since their drain and source terminals are DC-wise shorted and connected to the voltage VDD_{CG}. Thus, the original quality factor of the resonator is not significantly reduced. The supply voltage, VDD_{CG}, of the common-gate stage sets the operating point of the common-source transistors, $M3$ and $M4$. The gate voltage $V_{G,2}$ of the common-gate transistors $M1$ and $M2$ is derived from the input bias V_G.

B. Inductive coupled transformers and LO-buffer

The complete design integrates three stacked centre-tapped low-k inductive transformers of which $TF1$ and $TF3$ serve as baluns, as shown in Fig. 1. The output transformer is designed to transform the 50-Ω load to a high impedance

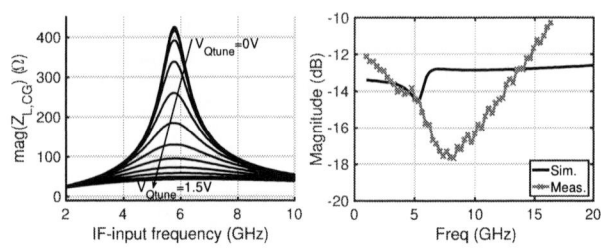

| (a) | (b) |

Fig. 2. (a) Simulated magnitude of $Z_{L,CG}$ for varying tuning voltages versus the IF input frequency, (b) Reflection coefficent of the differential IF input versus frequency @ $V_{Qtune} = 0\,\mathrm{V}$.

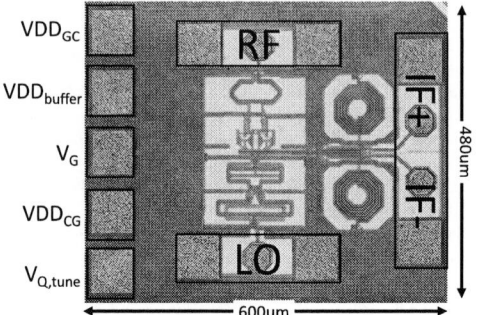

Fig. 3. Chipmicrograph of the fabricated mixer.

so that the Gilbert-quad can generate the desired output voltage swing. Additionally, it is matched over the broad RF operating band. Beside the physical dimensions of the transformers, an additional capacitance is added in shunt to the RF output and LO input pads introducing an additional design parameter to improve the matching bandwidth. Both baluns are implemented with respect to minimise the phase and amplitude deviations in order to improve the isolation between the ports.

The integrated LO-buffer is implemented as a differential transformer based amplifier. The input and output matching of the LO-buffer is employed by low-k transformers with high aspects ratios in order to reduce the occupied chip-area. The supply voltage of the buffer, VDD_{buffer}, is reused to set the DC-level of the amplified LO signal. Moreover, the transistor size of the LO-buffer is chosen similar to the Gilbert-quad in order to keep the impedance transformation ratio low and thereby reducing losses in the interstage matching. The input DC point of the buffer is set by the voltage, $V_{G,1}$, which is derived from the input bias voltage, V_G. Cross-coupled neutralisation capacitors, $C1$ and $C2$, are added to the LO-buffer to maximize its gain. The simulated peak gain is 7 dB at 55 GHz LO frequency.

III. EXPERIMENTAL RESULTS

The circuit characterisation is carried out on-wafer up to 67 GHz using Keysight's PNA-X network analyser. The LO input and the RF output are connected via ground-signal-ground (GSG) probing pads, as shown in Fig. 3.

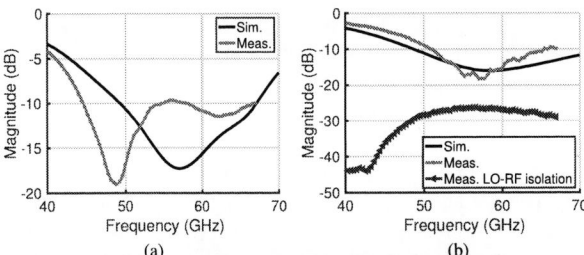

Fig. 4. Return loss characteristics at the LO- and RF-ports and isolation between LO- and RF-ports: (a) LO input port; (b) RF output port.

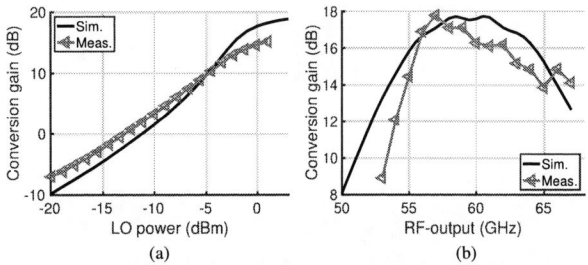

Fig. 5. Conv. gain vs.: (a) LO power @55 GHz; (b) LO- frequency @1 dBm.

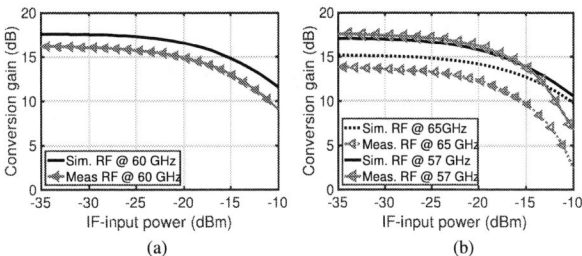

Fig. 6. Conversion gain over IF input power sweep for: (a) LO @ 55 GHz; (b) LO @ 52 and 60 GHz.

A differential GSSG probing pad is used for the IF connection which is generated by the PNA-X in a single-ended manner and fed through an external 180°-hybrid with an operating frequency range from 1 to 18 GHz. Fig. 2b shows the reflection coefficient of the differential IF input. The simulated magnitude of the insertion loss is below -12 dB up to 30 GHz. The measured reflection coefficient is characterised up to 18 GHz and it is below -10 dB. The results show that the measured and simulated reflection coefficients match well over a large bandwidth.

Fig. 4a and 4b present the measured and simulated reflection coefficients at the LO input, the RF output ports, and the isolation between LO- and RF-ports. The integrated LO-balun and the LO-buffer are designed to provide best matching at 55 GHz. The measured reflection coefficient shows a minor shift to lower frequencies due to the increased shunt capacitance in the tolerable region of process variation. Nevertheless, the frequency range from 45 to 67 GHz serves good matching with a return loss better than 9.8 dB. The mixer's output is matched to 50 Ω in order to provide maximum power transfer to the load. The measurement results show a good overlap to the simulated reflection coefficients where the -10 dB bandwidth spans between 50-70 GHz simulated and 52-67 GHz measured, respectively. The isolation between the LO- and RF-port is measured to be better than 26 dB in the band of interest although the LO-buffer is active and provides gain.

Fig. 5a shows the dependency of the conversion gain on LO power. The maximum LO power that can be delivered to the mixer up to 62 GHz is +1 dBm. Fig. 5b shows the conversion gain for different LO frequencies at a fixed LO power of +1 dBm resulting in different RFs. The maximum conversion gain of 17.6 dB is measured at an LO frequency of 52 GHz, a single-tone IF input at 5 GHz with a power level of -30 dBm, and 57 GHz RF output. The frequency shift in the LO input reflection coefficient, as shown in Fig. 4a, leads to a stronger power transfer at the lower LO frequencies which translates to higher conversion gain at lower frequencies, as shown in Fig. 5b. Nevertheless, the experimentally verified 3-dB bandwidth is 55-64.5 GHz where the above 10 dB conversion gain bandwidth can be given as 52.5-67 GHz which fully covers the targeted ISM/WiGiG bands.

The input-referred 1-dB compression point (P_{1dB}) is verified for different RF frequencies. As shown in Fig. 6a and Fig. 6b, the measured input-referred P_{1dB} is -20.5 dBm. At the P_{1dB} point the RF output power at 57 GHz is -2.9 dBm, whereas the saturated output power is slightly lower than -1 dBm. The input-referred 3rd-order intercept point (IIP3) is measured by using an external 50-GHz source with limited output power since the two internal sources of the PNA-X are combined in order to generate the two-tone IF input with a frequency offset of 10 MHz. Fig. 7a and Fig. 7b show the output spectra at different input power levels, i.e. -27 dBm and -26 dBm. In theory the IIP3 should be measured for input power levels much smaller than the P_{1dB}. However, due to the noise floor of the measured spectra the intermodulation products become reliably detectable at the given input power levels. By extrapolation, the IIP3, is calculated as -12 dBm, whereas in this particular case the gain is 14.2 dB and the input-referred P_{1dB} is -21 dBm which matches well with simulations.

The IF-bandwidth and the maximum achievable conversion gain are tuned by the variable quality factor of the inductive peaking stage in the interstage. The voltage, V_{Qtune}, is varied from 0 to 1.5 V and the conversion gain over the RF is plotted in Fig. 8a. Here, the LO frequency is kept constant at 55 GHz and the IF is swept. The resulting RF varies with the swept IF. The centre frequency of the IF-band is simulated and designed to be slightly above 5 GHz. As designed and simulated, the measured curves depict the maximum gain at 5.5 GHz IF which translates into 60.5 GHz RF output. Fig. 8b shows the compiled measurement results illustrating the trade-off

251

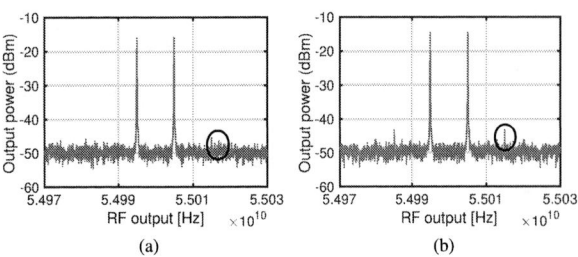

Fig. 7. RF-spectra of the two-tone measurement, P_{in}: (a) -27 dBm; (b) -26 dBm.

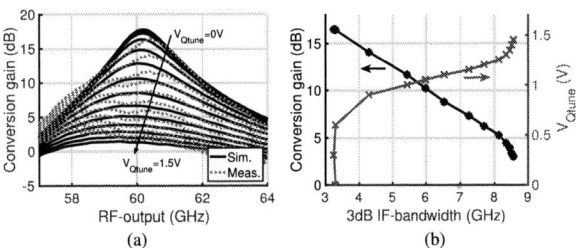

Fig. 8. Gain versus IF-bandwidth tuning: (a) Conversion gain over IF input frequency sweep for varying V_{Qtune} (LO @ 55 GHz); (b) Maximum conversion gain versus 3-dB IF-bandwidth over tuning range.

between conversion gain and achievable IF-bandwidth which can be tuned linearly for different control voltages, V_{Qtune}.

The comparison with state-of-the-art mixer circuits given in Table 1 shows that the proposed design exhibits excellent conversion gain, even though introducing a tunable and wide IF-bandwidth for standard CMOS. Similar conversion gain is achieved by employing processes like SiGe combined with additional amplification stages which leads to a considerably higher power consumption [9]. Furthermore, the occupied die area is the smallest while including the LO- and RF-baluns, the LO-buffer, pads, and tuneability.

Table 1. Comparison table of recently reported 60 GHz upconversion mixers.

Ref.	Process	Conv. gain (dB)	f_{IF} BW (GHz)	P_{LO} (dBm)	P_{DC} (mW)	P_{1dB} outp. ref. (dBm)	Area (mm²)
[7]	90 nm CMOS	-8	18-22	3	11	-19	0.45
[8]	65 nm CMOS	6.2	3-9.5	0	17.8	-5.8	0.41
[9]	180 nm SiGe	27	DC-10	n/a	495	+13	1.12
[10]	130 nm SiGe	-2.5	n/a	0	27	n/a	0.22
[11]	90 nm CMOS	0.78	DC-3.8	5	44.3	-11.7	0.74
[12]	90 nm CMOS	6	DC-1.4	n/a	12.1	n/a	0.52
This work	65 nm CMOS	**17.6**	**3.5-6.7** to **0.9-9**	1	w/o LO-buf.: 47.92, tot.:98.46	**-2.9**	**0.288**

IV. CONCLUSION

A high variable-gain and variable-bandwidth IF-to-mmW mixer in a 65 nm standard CMOS technology is implemented. The occupied die size is 0.28 mm² including the LO-buffer and baluns which makes the proposed mixer a promising candidate for fully integrated multistandard transmitters in standard CMOS. The integrated LO-buffer accomplishes high conversion gain at reasonable LO input drive level. The conversion gain versus bandwidth tuning can be controlled digitally in further integration. The ability to integrate the IF-to-mmW mixer on the same chip along with the baseband signal processing and the RF-DAC based transmitter omits the necessity of implementing circuits on special materials. Hence, the production cost and complexity of mobile communication devices can be reduced significantly. Furthermore, the IF and RF-bandwidth of the proposed mixer is fully compliant with the 60 GHz ISM/WiGig bands.

ACKNOWLEDGEMENT

This work has been funded by the German Research Foundation within the project of LP100-2.

REFERENCES

[1] C. Huang and W. Lin, "A radio transceiver architecture for coexistence of 4G-LTE and 5G systems used in mobile devices," IEEE MTT-S Int. Microw. Symp. (IMS), 2017.

[2] Y. Li, et al., "Eight-element MIMO antenna array for 5G/Sub-6GHz indoor micro wireless access points," 2018 Int. Works. on Antenna Technology (iWAT), 2018.

[3] G. R. Maccartney, et al., "Indoor Office Wideband Millimeter-Wave Propagation Measurements and Channel Models at 28 and 73 GHz for Ultra-Dense 5G Wireless Networks," in IEEE Access, vol. 3, pp. 2388-2424, 2015.

[4] A. Hemadeh, et al., "Millimeter-Wave Communications: Physical Channel Models, Design Considerations, Antenna Constructions, and Link-Budget," in IEEE Comm. Surv. & Tut., vol. 20, 2018.

[5] C. Huang and W. Lin, "A radio transceiver architecture for coexistence of 4G-LTE and 5G systems used in mobile devices," 2017 IEEE MTT-S Int. Microwave Symp. (IMS), 2017.

[6] E. Bechthum, et al., "A novel output transformer based highly linear RF-DAC architecture," 2013 Europ. Conf. on Circ. Theory and Design (ECCTD), 2013, pp. 1-4.

[7] A. Hamidian, S. E. Barbin and G. Boeck, "A wideband Gilbert cell up-converter in 90 nm CMOS for 60 GHz application," 2015 SBMO/IEEE MTT-S Int. Microwave and Optoelect. Conf. (IMOC), 2015.

[8] J. Lee, et al., "60GHz direct up-conversion mixer with wide IF bandwidth and high linearity in 65nm CMOS," 2017 IEEE Int. Symp. on Radio-Frequency Integration Techn. (RFIT), 2017.

[9] H. P. Forstner, et al., "A fully integrated homodyne upconverter MMIC in SiGe for 60 GHz wireless applications," IEEE 11th Topical Meeting on Silicon Monolithic Integrated Circ. in RF Sys., 2011.

[10] N. Mazor et al., "Highly Linear 60-GHz SiGe Downconversion /Upconversion Mixers," in IEEE Microwave and Wireless Components Letters, vol. 27, no. 4, pp. 401-403, 2017.

[11] M. Chou and H. Chiu, "A broadband 60 GHz CMOS up-converter design and analysis," 2016 IEEE Int. Works. on Electromagnetics: Applications and Student Innovation Competition (iWEM), 2016.

[12] Y. Lin, C. Wang, W. Wen and T. Tsai, "A 12.1 mW 50-67 GHz up-conversion mixer with 6 dB conversion gain and 30.7 dB LO-RF isolation in 90 nm CMOS," IEEE Radio and Wirel. Symp. (RWS), 2014.

A Derating-Rules Compliant Ka-Band GaN-on-Si Power Amplifier Designed for Highly Reliable Satellite Applications

F. Costanzo[#], L. Pace[#], P.E. Longhi[#1], W. Ciccognani[#], S. Colangeli[#], R. Leblanc[*], E. Limiti[#]

[#]Electronic Engineering Dept. University of Rome "Tor Vergata", Rome, Italy

[*]OMMIC SAS, Limeil Brèvannes, France

[1]longhi@ing.uniroma2.it

Abstract — **The present work shows a thermal-aware MMIC HPA design approach that could be adopted when designing GaN/Si amplifiers for space applications, where reliability is a prime concern. It is demonstrated how thermal requirements for space applications limit device performance in the case of GaN/Si technology. On the other hand, following a thermal load-pull approach, it is also demonstrated how, with proper selection of the load termination, GaN/Si technology is able to achieve competitive results at Ka-band also when enforcing space derating rules.**

Keywords — **Power Amplifiers, Space, Derating Rules, MMIC, RF, Ka-Band, Junction Temperature.**

I. INTRODUCTION

The upper part of the Ka-band, and in particular the spectrum from 32 to 36 GHz, is particularly appealing for the electronics for aerospace and defence community since several applications of interest are there allocated. A first example is earth imaging, and in particular detection of water vapor at 35-36 GHz. Slightly below, at 32-35 GHz we have high-resolution, close-range targeting radars, used in seekers or in Electronic Warfare systems. Whether spaceborne or airborne, all these applications require the generation of high RF power at the operating frequency in conjunction with enhanced circuit reliability – a fundamental aspect in mission critical systems – without excessively burdening the platform's DC power and thermal management system. From a circuit point of view these two aspects are related to a single crucial point: keeping device operating temperature as low as possible during operation. High efficiency is a key requirement for a High Power Amplifier (HPA) operating in space environment [1], since solar-cells and batteries are the only power supplies available on board. Likewise, there is a similar need for reduced power consumption in airborne environments, especially on unmanned platforms. Consequently, this poses a challenge to designers: devising circuits operating at high frequency under demanding efficiency constraints. Besides efficiency and output power, high reliability and controlled aging must be guaranteed for High Power Amplifiers (HPAs) adopted in spaceborne critical systems in order to withstand an adequate lifetime (10 years or more). To enhance reliability and ensure long-term operation, electronics systems for space and avionic environments must comply with derating rules [2], endorsed by national and international space agencies. Thus, circuits must be designed to operate at a reduced level of electrical and thermal stress as compared to more traditional terrestrial and consumer applications. The major source of stress for high-power active devices is self-heating, which especially in space or avionic applications is negatively impacted by the absence of efficient heat conduction and convection. Therefore, the active devices' maximum operating temperature must be reduced with respect to the technology's maximum rating referred to standard earth operation. This work provides several design solutions, or rather a whole design flow, that simultaneously accounts for RF performance while keeping under control the maximum operating temperature of the active devices.

II. MMIC REQUIREMENTS & TECHNOLOGY SELECTION

The realized circuit aims at demonstrating the capability of the selected technology and proposed design flow to provide adequate RF performance at Ka-Band, and in particular 31 – 35GHz, in conjunction with high reliability. The latter is quantified as limiting the active device's channel temperature to +160°C. In particular, it is assumed that, to achieve a low failure rate over a 20-year mission duration, a Mean Time to Failure (MTF) requirement of at least 108 hours would be reasonable. Then a recommendation for derating can be made, based on RF and DC life test results, that the simulated peak channel temperature should not exceed 160°C in electronic systems where high reliability is a must. The RF requirements can be summarised as follows: gain greater than 25 dB, 20 % Power Added Efficiency (PAE) and output saturated power of +35 dBm. Regarding the technology choice, Gallium Nitride (GaN) is already well established in HPAs for ground-based applications [1], and it has become an attractive option also in the space domain. Only European industrial-grade and space qualified technologies were considered, so our selection fell on OMMIC's D01GH. This process offers GaN-on-Si HEMTs with 100 nm channel length, featured by interesting overall performance in terms of frequency, noise [3] and, more important for the present application, a 3.3 W/mm maximum power density at 12 V drain bias.

3–4 April 2022, London, UK

III. DESIGN SOLUTIONS

The first and fundamental decision in the classical design flow of a HPA is determining the number of transistors and the total periphery of the final (output) stage. Typically, in single considering the maximum power density of the technology at the operating frequency as the fundamental input parameter for the decision. However, this cannot be the only input when the transistor's maximum channel temperature is also a critical requirement.

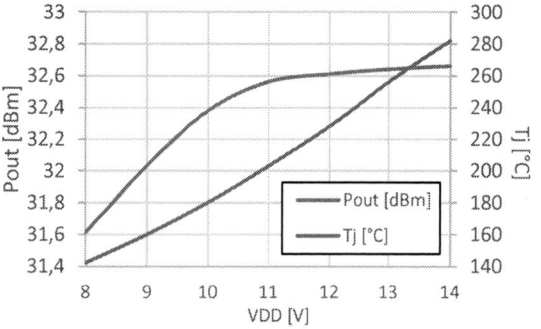

Fig. 1. 6x125um Pout and junction temperature Tj as a function of VDD.

Consequently, a trade-off between power density and operating channel temperature has to be accepted. The key factors affecting the junction temperature of the transistor mainly consist in the overall periphery of the transistor; both the gate and drain bias voltage value; and the extrinsic load presented to the active device. We will focus our attention on the final stage devices that are the most critical in terms of reliability. Via hole width is fixed at 20 μm, while its length can be increased arbitrarily to match the unitary gate width. The geometry used for this analysis has been selected with the following assumptions. The technology power density, in de-rated configuration to fulfil the maximum junction temperature operation, was initially set at 1.5 W/mm with the aim of a preliminary dimensioning the final stage of the HPA. Requested RF output power is more than 2 W. Also, we account for approximately 1 dB loss of generated RF power in the output matching network of the amplifier. Given these figures, the final stage periphery shall be at least 3 mm. Conservatively, a structure combining four 6 finger x 125 μm devices is selected to guarantee around 3W output power. The graph in Fig. 1 reports the output power of the single transistor and its channel temperature as a function of drain voltage V_{DD} at a fixed input power of +27 dBm at mid-band frequency. Base plate temperature is fixed at +80°C. The transistor is polarized in class AB corresponding to a gate voltage of −1.2 V. The junction temperature is estimated through:

$$T_j = T_{bottom} \exp\left(R_{th@T_{ref}} P_D / T_{ref}\right) \quad (1)$$

where T_{bottom} is the (absolute) base plate temperature. OMMIC PDK provides, for any device periphery, the value of $R_{th@TREF}$ at 20°C (293.15 K). The total thermal resistances ($R_{th@TREF}$ of the substrate evaluated at the reference temperature T_0^a (superscript 'a' denotes absolute temperatures in Kelvin) is:

$$R_{th@T_{ref}} = \frac{H}{A} \frac{1}{K_{th}(T_{ref})} \quad (2)$$

where H is the substrate thickness, A the area of the device footprint and $K_{th}(T_0^a)$ the thermal conductivity at T_0^a.

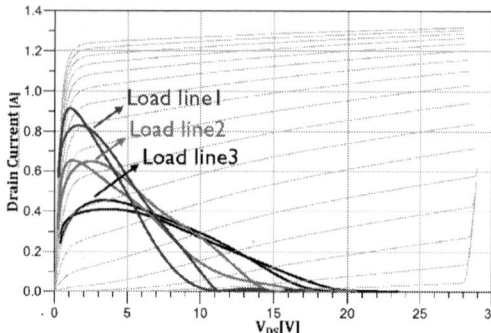

Fig. 2. Load lines corresponding to biases reported in Table 1 for a 6x125um device.

Table 1. Bias configuration for the three load lines reported in Fig. 2.

Load line	VGG (V)	VDD (V)	P_{OUT} (dBm)	Tj (°C)
1 - Blue	-1.2	9.0	32	160
2 - Red	-1.3	10.5	31	160
3 - Black	-1.4	12.0	30	160

Higher output power could be obtained in principle by increasing the drain voltage V_{DD}. A similar phenomenon occurs when V_{DD} is fixed and V_{GG} is swept. In this second case there is an adjustment of drain that has an impact on output power and channel temperature, although less dramatic than the case in which V_{DD} is swept. Graph in Fig. 1 was obtained for a fixed output termination. This is not the optimum case since output terminations should be re-tuned when V_{DD} changes. Fig. 2 shows the load lines for three V_{DD} values. V_{GG} is adjusted in order to modulate the drain current and obtain always 160°C channel temperature. Input signal is +27 dBm at mid-band frequency while backplate temperature is set at +80°C. Drain and gate bias configuration used for the three cases is reported in Table 1. By increasing V_{DD} we need to decrease V_{GG} to maintain low the dissipated power in absence of RF signal. Due to V_{DD} and by decreasing V_{GG} we obtain less gain and less output power. Moreover, by increasing V_{DD} and by decreasing V_{GG} we obtain an optimum load which moves towards the edge of the Smith Chart, as shown in Fig. 3.

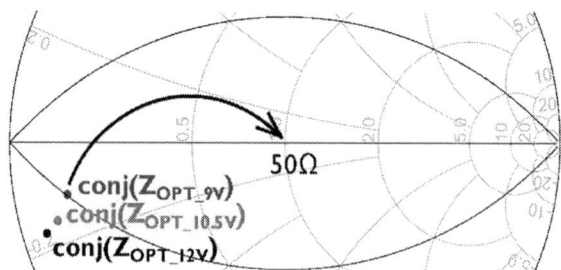

Fig. 3 Optimum load terminations as a function of bias point

Increasing V_{DD} his implies a higher complexity to match Z_{OPT} towards 50 Ω, and a greater insertion loss is expected by the combiner. Consequently, optimum engineering choice is to select a lower value of V_{DD} in conjunction with higher V_{GG} to simultaneously fulfil non-linear requirements and reliability aspects. A 4-stage topology, shown in Fig. 4, is adopted to obtain the prescribed 20 dB gain. The typical FET periphery scaling, increasing from input to output, is adopted. As stated before, the final stage periphery was set at 3mm, and we opted to realize the final stage with four separate transistors, thus combining the output of four 6x125 µm FETs. The input and inter-stage matching networks have been designed also to provide stability in-band and at lower frequencies. For this purpose, parallel R//C networks are applied to the FET's gate terminal.

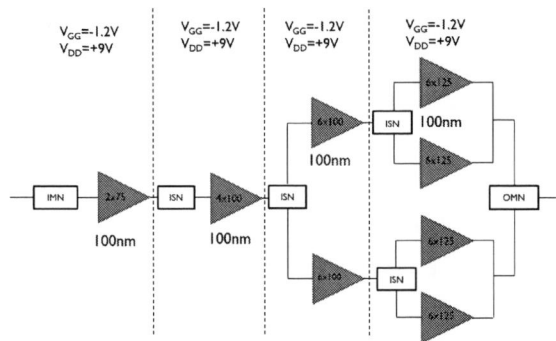

Fig. 4 Simplified block scheme of the designed HPA.

In addition, series R-C to ground have been inserted in the matching structures. After FETs stabilizing, we proceeded with the design of the output combination structure. The first step consisted in determining the optimum load impedance (Z_{OPT}) to be shown at the transistor's drain terminal that simultaneously fulfils the output power and maximum junction temperature specification. From a load pull analysis conducted at mid band frequency, the value of Z_{OPT} is found to be 12.6+13.1j Ω when the transistor is delivering +30 dBm output power. The Output combiner was then designed through an almost symmetrical structure so as to provide in-phase signal recombination, provide Z_{OPT} load to each FET when the output port is terminated on Z_0, and finally minimize insertion losses. The design then proceeded backwards with the synthesis of all the interstage and the input matching network.

Fig. 5 shows a microphotograph of the realized HPA. The structure has been slightly rendered asymmetrical to simplify an eventual integration in a multi-functional chip, thus positioning bias pads only on the upper side of the MMIC. This solution does not allow to show the same load to all the employed devices, but this effect has been minimized by an accurate positioning of decoupling capacitors.

IV. MMIC PERFORMANCE AND DISCUSSION

The design method here presented provides indication on the design of a power amplifier in which the junction temperature requirement is explicitly considered as a design

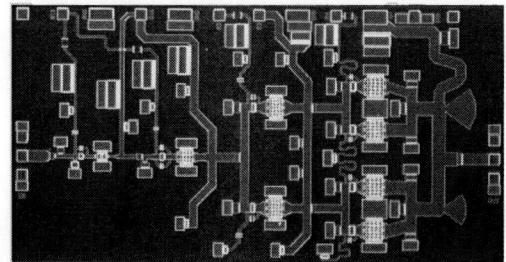

Fig. 5 Microphotograph of the designed MMIC. Size is 3.7x2 mm².

input. It is shown that lowering the V_{DD} helps obtain low channel operating temperature in conjunction with adequate output power level. A HPA is designed and tested to verify the feasibility of the proposed design solution. In Fig. 6 measured small signal (left) and non-linear (right) performance collected at 80°C backplate temperature are shown. A 27 dB gain was achieved in the whole design band. When the gain compression reaches 3 dB the designed MMIC shows a 35 dBm output power level, corresponding to a 11dBm available source power. A 20 % PAE is achieved, as well, being the dc power consumption of the chip around 16 W (+9.0 V by 1.78 A) at maximum PAE. All these three figures of interests show a very low spread all over the design band. The bottom graph of Fig. 6 also shows the Tj simulated behaviour of one of the last stage's devices, being the most critical one, obviously. As we can see, the temperature goes slightly above the 160°C reliability limit only at the band edges, thus assuring a more than satisfying overall performance despite working in derating conditions, showing the effectiveness of the presented thermal-aware approach and the selected technology. Table 2 reports a comparison between the presented MMIC and other GaN-based HPAs presented in literature. Despite adopting a drain bias level much lower than the values suggested by foundry for high power applications (i.e. 12-13 V), the presented MMIC still shows a more than satisfying performance while keeping the Tj below 160°C even in critical last-stage devices, contrarily to other HPAs shown in the same table where the temperature was pushed up to 200°C, thus ensuring to withstand a much higher operative life-time.

Table 2. Ka-band Power Amplifiers performance comparison.

REF	Freq (GHz)	PAE (%)	POUT (dBm)	VDD	Tech.	Max Tj / Tback (°C)
[4]	31-39	24	37	4	100nm GaAs	140
[5]	27-34	30	37	12	100nm GaN/Si	-/-
[6]	38-42	28	41	12	100nm GaN/Si	200 / 80
[7]	46-48	-	36	12	100nm GaN/Si	200 / 80
[8]	26-28	19	43	-	150nm GaN/SiC	-/-
This work	31-35	20	35	9	100nm GaN/Si	160 / 80

255

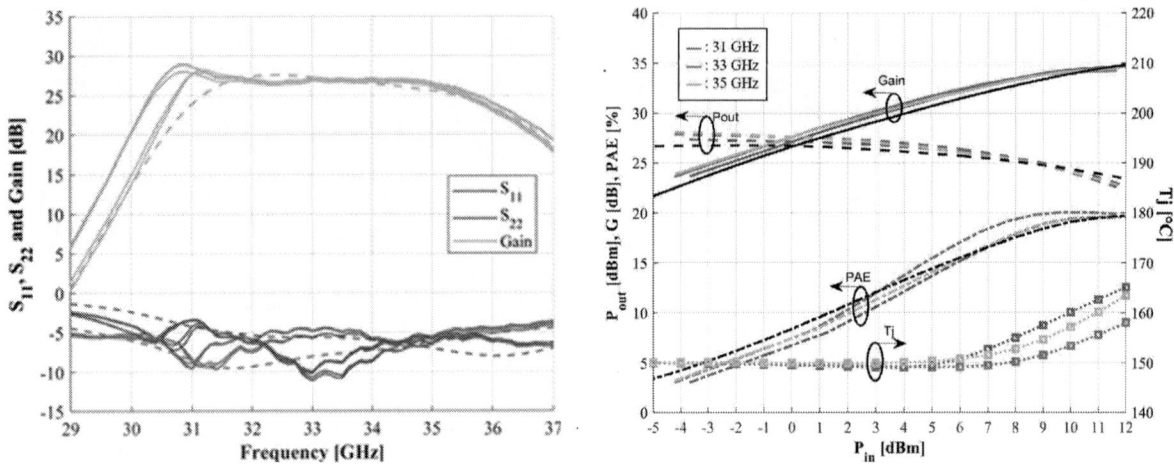

Fig. 6. Designed MMIC's S-Parameters (left) and non-linear performance (right). In right graph measured (solid lines) and simulated (dashed lines) performances of four chips are reported, while Tj behaviour in right graph is only simulated. Black lines report simulations where measured performance is given. NL simulations are given at 33 GHz.

V. CONCLUSIONS

This work shows a thermal-aware MMIC HPA design approach that could be adopted for the design of critical GaN/Si amplifiers for space environments. Despite it is demonstrated how derating rules strongly limit device performance requiring larger peripheries for the same targeted output power, it is also shown how with proper selection of the load termination, GaN/Si technology is able to achieve competitive results at Ka-band also when enforcing space derating rules. A HPA working in the 31-35 GHz band delivering 35 dBm output power has been designed and its performance have been shown also to demonstrate the effectiveness of the presented method.

ACKNOWLEDGMENT

Researchers have received funding from the European Union for project H2020-COMPET-2017 MiGANSOS No. 776322.

REFERENCES

[1] C. Ramella, A. Piacibello, R. Quaglia, V. Camarchia, and M. Pirola, "High efficiency power amplifiers for modern mobile communications: The load-modulation approach," Electron., vol. 6, no. 4, pp. 1–29, 2017, doi: 10.3390/electronics6040096.

[2] European Coorporation for Space Standarization, "Space product assurance Radiation hardness assurance - EEE components ECSS Secretariat ESA-ESTEC Requirements & Standards Division Noordwijk, The Netherlands," Ecss-Q-St-60-15C, no. October, pp. 1–32, 2012.

[3] L. Pace et al., "Design and Validation of 100 nm GaN-On-Si Ka-Band LNA Based on Custom Noise and Small Signal Models," Electronics, vol. 9, no. 1, p. 150, Jan. 2020, doi: 10.3390/electronics9010150.

[4] A. Alizadeh, M. Frounchi, and A. Medi, "On Design of Wideband Compact-Size Ka/Q-Band High-Power Amplifiers," IEEE Trans. Microw. Theory Tech., vol. 64, no. 6, pp. 1831–1842, Jun. 2016, doi: 10.1109/TMTT.2016.2554578.

[5] A. Gasmi et al., "10W power amplifier and 3W transmit/receive module with 3 dB NF in Ka band using a 100nm GaN/Si process,"

2017 IEEE Compd. Semicond. Integr. Circuit Symp. CSICS 2017, vol. 2017-Janua, pp. 1–4, 2017, doi: 10.1109/CSICS.2017.8240431.

[6] J. Moron, R. Leblanc, F. Lecourt, and P. Frijlink, "12W, 30% PAE, 40 GHz power amplifier MMIC using a commercially available GaN/Si process," in 2018 IEEE/MTT-S International Microwave Symposium - IMS, Jun. 2018, pp. 1457–1460, doi: 10.1109/MWSYM.2018.8439689.

[7] G. A. Joël Moron, Rémy Leblanc, Peter Frijlink, François Lecourt, Manuel Sigler, George Goussetis, "A Novel high-performance V-Band GaN MMIC HPA for the QV-LIFT project," in Ka and Broadband Communications Conference 2018, 2018, pp. 1–6.

[8] Y. Yamaguchi, J. Kamioka, M. Hangai, S. Shinjo, and K. Yamanaka, "A CW 20W Ka-band GaN high power MMIC amplifier with a gate pitch designed by using one-finger large signal models," in 2017 IEEE Compound Semiconductor Integrated Circuit Symposium (CSICS), Oct. 2017, pp. 1–4, doi: 10.1109/CSICS.2017.8240422.

Proceedings of the 16th European Microwave Integrated Circuits Conference

A 3.3 to 11.3 GHz Differential LNA with Slight Imbalance Active Balun in 0.15-μm GaAs pHEMT Process for Radio Astronomical Receiver

Ting-Hsuan Fan[#], Chau-Ching Chiong[*], and Huei Wang[#]

[#]Graduate Institute of Communication Engineering, National Taiwan University, Taipei 106, Taiwan

[*]Academia Sinica Institute of Astronomy and Astrophysics (ASIAA), Taipei, Taiwan

Abstract —This paper presents a fully-integrated broadband, low noise amplifier (LNA) with differential input and single-ended output for radio astronomical receiver. The inductive source degeneration is adopted to minimizing the noise figure, the negative feedback is applied for broadband design, and the fully on chip matrix balun is introduced behind the third-stage. The 3-dB bandwidth of the DLNA covers from 3.3 to 11.3 GHz. The measurement results demonstrate peak gain of 28.3 dB, average noise figure of 1.4 dB with DC power consumption of 85 mW. The chip area of the amplifier is 3×2 mm^2.

Keywords —active balun, differential low noise amplifier, GaAs pHEMT, Square Kilometre Array (SKA).

I. INTRODUCTION

The Square Kilometre Array (SKA) project is an on-going project with unprecedented sensitivity for radio astronomy. To receive and process the weak incoming signal from the universe, high sensitivity receiver is required. In general, dipole feed and built-in passive balun are usually designed and integrated together [1]. However, any 180 hybrid in front of low noise amplifier (LNA) will contribute noise to the system, thereby reducing the receiver sensitivity. To reduce the system noise, it is desired to place LNA as close as possible to the antenna feed. Therefore, LNA with differential input and power combining after signal amplification is an attractive choice for this type of receiver.

Passive baluns such as rat-race coupler or Marchand balun can be used as the bi-directional converters between differential and single-ended signals [2]. The disadvantages of passive baluns are inherently lossy, required large chip area at low frequency and limited bandwidth. On the other hand, active baluns demonstrate a wider bandwidth with little gain and phase imbalance. There are various types of active baluns. The single device balun [3] and the common-gate common-source (CGCS) balun [4] were proposed for narrowband applications due to the limited gain in common-gate cells. On the other hand, the matrix baluns have the potential of offering wideband performance with a small chip area [5]. Thus, the matrix balun is adopted in this design.

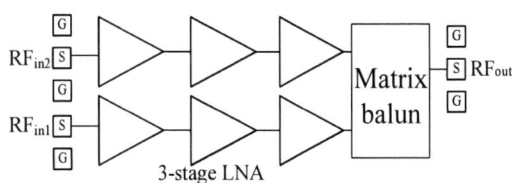

Fig. 1. Circuit block diagram of the proposed DLNA.

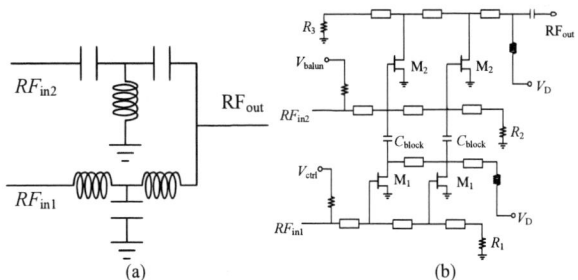

Fig. 2. Circuit schematic of the (a) conventional passive lumped L-C balun (b) proposed broadband matrix balun.

A differential LNA (DLNA) with slight imbalance matrix balun is presented in this paper. The proposed DLNA can cover the SKA band 5a (4.6 to 8.5 GHz) frequency ranges [6].

Compared with other published DLNAs, this work exhibits high gain, wide bandwidth and high common-mode rejection ratio (CMRR).

II. CIRCUIT DESIGN

The proposed DLNA is implemented in WIN's 0.15-μm GaAs pHEMT process with passivation layer. The block diagram of the DLNA is shown in Fig. 1. The DLNA is composed of two LNAs and a broadband active balun. Each LNA is with three stages to provide high gain. Inductive source degeneration is adopted to the first and the second stages for minimizing the noise figure. The input matching network transforms the 50-Ω load impedance to the optimal impedance for noise matching. In the first and the second

257

3–4 April 2022, London, UK

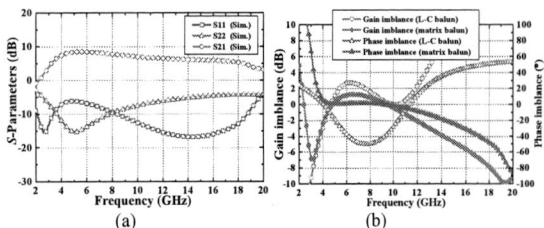

Fig. 3. The (a) simulated S-parameters of proposed balun and (b) simulated gain and phase imbalance of matrix balun and passive L-C balun.

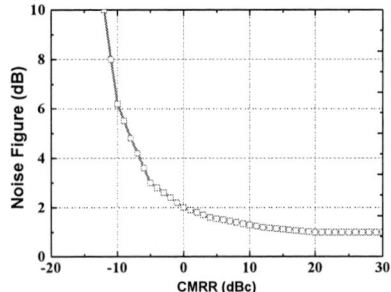

Fig. 4. The simulated noise figure versus CMRR of ideal DLNA.

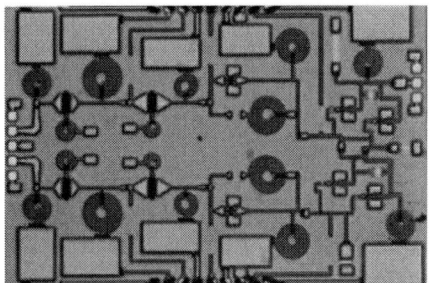

Fig. 5. Chip photograph with a chip size of 3×2 mm^2.

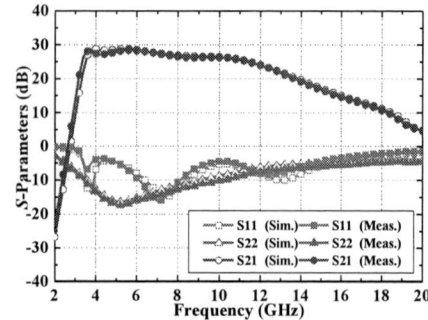

Fig. 6. Measured and simulated 2-port S-parameters of the DLNA.

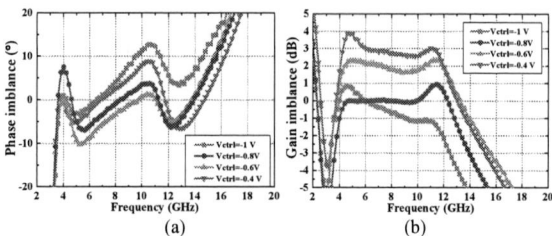

Fig. 7. The measured (a) phase and (b) gain imbalance of the DLNA.

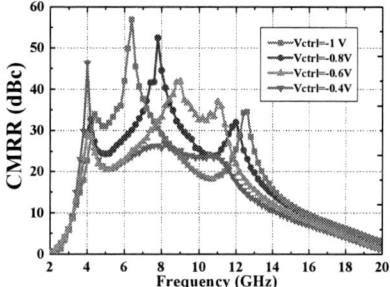

Fig. 8. The measured CMRR at different control voltage of the DLNA.

stage of the DLNA, the device size is 8 fingers with each gate width of 25 μm. In the third stage of the DLNA, the device size is 4 fingers with gate width of 25 μm. In order to achieve wider bandwidth, R-L-C feedback is applied to the third stages of the DLNA. Through R-L-C feedback, the gain is with positive slope in frequency response which can improve the flatness within the targeted bandwidth [7]. The optimized performance of the device occurs at the gate voltage bias of -1 V with the drain voltage of 1 V. As shown in Fig. 2 (b), the proposed matrix balun can be recognize as a six-port device with three open ports terminated by resisters R_1, R_2, and R_3. The other three are input/output ports, and denoted as RF_{in1}, RF_{in2} and RF_{out}. Theoretically, the relation between the input ports and output ports can be described by a six-port chain matrix representation. However, the full analytical expressions of six-port chain matrix over frequency are too complicated to calculate precisely for optimization purpose. Therefore, to simplify the design procedure of matrix balun, some design parameters shall be assigned first. In general, the matrix balun can be regarded as a matrix amplifier. We followed the design procedure in [5]. For frequency approach zero, the S-parameters can be expressed as [5]

$$S_{31} = \frac{18g_{m1}g_{m2}R_1R_2R_3Z_0^2}{(R_1+Z_0)(R_2+Z_0)(R_3+Z_0)} \quad (1)$$

$$S_{32} = \frac{-6g_{m2}R_2R_3Z_0}{(R_2+Z_0)(R_3+Z_0)} \quad (2)$$

where Z_0 denoted as the system impedance. RF_{in1}, RF_{in2} and RF_{out} are represented with port1, port2 and port3. The input signals from the two input ports will equally change the phase and the amplitude due to two common-source transistors M_2 in the matrix balun. This means that the gate and drain bias voltage (V_{balun} and V_D) and the size of M_2 transistors can be selected with high freedom. By sweeping the transistor parameters, the size and the bias condition of M_2 are chosen for flat frequency response and minimum power consumption. And then the size and the bias condition of M_1 can be derived by (1). The final size of M_1 and M_2 are both 2-finger with total 50-um gate width with $V_d = 1V$, $V_{balun} = -1V$ and $V_{ctrl} = -1V$. V_{ctrl} can be fine-tuned during the measurement for optimizing phase and amplitude imbalance. In the desired frequency from 4.6 to 8.5 GHz, the gain and phase imbalance are barely affected by

258

Table 1. Performance Comparison of Published DLNAs and Baluns

Ref.	Process	Topology	3-dB freq. (GHz)	BW (%)	Gain (dB)	P_{dc} (mW)	Noise Figure (dB)	Amp. Imbalance (dB)	Phase Imbalance (°)	Area (mm²)
[8]	0.13-μm CMOS	DLNA	22.9-26.4#	14.2	14.7	20.2	4.3	0.6	10*	0.82
[9]	0.18-μm SiGe BiCMOS	DLNA	17.5-25#	35.2	19.2	73.8	4	1	10.4	0.69
[10]	0.15-μm GaAs pHEMT	DLNA	4.3-16.3#	116	35.4	90	1.05	3.5*	48*	5
[11]	0.15-μm GaAs pHEMT	DLNA	1.5-3.7#	85	31.8	25	0.73	0.5*	30*	5
[12]	0.15-μm GaAs pHEMT	Balun	4.3-7*	48	-2	0	-	0.63	3.7	1
[5]	0.15-μm GaAs mHEMT	Balun	4-40	163	3.5	53	-	2	20	1.08
This work	0.15-μm GaAs pHEMT	DLNA	3.3-11.3#	110	28.3	85	1.4	1	7	6

* estimated from figure based on their papers or dissertations, # 3-dB bandwidth of differential-mode gains.

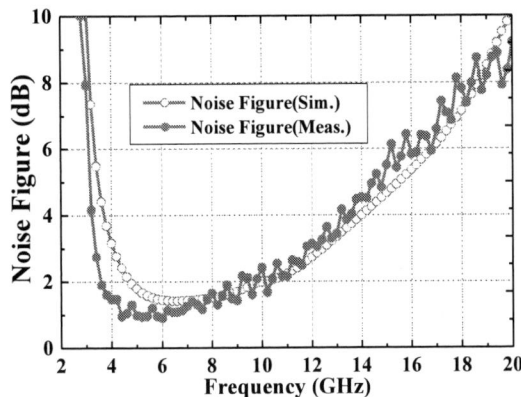

Fig. 9. The simulated and measured noise figure versus frequency.

stretching the transmission lines between transistors. The design is then optimized in simulation tool, and the open port termination resistors of $R_1 = 25$, $R_2 = 60$ Ω, and $R_3 = 330$ Ω are adopted for flat frequency response. Fig. 3(a) shows the simulated S-parameters of the matrix balun. The 3-dB small-signal gain bandwidth covers from 3.3 to 17.4 GHz. Higher gain or wider bandwidth can be achieved by adding more stages or decreasing. In [12], there is poor isolation between two output ports, it means that the two input signal will interfere with each other which in some applications can be a drawback. However, in this work, isolation is greater than 30 dB by cascading LNA in front of the matrix balun.

In order to demonstrate the advantage of the proposed matrix balun, as shown in Fig. 2 (a), a conventional passive lumped L-C balun consisting of 90° 3-element T-type high-pass phase shifter and low-pass phase shifter is also designed for comparison [10]. Fig. 3(b) shows the simulated gain and phase imbalance of the proposed matrix balun and the passive L-C balun. Within the target bandwidth from 4.6 to 8.5 GHz, passive L-C balun shows enormous phase (12° to 49°) and gain imbalance (0.1 to 2.8 dB). Moreover, it is well known that the poor CMRR is related to the higher common-mode gain due to the poor gain and phase imbalance of the DLNA. It will also degrade the noise performance of the system. Fig. 4 shows the simulated noise figure versus CMRR. The effect

of CMRR on noise figure was simulated by using an ideal DLNA with 10-dB gain and 1-dB noise figure. In conclusion, when the CMRR is worse than 20 dBc, the noise figure increases significantly.

III. MEASUREMENT RESULTS

The chip photograph of the proposed DLNA is shown in Fig. 5. The DLNA is measured through on-wafer probing with dc wire-bond to an external PCB board. The 2-port differential mixed-mode S-parameters are measured from 10 MHz to 20 GHz using Keysight Microwave Network Analyzer PNA-X N5245A. The measurement is carried out with mixed-mode with one differential port and one single-ended port. The network analyzer can calculate 2-port differential mixed-mode S-parameters based on 3-port measurements. The measured and simulated 2-port differential mixed-mode S-parameters of the proposed DLNA are shown in Fig. 6. The measurement results demonstrate 28.3 dB peak gain. The measured 3-dB small-signal gain bandwidth covers from 3.3 to 11.3 GHz. The measured gain imbalance and phase imbalance are shown in Fig. 7. When the Vctrl is chosen at -0.8 V. The measured gain imbalance is lower than 0.1 dB and the measured phase imbalance is within ± 5° in the desired operating frequency. Fig. 8 shows the measured CMRR which is higher than 25 dBc. Therefore, according to discussion mentioned above, the impact of CMRR on noise figure is not significant in this work. We followed the noise figure measurement procedure in [11]. According to the method, an external broadband passive balun with high return loss and low imbalances is connected in front of the device under test (DUT). Fig. 9 demonstrates an average noise figure of 1.4 dB.

This proposed DLNA with a broadband matrix balun is firstly to adjust gain and phase imbalance for better CMRR and resolve the disadvantages of on-chip passive balun such as high loss, enormous imbalance and limited bandwidth. Table I summarized the performance of the proposed DLNA and other published DLNAs and baluns in recent years. The proposed DLNA achieves high gain, wide bandwidth and good CMRR compared with the published HEMT LNAs in similar frequencies, and therefore it is suitable for radio astronomical receiver system.

IV. CONCLUSION

In this paper, a fully-integrated MMIC broadband DLNA fabricated in 0.15-μm GaAs pHEMT is presented. The proposed DLNA adopted inductive source degeneration to minimize the noise figure and R-L-C feedback to achieve wider bandwidth. The fully on-chip broadband matrix balun is designed with little gain and phase imbalance. Thus, the proposed DLNA exhibits the potential for application in the next-generation radio telescope systems.

ACKNOWLEDGMENT

This work is supported in part by research program from Academia Sinica Institute of Astronomy and Astrophysics (ASIAA), Taipei, Taiwan.

REFERENCES

[1] C. A. Balanis, "Linear wire antennas," in *Antenna Theory: Analysis and Design*, 3rd ed., USA: Wiley-Interscience, pp. 182-184.

[2] David M. Pozar, *Microwave engineering.*

[3] M. E. Goldfarb, J. B. Cole, and A. Platzker, "A novel MMIC biphase modulator with variable gain using enhancement-mode FET's suitable for 3 V wireless applications," in *IEEE Microw. Millimeter-Wave Monolithic Circuits Symp.*, Dig., May 1994, pp. 99–102.

[4] T. Hiraoka, T. Tokumitsu, and M. Akaike, "A minaturized broad-band MMIC frequency doubler," *IEEE Trans. Microw. Theory Tech.*, vol. 38, no. 12, pp. 1932–1937, Dec. 1990.

[5] M. Ferndahl and H. Vickes, "The combiner matrix balun, a transistor based differential to single-ended module for broadband applications," *IEEE MTT-S International Microwave Symposium*, Baltimore, MD, 2011, pp. 1-4.

[6] SKA website [Online] https://www.skatelescope.org/

[7] Y. Wang, C.-C. Chiong, J.-K. Nai and H. Wang, "A high gain broadband LNA in GaAs 0.15-μm pHEMT process using inductive feedback gain compensation for radio astronomy applications," *IEEE International Symposium on Radio- Frequency Integration Technology (RFIT)*, Sendai, Aug. 2015

[8] J. Yeh, C. Yang, H. Kuo and H. Chuang, "A 24-GHz transformer-based single-in differential-out CMOS low-noise amplifier," *IEEE Radio Frequency Integrated Circuits Symposium*, Boston, MA, 2009, pp. 299-302.

[9] J. Lee and C. Nguyen, "A K/Ka -Band Concurrent Dual-Band Single-Ended Input to Differential Output Low-Noise Amplifier Employing a Novel Transformer Feedback Dual-Band Load," *in IEEE Transactions on Circuits and Systems I: Regular Papers*, vol. 65, no. 9, pp. 2679-2690, Sept. 2018.

[10] Z. Jiang, Y. Chang, Y. Wang, C. Chiong and H. Wang, "High Gain Fully-Integrated Broadband Differential LNAs in 0.15-μm GaAs pHEMT Process Using R-L-C Feedback Gain Compensation for Radio Astronomical Receiver," *IEEE MTT-S International Microwave Symposium (IMS)*, Boston, MA, USA, 2019, pp. 999-100

[11] W.-C. Huang, C.-C. Chiong, and H. Wang, "A fully-integrated S-band differential LNA in 0.15-μm GaAs pHEMT for radio astronomical receiver," *IEEE Radio-Frequency Integration Technology (RFIT)*, Melbourne, Aug. 2018.

[12] C. Chang and Y. Lin, "On-chip transformer-coupled balun bandpass filter for 5-GHz applications," *IEEE MTT-S International Microwave Symposium*, Phoenix, AZ, 2015, pp. 1-4.

Benefits of AlGaN/GAN thermal ROM coupling with industrial non-linear transistor model

C. Chang, L. Brunel

United Monolithic Semiconductors SAS, France
{christophe.chang, laurent.brunel}@ums-rf.com

Abstract — **This work focuses on the coupling of a thermal Reduced Order Model (ROM) of a 0.15µm AlGaN/GaN HPA mounted in a measurement test fixture with a non-linear electro-thermal transistor model in order to evaluate the HPA junction temperature in operation.**

Keywords — **Reduced Order Model, GaN-based devices, thermal modelling, non-linear transistor modelling.**

I. INTRODUCTION

The junction temperature (Tj) knowledge of a MMIC is essential to ensure the circuit reliability and the performances within the lifetime defined for a given technology. [1]

In UMS Design Kit (DK), Tj is evaluated through a non-linear transistor model which uses an equivalent $R_{th}C_{th}$ network based on 3D Finite Element Method (FEM) thermal simulations results. At this stage, the analysis is limited to the transistor itself and is not suitable for large design since thermal crosstalk between transistors and thermal gradient in the assembly are not taken into account. Thus, a thermal simulation of the circuit within its environment (layout and assembly) is needed for each test condition (temperature and biasing) in order to evaluate maximum chip backside temperature which is used as input for transistor model allowing the junction temperature simulation.

UMS is willing to find a bridge between ANSYS thermal model and ADS environment to take into account the temperature effect during the design phase. CADFEM Company is proposing an ANSYS add-on (MOR inside ANSYS [2]) which translates a 3D FEM thermal model into a simplified model so called Reduced Order Model (ROM) in Spice format compatible with ADS software.

In this paper, a short description of UMS GH15 technology is given. Then thermal modelling methodology is described, and thermal ROM capabilities are discussed. A case study coupling the UMS GH15 model with a thermal ROM of a 20W class MMIC HPA mounted in a custom measurement test fixture is finally exposed.

II. TECHNOLOGY DESCRIPTION

GH15-10 technology [3] is based on AlGaN/GaN epitaxy on 4-inch SiC substrate. The transistor features a 150-nm T-shaped gate in combination with a Source-Terminated Field-Plate allowing to increase the drain voltage operation and, thus, to improve the RF performances. The substrate is thinned to 70 µm by means of a back-side process that ensures the compatibility with both soldering and gluing based assembly

processes. GH15-10 key parameters are summarized in the Table 1. UMS foundry provides a complete Design Kit available on Keysight ADS and Cadence AWR environments. Design kit library features thin-film resistors, self-inductances, via-holes, MIM capacitors (175pF/mm) models and linear and nonlinear transistor models suitable for noise, switch and amplifier designs.

The GH15 nonlinear Hot-FET model, dedicated for power amplifier (PA) application, uses an empirical model developed internally to ensure fast and accurate simulations. Moreover, it features electro-thermal capability providing the junction temperature and includes the trapping effects involved in the GaN technology for accurate large signal simulations.

Table 1 : GH15-10 technology main characteristics

Parameter	Typ. value
Cut-off frequency (ft)	40GHz
RF power density	3.5W/mm
IDSmax@VD=7 V, VG=2 V	1.4A/mm
GmMax@ VD=7 V	390mS/mm
Vpinch-off@VD=7V Ids=Idss/100	-3.2V
Drain-source voltage bias	20V
Leakage current@VD=7V	< 200µA/mm

III. UMS THERMAL MODELLING METHODOLOGY

A. Temperature assessment at transistor level

Junction temperature of AlGaN/GaN transistors has been assessed through FEM using ANSYS software thanks UMS background in this area. Based on a bibliographic study, a specific methodology has been developed for the thermal modelling of GaN devices: the heat sources location and dimensions have been determined through pure 2D physics-based simulations, and the interfacial thermal resistance between SiC and GaN layers is modelled by a thin layer so called TBR (Thermal Boundary Resistance) which thermal conductivity has been determined through Raman measurement [4].

Materials thermal property and its temperature dependence have been taken from the literature and assessed through several infrared (IR) and Raman experimental measurement. Indeed, IR and Raman measured temperatures have been compared with the simulated temperature values extracted from the ANSYS 3D model of the device. [5]

Thermal simulations have been performed for different devices topologies (number of gate fingers N, unitary gate width Wu) versus chip backside temperature (Tb) at a typical dissipated power of the technology. Then thermal resistance (R_{th}) values have been extracted to build R_{th} abacus as a function of N, Wu and Tb.

To represent properly the transistor thermal transient behaviour, two $R_{th}C_{th}$ Foster networks have been implemented into the GH15 DK transistor model. R_{th} extracted from ANSYS has been split in two parts composing the resistive part of each Foster network. The first cell represents the unitary gate self-heating and the second, the thermal coupling effect across the device. Transistor thermal capacitances C_{th} values are deduced from transient FEM simulations and are proportional to the transistor gate development. Thermal conductivity dependence of the GaN and the SiC substrate is taken into account thanks to a mathematical operation based on Kirchhoff transformation [6]. Based on this representation the non-linear transistor model is able to provide junction temperature value depending on the transistor geometry, the backside temperature (Tb) and heat dissipated power value.

B. Temperature assessment at circuit level

In order to evaluate accurately maximum peak junction temperature of a circuit, it is very important to take into account thermal crosstalk between adjacent transistors. However, temperature assessment of a complete MMIC composed of dozen transistors within its environment by FEM can be a time consuming process. Indeed, a very large number of elements would be needed to properly capture temperature gradient at the periphery of each gate fingers and evaluate Tj values. Therefore, UMS has developed a specific methodology for Tj evaluation at circuit level.

Thermal stack up is built into ANSYS software (radiator, PCB, package, die attach) with a 70 μm SiC thick block of the size of the chip. The heat dissipation is simplified thanks to imprints representing the transistor active area at the surface of the SiC substrate according to the circuit layout. As an example, a GH15 8x125μm transistor with 20μm gate pitch is represented by a dissipative area of 8x125x20 μm².

Temperature gradient across the structure is obtained by solving the thermal model with the following boundary conditions: a reference temperature is applied on the backside of the test structure, and the appropriate amount of power is dissipated on the corresponding area of each transistor. Depending on the bias point, dissipated power values are determined using electrical model. Maximum chip backside temperatures below each transistor of interest are then extracted from thermal model, and finally the junction temperatures values are determined using R_{th} abacus or simulated with the electrical model.

Although this methodology allows a precise evaluation of maximum peak junction temperature of a circuit within its environment, it requires a 3D FEM thermal simulation for each new biasing or environmental (temperature) conditions.

IV. REDUCED ORDER MODEL

A. MOR principle

Model Order Reduction is an area of mathematics which enables a formal approximation of the physical model and hence the generation of a compact model suitable for system level simulation as illustrated in the Fig 1.

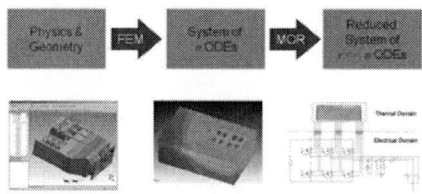

Fig 1: MOR principle illustration

The detail description of algorithm [6] based on the idea that high dimensional state vector (as the temperature vector) can be well approximated by lower dimensional subspace in which the original system can be projected. Performing MOR of FEM model can be directly performed by the software tool MOR inside ANSYS distributed by CADFEM.

B. MOR inside ANSYS software capabilities

The implementation of a thermal ROM in electrical simulation software should allow easy and fast electro-thermal analysis. Indeed, thanks to the high compatibility of Spice format, it might be easily implemented in any electrical simulation software. It allows taking into account thermal environment (assembly, circuit layout) and is a versatile feature that can be used either in DC (Rth) or in transient (Zth) in pulsed or radar mode; which is not possible using only FEM because of the huge computer resources requirements.

It offers a reduced time analysis since only one Finite Element Analysis (FEA) is needed to build the ROM that could be used in any conditions (temperature & biasing points). As there is no need for FEA, electro-thermal simulations can be performed in the electrical environment with the advantage to be faster than co-simulations. It might also be possible to build ROM library (package, test fixture) that could be easily coupled with the electrical model for fast overview of system thermal behaviour.

At last, ROM might handle 3D complex system such as System in Package (SiP), Wafer-level Packaging (WLP) considering a single ROM or by coupling different ROMs together.

C. Limitations

Despite plenty of advantages, ROM is a linear model which doesn't take into account material nonlinearities such as thermal conductivity versus temperature. However, it is possible to overcome this problematic by interpolation of several ROMs at different temperature or using mathematical function such as Kirchhoff function. Also, ROM depends on geometry (layout, assembly...), so any change requires building a new ROM.

V. ROM IMPLEMENTATION EXAMPLE

A. Thermal model description

The thermal model (Fig. 2) represents a 2 stages MMIC GH15 HPA brazed with 25 μm AuSn on a 150 μm thick molybdenum tab and 25 μm InPb on a 14 mm thick copper insert with a hole for thermocouple insertion. The chip is modelled as a 70 μm thick SiC block with the imprints of both active area of the transistors and passive elements according to circuit layout. The temperature has been fixed to 83°C at the backside of the measurement test fixture. Heat flows of 9W, 25.6W and 7.7W have been respectively applied on active area of stage 1, active area of stage 2 and passive elements.

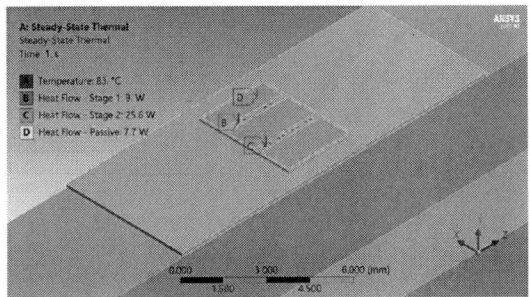

Fig. 2: Thermal model of GH15 HPA mounted in a test fixture

B. ROM implementation and validation

First step of ROM implementation process consists of validation of ROM model with respect to temperatures provided by ANSYS thermal model. Transient simulations have been performed in ADS Keysight environment. Same dissipated powers (43.3W) as the ANSYS model have been applied at the ROM input ports through 3 current sources corresponding respectively to the first and second amplifier stages and heat dissipated by the passives. On Fig. 3 are depicted the output backside temperature voltages compared to those given by the ANSYS model for 83°C test fixture backside temperature.

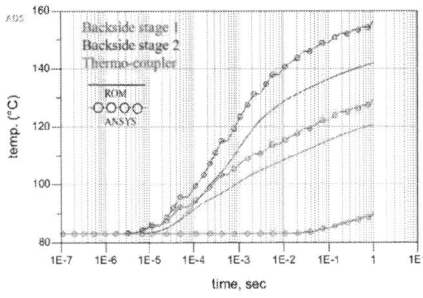

Fig. 3: ROM versus ANSYS temperatures

Transient ROM temperatures provide a fair prediction accuracy, but some discrepancies are visible and are more pronounced as far the temperature is increasing. This effect highlights the main limitation of the model reduction algorithm that can consider only linear system matrix. The main consequence is that material thermal conductivity is linear (set at 0°C by default). It explains why the ROM chip backside temperatures are lower than those given by the ANSYS model which takes into account the non-linearity of the SiC substrate thermal conductivity.

To overcome this restriction, Kirchhoff mathematical inverse transformation has been applied to the ROM output voltages. This mathematical transformation is well suited to linearize non-linear equation such as heat equations systems to a quasi-linear equivalent system.

Considering the substrate material thermal conductivity approximation given in (1), the "real" temperature T taking into account the non-linear material conductivity can be deduced by the inverse Kirchhoff transformation of a temperature T_{lin} deduced from the linear equation system resolution (2).

$$K(T) = K(T_{ref}) . \left(\frac{T}{T_{ref}}\right)^{-\alpha} \text{ with } \alpha > 0 \qquad (1)$$

$$T = T_{ref}\left[(1 + \alpha).\frac{T_{lin}}{T_{ref}} - \alpha\right]^{\left(\frac{1}{1+\alpha}\right)} \qquad (2)$$

ROM temperatures can be identified to solutions of a FEM linearized problem at 0°C reference temperature. By setting properly the α parameter and applying the transformation to the ROM output voltages through a sub-circuit as sketched in Fig. 4, the SiC backside transient temperatures (Fig. 5) are equivalent to those simulated in ANSYS.

Fig. 4 : ADS reduced thermal model representation

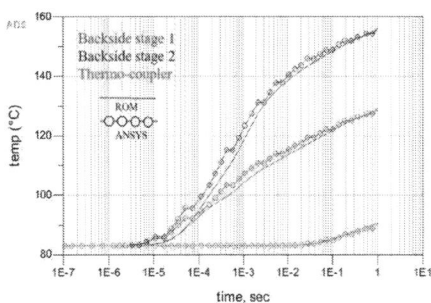

Fig. 5: ROM+Kirchhoff inverse transformed temperatures versus ANSYS temperatures

C. Electro-thermal model validation

Second step of the ROM implementation was the realization of the coupling of the reduced thermal model with GH15 UMS non-linear transistors in order to simulate the transistor junction temperatures. For this study, only the hottest transistors located in the middle of each PA stage have been considered.

UMS DK non-linear model thermal sub-circuit has been modified to realize the coupling with a ROM circuit. On the contrary to the actual DK thermal circuit using Foster thermal $R_{th}C_{th}$ network, a Cauer form has been chosen in order to represent the heat flux time transfer from the gate heat area to the transistor SiC substrate. The ADS schematic principle is illustrated in Fig. 6.

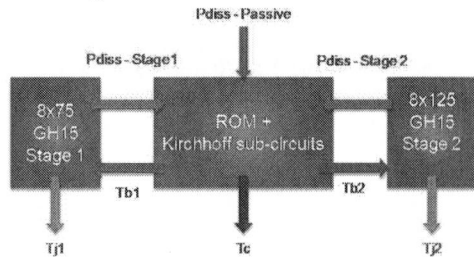

Fig. 6 : Thermal ROM & GH15 transistor model coupling principle

Prior to the transient simulations of the complete schematic, transistors bias voltages have been adjusted to provide the equivalent heat dissipated power as defined in the ANSYS thermal model. The simulated transient junction temperatures at the both stages are shown in the Fig. 7 for 83°C imposed below the test fixture. A very good accuracy of the junction temperature can be obtained by the transistor model coupled with the ROM of the test fixture.

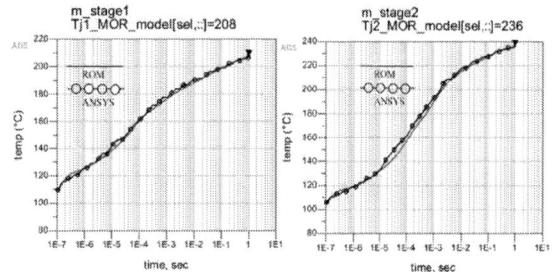

Fig. 7: ROM versus ANSYS junction temperatures

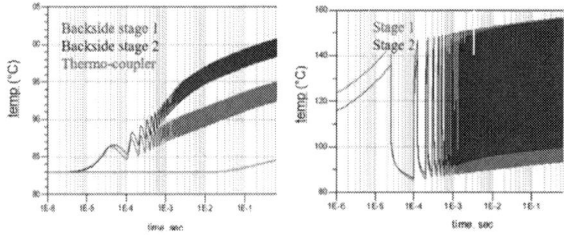

Fig. 8 : Transient ROM temperatures in pulsed mode (25μs, 10%)

Stage	T @25μs, 10% (°C)	Stage	T @25μs, 10% (°C)
Stage 1	149	Backside ST1	90
Stage 2	143	Backside ST2	92
		Thermo_coupler	84

Fig. 9 : Final peak ROM temperatures at time = 1 s

D. Junction temperature evaluation in pulsed mode

Previous paragraph has presented the method used to accurately realize the coupling of a ROM with an electro-thermal model. Thanks to this kind of representation, a MMIC end-user will benefit of time advantage of the electrical analysis and easy perform fast DoE to evaluate the maximum junction temperature of the circuit. In some cases such as pulsed mode thermal simulation, FEA isn't possible for material and time limitation. The use of the ROM is a solution to get the circuit temperatures. For example, a typical pulsed mode condition of 25 μs and 10 % of duty cycle for MMICs HPA have been simulated successfully with the ROM models. Fig. 8 shows the transient pulse behaviour of the thermo-coupler, SiC backside and junction temperatures and Fig. 9 table gathers the maximal temperatures after 1s duration.

VI. CONCLUSIONS & PERSPECTIVES

The coupling of a thermal ROM of GH15 HPA mounted in a custom measurement test fixture with a non-linear electro-thermal transistor model has been carried out. It allows fast MMIC electro-thermal simulations within electrical environment under different excitations (reference temperature, dissipated power, DC, pulsed, transient and radar modes). Once ROM has been validated by comparison with ANSYS simulations results, thermal analysis can be performed by the end-user without any new FEM simulations. Kirchhoff function allows accurate temperature prediction taking into account material nonlinearities versus temperature.

As a perspective, UMS plans to provide custom ROMs to its client through foundry service.

ACKNOWLEDGMENT

UMS would like to thank CADFEM for their support during this work.

REFERENCES

[1] R. J. Trew, D. S. Green, and J. B. Shealy. AlGaN/GaN HFET reliability. IEEE microwave magazine, June 2009, pp. 116-127

[2] E. B. Rudnyi and J. G. Korvink. Model Order Reduction for Large Scale Engineering Models Developed in ANSYS. Lecture Notes in Computer Science, v. 3732, pp. 349-356, 2006

[3] Di Giacomo-Brunel, V.; Byk, E.; Chang, C.; Grünenpütt, J.; Lambert, B.; Mouginot, G.; Sommer, D.; Jung, H.; Camiade, M.; Fellon, P.; et al. Industrial 0.15-μm AlGaN/GaN on SiC technology for applications up to Ka band. In Proceedings of the 2018 13th European Microwave Integrated Circuits Conference (EuMIC), Madrid, Spain, 23–25 September 2018; pp. 1–4.

[4] A. Sarua, H. Ji, K. P. Hilton, D. J. Wallis, M. J. Uren, T. Martin, and M. Kuball. Thermal Boundary Resistance Between GaN and Substrate in AlGaN/GaN Electronic Devices. IEEE TRANSACTIONS ON ELECTRON DEVICES, VOL. 54, NO. 12, DECEMBER 2007

[5] L. Baczkowski, J.C. Jacquet, O. Jardel, C. Gaquière, M. Moreau, et al. Thermal characterization using optical methods of AlGaN/GaN HEMTs on SiC substrate in RF operating conditions. IEEE Transactions on Electron Devices, Institute of Electrical and Electronics Engineers, 2015, 62, pp.3992 - 3998.

[6] T. Bechtold, E. B. Rudnyi, J. G. Korvink. Fast Simulation of Electro-Thermal MEMS: Efficient Dynamic Compact Models, Springer 2006, ISBN: 978-3- 540-34612-8

Gap in pagination due to withheld paper.

Pages 265-268

Proceedings of the 16th European Microwave Integrated Circuits Conference

A 300 GHz Frequency Doubler in Transferred Substrate InP DHBT Technology

Arsen Turhaner[#1], Maruf Hossain[*2], Mohamed Brahem[*3], Tom K. Johansen[#4]

[#]Department of Electrical Engineering, Technical University of Denmark, Denmark
[*]Ferdinand Braun Institut (FBH), Leibniz Institut für Höchstfrequenztechnik, Germany
{[1]aturh, [4]tkj}@elektro.dtu.dk, {[2]maruf.hossain, [3]mohamed.brahem}@fbh-berlin.de

Abstract — **A 300 GHz frequency doubler in transferred-substrate InP DHBT technology is presented. The frequency doubler has an unbalanced single ended topology and employs reflector networks for high efficiency. The experimental results show a saturated output power of -1.45 dBm at 296 GHz. The power consumption is only 9.9 mW leading to a record conversion efficiency of ~7% for transistor based sources in this frequency range.**

Keywords — **InP DHBT, THz circuits, frequency multiplier.**

I. INTRODUCTION

Today, many researches focus on the THz region of the spectrum using electronics techniques [1]. The inherent advantages of this part of the spectrum lead to emerging of applications at THz frequencies such as ultra-broadband short-range communication, high resolution radar and imaging, spectroscopy, and material characterization.

A variety of electronic solid-state devices have been used to design THz sources. Transistor based sources have a great potential for compact THz systems with functionalities such as amplification, frequency conversion, signal generation etc. Even antennas can be monolithically integrated on a single chip. As the frequency of operation approaches the fundamental limit of activity for the transistors, these THz sources are suffering from rather low efficiency [2-8]. There is thus a need for circuit configurations exhausting the potential of a given transistor technology.

In this paper, a high-efficiency frequency doubler design operating at 300 GHz is introduced along with the measurement results. The frequency doubler has an unbalanced single ended topology with reflector networks and would form the last stage of a multiplier chain. The organization of this paper is as follows. In section II, the transferred-substrate technology is briefly described. The circuit design is explained in detail in section III. In section IV, the measurement results are presented along with a discussion. Finally, section V concludes the paper.

II. TECHNOLOGY DESCRIPTION

High operating frequencies and several volts of breakdown voltages make the InP DHBTs popular for the high frequency and the high-power applications. To fully exploit the benefits

of the InP DHBTs for THz applications the extrinsic base-collector capacitance must be minimized. The approach followed at FBH is to employ a transferred-substrate (TS) process where the semiconductor part under the external base is removed. The transistors developed at FBH feature f_{max} of up to 530 GHz and simultaneous breakdown voltage of 5 V for a 0.4×6 μm^2 emitter area device [9]. Fig. 1 shows the schematic cross section of the FBH InP DHBT TS process. The process contains three electroplated gold-based metal layers (G2, G1 and GD) separated by BCB dielectrics for interconnects. Low-loss thin-film microstrip transmission lines can be formed between the 4.5 um thick top metal layer (G2) and the ground layer (GD). For circuit design, MIMs with specific capacitance of 0.3 fF/μm^2 and resistors with 25 Ω/sq. are available.

III. CIRCUIT DESIGN

One way to generate harmonics from a transistor is to reduce the angle of current conduction. For efficient generation of the second harmonic a class B operation is often used. Further enhancement of the second harmonic generation can be achieved by employing reflector networks [8] as shown in Fig. 2. These reflector networks can provide the optimum source and load impedances to the transistor. In order to find the optimum impedances for an efficient generation of second harmonic at 280 GHz using an available input power of 10 dBm, the optimization tool of the Keysight ADS is used. A transistor size of 0.5×6 um^2 is selected as a trade-off between output power and high-frequency operation. The transistor is

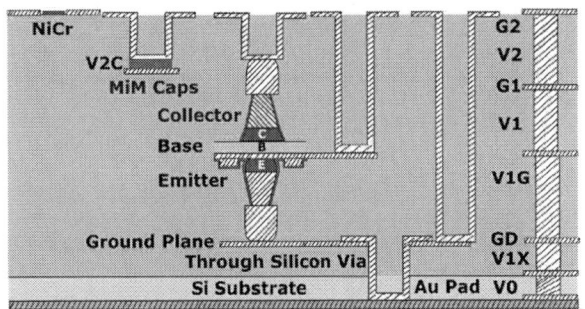

Fig. 1 Schematic cross sections of the InP DHBT TS process.

3–4 April 2022, London, UK

modelled using the FBH HBT large-signal model embedded in a parasitic network determined from 3D electromagnetic simulations [10]. Table 1 shows the optimum impedances and the output power for three different cases, 1) only the source impedance at the fundamental and the load impedance at the second harmonic are optimized, the rest are set to zero; 2) the load reactance at the fundamental is also included in the optimization; 3) the source reactance at the second harmonic is also included in the optimization. The last row demonstrates the results for the EM simulated final layout (see Fig. 4). In order to improve the efficiency and the suppression of the fundamental signal at the output, the source resistance at the second harmonic and the load resistance at the fundamental should be kept close to zero.

The optimum impedances found for the cases 2 and 3 are close to each other. Based on these cases, the input and the output matching/reflector networks are designed. The load reactance at the fundamental frequency given in the case 2 and 3 indicates that a long line is required at the output network. To get rid of that long line, the load impedance is tuned using iterative EM simulations. Otherwise, having a long line would introduce a high loss and occupy a large area. The final circuit schematic is shown in Fig. 3. The frequency doubler consists of DC-blocking capacitors, input and output matching/reflector networks and DC bias lines. The circuit is designed to operate at around 280 GHz.

The layout of the frequency doubler is seen in Fig. 4. Open-circuited stubs are placed at the input and the output which are quarter wavelength long at the frequency to be reflected, keeping the source resistance at the second harmonic and the load resistance at the fundamental close to zero. Capacitors on the bias lines filter out the high frequency signals and stabilize the DC bias voltages. Their values are chosen as 496 fF and 1984 fF. There is a 20 Ω resistor on the base bias line for low-frequency stabilization. DC-blocking capacitors are placed right in front of the RF pads. Thermal vias are placed close to the transistor for the cooling. In addition to this through substrate vias (TSVs) are added for mode suppression.

The full band simulation results of the frequency doubler are given in Fig. 5. Note that an available input power of 10 dBm is used. The base-emitter bias voltage is adjusted at each frequency step to maximize the second harmonic output

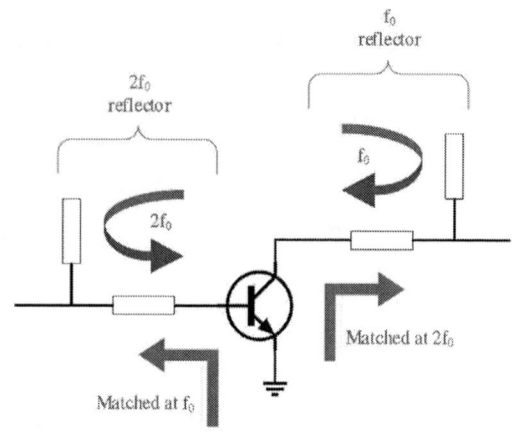

Fig. 2 Principle schematic of frequency doubler using reflector networks.

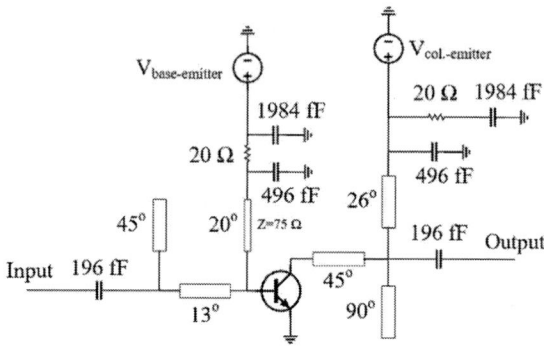

Fig. 3 Schematic of the frequency doubler (the electrical length of the transmission lines refers to 140 GHz).

power. The collector-emitter bias voltage is kept constant at 1.5 V. The second harmonic output power remains above -3.7 dBm for the full band. The 3 dB bandwidth is found approximately as 70 GHz, from 250 GHz to 320 GHz. Peak output power of 5 dBm is reached at 280 GHz as also stated in Table 1 at the last row. The fundamental suppression is more than 11.7 dB around the center frequency. However, it becomes much lower at frequencies away from the center frequency due to the matching/reflector networks being narrow band and the unbalanced topology of the circuit.

Table 1 Optimization of harmonic source and load impedances and the corresponding output power.

Cases	Source impedance (Ω)				Load impedance (Ω)				P_{out} (dBm)
	$Re\{Z_S(f_0)\}$	$Im\{Z_S(f_0)\}$	$Re\{Z_S(2f_0)\}$	$Im\{Z_S(2f_0)\}$	$Re\{Z_L(f_0)\}$	$Im\{Z_L(f_0)\}$	$Re\{Z_L(2f_0)\}$	$Im\{Z_L(2f_0)\}$	
Case 1	30.9	8.5	0	0	0	0	28.0	48.7	-3.5
Case 2	21.9	-1.6	0	0	0	165.9	39.5	12.4	7.2
Case 3	23.3	-1.4	0	17.7	0	197.7	44.4	10.1	7.5
After EM	22.2	7.1	2.1	20.0	6.2	56.0	62.2	-5.8	5.0

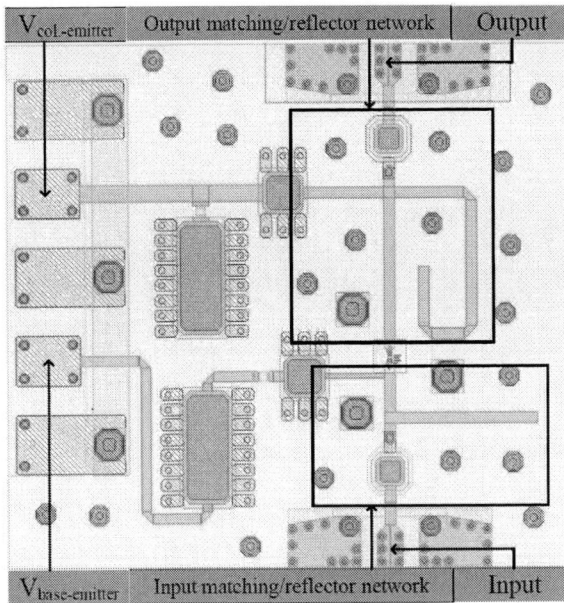

Fig. 4 The full layout of the frequency doubler.

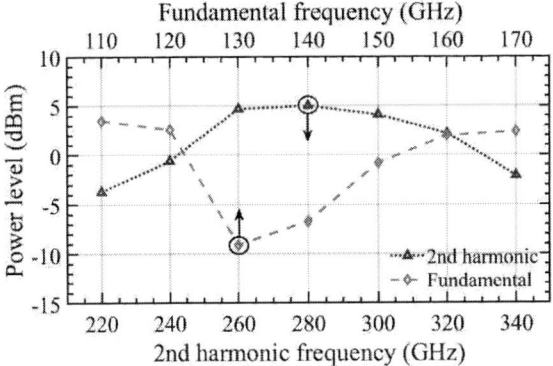

Fig. 5 The full band simulation of the second harmonic output power and the fundamental signal at the output.

IV. MEASUREMENT RESULTS

The chip photograph of the fabricated frequency doubler is shown in Fig. 6. The circuit was tested on-wafer using Formfactor probes and DC probes decoupled at the probe-tips. The input of the circuit is fed by a WR-5 module which includes a WR-10 ×6 multiplier, an amplifier, an attenuator, and a G-band frequency doubler from VDI with a WR-5 interface, followed by a ground-signal-ground (GSG) probe. In order to measure the output power of the circuit, a WR-3.4 GSG waveguide probe was connected to a WR3.4-to-WR10 taper, which is connected to the input of a power sensor and an Erickson PM4 power meter. The measurement setup is depicted in Fig. 7. Using that source, it was found that the circuit reaches its best performance at around 300 GHz, since the source power is not adequate at 140 GHz.

The small-signal characterization of the fabricated 0.5×6 um^2 InP DHBTs reveals a drop in peak f_{max} from approximately 456 GHz in the simulation model to 326 GHz. The cause of this drop is an increase in the base resistance. Therefore, measured second harmonic output power was approximately 4 dB lower than simulated with the original model at the frequencies of interest. The measurement results with respect to the frequency are given in Fig. 8, and compared with simulation results with an updated model. The available input power and the collector currents in these retro simulations are kept equal to those in the measurements. As it is seen in the figure, our retro-simulations with an updated model show a good agreement with the measurements. The fundamental suppression is higher at the lower frequencies. The fundamental output power depends on the input power. However, the difference between the available input power and the fundamental output power is slightly higher at the lower frequencies. The measured second harmonic output power reaches up to -1.45 dBm at 296 GHz. At this frequency and output power level, the circuit consumes only 9.9 mW of DC power. Hence, the dc-to-RF efficiency of the frequency doubler becomes ~7% using the definition given in [11].

Finally, Table 2 compares the performance of the frequency doubler to that of other transistor based frequency multipliers operating at around 300 GHz. As it is seen in the table, the output power of this work is highly competitive, and the efficiency of the proposed circuit is very high due to being a simple and standalone design, consuming very little DC power.

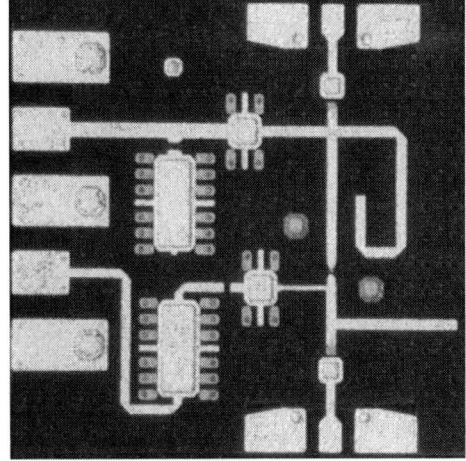

Fig. 6 The chip photograph of the frequency doubler. The chip size is 0.70 × 0.63 mm^2.

Fig. 7 The measurement setup.

Table 2 Transistor based frequency multipliers operating at around 300 GHz.

Reference	[2]	[3]	[4]	[5]	[6]	[7]	[8]	**This work**
Technology	130 nm SiGe HBT	130 nm SiGe BiCMOS	130 nm SiGe BiCMOS	120 nm SiGe HBT	35 nm mHEMT	800 nm TS InP DHBT	800 nm TS InP DHBT	**500 nm TS InP DHBT**
Topology	Amp. + Doubler	Tripler	Amp. + Doubler	VCO + Buffer + Doubler	Tripler	Quadrupler	Tripler	**Doubler**
Frequency (GHz)	325	280	270	290	324	325	303	**296**
$P_{out, peak}$ (dBm)	-3	3	-0.5	-1.7	-8.8	-7	-6.2	**-1.45**
P_{DC} (mW)	420	91.2	429	167	650	40	37.6	**9.9**
dc-to-RF efficiency	0.1%	2.2%	0.2%	0.4%	0.02%	0.5%	0.6%	**7.2%**

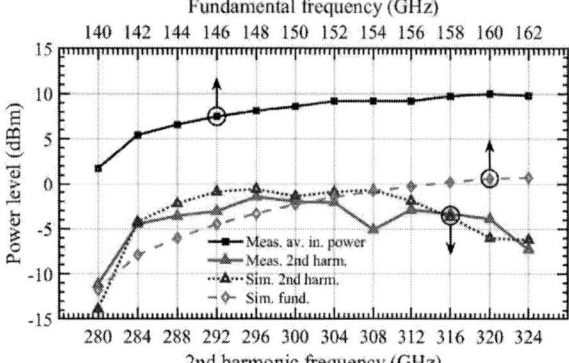

Fig. 8 The measurement of the available input power, the second harmonic output power, and the simulated output powers at the second harmonic and the fundamental with respect to frequency.

V. CONCLUSION

A 300 GHz frequency doubler implemented in transferred-substrate InP DHBT technology is presented. The circuit can deliver an output power of -1.45 dBm at 296 GHz. The circuit has a DC power consumption of only 9.9 mW and a dc-to-RF efficiency of ~7%.

ACKNOWLEDGMENT

This work was partly funded by the European Union's Horizon 2020 Research and Innovation Programme through the Convergence of Electronics and Photonics Technologies for Enabling Terahertz Applications (CELTA) Project under grant agreement no. 675683. Part of the work was funded also by the German BMBF within the "Forschungsfabrik Mikroelektronik Deutschland (FMD)" framework under ref. 16FMD02.

REFERENCES

[1] J.-S. Rieh, Introduction to Terahertz Electronics, Springer, 2021.

[2] E. Öjefors, B. Heinemann, and U. R. Pfeiffer, "Active 220- and 325-GHz frequency multiplier chains in an SiGe HBT technology," IEEE Transactions on Microwave Theory and Techniques, volume 59, number 5, pages 1311–1318, May 2011, issn: 0018-9480. doi: 10.1109/TMTT.2011.2114364.

[3] A. Turhaner, Y. Dong, V. Zhurbenko, and T. K. Johansen, "A 280-GHz self-mixing balanced frequency tripler in SiGe BiCMOS technology," IEEE Microwave and Wireless Components Letters, volume 30, number 10, pages 965–968, 2020. doi: 10.1109/LMWC.2020.3019589.

[4] F. Ahmed, M. Furqan, and A. Stelzer, "A 0.3-THz SiGe-based frequency doubler chip with 3-dB 50 GHz bandwidth and 17 dB peak conversion gain," in 2017 12th European Microwave Integrated Circuits Conference (EuMIC), October 2017, pages 134–137. doi: 10.23919/EuMIC.2017.8230678.

[5] S. P. Voinigescu et al., "A study of SiGe HBT signal sources in the 220–330-GHz range," in IEEE Journal of Solid-State Circuits, vol. 48, no. 9, pp. 2011-2021, Sept. 2013, doi: 10.1109/JSSC.2013.2265494.

[6] U. J. Lewark et al., "255 to 330 GHz active frequency tripler MMIC," in Proc. Integrated Nonlinear Microwave and Millimetre-Wave Circuits (INMMIC) Workshop, 2012.

[7] M. Hossain, K. Nosaeva, N. Weimann, V. Krozer, and W. Heinrich, "A 330 GHz active frequency quadrupler in InP DHBT transferred-substrate technology," in Proc. IEEE MTT-S Int. Microwave Symp. (IMS), San Francisco, CA, USA, May 2016, pp. 1–4.

[8] T. K. Johansen, M. Hossain, S. Boppel, R. Doerner, V. Krozer and W. Heinrich, "A 300 GHz active frequency tripler in transferred-substrate InP DHBT technology," 2019 14th European Microwave Integrated Circuits Conference (EuMIC), Paris, France, 2019, pp. 180-183, doi: 10.23919/EuMIC.2019.8909673

[9] N. G. Weimann, T. K. Johansen, D. Stoppel, M. Matalla, M. Brahem, K. Nosaeva, S. Boppel, N. Volkmer, I. Ostermay, V. Krozer, O. Ostinelli, and C. R. Bolognesi, "Transferred-Substrate InP/GaAsSb Heterojunction Bipolar Transistor Technology With f_{max}~0.53 GHz," in IEEE Trans. Electron Devices, Vol. 65, No. 9, pp. 3704 – 3710, Sept. 2018.

[10] T. K. Johansen, R. Doerner, N. Weimann, M. Hossain, V. Krozer, amd W. Heinrich, "EM simulation assisted parameter extraction for transferred-substrate InP HBT modeling," in International Journal of Microwave and Wireless Technologies 10, 700–708, 2018.

[11] S. A. Maas, Nonlinear Microwave and RF Circuits, 2nd edition. 685 Canton Street, Norwood, MA 02062: Artech House, 2003, page 482.

Proceedings of the 16th European Microwave Integrated Circuits Conference

55% Fractional-Bandwidth Doherty Power Amplifier in 130-nm SiGe for 5G mm-Wave Applications

Aniello Franzese[#], Nebojsa Maletic[#], Mohamed Eissa[#], Muh-Dey Wei[*], Renato Negra[*], Andrea Malignaggi[#]

[#]IHP – Leibniz-Institut für innovative Mikroelektronik, Germany
[*]Chair of High Frequency Electronics, RWTH Aachen, Germany
franzese@ihp-microelectronics.com

Abstract—This paper describes the design of a broadband Doherty power amplifier (DPA) implemented in a 130-nm SiGe BiCMOS technology for 5G applications (24.25-29.5 GHz band) with a compact size of 1.12 mm². Unlike other works at similar frequencies, the broadband amplifier employs an ultracompact single-section broadband lumped-element Wilkinson power splitter over the band 20-35 GHz. Current-combined amplifiers are employed as carrier and peaking amplifiers with different biasing, performing broadband operations while preserving efficiency and area. The amplifier has been validated through measurements, resulting in a 3-dB fractional bandwidth of 55% from 20 GHz to 35 GHz with a peak gain of 10.5 dB. Large-signal measurements show an output power 1-dB compression point (OP$_{1dB}$) of 15.4 dBm at 27 GHz, with a maximum power added efficiency (PAE) better than 20% over the band 23-31 GHz, while it is always better than 12% at 6-dB back-off. Furthermore, modulated signal measurements are reported for 16/64-QAM, showing gigabit-per-second data rates.

Keywords—Doherty, power amplifier, 5G, SiGe, BiCMOS, power added efficiency.

I. INTRODUCTION

With the fifth-generation (5G) radio, faster communications will be enabled leveraging on massive data rates provided by the possibilities of the high frequency spectrum, targeting the bands $n257$ and $n258$ (24.25-29.5 GHz). Achieving a steadily-higher amount of transmitted data requires complex modulation schemes [1], resulting in a high peak-to-average power ratio (PAPR), opening to a perennial trade-off between linearity and power efficiency on the power amplifier (PA) side. At lower frequencies, the DPA allowed the transmission of high PAPR signals, extending the linearity and increasing the efficiency at the power back-off (PBO) [2]. However, the DPA faces several challenges at mm-wave frequencies [3]. For this reason, several solutions for 5G applications have been proposed in the literature [1], [4]–[8]. Nevertheless, they do not explicitly target broadband capabilities, despite the work in [5] where varactor-loaded transmission lines (TLs) solution massively increase the size. Thus, this work aims to cover the 5G spectrum with an affordable silicon occupation preserving power efficiency. This is attained leveraging double parallel-cascode current-combined amplifiers and an ultracompact single-section lumped-element Wilkinson power splitter (WPS) that achieves broadband operations with precise phase and amplitude. In [9] and [10], the WPS acquires a compact form-factor exploiting the mutual inductance between the inductors used to synthesize the artificial TLs (ATLs).

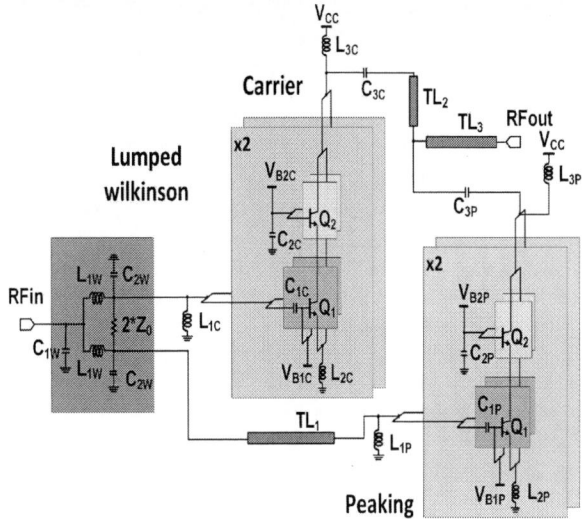

Fig. 1. Schematic of the designed Doherty PA.

However, the authors are shielding the inductors for safety reasons and exploiting the parasitic capacitance to avoid bulky capacitors. This work improves the bandwidth and insertion loss performance of the above mentioned WPSs, giving a valid alternative to coupler-based solutions, which could be designed only with metal shield removal.

II. CIRCUIT DESIGN

The schematic of the DPA is shown in Fig. 1 and it is a good compromise between area occupation and bandwidth, as shown in [11]. Indeed, load variations would affect heavily the bandwidth of a lumped output-impedance-inverter (OII) DPA. On the other hand, using double-inverter TLs on the peaking PA increases the area occupation. For this reason, in this design the OII is based on two TLs performing the active load modulation. Moreover, TL1 generates the required delay at the input of the peaking. The proposed DPA uses the broadband WPS, assuring compactness and precise phase and amplitude splitting for both carrier and peaking PAs.

A. Ultracompact single-section broadband Wilkinson

The aforementioned splitters do not assure precise amplitude balance, because they leverage on folded inductors. A metal shield underneath the inductor has been introduced to protect the WPS from substrate nonidealities, reduce magnetic

3–4 April 2022, London, UK

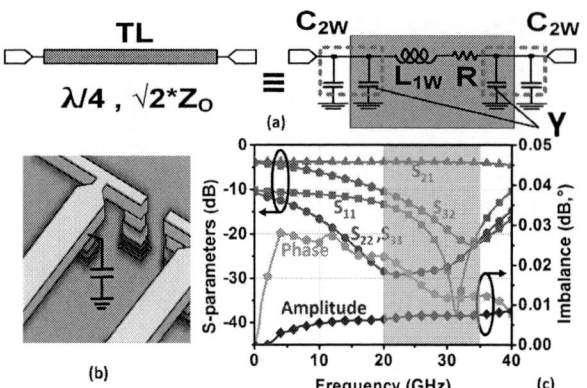

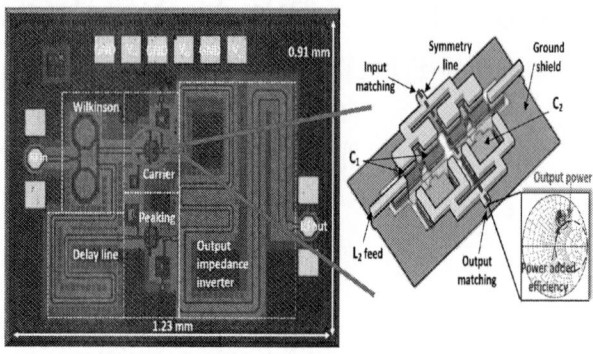

Fig. 2. (a) Transmission line and its lumped-component equivalent, (b) capacitor layout details and (c) S-parameters of the Wilkinson power splitter.

Fig. 3. Micrograph of the proposed power amplifier and its 3D core view.

coupling, and lower unwanted disturbance between the signals. On the other hand, it increases the shunt capacitances due to the closer ground plane. Nevertheless, this can be exploited as shown in Fig. 2a. Indeed, the TL finds an equivalent π-model, where shunt capacitances (pink rectangle), given by the ground closer to the inductors, are used to reduce the ATL shunt capacitors. Employing space underneath the interconnection lines as shown in Fig. 2b, further decreases both size and possible couplings. As a consequence, the capacitor C_{2W} in Fig. 1 is realized of two parallel capacitances, one given by the inductors and one lumped, as highlighted with a red dashed rectangle in Fig. 2a. The WPS shown in Fig. 1 can be synthesized with the following parameters [10]:

$$L_{1W} = \frac{\sqrt{2}Z_o}{\omega} \simeq 420\,pH, C_{2W} = \frac{1}{2\sqrt{2}\omega Z_o} \simeq 42\,fF, \quad (1)$$

and $C_{1W} = 2C_{2W}$. Now, L_{1W} can be EM simulated and the S-parameters [12] are used to calculate the $ABCD$ coefficients of the transmission matrix in (2), where $\underline{Z} = \frac{j\omega L_{1W}+R}{Z_o}$ and $\underline{Y}$ is the normalized admittance, extracted in (3).

$$\begin{bmatrix} A & B \\ C & D \end{bmatrix} = \begin{bmatrix} 1+\underline{Y}\,\underline{Z} & \underline{Z} \\ 2\underline{Y}+\underline{Y}^2\underline{Z} & 1+\underline{Y}\,\underline{Z} \end{bmatrix} \quad (2)$$

$$\underline{Y} = \frac{D-1}{\underline{Z}} = \frac{A-1}{\underline{Z}} \quad (3)$$

From (3), the Y value is obtained by dividing by Z_o. At this point, it becomes straightforward to calculate the lumped capacitor value from Y and C_{2W}. However, the interwinding capacitance and possible asymmetries should be added to the inductor model and thus the lumped capacitor value has to be tuned after employing EM simulations. This allows equal splitting, along with a broadband match, as shown in Fig. 2c. The simulated insertion loss is as low as 4 dB from 20 GHz to 35 GHz. Moreover, the simulated in-band return loss and isolation of the WPS are better than 10 dB with an area of $240 \times 100\,\mu m^2$.

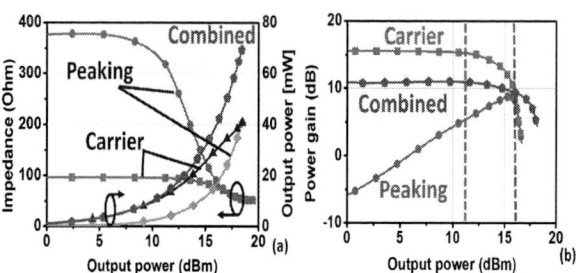

Fig. 4. Simulated (a) load modulation and output power and (b) power gain along with the carrier and combined OP$_{1dB}$s at 27 GHz.

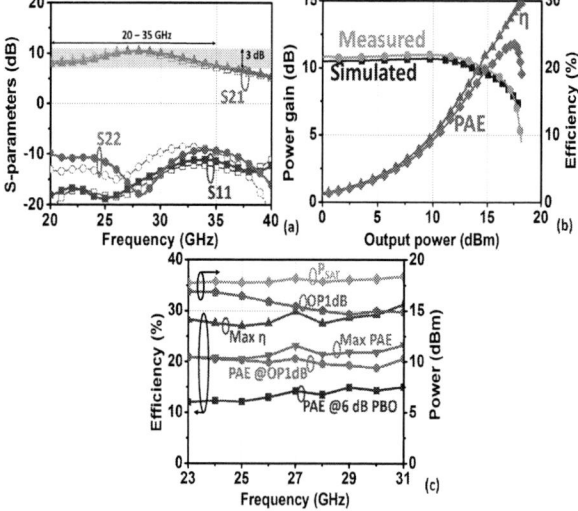

Fig. 5. Simulated (empty marker) and measured (solid marker) (a) S-parameters, (b) power gain along with efficiency measurements at 27 GHz and (c) power and efficiency across frequency.

B. PA core design

Both carrier and peaking PAs employ the same configuration, which is two parallel-cascode current-combined

274

Table 1. PERFORMANCE SUMMARY AND STATE-OF-THE-ART COMPARISON

	(Unit)	[1]	[4]	[5]	[6]	[7]	[8]	This Work
Technology		28nm bulk CMOS	130nm SiGe	130nm SiGe	0.15μm GaAS	45nm CMOS SOI	45nm CMOS SOI	130nm SiGe
Gain	(dB)	22	6.7	18.2	14.4	10	7	10
OP_{1dB}	(dBm)	16	17.5*	15.2	28.5	21.5	11*	15.4
P_{sat}	(dBm)	19.8	21.3	16.8	-	22.4	18	18.2
η_{max}	(%)	-	35*	29.4	-	-	33*	29.8
PAE_{max}	(%)	21	27*	20.3	37	40	24*	23.2
$PAE_{@P1dB}$	(%)	12.8	26*	19.5	37	39*	15*	20.6
$PAE_{@PBO}$**	(%)	13.5*	24.3	13.9	27	28	17	14.3
Area	(mm^2)	0.59	1.87	1.76	4.93	0.63	0.64	1.12
Supply	(V)	1	1.7	1.5	8***	2.4	2.5	3.3
Frequency	(GHz)	32	30	28	28	28	42	27
Bandwidth	(%)	25*	-	52	18*	-	-	55

* graphically estimated, ** at 6 dB, *** only power stage

slices. This results in an equal path length for all the transistor fingers as seen in the 3D view of Fig. 3. Each slice is comprised of two cascode amplifiers sharing the decoupling, C_2, on the basis of the common-base, ensuring compactness and lowering the output impedance transformation ratio (ITR). The load transformation networks (LTNs) have been optimized through load-pull simulations as trade-off between efficiency and output power. To avoid phase imbalance, the LTNs have been shielded with metal ground and the parasitic capacitance is exploited to lower the Q factor. Fig. 4a shows the simulated impedance and the output power for peaking, carrier, and combined amplifiers at the centre frequency. The output impedance of the carrier goes from $100\,\Omega$ to $50\,\Omega$, while the peaking follows the carrier behaviour starting from $370\,\Omega$. The reported output power shows how the peaking starts to deliver power, reaching the carrier at the saturation of the DPA. The power gain, reported in Fig. 4b, shows the linearity extension of the DPA with respect to the carrier. The OP_{1dB} is extended by more than $5\,dB$.

III. MEASUREMENT RESULTS

The DPA in Fig. 3 has been fabricated in IHP SG13S 130-nm technology occupying $0.91\,mm \times 1.23\,mm$ of silicon. S-parameters have been measured using a ZVA67 from R&S and they are presented in Fig. 5a with a 3-dB fractional bandwidth of 55%. The measured peak gain reported in Fig. 5b is $10.5\,dB$. Large-signal measurements from $23\,GHz$ to $31\,GHz$ are reported in Fig. 5c and present a collector efficiency (η) and PAE at $27\,GHz$ of 29.8% and 23.2%, respectively. The measured peak PAE at $6\,dB$ PBO over the band of interest is 15%. Moreover, an OP_{1dB} better than $15\,dBm$ is achieved in the band of interest as well as a saturation power greater than $17.5\,dBm$. Modulated signal measurements are performed with 16-/64-QAM signals at $27\,GHz$. The DPA's output is connected to a Keysight DSA-Z634A oscilloscope and the measured spectrum is shown for 400 MSym/s 16-QAM in Fig. 6a for both high back-off and $P_{avg} = 10\,dBm$, without digital predistortion. Fig. 6b reports the 200 MSym/s constellation for 16-/64-QAM with $P_{avg} =$

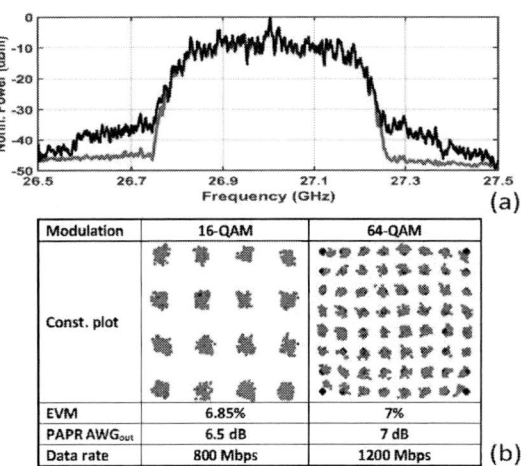

Fig. 6. 27 GHz (a) 400 MSym 16-QAM spectrum for high back-off (red) and 10 dBm P_{avg} (black), (b) 200 MSym 16/64-QAM measurements at 10 dBm P_{avg} output.

$10\,dBm$ and an error vector magnitude (EVM) of 6.85% / 7%. In-band linearity is limited by the setup in producing high symbol rates (400 MSym/s 64-QAM reports around 9% at the output of the signal generator). Table 1 reports state-of-the-art DPAs at mm-wave frequencies. The proposed DPA achieves the largest bandwidth with comparable performance to other wideband DPAs.

IV. SUMMARY

The proposed DPA is suitable for 5G mm-wave applications and exploits an ultracompact lumped-element broadband Wilkinson achieving a precise splitting in phase and amplitude. The splitter has been designed to exploit the parasitic capacitance of the inductor, reducing losses. The chosen amplifier topology is a compromise between bandwidth and area occupation, where both carrier and peaking amplifiers have been designed with a core based on two parallel-cascode current-combined slices, which ensures a broadband behaviour and high efficiency with 55% fractional bandwidth. The other performance of the designed circuit consists in a maximum η and PAE of 29.8% and 23.2%, respectively, at 27 GHz. Finally, modulated signal measurements have been carried with 16-/64-QAM, reporting an EVM of 6.85% / 7%.

ACKNOWLEDGMENT

The authors thank Mr. Thomas Mausolf for the help with measurements and IHP GmbH for the chip manufacture.

REFERENCES

[1] P. Indirayanti and P. Reynaert, "A 32 GHz 20 dBm-PSAT Transformer-based Doherty Power Amplifier for Multi-Gb/s 5G Applications in 28 nm bulk CMOS," in *2017 IEEE Radio Frequency Integrated Circuits Symposium (RFIC)*, 2017, pp. 45–48.

[2] P. Colantonio, F. Giannini, R. Giofrè, and L. Piazzon, "The AB-C Doherty power amplifier. Part I: Theory," *International Journal of RF and Microwave Computer-Aided Engineering*, vol. 19, no. 3, pp. 293–306, 2009. [Online]. Available: https://onlinelibrary.wiley.com/doi/abs/10.1002/mmce.20350

[3] P. M. Asbeck, "Will Doherty continue to rule for 5G?" in *2016 IEEE MTT-S International Microwave Symposium (IMS)*, 2016, pp. 1–4.

[4] M. Özen, N. Rostomyan, K. Aufinger, and C. Fager, "Efficient Millimeter Wave Doherty PA Design Based on a Low-Loss Combiner Synthesis Technique," *IEEE Microwave and Wireless Components Letters*, vol. 27, no. 12, pp. 1143–1145, Dec 2017.

[5] S. Hu, F. Wang, and H. Wang, "A 28-/37-/39-GHz Linear Doherty Power Amplifier in Silicon for 5G Applications," *IEEE Journal of Solid-State Circuits*, vol. 54, no. 6, pp. 1586–1599, June 2019.

[6] D. P. Nguyen, T. Pham, and A. Pham, "A 28-GHz Symmetrical Doherty Power Amplifier Using Stacked-FET Cells," *IEEE Transactions on Microwave Theory and Techniques*, vol. 66, no. 6, June 2018.

[7] N. Rostomyan, M. Özen, and P. Asbeck, "28 GHz Doherty Power Amplifier in CMOS SOI With 28% Back-Off PAE," *IEEE Microwave and Wireless Components Letters*, vol. 28, no. 5, May 2018.

[8] A. Agah, B. Hanafi, H. Dabag, P. Asbeck, L. Larson, and J. Buckwalter, "A 45GHz Doherty Power Amplifier with 23% PAE and 18dBm Output Power, in 45nm SOI CMOS," in *2012 IEEE/MTT-S International Microwave Symposium Digest*, June 2012, pp. 1–3.

[9] M. Love, M. Thian, F. van der Wilt, K. van Hartingsveldt, and K. Kianush, "Lumped-Element Wilkinson Power Combiners Using Reactively Compensated Star/Delta Coupled Coils in 28-nm Bulk CMOS," *IEEE Transactions on Microwave Theory and Techniques*, vol. 67, no. 5, pp. 1798–1811, 2019.

[10] S. Lee, M. Huang, Y. Youn, and H. Wang, "A 15 – 55 GHz Low-Loss Ultra-Compact Folded Inductor-Based Multi-Section Wilkinson Power Divider for Multi-Band 5G Applications," in *2019 IEEE MTT-S International Microwave Symposium (IMS)*, 2019, pp. 432–435.

[11] R. Quaglia and J. Lees, "Simplified analysis of the effect of load variation in common doherty power amplifier architectures," in *2019 IEEE Topical Conference on RF/Microwave Power Amplifiers for Radio and Wireless Applications (PAWR)*, 2019, pp. 1–3.

[12] B. Walker, "RLC Parameter Extraction Using the Transfer Matrix," July 2020. [Online]. Available: https://www.microwavejournal.com/articles/34236-rlc-parameter-extraction-using-the-transfer-matrix

Proceedings of the 16th European Microwave Integrated Circuits Conference

Full Octave Continuously Tunable SiGe Bipolar LC-VCO in K_u-Band

Christian Bredendiek[#], Klaus Aufinger[*], Nils Pohl[$#]

[#]Fraunhofer FHR, Fraunhofer Str. 20, D-53343 Wachtberg, Germany
[*]Infineon Technologies AG, Am Campeon 1-12, D-85579 Neubiberg, Germany
[$]Ruhr-Universität Bochum, D-44780, Germany
christian.bredendiek@fhr.fraunhofer.de

Abstract — **In this paper a signal generation concept is presented with frequency tuning ranges of more than one octave in the K_u-Band. On the MMIC a fundamental cross-coupled ultra-wideband K_u-Band LC-VCO is integrated along with a simple buffer and a frequency divider for PLL stabilization. The chip is fabricated in Infineon's BiCMOS production technology B11HFC which offers HBTs with an f_T of 250 GHz and f_{max} of 370 GHz. The chip has a power consumption of only 106 mW and uses an area of 0.52 mm^2. The VCO reaches an absolute frequency tuning range (FTR) of 12.1 GHz (relative FTR=75.1%) at a center frequency of 16.1 GHz. The phase noise at 1 MHz offset is as low as -104 dBc/Hz. The output level is around 0 dBm.**

Keywords — **Millimeter-wave Transceivers, Frequency Synthesis, MMICs, SiGe bipolar ICs, ultra-wideband.**

I. INTRODUCTION

With matured Silicon-Germanium hetero bipolar technologies, such as e.g.[1], readily available on the market many applications in industrial measurement scenarios can be exploited. This advancement heavily driven by the demand of the automotive industry in radar sensors for next generation driver assistance systems and self-driving cars already allows for the compact realization of high-precision industrial radar sensors such as [2],[3] and [4].

These sensors already feature compact, efficient, mono-static realization while still enabling high stability, precision and spatial resolution. Especially the spatial resolution, and hence the mandatory need for a high absolute bandwidth and highly linear frequency ramps in measurements with a FMCW radar, is very important for the evaluation of layer thicknesses. For the extrusion of multilayered plastics the high spatial resolution can be used to characterize the plastic and the thickness of its individual layers. To further improve this most sensor concepts reach for higher frequencies above 100 GHz to obtain larger absolute bandwidths as in [5],[6] and [7].

Here, however another approach is taken with the central signal generation part to further improve upon the frequency tuning range without using higher frequency bands above 100 GHz. An exemplary block diagram for such an ultra-wideband radar system is given in Fig. 1 of a commonly used mono-static radar sensor setup at E-Band for industrial purposes. The integration of most of the active mm-Wave parts have already been proven and demonstrated on a single transceiver MMIC in SiGe HBT technologies along with

the additional passive structures needed on a high-frequency front-end board. With this concept absolute bandwidths of up to 32 GHz in E-Band have already been demonstrated in [4].

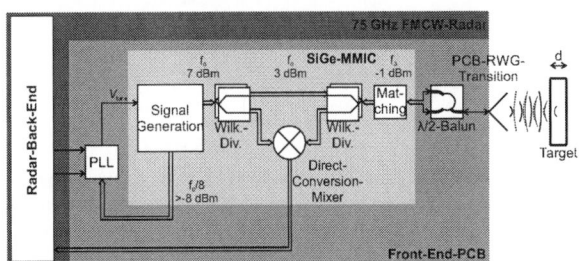

Fig. 1. Block diagram of a proposed 75 GHz FMCW radar system with a new approach on the signal generation compared to existing designs. Differential signal paths are drawn with double lines.

In order to stabilize the sensor in an elegant way with existing high performance off the shelf components, e.g. TIs LMX or ADs ADF series PLL chips with frequency ramp generation capabilities, a frequency output of typically lower than 15 GHz is needed. Also some sort of continuously tunable oscillator is always needed, such as a fundamental or push-push VCO with multiplier.

In Fig. 1 an intended level plan of such a mono-static system concept is also shown. The signal generation part has to deliver an output power of around 7 dBm to realize a TX output power of around -1 dBm at the PCB-interface while offering around 3 dBm to drive the LO-port of the direct-down-conversion mixer. The LO-TX-RX distribution can be done passively with for example Wilkinson-Dividers. The TX power has to be limited to around -1 dBm to keep the transceiver from self saturating the also integrated receive mixer due to strong reflections from the the mono-static concept. The power supply, digital signal processing and triggering of the fractional ramp-sweeps are done with a radar-back-end board to complete the realization of such an compact industrial ultra-wideband sensor.

II. MMIC DESIGN

The content of this chapter is presented by the block diagram in Fig. 2. All parts integrated on the MMIC from Fig. 2 are described in detail in the following part.

277

3−4 April 2022, London, UK

The design approach to achieve a higher absolute bandwidth with this signal generation concept is to sacrifice the efficiency and driving capabilities of a fundamental VCO at the intended frequency band by using a lower frequency VCO in combination with multipliers and amplifiers. Since the parasitic influences at lower frequencies are easier to handle a higher relative bandwidth can be achieved at lower bands. Additionally due to the reliance of this concept on multipliers and amplifiers we could use a VCO concept that has poor driving capabilities. The implementation of the VCO was heavily influenced by the investigations we conducted in [8]. This resulted in a fundamental K_u-Band VCO that is capable of a continuous frequency tuning range of more than one octave (>66% relative frequency tuning range).

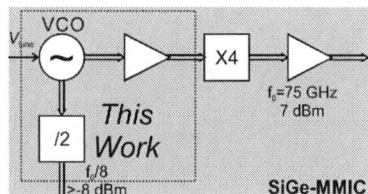

Fig. 2. Proposed signal generation concept shown in a block diagram for the proposed radar system. This work focuses on the signal source part in the K_u-Band on the left side of the figure.

There are two noteworthy aspects on the whole design methodology of the signal generation MMIC. The first is of course the pure differential design architecture which is easily realizable in silicon technologies. The second one is the entire DC-coupling of all circuit blocks so that only one additional biasing network had to be used. All other biasing was carried out by using emitter coupled pairs.

In Fig. 3 the chipfotograph of the presented MMIC is given with the highlighted building blocks from Fig. 2. The chip size is $930 \times 560\,\mu m^2$ and the highly symmetrical design due to the fully differential architecture can clearly be seen. The symmetry is only broken for the cross coupling of the VCO and in the needed crossings in the latches from the prescaler. It is also very obvious that the used area is mainly defined by the needed pad frame and the distance between pad rows needed to contact probes from three sides. The total area of the circuit blocks already including supply blocking caps could be as small as $470 \times 220\,\mu m^2$. For the implementation of the MMIC Infineon's B11HFC [1] automotive qualified production SiGe:C hetero bipolar technology with an f_T of 250 GHz and f_{max} of 370 GHz was used. The total current consumption of the presented MMIC from Fig. 3 is only 32 mA from a single 3.3 V supply.

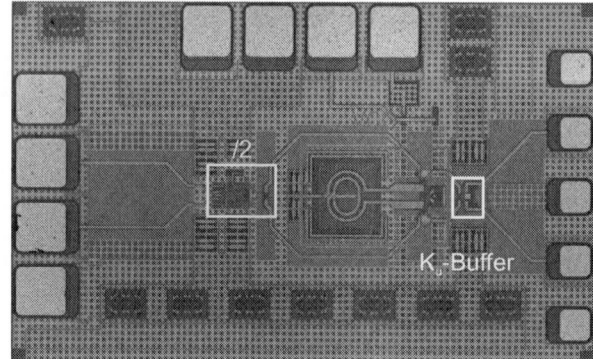

Fig. 3. Chipfotograph of the realized K_u-Band signal source. The chip size is $930 \times 560\,\mu m^2$

Two of the most appealing aspects on the design of the cross-coupled VCO are that is its inherent differential topology and its extremely easy to fulfill oscillation condition which relaxes the design complexity. It is differential since the positive feedback of the cross-coupled pair provides the necessary negative resistance for stable oscillation. The negative resistance can simply be obtained by considering the input impedance of the plain cross-coupled pair, which can be calculated to:

$$Z_{in,CC} = -\frac{2}{g_m} + \frac{2}{j\omega\left(C_{BE} + 4 \cdot C_{CB}\right)} \qquad (1)$$

A stable oscillation occurs when the losses of the loaded tank R_{Tank} are overcompensated by (1), which leads to

$$R_{Tank} < \frac{2}{g_m} \qquad (2)$$

at a oscillation frequency f_0 of

$$f_{0,CC} = \frac{1}{2\pi\sqrt{L_1\left(C_{BE} + 4 \cdot C_{CB} + C_{var}\right)}}. \qquad (3)$$

The realized topology of the cross-coupled LC-VCO for the K_u-Band is shown in Fig. 4. The classic cross-coupled pair formed by T_1 and T_2 is improved by using the decoupling capacitors $C_{1,2} = 150\,fF$ to remedy the otherwise limited voltage swing of only $2 \cdot V_{BE}$. But this requires additional biasing at the bases $V_{B1,B2}$ of the cross-coupled pair, where the only biasing network in the whole MMIC is used. Here this is realized with a simple resistor from positive supply, two symmetrical diodes connected to the bases, additional two symmetrical diodes connected to a resistor to negative supply (ground). Another improvement of the cross-coupled pair is the operation of the VCO in Class-A with a dedicated emitter degenerated current-source to control the negative resistance reasonably well and add some temperature stability.

278

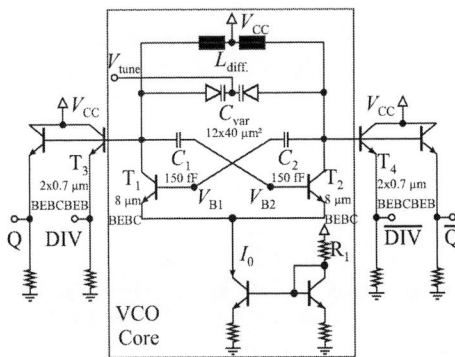

Fig. 4. Schematic of the fundamental K_u-Band LC-VCO.

As mentioned before one of the main drawbacks of this VCO architecture is its poor driving capability. Here, two HBTs with a shared physical collector node $T_{3,4}$ (BEBCBEB configuration) and two emitter stripes are used as emitter-followers. Effectively acting as four emitter-followers in order to directly connect the differential divider as well as the differential output buffer in a simple way.

The LC-tank of the VCO is realized with a center-tapped differential spiral inductor, which can clearly be seen in Fig. 3, and a differential pn-varactor offered by the used technology. The used two turn inductor is designed with the 2.5D simulator Sonnet and has an differential inductance of $L_{diff.} = 290\,pH$ and a quality factor of 18.9 at 16 GHz. The inductor with its area consumption of $130 \times 140\,\mu m^2$ dominates the area consumption of the VCO with its total size of $290 \times 140\,\mu m^2$. The VCO core with the differential varactor already included only consumes $160 \times 100\,\mu m^2$. The differential varactor offers a differential capacitance $C_{var} = 145\,fF$ at a reverse bias of 3.3 V ($V_{tune} = 6.6\,V$). It offers a capacitance variation of $\frac{C_{max}}{C_{min}} = 2.8$ when using it to its limit defined by its typical avalanche breakdown voltage of 7 V.

The core current I_0 is only 7.4 mA, biasing and the current mirror I_0 requires 1.6 mA totaling in 30 mW from a 3.3 V supply for the VCO core. The four additional emitter followers $T_{3,4}$ are using an additional current of 1 mA each. Resulting in a power consumption of 43 mW for the circuit given in Fig. 4.

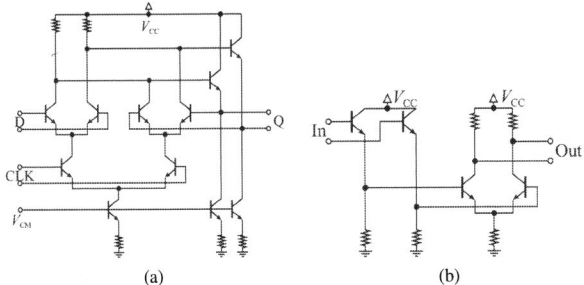

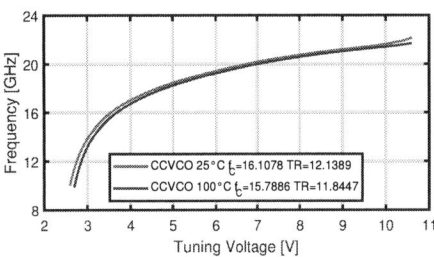

Fig. 5. (a) Schematic of the ECL-D-latches used in the static divider. (b) Schematic of the simple buffer used at the K_u-Band output as well as the divider.

The frequency divider is based on two D-type latches and the schematic of the used ECL-D-latches is shown in Fig. 5a. For the sample (D-input) and hold transistors of the latches as well as the differential Emitter-followers the available transistors from the PDK which offer a shared physical contact are used to reduce the parasitic capacitance in order to reach the highest frequency at a given current. One latch itself only uses a current of 1.1 mA. One additional Emitter-follower pair is used between the output DIV of the VCO and the clock input driving both latches. The voltage V_{CM} is supplied by a Emitter degenerated current mirror. The total current consumption for the divider with the input buffer (Emitter-follower) is 4.9 mA. The simulated maximum input frequency is 38 GHz after parasitic extraction of the layout.

The schematic of the simple buffer for the K_u-Band output and the divider are shown in Fig. 5b. Both share the same simple topology with a Emitter-follower followed by a differential pair with a resistor as a load. The current sources for the buffers are simple resistors which will not have a good temperature stability. Both buffers are also directly connected to the VCO and divider stage, respectively. The current consumption for the K_u-Band buffer is 8 mA and 5.6 mA for the buffer of the divider.

III. EXPERIMENTAL RESULTS

As a first experimental result the tuning curve of the realized ultra-wideband signal generation MMIC is shown in Fig. 6. The tuning characteristics are given for two different chuck temperatures of $\approx 25°C$ and $100°C$. The center frequency of the VCO is 16.1 GHz and it achieves a tuning range of 12.1 GHz. The tuning range relates to a relative frequency tuning range (rFTR) of 75.1%. The deviation due to temperature variation is quite low with only a small shift to lower frequencies for a chuck temperature of 100°C that can be observed. Compared to the realization in [8] this is a good improvement.

Fig. 6. Measurement of the tuning curve of the wideband VCO for two different temperatures ($\approx 25°C$ and $100°C$).

The second result is shown in Fig. 7 with the differential output power of the MMIC at the K_u-Band output. The output level of the VCO is roughly around 0 dBm. One interesting point that can be observed at high temperatures is the sudden drop in output power between 13 and 18 GHz. This occurs at the point where the tuning voltage V_{tune} equals the supply voltage V_{CC}. This needs to be investigated further by the authors.

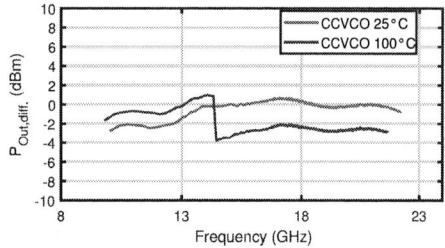

Fig. 7. Measurement of the differential output power at ≈25°C and 100°C.

In Fig. 8 the phase noise variation over the oscillation frequency at an offset frequency of 1 MHz is shown. The phase noise level is measured at the divider output and corrected for the frequency translation. The direct measurement at the K_u-Band output is also shown and matches the divider output very well. At low frequencies the highly optimized measurement script which was used with very small voltage steps to uncover load pull effects had problems with tracking the signal resulting in a much higher than actually present phase noise level.

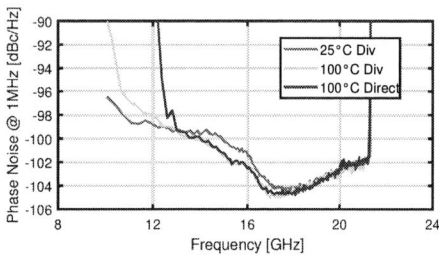

Fig. 8. Measurement of Phase Noise versus frequency taken at room temperature and 100°C.

As a last result in this paper the phase noise level versus the offset frequency is shown for the implemented signal generation MMIC at room temperature. The phase noise at the divider and an offset frequency of 1 MHz is as low as -111 dBc/Hz matching the -104 dBc from Fig.8 very well. The plot was taken at a frequency of 19.47 GHz with a tuning voltage of V_{tune} =6.3 V.

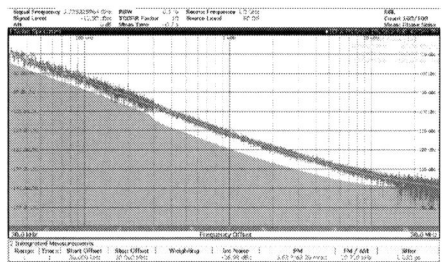

Fig. 9. Measurement of Phase Noise versus offset frequency taken at the divider output.

IV. CONCLUSION

We have presented the results for a signal generation MMIC in K_u-Band. The VCO can be tuned over a frequency range of more than one octave. To the best of the authors' knowledge this is the highest reported absolute continuous tuning range of 12.1 GHz (rFTR=75.1%) for SiGe LC-VCOs above 10 GHz. The phase noise level of -104 dBc/Hz at the used VCO core power of only 30 mW is good considering the extreme tuning range. The shown concept can be used to further increase the available bandwidth in E-Band radar systems.

ACKNOWLEDGMENT

The authors would like to thank Infineon Technologies AG and its staff members for fabricating the chips. We also want to thank Marcel van Delden and Florian Vogelsang from the Ruhr-University for their help in parts of the undertaken measurements.

REFERENCES

[1] J. Böck, K. Aufinger, S. Boguth, C. Dahl, H. Knapp, W. Liebl, D. Manger, T. F. Meister, A. Pribil, J. Wursthorn, R. Lachner, B. Heinemann, H. Rücker, A. Fox, R. Barth, G. Fischer, S. Marschmeyer, D. Schmidt, A. Trusch, and C. Wipf, "Sige hbt and bicmos process integration optimization within the dotseven project," in *2015 IEEE Bipolar/BiCMOS Circuits and Technology Meeting - BCTM*, Oct 2015, pp. 121–124.

[2] S. Gütgemann, C. Krebs, A. Küter, D. Nüßler, B. Fischer, and H. Krauthäuser, "Radar-based high precision thickness measurement for rolling mills," in *2018 15th European Radar Conference (EuRAD)*, 2018, pp. 122–125.

[3] N. Pohl, T. Jaeschke, and K. Aufinger, "An ultra-wideband 80 ghz fmcw radar system using a sige bipolar transceiver chip stabilized by a fractional-n pll synthesizer," *IEEE Transactions on Microwave Theory and Techniques*, vol. 60, no. 3, pp. 757–765, March 2012.

[4] C. Bredendiek, S. Hansen, G. Briese, and N. Pohl, "A full e-band single-channel sige transceiver mmic for monostatic fmcw radar systems," in *2020 15th European Microwave Integrated Circuits Conference (EuMIC)*, 2021, pp. 185–188.

[5] T. Jaeschke, C. Bredendiek, S. Küppers, and N. Pohl, "High-precision d-band fmcw-radar sensor based on a wideband sige-transceiver mmic," *IEEE Transactions on Microwave Theory and Techniques*, vol. 62, no. 12, pp. 3582–3597, Dec 2014.

[6] S. Thomas, C. Bredendiek, and N. Pohl, "A sige-based 240-ghz fmcw radar system for high-resolution measurements," *IEEE Transactions on Microwave Theory and Techniques*, vol. 67, no. 11, pp. 4599–4609, 2019.

[7] S. Hansen, C. Bredendiek, and N. Pohl, "D-band fmcw radar sensor for industrial wideband applications with fully-differential mmic-to-rwg interface in siw," in *2021 IEEE/MTT-S International Microwave Symposium (IMS)*, 2021.

[8] C. Bredendiek, K. Aufinger, and N. Pohl, "Investigation of integrated mmw-downconverter vcos in sige for offset-pll fmcw-transceivers," in *2020 IEEE 20th Topical Meeting on Silicon Monolithic Integrated Circuits in RF Systems (SiRF)*, 2020, pp. 43–46.

[9] ——, "Full waveguide e- and w-band fundamental vcos in sige:c technology for next generation fmcw radars sensors," in *2019 14th European Microwave Integrated Circuits Conference (EuMIC)*, Sep. 2019, pp. 148–151.

Proceedings of the 16th European Microwave Integrated Circuits Conference

An E-band Bidirectional PALNA in 0.13 μm SiGe BiCMOS Technology

R. Ahamed[#], M. Varonen[*], D. Parveg[*], M. Najmussadat[#], M. Kantanen[*], Y. Tawfik[#], K.A.I. Halonen[#]

[#]Department of Electronics and Nanoengineering, Aalto University, Finland

[*]VTT Technical Research Centre of Finland Ltd, Finland

raju.ahamed@aalto.fi, mikko.varonen@vtt.fi

Abstract — This paper presents an E-band bidirectional power amplifier and low-noise amplifier (PALNA) in a 0.13 μm SiGe BiCMOS technology. In the transmit mode, the LNA is isolated through the impedance matching network with enabler setting. In the receive mode, the input of the PA is isolated by matching network with enabler setting while the output of the PA is isolated by a differential switch to have minimal effect on the noise figure (NF) and input matching of the LNA. The measured PALNA achieves a peak gain of 24.3 dB at 72 GHz with a 3 dB bandwidth from 64 GHz to 85 GHz in the receive mode. The measured NF is 5.8 dB at 74 GHz. In the transmit mode, the differential PA results in 19.5 dB peak gain at 76 GHz and +9.2 dBm of saturated power at 74 GHz. The overall chip size is 1.25 mm^2. The PALNA consumes a DC power of 19 mW and 131 mW in the receive and transmit mode, respectively. The designed bidirectional PALNA is suitable for a half-duplex system with shared Tx/Rx antenna.

Keywords — BiCMOS, bidirectional, E-band, LNA, PALNA, PA, millimeter-wave, MMIC.

I. INTRODUCTION

Highly integrated beamforming transceivers have been demonstrated in [1], [2]. However, they are often limited by the size of antenna array when separate antennas are used for the transmitter (Tx) and the receiver (Rx) [3]. In time division duplex (TDD) communication system, a single antenna can be shared by both of the transmitter and the receiver. Using a single antenna for both the PA and the LNA increases integration and reduces the cost of the system [4], [5]. Front-end transmit-receive (T/R) switches (SW) are often used so that both the Tx and the Rx can share the same antenna. These additional high frequency switches add noise to the receiver and reduces the output power of the Tx while taking additional area and DC power. On the other hand, switchless PALNA is compact and consumes less DC power still maintaining the switching functionality between Tx and Rx [6].

To harvest a higher power, differential PA is an option that can ideally deliver 3-dB more power. However, a differential PA imposes challenge in the impedance transformation in the receive mode without a switch. In this case, a low-loss differential switch can be utilized at the output of the PA to have a minimal effect on the noise figure (NF) and input matching of the LNA.

In this paper, we present a bidirectional PALNA which implements a differential high-power PA and a single-ended LNA. A low-loss differential switch is utilized at the output of the PA to properly isolate the PA from the LNA in the receive mode. Switching of the LNA is done through the impedance matching network with enabler setting.

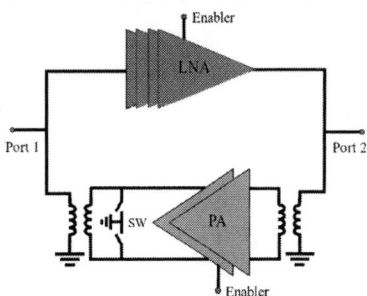

Fig. 1. Simplified block diagram of the PALNA.

II. PALNA DESIGN

A simple block diagram of the PALNA is shown in Fig.1. Fig.2 shows the schematic of the designed PALNA with the respective device sizing. The biasing circuits are omitted due to simplicity. In the receive mode, the LNA is active and the PA is inactive. In the transmit mode, the PA is active and deliver power to the antenna while the the LNA is inactive. The input of the differential PA and output of the LNA are connected through an impedance matching network. Similarly, the input of the LNA and output of the PA are connected through an impedance matching network. Switching of the LNA input, LNA output and the PA input is done through the impedance matching network with enabler settings. There is a differential switch at the output of the PA to properly isolate the PA from the LNA in the receive mode. As a result, there is negligible effect of the PA output on the LNA performance.

A. LNA design

The LNA is made of 4 cascaded common-emitter (CE) amplifier stages. The first stage is designed for minimum noise operation. Inductive degeneration is provided by the transmission line T_4 of 20 μm length for simultaneous noise and gain matching. The noise matching is achieved by the T_1, T_2 and C_1 and they transform the 50 Ω source impedance close to the optimum source impedance for minimum noise. The value of T_1, T_2 and C_1 are selected in such way that they can transform the off-impedance of the transistor M_1

3–4 April 2022, London, UK

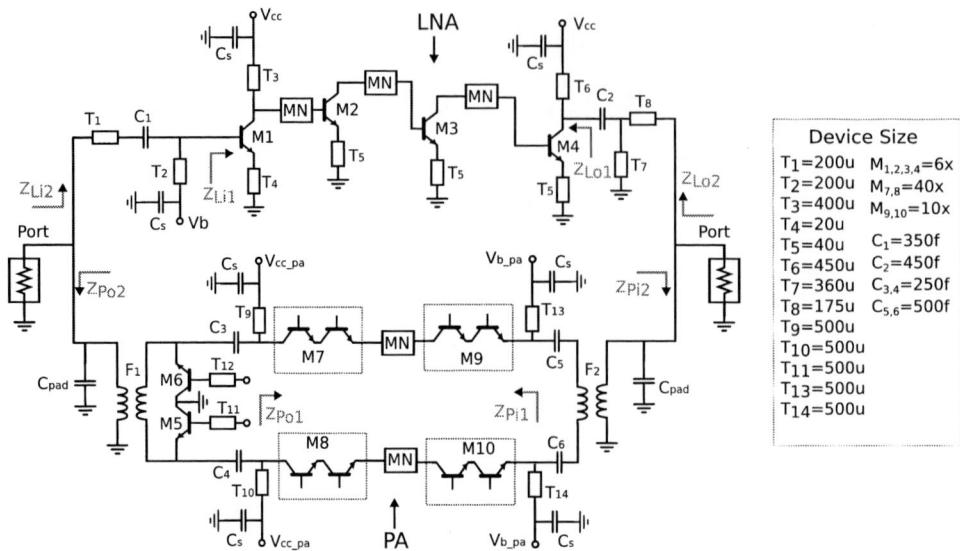

Fig. 2. Schematic of the designed E-band PALNA with associate device sizing. Biasing networks are ommited.

to a high impedance when the LNA is inactive. Second, third and fourth stage CE amplifiers are designed for high gain and conjugate matching is used between them. Emitter degeneration by transmission line is used also for gain stages for improving the stability. The output matching of the LNA is done by the T_6, T_7, T_7 and C_2 and the network can also transform off-impedance of the transistor M_4 to a high impedance when LNA is inactive.

In the receive mode, the LNA is active and the PA is inactive. The input and output impedance of the PALNA in receive mode is close to 50 Ω resulting a good input and output matching. In the transmit mode, when the transistors of LNA are turned off with a zero bias, the input matching network transform the off-impedance of the transistor Z_{Li1} to a high impedance Z_{Li2} (273-j17 at 76 GHz). Similarly, LNA output matching network transform the off-impedance of the transistor Z_{Lo1} to a high impedance Z_{Lo2}. This way the LNA is well isolated from the PA in the transmit mode. Fig.3 shows the respective impedance transformation in the smith chart for the frequency range of 71 GHz to 76 GHz.

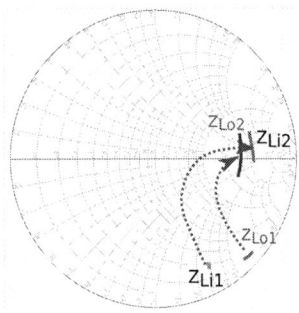

Fig. 3. Transformation of LNA input and output impedance in the transmit mode of the PALNA.

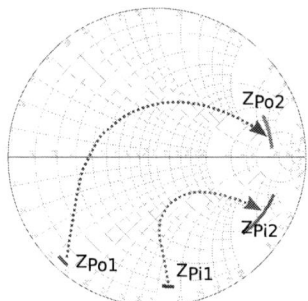

Fig. 4. Transformation of PA input and output impedance in the receive mode of the PALNA.

B. PA design

The differential topology is selected to deliver high output power from the PA. The differential PA consists of two cascode amplifier stages. The output stage has larger transistors with 40 multiplier and the input stage has smaller transistors with 10 multiplier. The cascode amplifier topology is chosen due to higher reverse isolation and higher power handling. Both of the stages are biased in class A operation. The input stage is optimized for gain and the output stage is optimized for power. The conversion between differential to single-ended signal is done by the transformer F_1 and F_2 . Transformers also resonate out the RF pad capacitance. The transformer also can serve as ESD protection.

The output matching elements of the PA are F_1, T_9, T_{10}, C_3 and C_4. They transform the parallel combination of the 50 Ω load and the off-impedance of the LNA Z_{Li2} to an impedance where the PA can deliver around 13 dBm of maximum power in the transmit mode. In the receive mode, the differential PA is inactive and the PA should show high impedance toward the LNA to have little impact on the LNA performance. But, unfortunately, the transformed off-impedance of the differential PA by the matching network alone is not sufficiently high. Consequently, it is affecting the LNA NF and input matching. To overcome this challenge, a differential switch utilizing reverse-saturated HBTs is used. The differential switch is capable of handling 3 dB more power than a typical single-ended switch. The switch is active in the receive mode. The differential switch is made of transistor $M5$, $M6$ with quarter-wave transmission lines T_{11} and T_{12} at the base and transformer F_1 that also acts as the balun in the transmit mode. The transistors $M5$ and $M6$ are turned on by applying 0.95 V at the bases. The on-state devices provide a low impedance and the transformer F_1 transform this low impedance to a high impedance Z_{Po2} (351+j121 at 76 GHz). This way the PA is well isolated from the LNA and have a minimal effect on the LNA. The differential switch draws a total DC current of 3.6 mA from 0.95 V supply. The switch is inactive in the transmit mode and the off-state capacitance of the transistors are resonated with the PA matching network resulting a little impact on the PA operation. The PA input matching network transform the off-impedance of the transistor Z_{Pi1} to a comparatively high impedance Z_{Pi2}. Fig.4 shows the respective impedance transformation in the smith chart.

C. Designed results of the PALNA

ADS momentum was used for EM simulations of the transmission lines, MIM capacitors and RF pads to obtain accurate modeling. RC extraction of the transistor was performed to take the high frequency parasitic effects into account. In the receive mode, the PALNA shows a peak gain of 27.5 dB, and a wideband noise performance with a minimum NF of 5.4 dB at 80 GHz. In the transmit mode, the PALNA shows a peak gain of 20.5 dB at 69 GHz with a peak saturated power of 12.5 dBm at 72 GHz.

III. EXPERIMENTAL RESULTS

The designed PALNA was fabricated in a 0.13 μm SiGe BiCMOS process. The die micrograph is shown in Fig.5. The total area of the chip is 1.25 mm^2.

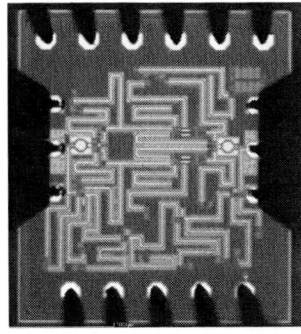

Fig. 5. Die micrograph of the PALNA.

S-parameters were measured over 30-100 GHz with an Agilent millimeter-wave PNA E8361C network analyzer using 200 μm pitch GSG on-wafer probes and LRRM calibration method. Measured S-parameters of PALNA in the receive mode is shown in Fig.6. The measurement results closely

follow the simulation results. It achieves a peak gain of 24.3 dB at 72 GHz with a 3-dB bandwidth from 64 GHz to 85 GHz. Both input and out matching are better than 10 dB from 67 GHz to 90 GHz.

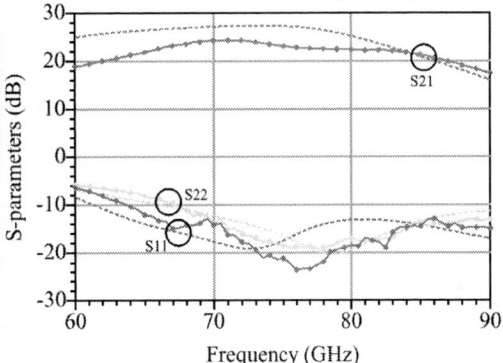

Fig. 6. Measured (solid) and simulated (dotted) s-parameters of PALNA in the receive mode.

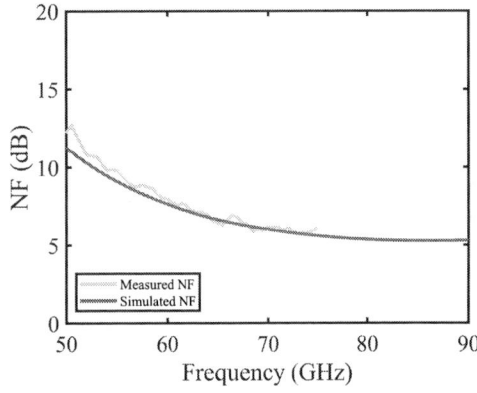

Fig. 7. Measured and simulated NF of the PALNA in the receive mode.

Noise figure was measured over 50-75 GHz using a noise diode and a noise receiver. The Noise receiver consists of a balanced mixer, LO-chain and Agilent N8973A noise figure analyzer. Frequency sweep of the measurement system is controlled by the noise figure analyzer via GPIB. Y-factor method is used to derive the noise figure. Fig.7 shows the measured and simulated noise figure of the PALNA. The measured NF and the simulated NF are very close to each other. The minimum measured NF is 5.8 dB at 74 GHz and it shows a very wideband noise characteristic as like as the simulation.

The measured and the simulated S-parameters of PALNA in the transmit mode is shown in the Fig.8. In the transmit mode, the differential PA results in 19.5 dB peak gain at 76 GHz and 3-dB bandwidth from 71.5 GHz to 79 GHz. Power characteristic of the PALNA is shown in Fig.9. The PA transmits +9.2 dBm of saturated power at 74 GHz. The measured saturated power is less than the simulated power. This is due to the fact that there is a low frequency oscillation found at 9.5 GHz. After rigorous analysis it is found that the

283

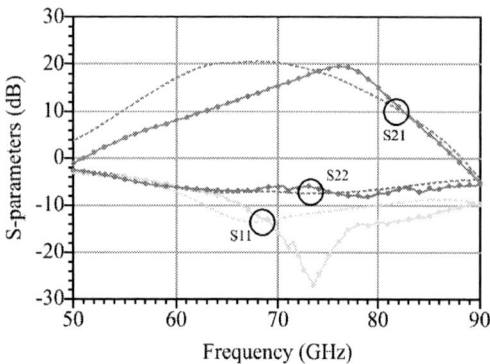

Fig. 8. Measured (solid) and simulated (dotted) *S*-parameters of PALNA in the transmit mode.

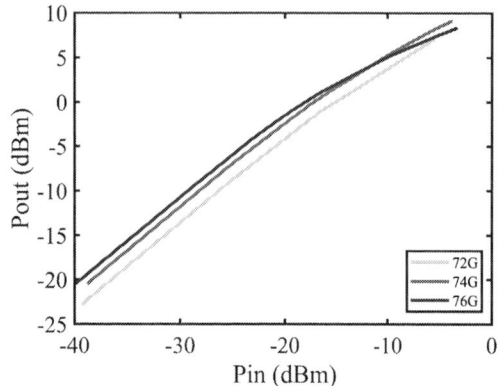

Fig. 9. Measured transmitted power of the PALNA.

oscillation comes from the DC supply loop as all the transistors are biased from the same supply. To kill the oscillation, the PA needs to operate at the lower bias with a 41 mA of collector current instead of 95 mA designed current. Consequently, the output power is less than the simulated power.

Table 1. Performance comparison to the published E-band PALNAs

	This work	[3]	[6]	[7]
Topology	Bidirectionl	Bidirectionl	Bidirectionl	Bidirectional
Process	130nm SiGe BiCMOS	90nm SiGe BiCMOS	90nm SiGe BiCMOS	130-nm SiGe:C BiCMOS
Fc (GHz)	72	70	78	60
Rx gain (dB)	24.3	21[a]	25[a]	17
NF (dB)	5.8	8.6	7.6	6.5 *
Rx FOM	18.6	9.2	6.5	5.1
Tx gain (dB)	19.5	23[b]	31[b]	16.5
P_{SAT} (dBm)	9.2	7.2	11	12
P_{DC} (mW)	131(PA) 19(LNA+sw)	85(PA) 21(LNA)	255(PA) 45(LNA)	130(PA) 50(LNA)

[a] Only LNA gain from simulation
[b] Only PA gain from simulation
* Simulated NF

Table 1 compares the measured performance of the designed E-band switchless PALNA to other published state-of-the-art switchless PALNAs in SiGe BiCMOS processes. The PALNA in this paper exhibits the lowest NF

and highest FOM in receive mode compared to other works. The FOM is defined as in (1) [7] where Gain and F are the magnitudes of S_{21} and noise figure, respectively. In the transmit mode, it provides a comparable results as well.

$$R_x FOM[GHz/mW] = \frac{Gain \times Frequency[GHz]}{(F-1) \times P_{DC}[mW]} \quad (1)$$

IV. CONCLUSION

The design and implementation of a bidirectional E-band PALNA has been presented in a 0.13 μm SiGe BiCMOS technology. The PALNA utilizes one differential switch at the output of the PA that enables proper isolation of PA resulting a high performance LNA without degrading the PA performance. The designed PALNA stands as the optimum choice between the switchless and the switched-PALNA specifically with differential PA. The measured PALNA achieves a peak gain of 24.3 dB at 72 GHz with a 3-dB bandwidth from 64 GHz to 85 GHz and a minimum NF of 5.8 dB in the receive mode. In the transmit mode, the differential PA achieves 19.5 dB peak gain at 76 GHz and +9.2 dBm of saturated power.

ACKNOWLEDGMENT

The authors would like to thank the Academy of Finland for supporting this work through TeraCom project. The work of M. Varonen was supported through the Academy of Finland Research Fellow project MIDERI (decision no 310234).

REFERENCES

[1] A. Natarajan, A. Komijani, G. Xiang, A. Babakhani, A. Hajimiri, "A 77-GHz Phased-Array Transceiver With On-Chip Antennas in Silicon: Transmitter and Local LO-Path Phase Shifting," *IEEE Journal of Solid-State Circuits*, vol.41, no.12, pp.2807–2819, Dec. 2006.

[2] B. Ku, O. Inac, M. Chang, Y. Hyun-Ho Yang, G.M. Rebeiz, "A High-Linearity 76–85-GHz 16-Element 8-Transmit/8-Receive Phased-Array Chip With High Isolation and Flip-Chip Packaging," *IEEE Transactions on Microwave Theory and Techniques*, vol.62, no.10, pp.2337–2356, Oct. 2014.

[3] T. Kijsanayotin, J. Li and J. F. Buckwalter, "A 70 GHz Bidirectional Front-End for a Half-Duplex Transceiver in 90-nm SiGe BiCMOS," *2015 IEEE Compound Semiconductor Integrated Circuit Symposium (CSICS)*, New Orleans, LA, 2015, pp. 1-4.

[4] I. Kallfass, S. Diebold, H. Massler, S. Koch, M. Seelmann-Eggebert and A. Leuther, "Multiple-Throw Millimeter-Wave FET Switches for Frequencies from 60 up to 120 GHz," *2008 European Microwave Integrated Circuit Conference*, 2008, pp. 426-429, doi: 10.1109/EMICC.2008.4772320.

[5] B. Kim, J. Jang, C. Kim and S. Hong, "Integration of SPDT Antenna Switch With CMOS Power Amplifier and LNA for FMICW Radar Front End," in *IEEE Transactions on Microwave Theory and Techniques*, vol. 66, no. 11, pp. 5087-5094, Nov. 2018, doi: 10.1109/TMTT.2018.2860970.

[6] P. Wu, T. Kijsanayotin and J. F. Buckwalter, "A 71–86-GHz Switchless Asymmetric Bidirectional Transceiver in a 90-nm SiGe BiCMOS," in *IEEE Transactions on Microwave Theory and Techniques*, vol. 64, no. 12, pp. 4262-4273, Dec. 2016, doi: 10.1109/TMTT.2016.2608891.

[7] A. Gadallah, M. H. Eissa, D. Kissinger and A. Malignaggi, "A V-Band Miniaturized Bidirectional Switchless PALNA in SiGe:C BiCMOS Technology," in *IEEE Microwave and Wireless Components Letters*, vol. 30, no. 8, pp. 786-789, Aug. 2020, doi: 10.1109/LMWC.2020.3005211.

A Ka-Band 40 W Output Power and 30 % PAE GaN MMIC Power Amplifier for Satellite Communication

Keigo Nakatani, Yutaro Yamaguchi, Masaomi Tsuru

Information Technology R&D Center, Mitsubishi Electric Corporation
Ofuna 5-1-1, 247-8501 Kamakura, Japan
Nakatani.Keigo@cj.MitsubishiElectric.co.jp

Abstract — This paper presents a Ka-band 40 W power and 30 % PAE GaN power amplifier MMIC using 0.15 μm gate-length GaN on SiC transistor with ISV structure. The fabricated 3-stage single-ended GaN power amplifier MMIC achieves 45.0 - 46.2 dBm output power and 23-30 % PAE over the 26 - 31 GHz band (FBW=17.5 %) under pulsed condition.

Keywords — Ka-band, GaN, power amplifiers, MMIC.

I. INTRODUCTION

In recent years, satellite communication (SATCOM) system has gradually increased frequency in response to requirement of more high data-rate and large capacity. In particular, Ka-band applications are hot topic for amount of communication systems, such as existing SATCOM, Military, high-resolution radar system and 5G mm-wave applications. Besides 5G/6G mm-wave applications will be driving change across the SATCOM system, through new satellites communication service using their high throughput. Operating power amplifier in Ka-band has problems to be solved due to increasing frequency channel video bandwidth requirements and high linear operation to realize multi-carrier systems. TWTAs (Traveling Wave Tube Amplifier) were generally used as microwave transmitter in Ka-band. Although TWTAs achieves high power and high efficiency in Ka-band, it is much hard to satisfy high linear operation with wideband characteristics. Therefore, solid state power amplifiers using GaN (Gallium Nitride) MMIC (Microwave Monoclinic Integrated Circuit) are one of the significant technologies to replace TWTAs in Ka-band.

GaN is suitable for mm-wave power amplifier with high efficiency and wideband characteristic due to high output impedance and low parasitic capacitance. Additionally, GaN has high output power density thus it produces good benefits of power amplifier module for chips size and cost efficiency. Therefore, high output power and high efficiency GaN power amplifier MMICs have been already reported in [1]-[8].

This paper presents a Ka-Band 40 W power and 30 % PAE 3-stage single-ended GaN power amplifier MMIC fabricated by Mitsubishi's 0.15 μm gate length GaN process on 50 μm thickness SiC substrate with ISV (Individual Source Via) structure. Fig.1 shows chip photograph of Ka-band 40 W GaN power amplifier MMIC. The fabricated 3-stage single-ended GaN power amplifier MMIC achieves 45.0 - 46.2 dBm output power and 23-30 % PAE over the 26 - 31 GHz band

(FBW=17.5 %) under pulsed condition. To the best of author's knowledge, that high output power and PAE one of the state-of-the art for GaN MMIC power amplifier in Ka-band.

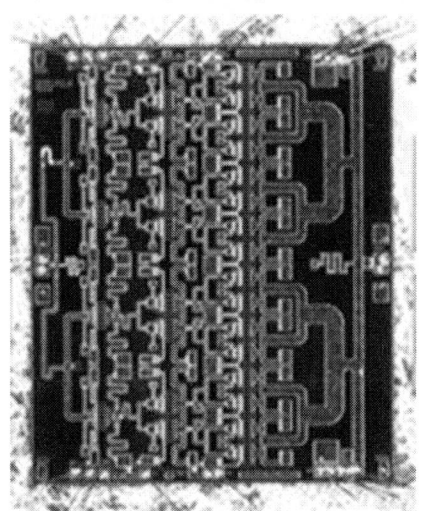

Fig. 1. Chip photograph of Ka-band 40 W GaN power amplifier MMIC. The die size is 3.75mm x 4.35mm.

II. DEVICE PERFORMANCES

Figure. 2. shows the measured input output characteristics of a unit cell transistor using load-pull measurement system at 28 GHz with optimized load impedance pre-matching circuit. The transistor was fabricated by Mitsubishi's 0.15 μm gate length process on 50 μm thickness SiC substrate. At 28 GHz, an 8-finger, 52 μm gate width transistor biased at Vd = 24 V, Id = 50 mA/mm typically achieves 3.8 W/mm output power density with 9.0 dB power gain and 50% PAE in efficiency load pull tuned under CW operation.

Figure. 3. shows small signal MAG (Maximum Available Gain) and MSG (Maximum Stable Gain) of 8-finger, 80 μm gate width transistors utilizing ISV and OSV (Outside Source Via) structure using scaling Angelov-GaN model. 0.64 mm gate width transistor was selected for the power amplifier design to get highest output power in Ka-band. OSV structure generally used in GaN-HEMT device structure in low frequency band within external source vias connected to each transistor source pad using fine metal lines and air bridges. It becomes difficult to operate in higher frequency band due to increase of parasitic elements. A calculated MSG is decreasing above 22 GHz. It is expected to be further reduced by the input matching circuit loss. Since, source pad was directly connected GND vias, source inductance is obvious decreasing and improves MAG frequency limit, ISV structure is suitable for transistor which operating in high frequency band.

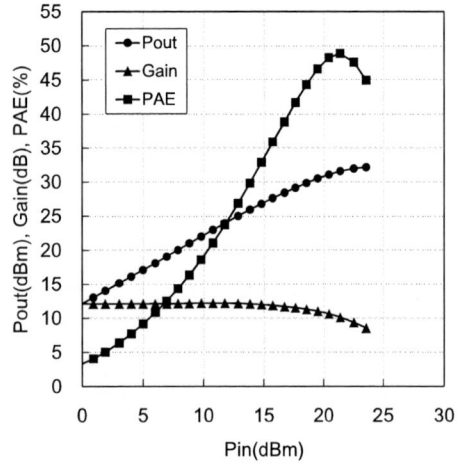

Fig. 2. Measured input output characteristics of a unit cell transistor at 28 GHz, 24 V, 50 mA/mm bias condition. The size of unit FET cell is 8-finger, 52 μm gate width.

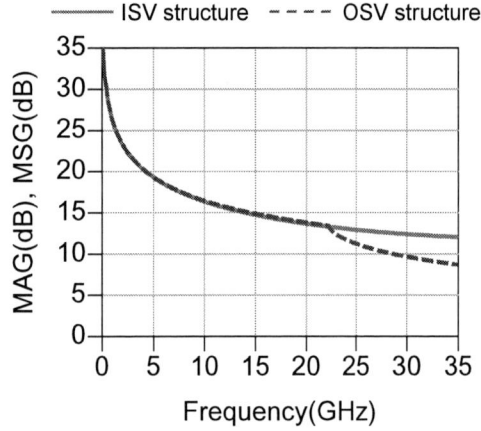

Fig. 3. Small signal MAG and MSG of 8-finger, 80 μm gate width transistors utilizing ISV and OSV structure using scaling Angelov-GaN model.

III. Power Amplifier Design

Figure. 4. shows Block diagram of 3-stage power amplifier MMIC. The amplifier consists of 3-stage Class-AB amplifiers. The gate width of unit FET cells are 8-finger, 80 μm gate width for each stage. 16-FET cells are combined for the final stage. 8-FET cells and 4-FET cells are used for the 2nd and the 1st driver stage. Linear gain is approximately 8 – 8.5 dB for each amplifier stage, total gain is 25 dB including

interstage matching networks (ISMN). ISMN used wideband BPF impedance matching technique utilizing considering parasitic elements of driver stage and final stage [9].

Figure. 5. Shows equivalent circuit of output matching network (OMN) for a unit cell transistor output section. OMN consists of TL_1, shunt inductors (L_1, L_2) and quoter wavelength transmission lines (TL_2, TL_3, TL_4). TL_1 and L_1 compensated transistor output impedance (Z_{FET}) with calculated by output resistance (R_o) and parasitic drain source capacitance of transistor (C_{out}) to realize optimum impedance (Z_1) and low loss circuit condition. TL_2, TL_3 and TL_4 used transforming Z_1 to Z_2, Z_3 and Z_{load} even as combining 16-FET cells. Fig. 6. shows simulation results of OMN. Return Loss (RL) approximately less than -17 dB and circuit loss of 0.55 - 0.65 dB are realized in 27 – 31GHz.

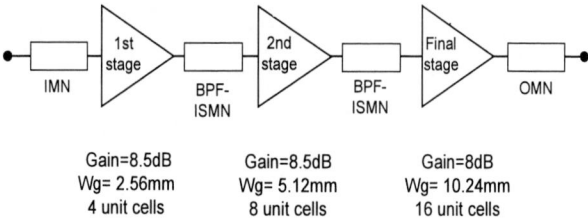

Fig. 4. Block diagram of 3-stage power amplifier MMIC

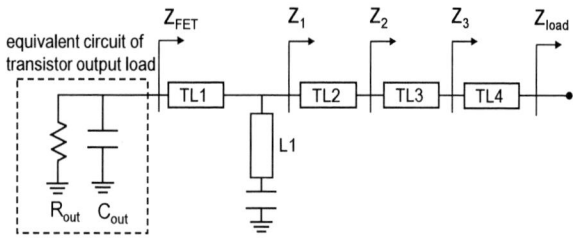

Fig. 5. Equivalent circuit of output matching network for a unit cell transistor output section

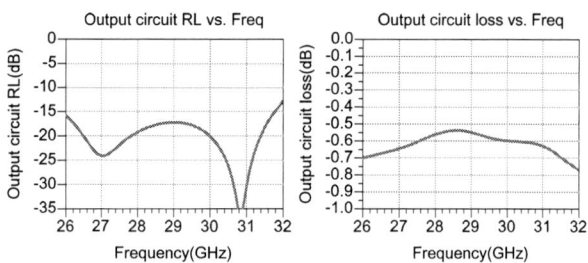

Fig. 6. Simulation results of OMN

IV. Measurement Results

Figure. 7 shows photograph of a GaN power amplifier MMIC on measurement test fixture. MMIC chip is soldered to a CuMoCu carrier plate and a copper heat sink is used. Fig. 8. shows the measurement results of S-parameters at V_d = 26 V, Id = 448 mA (I_{dq} = 25 mA/mm). The obtained small-signal gain is 20 - 25 dB in 26 - 31 GHz although matching frequency shifted to lower frequency band. Fig. 9. shows measurement

286

results of 26 GHz input-output characteristics at Vd = 26 V, Id = 448 mA. carrier plate backside temperature is 25 deg.C. The pulse condition is 1 mS and 10 % duty cycle. The measured maximum output power of 46.2 dBm, peak PAE of 30.2 % and G_p of 18 dB are obtained. Fig. 10. shows the measured maximum output power, G_p and peak PAE versus frequency under pulsed operation. The measured maximum output power of 45.0 - 46.2 dBm (31.6 – 41.7 W) and peak PAE of 23 – 30 % and G_p of 17.4 – 20.7 dB are obtained over 26 – 31 GHz (FBW = 17.5 %).

Fig. 7. Photograph of GaN power amplifier MMIC on measurement test fixture. MMIC chip is soldered to a CuMoCu carrier plate and a copper heat sink is used.

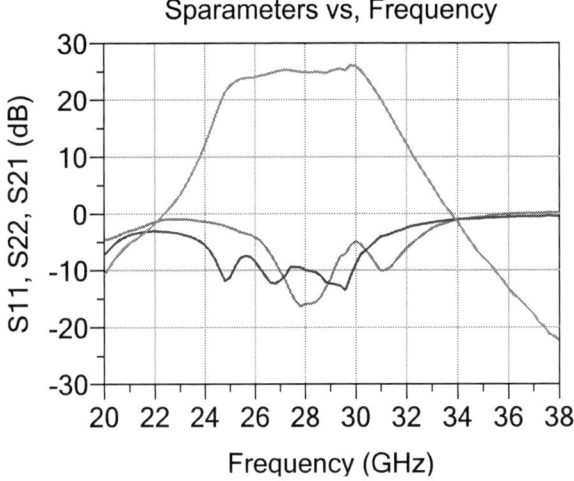

Fig. 8. Measurement results of S-parameters at Vd = 24 V, Id = 448 mA

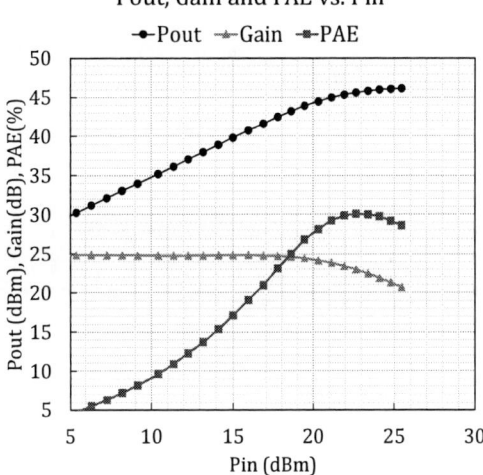

Fig. 9. Measurement results of 26GHz input-output characteristics at Vd = 26 V, Id = 448 mA, carrier plate backside temperature is 25 deg.C. The pulse condition is 1 mS and 10 % duty cycle.

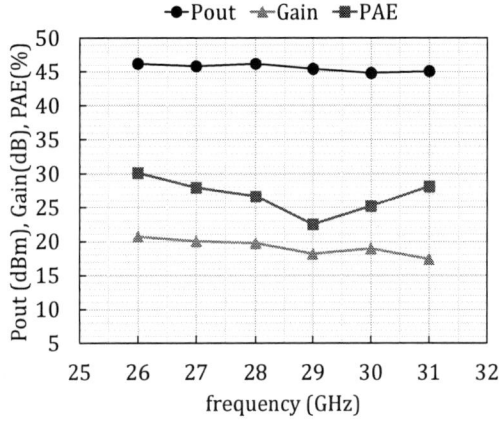

Fig. 10. Measurement results of frequency characteristics at Vd = 24 V, Id = 448 mA in 26 – 31 GHz. carrier plate backside temperature is 25 deg.C. The pulse condition is 1 mS and 10 % duty cycle.

V. CONCLUSION

In summary, a Ka-Band 40 W output power and 30 % PAE GaN power amplifier MMIC fabricated by Mitsubishi's 0.15 μm gate length GaN process has been reported. 3-stage single-ended amplifier consists of BPF ISMN and low loss wideband output matching circuit. The experimental result shows maximum output power of 45.0 - 46.2 dBm, peak PAE of 23 - 30 % with 17.4 – 20.7 dB power gain in 26 – 31 GHz band (FBW = 17.5 %). To the best of author's knowledge, that bandwidth with high output power is state-of-the art for GaN MMIC power amplifier in Ka-band as shown Table 1.

Table 1. Ka-Band Gan Power Amplifier MMIC Benchmarks.

Ref (#)	Process	Frequency (GHz)	FBW (%)	Stages (#)	Gain (dB)	Power (dBm)	PAE (%)	Die size (mm²)	Pd (W/mm)
[1]	GaN on Si	29-33	12.9	3	27	40.5	35	15.8	0.71
[2]	GaN on SiC	29-31	6.7	2	10	42.6	17.4	22.0	0.83
[3]	GaN on SiC	26.5-29	9.0	3	24	39.6	29.5	11.2	0.81
[4]	GaN on SiC	25.5-27	5.7	2	15.6	37.2	32	9.0	0.58
[5]	GaN on SiC	28-31	10.2	3	28	38.1	33	5.6	1.15
[6]	GaN on SiC	26-28	7.4	2	12	43.4	19.8	23.6	0.93
[7]	GaN on SiC	27-31	13.8	3	27	42.2	20	24.8	0.67
[8]	GaN on SiC	26-30	14.3	2	20	46	36	13.5	2.94
[9]	GaN on SiC	26.5-31	15.7	3	30	46.2	25.9	17.6	2.37
This Work	GaN on SiC	23-31	17.5	3	25	46.2	30.2	16.3	2.56

REFERENCES

[1] A. Gasmi, M. E. Kaamouchi, J. Poulain, B. Wroblewski, F. Lecourt, G. Dagher, P. Frijlink, and R. Leblanc, "10W power amplifier and 3W transmit/receive module with 3 dB NF in Ka band using a 100nm GaN/Si process," *in 2017 IEEE Compound Semiconductor Integrated Circuit Symposium (CSICS), Oct 2017, pp. 1–4.*

[2] K. Takagi, C. Y. Ng, H. Sakurai, and K. Matstushita, "GaN MMIC for Ka-Band with 18W," *in 2015 IEEE Compound Semiconductor Integrated Circuit Symposium (CSICS), Oct 2015, pp. 1–4.*

[3] Y. Noh, Y. Choi, I. Yom, "Ka-band GaN power amplifier MMIC chipset for satellite and 5G cellular communications," *in 2015 IEEE 4th Asia-Pacific Conference on Antennas and Propagation (APCAP), June 2015.*

[4] S. Samis, C. Friesicke, P. Feuerschütz, R. Lozar, T. Maier , P. Brückner, R. Quay and A. F. Jacob, "A 5 W AlGaN/GaN power amplifier MMIC for 25–27 GHz downlink applications," *in 2018 11th German Microwave Conference (GeMiC), Mar 2018.*

[5] C. F. Campbell, Y. Liu, M. Y. Kao, and S. Nayak, "High efficiency Ka-band Gallium Nitride power amplifier MMICs," *in 2013 IEEE International Conference on Microwaves, Communications, Antennas and Electronic Systems (COMCAS 2013), Oct 2013, pp. 1–5.*

[6] Y. Yamaguchi, J. Kamioka, M. Hangai, S. Shinjo, and K. Yamanaka, "A CW 20W Ka-band GaN high power MMIC amplifier with a gate pitch designed by using one-finger large signal models," *in 2017 IEEE Compound Semiconductor Integrated Circuit Symposium (CSICS), Oct 2017, pp. 1–4.*

[7] K. Nakatani, Y. Yamaguchi, M. Hangai, and S. Shinjo, "A Ka-Band CW 15.5W 15.6% Fractional Bandwidth GaN Power Amplifier MMIC Using Wideband BPF Inter-stage matching network," *in 2019 IEEE BiCMOS and Compound Semiconductor Integrated Circuits and Technology (BCICTS2019). Nov 2019.*

[8] S. Din, M. Wojtowicz, and M. Siddiqui, "High power and high efficiency Ka band power amplifier," *in 2015 IEEE MTT-S International Microwave Symposium, May 2015, pp. 1–4.*

[9] M. Roberg et al., "40 W Ka-Band Single and Dual Output GaN MMIC Power Amplifiers on SiC," *in Comp. Semicond. Integr. Circuits Techn. Symp. (BCICTS). IEEE, 2018, pp. 140–143.*

Probabilistic Poly Harmonic Distortion Model

Anna Davis Manjaly, Justin King

Department of Electronic and Electrical Engineering, Trinity College Dublin, Ireland
{manjalya, justin.king}@tcd.ie

Abstract — **This paper presents a probabilistic, nonlinear, frequency-domain behavioural model for RF transistor devices. The model is built on the poly harmonic distortion (PHD) framework, although the method itself is general and can be applied to many other modelling procedures. The key difference between this and existing models lies in the probabilistic prediction; rather than provide a single predictive output, the output of the proposed model is a probability distribution, with corresponding mean and variance distribution parameters that depend on the input independent variable. The model is first validated using simulated data to demonstrate the key theoretical principles. The model is then extracted from real-world experimentally measured load pull waveform data.**

Keywords — **Probabilistic model, Bayesian Approach.**

I. INTRODUCTION

The demand for higher data rates and connectivity is increasing with each new generation of the telecommunications network. Often relying on newer device technologies, such advancements require device models that are accurate with a short development time. Behavioural models, or 'black-box' models are well-suited to this task. These models do not require the details about the internal set-up of the device, and hence can be extracted quickly and accurately directly from measured device data.

However, all experimentally measured data are subject to error. These errors may be categorised into systematic and random errors. Systematic errors are those issues that can be identified and accounted for, e.g., the effect of cables in an RF system can be removed by calibration. Random errors, on the other hand, cannot be predicted. Such errors can be the result of thermal noise inside measurement equipment, for example. This work aims to quantify these random uncertainties using a Bayesian probabilistic approach.

There are existing studies that have quantified the uncertainties in the extraction of model parameters of behavioural models such as Scattering parameter (S-parameter) Modelling, X-parameter Modelling etc… [1],[3]. S-parameters, effectively the most used behavioural model in the RF field, are often applied to model the behaviour of linear systems or systems which behave linearly when driven by the signals with small amplitudes. Mubarak et al, in [1], have followed a physical paradigm for the analysis of uncertainties in Vector Network Analyzer S-parameter measurements. However, S-parameters are limited to the situations described above. X-parameter, on the other hand, are an extension of S-parameter formalism and are capable of modelling non-linear systems.

The X-parameter approach generally assumes a single dominant tone behaves non-linearly and the remaining harmonics behave linearly [2]. Basic X-parameters can provide good model prediction for nonlinear systems under matched or nearly matched load conditions. In [3], a probabilistic approach is applied to X-parameters, which gives a credible region for each X-parameter. In [4], the Quadratic Poly-Harmonic Distortion (QPHD) Model is introduced which can provide good prediction for nonlinear systems even in unmatched load conditions.

This work introduces a new probability model for scattered waves. A simple QPHD model is used to relate the scattered wave to the incident waves. In this work, the incident waves and the corresponding scattered waves are measured, and a Bayesian statistical analysis of these measurements is carried out. In this proposed technique, knowledge about the cause or origin of the uncertainties in the measurements or extracted model parameters is not required. Instead, a probability distribution for the scattered waves is inferred from measured data, using Bayesian inference methods.

This paper is divided into four sections. Section II presents the theory used in this work. In Section III, we provide theoretical validation of the proposed Bayesian model and extraction process. In Section IV, a probabilistic model of the scattered waves for a transistor is extracted and validated. Section V is the conclusion. Frequency domain quantities are represented by uppercase italic letters and the vectors are represented by bold uppercase letters throughout the paper.

II. THEORY

The Poly Harmonic Distortion (PHD) method is introduced in [5] as a method to extend the scattering formalism i.e., S-parameter models, to the non-linear regime. In Fig. 1, the waves incident on port p of the Device Under Test (DUT) at frequency q are denoted by A_{pq} and the waves scattered from port i of the DUT at frequency k are denoted by B_{ik}.

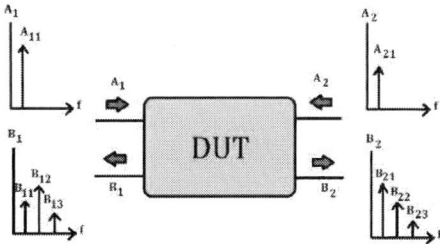

Fig 1. 2-port Device Under Test (DUT)

The pseudowaves A and B are defined as simple linear combinations of the port voltage V and current I at respective ports

$$A_{pq} = \frac{V_{pq} + Z_o I_{pq}}{2\sqrt{Z_0}}, \qquad (1)$$

$$B_{ik} = \frac{V_{ik} - Z_o I_{ik}}{2\sqrt{Z_0}}, \qquad (2)$$

where Z_0 is the characteristic impedance. In the PHD formalism, the scattered waves are related to the incident waves via Describing Functions (F_{ik}). Each scattered wave has a separate describing function [4]

$$B_{ik} = F_{ik}(A_{11}, A_{12}, A_{13}, A_{14}, \dots, A_{21}, A_{22} \dots). \qquad (3)$$

In this work, a polynomial describing function is used to model the nominal value of the scattered wave B_{ik} versus the incident waves A_{pq}. Two separate models are needed for the real and imaginary parts of the scattered wave. The suffixes R and I indicate the real and imaginary parts, respectively. For example, the scattered wave $B_{21,R}$, the dominant incident wave at port-1 as A_{11} and perturbations are caused by A_{21}. For simplicity, we have considered only the fundamental frequency for a two-port device. The scattered wave $B_{21,R}$ is represented as a function of two independent variables $A_{21,R}$ and $A_{21,I}$ as follows

$$B_{21,R} = w_0 + w_{1R}A_{21,R} + w_{1I}A_{21,I} + w_{2R}A_{21,R}^2 + w_{2I}A_{21,I}^2, \quad (4)$$

where $w_0, w_{1R}, w_{1I}, w_{2R}$ and w_{2I} are the model parameters, which are real numbers, and $A_{21,R}$ and $A_{21,I}$ are the real and imaginary parts of the incident wave at port 2. Since quadratic terms are used in (4), this model can be called a simplified Quadratic PHD (QPHD) model (since cross terms are omitted) [4]. The parameter w_0 can be calculated directly from the Large Signal Operating Point (LSOP) where only the dominant incident tone (A_{11} in this example) is present. The quantity $\mathbf{W}$ represents the weight vector excluding w_0, i.e.

$$\mathbf{W} = [w_{1R} \ w_{1I} \ w_{2R} \ w_{2I}].$$

A prior distribution is assumed for the model parameters $\mathbf{W}$ [6]. In this work, a distribution with zero mean and inverse variance $\alpha\mathbf{I}$ (i.e., isotropic) is taken as this prior distribution

$$p(\mathbf{W}|\alpha) = \mathcal{N}(\mathbf{W}|0, \alpha^{-1}\mathbf{I}), \qquad (5)$$

where $\mathbf{I}$ is the identity matrix, and $\mathcal{N}$ represents a multivariate Gaussian distribution. According to Bayes theorem, the posterior distribution of these weight parameters is proportional to the product of the prior distribution and a likelihood function (which we assume to be Gaussian also),

$$\underbrace{p\big(\mathbf{W}|B_{21,R_t}, A_{21,R_t}, A_{21,I_t}, \alpha, \beta\big)}_{\text{posterior}}$$
$$\propto \underbrace{p\big(B_{21,R_t}|\mathbf{W}, A_{21,R_t}, A_{21,I_t}, \beta\big)}_{\text{likelihood}} \underbrace{p(\mathbf{W}|\alpha)}_{\text{prior}}, \quad (6)$$

where B_{21,R_t}, A_{21,R_t} and A_{21,I_t} are the input training data, which are column vectors, and β is the estimated inverse variance of the data noise, respectively. Since the prior and likelihood distributions are Gaussian, the posterior distribution will also be Gaussian [6]. The quantity β is estimated using the maximum likelihood method and a relatively small value is given to α, i.e. we are assuming a broad prior distribution, which will be updated using the training data.

The mean m_N and variance S_N of the Gaussian posterior distribution of these model parameters can be computed from training data as follows

$$m_N = \beta S_N \boldsymbol{\phi}\big(A_{21,R_t}, A_{21,I_t}\big)B_{21,R_t}, \qquad (7a)$$

$$S_N^{-1} = \alpha\mathbf{I} + \beta\boldsymbol{\phi}\big(A_{21,R_t}, A_{21,I_t}\big)\boldsymbol{\phi}\big(A_{21,R_t}, A_{21,I_t}\big)^{\mathrm{T}}, \qquad (7b)$$

where N is the number of training samples and $\boldsymbol{\phi}(x, y)$ is the vector of basis functions, in this case given by

$$\boldsymbol{\phi}(x, y) = [x \ y \ x^2 \ y^2].$$

The predictive (posterior) distribution for $B_{21,R}$ will also be a Gaussian distribution

$$p\big(B_{21,R}|B_{21,R_t}, \alpha, \beta\big) = \mathcal{N}\big(B_{21,R}|m, v\big), \qquad (8)$$

where the mean m and variance v for the scattered waves can be computed from the training data as follows

$$m = \beta \ \boldsymbol{\phi}\big(A_{21,R}, A_{21,I}\big)\mathbf{S}\boldsymbol{\phi}\big(A_{21,R_t}, A_{21,I_t}\big)^T B_{21,R_t} \qquad (9a)$$

$$v = \beta^{-1} + \boldsymbol{\phi}\big(A_{21,R}, A_{21,I}\big)\mathbf{S} \ \boldsymbol{\phi}\big(A_{21,R}, A_{21,I}A_{21}\big)^T \qquad (9b)$$

$$\mathbf{S}^{-1} = \alpha\mathbf{I} + \beta\boldsymbol{\phi}\big(A_{21,R_t}, A_{21,I_t}\big)^T\boldsymbol{\phi}\big(A_{21,R_t}, A_{21,I_t}\big). \qquad (9c)$$

III. Validation of Modelling Approach

For validation of the proposed theory, we considered 500 complex-valued inputs A_{21} and computed the corresponding output values for $B_{21,R}$, the real part of the scattered wave phasor, using known model parameters. Gaussian noise with mean 0 and standard deviation $0.01 \ W^{1/2}$ is added to the scattered wave output.

Next, a prior probability distribution is assumed for the model parameters w_{1R}, w_{1I}, w_{2R} and w_{2I} – in this case a four-dimensional Gaussian distribution, as in (5). The Gaussian distribution is set to have zero mean and a precision parameter α of 0.005. For illustration purposes, the sequential update of the posterior distribution of two of the model parameters (w_{2R} and w_{2I}) is shown in Fig. 2.

Validation proceeds by observing the sequential update in the posterior model parameters distribution, after each 'prior' is updated via the likelihood function. After the first Bayesian iteration, using just five (assumed to be noisy) training data points $(A_{21_{t,n}}, B_{21_{t,n}})|_{n=1,\dots,5}$, the variance of the posterior distribution of model parameters has reduced considerably. The posterior distribution is estimated from the prior distribution

290

and likelihood function using Bayes theorem – see [6] for full details. In the second iteration, the posterior distribution of the first stage is now taken as the 'prior' distribution, and the likelihood function is calculated using another 15 data points $(A_{21_{t,n}}, B_{21_{t,n}})|_{n=6,\ldots,20}$. As can be seen, the variance of the posterior distribution of the second stage has further reduced compared to that from the first stage. Note that this *sequential* updating of the prior distribution is carried out here only for demonstrative purposes. This is not necessary–all available training data can be incorporated simultaneously, according to (7). To show this, the final plot in Fig. 2 shows the effect of a total of 500 training data points applied to the Bayesian model.

It is seen that the posterior distribution of model parameters has very small variance. For the first time, a PHD model contains information regarding its own certainty.

Table 1 shows the mean of the posterior model parameters distribution when considering 5, 20 and 500 data points.

Table 1. Mean of posterior Distributions of model parameters

No. Data Points	w_{1R}	w_{1I}	w_{2R}	w_{2I}
Actual Values	-0.4865	0.226	0.0018	-0.0464
5	-0.4938	0.2092	-0.0031	-0.0410
20	-0.4939	0.2286	-0.0005	-0.0447
500	-0.4872	0.2258	0.0018	-0.0465

IV. EXPERIMENTS & RESULTS

In this work, a 10 W RF GaN transistor (CGH40010F) is used as the Device Under Test (DUT). The scattered waves B_{21} at port-2 and incident waves A_{21} at port-2 are measured at various load impedances, at a fundamental frequency of 2 GHz. A total of 649 A_{21} values and corresponding B_{21} values are measured in this experiment. These measured data are divided into two subsets. The first subset contains 576 data points for training the model and the second subset contains 73 data points, to be used for model validation. A Gaussian distribution with zero mean and precision parameter $\alpha = 0.005$ is considered for the prior model parameters distribution.

Fig. 3 shows the probabilistic model for the B_{21} values corresponding to the A_{21} values in the training data subset. The red filled circle is the mean of the proposed probabilistic model (which depends on A_{21}), and the filled green circles are the experimentally measured data. Note that since we are assuming the data contain noise, we do not expect the model and measured data to be completely coincident in this plot. For a Gaussian model, the region extending to two standard deviations away from the predicted mean B_{21} value corresponds to a 95% credible area.

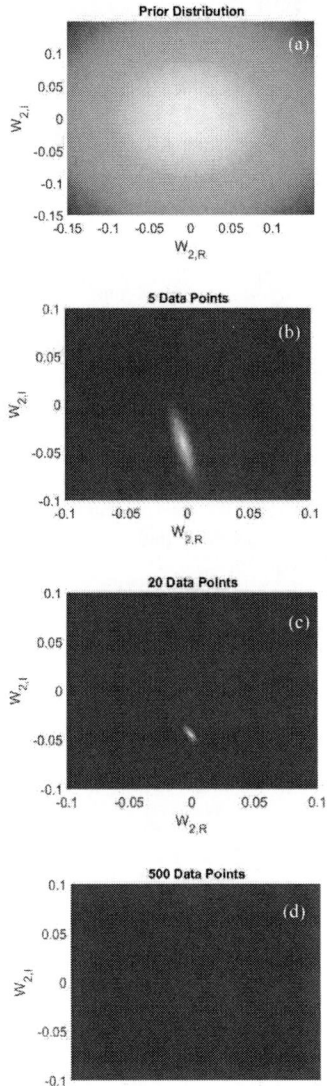

Fig. 2. Illustration of sequential Bayesian learning: Posterior Distribution of W_{2R} and W_{2I} estimated using (a) Prior Distribution (b) 5 Data points (c) 10 Data points (d) 500 Data points.

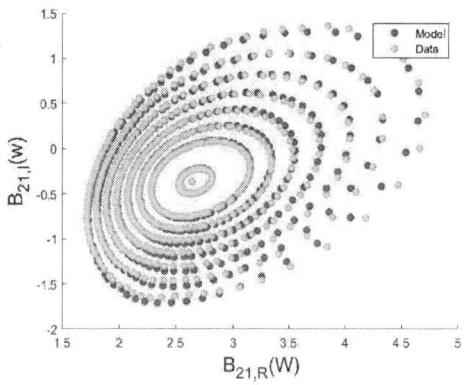

Fig. 3. Probabilistic Model for B_{21}

Fig. 4 shows the probabilistic model obtained for B_{21} values corresponding to the withheld A_{21} values i.e. these measurements are not part of the training data and are therefore unknown to the model. Fig. 5 shows the sequential updates of the posterior distribution of the model parameters w_{2R} and w_{2I}.

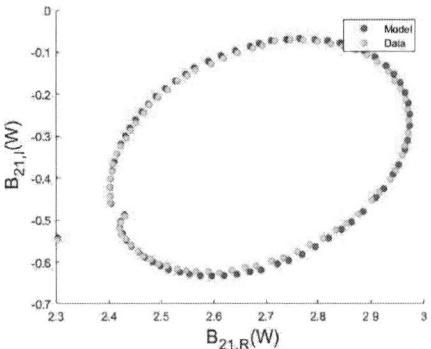

Fig. 4 Probabilistic Model for new test points (new A_{21} values)

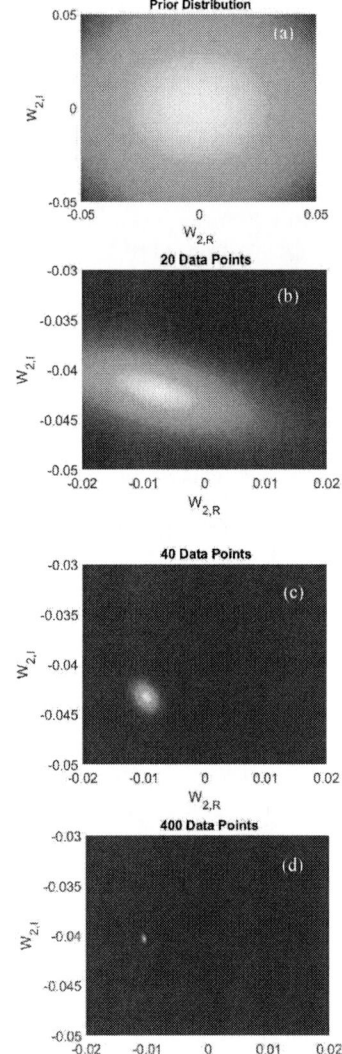

Fig.5 Illustration of sequential Bayesian learning: Posterior Distribution of W_{2R} and W_{2I} estimated using (a) Prior Distribution (b) 20 Data points (c) 40 Data points (d) 200 Data points.

It is observed that the variance of the posterior distribution of the model parameters is reduced by the sequential update of training data.

Fig. 6 shows the change in variance of a randomly selected data points of $B_{21,R}$ while estimating with 20, 40, 60,100,300 and 600 training data points, respectively. It is observed that the variance of each data point of $B_{21,R}$ reduces as the amount of training data increases.

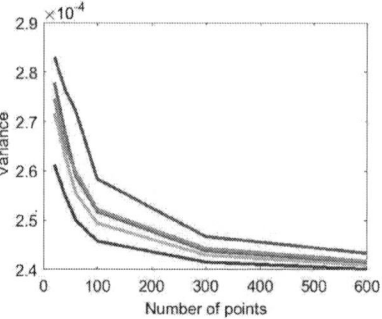

Fig. 6 Variance of single data point Vs Number of Data Points

V. CONCLUSION

This work provides a simple and convenient method to extract a probabilistic behavioural model for the scattered wave at each port of a DUT. This probabilistic model quantifies the uncertainties in the evaluation of the scattered waves without a requirement for knowledge of the causes of the uncertainties. The probabilistic model can be used to estimate the probabilistic values for important system parameters such as gain, efficiency etc. The variance of the probabilistic model can be reduced via further measurement data if higher confidence in the predicted output is required.

REFERENCES

[1] F. A. Mubarak and G. Rietveld, "Uncertainty Evaluation of Calibrated Vector Network Analyzers," in *IEEE Transactions on Microwave Theory and Techniques*, vol. 66, no. 2, pp. 1108-1120, Feb. 2018, doi: 10.1109/TMTT.2017.2756881.

[2] David E.Root, Jan Verspecht,.Jason Horn and ,Mihai Marcu, *X-parameters Characterization, Modeling and Design of Nonlinear RF and Microwave Components,* USA: Cambridge University Press, 2013.

[3] A. D. Manjaly, R. Sharma and J. King, "Probabilistic Behavioural Model Based on X-parameters," *2020 IEEE Asia-Pacific Microwave Conference (APMC),* 2020, pp. 119-121, doi: 10.1109/APMC47863.2020.9331530.

[4] B. Pichler, G. Magerl and H. Arthaber, "A Study on Quadratic PHD Models for Large Signal Applications," in *IEEE Transactions on Microwave Theory and Techniques*, vol. 67, no. 7, pp. 2514-2520, July 2019, doi: 10.1109/TMTT.2019.2915086.

[5] Verspecht and D. E. Root, "Polyharmonic distortion modeling," in *IEEE Microwave Magazine*, vol. 7, no. 3, pp. 44-57, June 2006, doi: 10.1109/MMW.2006.1638289.

[6] Christopher M. Bishop, *Pattern Recognition and Machine Learning,* Springer Science+Business Media, LLC, 233 Spring Street, New York, NY 10013, USA,2006 pp. 21-32,152-156. [Online]. Available: http:// www.springer.com.

Proceedings of the 16th European Microwave Integrated Circuits Conference

An S-band 34dBm Stacked-HBT Phase Driver in 0.25μm BiCMOS Technology for GaN-Based Phased-Array Radar Transmit Chain

J. Essing[1], A. Bossuet, R. Knight[2], A.P. de Hek[3], F.E. van Vliet[4]

TNO, the Hague, the Netherlands

{[1]jaap.essing, [2]rob.knight, [3]peter.dehek, [4]frank.vanvliet}@tno.nl

Abstract — This paper describes an integrated S-band phase-shifter driver-amplifier to drive Gallium-Nitride (GaN) based power amplifiers in a phased-array radar transmit chain. This phase driver is implemented in a 0.25μm SiGe BiCMOS process and packaged into a QFN5x5. By using device stacking, the phase driver achieves an output power of more than 32.5dBm (1.7W) across the band (2.6-3.4GHz), with a maximum of 34.2dBm (2.7W). The RMS phase error is less than 3.75° at a 6-bit phase resolution. This enables a low-cost highly-integrated transmitter front-end for phased-array radar applications.

Keywords — power amplifier, phase shifter, SiGe BiCMOS

I. INTRODUCTION

Front-ends for phased-array radar applications need to offer high-power and low-noise amplification for transmit and receive, respectively, transmit-receive switching and phase and amplitude control functionality for beamforming. A modern phased-array radar RF front-end configuration is shown in Fig. 1 [1]. The silicon (Si) receiver allows for digital beamforming by down-converting and conditioning the received signal such that it can be processed by the digital beamforming electronics. The GaN part provides the high-power and low-noise amplification and the transmit-receive switching functionality. The phase driver in the transmit chain needs to deliver phase-shifting functionality and sufficient output power to drive the GaN HPA. As the power generated by GaN HPAs is steadily increasing year on year, the required drive power needs to follow the same trend. To drive GaN HPAs having more than 100W output power [2], a drive level in the range of 1-3W is needed.

Although such phase driver can be implemented in a III-V compound technology as GaAs [1], a Si technology is becoming more favorable due to its low-cost and high integration capability [3]. However, the low breakdown voltage of silicon transistors complicates the realization of Watt-level output powers. In this work a stacked-HBT topology is used achieve the targeted output power in a 0.25μm SiGe BiCMOS process.

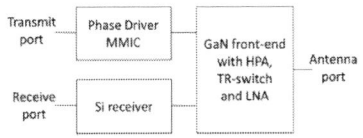

Fig. 1 Modern RF front-end for phased-array radar applications [1].

II. DESIGN

A. Architecture

The phase driver's block diagram is shown in Fig. 2, comprising in the RF-chain a vector modulator (VM) as phase-shifter and a two-stage driver amplifier (DRV). Moreover, a vector modulator offers gain control capabilities usable for future applications.

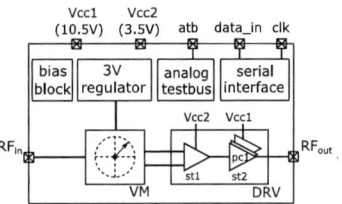

Fig. 2 Block diagram of phase-driver comprising vector modular (VM) and driver amplifier (DRV).

The single-ended RF input signal is converted to a differential signal at the VM's input and converted back to single-ended at the output of the driver's first stage (DRV-st1). The driver's second stage (DRV-st2) consists of sixteen power-combined power cells (pc) to achieve the targeted output power.

Two external supply voltages are applied, a 10.5V (V_{cc1}) and 3.5V (V_{cc2}) supply, the latter regulated on-chip down to 3V. The serial interface controls the RF blocks' quiescent currents (via the bias block), the VM's insertion phase and the analog test bus' (atb) read-out node.

B. Vector Modulator Design

The vector modulator comprises a transformer balun, a two-section RC poly-phase filter (ppf), two VM cores to control the gain in each of the two quadrature paths (I and Q) and two DACs (I and Q) to set the required gain values, as shown in Fig. 3 (left). The 5dB resistive attenuator is inserted to reduce the (loop) gain, making the system more robust against potential electrical instabilities. Moreover, it simplifies the input matching.

The VM cores' gain control is based on current steering the RF-current coming from the degenerated common-emitter pair to the RF output or the power supply via cascoded transistors, as also shown in Fig. 3 (right). Steering more current to the

3–4 April 2022, London, UK

power supply reduces the gain. The VM cores can switch their output polarity to cover all four quadrants, allowing for a phase control range of 360°. Each DAC output sets the base voltages of the VM core's cascoded transistors (Vgain, VP, VN) to obtain a specific gain (phase) setting. The resolution of each DAC is 6 bits, including a sign-bit for the polarity change.

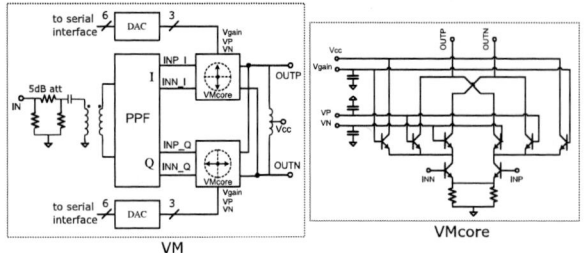

Fig. 3 VM's block diagram (left) and a VM core's circuit diagram (right).

The input impedance of a VM core will remain fairly constant versus the core's gain setting when using this architecture.

C. Driver Amplifier Design

The driver amplifier's second stage uses CPW transmission lines as 1:16 binary-tree splitter and combiner networks. These networks also include matching functionality, hereby using additional lumped reactive components. The combiner matching network is matched from 50Ω to the power cells' optimal load-line impedance. To increase its operating voltage and deliver 25dBm of output power per power cell, the cells are using a stacked-HBT topology. The benefits of a higher operating voltage are reduced currents and a reduced impedance matching ratio [4]. Three devices are stacked to increase the maximum voltage swing at the power cell's output, as shown in Fig. 4. Compensation capacitances at the device's base terminal (C_B) and across the device (C_{CE}) are added for the upper two devices ($Q_{2,3}$) to equalize all three the devices' voltages and (intrinsic) currents. This equalization maximizes the output power delivered by the stack. The design procedure in [4] is followed to achieve this. Although [4] focusses on stacked-FET design, the procedure is also applicable to stacked-HBT design as it allowed to incorporate (intrinsic) collector and emitter resistances.

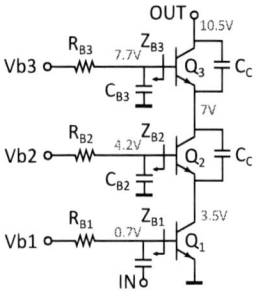

Fig. 4 Power cell of driver amplifier's second stage using a stack of three HBTs with the DC voltages indicated in blue.

The intrinsic base resistance was not included in the procedure, but its impact was considered negligible. The

external base resistors of the upper two devices ($R_{B2,3}$) need to be large-valued ($>10*X_{CB2,3}$) as $Z_{B2,3}$ needs to be capacitive at the fundamental frequency for proper voltage equalization. The most important difference with stacked-FET design is the base DC currents flowing when using HBT devices. The used extended high-voltage SiGe HBT devices have a high β of ~1500, resulting in small forward base currents. However, operating above their BV_{CEO} of 2.8V, to increase a device's output power, generates avalanche current that flows (partly) out of the base. This can cause non-negligible DC voltage-drop across the large-valued $R_{B2,3}$ resistors, which in turn changes the devices' V_{CE} voltages. Therefore, the devices' V_{CE} voltage is chosen to be 3.5V (only 0.7V above BV_{CEO}) to cause limited avalanche current and hence negligible DC voltage drop across $R_{B2,3}$ up to hard gain compression when using $R_{B2,3}=10*X_{CB2,3}$. This choice of V_{CE} is also beneficial for the electro-thermal stability of the CE-device (Q_1) compared to larger V_{CE} values. The resulting power cell's supply voltage V_{cc1} is therefore 10.5V (3*3.5V). The power cell's local bias circuit (not shown) presents low DC output resistances to the stack's V_B nodes ($V_{B1,2,3}$) to minimize further impact of the base currents.

The driver amplifier's first stage (see Fig. 2) is a common-emitter stage, converting at its output the differential signal back to a single-ended signal by using a transformer balun. This stage operates directly at V_{cc2} (3.5V) to give slightly more output power compared to 3V operation.

D. Thermal Analysis

The thermal behavior of the driver amplifier's second stage is investigated using COMSOL as the most power is dissipated in this stage. The simulated configuration includes the sixteen power cells in a QFN5x5 package, mounted on a PCB. This is shown in Fig. 5 (a) with a zoom-in on the power cells. Each power cell comprises the three stacked devices having each four emitter fingers. The steady-state junction temperatures (T_j) at each emitter finger along the indicated white line in Fig. 5 (a) are shown in Fig. 5 (b), hereby using the estimated average dissipated power for a duty-cycle of 15%.

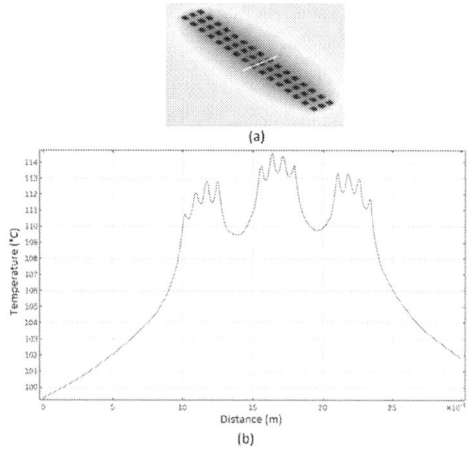

Fig. 5 (a) Configuration of the sixteen power cells used for thermal analysis. (b) Simulated steady-state junction temperatures at each emitter finger within one power cell along the indicated white line at 15% duty-cycle.

The simulated junction temperatures (T_j) remain below 115°C, which is far below the maximum allowable Tj of 160°C. Moreover, the impact of a temperature rise within the pulse is considered using a transient simulation in COMSOL. This resulted in an additional 6°C increase in T_j.

III. MEASUREMENT RESULTS

The chip is fabricated in a 0.25μm SiGe BiCMOS process with a peak f_t/f_{max} of 180/200 GHz and packaged into the QFN5x5, the latter mounted on a PCB for experimentation purposes. A photograph of the chip is shown in Fig. 6.

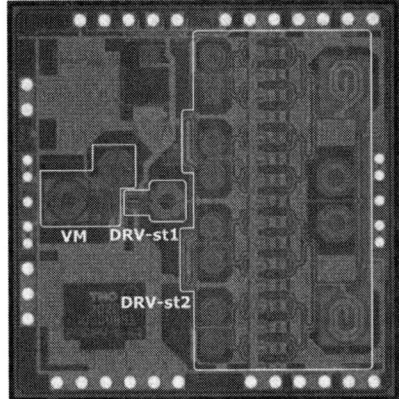

Fig. 6 Chip photograph of phase-driver (size: 2.3x2.3mm²).

The measurements are performed under pulsed-RF conditions using a pulse repetition frequency (PRF) of 100Hz and a pulse width (PW) of 100μs. This results in a duty-cycle of 1%. Measurements with a duty-cycle of 10% are also performed, showing similar results. A frequency step of 100MHz is used.

The measured small-signal gain for all 64 phase states is depicted in Fig. 7, showing an average peak gain of approximately 26.5dB, varying within a 1-1.5dB band across these states.

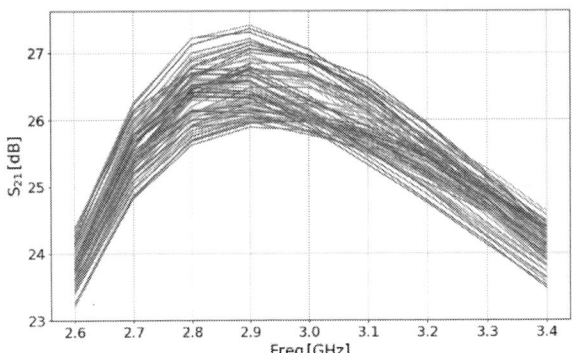

Fig. 7 Measured small-signal gain for all 64 phase states. V_{cc1}=10V, V_{cc2}=3.5V, PRF=100Hz and PW=100us.

The measured small-signal input match for all 64 phase states is depicted in Fig. 8, showing a return loss larger than 15dB.

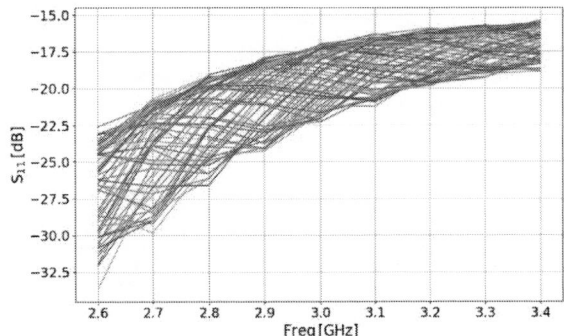

Fig. 8 Measured small-signal input match for all 64 phase states.

The measured small-signal relative insertion phase is depicted in Fig. 9. The RMS phase error is reported in Fig. 13. If strict monotonicity is required, the design should be adapted for this.

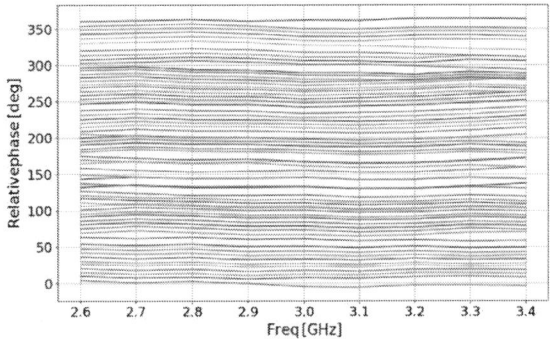

Fig. 9 Measured small-signal relative insertion phase for all 64 phase states.

The simulated and measured output power versus input power at five different frequencies are depicted in Fig. 10, showing good agreement and having a maximum measured saturated output power of 34.2Bm.

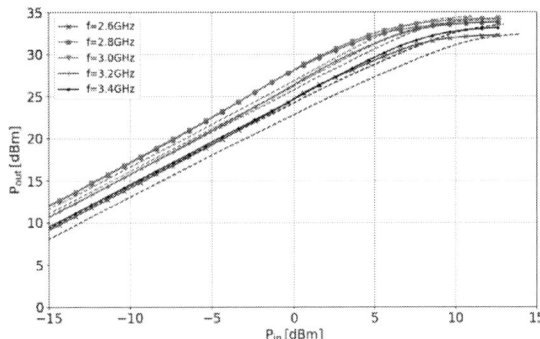

Fig. 10 Simulated (dashed) and measured (solid) output power versus input power for different frequencies at the 0° phase state. V_{cc1}=10V, V_{cc2}=3.5V, I_{ccq1}=340mA, PRF=100Hz and PW=100us.

The simulated and measured output power and PAE versus frequency at a fixed source power (P_S) of 11dBm are depicted in Fig. 11, showing good agreement. The maximum difference in PAE is approximately less than 3%.

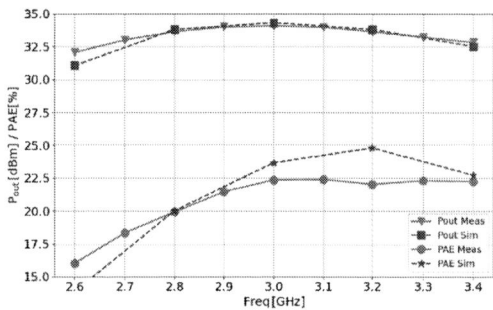

Fig. 11 Simulated (dashed) and measured (solid) output power (red) and PAE (blue) versus frequency at P_S=11dBm at the 0° phase state. V_{cc1}=10V, V_{cc2}=3.5V, PRF=100Hz and PW=100us.

The measured output power versus frequency for all 64 phase states at a source power of 10dBm is shown in Fig. 12. This depicts an output power variation smaller than 0.5dB across these states. The average large-signal gain across the states has here a maximum of approximately 23.5dB.

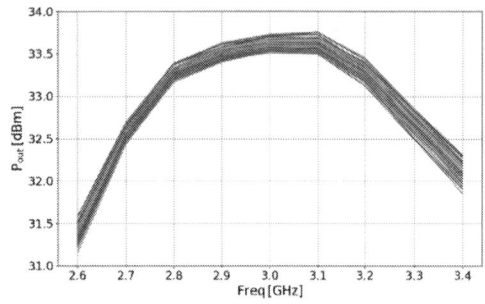

Fig. 12 Measured output power versus frequency for all 64 phase states at P_S=10dBm. V_{cc1}=10V, V_{cc2}=3.5V, PRF=100Hz and PW=100us.

The measured RMS phase and amplitude error versus frequency at a source power of 10dBm are shown in Fig. 13.

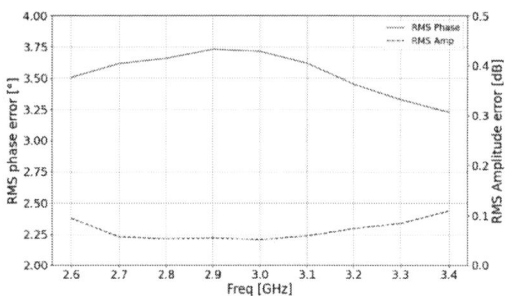

Fig. 13 Measured RMS phase error (solid) and RMS amplitude error (dashed) versus frequency at P_S=10dBm. V_{cc1}=10V, V_{cc2}=3.5V, PRF=100Hz and PW=100us.

These RMS errors are calculated with respect to their averages at this large-signal condition, which is at approximately 3dB gain compression. The RMS phase error and amplitude error are smaller than 3.75° and 0.11dB, respectively.

Table 1 summarizes the phase driver performance and compares it against published silicon-based (SiGe, CMOS)

power amplifiers. To the authors' best knowledge, the measured maximum output power of 34.2dBm is the highest reported output power for a SiGe or CMOS amplifier operating at 3GHz.

Table 1 Summary and comparison of measured performance.

	[5]	[6]	[7]	This work
Technology	SiGe	SiGe	CMOS	SiGe
freq. (GHz)	0.935	5	1.9	3
Psat,max (dBm)	34	33	35.3	34.2
PAE,max (%)	47	35.2	44.9	22.4
Gain (dB)	28	31.8	38	26.5
RMS phase error	NA*	NA*	NA*	<3.75**
RMS ampl error	NA*	NA*	NA*	<0.11**

*no phase-shifter (PA only), **at P_s=10dBm

IV. CONCLUSION

This paper presented an integrated S-band phase-shifter driver-amplifier to drive >100W GaN-based power amplifiers in a phased-array radar transmit chain. The design is implemented in a 0.25µm SiGe BiCMOS process and packaged into a QFN5x5. The phase driver realises a maximum output power of 34.2dBm by using device stacking, which is the highest reported output power for an SiGe/CMOS amplifier operating at 3GHz. The RMS phase error is less than 3.75° at a 6-bit phase resolution. This enables a low-cost highly-integrated transmitter front-end for phased-array radar applications.

ACKNOWLEDGMENT

This work is supported by the TKI HTSM Roadmap Electronics. The authors would like to thank NXP for their technical support and access to their technology.

REFERENCES

[1] M. V. Heijningen, J. Essing, and F. E. V. Vliet, "X-Band GaAs Phase Driver MMIC Optimized for GaN-Based Phased-Array Radar Transmit Chain," in 2018 13th European Microwave Integrated Circuits Conference (EuMIC), Sep. 2018, pp. 118–121.

[2] G. van der Bent, A. P. de Hek, F. E. van Vliet, and Z. Ouarch, "Single-Chip 100-Watt S-band Power Amplifier in 0.25 µm GaN HEMT MMIC Technology," in 2020 15th European Microwave Integrated Circuits Conference (EuMIC), Jan. 2021, pp. 21–24.

[3] C. Liu et al., "A Fully Integrated X-Band Phased-Array Transceiver in 0.13-µm SiGe BiCMOS Technology," IEEE Transactions on Microwave Theory and Techniques, vol. 64, no. 2, pp. 575–584, Feb. 2016.

[4] G. van der Bent, P. de Hek, and F. E. van Vliet, "Design Procedure for Integrated Microwave GaAs Stacked-FET High-Power Amplifiers," IEEE Transactions on Microwave Theory and Techniques, vol. 67, no. 9, pp. 3716–3731, Sep. 2019.

[5] G. Berretta, D. Cristaudo, and S. Scaccianoce, "cdma2000 PCS/cell SiGe HBT load insensitive power amplifiers," in 2005 IEEE Radio Frequency integrated Circuits (RFIC) Symposium - Digest of Papers, Jun. 2005, pp. 601–604.

[6] I. Ju, Y. Gong, and J. D. Cressler, "Highly Linear High-Power 802.11ac/ax WLAN SiGe HBT Power Amplifiers With a Compact 2nd-Harmonic-Shorted Four-Way Transformer and a Thermally Compensating Dynamic Bias Circuit," IEEE Journal of Solid-State Circuits, vol. 55, no. 9, pp. 2356–2370, Sep. 2020.

[7] H. Qian, Q. Liu, J. Silva-Martinez, and S. Hoyos, "A 35 dBm Output Power and 38 dB Linear Gain PA With 44.9% Peak PAE at 1.9 GHz in 40 nm CMOS," IEEE Journal of Solid-State Circuits, vol. 51, no. 3, pp. 587–597, Mar. 2016.

Proceedings of the 16th European Microwave Integrated Circuits Conference

A Phase Coherent DC-25 GHz 6-bit SiGe BiCMOS Step Attenuator with IP1dB >20 dBm

Hamza Kandis, Abdurrahman Burak, Cengizhan Kana, Melik Yazici, and Yasar Gurbuz

Faculty of Engineering and Natural Sciences, Sabanci University, Orhanli, Tuzla, 34956, Istanbul, Turkey

yasar@sabanciuniv.edu

Abstract — This paper presents a 6-bit step attenuator with high linearity and very low RMS amplitude and phase errors from DC to 25 GHz. The presented attenuator was designed with hybrid Π/T-type attenuation blocks. Isolated NMOS (iNMOS) devices were utilized instead of conventional NMOS transistors since iNMOS devices have fewer parasitics than its counterpart. This feature provides to keep the slope of the insertion loss (IL) constant and improves the power handling capability of the designed attenuator. This design was fabricated with IHP's 130 nm SiGe BiCMOS technology. The measured attenuator has a dynamic range of 23.96 dB with 0.38 dB fine steps. The RMS amplitude error of the designed attenuator is <0.38 dB from DC to 25 GHz. The maximum measured RMS phase error is 2.4° at 25 GHz. The measured $IP1_{dB}$ is > 20 dBm at 7.5 GHz and 12.5 GHz.

Keywords — Hybrid Pi/T-type, isolated NMOS (iNMOS), amplitude/phase control, high linearity, wideband.

I. INTRODUCTION

Many various applications such as high data rate wireless communication, wideband phased arrays with amplitude control and beamforming systems require wideband amplitude controllers [1]. Further, amplitude controllers can be used to limit the incoming power to the receiver chain and to adjust power levels precisely in a transmitter chain [2]. Although the main objective of an amplitude controller is controlling the signal level, this amplitude control must be provided without changing the phase of the signal.

Amplitude control in such systems can be achieved with attenuators or variable gain amplifiers (VGAs). Attenuators can provide the same functionality with better RF performances in terms of power handling capability and accurate amplitude/phase control with zero DC power consumption.

Most of the designed attenuators have continuous or step control. Among the step attenuators, switched path attenuators, distributed attenuators, and hybrid Π/T-type attenuators have been reported so far in the literature [3]–[5]. Hybrid Π/T-type attenuators provide to obtain more precise amplitude control, better phase compensation, and wider impedance matching than their counterparts. Therefore, hybrid Π/T-type attenuators were chosen as the topology.

In this paper, a phase coherent 6-bit SiGe BiCMOS hybrid Π/T-type step attenuator using iNMOS switches is reported. New design techniques are employed to flatten the gain response, widen the bandwidth, improve the linearity. The fabricated die provides linear-in-dB attenuation states with a low phase variation between DC to 25 GHz.

II. DC-25 GHz 6-BIT STEP ATTENUATOR DESIGN

A. Circuit Design

The block diagram of the designed attenuator is given in Fig. 1. The proposed attenuator was designed with six cascaded attenuation bits, which can cover the 23.96 dB attenuation range with 0.38 dB incremental states. One attenuation block was designed to compensate for the mismatches and variations that arise from the process. The attenuation capability of this block was selected as half of the LSB (in dB). The attenuation blocks were designed with hybrid Π/T-type topology due to exploiting the advantages in terms of RF performance, which were mentioned in the previous section. Besides, there is also a trade-off between using Π-type and T-type designs. Π-type design is a better choice for obtaining higher attenuation, since it has an extra path ground than T-type design. On the other hand, T-type design exhibits lower phase variation than Π-type design. Therefore, the smallest-in-dB four attenuation blocks and the compensation block were designed with T-type design whereas the remaining two attenuation blocks were designed with Π-type design. Also, the 12.16 dB attenuation block was realized by two cascaded 6.08 dB attenuation blocks due to the matching and linearity concerns. Furthermore, low-pass filter with R-L-R configuration was added to 3.04 dB, 6.08 dB and 12.16 dB attenuation blocks to further compensate the phase variation between the reference and attenuation states.

B. Comparison of NMOS and iNMOS as Series Switches

Conventional NMOS transistors include many parasitic components due to their physical structure. The IL increases as the frequency increases in attenuators that utilize conventional NMOS transistors. Thus, iNMOS transistors were used in this work. Fig. 2 shows almost two identical Π-type attenuation blocks when they are at the reference state. Fig. 2a has utilized iNMOS transistors as switches whereas Fig. 2b has utilized NMOS transistors. In order to compare the insertion loss of these networks, Y-parameters can be calculated. Then, these parameters can be converted into S-parameters to find the S_{21}, the insertion loss of each network.

Y-parameters of the Π-type attenuation blocks are given in (1). The admittances for Π-type attenuation block with iNMOS transistors can be calculated by using (2)-(5), whereas (6)-(8) can be used for the same block with NMOS transistors. S-parameter conversion can be realized by using Y to S conversion. Due to $sC_{sb}C_{db}$ term, (7) is larger than (4).

3–4 April 2022, London, UK

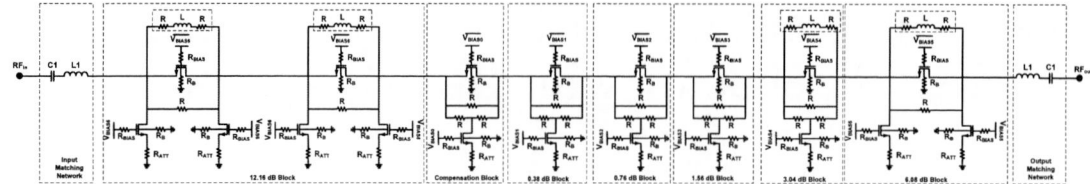

Fig. 1. The block diagram of the designed attenuator.

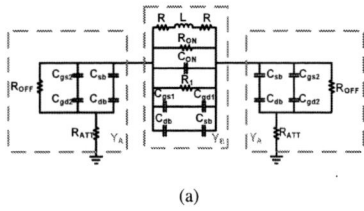

(a)

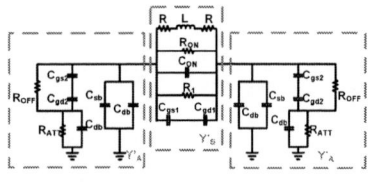

(b)

Fig. 2. Π-type attenuation blocks at the reference state with using (a) iNMOS, and (b) NMOS transistors.

Additionally, (5) is larger than (8) because of the contribution of the C_{db} and C_{sb} capacitors. Utilizing iNMOS transistors instead of NMOS transistors decreases IL since S_{21} is proportional to Y_B (Y_B') and inversely proportional to Y_A (Y_A').

$$Y = Y' = \begin{bmatrix} Y_{11} & Y_{12} \\ Y_{21} & Y_{22} \end{bmatrix} = \begin{bmatrix} Y_A + Y_B & -Y_B \\ -Y_B & Y_A + Y_B \end{bmatrix} \quad (1)$$

$$A = (C_{gs2} + C_{gd2})(C_{db} + C_{sb}) \quad (2)$$

$$B = s\left(\frac{C_{gs2} + C_{gd2}}{C_{gs2}C_{gd2}} + \frac{C_{db} + C_{sb}}{C_{db}C_{sb}}\right) \quad (3)$$

$$Y_A = \frac{A + B * R_{off}}{A(R_{off} + R_{ATT}) + B(R_{off}R_{ATT})} \quad (4)$$

$$Y_B = \frac{(2R + sL) + (R_{on}//R_1)}{(2R + sL)(R_{on}//R_1)} + \frac{sC_{gs1}C_{gd1}}{C_{gs1} + C_{gd1}} + \frac{sC_{db}C_{sb}}{C_{db} + C_{sb}} + sC_{on} \quad (5)$$

$$C = \frac{C_{gs2} + C_{gd2} + sC_{gs2}C_{gd2}R_{off}}{R_{off}(C_{gs2} + C_{gd2})} \quad (6)$$

$$Y_A' = \frac{(1 + sC_{db}R_{ATT})}{R_{ATT} + C + sC_{db}R_{ATT}C} + s(C_{sb} + C_{db}) \quad (7)$$

$$Y_B' = \frac{(2R + sL) + (R_{on}//R_1)}{(2R + sL)(R_{on}//R_1)} + \frac{sC_{gs1}C_{gd1}}{C_{gs1} + C_{gd1}} + sC_{on} \quad (8)$$

C. Design Steps for High Linearity Performance

The power handling capability of the conventional NMOS transistors decreases as the operating frequency increases. The physical structure of a conventional NMOS transistor includes parasitic components, and their effects become more dominant at higher frequencies. In order to improve the power handling capability of the designed attenuator, iNMOS transistors is a good choice, as for obtaining lower insertion loss. Isolation in the body terminal is achieved by including a deep n-well layer, which reduces the losses that arise from the substrate conductivity [6]. As analyzed in detail [7], the isolated body eliminates the parasitic capacitive effects, thus the slope of the IL is less dependent on the frequency. Fig. 3 shows the comparison between using NMOS and iNMOS transistors for reference and attenuation states. iNMOS transistor provides flatness in the attenuation states, which leads to a lower RMS amplitude error. Body floating technique was applied together with using iNMOS transistors to further increase the power handling capability of the attenuator [8]. The schematic models of a conventional NMOS and iNMOS for linearity analysis is given in Fig. 4. When the negative cycle of a high input power is applied, source-to-body and drain-to-body (D_{sb} and D_{db}) diodes will be turned on. Therefore, the current flow on these diodes will increase immediately, which will degrade the output power sooner. But, when an iNMOS transistor with body floating technique was utilized, the current flow on D_{sb} and D_{db} diodes will increase slower because of the high-valued resistor (R_b) placed between body and ground. Hence, utilizing iNMOS transistors with body floating technique improves the power handling capability of the design.

The linearity of the attenuator is affected the most in terms of RF performance due to bit sorting. Degraded linearity can cause suppression of the next stages in a chain or even breakdown of a whole system. To sustain the linearity, the largest attenuation bit was placed as the first block since this block has the highest IL than the others.

D. Design Steps for Wideband Performance

The transistor sizes were chosen at an optimum point for the IL, bandwidth, and linearity performance. As shown in Fig. 5, to design an attenuator with wider bandwidth, higher linearity, and lower IL, the width of the shunt transistor must

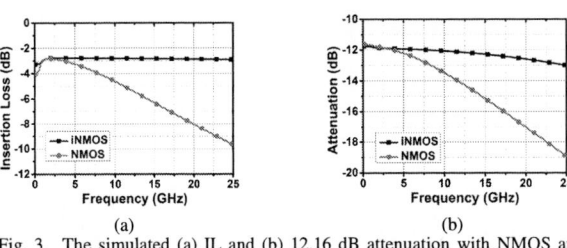

(a)   (b)

Fig. 3. The simulated (a) IL and (b) 12.16 dB attenuation with NMOS and iNMOS transistors.

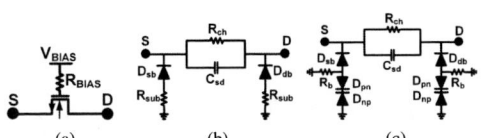

(a)   (b)   (c)

Fig. 4. (a) The bias connection of a MOS device; the schematic models of (b) NMOS and (c) iNMOS with body floating technique for linearity analysis.

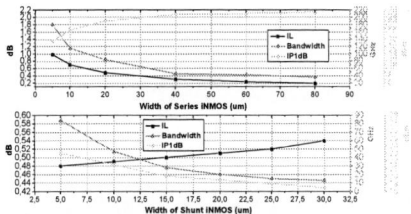

Fig. 5. The simulated linearity, IL and bandwidth performances of 6.08 dB block with different width of series and shunt iNMOS.

Fig. 6. The chip micrograph of the presented attenuator. The total chip size is 1.1 mm^2 (1.36 mm x 0.81 mm).

be the smallest. On the other hand, a shunt transistor with the smallest width leads to a higher Ron resistance. Thus, it becomes harder to obtain higher attenuation levels. Also, the optimum width of the series transistor was selected according to the simulations. The relation between the IL, bandwidth, and linearity with respect to the width of the series transistor was shown in Fig. 5. Since the bandwidth and linearity performance were prioritized, 20 μm width was chosen.

III. MEASUREMENT RESULTS AND DISCUSSION

The proposed attenuator was fabricated in IHP Microelectronics 130 nm SiGe BiCMOS technology. Fig. 6 depicts the die micrograph of the measured attenuator. Bias voltages were applied with Agilent E3631A DC power supply. On-chip inverters are utilized to reduce the number of DC pads. The S-parameters of the fabricated attenuator were measured by using Rohde & Schwarz (RS) ZVL67 network analyzer. The measured input return loss is better than 8.5 dB and the measured output return loss is better than 8.7 dB between DC-25 GHz, as shown in Fig. 7. The measured attenuator demonstrates state-of-the-art performance, where RMS amplitude error is <0.38 dB, and the measured maximum RMS phase error is 2.4° at 25 GHz as depicted in Fig. 8.

Linearity measurements were first performed with RS-ZVL67 network analyzer. Since, $IP1_{dB}$ point could not be measured with RS-ZVL67, input signals at 7.5 and 12.5 GHz were applied from Agilent E3267D PSG vector signal generator and then the corresponding output powers were measured with Agilent E4417A power meter when the attenuator was at the reference state. The measured results were given in Fig. 9. Although the $IP1_{dB}$ compression point could not be measured exactly, it is clear that the designed attenuator shows a state-of-the-art performance with a >20 dBm $IP1_{dB}$.

Fig. 10 (a) shows the real attenuation states and the IL of the designed attenuator. The measured IL <17.65 dB

between DC-25 GHz. Fluctuations were occurred during the measurement of the attenuation states due to the low power levels, which were also included in the RMS amplitude error. Fig. 10 (b) shows the relative attenuation states. It can be seen clearly that using iNMOS instead of conventional NMOS switches in the design of the attenuation blocks enables to achieve very flat states over a wide bandwidth.

The results of the presented wideband attenuator is given in Table 1 and compared with the similar works in the literature. The presented step attenuator exhibits the highest $IP1_{dB}$ with >20 dBm, and the widest bandwidth from DC to 25 GHz, and the lowest RMS phase error with < 2.4° with a compact chip size among the state-of-the-art attenuators. The IL is slightly higher than the other works, since their technologies offer transistors with lower R_{on} and C_{off} values. Moreover, the IL is also high due to the transistor size selection. As explained earlier, while we sacrifice the IL performance, we gain from the linearity and widen bandwidth. The beneficial effects of utilizing iNMOS transistors were proven in terms of high linearity and very low RMS amplitude and phase errors over a wide bandwidth.

IV. CONCLUSION

This paper presents a 6-bit step attenuator with a state-of-the-art performance, designed in 130 nm SiGe BiCMOS technology. The measured RMS amplitude error <0.38 dB were achieved with a measured RMS phase error of <2.4°. This work achieves the best $IP1_{dB}$, the widest bandwidth, and the lowest RMS phase error values in SiGe BiCMOS technology, to the best of the authors' knowledge.

ACKNOWLEGMENT

The authors would like to thank IHP Microelectronics for providing IC fabrication and technical support. This work is supported by The Scientific and Technological Research Council of Turkey under grants 120E345.

299

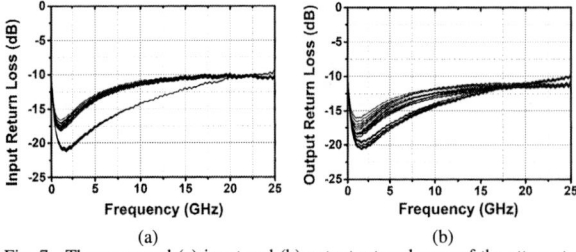

(a)   (b)

Fig. 7. The measured (a) input and (b) output return losses of the attenuator.

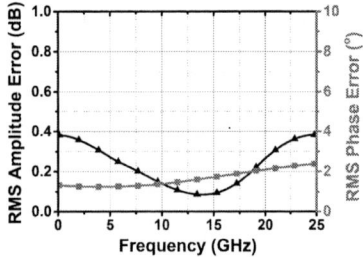

Fig. 8. The measured RMS amplitude and RMS phase errors of the attenuator.

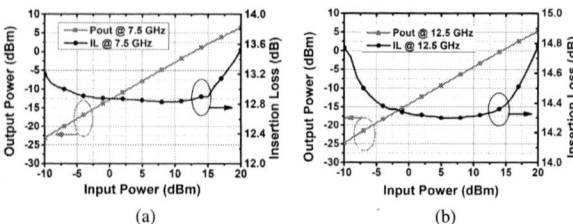

(a)   (b)

Fig. 9. The measured $IP1_{dB}$ compression point and corresponding IL when the input signal is (a) at 7.5 GHz and (b) at 12.5 GHz.

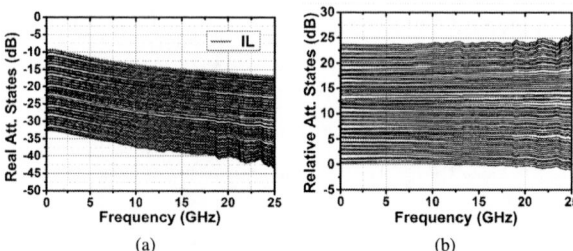

(a)   (b)

Fig. 10. The measured (a) attenuation states of the attenuator, and (b) relative attenuation states of the attenuator.

Table 1. Summary of the state-of-the-art step attenuators in the literature

References	[3]	[4]	[5]	This Work
Technology	0.18µm CMOS	0.18µm SOI CMOS	0.13µm SiGe BiCMOS	**0.13µm SiGe BiCMOS**
Topology	Hybrid Π/T-type	Switched path	Switched T-type	**Hybrid Π/T-type**
Frequency (GHz)	DC-14	DC-20	DC-20	**DC-25**
Number of Bits	6 LSB=0.5	6 LSB=1	6 LSB=0.5	**6 LSB=0.38**
Attenuation Range (dB)	31.5	31	31.5	**23.96**
Insertion Loss (dB)	3.7-10	3.1-7.6	1.7-7.2	**9.2-17.65**
Return Loss (dB)	>9	>12	>12	**>8.5**
RMS Amp. Error (dB)	0.5	0.5	0.37	**0.38**
RMS Phase Error (°)	4.2	2.5	4	**2.4**
$IP1_{dB}$ (dBm)	15	10	10	**>20**
Area (mm^2)	0.5*	0.63*	0.98	**1.1**

[1]Area without pads

REFERENCES

[1] Z. Zhang, N. Li, H. Gao, M. Li, S. Wang, Y.-C. Kuan, X. Yu, and Z. Xu, "A dc-32ghz 7-bit passive attenuator with capacitive compensation bandwidth extension technique in 55 nm cmos," pp. 1303–1306, 2020.

[2] H. Dogan, R. G. Meyer, and A. M. Niknejad, "Analysis and design of rf cmos attenuators," *IEEE Journal of Solid-State Circuits*, vol. 43, no. 10, pp. 2269–2283, Oct 2008.

[3] B. Ku and S. Hong, "6-bit cmos digital attenuators with low phase variations for x-band phased-array systems," *IEEE Transactions on Microwave Theory and Techniques*, vol. 58, no. 7, pp. 1651–1663, July 2010.

[4] M.-K. Cho, J.-G. Kim, and D. Baek, "A broadband digital step attenuator with low phase error and low insertion loss in 0.18-µm soi cmos technology," *ETRI Journal*, vol. 35, no. 4, pp. 638–643, 2013. [Online]. Available: https://onlinelibrary.wiley.com/doi/abs/10.4218/etrij.13.0112.0534

[5] I. Song, M. Cho, and J. D. Cressler, "Design and analysis of a low loss, wideband digital step attenuator with minimized amplitude and phase variations," *IEEE Journal of Solid-State Circuits*, vol. 53, no. 8, pp. 2202–2213, Aug 2018.

[6] X. J. Li and Y. P. Zhang, "Flipping the cmos switch," *IEEE Microwave Magazine*, vol. 11, no. 1, pp. 86–96, Feb 2010.

[7] M. Davulcu, C. Caliskan, I. Kalyoncu, M. Kaynak, and Y. Gurbuz, "7-bit sige-bicmos step attenuator for x-band phased-array radar applications," *IEEE Microwave and Wireless Components Letters*, vol. 26, no. 8, pp. 598–600, Aug 2016.

[8] Mei-Chao Yeh, Zuo-Min Tsai, Ren-Chieh Liu, K. . Lin, Ying-Tang Chang, and Huei Wang, "Design and analysis for a miniature cmos spdt switch using body-floating technique to improve power performance," *IEEE Transactions on Microwave Theory and Techniques*, vol. 54, no. 1, pp. 31–39, Jan 2006.

Proceedings of the 16th European Microwave Integrated Circuits Conference

A 26 GHz to 34 GHz Active Phase Shifter with Tunable Polyphase Filter for 5G Wireless Systems

Alok Sethi, Rehman Akbar, Mikko Hietanen, Timo Rahkonen, Aarno Pärssinen

University of Oulu, Finland
{firstname.lastname}@oulu.fi

Abstract — **This paper presents a vector summing Active Phase Shifter (APS) for 5G and 6G wireless systems. The circuit is fabricated using 45 nm CMOS SOI technology and exploits 10 bit control to set the desired amplitude and phase. The design uses a high-precision tunable polyphase filter (PPF) whose center frequency can be tuned from 23 GHz to 40 GHz. By utilizing a tunable PPF and independent gain tuning of the quadrature signals along with a simple calibration, a square phase constellation is achieved over the full operating band, resulting in an easy phase control with a mean resolution of 0.5 degree and 7 dB of amplitude control to facilitate amplitude tapering. The measured RMS phase and amplitude errors are as small as 0.6 degree and 0.08 dB, respectively. Active area occupied is 0.19 sq. mm, allowing a high degree of integration in a phased array system.**

Keywords — **phase shifter, vector summing, vector modulator, millimeter wave integrated circuits, 5G, 6G, CMOS SOI, polyphase filter.**

I. INTRODUCTION

The next generation of wireless systems, namely 5G and 6G, are transitioning to mmWave frequencies and require large analog beamforming systems to support the required link budget and system capacity. In order to achieve high spatial beam density, amplitude tapering schemes such as Dolph-Chebyshev, Taylor, and Bayliss are needed in a beamforming system [1]. Thus, a phase shifter with an inbuilt amplitude control is a good candidate for beamforming systems as it enables a large degree of integration and can implement a significant portion of the large dynamic range required for amplitude tapering.

There are two main ways to make phase shifters. First, by employing variable physical delay either using switchable filters [2] or changing transmission line characteristics [3]. Second, by variable summing of orthogonal vectors, namely vector-sum phase shifters (VSPS) [4]–[8]. Here, VSPS is used as it allows for higher phase resolution, inbuilt amplitude control and smaller die area.

In [6], a passive transformer based design is used for generating and summing the orthogonal vectors. Different weights for the vectors are implemented using two binary weighted switch arrays, with each having a seven bit binary control, resulting in a large but skewed constellation of phase points. This approach, even though provides good phase accuracy at one particular frequency, has a big calibration overhead and needs a complex search algorithm for operating over a large frequency band. Reference [7] uses a two stage

polyphase filter (PPF), which has a large insertion loss. A tunable RC-CR network is used in [8] to generate orthogonal vectors, however external analog voltage is used for changing the vectors weights.

In this work, an Active Phase Shifter (APS) is designed employing a single stage tunable PPF for generation of orthogonal vectors, with individual gain tuning of the vectors before summing them in a binary fashion. Through this approach, a near perfect square phase constellation is achieved, eliminating the need for any complex calibration or look-up table based phase mapping as required in [6]. As the manufactured IC targets a transmitter, the design decisions of utilizing a relatively large output buffer and not including a low noise amplifier are justified.

II. ACTIVE PHASE SHIFTER DESIGN

Block diagram of the APS is shown in Fig. 1. Differential RF signal is fed from a GSGSG probe and drives the tunable PPF that splits the signal into an I and a Q branch. From each branch, the signal then goes to a differential amplifier (DAMP) with tunable gain, followed by a quadrant switch and a variable gain amplifier (VGA).

The quadrant switch consists of cross-connected PMOS devices and is used to select the polarity of the signal that is going to the VGAs, thus selecting the quadrant of the phase constellation. The output current from both VGAs gets combined at the primary side of the current combiner, which is a transformer with a one turn primary coil and a two turn secondary coil. In the transformer, of the available eight metal layers, the primary coil uses the thickest two copper metals layers, named OB and OA, in the sixth and seventh layer of the metal stack. The secondary coil uses the fifth layer, named UA, of the metal stack. The coupling factor between the coils is around 0.3. The secondary coil of the transformer drives the output buffer which is connected to the output GSGSG pads.

Each VGA has four bits of control arranged in a binary weighted fashion, and each quadrant switch has one bit of control, thus in total there are ten control bits ($4 \times 2 + 1 \times 2$) for digitally regulating the amplitude and phase shift.

A. Tunable Polyphase filter

Accuracy of a VSPS depends on the accuracy of the quadrature generation. In this design, a PPF in a constant amplitude topology is used. A PPF is inherently narrow-band, providing perfect quadrature in a narrowband around its center

3–4 April 2022, London, UK

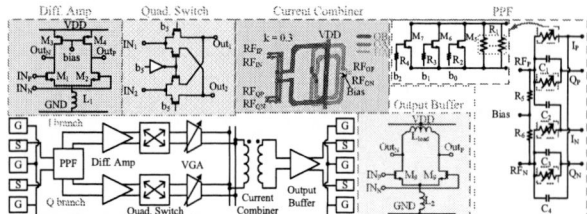

Fig. 1. Block diagram of the APS and schematics of various sub-blocks.

frequency. To make it operate over a wideband, tuning is required, which is achieved here by changing the overall resistance of the filter. In [9], a tunable PPF with continuous resistance tuning is used for generating narrowband quadrature LO signals. However, here the focus is on wideband modulated RF signals due to which attention is paid to the linearity of the filter. Further, due to the compressing nature of the LO buffers, higher amplitude imbalance can be tolerated, whereas for wideband modulated signals, both amplitude and phase mismatch play a more significant role.

In the tunable PPF, a fixed large resistor along with three binary weighted transistors that are operating in triode region, are used to provide variable resistance. The tuning bits can move the center frequency of the tunable PPF from 23 GHz to 40 GHz. The instantaneous bandwidth i.e., the bandwidth at a fixed digital configuration for which the amplitude and phase mismatch between the orthogonal basis vectors is less than 0.5 dB and 5° respectively, for the designed APS is in excess of 1.5 GHz and can be deduced from Fig. 5 and Fig. 6. Besides the used method, frequency tunability in the PPF could have been also achieved by using either switchable capacitor bank, varactors, or switchable resistor bank. However, dimensioning for those options given the required frequency range, would have resulted in a significant larger parasitic load compared to the selected solution of fixed large resistor with multiple MOS operating in triode region.

The schematic of the tunable PPF can be seen from Fig. 1. To reduce the mismatch related variation, R_1 is implemented by arranging four parallel resistors in a symmetric layout. Furthermore, M_{5-7} are made up of multiple unit sized transistors, with M_7 having the largest effective width and M_5 the smallest. Capacitors C_{1-4} are implemented using vertical natural capacitors (VNCAPs). Simulations showed that intra-chip mismatch effects do not cause noticeable changes in the PPF center frequency. Instead, the process variations caused 20–30 % variation in the center frequency compared to nominal, which can be partially compensated by the available tuning bits.

The NMOS transistors M_{5-7}, used for the resistance tuning have considerable amount of second order non-linearity when driven with large signals, which mixes further to IM3. This is fundamentally rising from the fact that the input signal directly modulates the V_{GS} and hence R_{on}. One way to minimize this effect is by allowing the gate voltage to float with the source voltage. This is achieved here by adding gate resistors R_{2-4}, which showed a significant improvement in the output IM3

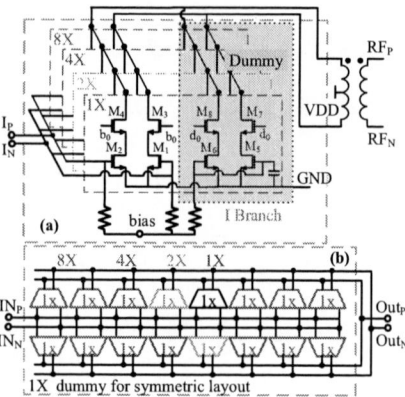

Fig. 2. Schematics and sketch of the layout arrangement of the VGA.

level in two-tone simulations.

B. Differential Amplifier

The differential signal after the PPF contains some common-mode because of inherent asymmetry in the PPF layout. Since the VGAs are a pseudo-differential structure, a strong common-mode signal will cause a reduction in the dynamic range and phase and amplitude errors in the output. Further, a tunability in gain is required to balance any mismatch between the I and Q branches. Perfectly matched I and Q branches assist in making a squared phase constellation, which significantly simplifies the process of controlling the phase shifter.

The schematic of the DAMP can be seen from Fig. 1. To provide tunability in the gain of the DAMP, M_{3-4} along with a variable bias source are used as a load. A tail inductor L_1, instead of a long channel NMOS, is used to improve the common-mode rejection ratio of the DAMP.

C. Variable Gain Amplifier

Schematic of the VGA can be seen from Fig. 2a. VGA consists of 16×2 pseudo-differential pairs with binary weighted on/off controls. Transistors M_{1-4} are part of the main tree and transistors M_{5-8} are part of the dummy tree. Transistors M_{3-4} and M_{7-8} are the controlling transistors for the main and the dummy trees, respectively. A complementary digital signal drives them i.e., when M_{3-4} are turned on M_{7-8} are turned off and vice-versa. This keeps the output impedance of the VGA constant, irrespective of the control word. Input transistors in both dummy and main trees are biased at the same potential, however RF signal is applied to input transistors of only the main tree.

In Fig. 2b, the layout of the VGA is sketched. Here, 1X represents the combination of a main and a dummy pseudo-differential pair, which acts as a unit cell in the layout of the VGA. Fifteen unit cells are required for implementing a four bit binary weighted control. The sixteenth unit cell is used for making a symmetrical layout and is electrically off.

302

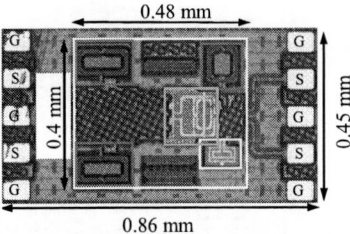

Fig. 3. Micrograph of the fabricated structure.

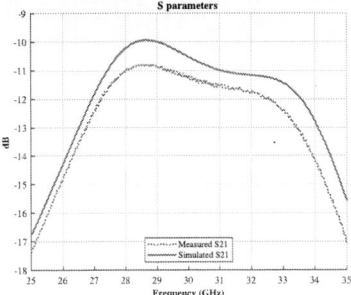

Fig. 4. Measured and simulated S21 parameter for the largest amplitude points.

D. Output Buffer

Schematic of the Output Buffer (OBUFF) can be seen from Fig. 1. OBUFF is a common source differential amplifier with an inductive load, L_{load} and a tail coil, L_2. L_{load} is a two turn coil and utilizes OB, OA, and UA. L_2 is also a two turn coil and is synthesized using the topmost Aluminum layer.

III. MEASUREMENT RESULTS

The APS integrated circuit is fabricated using 45 nm CMOS SOI technology and its micrograph is shown in Fig. 3. The dimensions including the input and output pads are 0.86 mm × 0.45 mm. Blocks in red, green, yellow and blue rectangles in the figure correspond to the tail inductor of the DAMP, current combining transformer, load inductor of the OBUFF, and tail inductor of the OBUFF, respectively. Core of the APS consumes a maximum DC power of 30 mW from a 1 V supply.

RF measurements are performed using Keysight PNA-X N5247A vector network analyzer along with Cascade Microtech Infinity series 67 GHz GSGSG probes. Calibration plane is set to the probe tips using an impedance standard substrate. Digital control is performed using on-chip registers. ESD protection is used for all the pads. A trivial calibration procedure is followed in which, first, the center frequency of the PPF is tuned to the desired point. Then the amplitude of the I and Q branches is matched using their respective DAMP gain control. The measured and simulated differential S21 curves are shown in Fig. 4 and they show a good match.

Measured normalized phase constellation showing all 1024 points at three different frequencies is shown in Fig. 5. The plots indicate that the designed electrical tuning in PPF and DAMPs is sufficient to generate a close to ideal square phase cloud. The red area represents a magnitude which is 7 dB

Table 1. Summary of the important performance parameters and comparison with related works

	This work	[2]	[6]	[7]	[8]
Technology	45 nm CMOS	28 nm CMOS	65 nm CMOS	130 nm SiGe	90 nm CMOS
Architecture	VSPS	STPS	VSPS	VSPS	VSPS
Quadrature Generator	Tunable PPF	-	Passive	Two stage PPF	Tunable RC-CR nw.
Frequency (GHz)	26–34	29–37	21–30	22–30	24–30 / 35–41
Resolution (°)	0.5	22.5	0.8	0.65	cont.[a]
RMS Phase error (°)	0.6–1.1	1.2–8.8	0.28–0.88	0.2	2–4 / 3.7–5.1
RMS Amp. error (dB)	0.08–0.24	1.1	0.4–0.5	0.2	0.1–0.6 / 0.1–0.35
IP_{1db} (dBm)	> 5.5	≈ 5	-	2	> −2.8 / > −3.7
P_{diss} (mW)[b]	30	-	-[c]	23	17.6
Area (mm^2)[d]	0.19	0.08	0.27[e]	0.48	0.34

[a] Measurements done using an external DAC.   [b] Core power
[c] Core consists of only switches   [d] Excluding the pads.
[e] Estimated.

lower than in the maximum gain circle. It has to be noted that the locus of constant amplitude will be a circle in such plots. Further, even though the phase resolution of the maximum amplitude points, i.e., the corner points of the phase constellation, is limited. It is still possible to use the corners to tune the strongest center tap of an amplitude tapered beamformer and rotate the rest of the taps accordingly. All the performance parameters are calculated on the black points in Fig. 5 i.e., out of 1024 points, around 700 points are used. The average phase resolution over the black points is 0.5°.

Below the phase constellations plots in Fig. 5, amplitude mismatch between I and Q vectors along with the phase difference from the ideal 90° between them i.e., $90 - \arctan(|A_Q|/|A_I|)$ is plotted. Here, A_Q and A_I are the measured complex vectors created by turning off all controlling transistors in the VGAs of I and Q branch, respectively.

Measured RMS phase and amplitude errors as a function of frequency are shown in Fig. 6. As can be seen, the maximum RMS phase error over the operating band is under 1.1°. Further, the RMS amplitude error is as small as 0.08 dB with the maximum being under 0.24 dB. Each curve in Fig. 6 corresponds to a calibrated configuration of the APS for different operating frequencies. Fig. 5 along with Fig. 6, also describe the achievable instantaneous bandwidth of the APS. Due to the limited maximum available differential power in the setup, it was not possible to push the APS to full compression, thus leading to some uncertainty in the reported IP_{1db} in the Table 1. Summary of the performance parameters and a comparison with other recent works is shown in Table 1. It can be seen from Table 1 that this work achieves highest resolution (an external DAC is required in [8]) with the lowest RMS amplitude error.

IV. CONCLUSION

This paper presents a vector summing based active phase shifter that can be accurately tuned to different center

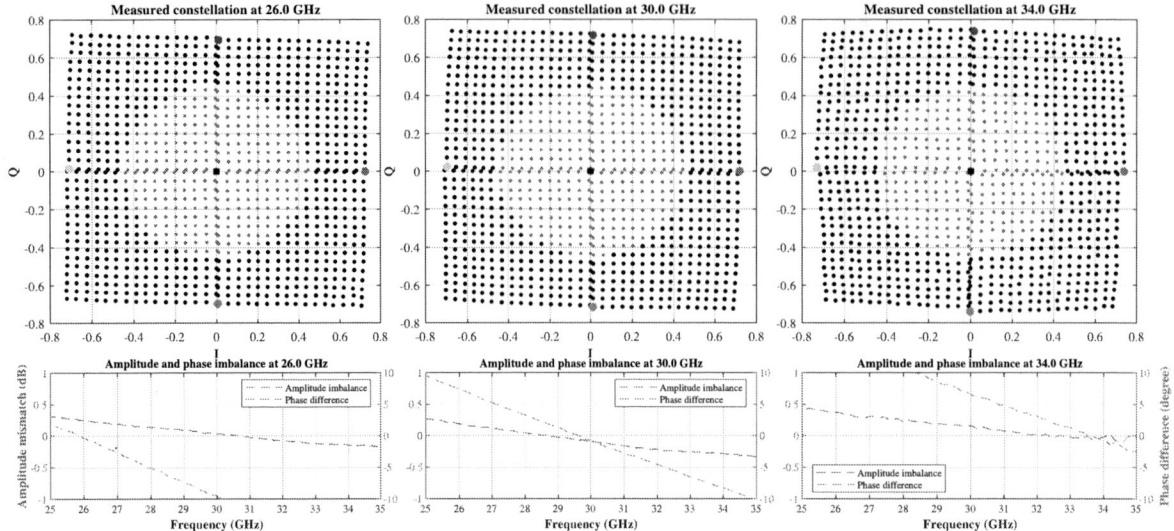

Fig. 5. Measured phase constellation along with measured amplitude mismatch in decibels and phase difference from the ideal 90° for the I and Q orthogonal basis vectors.

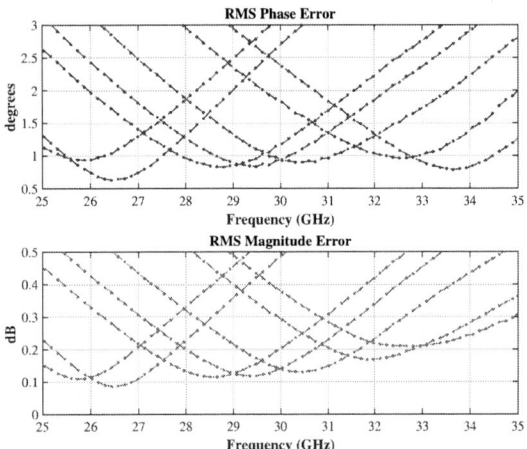

Fig. 6. Measured RMS phase and amplitude errors at different operating frequencies.

frequencies over the range of 26 GHz to 34 GHz. Using a simple calibration procedure involving a tunable PPF for frequency control and amplitude tuning via the DAMPs, the APS can generate a near ideal square phase constellation, resulting in an easily controllable phase shifter with 7 dB of amplitude control range and an average phase resolution of 0.5°. The measured RMS phase and amplitude error over the whole band was as small as 0.6° and 0.08 dB, respectively. The results were achieved in a small area of 0.19mm². This paper demonstrates the potential of this design to be a viable candidate for massive phased array systems utilized in area efficient 5G and future 6G applications.

Acknowledgment

Nokia is acknowledged for financing the project, GlobalFoundries for chip processing and Keysight for providing measurement equipment. This work was supported in part by the Academy of Finland 6Genesis Flagship (grant no. 318927).

References

[1] N. Fourikis, *Advanced Array Systems, Applications and RF Technologies*, 2000.

[2] M. Jung and B.-W. Min, "A Compact Ka-Band 4-bit Phase Shifter With Low Group Delay Deviation," *IEEE Microwave and Wireless Components Letters*, vol. 30, no. 4, pp. 414–416, apr 2020.

[3] W. H. Woods, A. Valdes-Garcia, H. Ding, and J. Rascoe, "CMOS millimeter wave phase shifter based on tunable transmission lines," in *Proceedings of the IEEE 2013 Custom Integrated Circuits Conference*, 2013, pp. 1–4.

[4] Y. Yu, K. Kang, C. Zhao, Q. Zheng, H. Liu, S. He, Y. Ban, L. L. Sun, and W. Hong, "A 60-GHz 19.8-mW current-reuse active phase shifter with tunable current-splitting technique in 90-nm CMOS," *IEEE Transactions on Microwave Theory and Techniques*, vol. 64, no. 5, pp. 1572–1584.

[5] D. Pepe and D. Zito, "Two mm-wave vector modulator active phase shifters with novel IQ generator in 28 nm FDSOI CMOS," *IEEE J. Solid-State Circuits*, vol. 52, no. 2, pp. 344–356, Feb. 2017.

[6] W. Zhu, W. Lv, B. Liao, Y. Zhu, Y. Dai, P. Li, L. Zhang, and Y. Wang, "A 21 to 30-GHz merged digital-controlled high resolution phase shifter-programmable gain amplifier with orthogonal phase and gain control for 5-G phase array application," in *2019 IEEE Radio Frequency Integrated Circuits Symposium (RFIC)*. IEEE, jun 2019.

[7] I. Kalyoncu, A. Burak, M. Kaynak, and Y. Gurbuz, "A 26-GHz vector modulator in 130-nm SiGe BiCMOS achieving monotonic 10-b phase resolution without calibration," in *2019 IEEE Radio Frequency Integrated Circuits Symposium (RFIC)*. IEEE, jun 2019.

[8] Y. Xiong, X. Zeng, and J. Li, "A frequency-reconfigurable CMOS active phase shifter for 5G mm-wave applications," *IEEE Transactions on Circuits and Systems II: Express Briefs*, vol. 67, pp. 1824–1828, 2020.

[9] F. Piri, M. Bassi, N. R. Lacaita, A. Mazzanti, and F. Svelto, "A PVT-Tolerant >40-db irr, 44% fractional-bandwidth ultra-wideband mm-wave quadrature LO generator for 5G networks in 55-nm CMOS," *IEEE Journal of Solid-State Circuits*, vol. 53, no. 12, pp. 3576–3586, 2018.

Proceedings of the 16th European Microwave Integrated Circuits Conference

A 25-50 GHz Digitally Controlled Phase-Shifter

Steeven Voisin[#*], Vincent Knopik[*], Eric Kerhervé[#]

[#] IMS Laboratory, Univ. Bordeaux INP, France

[*]STMicroelectronics, Crolles, France

steeven.voisin@st.com

Abstract — **This paper presents the design and the implementation of a phase shifter based on cascaded reflection type phase shifters (RTPS). The RTPSs are based on a 90° hybrid coupler and varactors digitally controlled. By cascading the RTPSs the phase shift range and the bandwidth are increased. To compensate the quality factor variation of the varactor at millimeter-wave frequencies, a simple compensation technique which requires no additional control is used. The proposed 7-bit controlled phase shifter achieves 216° of phase shift range with a fine step of 1.7° and an RMS phase error of 0.3°, the insertion losses are 10.9 dB on average, for a bandwidth of 25-50 GHz.**

Keywords — **RTPS, millimeter wave integrated circuits, phase shifter, Beamforming, 5G.**

I. INTRODUCTION

The phase shifter is an essential component of millimeter-wave frequency beamforming systems. Controlling the phase of the signal at each antenna allows the equivalent isotropic radiated power (EIRP) to be increased by directing the transmitted power in a specific direction.

The overall performance of these systems depends on the performance of each phase shifter. Several criteria are used to characterize the performance of the phase shifter: the phase resolution and accuracy, the insertion losses, the variation of losses according to the phase state and the power consumption. These criteria represent a real design challenge, particularly at millimeter-wave frequencies.

There are several phase shifter architectures, like the vector modulator (VM) [1] and the reflection-type phase shifter (RTPS) [2]-[5]. The former, based on active devices, provides gain but suffers from a poor linearity and a high power consumption. Inversely, the RTPS based on passive devices provides high linearity, bi-directionality, and no power consumption. In addition, it uses the advantage of integrated 90° hybrid couplers [6] which are very compact and have low losses at millimeter-wave frequencies, so it is possible to cascade them with a reduced impact on the surface area.

This paper presents a fully integrated phase-shifter consisting of three cascaded RTPSs. Each RTPS consists of an integrated 90° hybrid coupler and varactor as a reflective load. The digitally controlled phase shifter is designed for 25-50GHz, has a phase shift range of 215° with a step of 1.7°, the losses are 10.9 dB on average.

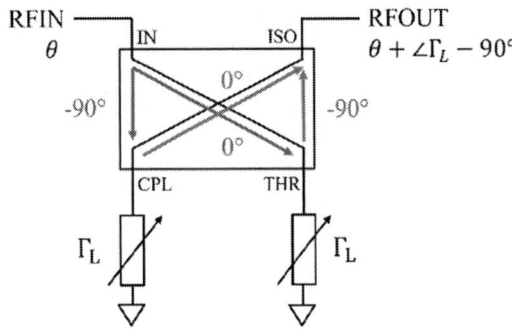

Fig. 1. Principle of RTPS with 90° hybrid coupler.

II. REFLECTION-TYPE PHASE SHIFTER DESIGN

Fig.1 shows the RTPS topology with a 90° hybrid coupler and a varactor as a reflective load. The input signal is split in two with a phase θ on the *THRU* port and a phase ($\theta - 90°$) on the *CPL* port. Then the signal is reflected by the varactor and is recombined in phase on the *ISO* port and out of phase on the *IN* port. The phase of the signal is modified by the phase of the reflection coefficient Γ_L of the varactor. The phase shift $\Delta\Phi$ between the output signal and the input signal can be written as in (1).

$$\Delta\Phi = \angle\Gamma_L - 90° \qquad (1)$$

The insertion losses of the phase shifter can be written as follows:

$$IL_{RTPS} = |\Gamma_L|_{dB} + 2IL_{hybrid} \qquad (2)$$

With IL_{hybrid} the insertion loss of the 90° hybrid coupler and $|\Gamma_L|$ the reflection loss of the varactor.

A. Hybrid Coupler

As can be seen from (1) and (2), the performance of the phase shifter depends on the performance of the 90° hybrid coupler, especially the insertion losses which count twice and the 90° phase shift between the two channels.

3–4 April 2022, London, UK

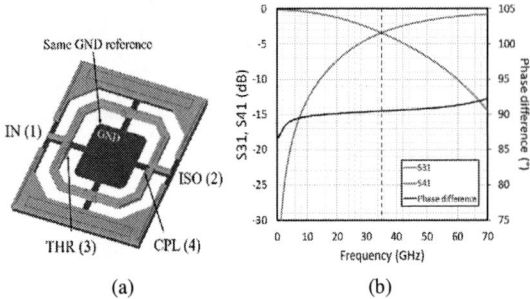

(a)   (b)

Fig. 2. (a) Layout and (b) EM simulation of 90° hybrid coupler

Fig. 2a shows the layout of the hybrid coupler. The *IN* port and the *ISO* port are respectively on the left and on the right side to be able to cascade them. The *THRU* port and the *CPL* port are placed inside the coupler, then both reflective loads can share the same reference ground which minimizes the phase difference between the two reflected signals. In addition, the phase shifter robustness to process variations is increased by placing the two loads close to each other.

Fig. 2b shows the EM simulation results of the 90° hybrid coupler. The coupler has a characteristic impedance of 40Ω, insertion losses of 0.41 dB and a phase difference of 90.5° at the resonant frequency of 35 GHz for a size of 135 x 141 μm² until the ground boundaries. The RTPS becomes very interesting at millimeter-wave frequencies with the compactness and the low losses of the 90° hybrid coupler.

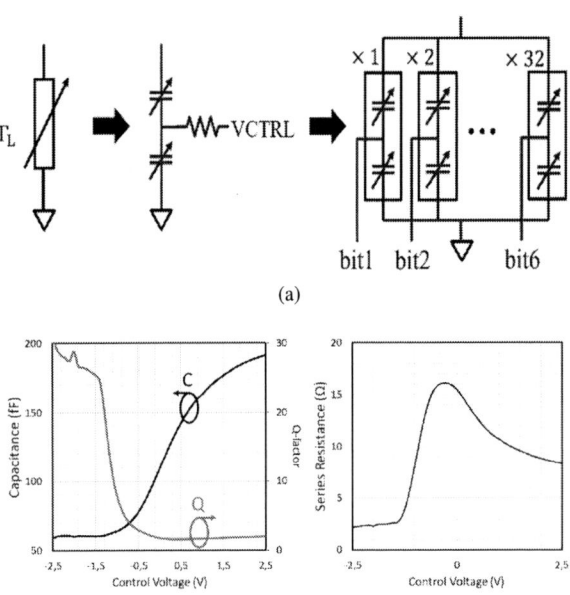

(a)

(b)   (c)

Fig. 3. (a) Elementary cell of varactor used for varactor bank as reflective load. (b) Measured Capacitance, *Q*-factor, and (c) equivalent series resistance of the 6bits-varactor versus the voltage control @40GHz

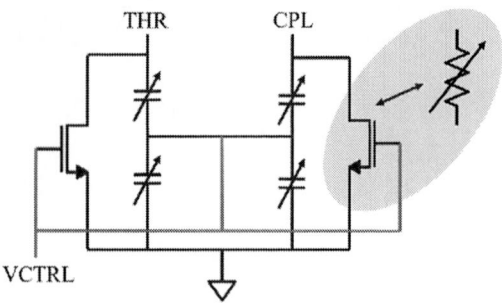

Fig. 4. Schematic of the compensation technique

B. Varactor

As shown in (1) the phase of the output signal can be controlled by changing the phase of the reflection coefficient of the reflective load. In the proposed RTPS the reflective load is based on a varactor. By changing the value of the varactor capacitance, the phase of the reflection can be changed as:

$$\angle\Gamma_{L} = 180° - \tan^{-1}\left(\frac{\frac{1}{C\omega}}{Z_0}\right) \tag{3}$$

To have a maximum phase shift range, the capacitance ratio *Cmax/Cmin* must be as high as possible. However, there is a trade-off between the quality factor *Q* and the varactor capacitance ratio. Indeed, increasing the width of the varactor *W* reduces the channel resistance but increases the overlapping capacitance and therefore reduces the capacitance ratio. Furthermore, the *Q*-factor is inversely proportional to the frequency, so this trade-off is even more challenging at millimeter-wave frequencies. The *Q*-factor will affect the insertion losses of the phase shifter and the ratio will affect the phase shift range.

The choice is to use a varactor with *W* = 1.2 μm and *L* = 0.7 μm as an elementary cell to maximize the capacitance ratio and reach a phase shift range of 60° without a strong *Q*-factor degradation. However, the design of the varactor is based on a tentative model which was optimistic about its performances: the *Q*-factor was over evaluated, and the ratio was under evaluated. In the section IV of this paper, the simulation results presented are based on the updated varactor model.

As shown in Fig. 3a two elementary varactors are implemented in series to form the basic cell of the reflective load. This allows to control the varactors voltage independently of each other. Thanks to that it is also possible to control the varactor in an "on/off" manner and thus, to control the phase shifter digitally. This has several advantages: it simplifies the interface with the DSP and removes the need of a high power consumption DAC. Furthermore, as shown in Fig. 3b, the capacitance value increases rapidly with a voltage around 0V. Using the varactor only at *Cmin* and *Cmax* makes it robust to potential process variations, and to frequency dependent behavior. This avoids adding correction block to a possible DAC. Furthermore, as shown in Fig. 3c, the equivalent series resistance of the varactor is also maximum at 0V and therefore the insertion losses are also maximum.

C. Compensation Technique

As shown in Fig. 3b, the Q-factor decreases with increasing varactor control voltage. Therefore, the insertion loss increases with the voltage. To compensate this variation in losses, a transistor is placed in parallel with the varactor (in Fig. 4). This transistor is equivalent to a voltage controllable resistance. When the transistor is ON the drain resistance is low and then the losses of the reflective load are increased. Inversely, when the transistor is OFF, its drain resistance is high, which does not affect the reflective load losses.

If the transistor is correctly sized, the gate of the transistor can be directly connected to the varactor control to simplify the control of the whole structure. In fact, the lower losses of the varactor at the $Cmin$ state are reduced to the losses at the $Cmax$ state by using a ON/OFF transistor in parallel equivalent to a variable resistance with no additional control. This reduces the variation of losses in the total phase shifter.

In addition, the transistor imposes a same voltage reference for both varactors connected in series. This guarantees the same potential difference on each varactor.

III. PHASE-SHIFTER IMPLEMENTATION

The proposed phase shifter is based on three RTPSs cascaded to initially reach 180° of phase shift range. Another block can be used to reach 360° of phase shift range like a VGA, because trying to do it with an RTPS is expensive in terms of insertion loss.

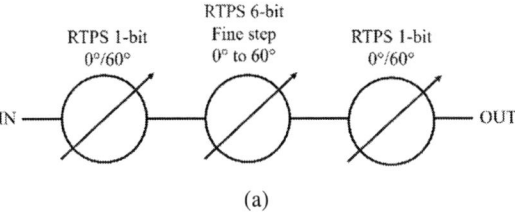

(a)

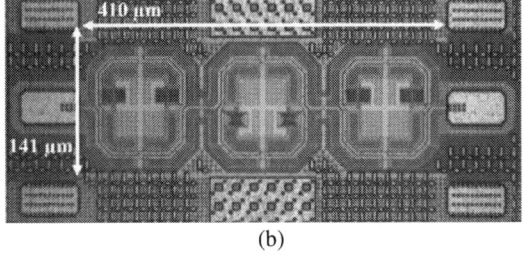

(b)

Fig. 5. (a) Schematic and (b) die photography of the proposed phase-shifter

Each RTPS covers 60° with the capacitance ratio of the varactors of 3. As shown in Fig. 5a one RTPS is controlled by 6 bits to have a fine step in phase and the other two are controlled by 1 bit to reach 180° phase. For the 6-bit RTPS the reflective load is composed of 6 elementary cells binary scaled. For the 1-bit RTPS the same number of elementary cells than

the 6-bit RTPS are used, i.e. 63 cells. Thus, the phase continuity is guaranteed when cascading the three RTPSs.

Each 1-bit RTPS insertion loss variation is compensated with a transistor connected to its control. For the 6-bit RTPS two transistors are used to compensate the loss variation and they are connected to bit 5 and bit 6 as they have the most important impact.

As shown in Fig. 5b each RTPS is flipped from the previous one and is surrounded by ground. Doing this, they do not talk to each other. Thus, we can optimize a single RTPS and cascade them without impacting the overall phase-shifter.

IV. MEASUREMENT RESULTS

The phase shifter was fabricated in 65nm CMOS SOI and was measured on wafer. A die photography is shown in Fig. 5b. The core area is only 141 × 410 µm².

The 6-bit RTPS was measured alone. The varactors are controlled in voltage with -2.5 V for the low level (coded 0) and 2.5 V for the high level (coded 1). Fig 6a shows the S-parameters of the 6-bit RTPS. The input and output reflection losses are less than -10dB from 34.5 GHz to 50 GHz and the insertion losses are of 3.3dB $\pm$ 1.8 dB without compensation. Fig. 6b shows the phase shift for each code. The phase shift range is 76.8°. The 6-bit RTPS alone was originally designed for a phase range of 60° with an insertion loss variation of $\pm$0.5 dB. As previously announced, this is due to the usage of a tentative model of the varactor. However, the results allow us to conclude on the proposed topology and the compensation technique functionality. The compensation transistor has been designed to compensate $\pm$0.5 dB for the 1-bit RTPSs and $\pm$0.3 dB for the 6-bit RTPS.

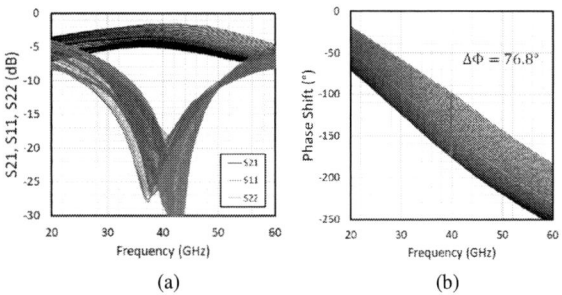

(a)   (b)

Fig. 6. Measured (a) S21, S11, and (b) phase shift of the 6-bit RTPS versus frequency for all codes.

Fig. 7 shows the S-parameters of the phase shifter. Thanks to the RTPS topology, the input and output impedances are close to 50 Ω and has low variation at 40 GHz. Moreover, by cascading 3 RTPSs and using a non-resonant reflective load, the bandwidth of the total phase shifter is greatly increased: S11 and S22 are less than -10 dB from 29 GHz to 51 GHz. Moreover, as seen in Fig. 7b, based on the average IL the 3dB-bandwidth starts at 23 GHz to reach 51 GHz. Also, as shown in Fig. 8b the phase shift range is relatively constant over 25 GHz to 50 GHz. This wideband operation is very interesting in 5G applications which aims to use several frequency bands in the Ka band (28GHz, 39GHz). As explained earlier in the paper, cascading the RTPSs does not impact each other. Therefore, the total losses correspond to the losses of the RTPS alone ×3 as do the phase shift range and the insertion loss variation. Then the insertion losses are of 9.5 dB on average with a variation of ±4.8 dB without compensation and are of 10.9 dB with a variation of ±3.5 dB with compensation.

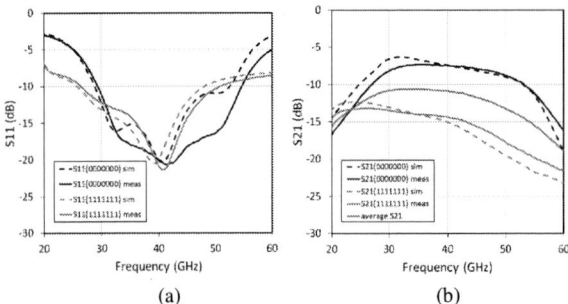

Fig. 7. Measured and Simulated (a) S11 and (b) S21 for the min code (0000000) and the max code (1111111) versus frequencies.

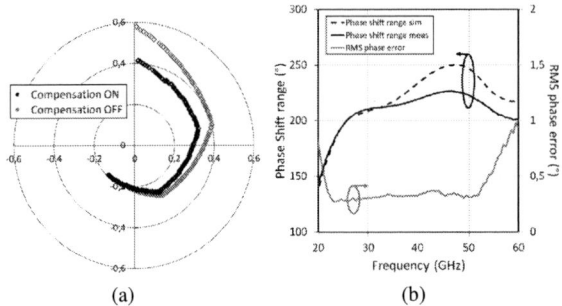

Fig. 8. (a) Polar plot of measured S21 versus with optimized code @40GHz and (b) Phase shift range and RMS phase error of compensated and optimized phase shifter versus frequencies.

Fig. 8a shows the polar plot of S21 at 40 GHz, the compensation reduces the variation of losses without impacting the phase resolution. Then the losses variations have been compensated from ±4.8 dB to ±3.5 dB that corresponds to 2× (±0.5 dB)+(±0.3 dB) as expected. The varactor can be optimized by reducing the length L, and then it can achieve 60° of phase shift range and reduce the insertion losses and the variation of losses.

The proposed phase shifter is controlled with 8 bits: 6 bits of control for the RTPS with a fine step and 2 bits for the other two RTPSs. By selecting the codes to have a 7-bit control with a constant step at 40 GHz, the phase shifter achieves a phase shift range of 216° with a step of 1.7° and an RMS phase error of 0.3° at 40 GHz. The phase shifter can be optimized for all frequencies and then the RMS phase error is less than 0.4° from 20GHz to 50GHz. This shows the very wide band property of the phase shifter not only in reflection and insertion losses but also in phase behavior.

Table 1 summarizes the different state-of-the-art digitally controlled RTPS at millimeter-wave frequencies. The proposed phase-shifter has the finest phase resolution with the lowest RMS phase error and the smallest occupied area. In addition,

by cascading three RTPS and using a non-resonant reflective load, the proposed phase shifter achieves the largest bandwidth.

Table 1. Performance summary and comparison.

Reference	**This Work**	[2]	[3]	[4]	[5]
Technology	**65 nm CMOS SOI**	SiGe 0.12 µm	45 nm CMOS SOI	45 nm CMOS	65 nm CMOS
Frequency (GHz)	**29-51**	57-66	29	28	28
Phase-shift range (°)	**216***	200	360	340	360
Bit control	**7 bits**	6 bits	6 bits	7 bits	5 bits
Phase resolution (°)	**1.7***	/	5.6	2.8	11.25°
RMS phase error (°)	**0.3***	2.6	0.3	/	0.3
IL (dB)	**10.9 ± 3.5***	8.2 ± 0.5	6.8 ± 0.1	9 ± 5	7.75 ± 0.3
Area (mm²)	**0.018**	0.28	0.065	0.18	0.16

*@40GHz

V. CONCLUSION

In this paper a phase shifter composed of three cascaded RTPS is presented. This phase shifter has the advantage of consuming almost zero power and having a high linearity. The main challenge is to design the varactor at these frequencies because of the trade-off between the capacitance ratio and the Q-factor. In the proposed circuit the goal is to achieve a 60° phase shift range per RTPS and cascading them to achieve 180° phase shift range. With the measurement the various key points of this structure are well validated: the possibility of cascading the RTPSs without coupling between them thanks to layout techniques, the simple compensation of loss variation without additional control and without impacting the phase behavior, and the increase of the phase shifter bandwidth by cascading RTPSs and using non-resonant reflective loads.

REFERENCES

[1] K. E. Drenkhahn, A. Gadallah, A. Franzese, C. Wagner and A. Malignaggi, "A V-Band Vector Modulator Based Phase Shifter in BiCMOS 0.13 µm SiGe Technology," *2020 15th European Microwave Integrated Circuits Conference (EuMIC)*, 2021, pp. 65-68

[2] R. B. Yishay and D. Elad, "A 57–66 GHz reflection-type phase shifter with near-constant insertion loss," *2016 IEEE MTT-S International Microwave Symposium (IMS)*, 2016, pp. 1-4.

[3] J. Xia, M. Farouk and S. Boumaiza, "Digitally-Assisted 27-33 GHz Reflection-Type Phase Shifter with Enhanced Accuracy and Low IL-Variation," *2019 IEEE Radio Frequency Integrated Circuits Symposium (RFIC)*, 2019, pp. 63-66.

[4] R. Bhattacharya, V. Aggarwal, A. Gupta, T. Kukal and S. Aniruddhan, "An 8-Channel Varactor-Less 28-GHz Front End With 7-Bit Resolution 340° RTPS for 5G RF Beamformers," in *IEEE Transactions on Circuits and Systems II: Express Briefs*, vol. 66, no. 12, pp. 1937-1941, Dec. 2019.

[5] R. Garg and A. S. Natarajan, "A 28-GHz Low-Power Phased-Array Receiver Front-End With 360° RTPS Phase Shift Range," in *IEEE Transactions on Microwave Theory and Techniques*, vol. 65, no. 11, pp. 4703-4714, Nov. 2017.

[6] V. Knopik, B. Moret and E. Kerherve, "Integrated scalable and tunable RF CMOS SOI quadrature hybrid coupler," *2017 12th European Microwave Integrated Circuits Conference (EuMIC)*, 2017, pp. 159-16.

Proceedings of the 16th European Microwave Integrated Circuits Conference

A 270 – 330 GHz Vector Modulator Phase Shifter in 130nm SiGe BiCMOS

Mohammad Hassan Montaseri[1], Sumit Pratap Singh[1], Markku Jokinen[1], Timo Rahkonen[2], Marko E. Leinonen[1], Aarno Pärssinen[1]

[1]Center for Wireless Communications, University of Oulu, Finland
[2]Circuits and Systems Lab, University of Oulu, Finland
{firstname.lastname}@oulu.fi

Abstract — **This paper presents a wideband vector modulator phase shifter measured at frequency bands reaching and ever surpassing the f_t of the technology for 6G beamforming for extremely high data rate applications. The circuit is designed based on 130nm SiGe BiCMOS technology with f_t / f_{max} of 300/500 GHz. The chip occupies a die area of 0.53×0.48 mm² and offers a total 360° of phase variation with a pad-to-pad gain of -10dB over its 270 – 330 GHz operational ranges.**

Keywords — **phased array, phase shifter, vector modulator, SiGe, HBT, BiCMOS, beamforming, 6G.**

I. INTRODUCTION

Due to demands for extremely high data rates, upper mm-wave bands up to THz region have become an attractive range for recently discussed 6G wireless communications systems. The envisioned properties and performance requirements of the 6G systems impose much more burden on the electronics design as the physical limitations of some of the available technologies are at the edge of reasonable performance. Thus, it is increasingly difficult to design an integrated circuit solution to meet the stringent requirements. This is due to the intrinsic behavior of the semiconductors at frequencies higher than $f_t/10$ as the key performance characteristics of the active components are heavily impacted at the mentioned bands. Atmospheric absorption and attenuation along with high path loss and hence limited link budget also add to the problem. One approach to address the issues is the phased array antenna solution, i.e. beamforming [1] – [6].

Phase shifters (PS) are the key building blocks of the beamforming systems utilized in long range connectivity and multiple-user multiple-input-multi-output (MIMO) systems. Demand for higher data rate envisioned by 6G wireless communications systems, and the need for beamforming for mobility require spacial scanning systems. Thus, the design of phase shifters is of great importance.

Since the vector modulators are in favor of gain and smoother phase variation, directly proportional to the control mechanism e.g. number of bits in case of DAC controlled ones, they have been widely utilized in almost every frequency band [1] – [6]. Compact size has been achieved at the cost of DC power consumption compared to their passive counterparts [7] – [9].

In this respect, this paper presents a 20% fractional bandwidth phase shifter with a center carrier frequency of

300 GHz, i.e. f_t of the technology. The rest of this paper is organized as follows. In section II, the designed vector modulator phase shifter (VMPS) circuitry is discussed. The measurement results are then shown in section III. The paper is concluded in section IV.

II. VECTOR MODULATOR PHASE SHIFTER CIRCUITRY

The schematic of the implemented vector modulator is shown in Fig. 1. It is based on the concept in [10] wherein the simulated results were presented without implementation. The input signal is first decomposed into two namely in-phase (I) and quadrature (Q) signal components each of which is then weighted passing through a variable gain amplifier (VGA). The weighted I and Q signals are summed at the output node of the VGAs to form the phase shifted version of the input signal.

The circuit consists of two Marchand baluns, one coupled-transmission-line based quadrature hybrid coupler, two Gilbert cells, and two 5-bit current steering digital-to-analog converters (DAC) equipped with serial-to-parallel shift registers. The Marchand baluns are utilized to convert the single-ended signal to their differential counterpart and vice versa at the input and output sides, respectively.

Core of the circuit is designed inherently to be fully differential. Marchand baluns at the input and output facilitate single ended measurements at the desired frequency bands of 220 – 330 GHz corresponding to the available equipment. Each balun has simulated loss of ~1dB that is included in the total measured gain later in this paper. The quadrature hybrid coupler for initial 90° phase shift, which is comprised of two quarter wave coupled transmission lines shielded with ground walls on the sides, generates the differential I and Q signal components for the input of the VGAs. Gilbert cells are used to realize the VGA circuitry with I and Q signals applied to the tail transistors and gain controlling signals to the differential pairs. Due to the simplicity of the Gilbert cell circuitry, it has primarily been the candidate for such high frequencies [1] – [6]. The inductance L_{cp} partly compensates the capacitances reside in the translinear loop [10]. The capacitance C_b ensures only DC signals at the base nodes of the differential pairs in addition to guarantee total DC current through transistors $Q_{b1} - Q_{b4}$. They also form a logarithmic operation to cancel out the antilog property of the base-emitters of the differential pairs. This yields more linear multiplication from digital control words to the output

3–4 April 2022, London, UK

constellation points over the frequency range. The weighting factors are determined by the two 5-bit current steering DACs in each branch. The weighed I and Q signals are then summed up at the output node of the Gilbert cells to form the phase shifted version of the input signal.

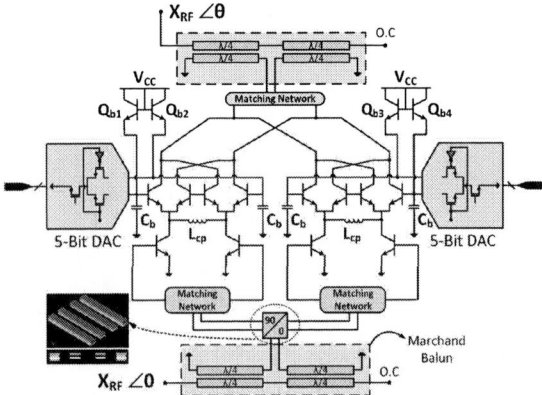

Fig. 1. Circuit schematic of the designed vector modulator phase shifter.

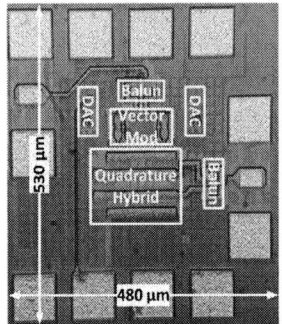

Fig. 2. Photograph of the designed vector modulator phase shifter.

III. MEASUREMENT RESULTS

The chip has been fabricated using 130nm SiGe BiCMOS process with f_t / f_{max} of 300/500 GHz. The microphotograph of the circuit is shown in Fig. 2. The chip dimensions are 0.53×0.48 mm^2, including the bias and signal pads. The active area of the phase shifter is only 0.179×0.13 mm^2. The chip draws 45 mA from 2.5 V of which 8 mA was burnt by the DACs, 7 mA burnt in the current mirror references, and 4 mA by the log/antilog circuitry. So, the net power consumed by the vector modulator phase shifter core is $2.5\ V \times 26\ mA = 65\ mW$.

The chip was measured on wafer using VDI WR3.4 extension kits connected to GSG infinity probes from Cascade Microtech. The probe tip short-open-load-thru (SOLT) calibration method on the P/N 138-357 calibration substrate was performed. An automatic measurement routine was developed based on MATLAB scripts consisting of the digital control codewords along with the commands for the Keysight PNA N5247B. The measurements were conducted in the time domain continues-wave (TDCW) mode. The schematic of the measurement setup is shown in Fig. 3.

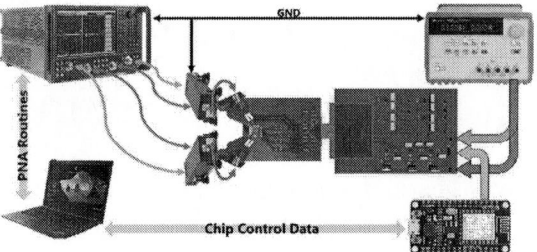

Fig. 3. Measurement setup block diagram.

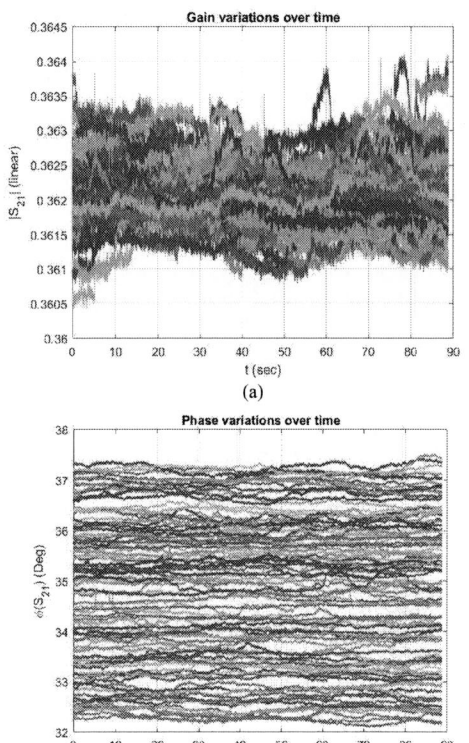

Fig. 4. The variation of the reference point (a) amplitude and (b) phase over time.

To extract the constellation points the gain S_{21} was measured in time domain. This is to prevent phase miss-locking in frequency domain. In other words, in frequency sweeping mode the internal/local oscillators of the PNA tend to lock to the reference frequency rather than phase. Therefore, the reference phase tends to vary randomly with each new frequency setting. To prevent this, for measurement purposes, the PNA was set to TDCW mode to ensure that the reference phase remain constant throughout the whole measurement sequence. Even in TDCW mode, the reference phase varies over the time due to time dependent variation of the measurement equipment systems properties. This also yields improper phase measurement. The variation of the reference point is shown in Fig. 4.

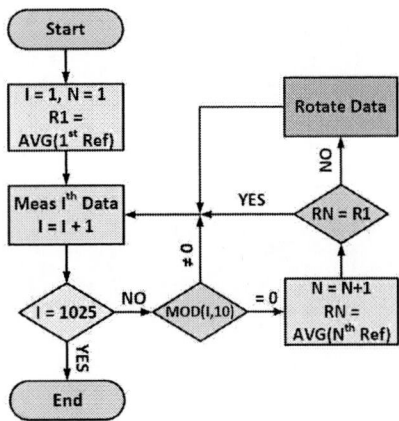

Fig. 5. The flowchart of the measurement routine.

To verify the measurement accuracy over time, the variation of the gain is shown in linear scale instead of logarithmic (Fig. 4(a)). Error seen from Fig. 4(a) the amplitude variation of the gain is quite negligible and limited close 0.5% on average. Hence, it can be ignored. Being on the order of $-2.5° \leq \Delta\phi \leq 2.5°$, the phase variation on the other hand varies in a larger scale with time (Fig. 4(b)). In phase shift measurements, a constellation point corresponding to the maximum gain was chosen as a reference, i.e. the code word was set to one of the outermost corners of the constellation diagram. Then, the automatic measurement routines were set to return to and measure the exact same reference point after every ten code words measurement sequence. The measured reference was then averaged. In case there existed a variation in the reference point, the phase of the measured group of gains were rotated accordingly. And then the rest of the measurement continued. This is shown in the flowchart diagram of Fig. 5.

Reference data point is defined in the measurements by averaging over 10001 samples and the rest of the constellation points were measured against that reference without averaging. Figs. 6 – 8 illustrate the measured constellation points at 270 GHz, 300 GHz, and 330 GHz, respectively. The presented circuit covers the whole plane, i.e. complete 360° of phase rotation for the whole band. In case of 330 GHz a hollow area was observed due to lack of points in origin. However, phase shifting was still operating properly.

Frequency response of the designed vector modulator phase shifter is shown in Fig. 9. The frequency response of the designed circuit is broad, i.e. ~100 GHz. Though, the constellation points start to be more distorted at frequencies below 260 GHz. Based on the plots, gain control in VMPS amplifiers tends to be more limited at some bands due to feedthrough. The more the gain can be controlled towards lower levels, i.e. more negative in dB scale, the more evenly constellation points are distributed towards the center of the constellation plane (Figs. 6 – 8). Limited gain range is clearly visible at 330GHZ in Fig. 8.

Table 1 summarizes recent papers on vector modulator phase shifters at frequencies above 100 GHz. It can be seen the presented vector modulator phase shifter offers the highest operating frequency with comparable performance.

Furthermore, the presented circuit is also the first phase shifter that has been measured at frequency bands of f_t and above.

Multistage cascade amplification has been utilized in [2], [3], and [6]. Therefore, the highest gain and dissipated DC power have been reported. The end resolution and phase error are directly proportional to the number of bits in the utilized DACs as well as calibration circuitry. Therefore, the higher the number of bits, the better the phase resolution and accuracy.

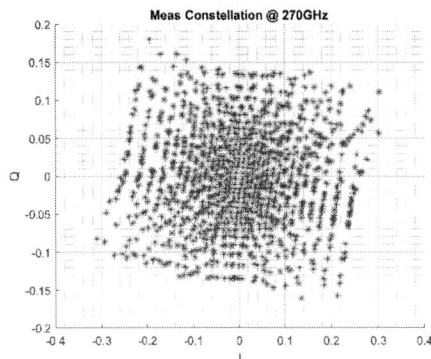

Fig. 6. Measured constellation points of the designed vector modulator phase shifter at 270 GHz.

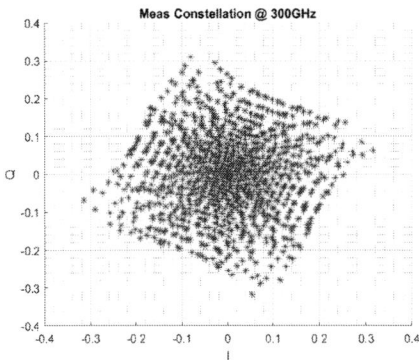

Fig. 7. Measured constellation points of the designed vector modulator phase shifter at 300 GHz.

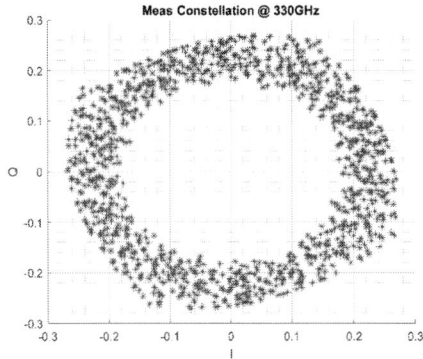

Fig. 8. Measured constellation points of the designed vector modulator phase shifter at 330 GHz.

Table 1. Comparison of the recently published vector modulator phase shifters

Ref	[1]	[2][a]	[3][a]	[4]	[5]	[6][b]	This work
Technology (nm)	SiGe 130	SiGe 130	SiGe 130	SiGe 120	InP 250	InP	SiGe 130
Frequency (GHz)	210	170	210	210	210	210	210
f_t/f_{max}	300/450	300/450	250/340	260/340	370/650	375/Nan	300/500
Gain (dB)	-9.5	16[c]	15[c]	-5.8	-10	10[c]	-10
Resolution (Deg)	15	5.625	Analog Control	11.25	22.5	Nan	11.25
RMS Phase Error (Deg)	Nan	Nan	Analog Control	2.2	10.2	Nan	0.98
Input P_{1dB}	Nan	-6.2	-9.3	Nan	-0.7	Nan	+1[d]
DC P_{Diss}	30[e]	330[f]	225[f]	30	42	Nan	65
Area (mm²)	0.075	1.966[g]	0.1	0.405[g]	0.23[g]	0.375[g]	0.023

a) Only TX chain is reported here. Original paper reports both TX and RX. b) The design covers only the first quarter of the whole plane, i.e. maximum phase shift of 90°. c) Amplifier embedded front-end. d) Based on simulations. The available equipment did not offer enough drive for the chip to enter saturation. e) The authors defined a range from 0 mA to 012.5mA. f) PA power consumption is included. g) The given dimensions include all the passives, i.e. baluns, etc., pads, and/or extra amplifier circuits.

So, in case of analog control, i.e. [1] and [3], the phase resolution can be the best of all, and RMS phase error marginally approaches zero. In this paper the compression point is reported only based on simulation. This is because the utilized VDI frequency extender cannot provide enough power to compress the phase shifter. The saturated output power of the VDI WR3.4 is measured to be ~3 dBm [11]. Passing this through the ~6.5dB insertion loss probes, the authors managed to drive the chip with −3.5dBm signal power which is insufficient to compress.

IV. CONCLUSION

A wideband vector modulator phase shifter with a 60 GHz (20% fractional) bandwidth was presented in this paper. The presented circuit was measured at frequencies 270 − 330 GHz. The circuit is the only vector modulator phase shifter which has been measured to be operational at frequencies above the unity current gain of the technology, i.e. 300 GHz. The designed circuit offers full 360° signal rotation with a −10 dB gain. Compensation technique for translinear loop improved the measured constellation of the phase shifter. To extract the constellation points a statistical processing was performed using multiple reference point scanning.

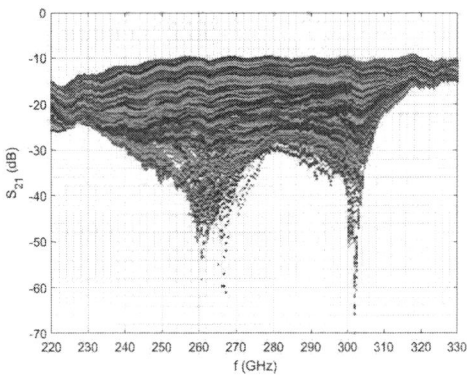

Fig. 9. Measured frequency response, i.e. S_{21}, of the designed vector modulator phase shifter at the band of 220 − 330 GHz.

ACKNOWLEDGMENT

This work was supported by the Academy of Finland 6Genesis Flagship (grant no. 318927). The authors would like to express their gratitude towards Mr. Mostafa Jafari-Nokandi and Mr. Joh-Johan Toivanen for their help and support in setting up the measurement equipment.

REFERENCES

[1] P.V. Testa, C. Carta, F. Ellinger, "A 140-210 GHz Low-Power Vector Modulator Phase Shifter in 130 nm SiGe BiCMOS Technology," IEEE APMC, Nov 2018.

[2] M. Elkhouly, M.J. Holyoak, D. Hendry, M. Zierdt, A. Singh, M. Sayginer, S. Shahramian, Y. Baeyens, "D-band Phased-Array TX and RX Front Ends Utilizing Radio-on-Glass Technology", IEEE RFIC, Aug 2020.

[3] E. Öztürk, H. J. Ng, W. Winkler, and D. Kissinger, "0.1mm² SiGe BiCMOS RX / TX channel front-ends for 120 GHz phased array radar systems," in Proc. IEEE SiRF, Jan. 2017, pp. 50–53.

[4] R. B. Yishay and D. Elad, "D-band 360 phase shifter with uniform insertion loss," in Proc. IEEE/MTT-S Int. Microwave Symp. - IMS, Jun. 2018, pp. 868–870.

[5] Y. Kim, S. Kim, I. Lee, M. Urteaga and S. Jeon," A 220-320-GHz Vector-Sum Phase Shifter Using Single Gilbert-Cell Structure with Lossy Output Matching," IEEE Trans. on Microw. Theory and Tech., vol. 63, no. 1, pp. 256-265, Jan. 2015.

[6] H. G. Yu, K. J. Lee, and M. Kim, "300 GHz vector-sum phase shifter using InP DHBT amplifiers," IET Electron. Lett., vol. 49, no. 4, pp. 263–264, Feb. 2013.

[7] KJ. Koh, J.W. May, and G.M. Rebeiz, "A Q-band (40–45 GHz) 16-element phased-array transmitter in 0.18-µm SiGe BiCMOS technology," IEEE Radio Frequency Integrated Circuits Symposium, June 2008.

[8] K. Kibaroglu, M. Sayginer, and G.M. Rebeiz, "A low-cost scalable 32-element 28-GHz phased array transceiver for 5G communication links based on a 2 × 2 beamformer flip-chip unit cell," IEEE J. solid-state circuits, vol. 53, no. 5, pp. 1260-1274, May 2018.

[9] Y. Yeh, B. Walker, E. Balboni, and B. Floyd, "A 28-GHz Phased-Array Receiver Front End with Dual-Vector Distributed Beamforming", IEEE J. Solid-State Circuits, vol. 52, no. 5, pp. 1230-1244, May 2017.

[10] M. Montaseri, M. Jafari-Nokandi, A. Pärssinen, and T. Rahkonen, "Analysis of HBT Vector Modulator Phase Shifters Based on Gilbert Cell for sub-THz Regimes", IEEE ISCAS, Oct 2020.

[11] M.E. Leinonen, K. Nevala, N. Tervo, A. Pärssinen, "Linearity Measurement of 6G Receiver with One Transmission Frequency Extender Operating at 330 GHz", IEEE RWW, Jan 2021.

Gap in pagination due to unavailable papers.

Pages 313-340

Optoelectronic Millimeter-Wave Integrated Circuits Fabricated in Pure Silicon-Based Technologies

Uroschanit Yodprasit [#1], Wolfgang Winkler [#2]

[#] Silicon Radar GmbH, Frankfurt (Oder), Germany

{[1]yodprasit, [2]winkler}@siliconradar.com

Abstract — **The design and implementations of optoelectronic millimeter-wave circuits integrated in silicon technologies are presented in this paper. For demonstrating the feasibility and the performance of the utilized technologies, a low-noise amplifier (LNA) with optical input and an optically synchronized voltage-controlled oscillator (VCO) have been implemented. The chips are fabricated using 0.25 μm SiGe:C BiCMOS electronic-photonic integrated circuit (EPIC) technology with f_T =190 GHz. Operated from a 3.3-V supply voltage, the LNA feature a wide bandwidth of around 50 GHz while consuming 80 mW power. The VCO features a free-running tuning range from 58.5 GHz to 65.7 GHz and has a phase noise of -85 dBc/Hz at 1-MHz offset at 58.5 GHz. It is shown that the oscillator performance can be increased considerably by injecting an optical signal to stabilize the oscillator.**

Keywords — **photonics, optoelectronics, photo detectors, low-noise amplifiers, voltage-controlled oscillators, injection locking**

I. INTRODUCTION

Optoelectronic circuits are recently playing essential role in many applications. Photonics provides numerous advantages in high-speed data communication and radar compared to wireless or copper-based interconnects such as higher signal integrity, lower signal loss and lower interferences. Photonics is demonstrated to enable a new generation of miniaturized, heterogeneous, and distributed radars, i.e., future radars on chip with different features, working in different radio spectral regions, and organized in spatially distributed sensors for the enhanced detection of a wider range of target properties. In these systems, the use of photonics assures benefits in terms of frequency flexibility, accuracy, and computational load reduction [1].

Since most of today's signal processing technologies are still in electrical domain, optical data routed via optical fibers need to be converted to electrical signal for subsequent signal processing steps. However, in order to handle high-speed optical signals that can have a clock rate of more than 50 GHz, ultra-wideband electronic circuits operating at millimeter-wave frequencies are mandatory.

In this paper, the designs of an optoelectronic 50 GHz wideband LNA and a 60 GHz optically injection-locked VCO for radar clock distribution applications are described. The corresponding test structures (i.e., circuits without photo detectors) were employed to characterize the electrical-only characteristics of the LNA and the VCO. The electro-optical characteristics of the complete structures consisting of photo detectors as optical receiving elements were also verified experimentally.

II. IHP SG25H4/H5 EPIC SiGe TECHNOLOGIES

Silicon Photonics is a key technology for the development of high performance communication and innovative radar systems. One advantage is the use of established silicon processes, which enables high scale co-integration and low cost production. Hybrid or monolithic co-integration techniques are most commonly used. The hybrid integration approach offers a flexible choice of photonic and electronic technologies to be combined. Here, the interconnection between the two chips is achieved via bond pads and various bonding techniques. However, this comes with the disadvantage of introducing additional parasitic elements into the circuit design, which reduce the overall circuit performance.

The monolithic co-integration approach allows photonic and electronic components to be fabricated on the same chip. This enables the shortest possible connection between optics and electronics and consequently reduces the parasitic influence of interconnects. This results in a higher level of integration as well as higher data rates, lower power consumption, lower noise and a smaller chip size.

The monolithic co-integration approach was used to implement the 0.25μm photonic SiGe BiCMOS processes SG25H4 EPIC and SG25H5 EPIC (cross section shown in Fig. 1) combining 0.25μm CMOS, high-speed hetero-junction bipolar transistors (HBT) with f_T/f_{max}=180/220 GHz (H4) or f_T/f_{max}=220/290 GHz (H5) and photonic devices for optical C- and O-band applications. The processes offer a waveguide-based germanium photodiode (PD) with a 3-dB electro-optical bandwidth of about 65 GHz and capacitance of 10 fF at -2 V bias. The combination of a responsivity of 0.7 A/W in the C-band and a dark current lower than 100 nA ensures a good sensitivity of the photo detector [2].

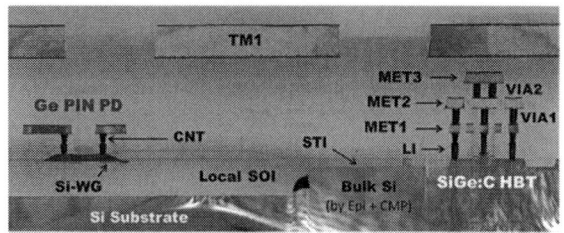

Fig. 1. Cross-section of IHP's EPIC technology platform (image of the first generation EPIC technology with 30 GHz photo detector). Locally reconstructed bulk Silicon by using epi+CMP enables frontend-of-line integration approach of SiGe electronics with state-of-the-art SOI-based photonics.

III. Optoelectronic 50 GHz Wideband LNA

The proposed optoelectronic wideband LNA is intended to be utilized as the front-end of an optical receiver, of which the structure is shown in Fig. 2. The receiver receives an optical input signal from an optical fiber. The on-chip photo detector consisting of an optical waveguide and a photodiode converts the optical signal into an electrical current.

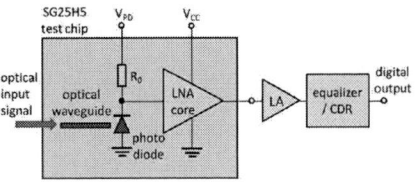

Fig. 2.   Structure of a conventional optical receiver (highlighted part is the test chip implemented in this work).

Together with the input impedance of the LNA, this generates an input voltage for the LNA core. The amplified voltage from the LNA is then fed into a limiting amplifier (LA) and an equalizer and/or a clock-data recovery (CDR) circuit for subsequent digital signal processing.

The optoelectronic LNA proposed in this paper is highlighted in Fig. 2. The LNA core is a three-stage cascode distributed amplifier [3], of which the schematic diagram is shown in Fig. 3. The distributed amplifier structure has been selected for this application because of its ultra-wideband characteristic. The cascode stage additionally improves the gain and the isolation of each stage.

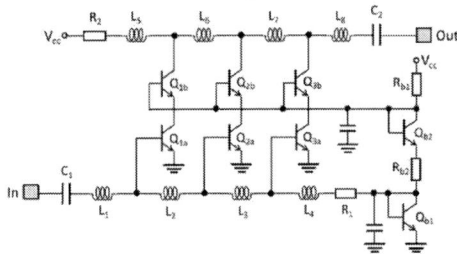

Fig. 3.   Three-stage cascode distributed amplifier as the LNA core.

The inductors L_1 to L_8 and the termination resistors R_1 and R_2 in the base line and in the collector line were optimized for wide bandwidth of greater than 50 GHz. The capacitors C_1 and C_2 serve as DC blocking capacitors and were taken into the bandwidth optimization process. The input of the LNA is connected to a photo detector comprising of a photodiode and an optical waveguide. From Fig. 2, the photodiode is biased through the resistor R_0 and the DC voltage V_{PD} in order to set the optimum sensitivity. The input and the output sections are, respectively, matched to 50-Ohm source and load. The bias voltages for the driver transistors (Q_{1a} to Q_{3a}) and those of the cascode transistors (Q_{1b} to Q_{3b}) are generated by a bias network consisting of Q_{b1} and Q_{b2}.

IV. Optically Injection-Locked 60-GHz VCO

The overview of the optically injection locked VCO is shown in Fig. 4. It consists of an optical waveguide, a photo diode, an oscillator core and an output buffer.

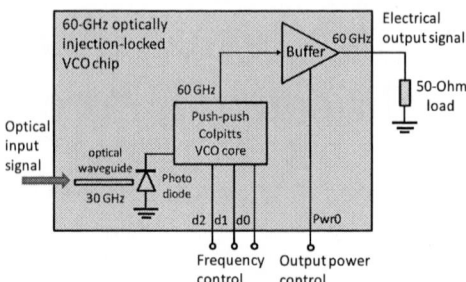

Fig. 4.   Structure of the proposed 60-GHz optically injection-locked VCO.

A. Colpitts push-push VCO core

The circuit diagram of the Colpitts VCO core is shown in Fig. 5. Two common-collector transistors $Q_{1a,b}$ form the gain stage to sustain an oscillation at around 30 GHz. The free-running oscillation frequency is defined by the inductors $L_{1a,b}$ and the effective capacitance of the capacitors $C_{1a,b}$ and $C_{2a,b}$ and the three varactors. The frequency tuning of the VCO depends on these varactors, in which the capacitances are varied by the voltages d_0 to d_2. The inductors $L_{2a,b}$, the resistor $R_{1a,b}$ and the capacitor $C_{3a,b}$ are used for biasing the emitter of the gain stage and also for serving as an injection input, receiving an injection current from the photodiode PD_{1a}.

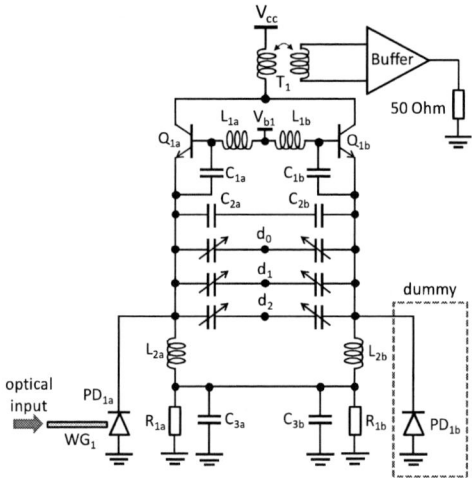

Fig. 5.   Detailed schematic diagram of the Colpitts VCO core.

Note that the photodiode PD_{1b} acts only as a dummy to preserve a good symmetry. The common-collector node generates a frequency doubled from the free-running frequency (i.e., thus around 60 GHz) by employing the nonlinearity of the transistors. The transformer T_1 is then used to convert this single-ended 60-GHz output signal for the differential input of the output buffer.

B. The VCO output buffer

A buffer amplifier is connected to the VCO core to drive an off-chip 50-Ohm load. The schematic of the output buffer is shown in Fig. 6. The circuit topology is based on a cascode differential amplifier with inductive load.

342

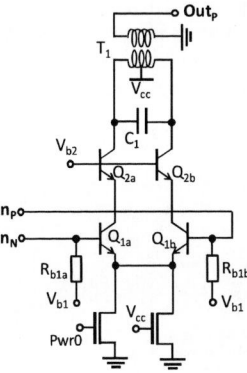

Fig. 6.   Detailed schematic diagram of the output buffer.

The transistors $Q_{1a,b}$ receive the output from the VCO core, while the cascode transistors $Q_{2a,b}$ increase the gain and provide a good isolation. The output Out_p has a DC ground provided through the secondary turn of the transformer T_1 for ESD protection. The capacitor C_1 is optimized together with the inductance of T_1 for maximizing the output power. The output power of the buffer can be adjusted by varying the voltage Pwr0.

V.   FABRICATION AND MEASUREMENT

The proposed optoelectronic wideband LNA is fabricated using IHP SG25H5 EPIC SiGe technology, while the optically injection-locked VCO is fabricated in SG25H4 technology [2]. The microphotographs of the optoelectronic LNA and the optically injection-locked VCO are shown, respectively, in Fig. 7 (a) and (b), respectively. Note that the test chips without photo detectors (marked in the white boxes) are fabricated in separate chips for characterizing the electrical-only characteristics.

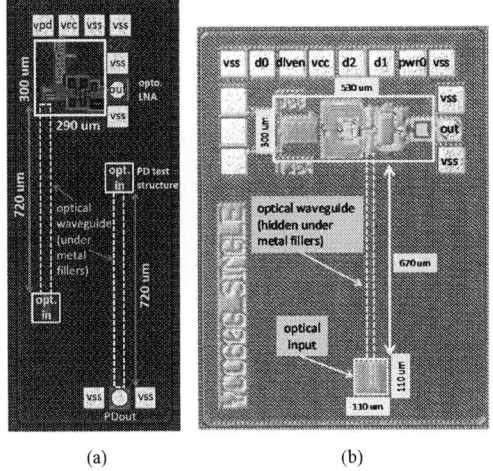

(a)   (b)

Fig. 7.   Chip microphotographs of (a) optoelectronic LNA and (b) optically injection-locked VCO

The characterizations of the LNA and those of the VCO were done on wafer and the measurements of each circuit are described as follows.

A. Optoelectronic 50-GHz wideband LNA

The optoelectronic LNA was characterized from wafers on a probe station. The test setup for the version with a photo detector is illustrated in Fig. 8. For the version without the photo detector employed for characterizing only the electrical characteristics, standard S-parameter measurement setup was applied.

Fig. 9 shows the measured power gain and the output matching characteristics in comparison to simulation results of the LNA test structure (i.e., without the photo detector). It can be seen that the measurement results match reasonably well to those from the simulations. Consuming 80 mW of power, the measured maximum gain is 13 dB and the 3-dB bandwidth is 50 GHz ranging from 7 GHz to 57 GHz. In comparison to simulation, the bandwidth is slightly reduced and the LNA reveals a narrower output matching bandwidth in the measurement accordingly.

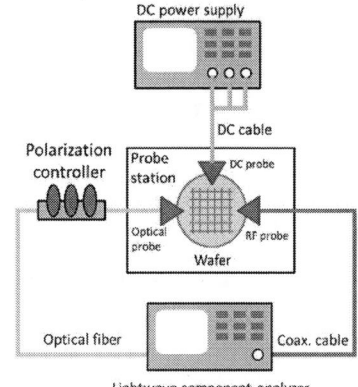

Fig. 8.   Measurement setup for the optoelectronic wideband LNA.

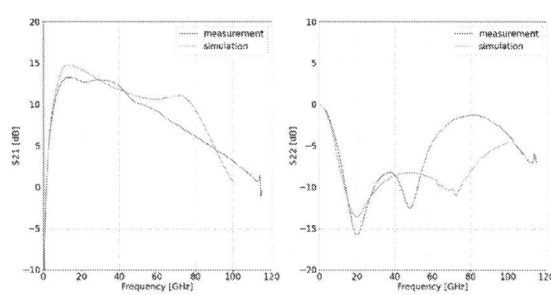

Fig. 9.   Measured S parameters in comparison with simulation results of the LNA test structure (i.e., without photo detector).

B. Optically injection-locked 60-GHz VCO

The test setup for the on-wafer measurement of the VCO is illustrated in Fig. 10. It consists of a Keysight tunable laser source TLS81980A (l = 1550 nm), two light polarization controllers, a MACH-40 intensity modulator, a bandpass filter, a Manlight Erbium-doped fiber amplifier (EDFA), an Anritsu MG3697C signal generator, a Keysight N4985A RF amplifier, an 8-channel DC power supply, a Rohde & Schwarz FSW signal and spectrum analyzer (from 2 Hz up to 67 GHz), DC and RF probes and a probe station.

343

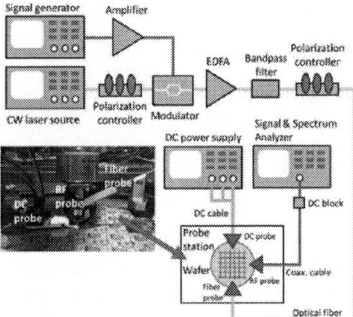

Fig. 10.   Measurement setup for the optically injection-locked VCO.

The oscillation frequency, the locking behavior and the phase noise characteristics were measured using a spectrum analyzer, while the output power was measured using a power meter. The optical input was produced by modulating a continuous laser source with an RF signal from a frequency generator for a required locking frequency. From three chip samples, the measured tuning range and the output power are shown in Fig. 11 (a) and (b), respectively.

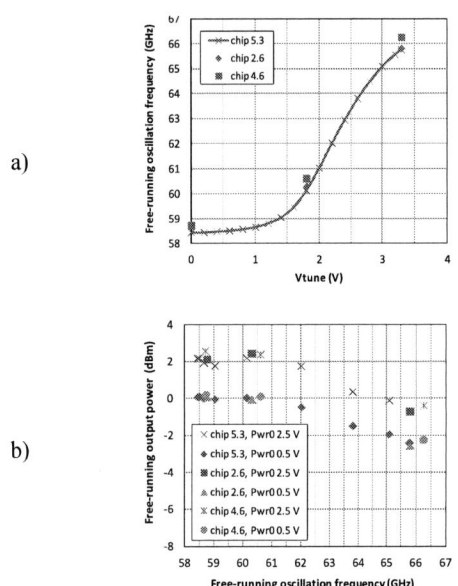

Fig. 11.   VCO test structure measurement from three chip samples (a) tuning range and (b) output power.

The tuning range was measured about 7.5 GHz and the VCO can deliver an adjustable output power (varying Pwr0 from 0.5 V to 2.5 V) from about 0 dBm to 2 dBm in the lower band. The output power starts to drop at the frequencies higher than 61 GHz. At 66 GHz, the output power can be varied between -2.2 dBm to -0.4 dBm.

The optically injection-locked VCO was measured with an injection of the optical input signal. The phase noise characteristics are shown in Fig. 12 (a) the free-running and (b) the optically injection-locked VCO at the upper limit of the locking range of 64 GHz together with (c) the free-running

and (d) the injection-locked phase noise of the VCO at the lower limit of the locking range of 58 GHz. For comparison, the phase noise curve for (e) the CW laser source and (f) the electrical signal generator at 32 GHz are also shown.

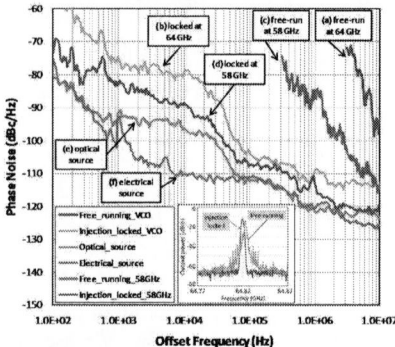

Fig. 12.   Phase noise characteristic of the VCO operating at 64 GHz (inset: output power spectrum in locked and free-running mode).

VI. CONCLUSION

An optoelectronic wideband 50-GHz LNA and an optically injection-locked 60-GHz VCO are designed and fabricated in SiGe EPIC technologies. The optoelectronic wideband LNA is suitable for serving as an optical front-end for clock distribution. The optically injection-locked VCO is suitable for implementing highly-stable millimeter-wave signal generators, where high-spectral-purity and high-stability clock distribution over large area are required. Both circuits can have application fields in radar or communication systems. Since the proposed circuits have been fabricated in Si-based technology, interfacing between the photonic parts and the electronic parts are simplified, enabling system-on-chip integration and reducing cost of the complete system.

ACKNOWLEDGMENT

The authors would like to thank Stefan Simon, Anna Peczek, Falk Korndörfer and Lars Zimmermann of IHP Microelectronics in Frankfurt (Oder), for measurements and fruitful discussions and for fabricating the chips.

REFERENCES

[1] Giovanni Serafino et al.," Toward a New Generation of Radar Systems Based on Microwave Photonic Technologies", IEEE JOURNAL OF LIGHTWAVE TECHN., VOL. 37, NO. 2, 2019, pp. 643-650

[2] D., Knoll. et al., "SiGe BiCMOS for Optoelectronics," ECS Transactions, vol.75, issue 8, pp. 121-139, Sept. 2016.

[3] U., Yodprasit et al., "12 GHz to 40 GHz 0.13-μm SiGe BiCMOS circuits for UWB 3D real-time OFDM MIMO imaging radar applications," 2018 11th German Microwave Conference (GeMiC), Freiburg, 2018, pp. 339-342.

[4] M. H. Eissa et al., "A wideband monolithically integrated photonic receiver in 0.25-μm SiGe:C BiCMOS technology," ESSCIRC Conference 2016: 42nd European Solid-State Circuits Conference, Lausanne, 2016, pp. 487-490.

[5] K., Balakier et al., "Integrated Semiconductor Laser Optical Phase Lock Loops," IEEE Journal of Selected Topics in Quantum Electronics, vol.24, issue 1, pp. 351-357, Jan. 2018.

[6] N., Deparis, et al., "A 2pJ/bit Pulsed ILO UWB Transmitter at 60 GHz in 65-nm CMOS-SOI," IEEE ICUWB Conf., pp. 113-117, 2009

Gap in pagination due to unavailable papers.

Pages 345-369

Microwave sensing using metal-insulator-metal diodes based on 4-nm-thick hafnium oxide

Martino Aldrigo[#], Mircea Dragoman[#], Sergiu Iordanescu[#], Mazen Al Shanawani[*], George Deligeorgis[$]

[#]IMT-Bucharest, L4, Voluntari (Ilfov), Romania

[*]DEI, Faculty of Engineering, University of Bologna, Bologna, Italy

[$]Foundation for Research and Technology Hellas, Institute of Electronics Structure and Laser (FORTH–IESL), Heraklion, Greece

{martino.aldrigo; mircea.dragoman; sergiu.iordanescu}@imt.ro, mazen.shanawani@unibo.it, deligeo@physics.uoc.gr

Abstract — In this paper, we present a state-of-the-art metal-insulator-metal (MIM) diode, in which the bottom electrode is made of platinum and the top one of titanium, while the in-between oxide is an atomic layer deposition (ALD)-deposited hafnium oxide (HfO_2) layer with a thickness less than 4 nm, and a footprint area of few square microns. Provided that such devices are suitable for detection and energy harvesting systems at frequencies where classical Schottky diodes fail to comply with the desired specifications of high cut-off frequency and monolithic integration, the proposed MIM demonstrated to be an excellent detector up to 40 GHz, with a maximum current density of over 2×10^3 A/cm^2, a measured voltage responsivity of about 500 mV/mW at 10 and 24 GHz, and a minimum noise equivalent power of only 160 pW/$\sqrt{Hz}$. As such, the HfO_2-based MIM diode could be used as an efficient low-voltage microwave sensor to detect radiofrequency signals in the upcoming smart 5G transceivers.

Keywords — Microwave sensors, diodes, nanoscale devices, hafnium compounds.

I. INTRODUCTION

Metal-insulator-metal (MIM) diodes have been gaining an increasing interest in the last years, thanks to the fact that they are tunnelling devices with a cut-off frequency limited only by the tunnelling time, i.e., in the order of just few femtoseconds. In order to achieve such high cut-off frequencies, both insulator's and metals' roughness must be in the Ångstrom range [1]. If this requirement is not met, the quantum tunnelling effect degrades and very low current values can be expected. The history of MIM diodes dates back to the previous century, when J. G. Simmons [2] and A. Sanchez *et al.* [3] proposed the first analytical models to describe the physics and working principle of such devices. After that, more complex models were developed to take into account the thermionic emission [4] and the Frenkel Poole emission (FPE) [5]. The typical signature of a MIM diode is its nonlinear current-voltage dependence due to quantum tunnelling, with a tunnelling time between 10^{-12} s and 10^{-15} s (which translates into a working frequency of tens or even hundreds of THz). As a consequence, such devices have been the subject of widespread research in the field of optical/solar rectennas [6],[7] and, more generally, high-frequency energy-harvesting [8]-[9]. In this respect, a promising oxide material for high-frequency applications is hafnium oxide (HfO_2), a high-κ dielectric largely exploited as gate oxide in FETs but also used in millimetre-wave energy harvesters with outstanding performance [10]. This material

can be grown via atomic layer deposition (ALD) techniques in nanometric scale thin films (i.e., 2–6 nm) with a very low roughness (i.e., 0.2–0.3 nm). Moreover, a CMOS-compatible fabrication process would be highly desirable for large-scale, low-cost mass production. "Smart" transceivers with autonomous reconfiguration capabilities represent a target for adaptive wireless systems. In this sense, an electromagnetic (EM) power level sensor could be beneficial to monitor the RF input power level and interference when the system is idle. This can provide information about incoming RF power density (critical to establish the noise floor), implement wake-up functions (super-regenerative receiver) in low-power multiple-input multiple-output (MIMO) mobile applications, and realise health monitoring if RF leakage of adjacent emitting modules occurs.

The aim of this manuscript is to present a state-of-the-art MIM diode, in which the bottom electrode is made of platinum (Pt) and the top one of titanium (Ti), while the insulating oxide is an ALD-deposited HfO_2 layer with a thickness below 4 nm, and an active area of few square microns. The device performs excellently up to the millimetre-wave range and, thanks to its high responsivity and low noise equivalent power, it meets the requirements of an EM interference sensor with CMOS-compatible characteristics.

The paper is organised as follows: Section II presents the theoretical modelling of the proposed HfO_2-based MIM diode, together with the final microwave layout of the sensor, suitable for on-wafer measurements; Section III describes the fabrication of the MIM diodes; Section IV is dedicated to the full DC and microwave characterisation of the devices, with a discussion about the results and future perspective. Finally, some conclusions are provided in Section V.

II. THEORETICAL MODELLING AND MICROWAVE LAYOUT OF THE HAFNIUM OXIDE BASED MIM DIODE

A. Theoretical Modelling

It is well known that, in a MIM diode, the thin insulator (oxide) layer between the two electrodes represents a potential barrier, with the electrons tunnelling through this barrier in about 10^{-12}–10^{-15} s. The tunnelling probability of electrons falls exponentially with insulator thickness. Moreover, thermally excited electrons can be emitted above the barrier as well. Possible solutions for increasing nonlinearity and asymmetry are using 1) different metals for the bottom and top electrodes (as in the proposed device) and 2) an asymmetric insulator in-between the two metals (which is technologically challenging and high-cost when combined with a thin thickness to enhance tunnelling at low power levels). A benchmarking physical model of the total current density J_{tot} accounts for both thermionic emission-limited J_{th} and tunnel-limited J_{tun} contributions, including the image force impact on the potential barrier. The "composite Simmons model" is described as follows:

$$J_{th} = A_{th}T^2 e^{-\varphi_1'/k_B T}\left(1 - e^{-qV/k_B T}\right) \tag{1}$$

$$J_{tun} = J_0\left[\hat{\varphi}e^{-A_{tun}\sqrt{\hat{\varphi}}} - (\hat{\varphi} + qV)e^{-A_{tun}\sqrt{\hat{\varphi}+qV}}\right] \tag{2}$$

$$J_{tot} = J_{th} + J_{tun}. \tag{3}$$

In the above equations, $J_0 = q/[2\pi h(\alpha\Delta s)^2]$, $A_{tun} = (4\pi\alpha\Delta s/h)\sqrt{2m}$ and $\hat{\varphi} = \frac{1}{\Delta s}\int_{s_1}^{s_2}\varphi(x)dx$. $A_{th} = 4\pi mqk_B^2/h^3$ is Richardson's thermal constant (where m is electron's mass, q is electron's charge, k_B is Boltzmann's constant, and h is Planck's constant), T is the temperature, φ_1' is the maximum barrier height above the negatively biased electrode, V is the voltage applied to diode's electrodes, α is a correction factor, $\Delta s = s_2 - s_1$ is the tunnelling distance (s_1 and s_2 are the positions where the potential energy barrier intersects the Fermi level of the negatively biased electrode), $\hat{\varphi}$ is the mean barrier height above the Fermi level of the negatively biased electrode, $\varphi(x) = \varphi_1 + (\Delta\Phi - qV)(x/t)$ is the potential energy between the electrodes, x defines the position coordinate inside the insulator, and t is insulator's thickness. Φ_{m1} and Φ_{m2} are the metal work functions of the top and bottom electrodes respectively, while φ_1 and φ_2 are the metal-insulator barrier heights ($\varphi_i = \Phi_{mi} - \chi_{ins}$, where χ_{ins} is insulator's electron affinity), then $\Delta\Phi = \Phi_{m1} - \Phi_{m2}$. Finally, $V_{tot} = \varphi(x) + (-0.288t)/[\varepsilon x(t-x)]$ is the total potential in the insulator (with permittivity ε), which takes into account the image potential between the two electrodes. The image potential has an important impact on the accurate modelling of thermionic emission-limited current, since it lowers the potential barrier. A schematic cross section of the MIM diode is shown in Fig. 1.

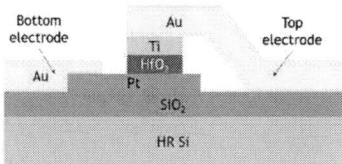

Fig. 1. Schematic cross section of the proposed hafnium oxide based MIM diode on HR Si/SiO₂ substrate.

In the present case of study, $\Phi_{Pt} = 5.65$ eV, $\Phi_{Ti} = 4.33$ eV, $\varepsilon \approx 20$, $\chi_{HfO2} = 2 \pm 0.25$ eV, and $t \approx 4$ nm. The contact area (i.e., MIM diode's area) is between 1×1 μm² and 3.5×3.5 μm². Different combinations of metals for the bottom and top electrodes are possible, but this entails a long parametrisation and optimisation process. For CMOS compatibility of the envisaged MIM-based RF sensor, the substrate is a high-resistivity silicon/silicon oxide (HR Si/SiO₂) wafer. An accurate implementation of (1)-(3) in MATLAB® gives an

estimation of the I-V and J-V curves, as displayed in Fig. 2. As J_{th} depends on the maximum potential barrier height, the relatively high barrier values in this example suppress the thermionic contribution. Therefore, the I-V curve is mainly controlled by the tunnelling phenomenon. We stress here that we chose the voltage range [-1,1] V because a MIM diode cannot work at high voltages due to the very thin oxide, in order to avoid junction destruction due to excess current.

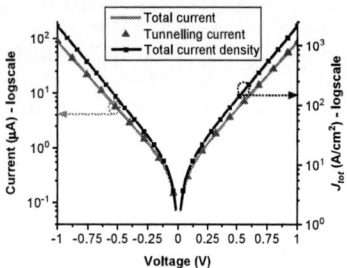

Fig. 2. Left vertical axis. comparison between the total current through an HfO₂ MIM diode (solid red curve) and the tunnelling current (blue triangles). The thermionic contribution is not displayed, as it is several orders of magnitude smaller. Right vertical axis: total current density J_{tot}.

B. Microwave Layout

In order to have a measurable device, we need to provide a layout which is suitable for on-wafer characterisation using standard coplanar waveguide (CPW) probe tips. Hence, the MIM diode needs to be embedded into a CPW line with a characteristic impedance of 50 Ω on HR Si/SiO₂ substrate. This way, the final geometry has a twofold purpose: 1) allows measuring the diodes in DC in a straightforward way, using a semiconductor characterisation system (SCS); 2) the diode can be connected directly to a microwave generator to provide the desired high-frequency signal at the input port, and to read the detected signal at the output port through an oscilloscope. Using CPW probe tips with high-quality coaxial cables is a guarantee that the contact resistance is very low, the influence of parasitic effects is reduced, and the supplementary attenuation in the chain generator-diode can be quantified precisely. The resulting layout is shown in Fig. 3.

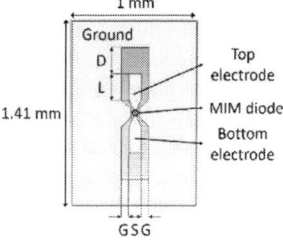

Fig. 3. Top view of the CPW-based device for on-wafer measurements.

As shown in Fig. 3, the device is only 1×1.41 mm², G = 60 μm and S = 100 μm (to ensure 50 Ω), L = 200 μm (to minimise parasitic inductance and series resistance of the lead lines), and D = 200 μm (to diminish the open-end capacitive effects between lead lines and ground). For the fabrication of the MIM diodes, a 525-μm-thick HR Si was used, on top of which a 300-nm-thick SiO$_2$ was thermally grown. The CPW structure was realized using a gold layer of 500 nm to limit skin depth effects at high frequencies.

III. FABRICATION PROCESS OF THE RF SENSORS

To fabricate the RF sensors, the bottom electrode was first defined by photolithography, a 20 nm Cr/100 nm Pt layer was deposited be e-gun evaporation, and the excess metal was lifted-off. After that, a layer of 100-nm-thick Si$_3$N$_4$ was deposited by plasma enhanced chemical vapour deposition (PECVD) everywhere, and an area around the actual MIM contact was defined by removing the Si$_3$N$_4$ over the remaining areas. This also exposed the edge of the bottom electrode. To define the active area of the contact, near the centre of the Si$_3$N$_4$ spacer, e-beam lithography was used (30 kV in a RAITH/JEOL 7000F system) and plasma etching of the Si$_3$N$_4$ using CF$_4$/O$_2$ plasma was performed. Subsequently, ALD of a HfO$_2$ thin film (< 4 nm) was carried out using a PicoSun R200 Advanced ALD rector with Tetrakis(ethylmethylamido)-hafnium (TEMAHf) and H$_2$O as oxidiser at 250°C. The HfO$_2$ layer was patterned using Cl$_2$/BCl$_3$ plasma etching. The top electrode was defined using a lift-off process and evaporation of 100 nm Ti/200 nm Au, thus creating the asymmetric MIM structure. Finally, a thick Au (500 nm) layer was deposited to define the CPW structure. This completed the fabrication process. Some optical and SEM pictures of the resulting devices are shown in Fig. 4.

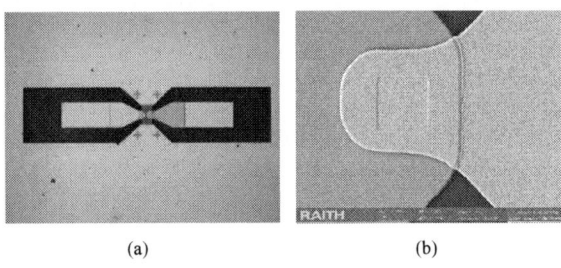

(a)   (b)

Fig. 4. (a) Optical picture of a generic RF sensor with the MIM diode integrated with a CPW line and (b) SEM magnification of the contact area. In Fig. 4b, the central square corresponds to the location of the HfO$_2$ thin film.

IV. DC AND MICROWAVE CHARACTERISATION OF THE HAFNIUM OXIDE BASED RF SENSORS

A. DC Characterisation

For the DC characterisation of the fabricated MIM diodes, we performed measurements at room temperature ($T = 290$ K) using a Keithley SCS 4200 station. The significant performance indicators of a MIM diode are the differential resistance R_D (in Ω), the noise equivalent power NEP (in pW/√Hz), the nonlinearity χ, and the sensitivity γ (in V^{-1}), defined as follows:

$$R_D = 1/(\partial I/\partial V) \tag{4}$$
$$NEP = \sqrt{4k_B T R_{D0}}/\beta \tag{5}$$
$$\chi = (\partial I/\partial V)/(I/V) \tag{6}$$
$$\gamma = (\partial^2 I/\partial V^2)/(\partial I/\partial V). \tag{7}$$

In (4), $R_{D0} = R_D(V = 0)$ and, in (5), β is the responsivity. In Figs. 5a-b and 6, we show a comparison between the

predicted I-V characteristic and the measured one for the case of a contact area of 2.5×2.5 μm^2 (which shows a nice agreement), and the extracted R_D-V and NEP-V curves, respectively. One can see the following: 1) the modelled I-V curve (using the composite Simmons model) overestimates the total current for absolute voltage values greater than 0.6 V, this may be attributed to uncertainties in the oxide thickness or defect assisted tunnelling in the dielectric that is not taken into account by the Simmons model, and to fabrication tolerances; 2) the extracted R_D is in the order of tens of kΩ (the maximum value being around 530 kΩ), which is consistent with the typical differential resistance of a MIM diode; 3) the maximum NEP is only 160 pW/√Hz, which is a very low value, optimal for low-noise microwave sensing of incoming EM interference.

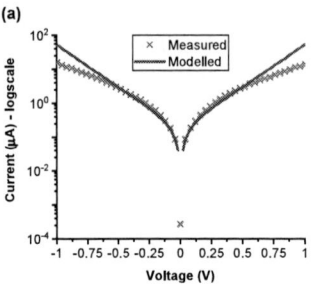

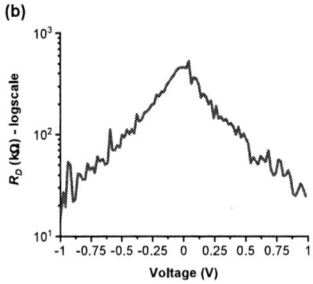

Fig. 5. (a) Measured (red crosses) and modelled (solid blue curve) I-V characteristics; (b) extracted R_D.

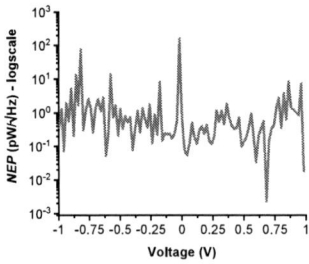

Fig. 6. Extracted NEP of the proposed HfO$_2$ based MIM diode.

Finally, $\chi_{max} \approx 4$ and $\gamma_{max} \approx 22$ V^{-1}, hence excellent values that make the proposed HfO$_2$ based MIM diode an ideal candidate for microwave sensing in wireless systems.

B. Microwave Characterisation

The microwave characterisation of the fabricated MIM diodes was carried out using an Agilent analogue signal generator, connected through a bias tee to one port of the device shown in Fig. 4a, whereas the other port was connected through a load resistor to a Tektronix oscilloscope. The load resistor had a value of 100 kΩ, very close to R_D, thus enhancing the output signal. The input signal was a microwave sinusoidal carrier modulated by a square signal of frequency f_{AM} = 1 kHz. The best results for the detector were obtained by using a small bias voltage of ±0.5 V, in order to move the operating point of the diode in the nonlinear region of its I-V characteristic. First, in Fig. 7 we show a screenshot of the detected voltage at 10 GHz (a frequency of interest for radar applications), in which one can notice how the pulses of the AM signal are well shaped. We also repeated the same measurements having in mind, as potential application, an indoor localisation system based on a Frequency-Modulated Continuous-Wave (FM-CW) radar in the 24 GHz (24–24.25 GHz) ISM band. The detected voltage maintained its shape with a similar dynamic range.

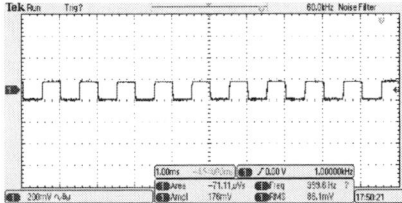

Fig. 7. Screenshot of the detected voltage at 10 GHz.

Finally, we measured the detected voltage as a function of the input power P_{IN}, at 10 and 24 GHz, thus obtaining the curves for the voltage responsivity R_V (in mV/mW), displayed in Fig. 8 for both cases of an applied bias voltage of -0.5 and 0.5 V. This figure shows that the maximum responsivity $R_{V,max}$ is about 500 mV/mW at both frequencies of interest (hence, an outstanding value for such devices based on nanoscale oxides) for an input power level of just -14 dBm. We stress here that we can expect even higher values for R_V if using a lock-in amplifier (which was not available in our setup). The maximum value of P_{IN} (between -2 and 2 dBm) was limited by power-induced nonlinear effects, resulting in strong oscillations and noise of the detected voltage.

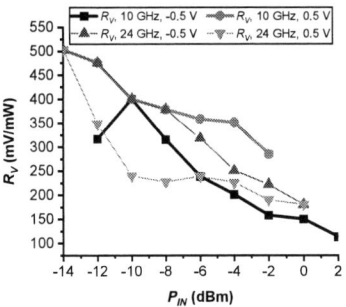

Fig. 8. Measured voltage responsivity of the proposed HfO$_2$ based MIM diode at 10 (black and red curves) and 24 GHz (blue and pink curves), for an applied bias voltage of -0.5 V and 0.5 V.

It can be also possible to estimate the cut-off frequency f_c of the proposed HfO$_2$ based MIM diode. Using an approach similar to the one in [11], we can write $f_c = 1/(2\pi RC)$, where

$R \approx 1$ Ω is the maximum (estimated) resistance of the lead lines (i.e., the CPW pads) and $C \approx 50\div550$ fF is the contact area's capacitance. With this data, f_c is between 300 GHz and 3 THz, depending on the contact area. Hence, the proposed HfO$_2$ based MIM diode is suitable for applications well beyond the millimetre-wave range.

V. CONCLUSION

In this manuscript, a state-of-the-art HfO$_2$ based MIM diode for microwave sensing has been presented. Starting from rigorous physical and computational modelling, the device was integrated with a CPW line for on-wafer DC and high-frequency characterisation. With a maximum current density of over 2×10^3 A/cm^2, a measured voltage responsivity of over 500 mV/mW at 10 and 24 GHz for an input power level as low as -14 dBm, and a minimum noise equivalent power of only 160 pW/$\sqrt{}$Hz, the proposed MIM diode demonstrated to be an excellent detector up to 40 GHz, with a cut-off frequency between 300 GHz and 3 THz.

ACKNOWLEDGMENT

This work was supported in part by the European Project H2020 ICT-07-201 "NANOSMART" under Grant No. 825430 and in part by two grants of the Romanian Ministry of Research, Innovation and Digitalization, CCCDI-UEFISCDI, under Project PN-III-P3-3.6-H2020-2020-0073 and Project PN-III-P2-2.1-PED-2019-0052, within PNCDI III.

REFERENCES

[1] E. W. Cowell III, N. Alimardani, C. C. Knutson, J. F. Conley Jr., D. A. Keszler, B. J. Gibbons, and J. F. Wager, "Advancing MIM electronics: amorphous metal electrodes," *Advanced Materials*, no. 23, pp. 74–78, 2011.

[2] J. G. Simmons, "Electric Tunnel Effect between Dissimilar Electrodes Separated by a Thin Insulating Film," *J. Appl. Phys.*, vol. 34, 2581, 1963.

[3] A. Sanchez, C. F. Davis, K. C. Liu, and A. Javan, "The MOM tunneling diode: theoretical estimate of its performance at microwave and infrared frequencies", *J. Appl. Phys.*, vol. 49, 5270, 1978.

[4] T. O'Regan, M. Chin, C. Tan, and A. Birdwell, "Modeling, Fabrication, and Electrical Testing of Metal-Insulator-Metal Diode", ARL-TN-0464 (2011).

[5] M. L. Chin, "Planar metal–insulator–metal diodes based on the Nb/Nb$_2$O$_5$/X material system," *Journal of Vacuum Science & Technology B*, vol. 31, 051204, 2013.

[6] S. Grover, and G. Moddel, "Applicability of Metal/Insulator/Metal (MIM) Diodes to Solar Rectennas," *IEEE J. Photovolt.*, vol. 1, no. 1, pp. 78–83, Jul. 2011.

[7] D. Matsuura, M. Shimizu, and H. Yugami, "High-current density and high asymmetry MIIM diode based on oxygen-non-stoichiometry controlled homointerface structure for optical rectena," *Sci. Rep.*, vol. 9, 19639, 2019.

[8] K. Bhatt, S. Kumar, and C. C. Tripathi, "Highly sensitive Al/Al$_2$O$_3$/Ag MIM diode for energy harvesting applications," *International Journal of Electronics and Communications*, vol. 111, 152925, Nov. 2019.

[9] S. Shriwastava, and C. C. Tripathi, "Metal–Insulator–Metal Diodes: A Potential High Frequency Rectifier for Rectenna Application," *Journal of Elec. Materi.*, vol. 48, pp. 2635–2652, 2019.

[10] M. Aldrigo, et al., "Harvesting Electromagnetic Energy in the V-Band Using a Rectenna Formed by a Bow Tie Integrated With a 6-nm-Thick Au/HfO$_2$/Pt Metal–Insulator–Metal Diode," *IEEE Trans. Electron Devices*, vol. 65, no. 7, pp. 2973–2980, Jul. 2018.

[11] M. Bareiß, et al., "High-Yield Transfer Printing of Metal–Insulator–Metal Nanodiodes," *ACS Nano*, vol. 6, no. 3, pp. 2853–2859, 2012.

Automatic Nonlinear Nonquasi-Static Diode Model Extraction from Large-Signal Measurements

A. García-Luque[#1], T. M. Martín-Guerrero[#2], A. Santarelli[*3], C. Camacho-Peñalosa[#4]

[#]Instituto Universitario de Investigación en Telecomunicación, Universidad de Málaga-Andalucía Tech, 29010 Málaga, Spain
[*]Department of Electrical, Electronic and Information Engineering (DEI), University of Bologna, 40126 Bologna, Italy
{[1]agl, [2]teresa, [4]ccp}@ic.uma.es, [3]alberto.santarelli@unibo.it

Abstract— An automatic nonlinear nonquasi-static (NQS) model extraction method is proposed. The procedure is based on a NQS charge that depends on two quasi-static (QS) functions: electric charge and delay. The use of Nonlinear Function Sampling (NFS) operator leads to a crucial link between real and imaginary parts of the spectral coefficients of the QS state functions. This approach is used to build-in a CAD (Computer-Aided Design) model that takes into account these NQS effects for a diode. The excellent numerical results allow to be optimistic in the integration of extracted models into CAD tools looking for emulating diode experimental behaviour.

Keywords— Diode, Nonquasi-static Nonlinear Modelling, Large-Signal, Convolution, and Nonlinear Function Sampling.

I. INTRODUCTION & MOTIVATION

Modern RF architectures (embracing the so-called *5G New Radio*) must respond successfully to a massive demand by guaranteeing small power consumption, spectral efficiency, high signal quality, lowest cost, automation and maximum integration [1].

The proliferation of novel AlGaN/GaN active components, the development of new CAD tools and the availability of powerful nonlinear measurement set-ups, such as Nonlinear Vector Network Analyzers (NVNAs), has completely shaken this sector up tending towards the development of SDR (Software Defined Radio) and modern reconfigurable antennas.

In this context, new co-design and manufacturing strategies are needed in order to emulate experimental responses, including the efficient modelling of active devices, which operate in conditions where non-linearity plays a main role.

The QS approach (state functions exclusively dependent of instantaneous junction voltages), which offers a good compromise between empirical and physical behaviour, speeds up modelling tasks and allows easy integration into CAD tools, is often no longer valid in modern high-frequency prototypes.

As a consequence of this demanding frequency increase, the finite times required for the electric charges to redistribute according to voltage changes need to be considered. These NQS phenomena characterisation has always been considered a significant challenge and, since years ago, several groups have suggested simplified NQS approaches to improve HBTs (Heterojunction Bipolar Transistors) [2] and FETs (Field-Effect Transistors) modelling [3].

Historically, all these works have in common that a large-signal dynamic model of electron devices can be built by considering a charge-based definition where the intrinsic port current $i(t)$ is the sum of a conduction $i_{con}(t)$ and a displacement $i_{dis}(t)$ component (nonlinear current and charge source, respectively):

$$i(t) = i_{con}(t) + i_{dis}(t) = i_{con}(v(t)) + \frac{dq_{NQS}(t)}{dt}, \quad (1)$$

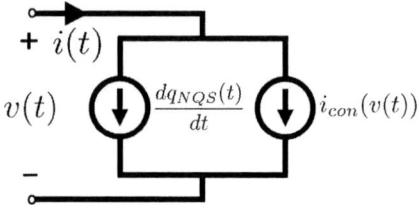

Fig. 1. NQS intrinsic equivalent one-port model adopted.

whose $q_{NQS}(t)$ is the port charge. One of the original NQS first-order approaches [3], [4], proposed for the displacement current of voltage-controlled devices, is adopted in this work:

$$\frac{dq_{NQS}(t)}{dt} = \frac{dq_{QS}(v(t))}{dt} - \frac{d[\tau_{QS}(v(t))i_{dis}(t)]}{dt}. \quad (2)$$

According with last definitions, it is considered the NQS diode model topology shown in Fig. 1, where:

- The intrinsic diode is characterised by a total current $i(t)$ and a junction voltage $v(t)$. Three QS functions are highlighted here (i.e., with dependence on instantaneous $v(t)$ only): $i_{con}(v(t))$ (conduction current component), $\tau_{QS}(v(t))$ (finite electric charge redistribution time) and $q_{QS}(v(t))$ (QS port electric charge).
- The displacement current $i_{dis}(t)$ is derived from the instantaneous NQS port charge $q_{NQS}(t)$, which relies on both QS functions: $\tau_{QS}(v(t))$ and $q_{QS}(v(t))$.

A technique for the extraction of the QS charge and delay functions in (2) is proposed in the next section.

II. NQS CHARGE EXTRACTION TECHNIQUE

A. Preliminary Definitions & Considerations

Focusing on (2), the goal of this work is to formulate an automatic extraction method to obtain consistent NQS models from large-signal device measurements. By integrating the displacement current (2), the $q_{NQS}(t)$ is obtained [5] as:

$$q_{NQS}(t) = q_{QS}(v(t)) - \tau_{QS}(v(t))\frac{dq_{NQS}(t)}{dt}. \quad (3)$$

3–4 April 2022, London, UK

This formulation is somehow equivalent to a relaxation time approach. Rearranging elements in (3), it is finally achieved:

$$i_{dis}(t) = \frac{dq_{NQS}(t)}{dt} = -\frac{q_{NQS}(t) - q_{QS}(v(t))}{\tau_{QS}(v(t))}. \quad (4)$$

By considering a periodic regime ($T = 1/f_0$), any signal involved and time-function in (3) can be expanded in Fourier series. With reference to a generic $x(t)$ function and by limiting the summation to N harmonics:

$$x(t) = \sum_{k=-N}^{N} X_k e^{(j\omega_k t)} =$$

$$= X_0 + \sum_{k \geq 1}^{N} \left[2\Re e[X_k] \cos(\omega_k t) - 2\Im m[X_k] \sin(\omega_k t) \right], \quad (5)$$

where $\omega_k = k(2\pi f_0)$, with $k = 0$ [DC], $\pm 1, ..., \pm N$, and X_k are the complex spectral Fourier coefficients.

The extraction procedure outlined in Fig. 2 aims at modelling the NQS charge (3) by extracting the unknown QS functions $\tau_{QS}(v(t))$ and $q_{QS}(v(t))$ from $i(t)$ and $v(t)$ measurements at intrinsic diode terminals, obtained with a high-frequency single-tone (f_0) large-signal characterisation, provided that the conduction current $i_{cond}(t)$ is preliminary de-embedded from $i(t)$ waveforms.

Different methods for obtaining the conduction current have been proposed. For instance, in [6] conduction and displacement current components are separated from measurements taken at frequencies where NQS effects are negligible, while in [7] conduction current is obtained from low-frequency characterisations. Once the $i_{con}(v(t))$ (along with any I_{con_k}) is known, the NQS charge coefficients Q_{NQS_k} can be evaluated from (1) expressed in spectral terms:

$$I_k = I_{con_k} + j\omega_k Q_{NQS_k}. \quad (6)$$

Before the formulation, the *mean relative error* $\overline{\delta}$, which will be useful to validate the extraction, is introduced:

$$\overline{\delta}[\%] = mean \left(\left| \frac{extracted\ value - expected}{expected} \right| * 100 \right) \quad (7)$$

B. Convolution Procedure

The periodic and limited-bandwidth functions in (3) can be turned into frequency domain according to:

$$Q_{NQS_k} = Q_{QS_k} - T_{QS_k} * dQ_{NQS_k}, \quad (8)$$

where these variables symbolise vectors of spectral coefficients of $q_{NQS}(t)$, $q_{QS}(t)$, $\tau_{QS}(t)$ and $dq_{NQS}(t)/dt$, respectively, item $*$ stands for the *convolution* operator and each vector follows this example size: $Q_{QS_k} = [Q_{QS_{-N}}, ..., Q_{QS_{-1}}, Q_{QS_0}, Q_{QS_1}, ..., Q_{QS_N}]$. The *convolution* operator in (8) can be expanded as:

$$Q_{NQS_k} = Q_{QS_k} - \underbrace{\sum_{l=-N}^{l=N} T_{QS_l} dQ_{NQS_{k-l}}}_{T_{QS_k} * dQ_{NQS_k}} \quad (9)$$

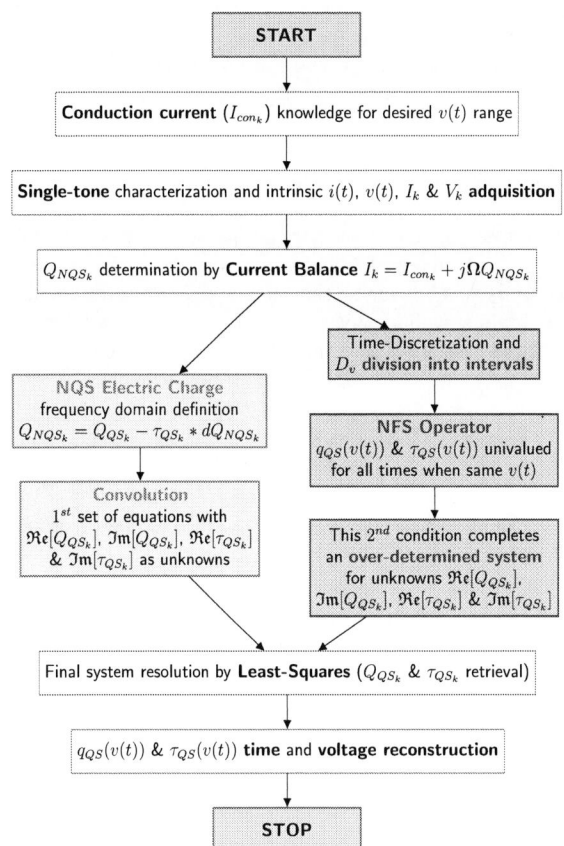

Fig. 2. Complete step-by-step NQS extraction procedure for one-port devices based on two key factors: *Convolution* (brown) and *NFS Operator* (blue).

Since coefficients with subscripts l and $-l$ are complex conjugates, by separating real and imaginary parts in (8), it is obtained:

$$\Re e[Q_{NQS_k}] + j\Im m[Q_{NQS_k}] =$$
$$\Re e[Q_{QS_k}] + j\Im m[Q_{QS_k}]$$
$$-(\Re e[T_{QS_0}] + j\Im m[T_{QS_0}])(\Re e[dQ_{NQS_k}] + j\Im m[dQ_{NQS_k}])$$
$$-\left(\sum_{l=1}^{l=N} \Re e[T_{QS_l}](A) + \Im m[T_{QS_l}](B) \right)$$
$$-j\left(\sum_{l=1}^{l=N} \Re e[T_{QS_l}](C) + \Im m[T_{QS_l}](D) \right), \quad (10)$$

where $A = \Re e[dQ_{NQS_{k-l}}] + \Re e[dQ_{NQS_{k+l}}]$, $B = -\Im m[dQ_{NQS_{k-l}}] + \Im m[dQ_{NQS_{k+l}}]$, $C = \Im m[dQ_{NQS_{k-l}}] + \Im m[dQ_{NQS_{k+l}}]$, and $D = \Re e[dQ_{NQS_{k-l}}] - \Re e[dQ_{NQS_{k+l}}]$.

Relations (10) represent a first set of $N+1$ equations in terms of the $4(N+1)$ unknown spectral coefficients: $\Re e[Q_{QS_k}]$, $\Im m[Q_{QS_k}]$, $\Re e[T_{QS_k}]$ and $\Im m[T_{QS_k}]$ for $k = 0, 1, 2, ..., N$.

C. Introducing the NFS Operator

Given a generic QS quantity $Z(v(t))$, known through $Z(t)$ and $v(t)$ waveforms, the *NFS operator* leads to extract the

375

nonlinear algebraic function $Z(v)$ out of the waveforms time samples. Application of the *NFS operator* will set up extra relationships between real and imaginary parts of Q_{QS_k} and T_{QS_k} allowing extraction of these unknowns, by means of standard least square solution algorithms of over-determined equation systems.

According to the example in Fig. 3 for a generic QS quantity $Z(v(t))$, the one-dimensional control voltage domain D_v (with excursion between v_{min} and v_{max}) is divided into several small intervals Δ_v of equal voltage length. Provided that Δ_v is small enough, each sample of $Z(t)$, corresponding to $v(t)$ falling within a given Δ_v interval, can be approximated by a unique value and associated with the average voltage value in that interval. Let us adopt the following notation:

- Let v_j (with $j = 1, 2, ..., N_v$) be the values of control voltage v, where the N_v samples of the functions: $q_{QS}(v)$ and $\tau_{QS}(v)$ will be evaluated
- Let t_j^p (with $p = 1, 2...P_j$) be each of the P_j time samples corresponding to $v(t_j^p)$ in the Δ_v interval around the v_j level.

On this basis, the following relation for $q_{QS}(v(t))$ is held verified for each voltage value v_j, (same for $\tau_{QS}(v(t))$):

$$q_{QS}(v_j) \approx q_{QS}(v(t_j^p)) \approx \frac{1}{P_j} \sum_p q_{QS}(v(t_j^p)). \quad (11)$$

By expanding $q_{QS}(v(t))$ in Fourier series as in (5) and applying the *NFS operator* as in (11), it is obtained (DC excluded, the same holding for T_{QS_k}):

$$\sum_{k=1}^{k=N} \left[2\Re[Q_{QS_k}]cos(\omega_k t_j^p) - 2\Im[Q_{QS_k}]sin(\omega_k t_j^p) \right] \approx$$
$$\sum_{k=1}^{k=N} \left[2\Re[Q_{QS_k}]\langle cos(\omega_k t_j)\rangle_p - 2\Im[Q_{QS_k}]\langle sin(\omega_k t_j)\rangle_p \right], \quad (12)$$

where the compact notation used for the *sin* and *cos* terms has the following meaning (same for the *cos* terms):

$$\langle sin(\omega_k t_j)\rangle_p \equiv \frac{1}{P_j} \sum_p sin(\omega_k t_j^p). \quad (13)$$

Taking into account very small voltage intervals Δ_v, the deviations between actual and averaged trigonometric function samples can be considered almost negligible (same holding for the *cos* terms), i.e.:

$$\underbrace{\Delta sin_k(t_j^p)}_{ERROR} = \underbrace{sin(\omega_k t_j^p)}_{ACTUAL} - \underbrace{\langle sin(\omega_k t_j)\rangle_p}_{AVERAGE} \approx 0. \quad (14)$$

Due to this reason, (12) can be finally rewritten as:

$$\sum_{k=1}^{k=N} \left[2\Re[Q_{QS_k}]\Delta cos(\omega_k t_j^p) - 2\Im[Q_{QS_k}]\Delta sin(\omega_k t_j^p) \right] \approx 0 \quad (15)$$

A similar relation can be stated for T_{QS_k}. To sum up, spectral coefficients of QS charge and delay must fulfil an over-determined system of equations given by:

- $N + 1$ *convolution* equations (10)

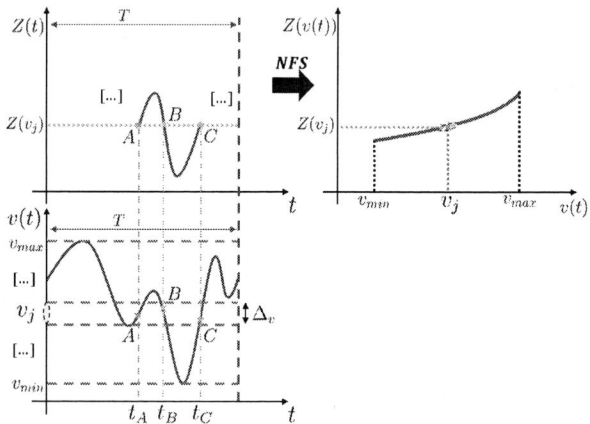

Fig. 3. Samples A, B and C of the periodic $v(t)$ are allocated into the yellow interval (v_j average) and offer same $Z(t)$. In voltage domain, *NFS* allows a single v_j value (X axis) and a common $Z(v_j)$ (Y axis) for all these ones.

- $2N_v$ *NFS*-based equations (15), corresponding to $q_{QS}(v(t))$ and $\tau_{QS}(v(t))$ and each voltage value v_j with $j = 1, 2, ..., N_v$.

III. NUMERICAL EXTRACTION EXAMPLE

A preliminary validation of the proposed technique is given here in terms of a numerical MatLab™ test. A $q_{NQS}(t)$ waveform is analytically calculated by exploiting literature-based expressions for QS functions $q_{QS}(v(t))$ and $\tau_{QS}(v(t))$ and by assuming $v(t)$ as a known non-sinusoidal periodic waveform. The suggested method is then used for function reconstruction and validation.

In particular, a $q_{QS}(v(t))$ diode foundry model was supposed from [8] along with a Schottky Gate FET $\tau_{QS}(v(t))$ from [9]. This $\tau_{QS}(v(t))$ corresponds to about 30% of the signal period duration $T = 1/f_0$. In addition, the intrinsic $v(t)$ has been defined by the following periodic signal (emulating large-signal experimental conditions):

$$v(t) = 0.93 + 0.49[sin(w_0 t + \pi) + 0.3cos(w_0 t + \pi/3)$$
$$+ 0.1sin(2w_0 t) + 0.25cos(2w_0 t) + 0.15sin(3w_0 t)]. \quad (16)$$

As a first-step, the NQS charge was computed from both QS functions using (8) in vector form:

$$Q_{NQS_k} = [\mathbf{I_D} + j\mathbf{T_c}\mathbf{\Omega}]^{-1} Q_{QS_k} \quad (17)$$

where $\mathbf{I_D}$ being the identity matrix, $\mathbf{\Omega}$ a diagonal-matrix of harmonic frequencies ($k2\pi f_0$ for $k = 0, \pm 1, ..., \pm N$) and $\mathbf{T_c}$ a matrix which performs the *convolution* procedure.

Preliminary voltage-domain curves and waveforms computed from the starting data are plotted in Fig. 4 for $f_0 = 2$ GHz, $M = 10000$ time-samples and $N = 15$.

With $dQ_{NQS_k} = j\mathbf{\Omega}Q_{NQS_k}$, the over-determined system from (10) and (15) (including the equations for T_{QS_k}) was built. The Δ_v intervals and the j integer (interval-ID) corresponding to a generic v waveform value were evaluated by means of:

$$\Delta_v = \frac{(v_{max} - v_{min})}{N_v} \qquad j = \left\lceil \frac{v - v_{min}}{\Delta_v} \right\rceil \quad (18)$$

where $\lceil\ \rceil$ indicates rounding to the upper integer. The over-determined system was solved by standard *Least-Squares* algorithms achieving full extraction of $q_{QS}(v(t))$ and $\tau_{QS}(v(t))$, which are compared versus the original ones.

Fig. 5 shows the extraction results at $f_0 = 2$ GHz for different choices of N_v. $M = 10000$ time-samples and $N = 15$ harmonics were considered for input data. It can be seen how the number of voltage intervals (N_v) affects the extraction accuracy. Good results (i.e. with small $\bar{\delta}$) are achieved with sufficiently high N_v numbers ($\Delta_v = 1.6e - 6$ V).

The test was repeated at different frequencies. Fig. 6 displays results for $f_0 = 1$ GHz and $f_0 = 4$ GHz. By increasing the frequency, NQS effects are more evident and better accuracies are achieved at lower N_v numbers.

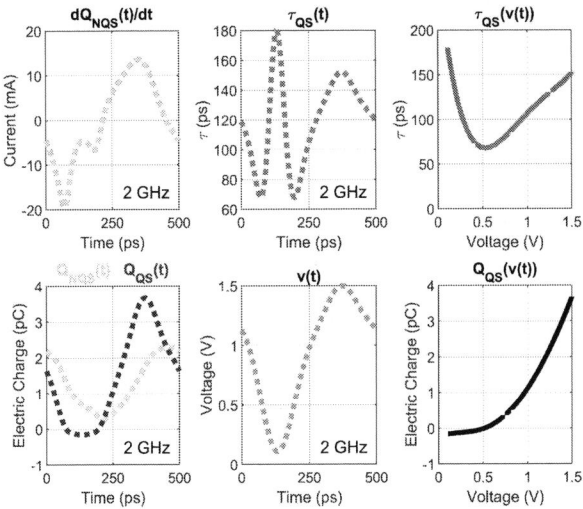

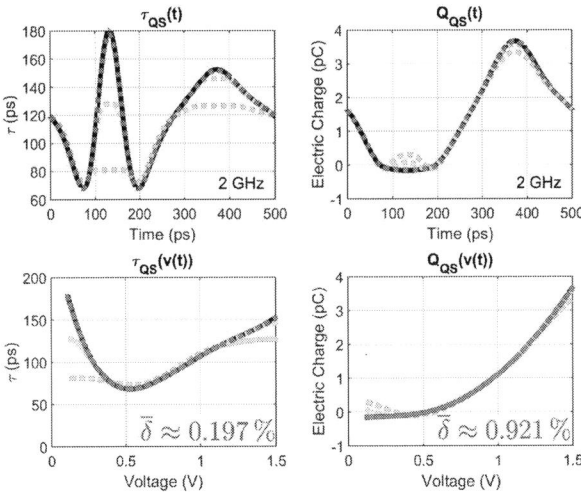

Fig. 4. Previous extraction voltage curves and waveforms ($f_0 = 2$ GHz).

Fig. 5. QS extracted functions for $f_0 = 2$ GHz with $M = 10000$ samples and $N_v = 10000$ (green), 50000 (cyan), 900000 (pink) and originals (black).

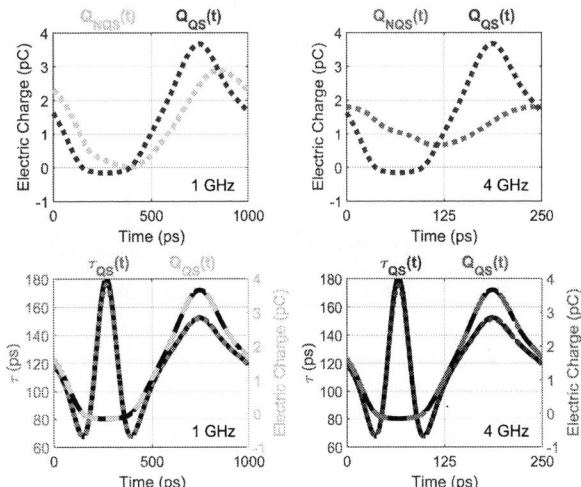

Fig. 6. Results with $M = 10000$ samples for $f_0 = 1$ GHz ($N_v = 1000000$) and $f_0 = 4$ GHz ($N_v = 700000$). Extracted (color) and originals (black).

IV. CONCLUSION

An automatic extraction method for NQS diodes has been presented. The technique relies on a NQS charge model using quasi-static functions of charge and relaxation time. The procedure, which involves *convolution* and *NFS*-based equations, has been preliminarily tested with a numerical case and allows optimism for an upcoming CAD tool integration looking for accurate validation under large-signal conditions.

ACKNOWLEDGMENT

This work has been supported by the I Plan Propio de Investigación, Transferencia y Divulgación Científica de la Universidad de Málaga under the PhD Fellowship code-401.

REFERENCES

[1] Qualcomm™, "Future of 5G." Feb. 2020, [Accessed: 03.16.2021]. [Online]. Available: https://www.qualcomm.com/news/media-center

[2] A. Ouslimani et al., "Nonquasi-Static Large Signal Transient Model for Heterojunction Bipolar Transistors," in *1999 IEEE MTT-S International Microwave Symposium Digest*, vol. 2, 1999, pp. 427–430 vol.2.

[3] P. Roblin et al., "Improved Small-Signal Equivalent Circuit Model and Large-Signal State Equations for MOSFET/MODFET Wave Equation," *IEEE Trans. Electron Devices*, vol. 38, no. 8, pp. 1706–1718, 1991.

[4] R. Daniels et al., "A Universal Large/Small Signal 3-Terminal FET Model using a Non-Quasi-Static Charge-Based Approach," *IEEE Trans. Electron Devices*, vol. 40, no. 10, pp. 1723–1729, 1993.

[5] M. Foisy et al., "Large Signal Relaxation Time Model for HEMTs," in *IEEE MTT-S Microwave Symposium Digest*, 1992, pp. 251–254 vol.1.

[6] T. M. Martín-Guerrero et al., "Automatic Extraction of Measurement Based Large Signal FET Models by Nonlinear Function Sampling," *IEEE Trans. Microw. Theory Techn*, vol. 68, no. 5, pp. 1627–1636, 2020.

[7] V. Vadalà et al., "A New Dynamic-Bias Measurement Setup for Nonlinear Transistor Model Identification," *IEEE Trans. Microw. Theory Techn*, vol. 65, no. 1, pp. 218–228, 2017.

[8] Keysight Tech.™, "Custom Modeling with Symbolically Defined Devices," Jan. 2008, [Accessed: 03.15.2021]. [Online]. Available: edadocs.software.keysight.com/display/ads2008U1/User-Defined+Models

[9] T. Suemitsu, "An Intrinsic Delay Extraction Method for Schottky Gate FETs," *IEEE Electron Device Letters*, vol. 25, no. 10, pp. 669–671, 2004.

Proceedings of the 16th European Microwave Integrated Circuits Conference

Compact GaN RF-Switches for Power Applications

Samira Driad[#], Charles Teyssandier[#], Laurent Caille[#], Christophe Chang[#], Laurent Brunel[#], Benoit Lambert[#], Hermann Stieglauer[*], Valeria Brunel[#]

[#]UMS SAS, France
[*]UMS GmbH, Germany

{samira.driad, charles.teyssandier, laurent.caille, christophe.chang, laurent.brunel, benoit.lambert}@ums-rf.com,
{hermann.stieglauer, valeria.brunel}@ums-rf.com

Abstract — Transistors with large gate-to-gate distance (gate pitch) are commonly used for High Power Amplifier (HPA) to keep the channel temperature below the critical level. However, the large pitch in switching application shows to prone insertion losses that could be penalizing. In addition, a large gate to drain distance leads to high breakdown voltage but could affect the On-resistance (Ron). This work investigates the best trade-off for transistor gate pitch and gate-to-drain dimensions to optimize the performances while keeping promising power handling capability.

Keywords — RF-GaN switch, insertion losses, high isolation, device electrical modelling.

I. Introduction

Radio Frequency (RF) switches are crucial components for wireless communication systems and radars [1], [2]. In addition, high switching power is important for the overall systems: emitters, drivers, phase shifters etc. To achieve high system's performances, the RF-switch should combine: low insertion losses, high isolation in a broad frequency range with low DC power consumption.

GaN FET technologies based RF-switches are very attractive due to several unique features comparing to other technologies ([3], [4], and [5]). Combining the known advantages of the AlGaN/GaN hetero-junction (high electron mobility and high voltage) with the compact design of the transistor by reducing the gate pitch and the gate to drain (GD) distance, we have recently demonstrated an effective improvement in the insertion losses and the isolation with promising power handling capabilities.

In this paper, after a quick description of UMS GH25-10 technology and a reminder of the switching operation mode, an overview of the RF-characterization at ambient and at high temperature is presented. The insertion losses, the breakdown voltage and power handling versus the geometry of the transistor are evaluated to highlight the performances versus the topology. The associated electro-thermal model is also implemented and successfully validated with the measurement.

II. GH25-10 Switch technology and Operating Mode

A. Technology Description

The GH25-10 technology is based on an AlGaN/GaN epitaxy on -SiC substrate. The AlGaN/GaN High Electron Mobility Transistor (HEMT) fabrication starts with standard ohmic contacts. Then, after device isolation, the T-shaped gates of 0.25μm footprint are defined and passivated. After

that, the front-side process is completed by passive elements in order to allow MMIC designs. BCB coating is added as a mechanical protection. The wafer is finally thinned down and followed by a via-hole process for source grounding

The UMS industrial GH25-10 is also space evaluated.

B. RF-Switch Operating Mode

The RF-FET switches are based on the Field-Effect Transistors used as the switching elements. GaN transistor in switch configuration is a three-port device with a centered gate layout, symmetric with respect to source and drain. This symmetry is practically the only difference with respect to a conventional transistor for PA applications. The drain–source channel is controlled by the gate voltage VGS:

- When VGS= 0 V, the switch is ON: the switch is determined by the ON-state resistance between drain and source and by the drain–source parasitic and capacitive effects, whose influence increases with frequency
- When VGS < VTH, the switch is OFF, and its isolation is limited by the depletion and parasitic capacitances.

Then, basically, the equivalent circuit is based on the resistance in parallel with the off-capacitor as shown the Fig 1.

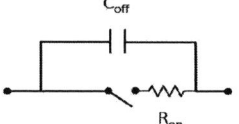

Fig. 1. Basic switch equivalent circuit.

Ron gives the insertion losses at the On-state mode. Then, the lower is Ron, the lower the insertion losses are. The Coff is related to the isolation level: the lower is the Coff, the higher is the switch isolation. The main Figure of the Merit (FoM) of the RF-switch is defined as in the following:

$$FoM = \frac{1}{Ron * Coff} \quad (1)$$

III. Methodology and Characterisation

A. Methodology

The topology of the FET-switches should be optimised to meet good isolation with low insertion losses while saving a good power handling. In this work, in addition to the standard switch topology directly driven by the one for HPA, some

3–4 April 2022, London, UK

topologies are designed to find the compromise in terms of gate to gate (GG) pitch and gate to drain/source spacing. The so-called standard topology has a GG pitch of 40 μm and a gate to drain/source (GD=GS) spacing of 1.7 μm. The pitch is then reduced to 25 μm for the compact topologies with various gate to drain/source spacing: 1 μm, 1.2 μm, 1.4 μm, 1.7 μm, 2 μm. Fig. 2 reports the standard and the downsized cells in terms of GG-pitch used for this study: the transistors are in series switch configuration and present an integrated large series resistor (typically a few kilo-ohm) at the gate terminal as RF choke. Both have here GD=GS of 1.7μm.

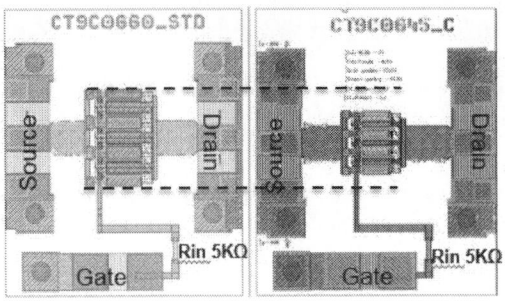

Fig.2. (left) standard (STD) RF Switches with GG-pitch of 40μm and, (right) compact RF-Switches with GG-pitch of 25μm and both with GD=GS=1.7μm.

B. S-Parameter Characterization

On-wafer S-Parameter measurements are carried out up to 20 GHz at the ambient temperature at Vds=0 V, Vgs= -40 V to Vgs= 1 V. Then, the On-resistance and the Off-capacitance are extracted @ 2GHz respectively according to the following equations (2) and (3):

$$Coff\left(\frac{pF}{mm}\right) = [((Im(Y22) + Im(Y12))/\omega]/W \qquad (2)$$

$$Ron\ (\Omega. mm) = \left[\frac{1}{(Real\ (Y22) + Real\ (Y12))}\right] * W \qquad (3)$$

where ω=2*π*f, f is the frequency, W is the total gate width.

Fig 3 shows the Off-capacitance (Coff) results obtained on both: the compact gate pitch switch CT9C00660_C (25μm) and the standard switch CT9C0660_Std (40μm).

The result shows that the reduction of the gate pitch from 40 μm to 25 μm allows the reduction of the Coff and the improvement of the FoM by 11% by considering the Ron of 2.29 Ω.mm for the gate-drain spacing of 1.7μm for both. As expected, the reduction of the pitch improves directly the switch isolation.

On the other hand, the S-parameters performed on the compact switches with various gate-drains spacing allow the extraction of the On-resistance and the Coff at 2GHz.

Table.1 shows the evolution of the Off-capacitance and the On-resistance versus the gate to drain distance for the compact pitch of 25μm.

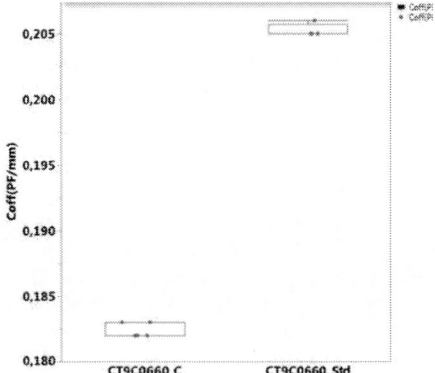

Fig. 3. Coff comparison between the standard gate pitch (40μm) and the downsized switch pitch of 25μm, gate to drain distance of 1.7μm for both.

Table 1. Ron and Coff of the downsized RF-Switches pitch versus the gate to drain distances @ 2GHz.

Device 8x100μm	GD=1μm	GD=1.2μm	GD=1.4μm	GD=1.7μm
Coff (PF/mm)	0.192	0.188	0.183	0.185
Ron (Ω.mm)	2.004	2.06	2.18	2.33

The results of the table.1 show that the GD distance has quite low impact on the Coff. Conversely, the Ron is sensitive since it increases when the distance GD increases. Consequently, the small GD distances are more relevant when looking for low insertion losses. The FoM of the compact switch versus the GD distances is given on the fig 4.

The smallest gate-drain spacing is more relevant in terms of the FoM. However, the small GD distance should impact the Breakdown voltage and the power handling capability of the switch.

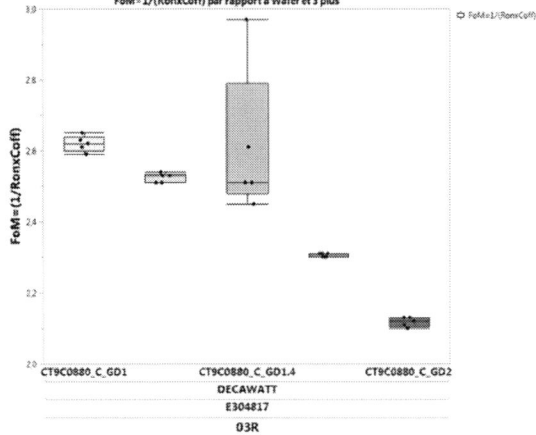

Fig.4. Figure of Merit of the 8x100μm RF-Switch versus GD distance with a GG-pitch of 25μm.

To find a compromise between the Figure of Merit (in terms of low losses with high isolation) and the breakdown voltage capability further DC measurements are performed and discussed in the next section.

C. DC Breakdown Voltage Measurement

DC measurements are carried out on the Keithley test bench at Vds=0 V and Vgs sweep from 0 V to -200 V with the step of 1 V in the temperature range of 25°C to 125°C by the step of 25°C. Then, the breakdown voltage is extracted at Id of 3 mA/mm. The measurement configuration and the main results @25°C versus the GD spacing (1, 1.2, 1.4, 1.7, and 2) µm are given in the following:

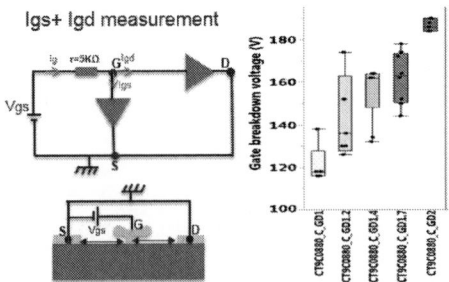

Fig. 5. (Left) Set-up configuration, (right) Breakdown voltage results versus the GD spacing extracted at 25°C and at 3mA/mm on 8x100-µm FET.

As expected the breakdown voltage extracted for Id of 3mA/mm increases with the increase of GD spacing. This is due to the better field distribution in the larger GD spacing allowing to reach over than 180V breakdown voltage for GD of 2µm (CT9C0880_C_GD2 devices) @ 25°C. However, the reverse leakage current increases with the temperature resulting in the degradation of the breakdown voltage.

The breakdown measurements are carried out in the range of 25°C to 125°C by the step of 25°C. The medium values obtained on the measured devices are taken into account. From 25°C to 125°C, the decrease of the breakdown voltage starts to be significant when the temperature exceeds 75°C with an increase in the leakage dispersion between devices. The minimum breakdown voltage obtained after 125°C is around 80V while it stills around 110V till 75°C.

The breakdown results obtained versus the temperature allow the definition of the maximum rating set parameters of the GH25-10 RF-switches as shown in the following table for both the compact and the standard GG-pitch with respectively GD spacing of 1µm and 1.7µm, table .2.

Table 2. Maximum rating of the GH25-10 RF-FET-Switches.

Devices	Parameters	voltage	ROR	AMR
CT9C0880 (GG40,GD1.7)µm	Vgd/s DC+RF	Vgs/d=-30V	-60V	-80V
CT9C0880 (GG25,GD1)µm	Vgd/s DC+RF	Vgs/d=-30V	-55V	-60V

D. Power Handling Capability

Maximum power handling for RF switches is the power level that will correspond to the onset of gain compression. Once the calculated power handling is exceeded, the insertion loss of the switch starts to go up.

The RF power handling for an off-state FET-switch is maximized for a symmetric topology with Vds= 0 V, for the gate bias voltage centred between the breakdown and the pinch-off voltage.

$$Pmax_{off}(W) = \frac{(VBD - VP)^2}{(2 * Z0)} \qquad (4)$$

Z_0: is Z_{load}=50 Ω, Vp= Vgs at Ids=100 mA/mm, V_{BD} is the measured breakdown voltage.

Based on the equation "(4)", the results show that from 1 µm to 1.7 µm of gate-drain spacing, the theoretical $Pmax_{off}$ increases from 108 W to 189 W with a ratio of 54%, showing that the 1.7 µm GD spacing is needed for high power applications. However, the power handling for the smallest GD spacing (1 µm) stills relevant for applications that require low insertion losses. For safe estimation of the switch power handling, the defined maximum rating shown in table.2 should be taken into account. The UMS RF-FET-switches are available with both 1 µm and 1.7 µm gate-drain spacing when high power handling is critical.

IV. NONLINEAR MODEL VALIDATION

GH25-10-UMS technology provides two cold FET transistor topologies for switch applications: TZ9C transistor for the parallel switch and CT9C for serial switch. Both in standard topology (pitch of 40 µm, GD spacing of 1.7 µm) and in the compact switch topology (gate-pitch of 25 µm, GD spacing 1µm and 1.7µm). Consequently, two new compact topologies models are available in the GH25_UMS PDK. Then the switch: GH25CNSSW_10 for serial and GH25CNCF for parallel configurations.

The UMS nonlinear Cold-FET models contained into the Design Kit is multi-bias and scalable in terms of the number of fingers and the gate width as follows:

Table 3. Validity domain of the GH25-10 model.

	Gate width
	30 µm -100 µm (up to 150 µm for 1 finger topology)
Number of gate finger	1,2 4, 6,8
Gate-drain spacing	1 µm or 1.7 µm
Gate to gate pitch	25 µm or 40 µm
Frequency	0.5 to 30 GHz

The gate voltage bias can be chosen from -25 V to 1.8 V, in the temperature range of 25 °C to 100 °C. The cold-FET nonlinear model relies on both S parameters and pulsed I-V measurements, done at Vds = 0 V and at different gate voltage quiescent points. The nonlinear current sources Ids, Igd (gate-drain diode) and Igs (gate-source diode) are deduced from pulsed I-V characteristics, obtained by excursion around the bias point using very short pulses (≤1 µS).

The model validation of the compact switch is shown on the following figures. The insertion losses and the isolation are given on the figure 6 for the switch compact with GG pitch of 25µm and GD=1µm.

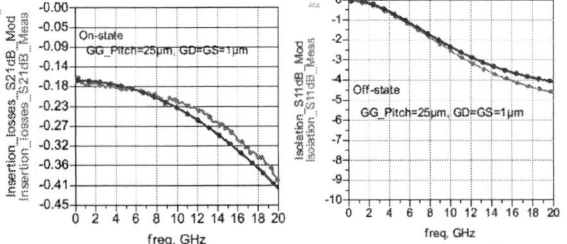

Fig. 6. (Blue line) model, (Red line) measurement: (left) insertion loss at on state, (right) isolation versus frequency, on 8x100µm switch-FET width.

We can see the good switch performance obtained and the good agreement between the nonlinear electrical model and the measurement. On the both following charts the measured Coff of the compact pitch 25µm (red line) and the standard one of 40µm (green line) are plotted. The model is simulated first with pitch of 40µm on the left and then with the pitch of 25µm on the right

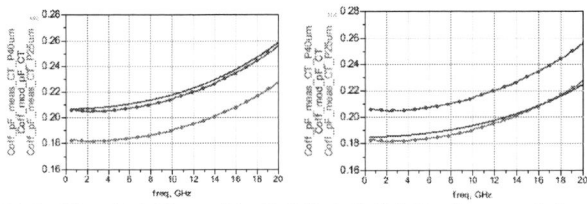

Fig.7. (Blue line) The model, (Red line) Coff (pF/mm) extracted from measurement @ 2 GHz of the compact switch of 20µm, (green line): standard pitch of 40µm. Both with GD spacing of 1.7µm.

We can see that the model reproduces accurately the behaviour of the Coff for both topologies since the simulation is in a good agreement on the left with the standard pitch and also on the right with the compact gate pitch.

The model accuracy is also well achieved at the On-state for both gate to drain spacing of 1.7 µm and 1 µm as shown on fig8.

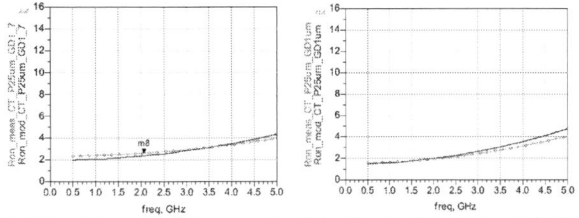

Fig.8. Ron (ohm.mm) of the compact switch with gate pitch of 25µm @2GHz, (left): GD spacing of 1.7µm. (right): GD spacing of 1µm. (Blue line) the model, (Red line) measurement.

V. CONCLUSION

This paper deals with the investigation of the optimum HEMT GaN topologies to reach the best trade-off between the low insertion losses, the high isolation and the high breakdown voltage to allow a good power handling capabilities.

After device fabrication measurements are carried out to evaluate the device's performance and select the relevant topology regarding the targeted applications. The evaluation of the new compacted devices demonstrates state of the art performances with regard to insertion loss, bandwidth and RF power handling capability.

The results presented here strongly suggest that GaN transistor technology is well suited to high power RF switching applications. In addition to good technology performance, an electrical model was developed to allow designing with GH25-10 cold FET technology. The validation of the model carried out by comparison with measurement shows a good agreement and relevant accuracy level.

REFERENCES

[1] Hangai, M., Nishino, T., Kamo, Y. and Miyazaki, M., "An Sband100W GaN Protection Switch," IEEE MTT-S Int.Microwave Symp. Dig., 2007, pp. 1389-1392

[2] Ishida, H., Hirose, Y., Murata, T., Ikeda, Y., Matsuno, T., Inoue,K., Uemoto, Y., Tanaka, T., Egawa, T., and Ueda, D., "A High-Power RF Switch IC Using AlGaN/GaN HFETs With Single-Stage Configuration IEEE Transaction on Electronic Devices, vol. 52, no. 8, pp. 1893- 1899, August 2005.

[3] Campbell, C. F., and Dumka, D. C., "Wideband High Power GaN on SiC SPDT Switch MMICs," IEEE Mtt-S Int Microwave Symp. Dig., 2010, pp. 145-148Microwave Symp. Dig., 2007, pp. 1389-1392

[4] Ma, B. Y., Boutros, K. S., Hacker, J. B., and Nagy, G., "High PowerAlGaN/GaN Ku-Band MMIC SPDT Switch and DesignConsideration," IEEE Mtt-S Int. Microwave Symp. Dig., 2008, pp. 1473-1476

[5] S. Shukla and J. Kitchen, "GaN-on-Si switched mode RF power amplifiers for non-constant envelope signals," 2017 IEEE Topical Conference on RF/Microwave Power Amplifiers for Radio and Wireless Applications (PAWR), 2017, pp. 88-91, doi: 10.1109/PAWR.2017.7875581.

Proceedings of the 16th European Microwave Integrated Circuits Conference

Analysis of RF Stress Influence on Large-Signal Performance of 22nm FDSOI CMOS Transistors utilizing Waveform Measurement

Dang Khoa Huynh[†*], Quang Huy Le[†*], Steffen Lehmann[#], Zhixing Zhao[#], Germain Bossu[#], Wafa Arfaoui[#], Defu Wang[†], Thomas Kämpfe[†], Matthias Rudolph[*]

[†]Center Nanoelectronic Technologies, Fraunhofer IPMS, Germany
[*]Ulrich-L.-Rohde Chair of RF and Microwave Techniques, Brandenburg University of Technology, Germany
[#]GlobalFoundries, Germany
dang.khoa.huynh@ipms.fraunhofer.de

Abstract — The following study employs RF waveform engineering to monitor degradation in 22nm FDSOI transistor at high-frequency region. The current and voltage waveforms are measured, reconstructed, and de-embedded to the device's intrinsic during large-signal CW RF stress testing. This technique provides extra information on device performance compared with standard DC and RF figures of Merits degradation. With clear pictures of where on the output IV plane the degradation is occurring, device designers can get an insight into the degradation behavior limiting RF performance. It is observed that devices show a different behavior under RF stress in comparison to DC-stress-induced degradation.

Keywords — hot-carrier, RF, reliability, large-signal, FDSOI, 22FDX®, 5G.

I. INTRODUCTION

Fully depleted silicon-on-insulator (FDSOI) technology has been matured over the past years to enhance high-frequency performance. Therefore, FDSOI has become more attractive as a cost-effective solution for automotive and 5G application. With sophisticated modulation schemes, the transistor devices operating as power amplifiers (PAs) are often subjected to high-peak voltage and current far exceeding the nominal value. Hence, it has brought more challenges to the reliability issues. Such stress of FDSOI devices conducting in practical RF environment has been presented in [1].

In this work, a new RF reliability analysis method for monitoring degradation of RF transistors is presented. Utilizing vector network analyzer (VNA)-based harmonic load-pull measurement, the RF waveforms presented to devices during large-signal continuous wave (CW) RF stress testing can be monitored to emulate practical PAs environment. This gives the ability to get a unique insight into the degradation behavior.

II. RF RELIABILITY METHODOLOGY

The devices under test (DUTs) are the thin-oxide (SG) super-low threshold voltage (SLVT) p-channel FDSOI with total width $W_{tot} = 16$ μm. The test structures are provided in common-source configuration with back-gate tied to ground and embedded in ground-signal-ground fixtures (GSG).

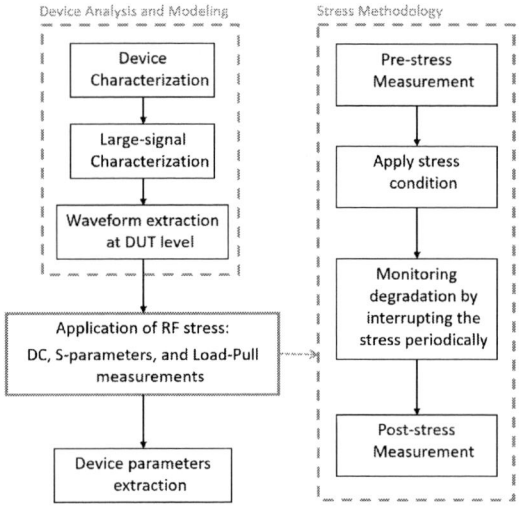

Fig. 1. Diagram of the proposed methodology to address reliability in the RF domain.

A. Measurement Setup

The diagram of Fig. 1 illustrates the methodology developed to address reliability in the RF domain. The device is first characterized by a set of DC, S-parameters, and load-pull measurements. An empirical large-signal model can be extracted from those sets of measurement data. Because reliability analysis is often conducted at high power and high bias conditions, load-pull measurement risks pre-stressing the devices when sweeping impedances to find the optimum points. Therefore, the extracted model is used to determine the source and load conditions for the RF stress applied to a fresh device.

Fig 2 shows the vector-receiver load-pull system used in this study. It consists of a Keysight N5247B VNA and Maury automated impedance tuners. Because of narrow bandwidth behavior, the VNA measurement systems lose phase relationships when the local oscillator (LO) is frequency shifted from the fundamental tones to the harmonic tones. In this system, one comb generator is used to calibrate the

3–4 April 2022, London, UK

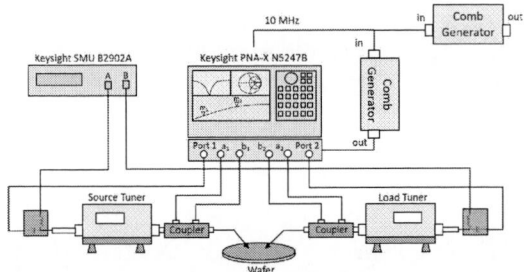

Fig. 2. Vector-receiver load-pull measurement setup.

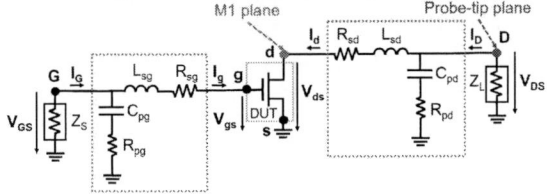

Fig. 3. Simplified fixture parasitic model.

absolute phase at the measurement plane, while the other serves as a phase reference once the LO has been shifted [2]. This setup enables measurement of the calibrated incident and reflected power waves at the different harmonic tones. We can get the RF waveforms versus different power levels and impedances directly at the extrinsic planes, by calculating the combination of those power waves.

B. Large-Signal Waveform De-embedding

De-embedding those large-signal power waves is more complex than in small-signal cases because the measurement is taken in non-50Ω environment. The equation to transform the vector harmonic power waves a1, a2, b1, b2 measured by the load-pull setup into equivalent voltage and current vectors are given as follows [3]:

$$V = V_{DC} + \sum_{i=1}^{n} \sqrt{Z_0}(a_i + b_i) \quad (1)$$

$$I = I_{DC} + \sum_{i=1}^{n} \sqrt{1/Z_0}(a_i - b_i) \quad (2)$$

where V_{DC} and I_{DC} are measured DC voltage and current, i is an index for a harmonic number.

Once converted these voltage and current waveforms can be de-embedded to the metal layer M1 using a simplified fixture parasitic model and the following equations [2]:

$$I_g = I_G - \frac{V_{GS}}{Z_{Cpg}} \quad (3)$$

$$I_d = I_D - \frac{V_{DS}}{Z_{Cpd}} \quad (4)$$

$$V_{gs} = V_{GS} - I_g(R_{sg} + Z_{Lsg}) \quad (5)$$

$$V_{ds} = V_{DS} - I_d(R_{sd} + Z_{Lsd}) \quad (6)$$

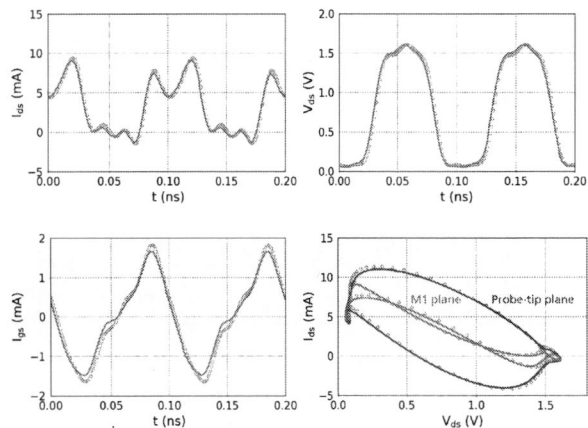

Fig. 4. Waveforms reconstructed/de-embedded from measurement (symbol) and simulation (solid line) at $f_0 = 10$ GHz, $P_{avs} = -5$ dBm ($V_{DS} = 0.8$ V, $I_{DS} = 2$) mA.

The fixture parasitic components depicted in Fig. 3 can be extracted from S-parameters measurement of the open and short structures respectively [4]. For validation, the de-embedded waveforms using this simplified fixture parasitic model are checked again with the large-signal model [5]. As depicted in Fig. 4, they show good agreement with the simulation result. It can be seen that the parasitic parallel capacitances cause a negative output current with a bigger swing, which is not the case. Therefore, the de-embedded waveforms reflect more correctly to the device's nonlinear behavior, especially under stress conditions.

III. EXPERIMENT

A. Reliability Measurement Under DC and RF Stresses

In this section, three experiments were done to compare the degradation under different stresses. For DC stress, the device was biased at high voltage with $V_{GS} = V_{DS} = -1.4V$ [4]. Therefore, the current density and the lateral electrical field increase to expedite the hot-carrier effect.

For the RF stress, the input power was set to -9.5 dBm with class-A bias condition at 10 GHz fundamental frequency. The source and load impedances were tuned for V_{GS} and V_{DS} peak voltage at about -1.4 V and -1.8 V. In these conditions, the distortion of the drain-voltage waveform did not occur as the power was smaller than P1dB (Fig. 5). The drain-voltage signal was sinusoidal with rms values equal to -1.29 V and -1.62 V respectively. Under this high-voltage and high-current condition, more carriers with enough energy induced more amount of carrier trapping and interface traps for the oxide/silicon interface. The gate voltage swing was amplified and out of phase to the drain signal and increased the electric field at the gate oxide.

B. Results and Discussion

Fig. 6 shows the comparison of linear and saturation current degradation after DC and RF stresses. After 10000 seconds of stress, the degree of degradation under DC stress is higher than the one under RF stress condition. Although the RF

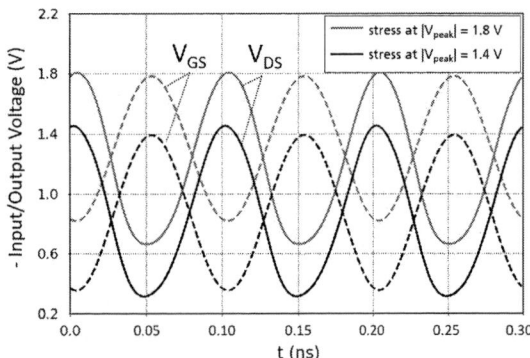

Fig. 5. Input V_{GS} (dashed lines) and output voltage V_{DS} (solid lines) waveforms reconstructed from measurement at $V_{peak} = -1.4V$ and $V_{peak} = -1.8V$ stress condition.

stress voltage swing has a higher peak (-1.8 V) than DC stress voltage (-1.4 V), the device shows to be more resilient during high-frequency operation. Moreover, the degrading factor starts to roll down after 1000 seconds in the RF stress condition. As a consequence, there are two mechanisms that contribute to the degradation in this case. The degradation of power performances is also shown in Fig 7. Even after compensating for the threshold voltage shifted of the stressed device, the performance cannot be recovered.

Fig 8 shows the transistor waveform before and after stress. The initial waveform (fresh) is the blue line while the red line depicts the device after stress. The input voltage is not shown since it was unchanged throughout the stress period. There is only a very slight change visible in the RF gate current over the stress period, suggesting that the effect on the RF displacement current passing through the input capacitance is negligible. The RF output current reduces over time but only at the high current end. In contrast, the RF output voltage waveform can be seen to be degrading gradually at both ends of its swing. It is more visible in lower drain current bias condition.

Plotting the output current and voltage waveforms against one another results in the device dynamic load lines, which have been overlaid onto DC-IV measurements taken before and after the stress measurements in Fig. 9. Using these graphs, the device's degradation behavior can be monitor easier. The slope of the drain current after stress became much larger. This characteristic is similar to the channel length modulation. It is shown that the on-resistance decreases after stress. This is the main issue that causes a decrease in power performance after stress as it limits the drain voltage swing. Because the threshold voltage (V_T) increases after stress resulted in drain current decreasing, measurement at new input voltage with V_T compensation taking into account ($V_{GS} - \Delta V_T$) can increase the maximum voltage swing to the same value as before. The dynamic load lines of pre-stress and after V_T correction measurement are similar at the low current region. But it is still smaller at high current region. As a result, the device's performance can not be recovered to the point before stress (as

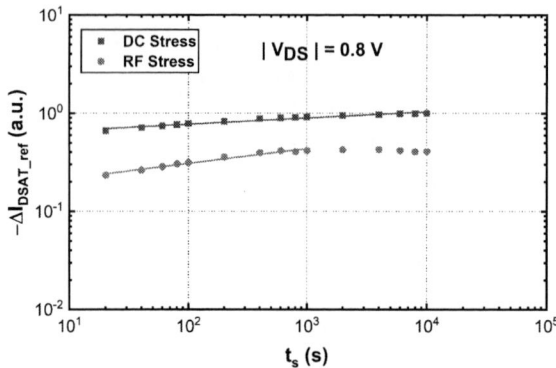

(a) Saturation current shift measured at $|V_{DS}| = 0.8$ V, $|I_{DS}| = 2$ mA

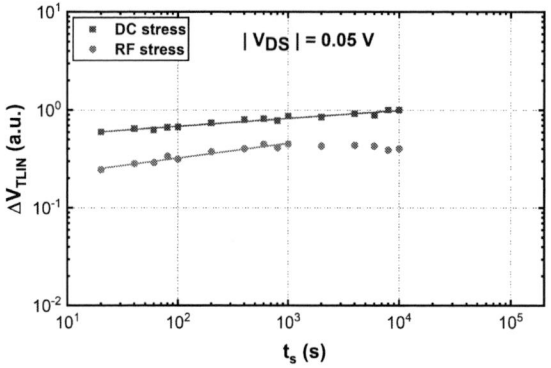

(b) Threshold voltage (linear region) shift measured at $|V_{DS}| = 0.05$ V
Fig. 6. Comparison of degradation under DC (blue symbol) and RF (red symbol) condition as a function of stress time

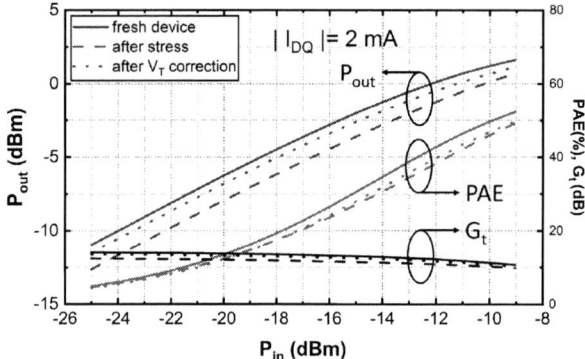

Fig. 7. Output power, Transducer Gain (Gt) and PAE of fresh and stressed devices. The performance is measured at reference quiescent points $|V_{DQ}| = 0.8V$, $|I_{DQ}| = 2mA$

depicted in Fig. 7) indicating permanent damage has caused the degradation.

The devices' dynamic input characteristics depicted in Fig. 10 shows the input RF current vs voltage. The graphs highlight change in current flow in relation to the input voltage cycle. The degradation is more visible in the high current region with changes in the slopes resulted in the increase of threshold voltage.

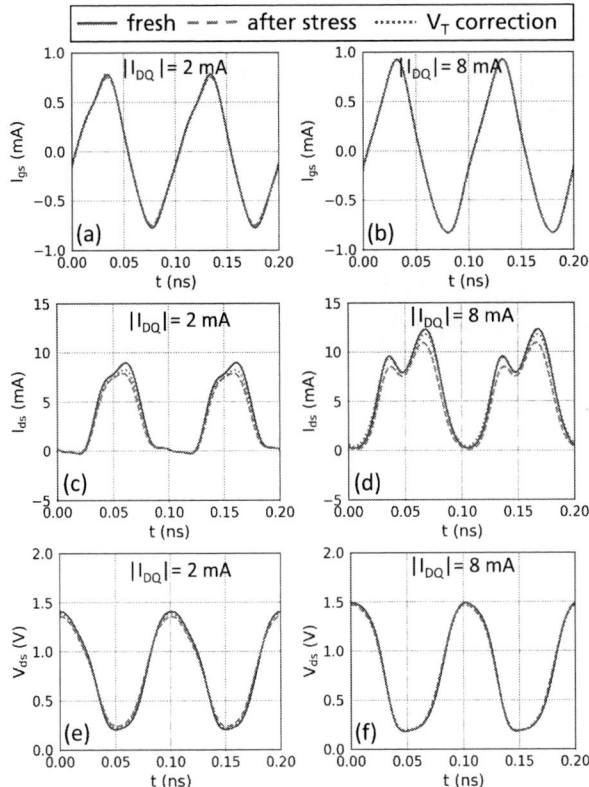

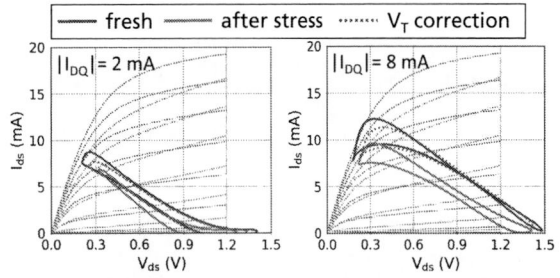

(a) Dynamic load lines after DC stress at $|V_{stress}| = 1.4V$

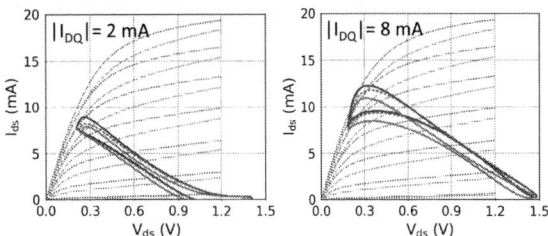

(b) Dynamic load lines after RF stress at $|V_{peak}| = 1.8V$

Fig. 9. Dynamic load lines and DC-IVs ($|V_{GS}| = 0$ V to 1.2 V in 1.5 V step) before, after stress, and after V_T correction measured at reference quiescent points $|V_{DQ}| = 0.8V$, $|I_{DQ}| = 2mA$.

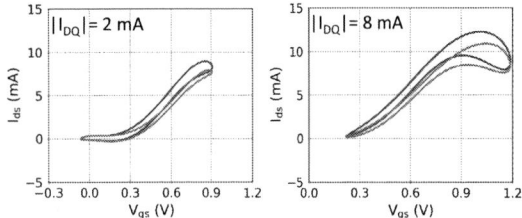

Fig. 10. Dynamic transfer characteristics measured before (blue line) and after stress period (red line) measured at reference quiescent points $|V_{DQ}| = 0.8V$, $|I_{DQ}| = 2mA$.

Fig. 8. Reconstructed RF waveforms of fresh device, after stress and after threshold voltage correction showing from top: (a)-(b) input current, (c)-(d) output current and (e)-(f) output voltage waveforms after 10000 seconds stress period in Class A at $|V_{peak}| = 1.8$ V, $f_0 = 10$ GHz. The RF waveforms are measured at reference quiescent points $|V_{DQ}| = 0.8$ V, $|I_{DQ}| = [2; 8]$ mA

IV. CONCLUSION

In this paper, the effects of RF stress on large-signal performance in comparison to DC have been presented. From the experimental results of the DC I–V curve, the threshold voltage is increased, and the slope of drain current became much larger which was similar to the channel length modulation of MOSFET. It is caused by localized damage near the drain region and the reduction of effective channel length in stressed devices. As the result, the power performance is degraded due to it limits the drain voltage swing and decreases the maximum peak drain current. However, the stress shows little effect on the displacement current as the input current waveform remains unchanged. It can be observed that devices degrade differently under RF stress conditions compared to DC. Hence, applying DC stress only would not guarantee strong validity to analyze the device's behavior at RF operation conditions. In particular, it also can underestimate the effect of voltage/current swing on large-signal performance.

ACKNOWLEDGMENT

The authors would like to thank GlobalFoundries Dresden for providing the test structures. This work is funded by project BEYOND5 from the ECSEL Joint Undertaking and project WIN FDSOI from IPCEI.

REFERENCES

[1] Q. H. Le, D. K. Huynh, D. Wang, T. Kämpfe, and S. Lehmann, "DC-110 GHz characterization of 22FDX® FDSOI transistors for 5G transmitter front-end," in *ESSDERC 2019 - 49th European Solid-State Device Research Conference (ESSDERC)*, Sep. 2019, pp. 218–221.

[2] C. Charbonniaud and T. Gasseling, "VNA based load pull harmonic measurement de-embedding dedicated to waveform engineering," in *2015 IEEE International Conference on Microwaves, Communications, Antennas and Electronic Systems (COMCAS)*, 2015, pp. 1–6.

[3] D. Barataud, A. Mallet, M. Campovecchio, J. M. Nebus, J. P. Villotte, and J. Verspecht, "Measurements of time domain voltage/current waveforms at R.F. and microwave frequencies for the characterization of nonlinear devices," in *IMTC/98 Conference Proceedings. IEEE Instrumentation and Measurement Technology Conference. Where Instrumentation is Going (Cat. No.98CH36222)*, vol. 2, 1998, pp. 1006–1010 vol.2.

[4] D. K. Huynh, Q. H. Le, P. Duhan, D. Wang, T. Kämpfe, and M. Rudolph, "Analysis of hot-carrier degradation in 22nm FDSOI transistors using RF small-signal characteristics," in *2020 German Microwave Conference (GeMiC)*, 2020, pp. 244–247.

[5] Q. H. Le, D. K. Huynh, S. Lehmann, Z. Zhao, T. Kämpfe, and M. Rudolph, "Empirical Large-Signal Modeling of mm-Wave FDSOI CMOS Based on Angelov Model," *IEEE Transactions on Electron Devices*, vol. 68, no. 4, pp. 1446–1453, 2021.

Proceedings of the 16th European Microwave Integrated Circuits Conference

Towards an Excitable Microwave Spike Generator for Future Neuromorphic Computing

Qusay Al-Taai[1], Razvan Morariu[1], Jue Wang[1], Abdullah Al-Khalidi[1], Ali Al-Moathin[1], Bruno Romeira[2], José Figueiredo[3], Edward Wasige[1]

[1]High Frequency Electronics Group, James Watt School of Engineering, University of Glasgow, UK

[2]International Iberian Nanotechnology Laboratory, Portugal

[3]Centra-Ciências and Departamento de Física, Faculdade de Ciências, Universidade de Lisboa, Campo Grande, Portugal

qusayraghibali.al-taai@glasgow.ac.uk

Abstract—This paper describes the systematic approach to develop low power consumption excitable neuromorphic spike generators using nano-sized resonant tunnelling diode (RTD), including fabrication, characterization and device modelling and spike circuit simulation. The fabrication process of nano sized RTDs has been developed and devices exhibit peak currents of up to 100 µA. The energy efficiency of the RTD spike generator can reach as low as 0.09 fJ per spike. An accurate small signal model of nano RTD has also been developed and is described. This nano-RTD technology could underpin the development of energy efficient neuromorphic computing in the very near future.

Keywords—Resonant tunnelling diode, RTD, neuromorphics, ultrafast electronics, spike generator.

I. INTRODUCTION

The recent rise of Artificial Intelligence (AI) systems powered by computers that can learn without the need for explicit instructions is transforming our digital economy and our society as a whole. Computational deep neural network models are inspired by information processing in the human brain. However, today's computing hardware, based on von Neumann architectures, is inefficient at implementing these neural networks largely because of the high power consumption per unit area required, typically >10 W/cm² compared to around 0.01 W/cm² for the human brain [1]. As such, research for new low energy computing paradigms is underway.

The nanometre-sized RTD (nano-RTD) is a nanoelectronic structure that can be easily integrated with a photodiode on a chip working at room temperature. Its unique current-voltage characteristic exhibits a negative differential resistance (NDR) region, and so it can be used in the design of excitable spike (pulsed) generators to produce short electrical and optical pulses mimicking the spiking behaviour of biological neurons thus making it one of the target candidates for such neuromorphic applications [2]. RTD has attracted a lot of research interest for terahertz (THz) [3] and logic applications [4], but here we propose a novel application in neuromorphic computing.

In this paper, we report on the systematic approach to develop low power consumption RTD technology, including device fabrication, characterization, accurate device modelling process and spike generator circuit simulation. The results are expected to pave the way for building low power consumption excitable electrical spike generators operating in the microwave

frequency range and aimed at the currently fast evolving neuromorphic computing applications.

II. DEVICE DESIGN AND FABRICATION

A. Epitaxial structure design

In this section we describe the design of the epitaxial RTD wafers. It was grown on a semi-insulating InP substrate by molecular beam epitaxy (MBE). The double barrier quantum well (DBQW) structure of RTD comprises a 5.7 nm $In_{0.53}Ga_{0.47}As$ quantum well, 1.7 nm AlAs barriers, with 100 nm thick lightly doped InAlGaAs spacers. The collector and emitter layers are made of highly Si doped $In_{0.53}Ga_{0.47}As$ layers. The details of the epitaxial layers structure of 1000A wafer are shown in Table 1.

Table 1. Structure Specification of 1000A wafer.

Layer	Type	Doping level	Material	Thickness (Å)	Description
12	N++	2.0e19	InGaAs	1000	Contact
11	N+	2.0e18	InAlAs	1000	Collector
10	N-	2.0e16	InAlGaAs	1000	Spacer
9	I	---	InGaAs	20	---
8	I	---	AlAs	17	Barrier
7	I	---	InGaAs	57	Well
6	I	---	AlAs	17	Barrier
5	I	---	InGaAs	20	---
4	N-	2.0e16	InGaAs	200	Spacer
3	N+	2.0e18	InAlAs	1000	Emitter
2	N++	2.0e19	InGaAs	5000	Contact
1	I	---	In(x)AlAs	1000	---
Substrate			InP		

B. Design and Fabrication

We designed and fabricated mesa structures with circular geometry with diameters of 500, 700, 800 nm. The fabrication process starts with metal evaporation of Ti/Pd/Au to form the top contact of the RTD. Dry etching InGaAs with $Cl_2/CH_4/H_2$ gases at 60˚C to define the RTD mesa. The bottom contact metal (Ti/Pd/Au) was then evaporated. Low dielectric constant polymer Benzocyclobutene (BCB) was used for device passivation, followed by an etch back process. Finally, bond-pad metallisation was carried out using Ti/Au. Fig.1 shows the SEM images for different stages of 500 nm nano-RTD fabricated devices.

386

3−4 April 2022, London, UK

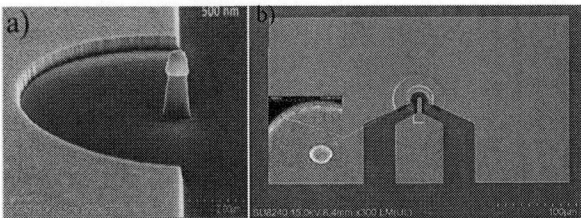

Fig. 1. SEM images of a fabricated 500 nm RTD a) top contact/mesa and bottom contact, and b) the bond pads of completed device; the inset showing the bond pad over the 500 nm top contact.

III. MEASUREMENTS AND RESULTS

A. Static current-voltage (I-V) characterization of RTDs

The I-V characterisation of the fabricated devices was done using the Keysight's semiconductor parameter analyser (B1500). The measured characteristics of the fabricated nano-RTD devices with diameters of 500, 700, and 800 nm are shown in Fig. 2. The peak DC power consumption of 500 nm RTD is 64.8 µW. For the 500 nm diameter device, the energy consumption per cycle can reach as low as 0.09 fJ if it is employed in an oscillatory circuit.

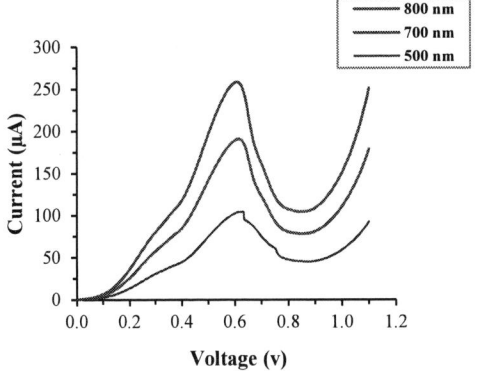

Fig. 2. I-V characteristics of 500, 700 & 800 nm RTD device.

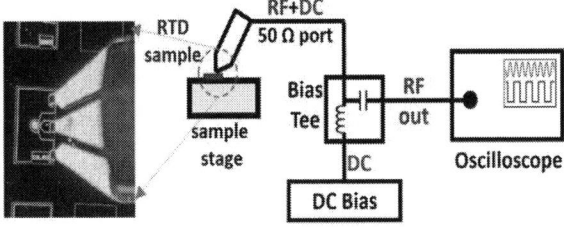

Fig. 3. Schematic circuit for the self-oscillation measurement.

To assess the spiking behaviour of the devices, an RTD device of 500 nm mesa was used. Fig.3. shows the setup for this measurement. While Fig.4 shows the measurement equivalent circuit. Here, R_b/L_b represent the resistance and inductance of the biasing cable. $L_{bias-tee}$ and $C_{bias-tee}$ denote the inductance and capacitance of the bias-tee while R_L represent the input impedance of the oscilloscope.

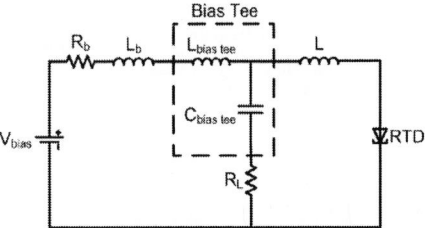

Fig. 4. The measurement equivalent circuit of self-oscillation measurement.

Relaxation oscillations were observed on the oscilloscope. When the RTD was biased at 0.6 V in the NDR region near the peak voltage, Vp, the time domain pulse was as shown in Fig. 5. The number of spikes can be controlled by adjusting the applied biasing voltage. It is worth to note that phase of the spike is opposite depending on the bias position. The results are consistent with our simulation presented in section D.

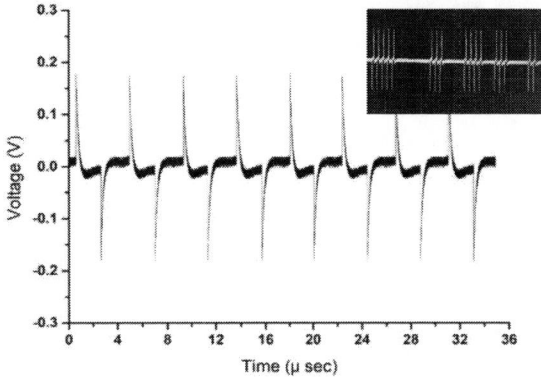

Fig. 5. Experimental traces of output spikes generated by 500 nm nano-RTD device when biasing around peak point. The inset presents the number of spikes during simply adjustment of biasing voltage.

Fig.6 shows another set of experimental results of generated spikes using an RTD-oscillator [5]. The oscillator was driven by pseudo-random bit sequence (PRBS) with an amplitude of 1V, biased at -0.59 V close to peak voltage V_p, and the frequency of the PRBS was set at 500 MHz. The resultant spiking behaviour is shown in Fig.7. The amplitude of the spikes is around 2.5 mV, and the spike duration is 450 psec. The amplitude of spikes is rather low (~2mV), which is attenuated due to the shunt circuit of the RTD-oscillator circuit. In this experiment, we find that when applying an electrical signal and biasing the RTD in either first or second PDR regions slightly below/above the peak/valley voltage, we found that the RTD responds to the external signal and generate spikes; only if the PRBS signal exceeds a threshold point, which mainly depends on the biasing value. Two main factors characterise spiking behaviour namely the biasing positions and the amplitude of PRBS input signal. In addition, the control of the pulse shape can be achieved via several parameters, such as the adjusting between time constants associated to the capacitive part of the circuit (RC time) and the inductive part (inductance associated time constant).

387

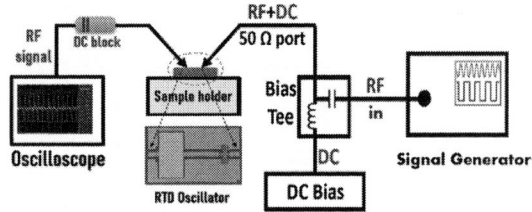

Fig.6, Schematic circuit for the spike measurements.

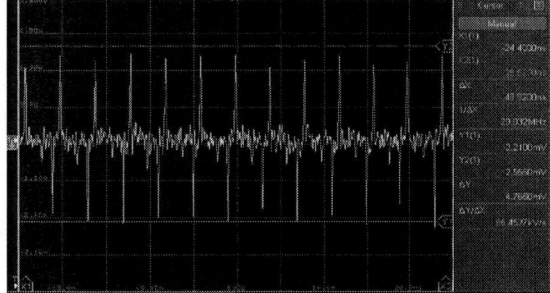

Fig. 7. Experimental traces of output spikes generated by an RTD driven by pulse generator. The input bias voltage the PRBS square wave amplitude was set to 1V, the RTD oscillator was biased at 0.56V, the frequency was set at 500 MHz.

B. High frequency characterisation of nano-RTDs

Here, a vector network analyzer (VNA) was used. The calibration of the VNA was done using the short-open-load-through (SOLT) technique with a port power of −17 dBm. The frequency range was 10 MHz–110 GHz. Fig. 8. shows the measured S-parameters (for a bias point in the NDR region) of 500 nm RTD which reflect the stability of nano RTD devices.

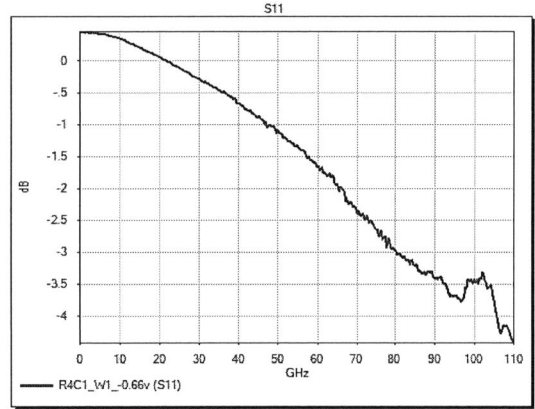

Fig. 8. S-parameters (S11) of 500 nm RTD device.

C. Small-signal modelling of nano RTD devices

The small signal equivalent circuit model of the RTD, first introduced in [6] consists of the metal-semiconductor contact and access resistance R_s, in series with the parallel combination of the device self-capacitance C_n together with the device conductance G_n which models the device current–voltage characteristics, and the quantum well inductance L_{qw}, which

models the charging and discharging effects of the quantum well [7]. The complete device circuit, as shown in Fig. 9, is completed by the extrinsic elements C_p and L_p, which model the parasitic components introduced by the metallic bond-pads [8].

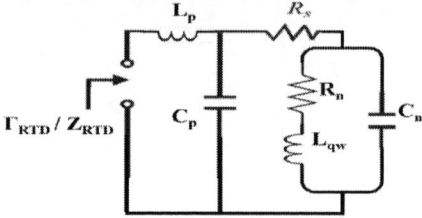

Fig. 9. Small-signal equivalent circuit of an RTD device comprises R_s R_n, L_{qw} and C_n. The bond-pad is modelled by parasitic elements L_p and C_p.

In order to accurately determine the realised device equivalent circuit elements, the acquired S-parameter data was first converted to real and imaginary Z-parameters and fitted using the proposed model over the entire frequency range using a direct optimization procedure [9].

Good agreement between measurement and simulation was obtained across the entire nano RTD device bias range for frequencies up to 110 GHz. This is illustrated by the graphs in Fig. 10, with only two bias points results are shown, i.e. 0.4V in 1st PDR region and 0.65V in NDR region.

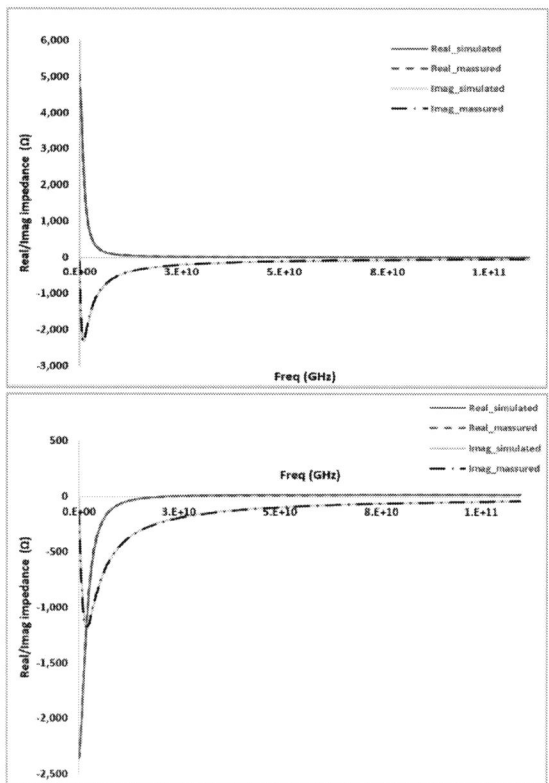

Fig. 10. Graphs of real and imaginary parts of Z-parameters, the dashed lines are for measured while the sold lines for simulated parameters for a 500 nm diameter nano-RTD in the NDR region at a) 0.4V and b) 0.65V.

A summary of the extracted frequency independent RTD parameters corresponding to the three regions of interest is given in Table 2.

Table 2. Summary of parameter extraction

Bias (V)	L_p (pH)	C_p (fF)	R_s (Ω)	G_n (mΩ^{-1})	C_n (fF)	L_{qw} (nH)
0.4	9.35	0.1	22	0.639	33.3	2
0.65	9.35	0.1	22	-1.563	33	-2.2
1	9.35	0.1	22	0.505	33.7	2.7

As predicted by the measured device DC characteristics, the differential conductance G_n becomes negative in the NDR region, with its relatively low magnitude attributed to the designed sub-micron device active area. Furthermore, a bias dependency of the device self-capacitance was observed, with a peak value obtained within the NDR region. This observation is consistent with the predicted behaviour described in [10], and is attributed to the quantum capacitance component, which models the charge variation through the double-barrier quantum-well structure.

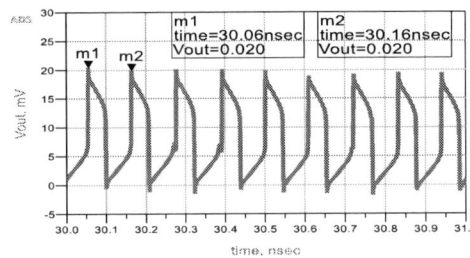

Fig. 11 Simulated relaxation oscillations with a 500 nm diameter RTD device showing that 10 GHz switching oscillations can be achieved. The resonating inductance was 34 nH.

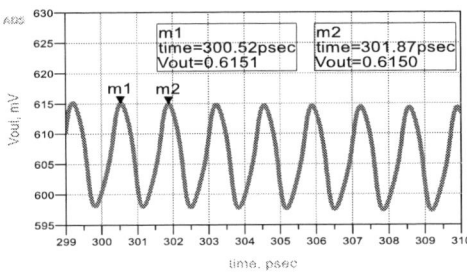

Fig. 12 Simulated sinusoidal oscillations with a 500 nm diameter RTD device showing that terahertz frequencies can be generate. The resonating inductance here was 100 pH.

D. Relaxation and sinusoidal nanopillar RTD oscillators

Using the measured device characteristics and extracted device self-capacitances, preliminary RTD oscillator simulations have been carried out and show that nano-sized RTDs can support both relaxation and sinusoidal oscillations. Fig. 11 shows simulation oscillator results of a 500 nm diameter RTD resonated with a 34 nH inductor and driving a 700-Ω load, while Fig. 12 shows similar results for 100 pH resonating

inductance. In the former case, relaxation or switching oscillations are the result while in the latter case, sinusoidal oscillations occur. Through the careful design and driving of the relaxation oscillator, an excitable pulse generator required for neuromorphic computing could be realised.

IV. CONCLUSION

The fabrication and characterization of nano-RTD devices has been described in this paper. In addition, an accurate small signal model of the device has also been described. The devices have been used in experimental demonstration of relation oscillators and in simulated sinusoidal oscillators. The power consumption can be as low as 0.09 fJ per cycle. The RTD which includes a light absorption layer, can be optically controlled. This would extend our research into the optical domain. Further work in this regard is ongoing.

ACKNOWLEDGMENT

The authors would like to thank the staff of James Watt Nanofabrication Centre (JWNC) for assistance in device fabrication. The work was funded by the Horizon 2020 FET-OPEN ChipAI project, grant agreement 828841.

REFERENCES

[1] P. A. Merolla, J. V. Arthur, R. Alvarez-Icaza, A. S. Cassidy, J. Sawada, F. Akopyan, B. L. Jackson, N. Imam, C. Guo, Y. Nakamura, B. Brezzo, I. Vo, S. K. Esser, R. Appuswamy, B. Taba, A. Amir, M. D. Flickner, W. P. Risk, R. Manohar, and D. S. Modha, "A million spiking-neuron integrated circuit with a scalable communication network and interface," *Science,* vol. 345, no. 6197, pp. 668-673, 2014.

[2] B. Romeira, J. Javaloyes, C. N. Ironside, J. M. L. Figueiredo, S. Balle, and O. Piro, "Excitability and optical pulse generation in semiconductor lasers driven by resonant tunneling diode photo-detectors," *Optics Express,* vol. 21, no. 18, pp. 20931-20940, 2013.

[3] S. Suzuki, M. Shiraishi, H. Shibayama, and M. Asada, "High-Power Operation of Terahertz Oscillators With Resonant Tunneling Diodes Using Impedance-Matched Antennas and Array Configuration," *IEEE Journal of Selected Topics in Quantum Electronics,* vol. 19, no. 1, pp. 8500108-8500108, 2013.

[4] K. Arzi, S. Suzuki, A. Rennings, D. Erni, N. Weimann, M. Asada, and W. Prost, "Subharmonic Injection Locking for Phase and Frequency Control of RTD-Based THz Oscillator," *IEEE Transactions on Terahertz Science and Technology,* vol. 10, no. 2, pp. 221-224, 2020.

[5] J. Wang, Wang, l., Li, C., Patarata Romeira, B.M., Wasige, E., "28 GHz MMIC resonant tunnelling diode oscillator of around 1mW output power," *Electronics Letters,* vol. 49, pp. 3, 2013.

[6] E. R. Brown, C. D. Parker, and T. C. L. G. Sollner, "Effect of quasibound‐state lifetime on the oscillation power of resonant tunneling diodes," *Applied Physics Letters,* vol. 54, no. 10, pp. 934-936, 1989.

[7] R. Lake, and Y. Junjie, "A physics based model for the RTD quantum capacitance," *IEEE Transactions on Electron Devices,* vol. 50, no. 3, pp. 785-789, 2003.

[8] J. Wang, A. Al-Khalidi, K. Alharbi, A. Ofiare, H. Zhou, E. Wasige, and J. Figueiredo, "High performance resonant tunneling diode oscillators as terahertz sources.", 46th European Microwave Conference (EuMC), pp. 341-344, 2016.

[9] R. Morariu, J. Wang, A. C. Cornescu, A. Al-Khalidi, A. Ofiare, J. M. L. Figueiredo, and E. Wasige, "Accurate Small-Signal Equivalent Circuit Modeling of Resonant Tunneling Diodes to 110 GHz," *IEEE Transactions on Microwave Theory and Techniques,* vol. 67, no. 11, pp. 4332-4340, 2019.

[10] R. Lake, and J. Yang, "A physics based model for the RTD quantum capacitance," *IEEE Transactions on Electron Devices,* vol. 50, no. 3, pp. 785-789, 2003.

Numerical and Experimental Investigations of Self-mixing Effect of a Planar Gunn Diode Oscillator

Ming Yan Zhong[1], David R. S. Cumming[2], and Chong Li[3]

James Watt School of Engineering, University of Glasgow, UK

[1]m.zhong.1@research.gla.ac.uk, {[2]davd.cumming.2, [3]chong.li}@glasgow.ac.uk

Abstract — In this paper, we investigate the self-mixing effect of planar Gunn diode oscillators numerically and experimentally. The simulation shows a 4 µm long GaAs-based heterostructure planar Gunn diode can generate not only an oscillation frequency of 27.5 GHz but also a down-converted signal of 2.5 GHz when an external signal of 30 GHz is injected to the diode. The conversion loss varies between 14 dB and 20 dB depending on the biasing condition as well as the amplitude and frequency of the input signal. Experiments confirmed the simulation results. The self-mixing effect of planar Gunn oscillators show a great potential for simplifying RF frontends for millimeter-wave applications such as 5G and beyond communications, radar, and imaging systems.

Keywords — Gunn diode, self-mixing effect, heterodyne, down-conversion.

I. INTRODUCTION

Millimetre-wave (mm-wave) refers to the frequency band in the electromagnetic spectrum from 30 GHz to 300 GHz. It has been applied in many fields in the recent years such as wireless communications, security imaging and remote sensing, radio astronomy, and earth science [1]. Compared with microwave, mm-wave has shorter wavelength, meaning it is suitable for dense communication networks, such as personal area networks (PLANs) to enhance spectrum utilization and data rate [2]. As the core of mm-wave systems, the front-end module, which generally contains local oscillators (LOs), mixers, amplifiers, and filters, becomes more and more challenging due to smaller size and lower efficiency. New devices and front-end architectures with higher efficiency and simplicity are always desirable. For example, in a heterodyne system, the incident signal (RF) is mixed, also called down-converted, with a defined local oscillation signal to an intermediate frequency (IF) signal, that always requires an external mixer and a local oscillator (LO) [3]. At mm-wave frequencies, it's extremely challenging to obtain high performance LOs and mixers using a single semiconductor technology. The separation of mixers and local oscillators requires bulky and costly hybrid integration. Fortunately, planar Gunn diodes could be a game changer because it has a heterostructure structure that is compatible with high electron mobility transistors (HEMTs). Plus their intrinsic nonlinearities in conductance and capacitances make them ideal as mixers like other nonlinear devices [4]. Thus, fully integrated monolithic mm-wave frontends with all active devices fabricated on a single substrate using the HEMT-like planar Gunn diode technology is possible [5] [6] [7].

In this paper, we'll investigate the self-mixing effect of a planar Gunn diode using a commercially available physics-based numerical modelling tool. The DC characteristics, time-domain waveforms and the spectrum of self-oscillation and self-mixing effect of the device are successfully predicted by the numerical tool. Experimental results further confirm and validate the simulation results, indicating planar Gunn diodes have the potential to simply mm-wave frontends for many applications.

II. THE PLANAR GUNN DIODE

A. Device Structure

A planar Gunn diode having a GaAs channel that is sandwiched by two silicon δ-doped $Al_{0.23}Ga_{0.77}As$ layers was investigated in this study. Fig.1 illustrates a scanning electron microscopy (SEM) image of the fabricated diode, constructed in a coplanar waveguide (CPW) test structure, and the cross-sectional view of its epitaxial layers. The anode-cathode distance, L_{ac}, and the velocity of Gunn domains, v_{domain}, determine the oscillation of frequency of the diode. The relationship is approximately as $f_{osc}=(L_{ac}-L_{dead})/v_{domain}$ if the dead-zone L_{dead} is considered [5]. For a given material structure as shown in Fig. 1 and a fixed L_{ac}, f_{osc} varies as change of bias voltage. This is because the domain velocity varies as bias voltage changes.

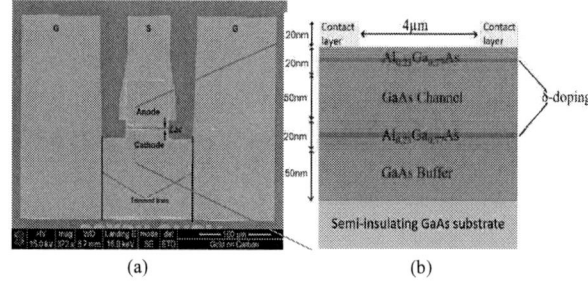

Fig. 1. (a) Scanning electron micrograph of a typical planar Gunn diode oscillator with coplanar waveguide (CPW) test structure. (b) Illustration of the epitaxial layers of the Gunn diode used in this study.

B. Numerical model in COMSOL

The Gunn diode is modelled using a commercial software COMSOL. COMSOL solves carrier transport equations, Poisson's equation and the carrier continuity equations using the finite element method (FEM). A 2D model was constructed to save computing time. The anode and cathode electrodes are modelled with good Ohmic contact that has a metal contact boundary condition and a 20 nm highly doped GaAs with carrier concentration of 2×10^{20} cm^{-3} underneath. Two Al$_{0.23}$Ga$_{0.77}$As barrier layers are lightly doped with electron concentration of 5×10^{16} cm^{-3}. The GaAs channel and buffer are modelled unintentionally n-doped with electron concentration of 10^2 cm^{-3}. Gunn Effect occurs under high electric field however COMSOL has no such built-in mobility model, a user defined high-field mobility model was used [9]:

$$\mu_h = \frac{\mu_0 + (\frac{V_{sat}}{E})(\frac{E}{E_0})^4}{1+(\frac{E}{E_0})^4} \qquad (1)$$

where μ_0 is the low field mobility of GaAs, v_{sat} is the saturation velocity, E_0 is the reference electric field and E is the dependence electric field. When a device is biased with a voltage greater than the threshold voltage i.e., $E > E_0$, the electron transfer effect will occur and result in reduction of overall mobility, and therefore lowered current. The built-in continuity/heterojunction condition is used with continuous quasi-Fermi level model. In addition, trap-assisted recombination condition is used with Shockley-Read-Hall model and the electron/hole lifetime is 0.1 µs. Fig. 2 illustrates the band structures and electron concentration distribution of the diode when the bias voltage is 0 V. A triangle quantum well and a 2DEG are visible at the interface between the top Al$_{0.23}$Ga$_{0.77}$As barrier and the GaAs channel. Other material properties used in the modelling are shown in Table 1.

Table 1. Summary of other properties of the materials used in the simulation.

Parameters	GaAs	AlGaAs
Relative permittivity	12.9	12.2
Bandgap (eV)	1.414	1.71
Affinity (eV)	4.07	3.82
Effective conduction band density of states (cm^{-3})	4.7×10^{17}	5.9×10^{17}
Low field mobility (cm$^2\cdot$V^{-1}s^{-1})	8500	4000
Electron saturation velocity (cm$\cdot$s^{-1})	1×10^7	0.8×10^7

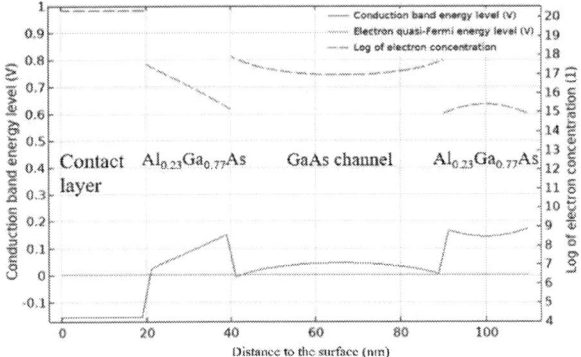

Fig. 2. Band structures and electron distribution of a 4 µm planar Gunn diode.

C. Device fabrication

The anode and cathode are defined by using electron beam lithography (EBL). Ohmic contacts are formed by evaporating Pd/Ge/Au/Pd/Au and annealed at 400 °C for 60 seconds. Additionally, 1:1:10 H$_2$O$_2$:H$_2$O:H$_2$SO$_4$ was used to etch the mesa for 90s. 200 nm gold was evaporated to form CPW pads for RF and DC measurements. At last, the unwanted GaAs cap between electrodes were etched away by dipping in 3:1 citric acid:H$_2$O$_2$ solution for 20 seconds. More detailed fabrication process can be found elsewhere [11].

III. EXPERIMENTAL AND NUMERICAL RESULTS

A. Stationary simulation

Stationary study provides IV characteristics of the diode. By using auxiliary sweep function, the IV curve of a 4 µm device is derived and plotted against the experimental results as shown in Fig.3. When the bias voltage reaches around 3V, the Gunn diode starts exhibiting negative differential resistance or NDR where the current decrease as the voltage increase. The slight discrepency between the measured and simulated resutls is probably due to the self heating which would be considered in the future study.

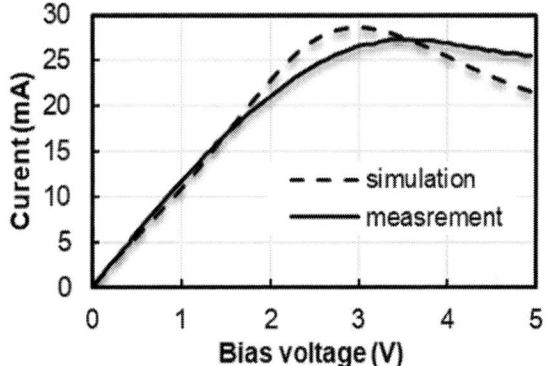

Fig. 3. Measured and simulated I-V characteristics of a 4 µm planar Gunn diode.

B. Modelling the self-oscillation of Gunn diode

Self-oscillation simulation of the Gunn diode is modelled using COMSOL's time-dependent solver. A ramp up function for bias voltage was used and the slope was set to be 1×10^{10} which means the bias voltage ramps up to 4.5 V in 100 ps. Fig. 4 shows the simulated waveform in time domain and its Fourier Transform in frequency domain. Note the amplitude shown here is current. The diode oscillated at 27.5 GHz at 4.5 V bias voltage, which marched the experimental results very well.

C. The self-mixing effect

Gunn diodes have intrinsic nonlinearities in their conductance and capacitance that make them possible to mix with incoming signals in addition to their self-oscillations. In COMSOL, we deliberately injected a 30 GHz (f_{RF}) sinusoidal signal with an amplitude of 10 mV between the anode and cathode when the bias voltage was set at 4.5 V and the corresponding waveform and its spectrum are shown in Fig. 5 where both self-oscillation (f_{LO}=5 GHz) and self-mixing effect (f_{RF} - f_{LO} =30 GHz-27.5 GHz = 2.5 GHz) are seen. The 2nd order mixing ($2f_{LO}$-f_{RF}=25 GHz) is also visible in Fig. 5b.

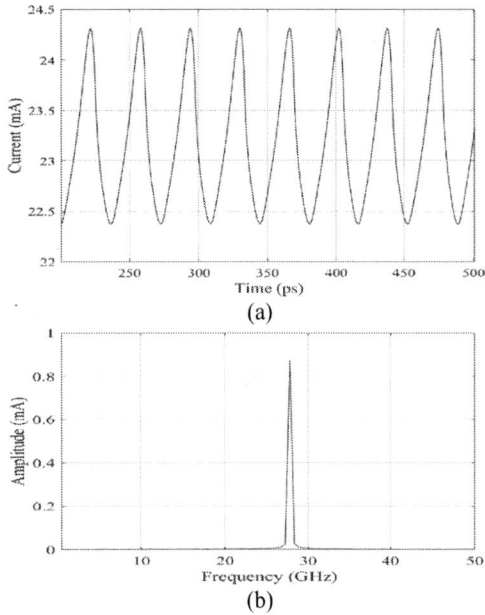

(a)

(b)

Fig. 4. (a) waveform in time domain and (b) the corresponding spectrum modelled in COMSOL.

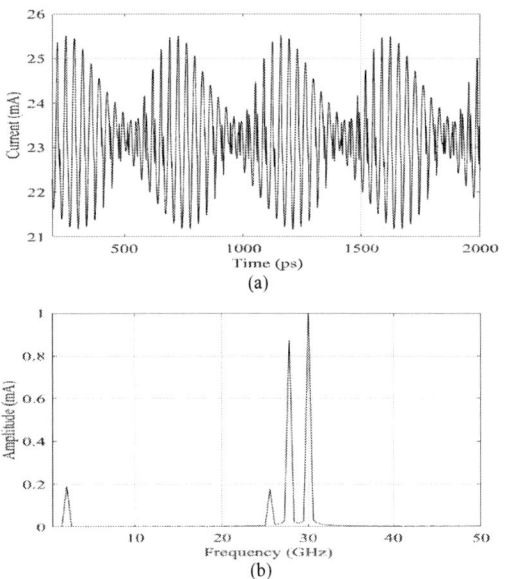

(a)

(b)

Fig. 5. Simulated self-mixing effect of a planar Gunn diode when it is biased at 4V (a) waveform in time domain and (b) its frequency spectrum.

To verify the simulation results, we trimmed the side grounds of the coplanar waveguide and converted the one-port Gunn diode to be a two-port device (Fig.1a). The trimmed diode was measured in a setup shown in Fig. 6. The diode was biased through two bias-tees (Anritsu 250V). The incident signal was generated by an external signal generator (Wiltron 68369B) to the cathode side of the Gunn diode through a Formfactor's ground-signal-ground (GSG) ACP probe. A spectrum analyzer (Agilent 4448A) was connected to the anode side to measure the output of the diode. Fig. 7 shows a snapshot of the spectrum analyzer where the self-oscillation frequency of 27.5 GHz, incident signal of 30 GHz and the down-mixed signal of 2.5 GHz are shown. Note the diode was biased at 4.22 V and the DC current was 24.5 mA [5]. Discrepancy on bias voltage and current between the simulation and experiment is due to system loss which include probes, cables, and bias tees. The measured conversion loss of target planar diode is around 20±2.5 dB [5].

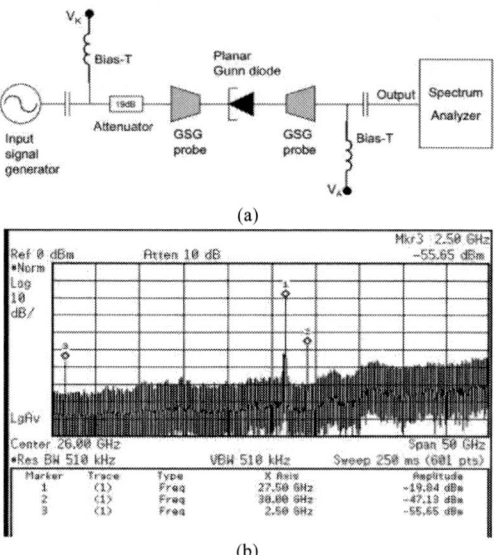

(a)

(b)

Fig. 6. (a) Illustration of the measurement setup and (b) the spectrum shown on the spectrum analyser, where 1, 2 and 3 indicate the self-oscillation signal frequency of the diode, the incident RF signal and the IF signal.

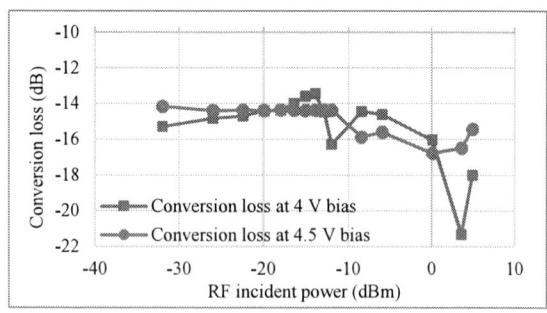

Fig. 7. Conversion loss (dB) VS amplitude of the incident wave (V) at 4.0 V and 4.5 V.

We numerically investigated the relationship between the power of the incident wave and the conversion loss. The power was varied from − 32 dBm to 5 dBm for two biasing conditions 4.0 V and 4.5 V and the results are shown in Fig. 7. One can see the conversion loss remain almost constant while the power of the input signal small however once the input signal's strength is increased. The conversion loss decreases at 4.0 V but tends to be stable at 4.5 V.

Furthermore, we investigated the linearity of the IF output power against the RF input power and the results are shown in Fig. 8. From this we can see that the Gunn diodes has a linear relationship between the input power and down-converted IF signal.

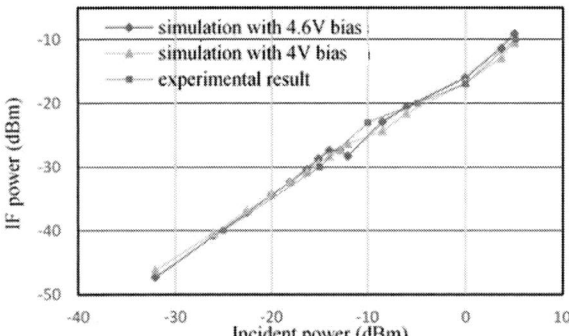

Fig. 8. Relationship with RF power and IF power

Both simulation and experimental results indicate that the conversion loss of the diode is around 20 dB at 30 GHz that is slightly higher than other nonlinear devices such as barrier diode (SBD) and field effect transistor at similar frequency range [12] [13]; however planar Gunn diodes have the advantage of self-mixing that prevents from using external local oscillators. The demonstrated planar Gunn diode has comparable conversion loss to other self-oscillating devices such as resonant tunnelling diodes at the frequency of interest, as shown in Table 2.

Table 2. Performance comparison of self-oscillating mixers

Frequency (GHz)	Conversion loss (dB)	LO power (dBm)	Ref.
11	10	-40	[14]
87	25	-53	[15]
350	10	-11	[16]
30	20	-20	This paper

IV. CONCLUSIONS

In this paper we have demonstrated the self-mixing effect of a planar Gunn diode using both numerical and experimental approaches. A 4 μm AlGaAs/GaAs HEMT-like planar Gunn diode generates a 27.5 GHz oscillation and down-converts a 30 GHz RF signal to 2.5 GHz. Approximately 20 dB conversion loss was obtained. Future work can be concentrated on improving the conversion efficiency using more compact design of the device, implementing impedance matching circuit

and other materials e.g. InGaAs [17]. Nevertheless, this work shows that Gunn diodes have the potential to be used as self-oscillating mixers to simply RF front-ends in novel mm-wave systems.

REFERENCES

[1] K. C. Huang and Z. C. Wang, *Millimeter wave communication systems*, John Wiley & Sons, 2011.
[2] D. P. Jaco and S. Saurabh, *Millimeter-wave power amplifiers*, Springer, 2017.
[3] R. A. Lewis, "A review of terahertz detectors," *Journal of Physics D: Applied Physics.*, vol. 52, 2019.
[4] A. Khalid, N. J. Pilgrim, G. M. Dunn, M. C. Holland, C. R. Stanley, I. G. Thayne, and D. R. S. Cumming, "A planar Gunn diode operating above 100 GHz." *IEEE Electron Device Lett.*, vol.28, pp. 849-851, Oct. 2007.
[5] L. B. Lok, C. Li, A. Khalid, N. J. Pilgrim, G. M. Dunn, and D. R. S. Cumming, "Demonstration of the self-mixing effect with a planar Gunn diode at millimeter-wave frequency," in *IRMMW-THz 2010 - 35th Int. Conf. Infrared, Millimeter, Terahertz Waves, Conf. Guid.*, pp. 9–10, 2010.
[6] M. J. Lazarus, E. D. Bullimore, and S. Novak, "A sensitive millimeter wave self-oscillating Gunn diode mixer." *Proceedings of the IEEE*, vol. 59, pp. 812-4, May 1971.
[7] J. Krasavin and H. Hinrikus, "Performance and optimization of Gunn self-oscillating mixer." *IEEE Microwave Guided Wave Letts.*, vol. 5, pp. 177-9, Jun. 1995.
[8] M. Ali, "Design and simulation of planar electronic nanodevices for terahertz and memory applications," Ph.D. thesis, Univ. of Manchester, p. 158, 2013.
[9] M. Pokorný and Z. Raida, "Multi-physical model of Gunn diode," in *17th Int. Conf. Microwaves, Radar Wirel. Commun*, pp. 2–5, 2008.
[10] G. Rieke, *Detection of Light*, 2nd ed., Cambridge University Press, pp. 276-300, 2002.
[11] C. Li, "Design and characterisation of millimetre-wave planar Gunn diode and integrated circuits," Ph.D. thesis, Univ. of Glasgow, p. 97, 2012.
[12] A. Rogalski and F. Sizov, "Terahertz detectors and focal plane arrays," *Opto-Electronics Review*, vol. 19, no. 3, pp. 346-404, 2011.
[13] D. Glaab, S. Boppel, A. Lisauskas, U. Pfeiffer, E. Öjefors, and H. G. Roskos, "Terahertz heterodyne detection with silicon field-effect transistors," *Applied Physics Letters*, vol. 96, no. 4, pp. 042106-042106-3, 2010.
[14] G. Millington, R. E. Miles, R. D. Pollard, D. P. Steenson, and J. M. Chamberlain, "A resonant tunnelling diode self-oscillating mixer with conversion gain," in *IEEE Microwave and Guided Wave Letters*, vol. 1, no. 11, pp. 320-321, 1991.
[15] G. Millington, R. E. Miles, R. D. Pollard, D. P. Steenson, J. M. Chamberlain, and M. Henini, "Millimetre-wave self-oscillating mixers using resonant tunnelling devices," *23rd European Microwave Conference*, pp. 255-256, 1993.
[16] Y. Nishida, N. Nishigami, S. Diebold, J. Kim, M. Fujita, and T. Nagatsuma "Terahertz coherent receiver using a single resonant tunnelling diode," *Scientific reports*, vol. 9, no. 1, pp 1-9, 2019.
[17] C. Li, A. Khalid, S. H. Paluchowski Caldwell, M. C. Holland, G. M. Dunn, I. G. Thayne, and D. R. S. Cumming, "Design, fabrication and characterization of In0.23Ga0.77As-channel planar Gunn diodes for millimeter wave applications," *Solid State Electron.*, vol. 64, no. 1, pp. 67–72, Oct. 2011.

A

Acri, Giuseppe 313
Ahamed, R. 281
Ahmed, Ahmed S.H. 129
Akbar, Rehman 301
Aldrigo, Martino 370
Algani, C. 422
Alian, AliReza 30
Al-Khalidi, Abdullah 386
Al-Moathin, Ali 386
Al Shanawani, Mazen 370
Al-Taai, Qusay 386
Altabas, Jose A. 173
Ambacher, Oliver 144
Amendola, G. 113
Amiaud, A.-C. 152
Amschl, Dominik 76
Andersson, Kristoffer 51, 108
Ando, Yuji 193
Andrei, Cristina 133
Aniel, Frédéric 313
Arfaoui, Wafa 382
Ashley, Andrea 329
Aufinger, Klaus 39, 277
Azevedo Goncalves, Joao Carlos 125

B

Babenko, Akim 354
Bao, Mingquan 51, 108
Barakat, Adel 26, 317
Barataud, Denis 9, 362
Bar-Helmer, Noam 213
Barton, Taylor 418
Bekker, Elizabeth 345
Beleniotis, Petros 63
Ben-Sassi, M. 9
Ben Yishay, Roee 47
Berglund, Bo 402
Berroth, Manfred 169, 177
Bettoumi, Ines 398
Bhutani, Akanksha 345
Bierbuesse, David 249
Biondi, Andrea 92
Biswas, Debaleen 193
Blampey, Benjamin 80
Blanck, Hervé 233
Błaszczuk, B. 43
Blondy, Pierre 398
Boccia, L. 113
Bonani, F. 189
Bose, Arijit 193
Bossu, Germain 382
Bossuet, Alice 80, 293
Bouvot, Simon 125

Bowers, Steven M. 229
Brahem, Mohamed 269
Braun, Tobias T. 205
Bredendiek, Christian 205, 277
Brezza, E. 1
Brunel, Laurent 261, 378
Brunel, Valeria 378
Brunner, Frank 185
Buballa, Frowin 156
Buisman, Koen 51
Bunea, Alina-Cristina 321
Burak, Abdurrahman 217, 297
Byk, Estelle 233

C

Cabria, L. 88
Caille, Laurent 378
Çalışkan, Can 217
Callet, Guillaume 233
Camacho-Peñalosa, Carlos 374
Camiade, Marc 233
Campovecchio, Michel 362
Cariani, Luca 92
Carta, Corrado 136, 181, 221
Carvalho, Nuno Borges 5
Caspar, C. 169
Catoggio, E. 189
Centurelli, F. 113
Chang, C. 261, 378
Chartier, Sébastien 121, 209
Chaudhary, Sanket 5
Chen, Baichuan 26, 317
Chen, Chun-Nien 55
Chen, Jr-Tai 75
Chen, Peng 410
Chen, Zhi-Yong 410
Chevalier, Pascal 113, 168
Chevtchenko, Serguei 63
Chiong, Chau-Ching 257
Choi, J.H. 169
Ciccognani, W. 253
Clifton, John Christopher .... (NA), (NA)

Colangeli, S. 253
Colantonio, P. 88
Collaert, Nadine .... 30, (NA)
Cooman, Adam 5
Coquillas, B. 152
Corsi, Jordan 313
Costanzo, F. 253
Cumming, David R.S. 390

EuMIC 2021 Brief Author Index

D

Dan, Iulia	117
Danneville, François	125
Davis Manjaly, Anna	289
Deborgies, François	92
de Hek, A.P.	160, 293
Dehos, Cédric	80
Deligeorgis, George	370
Demenitroux, Wilfried	362
Denis, Pierre	233
Der, Adam	418
Dietz, Marco	39
Di Martino, Stefano	76
Dinari, Mehdi	353
Dinescu, Adrian	321
Divay, Alexis	80
Doerner, Ralf	140, 185, 225, 237
Donati Guerrieri, S.	189
Dragoman, Mircea	370
Driad, Samira	378
Du, Xuan-Quang	177
Dubois, Emmanuel	125
Ducournau, Guillaume	1, 125, 313
Duffy, Maxwell R.	358
Dupouy, Emmanuel	197
Dyskin, Aleksey	209

E

Eissa, Mohamed	273
Elad, Danny	47
ElKashlan, Rana	30
Ellinger, Frank	100, 136, 181, 221
Eltayeb, Mathani	59
Ermolov, Vladimir	337
Essing, J.	293

F

Fager, Christian	51, 402
Fakhfakh, Seifeddine	233
Fan, Ting-Hsuan	257
Farid, Ali	129
Favede, Laurent	233
Ferrari, Philippe	313
Figueiredo, José	386
Fiorese, Victor	125
Fonte, A.	113
Franzese, Aniello	273
Freund, R.	169
Friesicke, C.	164
Fritsche, David	136

G

Gaillard, Florent 197
Gallardo, Omar 173
Gaquière, Christophe 1, 125
García-Luque, A. 374
Garmash, Sergey 265
Gatzastras, Athanasios 121
Gauthier, A. 1
Gerfers, Friedel 156
Ghaleb, Hatem 100
Ghione, G. 189
Gibiino, Gian Piero 59
Giofrè, R. 88
Giroto de Oliveira, Lucas 345
Gloria, Daniel 1, 125
Gorman, Melissa C. 148
Grimault-Jacquin, Anne-Sophie 313
Grötsch, Christopher 117
Grözing, Markus 169, 177
Gurbuz, Yasar 217, 297

H

Haapalinna, Atte 325
Hadipour, Kambiz 76
Hagelauer, Amelie 39
Halonen, K.A.I. 281
Hanay, Oner 249
Hansryd, Jonas 108
Haque, Sanaul 185
Heimlich, Michael C. 148, 245
Heinrich, Wolfgang 140, 156, 225, 237
Heinz, Felix 144
Hellepee, C. 9
Hietanen, Mikko 301
Hilt, Oliver 185
Hoffmann, Thomas 156
Holmberg, Heikki 325
Honjo, Kazuhiko 241
Hossain, M. 140, 237, 269
Hou, Rui 402
Hujanen, Arto 325
Hüssen, Lukas 406
Huynh, Dang Khoa 17, 22, 382

I

Iordanescu, Sergiu 370
Ishikawa, Ryo 241
Issakov, Vadim 112
Itcia, E. 152
Iupikov, Oleg 349
Ivashina, Marianna V. 349
Iyer, Vinay 229

J

Jacob, Arne F. 164
Jahan, Nusrat 26
Jameson, Samuel 213
Jensen, Jesper B. 173
Johansen, Tom K. 140, 173, 225, 237, 269
John, Laurenz 117
Jokinen, Markku 309
Jonsson, Rolf 51, 108
Joram, Niko 100
Jordão, Marina 5
Jungnickel, V. 169

K

Kahmen, Gerhard 21, 71, 96
Kallfass, Ingmar 117, 121, 209
Kämpfe, Thomas 17, 22, 136, 382
Kana, Cengizhan 297

Kandis, Hamza 297
Kantanen, Mikko 281
Karman, S. 113
Katz, Oded 47
Kaule, Evelyne 133
Kawamura, Yoshifumi 84
Kerhervé, Eric 152, 305
Khaled, Ahmad 30
Khandelwal, S. 13
Kikuchi, K. 13
King, Justin 289
Kiryukhina, Kateryna 398
Kishchinsky, Andrei 265
Klemm, M. 43
Knauder, Daniel 76
Knight, R. 293
Knopik, Vincent 305
Koelpin, Alexander 71, 96
Kolb, Katharina 39
Krozer, Viktor 140, 225, 237
Kruglov, Dmitrii 349
Kucharski, M. 43
Kupska, Monika 173
Kursu, Olli 104
Kuwata, Eigo 84

L

Lahbib, I. 9
Lambert, Benoit 378
Lamminen, Antti 337
Lasser, Gregor 358, 366
Le, Quang Huy 17, 22, 136, 382
Leblanc, Rémy 51, 253
Leclerc, E. 422
Lehmann, Steffen 17, 22, 382
Leinonen, Marko E. 309
Lépilliet, Sylvie 1, 125
Leuther, Arnulf 121, 144
Levantino, S. 113
Li, Chong 67, 390
Li, Mu-Heng 55
Li, Songhui 136, 221
Li, Teng 345
Limiti, E. 253
Linnhoff, Sebastian 156
Liu, Bo 67
Liu, Rui-Jia 410
Londhe, Sumeet 213
Longhi, P.E. 253
Lopez, M. 88
Louis, Bruno 152

M

Ma, Qiang 193
Maaskant, Rob 349
Mahmud, Mir Hassan 217
Mahon, Simon J. 148, 245
Maier, T. 164
Maiwald, Tim 39
Maletic, Nebojsa 273
Malignaggi, Andrea 273
Malko, Aleksandra 233
Mallet-Guy, Benoît 353
Malmqvist, Robert 51, 108
Mancuso, Yves 353
Mansour, Raafat R. 34
Margalef-Rovira, M. 1
Martineau, Baudouin 80
Martín-Guerrero, Teresa M. 374
Massler, Hermann 121
Maye, C. 1
Mazzanti, A. 113
Merlet, T. 152
Montaseri, Mohammad Hassan 309
Morandini, Yvan 80
Morariu, Razvan 386
Morris, Kevin .... (NA)
Moulin, Maxime 313
Musch, Thomas 394
Mustacchio, C. 113

N

Najmussadat, M. 281
Nakatani, Keigo 285
Nallatamby, Jean-Christophe 197
Nayak, Anurag 17
Neculoiu, Dan 321
Negra, Renato 249, 273, 406
Neveux, G. 9
Ng, Herman Jalli 71, 96
Nogales, Connor 358, 366
Nonet, Olivier 362

O

Oishi, Toshiyuki 201
Okada, E. 1
Olavsbråten, Morten 59
Otsuka, Tomohiro 201
Ozdol, Ali Bahadir 217
Ozkan, Tahsin Alper 217

P

Pace, L. 253
Pallotta, A. 113
Parkkinen, Katja 325
Pärssinen, Aarno 104, 301, 309
Parvais, Bertrand 30
Parveg, Dristy 281
Passerieux, D. 9
Pech, D. 169
Peralagu, Uthayasankaran 30
Perez-Cisneros, Jose-Ramon 402
Petricli, I. 113
Piesiewicz, R. 43
Pinto, Mauricio 354
Pires, Sergio C. 5
Pistono, Emmanuel 313
Ploneis, Frederic 362
Podevin, Florence 313
Pohl, Nils 205, 277, 333
Pokharel, Ramesh 26, 317
Popović, Zoya .... (NA), 354, 358, 366
Potschka, Julian 39
Psychogiou, Dimitra 329
Puig, Olivier 398
Pursula, Pekka 337
Putcha, Vamsi 30

Q

Quay, Rüdiger 164

R

Radchenko, Alexey 265
Rahkonen, Timo 104, 301, 309
Raja, P. Vigneshwara 197
Rantakari, Pekka 325
Redois, S. 152
Resca, Davide 92
Riedmueller, Sandra 233
Rieß, Vincent 100
Rocchi, Marc 38
Rodriguez, Raul 30
Rodwell, Mark 129
Romeira, Bruno 386
Ropers, G. 169
Roussel, L. 152
Rücker, Holger 333
Rudolph, Matthias .. 17, 22, 63, 133, 185, 225, 382

S

Saad, Paul 402
Saarilahti, Jaakko 337
Saijets, Jan 325
Saleh, H. 9
Samanta, Kamal K. ....(NA)
Samis, S. 164
Samori, C. 113
Santarelli, Alberto 59, 374
Scappaviva, Francesco 92
Schmidt, C. 169
Schnieder, Frank 185
Schoepfel, Jan 205
Schostak, J. 169
Sear, William 418
Seidel, Andres 221
Seifert, S. 237
Seo, Munkyo 129
Seshimo, Takuya 241
Sethi, Alok 301
Seyyedrezaei, Seyyedmohsen 181
Sheth, Jay 229
Shivan, T. 237
Shrestha, Amit 225
Silva Valdecasa, Guillermo 173
Singh, Sumit Pratap 309
Singh, Tejinder 34
Snowden, Christopher ....(NA)

Socher, Eran 213
Sommet, Raphaël 197
Sonnenberg, Timothy 354
Sow, O. 9
Soylu, Utku 129
Squartecchia, Michele 173
Starke, David 333
Stieglauer, Hermann 378
Sutbas, Batuhan 71, 96
Szilagyi, Laszlo 136, 221

T

Takahashi, Hidemasa 193
Takayama, Yoichiro 241
Tamminen, Aleksi 337
Tannert, Tobias 169, 177
Tawfik, Y. 281
Tesolin, F. 113
Tessmann, Axel 117
Testa, Paolo Valerio 221
Teyssandier, Charles 378
Thapa, Samundra K. 26, 317
Thayyil, Manu Viswambharan 181
Thome, Fabian 144
Tommasino, Pasquale 113
Torii, Takuma 84
Traversa, Antonio 113
Trifiletti, Alessandro 113
Tsuchiya, Yoichi 193
Tsuru, Masaomi 84, 201, 285
Turhaner, Arsen 269

U

U'Ren, Greg D. 9

V

Vähä-Heikkilä, Tauno 325
van Delden, Marcel 205, 394
Vanden Bossche, Marc 5
van der Bent, G. 160
van Vliet, F.E. 160, 293
Varonen, Mikko 281
Verploegh, Shane 354
Vincent, Loïc 313
Vitulli, Francesco 92
Vogelsang, Florian 333
Voisin, Steeven 305

W

Wagner, Sandrine 117
Wakejima, Akio 193
Walther, Bent 394
Wambacq, Piet 30
Wang, Defu 136, 382
Wang, Huei 55, 257
Wang, Jing 67
Wang, Jue 386
Wang, Yunshan 55
Wasige, Edward 386
Watkins, Gavin T. 414
Wei, Muh-Dey 273
Weigel, Robert 39
Weikle, Robert M. 229
Wentzel, Andreas 156, 225
Wessel, Jan 71, 96
Widmann, Daniel 177
Winkler, Wolfgang 341
Wittemeier, Jonathan 333
Wörmann, Janis 209
Wünsch, S. 169

X

Xia, Jing 410
Xie, Linli 229
Xu, Xin 136, 221
Xue, Li-Yuan 67

Y

Yacoub, Hady 140, 225, 237
Yamaguchi, Yutaro 201, 285
Yamamoto, H. 13
Yazici, Melik 217, 297
Yodprasit, Uroschanit 341
Yu, Chao 410

Z

Zerounian, Nicolas 313
Zhang, Linsheng 229
Zhang, Lv 410
Zhao, Zhixing 17, 22, 382
Zhong, Ming Yan 390
Zhou, Han 402
Zhu, Xiao-Wei 410
Zhu, Yu 100
Zwick, Thomas 345

IEEE
445 Hoes Lane
Piscataway, NJ 08854-4141

ISBN 978-1-6654-4722-5